Handbook of Experimental Pharmacology

Volume 154/II

Editorial Board

G.V.R. Born, London
M. Eichelbaum, Stuttgart
D. Ganten, Berlin
H. Herken, Berlin
F. Hofmann, München
L. Limbird, Nashville, TN
W. Rosenthal, Berlin
G. Rubanyi, Richmond, CA
K. Starke, Freiburg i. Br.

Springer
*Berlin
Heidelberg
New York
Barcelona
Hong Kong
London
Milan
Paris
Tokyo*

Dopamine in the CNS II

Contributors

E. Aquas, J. Bierbrauer, D. Boulay, S.B. Caine, C. Cepeda,
A. Chéramy, M.-F. Chesselet, G.U. Corsini, R. Depoortere,
M. Diana, G. Di Chiara, B.J. Everitt, J. Glowinski, L.H. Gold,
L. Hilwerling, D. Joel, M.L. Kemel, C. Konradi, G.F. Koob,
M.S. Levine, J.E. Leysen, R. Maggio, M. Morelli, E. Ongini,
G. Perrault, T.W. Robbins, D.J. Sanger, J.M. Tepper, F. Vaglini,
I. Weiner, P. Willner

Editor:
Gaetano Di Chiara

 Springer

Professor
GAETANO DI CHIARA
Department of Toxicology
University of Cagliari
Via Ospedale
09125 Cagliari
Italy
e-mail: diptoss@tin.it

With 36 Figures and 9 Tables

ISBN 3-540-42720-1 Springer-Verlag Berlin Heidelberg New York

Library of Congress Cataloging-in-Publication Data
Dopamine in the CNS / editor, Gaetano Di Chiara ; contributors, E. Aquas.
   p. cm. – (Handbook of experimental pharmacology ; 154)
  Includes bibliographical references and index.
  ISBN 3540427198 (hardcover : alk. paper) – ISBN 3540427201 (hardcover : alk. paper)
  1. Dopamine. 2. Central nervous system. 3. Dopaminergic mechanisms. 4. Dopaminergic neurons. I. Di Chiara, Gaetano. II. Aquas, E. III. Series.
QP905 .H3 vol. 154
[QP364.7]
615'.1
[612.8'042 2 21]

2002022665

This work is subject to copyright. All rights are reserved, whether the whole or part of the material is concerned, specifically the rights of translation, reprinting, re-use of illustrations, recitation, broadcasting, reproduction on microfilms or in any other way, and storage in data banks. Duplication of this publication or parts thereof is permitted only under the provisions of the German Copyright Law of September 9, 1965, in its current version, and permission for use must always be obtained from Springer-Verlag. Violations are liable for Prosecution under the German Copyright Law.

Springer-Verlag Berlin Heidelberg New York
a member of BertelsmannSpringer Science+Business Media GmbH

http://www.springer.de

© Springer-Verlag Berlin Heidelberg 2002

The use of general descriptive names, registered names, etc. in this publication does not imply, even in the absence of a specific statement, that such names are exempt from the relevant protective laws and regulations and free for general use.

Product liability: The publishers cannot guarantee the accuracy of any information about dosage and application contained in this book. In every individual case the user must check such information by consulting the relevant literature.

Cover design: design & production GmbH, Heidelberg
Typesetting: SNP Best-set Typesetter Ltd., Hong Kong

SPIN: 10835059   27/3020xv-5 4 3 2 1 0 – printed on acid-free paper

Preface

Dopamine, like Cinderella, has come a long way since its discovery. Initially regarded as a mere precursor of noradrenaline, dopamine has progressively gained its present status of a common target for major drug classes and a substrate for some basic functions and dysfunctions of the central nervous system (CNS). A tangible sign of this status is the fact that dopamine has been the main subject of the studies of the Nobel laureates of 2000, ARVID CARLSSON and PAUL GREENGARD, who also contribute to this book.

The understanding of the function of dopamine was initially marked by the discovery, made in the early 1960s by HORNYKIEWICZ, BIRKMAYER and their associates, that dopamine is lost in the putamen of parkinsonian patients and that the dopamine precursor, L-dopa, reverses their motor impairment. For many years the clinical success of L-dopa therapy was quoted as a unique example of rational therapy directly derived from basic pathophysiology. For the next 10 years, on the wing of this success, dopamine was regarded as the main substrate of basal ganglia functions and was assigned an essentially motor role.

In the early 1970s, studies on the effect of dopamine-receptor antagonists on responding for intracranial self-stimulation and for conventional and drug reinforcers initiated a new era in the understanding of the function of dopamine as related to the acquisition and expression of motivated responding.

This era has merged into the present one, characterized by the notion of dopamine as one of the arousal systems of the brain, modulating the coupling of the biological value of stimuli to patterns of approach behaviour and the acquisition and expression of Pavlovian influences on instrumental responding.

This notion of dopamine has shifted the interest from typically motor areas of the striatum to traditionally limbic ones such as the nucleus accumbens and its afferent areas, the prefrontal cortex, the hippocampal formation and the amygdala. Through these connections, the functional domain of dopamine now extends well into motivational and cognitive functions.

This long development has been marked at each critical step by the contribution of pharmacology: from the association between reserpine akinesia and dopamine depletion and its reversal by L-dopa in the late 1950s, to the blockade of dopamine-sensitive adenylate cyclase by neuroleptics in the early 1970s, to the involvement of the dopamine transporter in the action of cocaine in the 1980s. In no other field of science has pharmacology been as instrumental for the understanding of normal and pathological functions as in the case of dopamine research.

This book intends to provide a rather systematic account of the anatomy, physiology, neurochemistry, molecular biology and behavioural pharmacology of dopamine in the CNS. Nonetheless, the classic extrapyramidal function of dopamine and its role in the action of antiparkinsonian drugs has received relatively little attention here. One reason is that this topic has been the subject of a previous volume of this series. Another reason, however, is that in spite of their systematic layout, even these volumes cannot avoid being a reflection of the times, that is, of the current interests of the research on dopamine.

G. DI CHIARA

List of Contributors

ACQUAS, E., Department of Toxicology, University of Cagliari, Via Ospedale 72, 09125 Cagliari, Italy

BIERBRAUER, J., Klinik für Psychiatrie und Psychotherapie II der Johann Wolfgang Goethe-Universität Frankfurt, Germany
e-mail: j-bier@gmx.de

BOULAY, D., Sanofi-Synthélabo, 31 ave P. Vaillant-Couturier, 92220 Bagneux, France

CAINE, S.B., Alcohol and Drug Abuse Research Center, McLean Hospital, Harvard Medical School, Belmont, Massachusetts USA

CEPEDA, C., Mental Retardation Research Center, 760 Westwood Plaza NPI 58–258, University of California at Los Angeles, School of Medicine, Los Angeles, CA 90024, USA
e-mail: ccepeda@mednet.ucla.edu

CHÉRAMY, A., Chaire de Neuropharmacologie, INSERM U 114, Collège de France, 11, place Marcelin Berthelot, 75231 Paris Cedex 05, France

CHESSELET, M.-F., Department of Neurology, UCLA School of Medicine, 710 Westwood Plaza, Los Angeles, CA 90095-1769, USA
e-mail: mchessel@mednet.ucla.edu

CORSINI, G.U., Department of Neuroscience, Section of Pharmacology, University of Pisa, Via Roma 55, 56126 Pisa, Italy
e-mail: gcorsini@drugs.med.unipi.it

DEPOORTERE, R., Sanofi-Synthélabo, 31 ave P. Vaillant-Couturier, 92220 Bagneux, France
e-mail: ronan.depoortere@sanofi-synthelabo.com

DIANA, M., Department of Drug Sciences, University of Sassari, Via Muroni 23/a, 07100 Sassari, Italy

DI CHIARA, G., Department of Toxicology, University of Cagliari, Via Ospedale 72, 09125 Cagliari, Italy
e-mail: diptoss@tin.it

EVERITT, B.J., Department of Experimental Psychology, and MRC Co-operative in Brain, Behaviour and Neuropsychiatry, University of Cambridge, Downing Street, Cambridge CB2 3 EB, UK

GLOWINSKI, J., Chaire de Neuropharmacologie, INSERM U 114, Collège de France, 11, place Marcelin Berthelot, 75231 Paris Cedex 05, France
e-mail: marie-helene.levi@college-de-france.fr

GOLD, L.H., Department of Neuropharmacology, CVN-7, The Scripps Research Institute, 10550 North Torrey Pines Road, La Jolla, CA 92037, USA

HILWERLING, L., Klinik für Psychiatrie und Psychotherapie II der Johann Wolfgang Goethe-Universität Frankfurt, Germany

JOEL, D., Department of Psychology, Tel Aviv University, Ramat-Aviv, Tel Aviv 69978, Israel
e-mail: djoel@post.tau.ac.il

KEMEL, M.L., Chaire de Neuropharmacologie, INSERM U 114, Collège de France, 11, place Marcelin Berthelot, 75231 Paris Cedex 05, France

KONRADI, C., Molecular and Developmental Neuroscience Laboratory and Department of Psychiatry, Massachusetts General Hospital and Harvard Medical School, Boston, MA 02114, USA
e-mail: konradi@helix.mgh.harvard.edu

KOOB, G.F., Department of Neuropharmacology, CVN-7, The Scripps Research Institute, 10550 North Torrey Pines Road, La Jolla, CA 92037, USA
e-mail: gkoob@scripps.edu

LEVINE, M.S., Mental Retardation Research Center, 760 Westwood Plaza NPI 58–258, University of California at Los Angeles, School of Medicine, Los Angeles, CA 90024, USA
e-mail: mlevine@mednet.ucla.edu

LEYSEN, J.E., Janssen Pharmaceutica n.v., Turnhoutseweg 30, 2340 Beerse, Belgium
e-mail: jleysen2@janbe.jnj.com

MAGGIO, R., Department of Neuroscience, Section of Pharmacology, University of Pisa, Via Roma 55, 56126 Pisa, Italy

MORELLI, M., Department of Toxicology, University of Cagliari, Palazzo delle Scienze, Via Ospedale 72, 09124, Cagliari, Italy
e-mail: micmor@tin.it

ONGINI, E., Nicox Research Institute, Via Ariosto 21, 20091 Bresso, Milano, Italy
e-mail: Ongini@nicox.com

List of Contributors

PERRAULT, G., Sanofi-Synthélabo, 31 ave P. Vaillant-Couturier, 92220 Bagneux, France

ROBBINS, T.W., Department of Experimental Psychology, and MRC Co-operative in Brain, Behaviour and Neuropsychiatry, University of Cambridge, Downing Street, Cambridge CB2 3 EB, UK
e-mail: t.robbins@psychol.cam.ac.uk

SANGER, D.J., Sanofi-Synthélabo, 31 ave P. Vaillant-Couturier, 92220 Bagneux, France

TEPPER, J.M., Center for Molecular and Behavioral Neuroscience, Rutgers, The State University of New Jersey, 197 University Avenue, Newark, NJ 07102, USA

VAGLINI, F., Department of Neuroscience, Section of Pharmacology, University of Pisa, Via Roma 55, 56126 Pisa, Italy

WEINER, I., Department of Psychology, Tel Aviv University, Ramat-Aviv, Tel Aviv 69978, Israel
e-mail: weiner@post.tau.ac.il

WILLNER, P., Centre for Substance Abuse Research, University of Wales Swansea, Swansea SA2 8PP, UK
e-mail: p.willner@swansea.ac.uk

Contents

CHAPTER 13

Electrophysiological Pharmacology of Mesencephalic Dopaminergic Neurons
M. Diana and J.M. Tepper. With 6 Figures 1

A. Introduction 1
B. Anatomical Organization 2
C. Basic Electrophysiological Properties 4
    I. Extracellular Recordings 4
    II. Intracellular Recordings 7
D. Afferents to Dopaminergic Neurons 8
    I. GABAergic Afferents 8
    II. Glutamatergic Afferents 14
    III. Cholinergic Afferents 19
    IV. Monoaminergic Afferents 20
E. Autoreceptor-Mediated Effects on Dopaminergic Neurons 22
    I. Somatodendritic Autoreceptors 22
    II. Axon Terminal Autoreceptors 24
    III. Are D_2 Autoreceptors Different from Other
        D_2 Receptors? 25
    IV. Are Autoreceptors Ubiquitous Among Dopaminergic
        Neurons? 26
    V. What Are the Physiological Roles of Autoreceptors? ... 28
F. Miscellaneous Neuropharmacology 31
    I. Gamma-Hydroxybutyric Acid 31
    II. Glycine 31
    III. Neuropeptides 31
G. Acute and Chronic Effects of Antipsychotics on
   Dopaminergic Neurons 33

The first 12 chapters of this monograph are found in the companion volume (HEP 154/I). Its contents are reprinted immediately after those of the present volume.

	I. Differences Between Effects of Typical and Atypical Antipsychotics	33
	II. Effects of Chronic Antipsychotic Drug Administration – The Depolarization Block Hypothesis	34
H.	Dopaminergic Neurons and Drugs of Abuse: Acute and Chronic Studies	36
	I. Acute Effects of Drugs of Abuse on Dopaminergic Neurons	36
	II. Chronic Effects of Drugs of Abuse on Dopaminergic Neurons	38
	III. Withdrawal Following Chronic Administration	40
I.	Conclusions	42
	References	43

CHAPTER 14

Presynaptic Regulation of Dopamine Release
J. GLOWINSKI, A. CHÉRAMY, and M.-L. KEMEL. With 4 Figures 63

A.	Introduction	63
B.	Interactions Between Heteroreceptors or Heteroreceptors and D_2 Autoreceptors Present on Dopaminergic Nerve Terminals	64
C.	Role of Diffusible Messengers in the Presynaptic Control of Striatal Dopaminergic Transmission	67
D.	Local Circuits Involved in the Control of DA Transmission in Striatal Compartments	70
	I. Similarities and Differences in the Presynaptic Regulation of DA Release in Striatal Compartments	72
	II. The GABA- and Dynorphin-Dependent Inhibitions of DA Transmission Triggered by Acetylcholine Occur in Two Distinct Matrix Territories	73
	III. NMDA-Dependent Local Inhibitory Circuits of DA Transmission Occur in Both Striatal Compartments and Involve GABA and Dynorphin	74
	IV. Facilitation by DA of the NMDA-Sensitive Local Inhibitory Circuits Involved in the Presynaptic Regulation of DA Release in Striatal Compartments	74
E.	Conclusions	78
	References	78

CHAPTER 15

Dopamine – Acetylcholine Interactions
E. ACQUAS and G. DI CHIARA. With 9 Figures 85

A.	Introduction	85
B.	Dopamine – Acetylcholine Interactions in the Basal Ganglia	85

I. Early Studies	86
II. Direct D_1 Receptor-Mediated Facilitation of Striatal Acetylcholine Transmission	89
III. Separate Transduction Pathways for D_1 and D_2 Receptor-Mediated Influences on Acetylcholine Transmission	92
IV. Independent Gating of Input to Striatal Acetylcholine Neurons by Dopamine Receptor Subtypes	92
V. Relative Role of D_1 and D_2 Receptors in the Control of Striatal Acetylcholine Function	95
VI. Nicotinic Receptors and Dopamine Neurons	98
VII. Actions of Nicotine on Dopamine Function	98
VIII. Mechanism of Nicotine Actions on Dopamine Function	100
C. Dopamine – Acetylcholine Interactions Outside the Basal Ganglia	103
I. Dopaminergic Regulation of Cortical and Hippocampal Acetylcholine Transmission	104
References	105

CHAPTER 16

Dopamine – Glutamate Interactions
C. KONRADI, C. CEPEDA, and M.S. LEVINE. With 2 Figures 117

A. Introduction	117
B. Neuropharmacological Interactions	117
I. Dopamine and Glutamate Act Within the Same Neuronal Circuits	117
II. Dopamine and Glutamate Receptors in the Striatum	118
III. Reciprocal Release Regulation of Dopamine and Glutamate by Dopamine Receptors and Ionotropic Glutamate Receptors	119
IV. Reciprocal Regulation of Receptor Synthesis	120
V. Glutamate Regulates the Synthesis of Dopamine in Striatal Synaptosomes	120
VI. Glutamate and Dopamine Are Co-released from Dopamine Neurons	121
C. Intraneuronal Interactions	121
I. Dopamine Receptors and NMDA Receptors Cooperatively Modulate Gene Expression	121
D. Electrophysiological Interactions	122
I. Striatal Organization	122
II. Dopamine Modulates Glutamate Inputs	124
E. Interaction of the Dopamine and Glutamate Neurotransmitter Systems in Other Brain Areas	126

F. Functional Consequences of the Interaction of the Glutamate and Dopamine Systems 127
References 128

CHAPTER 17

Dopamine – Adenosine Interactions
M. MORELLI, E. ACQUAS, and E. ONGINI. With 2 Figures 135

A. Adenosine in the CNS 135
    I. Receptor Distribution 135
    II. Adenosine in CNS Pathology 137
B. Pharmacology of Adenosine Receptors 138
C. Adenosine – Dopamine Interactions 139
    I. Dopamine D_1 and Adenosine Receptors 140
    II. Dopamine D_2 and Adenosine Receptors 141
    III. Modulation of Dopamine Release 143
D. Therapeutic Implications 144
References 145

CHAPTER 18

Dopamine – GABA Interactions
M.-F. CHESSELET 151

A. Introduction 151
B. The Anatomical Relationship Between Nigrostriatal Dopamine and GABAergic Neurons 151
    I. Substantia Nigra 151
    II. Striatum 152
    III. Other Basal Ganglia Regions 152
C. Functional Interactions Between GABA and Nigrostriatal Dopaminergic Neurons 153
    I. Striatum 153
        1. Effects of GABA on Dopaminergic Neurons 153
        2. Effects of Dopamine on GABAergic Output Neurons 153
        3. Dopaminergic Regulation of Striatal GABAergic Interneurons 155
        4. Dopaminergic Regulation of Striatal GABA Release ... 155
    II. Dopamine – GABA Interactions in the Globus Pallidus ... 156
    III. GABA – DA Interactions in the Internal Pallidum 158
    IV. DA – GABA Interactions in the Substantia Nigra 159
    V. Functional Implications of GABA – DA Interactions Within the Basal Ganglia 160

Contents   XV

D. DA – GABA Interactions in the Mesolimbic Pathway   161
E. DA – GABA Interactions in the Mesocortical System   163
References   165

CHAPTER 19

Dopamine – Its Role in Behaviour and Cognition in Experimental Animals and Humans
T.W. ROBBINS and B.J. EVERITT. With 1 Figure   173

A. Introduction   173
B. Electrophysiological and Neurocomputational Approaches   174
C. Neuropharmacological Evidence for a Role for Dopamine in
   Learning   176
     I. Overview of Results from In Vivo Monitoring Studies   176
    II. Psychopharmacological Evidence of Specific Actions of
      Dopamine on Learning and Memory   178
   III. The Possible Complication of a Role for Dopamine in
      Attentional Function   182
   IV. Models of ADHD   185
D. Working Memory   186
     I. Problems of Interpretation of the Role of the PFC in
      Working Memory   190
E. Evidence for a Role for Dopamine in Cognition in Humans   192
     I. Dopamine and Cognition in Clinical Disorders:
      Parkinson's Disease, Schizophrenia, Acute Brain Injury
      and ADHD   193
    II. Effects of Dopaminergic Drugs on Cognition in Normal
      Human Volunteers   196
F. Conclusions and Future Directions   199
References   203

CHAPTER 20

Molecular Knockout Approach to the Study of Brain Dopamine Function
G.F. KOOB, S.B. CAINE, and L.H. GOLD. With 6 Figures   213

A. Introduction   213
B. Limitations of the Knockout Approach: Compensation and
   Epistasis   216
C. Overview of the Midbrain Dopamine System in Motor Behavior
   and Reward   218
D. Overview of the Dopamine Receptor Subtypes in Motor
   Behavior and Reward   219
E. D_1 Receptor Knockouts   220

F. D_2 Receptor Knockouts	222
G. D_3 Receptor Knockouts	224
H. Knockout of the Dopamine Transporter	225
I. Knockout of Tyrosine Hydroxylase Gene	227
J. Other Knockouts	229
K. Summary and Conclusions: What We Know That We Did Not Know Before Knockouts	231
References	232

CHAPTER 21

Behavioural Pharmacology of Dopamine D_2 and D_3 Receptors: Use of the Knock-out Mice Approach
R. Depoortere, D. Boulay, G. Perrault, and D.J. Sanger.
With 4 Figures

	239
A. Introduction	239
B. Behavioural Pharmacology of DA D_2/D_3 Receptor Agonists	240
C. Correlational Studies Using DA D_2/D_3 Receptor Agonists	241
D. Behavioural Pharmacology of DA D_2/D_3 Receptor Antagonists	242
E. Dopamine D_3 Receptor Knock-out Mice	245
I. Analysis of the Phenotype of D_3 Receptor Knock-out Mice	246
II. Effects of DA Receptor Ligands in D_3 Receptor Knock-out Mice	246
1. DA D_2/D_3 Receptor Agonists	247
2. DA D_2/D_3 Receptor Antagonists	247
3. Psychostimulants	247
F. Dopamine D_2 Receptor Knock-out Mice	247
I. Analysis of the Phenotype of D_2 Receptor Knock-out Mice	248
II. Effects of DA Receptor Ligands in D_2 Receptor Knock-out Mice	248
1. DA D_2/D_3 Receptor Agonists	249
2. DA D_2/D_3 Receptor Antagonists	249
3. Psychotropic Agents	249
G. Direct Comparison Between D_2 and D_3 Receptor Knock-out Mice	250
I. Comparison of Avoidance Behaviour of D_2 and D_3 Receptor Knock-out Mice	251
II. Comparison of Effects of DA Receptor Ligands in D_2 and D_3 Receptor Knock-out Mice	252
1. Psychotropic Agents	252
2. DA Receptor Antagonists	255

H. Conclusions	256
References	257

CHAPTER 22

Dopamine and Reward
G. DI CHIARA ... 265

A. Introduction	265
B. Terminology	266
I. Reward, Reinforcer, Incentive	267
II. Motivation and Instrumental Responding	268
1. Incentive-Motivational Responding	269
2. Instrumental Responding	270
C. Early Studies: The Original and the Revised Anhedonia Hypothesis	272
D. Testing the Original Anhedonia Hypothesis	274
I. Sweet Reward	274
II. Operant Responding for Sweet Reward	280
E. The Motor Deficit Issue	281
F. Response-Reinforcement Functions	285
I. Reward Summation Studies	285
II. Intensity-Threshold Studies	286
III. Response-Reinforcement Matching Studies	287
G. Dissociating Reinforcement from Incentive-Motivation and Performance	289
H. Incentive Accounts of the Role of Dopamine in Behaviour	291
I. Stimulus-Bound Incentive Role of Dopamine?	293
II. Dopamine and Incentive Arousal	294
III. Incentive Role of Drug-Stimulated Dopamine Transmission	295
I. Associative Learning Accounts	296
I. Pavlovian Incentive Learning	296
II. Place-Conditioning Studies	301
J. An Interpretative Framework of the Role of Dopamine in Reward	304
References	309

CHAPTER 23

Molecular and Cellular Events Regulating Dopamine Neuron Survival
G.U. CORSINI, R. MAGGIO, and F. VAGLINI ... 321

A. Introduction	321
B. Mechanisms of DA Cell Death	322
C. Extraneuronal Events	323

	I.	Noradrenergic System	323
		1. NE in Experimental Parkinsonism	325
	II.	Excitatory Amino Acids	327
		1. Excitotoxicity in PD	327
		2. Excitotoxicity in Experimental Parkinsonism	329
		3. The MPTP Model	329
		a) Species Differences in MPTP Toxicity	331
		b) MPP$^+$ Kinetics	332
		c) Excitotoxicity in the MPTP Model	335
		4. Methamphetamine Toxicity	336
		5. 6-OHDA Toxicity	337
		6. Conclusions on Excitotoxicity	339
	III.	Neurotrophic Factors	340
D.	Intraneuronal Events	343
	I.	Oxidative Stress	343
	II.	Nitric Oxide	345
	III.	Apoptosis and Mitochondria	346
	IV.	Cytochrome P450 System	347
		1. Cytochrome P450 in the CNS	348
		2. P450 System and DA Neurons	349
		a) CYP 2D6	350
		b) CYP 2E1	351
		3. The P450 System in PD	351
		4. P450 in Experimental Parkinsonism	353
E.	Toxicity of Dopamine	355
	I.	DA and Apoptosis	358
F.	Conclusions About the Pathogenesis of PD	359
References	362

CHAPTER 24

Dopamine and Depression
P. WILLNER 387

A.	Introduction	387
B.	DA Function in Affective Disorders	389
	I. DA Turnover	389
	II. DA Receptors	391
	III. Neuroendocrine Studies	392
	IV. Summary	393
C.	Mood Effects of DA Agonists and Antagonists	393
	I. Psychostimulants	393
	II. DA-Active Antidepressants	394
	III. Neuroleptic-Induced Depression	396
	IV. Parkinson's Disease	397
	V. Neuroleptics as Antidepressants	397

Contents

VI. Summary	399
D. Dopaminergic Consequences of Antidepressant Treatment	399
I. DA autoreceptor Desensitization	399
II. Sensitization of D_2/D_3 Receptors	400
III. Clinical Evidence	401
IV. Summary	402
E. Dopaminergic Mechanisms in Animal Models of Depression	402
I. D_2/D_3 Receptor Sensitization as a Mechanism of Antidepressant Action	402
II. Clinical Evidence	404
III. Reciprocal Changes in DA Responses to Reward and Stress	404
IV. Summary	405
F. Conclusions	406
I. Limitations of the Dopamine Hypothesis	406
II. Syndromes or Symptoms?	407
III. The Wider Picture	408
References	409

CHAPTER 25

Dopamine in Schizophrenia
Dysfunctional Information Processing in Basal Ganglia – Thalamocortical Split Circuits

I. WEINER and D. JOEL. With 1 Figure	417
A. The Dopamine Hypothesis of Schizophrenia	417
B. Schizophrenia as a Dopamine-Dependent Dysfunctional Information Processing in Basal Ganglia – Thalamocortical Circuits	423
I. Circuit Models of Schizophrenia	425
II. The Split Circuit Model of Schizophrenia	428
1. Striatum as a Contention Scheduling Device	430
2. The Interaction Between the Striatum and the Frontal Cortex	432
3. Contention Scheduling of Goals by the Limbic Striatum	433
4. The Role of Tonic and Phasic DA in the Contention Scheduling of Goals	433
a) Tonic and Phasic DA Release	434
b) The Establishment of Goals in the Limbic Striatum	436
c) Goal Selection	436
d) Goal Maintenance and Energizing	437
e) Switching Between Goals	438
5. The Translation of Goals to Behavior	439

6. Schizophrenia 440
   a) Fronto-temporo-limbic Cortical Dysfunction and
      Dysregulation of Tonic and Phasic DA Transmission
      in Schizophrenia 442
   b) The Consequences of Fronto-temporo-limbic
      Cortical Dysfunction: Disrupted Establishment
      of Goals 442
   c) The Consequences of Dysregulation of the DA
      Input to the Limbic Striatum 444
      α) Reduced Tonic DA: Goal Selection, Activation
         and Maintenance 444
      β) Abnormal Phasic DA Release:
         Learning and Switching 446
   d) Summary: Phasic and Tonic DA Dysregulation and
      Schizophrenia Symptoms 449
References 451

CHAPTER 26

Atypical Antipsychotics
J.E. LEYSEN. With 1 Figure 473

A. Introduction 473
B. Receptor Binding Profile of Antipsychotics 474
C. Interaction with Dopamine Receptors 481
D. Interaction with 5-HT_2 and Other 5-HT Receptors . 483
E. Interaction with Various Biogenic Amine Receptors . 484
F. Future Antipsychotics 485
G. Conclusions 486
Abbreviations 487
References 487

CHAPTER 27

Sleep and Wake Cycle
J. BIERBRAUER and L. HILWERLING 491

A. Introduction 491
B. Dopaminergic Action in Sleep 492
   I. D_2 Antagonists 492
   II. D_2 Agonists 493
   III. D_1 Antagonists 494
   IV. D_1 Agonists 494
   V. More Specific Studies 494
   VI. Catecholaminergic Pathway Modulation 495
   VII. Temperature Regulation 495

C. Pharmacological Interactions	496
I. Serotonin	496
II. Adrenergic System	497
III. Acetylcholine	498
IV. Histamine	498
1. H_1 Receptor	499
2. H_2 Receptor	499
3. H_3 Receptor	499
V. GABAergic System	499
D. Summary	500
References	501
Subject Index	507

Contents of Companion Volume 154/I

Dopamine in the CNS I

CHAPTER 1
Brain Dopamine – A Historical Perspective
O. HORNYKIEWICZ 1

CHAPTER 2
Birth of Dopamine: A Cinderella Saga
A. CARLSSON 23

CHAPTER 3
The Place of the Dopamine Neurons Within the Organization of the Forebrain
S.N. HABER 43

CHAPTER 4
Synaptology of Dopamine Neurons
S.R. SESACK 63

CHAPTER 5
D_1-Like Dopamine Receptors: Molecular Biology and Pharmacology
H.B. NIZNIK, K.S. SUGAMORI, J.J. CLIFFORD, and J.L. WADDINGTON 121

CHAPTER 6
Understanding the Function of the Dopamine D_2 Receptor: A Knockout Animal Approach
S. TAN, B. HERMANN, C. IACCARINO, M. OMORI, A. USIELLO, and
E. BORRELLI 159

CHAPTER 7
The Dopamine D_3 Receptor and Its Implications in Neuropsychiatric Disorders and Their Treatments
P. SOKOLOFF and J.-C. SCHWARTZ 185

CHAPTER 8
Dopamine D$_4$ Receptors: Molecular Biology and Pharmacology
O. CIVELLI 223

CHAPTER 9
Signal Transduction by Dopamine D$_1$ Receptors
J.-A. GIRAULT and P. GREENGARD 235

CHAPTER 10
The Dopamine Transporter: Molecular Biology, Pharmacology and Genetics
C. PIFL and M.G. CARON 257

CHAPTER 11
Cellular Actions of Dopamine
D.J. SURMEIER and P. CALABRESI 299

CHAPTER 12
Dopamine and Gene Expression
E.J. NESTLER 321

Subject Index 339

CHAPTER 13

Electrophysiological Pharmacology of Mesencephalic Dopaminergic Neurons

M. DIANA and J.M. TEPPER

We dedicate this chapter to the memory of Dr. Stephen J. Young, mentor, colleague and friend. For decades Steve contributed tirelessly and selflessly to the advancement of the science of countless students, colleagues and scientists around the world. His presence is sorely missed.

A. Introduction

In spite of the fact that actions of dopamine, as a neurotransmitter in its own right, were foreseen as early as the 1930s (BLASCHKO 1939) and explicitly postulated in the 1950s (CARLSON et al. 1958), it took over a decade more to begin to explore the electrophysiological features, characteristics, and responsiveness to drugs of central dopaminergic neurons (BUNNEY et al. 1973b; GROVES et al. 1975). In the 1960s much effort was employed attempting to map the location of catecholamine neurons in the mammalian central nervous system. The use of the histofluorescence technique (FALCK et al. 1962) coupled with lesion experiments enabled anatomists to locate dopaminergic cell bodies in the mesencephalon (ANDEN et al. 1964; BERTLER et al. 1964). Subsequent work (DAHLSTROM and FUXE 1964; ANDEN et al. 1965; UNGERSTEDT 1971) refined and extended those initial and pioneering findings and formed the basis for modern anatomical (see SESACK this volume for an updated view), biochemical, and electrophysiological investigation of central dopaminergic neurons.

Physiological studies of central dopaminergic neurons began with in vivo extracellular recordings which described the basic electrophysiological and pharmacological properties of mesencephalic dopaminergic neurons (BUNNEY et al. 1973a,b). From the very beginning, the unusually long duration action potential, the persistent low frequency of spontaneous discharge, including unusually low frequency burst firing and slow conduction velocity (DENIAU et al. 1978; GUYENET and AGHAJANIAN 1978), together with inhibitory responses to dopamine and dopamine agonists such as apomorphine and amphetamine (BUNNEY et al. 1973a,b; GROVES et al. 1975) have been unanimously recognized as the extracellular, electrophysiological "fingerprint" of dopamine-containing neurons in the midbrain.

There are several compelling reasons for studying central dopaminergic systems over and above their uniqueness and intrinsically interesting properties. Chief among them is the central role that they play in mediating the effects of antipsychotic drugs, and in the neurobiology of many psychotropic drugs, drug abuse, and addiction. In this chapter we review some of the principal aspects of the neurobiology of dopaminergic neurons as they relate to the pharmacology of psychotherapeutic drugs and drugs of abuse. Electrophysiological studies of dopaminergic neurons have provided important evidence implicating these cells as components of systems of fundamental importance in normal CNS functioning as well as in various pathological conditions including degenerative disorders such as Parkinson's disease, schizophrenia, and drug addiction. Controversy and disagreement with respect to the interpretation of data is common in the scientific literature, and the literature on the neurophysiology and neuropharmacology of dopaminergic neurons is no exception. Where relevant, we will point out some of the current areas of contention and discuss them in light of recent findings.

B. Anatomical Organization

Although some dopaminergic neurons are located elsewhere in the brain (i.e., tuberoinfundibular dopaminergic neurons that regulate the release of prolactin from the anterior pituitary gland; MOORE et al. 1987 and in the retina where they regulate receptive field size by altering the conductance of electrotonic synapses e.g., TERANISHI et al. 1983), most of the dopaminergic neurons in the central nervous system are located in the midbrain. In the present chapter, we will focus on the dopaminergic pathways originating in the mesencephalon which have been most extensively studied and whose function has been most convincingly linked to human psycho- and neuropathology. Although the topography of their inputs and outputs differs somewhat, the mesencephalic dopaminergic neurons exist for the most part as a single continuous and contiguous group of cells, and the axon of many of these neurons collateralizes to one or more additional target structures (FALLON 1981). However, historically the midbrain dopaminergic cell groups and their projections have been functionally subdivided into three systems: the nigrostriatal, mesolimbic, and mesocortical dopaminergic systems.

Most of the cell bodies of origin of the nigrostriatal dopaminergic system are located in the substantia nigra pars compacta (A9 in the terminology of DAHLSTROM and FUXE 1964) with the remainder being located in the pars reticulata. The neurons are medium to large sized, multipolar, fusiform, or polygonal in shape and emit 3–5 large, rapidly tapering smooth dendrites. There is no local axon collateral arborization within the substantia nigra (JURASKA et al. 1977; TEPPER et al. 1987b). These neurons send their axons anterior and rostral to the neostriatum where they form Gray's type II symmetrical synapses, mainly on the dendrites or the necks of the dendritic spines of the striatal medium spiny projection neurons (PICKEL et al. 1981; FREUND et al. 1984) (See Fig. 1).

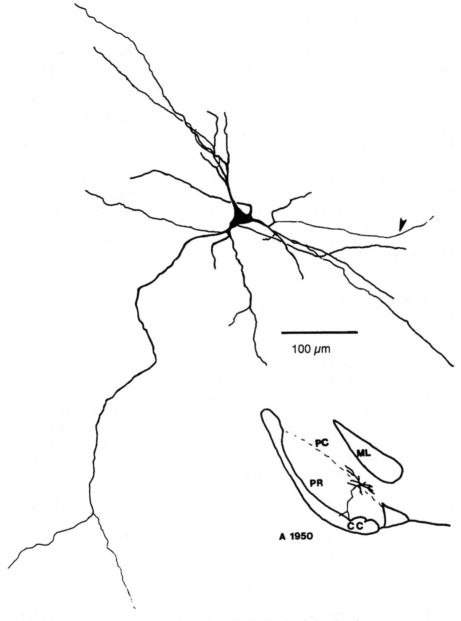

Fig. 1. Drawing tube reconstruction of an HRP-filled substantia nigra pars compacta neuron that was antidromically activated from both ipsilateral globus pallidus and neostriatum. The *inset* is drawn approximately to scale to illustrate the location of the dendritic arborization of the neuron within substantia nigra. The coordinates refer to the location of the coronal section from the atlas of KONIG and KLIPPEL (1963). The *arrow* points to the proximal portion of the axon, which emerges from a dendrite. PC, pars compacta, PR, pars reticulata, ML, medial lemniscus. (Reproduced from TEPPER et al. 1987b with permission of the publishers)

Most of the cells of origin of the mesolimbic dopaminergic system are located medial to the main body of the substantia nigra pars compacta in the ventral tegmental area (A10 in the terminology of DAHLSTROM and FUXE 1964) and medial substantia nigra. These neurons project to the ventral part of the striatal complex, including the nucleus accumbens (both core and shell) and the olfactory tubercle.

The mesocortical dopaminergic projection arises from the mediodorsal, most parts of the pars compacta and ventral tegmental areas (VTAs) and innervates the prefrontal, cingulate, perirhinal, and entorhinal cortices in a loosely topographical manner (for review see FALLON and LAUGHLIN 1995).

The most caudal, lateral, and superior extension of the midbrain dopaminergic cell group, and the smallest of the three cell groups, is termed the retrorubral field (A8 in the terminology of DAHLSTROM and FUXE 1964) and innervates largely striatal regions. For a more detailed description of the anatomical organization of mesencephalic dopaminergic neurons in rat, the reader is referred to other chapters in this volume and to the excellent review by FALLON and LAUGHLIN (1995).

C. Basic Electrophysiological Properties

I. Extracellular Recordings

In in vivo extracellular recordings from anesthetized adult rats, midbrain dopaminergic neurons fire spontaneously at slow rates, averaging around 4 spikes per second (BUNNEY et al. 1973b; DENIAU et al. 1978; GUYENET and AGHAJANIAN 1978; BUNNEY 1979; TEPPER et al. 1982). Dopaminergic neurons exhibit three distinct modes or patterns of firing. The most common pattern of activity in vivo is a random, or occasional mode of firing characterized by an initial, prolonged trough in the autocorrelation function representing a long post-firing inhibition. The next most common firing pattern is a very regular, pacemaker-like firing, characterized by very regular interspike intervals with a low coefficient of variation, and a lack of bursting. The third and least common mode of firing is bursty firing, characterized by stereotyped bursts of 2–8 action potentials in which the first intraburst interspike interval is around 60 ms, followed by progressively increasing interspike intervals and progressively decreasing spike amplitudes (WILSON et al. 1977; GRACE and BUNNEY 1984a,b; TEPPER et al. 1995). In anesthetized, unanesthetized, and freely moving rats (FREEMAN et al. 1985; DIANA et al. 1989), dopaminergic neurons often switch between different firing modes, and these firing patterns can best be thought of as a existing along a continuum, with the pacemaker-like firing on one end and bursty firing on the other (Fig. 2). The bursty mode of firing has generated particular interest as action potentials fired in bursts have been linked to an increased overflow of dopamine in terminal areas compared to an equal number of evenly spaced action potentials (GONON 1988) which could alter dopaminergic neurotransmission in axonal terminal fields qualitatively

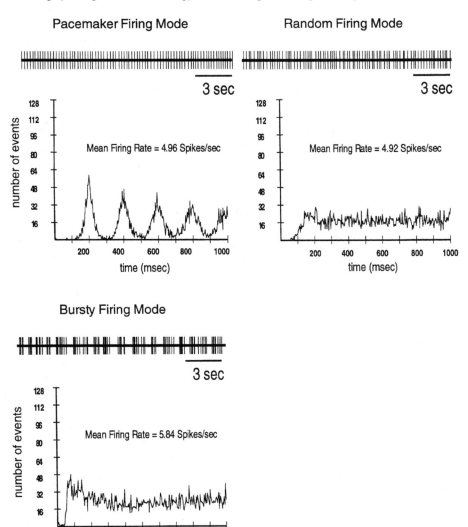

Fig. 2. Autocorrelograms of representative neurons exhibiting the three firing modes of dopaminergic neurons in vivo. Above each autocorrelogram is the first approximately 15 s of the spike train used to create the autocorrelogram. Bin width = 3 ms. (Reproduced from TEPPER et al. 1995 with permission of the publishers)

as well as quantitatively (e.g., GONON 1997), and which may play a role in the dendritic release of dopamine (BJORKLUND and LINDVALL 1975; GROVES et al. 1975; CHERAMY et al. 1981) as well.

Anesthesia affects the expression of the three firing patterns and their responsiveness to drugs (MEREU et al. 1984b; KELLAND et al. 1990a). Although all three firing patterns are expressed in unanesthetized freely moving or immobilized preparations, burst firing is more common in unanesthetized rats than under any anesthetic (WILSON et al. 1977; FREEMAN et al. 1985; DIANA et al. 1989; KELLAND et al. 1990a). Different anesthetics also differentially affect the distribution of firing patterns; burst firing is expressed least under urethane, is intermediate under chloral hydrate, and is expressed most under ketamine anesthesia with an incidence almost equal to that observed in unanesthetized preparations (KELLAND et al. 1990a).

The extracellularly recorded action potential of midbrain dopaminergic neurons is of unusually long duration, almost always greater than 2 ms and sometimes as much as 5 ms depending on the level of depolarization of the neuron, and often displays a notch or inflection on the initial rising phase termed an initial segment-somatodendritic (IS-SD) break (BUNNEY et al. 1973b; GUYENET and AGHAJANIAN 1978; GRACE and BUNNEY 1983b) by analogy to a similar phenomenon in spinal motoneurons (COOMBS et al. 1957; ECCLES 1957).

Early studies using antidromic activation of mesencephalic dopaminergic neurons from terminal fields in striatum revealed that these neurons have very slow conduction velocities (~0.5 m/s in rat; DENIAU et al. 1978; GUYENET and AGHAJANIAN 1978) consistent with their thin (less than 1 μm) and unmyelinated nature (TEPPER et al. 1987b). Most of the time (64%; TRENT and TEPPER 1991) the antidromic response consists of a small spike, assumed to be an initial segment (IS) spike (COOMBS et al. 1957; ECCLES 1957; GUYENET and AGHAJANIAN 1978). Multiple discrete antidromic latencies are often present, presumably reflecting the highly branched nature of the terminal field, giving rise to multiple sites of initiation of the antidromic spike (COLLINGRIDGE et al. 1980; TEPPER et al. 1984a).

Although many of the early extracellular recording studies focused on dopaminergic neurons in substantia nigra, the majority of subsequent studies revealed that with a few exceptions, VTA neurons exhibit electrophysiological and pharmacological properties that are similar or identical to those of substantia nigra dopaminergic neurons in most ways (e.g., BUNNEY 1979; WANG 1981a–c; FREEMAN et al. 1985; MEREU et al. 1985; FREEMAN and BUNNEY 1987; CLARK and CHIODO 1988).

The most commonly reported difference between A9 and A10 dopaminergic neurons has to do with the pattern and rate of spontaneous activity in vivo. Although A10 neurons exhibit the same range of firing patterns as A9 neurons, many studies report that the incidence of burst firing is greater among VTA neurons than substantia nigra pars compacta neurons (GRENHOFF et al. 1986, 1988; CHARLETY et al. 1991). Interestingly, it does not appear as if the characteristics of the burst firing are different; most of the burst parameters are the same among A9 and A10 neurons, but the proportion of A10 neurons firing in the bursty mode is greater (CHIODO et al. 1984; GRENHOFF et al. 1986, 1988; CHARLETY et al. 1991). Despite this consistent difference, the mean firing rates of A9 and A10 dopaminergic neurons are usually reported to be about the same (e.g., WANG 1981a,b; GRENHOFF et al. 1986, 1988; FREEMAN and BUNNEY 1987; GARIANO et al. 1989b; SHEPARD and BUNNEY 1988; CHARLETY et al. 1991; but see also CHIODO et al. 1984). One reason put forth for the difference in proportion of burst firing neurons is a difference in autoreceptor number and/or sensitivity (CHIODO et al. 1984), but for reasons discussed below (see discussion in Sect. E.IV) this does not seem the most likely explanation. Rather, as suggested previously (e.g., GRENHOFF et al. 1988) a difference in afferent inputs may be responsible. Various afferents to midbrain dopaminergic neurons and the effects they have on firing rate and pattern are discussed below (see Sect. D). In that context, it is interesting to note that one of the most striking qualitative differences between A9 and A10 neurons is that dopaminergic neurons in the VTA appear to receive a significantly greater number of glutamatergic asymmetric, presumably excitatory, synaptic contacts than those in the substantia nigra (SMITH et al. 1996).

II. Intracellular Recordings

The first data from intracellular recordings from identified rat dopaminergic neurons were published by GRACE and BUNNEY in a memorable series of papers in the early 1980s (GRACE and BUNNEY 1980, 1983a,b, 1984a,b). This accomplishment was rendered even more impressive by the fact that these were in vivo recordings from the substantia nigra, a structure deep in the midbrain where the dopaminergic neurons are situated in a layer only a few cells thick. These recordings verified that the unusually long duration action potential was not an artifact of damage or extracellular recording. The action potential had an inflection that, upon digital differentiation, was virtually identical to the IS-SD break previously noted in extracellular recordings. Furthermore, the small antidromic spike observed extracellularly could be seen intracellularly and converted to a full spike by injecting depolarizing current, consistent with its tentative extracellular identification as an IS spike. Spontaneous spikes were seen to arise from a slow depolarization and were followed by large amplitude, long-lasting spike afterhyperpolarizations. Application of hyperpolarizing current pulses revealed a slowly developing inward rectification, and the episodes of slow-burst firing first seen with extracellular recordings were observed to occur superimposed upon large spontaneous depolarizations (GRACE and BUNNEY 1980, 1983a,b, 1984a,b).

Subsequent in vitro recordings revealed that the long, slow afterhyperpolarization was due to a calcium-activated potassium conductance and that the slowly developing inward rectification was blocked by tetraethylammonium (TEA), suggesting its mediation by I_h (KITA et al. 1986). The slow afterhyperpolarization is very sensitive to apamin, and plays a significant role in regulating the firing pattern of dopaminergic neurons (SHEPARD and BUNNEY 1988; PING and SHEPARD 1996). A number of pharmacologically and electrophysiologically distinct low- and high-threshold calcium conductances have been identified in midbrain dopaminergic neurons (e.g., LLINÁS et al. 1984; NEDERGAARD et al. 1988, 1993; NEDERGAARD and GREENFIELD 1992; KANG and KITAI 1993a,b; CARDOZO and BEAN 1995; GALARRAGA and BARGAS 1995; WILSON and CALLAWAY 2000). Dopaminergic neurons also exhibit several different types of voltage-dependent potassium channels (SILVA et al. 1990). A transient, 4-aminopyridine (4-AP)-sensitive, TEA-insensitive A-current that is largely inactivated at the most stable subthreshold membrane potentials is expressed, as is a sustained outward current and at least two different types of calcium-activated potassium current (SILVA et al. 1990; CARDOZO and BEAN 1995), plus the inwardly rectifying I_h mentioned above. Although the conductances responsible for the bursty and random firing patterns have not yet been identified conclusively, it appears that the pacemaker firing pattern emerges as a result of an intrinsic membrane potential oscillation, resulting from a low threshold, non-inactivating calcium conductance, and a calcium-activated potassium conductance (HARRIS et al. 1989; YUNG et al. 1991; NEDERGAARD and GREENFIELD 1992; KANG and KITAI 1993a,b; WILSON and CALLAWAY 2000). A single action potential is fired at the peak of the oscillation and the resulting calcium-dependent spike afterhyperpolarization is sufficient to prevent any further spiking. Although results from early studies suggested that the dopaminergic cell bodies were electrically inexcitable (GRACE and BUNNEY 1983b), excised patch clamp recordings from the soma and dendrites of dopaminergic neurons have revealed voltage-gated inward and outward currents underlying active propagation of spikes in the soma and dendrites of these neurons (HAUSSER et al. 1995).

The biggest difference between dopaminergic neurons recorded in vivo and in vitro is the absence of the random or bursty firing patterns in the slice preparation, likely due to the loss of afferents in the slice (GRACE 1987; LACEY et al. 1989; but see also MEREU et al. 1997). Another difference is the higher input resistance observed in vitro (70–250MΩ; KITA et al. 1986) compared to in vivo (18–35MΩ; GRACE and BUNNEY 1983a) also presumably due to the reduced number of functional afferents in the slice preparation (Fig. 3).

D. Afferents to Dopaminergic Neurons

I. GABAergic Afferents

The vast majority of afferent boutons synapsing on dopaminergic perikarya and dendrites in substantia nigra, perhaps as much as 70%–90%, are γ-aminobutyric acid (GABA)ergic. Most of the GABAergic input originates from the striatum, globus pallidus, and the pars reticulata of the substantia nigra (RIBAK et al. 1976, 1980; SOMOGYI et al. 1981; NITSCH and RIESENBERG

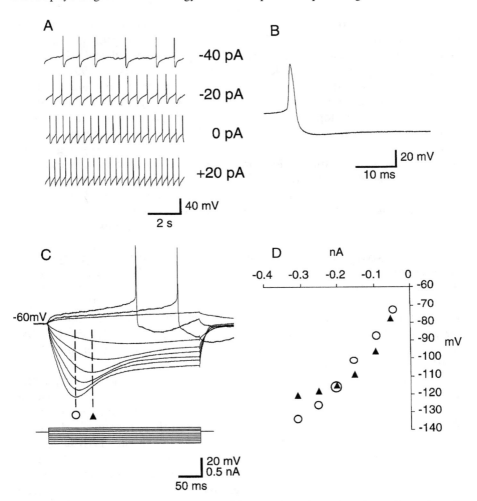

Fig. 3A–D. Electrophysiological identification of substantia nigra dopaminergic neurons in vitro. **A** Spontaneously active dopaminergic neuron firing in the typical pacemaker-like mode seen in vitro. Constant current injection of hyper- and depolarizing pulses manipulated pacemaker-like firing between 0.8 and 4 Hz. Action potential amplitudes are truncated due to aliasing. **B** Action potentials were of long duration (>2 ms) and exhibited large afterhyperpolarizations. **C** Intracellular injection of current pulses revealed a slow depolarizing ramp potential in the depolarizing direction and a strong time-dependent inward rectification when the membrane was hyperpolarized. **D** Current–Voltage plots show nearly linear slope and minimal inward rectification at the onset of hyperpolarizing current pulses (*open circles*) and a much more pronounced slowly activating inward rectification when I_h begins to activate after about 100 ms (*solid triangles*). (Reproduced from IRIBE et al. 1999 with permission of the publishers)

1988; SMITH and BOLAM 1989; TEPPER et al. 1995). Dopaminergic neurons express both of the two principal subtypes of GABA receptor, GABA$_A$ and GABA$_B$ receptors, and are quite effectively hyperpolarized by bath application of GABA$_A$- or GABA$_B$-selective agonists in vitro (LACEY 1993).

There is a massive GABAergic input to the substantia nigra from the neostriatum, both the dorsal and ventral parts. Although most of these fibers synapse on the non-dopaminergic neurons in the pars reticulata (GROFOVA and RINVIK 1970), there are monosynaptic inputs to dopaminergic neurons (SOMOGYI et al. 1981; BOLAM and SMITH 1990). Early in vivo recording studies showed that striatal stimulation produces monosynaptic inhibitory postsynaptic potentials (IPSPs) that could be blocked by picrotoxin in substantia nigra, thus suggesting that striatonigral inhibition was mediated by GABA$_A$ receptors; however, the neurons were not identified in these studies and appear to have been pars reticulata GABAergic neurons (PRECHT and YOSHIDA 1971; YOSHIDA and PRECHT 1971).

Later in vivo intracellular recording studies from identified substantia nigra dopaminergic and non-dopaminergic neurons also revealed a monosynaptic inhibitory postsynaptic potential evoked by striatal stimulation that is also mediated by a GABA$_A$ receptor (GRACE and BUNNEY 1985), and the striatal-induced inhibition of antidromically identified nigrostriatal dopaminergic neurons recorded extracellularly in vivo is abolished by the GABA$_A$ receptor antagonist, bicuculline, but not by the GABA$_B$ receptor antagonist, CGP-55845 A (PALADINI et al. 1999a).

In contrast, in vitro studies show that both GABA$_A$ and GABA$_B$ IPSPs are elicited in substantia nigra and VTA dopaminergic neurons following stimulation of various places within the slice (HAUSSER and YUNG 1994), although it is difficult to be certain of the origin of these responses. However, activation of D$_1$ receptors in substantia nigra has been shown to selectively facilitate GABA$_B$ responses elicited by high frequency trains of stimuli delivered locally to dopaminergic neurons in vitro (CAMERON and WILLIAMS 1993). Since only the striatonigral afferents to nigra are known to express D$_1$ receptors (HARRISON et al. 1990), these data suggest that at least some of the GABA$_B$ IPSPs are mediated via the striatonigral pathway (CAMERON and WILLIAMS 1993). One possible explanation for the different results obtained in vivo and in vitro is that most of the in vivo studies used single-pulse stimuli, whereas CAMERON and WILLIAMS (1993) used trains. However, attempts to evoke GABA$_B$-mediated responses in vivo by stimulating the striatum with high frequency trains similar to those used in vitro were unsuccessful (PALADINI et al. 1999a). It is also possible that for some reason the stimulus-evoked release of GABA has better access to GABA$_B$ receptors in the slice preparation than it does in vivo, perhaps because of reduced GABA uptake, or because the stimulation in vitro causes activation of a population of GABAergic afferents that is not activated in vivo. Along these lines it is interesting to note that spontaneous miniature IPSPs in dopaminergic neurons appear to be exclusively GABA$_A$-mediated (HAUSSER and YUNG 1994).

Although the origin of the GABA$_B$ responses in vitro remain unclear, the bulk of the data suggest that in vivo, striatal GABAergic inhibition of dopaminergic neurons is mediated largely or exclusively by GABA$_A$ receptors.

There is also a significant input to substantia nigra from globus pallidus. Although the pallidal projection also appears to terminate preferentially on non-dopaminergic neurons of the substantia nigra pars reticulata (SMITH and BOLAM 1989), there is also a significant projection to pars compacta (HATTORI et al. 1975). Stimulation of the globus pallidus elicits monosynaptic IPSPs in dopaminergic neurons in vivo (TEPPER et al. 1987b), and like striatal-evoked inhibition, inhibition of nigrostriatal neurons evoked by electrical stimulation of the globus pallidus can be completely blocked by GABA$_A$, but not GABA$_B$ antagonists (PALADINI et al. 1999a).

The third major GABAergic input to dopaminergic neurons arises from axon collaterals of pars reticulata neurons. GRACE and colleagues (GRACE and BUNNEY 1979, 1985; GRACE et al. 1980) provided an important clue to understanding synaptic responses in substantia nigra by showing that there is a reciprocal relation between the spontaneous firing of non-dopaminergic neurons in the pars reticulata and dopaminergic neurons of the pars compacta. A second important finding was that very low intensity stimulation of neostriatum produced excitation of dopaminergic neurons (GRACE and BUNNEY 1985). These data were interpreted to indicate that there exists a monosynaptic pathway between a population of GABAergic neurons in pars reticulata and dopaminergic neurons in pars compacta.

The pars reticulata neuron observed to fire reciprocally with dopaminergic neurons in vivo in extracellular recordings was not identified in the first studies except to note that the neurons fired between 15 and 40Hz, exhibited brief-duration (~0.5ms) spikes, were excited by tail pinch, were more sensitive to inhibition by GABA than dopaminergic neurons, could not be antidromically activated from thalamus, and comprised a subpopulation of non-dopaminergic pars reticulata neurons (GRACE and BUNNEY 1979; GRACE et al. 1980). However, subsequent reports tentatively identified the neuron as an interneuron (e.g., GRACE and BUNNEY 1985, 1986; SMITH and GRACE 1992; GRACE et al. 1997). This suggestion of a class of pars reticulata interneurons that mediate a number of indirect effects on dopaminergic neurons has by now been generally accepted and is widely cited by a number of physiologists and pharmacologists (e.g., MEREU and GESSA 1985; JOHNSON and NORTH 1992; SANTIAGO and WESTERINK 1992; ZHANG et al. 1992, 1993). However, although suggested on the basis of Golgi staining studies (e.g., SCHWYN and FOX 1974; JURASKA et al. 1977; FRANCOIS et al. 1979) the existence of one or more classes of nigral interneurons has never been conclusively identified, an admittedly difficult task.

Pars reticulata projection neurons that send their main axons to tectum or thalamus issue axon collaterals within both substantia nigra pars reticulata and pars compacta (DENIAU et al. 1982; GROFOVA et al. 1982). These collaterals synapse on other non-dopaminergic pars reticulata neurons (DENIAU et al.

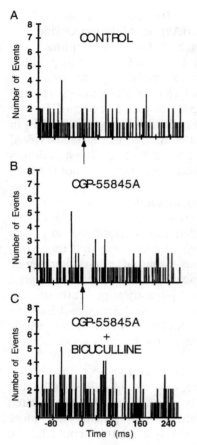

Fig. 4A–C. Presynaptic inhibitory $GABA_B$ receptors present on the terminals of local collaterals of pars reticulata nigrothalamic neurons are responsible for masking the inhibitory effects of antidromic activation of nigrothalamic neurons on dopaminergic neurons. The presynaptic inhibition is unmasked by local application of the selective $GABA_B$ receptor antagonist, CGP-55845 A. **A** Stimulation of thalamus (1.0 mA) fails to affect the firing of a nigrostriatal dopaminergic neuron. **B** Application of CGP-55845 A reveals an inhibition (suppression to 0% of control for 24 ms duration). **C** Application of bicuculline together with CGP-55845 A abolishes the unmasked inhibition. Peri stimulus time histograms (PSTH) consist of 100 trials each with 2-ms bin width. (Reproduced from Paladini et al. 1999a with permission of the publishers)

1982) as well as on dopaminergic neurons (Tepper et al. 2002). When these pars reticulata neurons are selectively activated antidromically by electrical stimulation of the thalamus or tectum, most dopaminergic neurons are inhibited (Tepper et al. 1995). This inhibition is blocked by the selective $GABA_A$ receptor antagonist, bicuculline, but not by the selective $GABA_B$ receptor antagonists, 2-hydroxysaclofen or CGP-55845 A (Tepper et al. 1995; Paladini et al. 1999a). Thus, pars reticulata GABAergic projection neurons provide an important monosynaptic GABAergic input to nigral dopaminergic neurons.

In contrast to $GABA_A$ receptor blockade, $GABA_B$ receptor blockade not only failed to block inhibition elicited by electrical stimulation of striatal, pallidal, or nigral reticulata afferents, but rather potentiated it (Paladini et al. 1999a), as shown in the example in Fig. 4. This is likely due to the presence of inhibitory presynaptic $GABA_B$ receptors on the terminals of GABAergic afferents to the dopaminergic neurons. These presynaptic receptors serve to inhibit evoked release of GABA (GIRALT et al. 1990) and reduce IPSP/C amplitude (HAUSSER and YUNG 1994; SHEN and JOHNSON 1997). There is apparently enough endogenous GABA in the substantia nigra in vivo to activate these autoreceptors such that when they are blocked by local application of GABA$_B$ antagonists, GABA release is enhanced and the postsynaptic GABA$_A$-mediated inhibition is increased (PALADINI et al. 1999a).

In addition to their inhibitory effects on the rate of spontaneous activity, the GABAergic inputs contribute significantly to the regulation of the firing pattern of midbrain dopaminergic neurons. Local application of the GABA$_A$ receptor antagonists, bicuculline or picrotoxin, causes dopaminergic neurons to switch to the bursty firing pattern (TEPPER et al. 1995; PALADINI and TEPPER 1999). The transition is quite robust, and is independent of the baseline firing rate, firing pattern, or the change in firing rate due to application of the drug, suggesting that it is not due simply to increased depolarization and/or firing rate caused by blocking GABA$_A$ receptors. The effect is specific to blocking GABA$_A$ receptors; blockade of GABA$_B$ receptors with 2-OH-saclofen or CGP-55845 A produces a slight but consistent and statistically significant reduction in firing rate and regularization of the firing pattern (TEPPER et al. 1995; PALADINI and TEPPER 1999). This latter effect appears due to increased GABA release as a result of blockade of the presynaptic GABA$_B$ receptors discussed above. This results in increased stimulation of postsynaptic GABA$_A$ receptors on dopaminergic neurons and decreased burst firing, probably due to the GABA$_A$-mediated decrease in input resistance (CANAVIER 1999; PALADINI et al. 1999b). Subsequent experiments revealed that a significant source of the GABAergic input that was blocked by bicuculline or picrotoxin resulting in burst firing was the pars reticulata, and that the reticulata efferents could be effectively modulated by output from the globus pallidus (CELADA et al. 1999). Thus, increased activity in pallidum led to inhibition of reticulata GABAergic projection neurons and disinhibition of nigrostriatal dopaminergic neurons resulting in burst firing. Conversely, decreased activity in pallidum led to increased firing of reticulata neurons and the abolition of burst firing in dopaminergic neurons (CELADA et al. 1999). Although the mechanism or mechanisms underlying endogenous burst firing in dopaminergic neurons are incompletely understood (see below), it is clear that GABAergic afferents, acting at postsynaptic GABA$_A$ receptors on dopaminergic neurons can modulate the firing pattern of these neurons in vivo in an extremely powerful and consistent manner.

The roles and physiological significance of postsynaptic GABA$_B$ receptors on mesencephalic dopaminergic neurons are less clear. The receptors are certainly present, and dopaminergic neurons respond to selective GABA$_B$ agonists in vitro with a large conductance increase to potassium and a hyperpolarization (LACEY et al. 1988; LACEY 1993), and local electrical stimulation in slices of substantia nigra can elicit GABA$_B$ IPSPs or IPSCs (e.g., SUGITA et al.1992; CAMERON and WILLIAMS 1993). On the other hand, neither the striatal, pallidal, nor pars reticulata inputs appear to stimulate GABA$_B$ receptors on dopaminergic neurons in vivo to any significant degree as discussed above (PALADINI et al. 1999a), so the source(s) of the input to GABA$_B$ postsynaptic receptors remains unclear. In vivo, application of the GABA$_B$ agonist, baclofen, reduces dopaminergic neuron firing rate and leads to a regularization of firing pattern (e.g., ENGBERG et al. 1993). However, although intravenous administration of the selective GABA antagonist, CGP35348, antagonized the effects of baclofen, it was without effect on firing rate or firing pattern when given alone, suggesting that the receptor was not effectively stimulated in vivo under the conditions of the experiment, consistent with the results of TEPPER et al. (1995) and PALADINI and TEPPER (1999). On the other hand, in a more recent study, SCH 50911, a novel GABA$_B$ antagonist, was shown to increase the firing rate and burstiness of dopaminergic neurons when administered intravenously, suggesting that the postsynaptic GABA$_B$ receptors were effectively stimulated by endogenous GABA (ERHARDT et al. 1999). GABA, as well as GABA$_B$ agonists and antagonists will act both on pre- and postsynaptic receptors, and it is likely that methodological differences, possibly differences in the potencies and/or tissue distribution of the different GABA$_B$ antagonists, accounts for these discrepancies by altering the balance of effects on the pre- and postsynaptic GABA$_B$ receptors. Thus at present, the source(s) of inputs that activate GABA$_B$ receptors as well as the physiological significance of GABA$_B$ receptor activation in midbrain dopaminergic neurons remain to be determined.

II. Glutamatergic Afferents

The best characterized glutamatergic (i.e., excitatory amino acid) afferents to substantia nigra arise from the frontal cortex (USUNOFF et al. 1982; USUNOFF 1984; SESACK and PICKEL 1992; NAITO and KITA 1994), subthalamic nucleus (STN; CHANG et al. 1984; KITA and KITAI 1987; DAMLAMA and TEPPER 1993) and pedunculopontine nucleus (PPN), which also sends cholinergic afferents to substantia nigra (MOON-EDELY and GRAYBIEL 1983; SUGIMOTO and HATTORI 1984; CLARKE et al. 1987; RYE et al. 1987; GOULD et al. 1989; DAMLAMA and TEPPER 1993; FUTAMI et al. 1995). Midbrain dopaminergic neurons express both N-methyl-D-aspartate (NMDA) and non-NMDA glutamate receptors (MEREU et al. 1991) and respond to local application of glutamate in vivo with an increase in spontaneous firing rate (SCARNATI and PACITTI 1982). As the principal mediators of excitatory synaptic transmission in substantia nigra, these afferents have been the subject of considerable study. Moreover, glutamate application induces an increase in burstiness in dopaminergic neurons (GRACE and BUNNEY 1984b; OVERTON and CLARK 1992, 1997) as does intracellular loading with calcium (GRACE and BUNNEY 1984b), and the incidence of spontaneous burst firing has been reported to be decreased by NMDA antagonists (CHERGUI et al. 1993). In addition, stimulation of NMDA receptors on dopaminergic neurons in vitro produces a stereotyped form of a calcium-independent rhythmic burst firing that appears to be dependent on sodium influx through the NMDA channel and the operation of an electrogenic sodium pump (JOHNSON et al. 1992). Thus, it is as a potential mechanism for inducing burst firing that the glutamatergic afferents, especially those originating in frontal and prefrontal cortex, have received special interest (OVERTON and CLARK 1997).

Glutamate also acts on dopaminergic neurons through metabotropic receptors which are divided into eight subgroups (DE BLASI et al. 2001). Although it is unclear if all these subgroups are present on dopaminergic neurons (BONCI et al. 1997) there have been reports describing the action of metabotropic glutamate receptor agonists on the electrophysiological properties of dopaminergic neurons in vitro and in vivo. In vitro intracellular recordings studies obtained from rats slices, have reported that stimulation of metabotropic glutamate receptors with Trans-1-amino-cyclopentane-1,3-dicarboxylate (t-ACPD), a selective agonist for the R1 subtype of the metabotropic glutamate receptor, produces a depolarization (MERCURI et al. 1992) and a sustained increase in firing rate (MERCURI et al. 1993). This depolarization seems to be mediated by a cation-mediated inward current independent of calcium mobilization (GUATTEO et al. 1999). In contrast, other studies have reported an IPSP after stimulation of mGluR1 (FIORILLO and WILLIAMS 1998) and a blockade of this effect by amphetamine (PALADINI et al. 2001). Furthermore, in the only published study on the role of metabotropic glutamate receptors on dopaminergic neurons in vivo (MELTZER et al. 1997), an inhibition followed by excitation of firing rate was reported after microiontophoretic application of 1-aminocyclopentane-1,3-dicarboxylate (1 S,3R-ACPD), a putative metabotropic glutamate receptor selective agonist and both these effects were antagonized by application of the metabotropic glutamate receptor antagonist (S)-4-carboxy-phenylglycine. These findings would imply that glutamate is not solely an excitatory neurotransmitter in the midbrain but that its actions have to be viewed in a broader sense. At present is unclear if the metabotropic glutamate receptor-mediated IPSP is due to the particular stimulating conditions employed (FIORILLO and WILLIAMS 1998) or really represents an effect of physiological importance. If the latter turns out to be the case, it will add considerably to the role of glutamate on the regulation of dopaminergic neurons and their response to drugs.

In the first report to implicate cortex (frontal and anterior cingulate) in the elicitation of bursting in nigrostriatal neurons, cortical stimulation in urethane-anesthetized rats was shown to elicit burst discharges that closely resembled spontaneous bursts (GARIANO and GROVES 1988). However, this response occurred only in a very small proportion of nigral dopaminergic neurons (5%), at a latency of over 200ms, and was preceded by a substantial inhibition of firing (NAKAMURA et al. 1979; GARIANO and GROVES 1988). No attempts to block the bursts with glutamate antagonists were made and given the long latency, mediation by a monosynaptic glutamatergic input from cortex seemed unlikely. Soon after, inactivating the prefrontal cortex by local cooling was shown to abolish bursting and induce pacemaker-like firing in dopaminergic neurons (SVENSSON and TUNG 1989). On the other hand, lesions of medial prefrontal cortex were largely without effect on the spontaneous activity of substantia nigra dopaminergic neurons, although there was a significant reduction in the number of VTA neurons encountered per track (SHIM et al. 1996), consistent with a greater innervation of VTA dopaminergic neurons by glutamatergic afferents compared to substantia nigra (SMITH et al. 1996). Interestingly, the prefrontal lesions were associated with a slight increase in the spontaneous firing rate of substantia nigra dopaminergic neurons (SHIM et al. 1996), perhaps due to the preferential site of termination of corticonigral afferents on GABAergic pars reticulata neurons thereby activating feed-forward inhibition onto the dopaminergic neurons (HAJOS and GREENFIELD 1994; TEPPER et al. 1995). Subsequent studies replicated the finding of initial inhibition followed by extremely long latency burst responses after frontal cortical stimulation. They showed that the burst response could be blocked by NMDA but not by non-NMDA antagonists (see OVERTON and CLARK 1997 for review), providing strong evidence for a role of the glutamatergic corticonigral projection in the modulation of dopaminergic neuron firing pattern.

Reports of the effects of STN stimulation on the activity of substantia nigra dopaminergic neurons in vivo have been, perhaps surprisingly, more contradictory. In the earliest report, electrical stimulation of the subthalamic nucleus was found to be excitatory to dopaminergic and non-dopaminergic nigral neurons (HAMMOND et al. 1978). In a subsequent study that used local infusions of bicuculline to stimulate the subthalamic nucleus pharmacologically, approximately equal numbers of excitatory and inhibitory responses were found among dopaminergic neurons, although almost all of the non-dopaminergic neurons in pars reticulata were excited (ROBLEDO and FÉGER 1990). More recently, biphasic effects of electrical or pharmacological stimulation of subthalamic nucleus on nigral dopaminergic neurons were again reported, with an initial inhibition predominant following electrical stimulation that was followed in 35% of the neurons by a burst-like response (SMITH and GRACE 1992). Pharmacological activation of the subthalamic nucleus by bicuculline infusion led to an initial decrease in firing rate and the incidence of burst firing with the opposite biphasic effects following inactivation of the subthalamic nucleus with muscimol (SMITH and GRACE 1992). In another study, local infusions of GABA or bicuculline into subthalamic nucleus produced decreases and increases in firing rate and burst firing in nigral dopaminergic neurons, but these effects were observed in only about half of the neurons, with the other half showing the opposite effects (CHERGUI et al. 1994).

The STN-evoked inhibitory responses seen in the in vivo studies are almost certainly an indirect effect, resulting from subthalamic stimulation-induced activation of GABAergic axons or neurons synaptically activated by the stimulus. In vitro studies revealed that the depolarizing response seen in response to subthalamic stimuli in dopaminergic neurons (NAKANISHI et al. 1987) was composed of a nearly superimposed monosynaptic excitatory post-synaptic potential (EPSP) comprising both NMDA and non-NMDA components, and a monosynaptic and/or polysynaptic GABA$_A$-mediated IPSP (IRIBE et al. 1999). The monosynaptic IPSP arose from stimulation of descending GABAergic striatonigral and/or pallidonigral fibers and was eliminated by hemisection of the brain anterior to the subthalamic nucleus several days before the in vitro recordings. In some cases, however, an IPSP remained after the hemisection that could be abolished with bicuculline or 6-cyano-7-nitroquinoxaline-2,3-dione (CNQX; Fig. 5). The latter effect indicates that the IPSP arose from glutamatergic excitation of a GABAergic neuron whose subthalamic input and outputs to dopaminergic neurons remained intact in the slice preparation, most likely the pars reticulata GABAergic projection neurons (TEPPER et al. 1995; IRIBE et al. 1999).

Stimulation of the PPN in vivo induces short latency excitation in a significant fraction of nigral dopaminergic neurons (SCARNATI et al. 1984). In brain slices, stimulation of the PPN produces monosynaptic EPSPs that consist of both glutamatergic and cholinergic components that appear to converge on single dopaminergic neurons (FUTAMI et al. 1995). The pharmacology of the glutamatergic component is not well established; however, in one extracellular recording study, NMDA-selective antagonists were ineffective at blocking excitatory effects of pedunculopontine stimulation which were blocked by broad spectrum glutamate antagonists, suggesting that in vivo the predominant effect may be mediated principally by non-NMDA glutamate receptors (DiLORETO et al. 1992). Compared to the subthalamic nucleus and prefrontal cortex, inhibitory responses are relatively rare with pedunculopontine stimulation. This may be because a larger proportion of pedunculopontine afferents terminate on dopaminergic neurons and dendrites as opposed to pars reticulata GABAergic neurons. For example, only about 10% of subthalamic afferents terminate on tyrosine hydroxylase-positive cells and dendrites in substantia nigra, the remainder synapsing on non-dopaminergic pars reticulata neurons, whereas almost 38% of boutons originating in the pedunculopontine nucleus synapse on dopaminergic dendrites (DAMLAMA 1994). Thus, the balance of input is shifted more towards the monosynaptic pedunculopontine–dopaminergic neuron pathway than the disynaptic pathway through pars reticulata (IRIBE et al. 1999). Thus, although not yet as well studied as the subthalamic afferents, the excitatory input from the pedunculopontine nucleus may prove to be at least equally important as a source of monosynaptic excitation of dopaminergic neurons.

Although there are many reports that NMDA agonists elicit burst firing in dopaminergic neurons in vivo and in vitro (GRACE and BUNNEY 1984b; JOHNSON et al. 1992; OVERTON and CLARK 1992), and that kynurenate, a broad-spectrum excitatory amino acid antagonist, inhibits burst firing (CHARLETY et al. 1991), there are other reports that NMDA or *l*-glutamate, acting through NMDA receptors as demonstrated by blockade of their effects with selective NMDA antagonists, produced increases in midbrain dopaminergic neuron firing rate without significantly increasing bursting in vitro (e.g., SEUTIN et al. 1990; WANG and FRENCH 1993; CONNELLY and SHEPARD 1997). In addition, non-

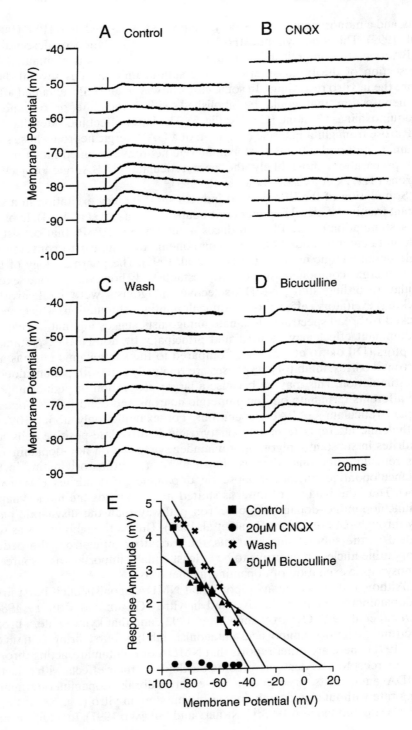

NMDA, mGluR1 agonists have been reported to induce burst firing in dopaminergic neurons (MELTZER et al. 1997), even in the presence of NMDA receptor antagonists (ZHANG et al. 1994). Blockade of the long-lasting spike afterhyperpolarization by apamin also induces burst firing in vitro (SHEPARD and BUNNEY 1988). Finally, rhythmic burst firing induced by NMDA or NMDA plus apamin in vitro is abolished by GABA$_A$ receptor agonists (PALADINI et al. 1999b), suggesting that in vivo, NMDA-related burst firing may be controlled or gated in a permissive fashion depending on the level of activity in GABAergic afferents.

There is little doubt that the glutamatergic afferents to dopaminergic neurons are the most important source of their excitatory input. However, while it is virtually certain that glutamatergic inputs play an important role in the modulation of dopamine neuron firing pattern (OVERTON and CLARK 1997), it is probably not the case that NMDA receptor stimulation of dopaminergic neurons is exclusively or perhaps even primarily responsible for evoking bursty firing in vivo. There is also good evidence that dopaminergic neuron firing pattern is modulated to an important extent by other transmitter/receptor systems including GABAergic (TEPPER et al. 1995; CELADA et al. 1999; PALADINI and TEPPER 1999), cholinergic (GRENHOFF et al. 1986; FUTAMI et al. 1995; KITAI et al. 1999), and non-NMDA glutamatergic systems (ZHANG et al. 1994; MELTZER et al. 1997).

III. Cholinergic Afferents

The substantia nigra is rich in acetylcholinesterase, and choline acetyltransferase-positive synapses are made onto the dendrites of dopaminergic neurons (BENINATO and SPENCER 1988). The principal source of the cholinergic input is likely the pedunculopontine and laterodorsal tegmental nuclei (GOULD et al. 1989; DAMLAMA and TEPPER 1993). A number of nicotinic receptor subunits are expressed by mesencephalic dopaminergic neurons including α3, α4, α5, α7, β2, and β3 (SORENSON et al. 1998), and bath application of nicotine produces an inward current and depolarization that exhibits

◄
Fig. 5A–E. The IPSP component of the subthalamic nucleus-evoked depolarizing postsynaptic potential (DPSP) in some dopaminergic neurons is polysynaptic. Under control conditions, subthalamic stimulation produced a DPSP with a reversal potential of −38.8mV (**A, E**) indicating that it is composed of an EPSP and near simultaneous IPSP. Addition of CNQX to the bath completely abolished both components of the DPSP (**B, E**) indicating that the IPSP resulted from glutamate-dependent synaptic activation of an inhibitory neuron whose inputs and outputs remained intact in the slice. After a 1-h wash, the DPSP returned and still exhibited a hyperpolarized reversal potential as before drug application (**C, E**). Subsequent application of bicuculline shifted the reversal potential in the positive direction to 12.6mV (**D, E**) showing that the IPSP component of the DPSP was GABA$_A$-mediated. Traces in A–D are each the average of four single sweeps. (Reproduced from IRIBE et al. 1999 with permission of the publishers)

desensitization with a time course of tens of seconds (CALABRESI et al. 1989; SORENSON et al. 1998). The response is sensitive to κ-bungarotoxin but not α-bungarotoxin and is thus more similar to the nicotinic response seen at peripheral autonomic ganglia than at the neuromuscular junction (CALABRESI et al. 1989). In vivo, local or systemic administration of nicotine agonists produces excitation of nigrostriatal (LICHTENSTEIGER et al. 1982) and VTA dopaminergic neurons (MEREU et al. 1987) together with an increment in burst firing of dopaminergic neurons (GRENHOFF et al. 1986). It is interesting to note that the increase in firing rate and increase in burst firing were only poorly correlated, suggesting a possible nicotinic effect on firing pattern independent of its effect on firing rate (GRENHOFF et al. 1986).

Dopaminergic neurons also express muscarinic receptors, and are depolarized by muscarinic agonists in vitro with a pharmacological profile resembling that of the M_1 receptor, although the mechanism of the response appears different from that of the classic m-current closure of potassium channels (LACEY 1993). In addition to these postsynaptic actions, acetylcholine (ACh) acts presynaptically in substantia nigra to inhibit release of GABA from GABAergic afferents through an M_3 receptor (GRILLNER et al. 2000).

Stimulation of the pedunculopontine nucleus in vivo produces mostly excitation of dopaminergic neurons at short latencies ranging from 3 to 5 ms (SCARNATI et al. 1984), consistent with the conduction time of cholinergic neurons from the pedunculopontine nucleus to the substantia nigra (FUTAMI et al. 1995; TAKAKUSAKI et al. 1996). The EPSP that underlies the excitation seen extracellularly in vivo is composed of both nicotinic and pirenzepine-sensitive muscarinic components (FUTAMI et al. 1995). Pedunculopontine stimulation also produces burst firing in nigral dopaminergic neurons in vivo (LOKWAN et al. 1999). The bursts observed were brief (averaging two spikes) and occurred at extremely long latency (~100 ms). As no antagonists were tested, the transmitter and receptor underlying the evoked bursts remains to be determined. The bursting could be glutamate-mediated as suggested by the authors, cholinergic, or might depend on an interaction of the two transmitter systems (e.g., FUTAMI et al. 1995; KITAI et al. 1999).

IV. Monoaminergic Afferents

A projection from the dorsal raphé nucleus to the substantia nigra has been described on the basis of anatomical, electrophysiological, and pharmacological bases. Retrograde and anterograde tract tracing studies both reveal a significant input to substantia nigra and VTA from the dorsal raphé nucleus (FIBIGER and MILLER 1977; CORVAJA et al. 1993), and the ventral regions of the substantia nigra and VTA are rich in serotonergic axons and boutons that make asymmetric synapses onto both dopaminergic and non-dopaminergic dendrites (HERVÉ et al. 1987; MORI et al. 1987; CORVAJA et al. 1993). In early studies, stimulation of the dorsal raphé was shown to inhibit the firing of both pars compacta (dopaminergic) and pars reticulata (non-dopaminergic) neurons in vivo (DRAY et al. 1976; FIBIGER and MILLER 1977), effects that were abolished by depletion of serotonin (FIBIGER and MILLER 1977). A later study revealed more modest effects, with dorsal raphé stimulation exerting modest inhibitory effects only on dopaminergic neurons firing at less than 4Hz; more rapidly firing neurons were unaffected (KELLAND et al. 1990b). 5-Hydroxytryptamine (5HT)$_{1A}$ agonists exerted effects consistent with this, leading at high doses to excitation of slowly firing cells without affecting more rapidly firing neurons, while 5HT$_{1B}$ agonists were without effect (KELLAND et al. 1990b).

These inhibitory effects of serotonin are difficult to reconcile with the asymmetric synapses made by dorsal raphé neurons on dopaminergic dendrites, which are usually associated with excitatory synaptic actions. Furthermore, serotonin has been found to enhance the release of dopamine from substantia nigra in vivo (GLOWINSKI and CHERAMY 1981) and the VTA in vitro (BEART and McDONALD 1982). In vitro, serotonin has been found to facilitate a dendritic calcium conductance (NEDERGAARD et al. 1988), and produces a clear depolarization and excitation of substantia nigra dopaminergic neurons (NEDERGAARD et al. 1991). These effects are mediated postsynaptically, but not by 5HT$_{1A}$ or 5HT$_2$ receptors. These data also seem inconsistent with a classical inhibitory action of serotonin on mesencephalic dopaminergic neurons.

Perhaps some of the discrepancy can be resolved by data showing that stimulation of the dorsal raphé with short trains of pulses reduces the dendritic excitability of dopaminergic dendrites, as measured by somatodendritic invasion of antidromic spikes (TRENT and TEPPER 1991). The depression in dendritic excitability was unrelated to changes in the mean firing rate or to the strength or duration of neostriatal-evoked inhibition. This effect was abolished by depletion of serotonin with para-chlorophenylalanine for 3 days prior to recording and could be reinstated by administration of 5 hydroxytryptophan and was also blocked by systemic administration of the non-specific serotonin antagonist, metergoline, indicating that it was serotonergic in nature. In addition, the depression in dendritic excitability could be, perhaps surprisingly, also blocked by haloperidol. These data were interpreted to indicate that the raphé inputs to nigral dopaminergic dendrites produced a local depolarization that resulted in local release of dopamine that subsequently activated somatodendritic autoreceptors which led to a local hyperpolarization of the dendrites and a reduction in dendritic excitability, without grossly affecting the firing rate of the neuron as a whole (TRENT and TEPPER 1991). This interpretation is consistent with the asymmetric character and location of the serotonergic synapses on the dopaminergic neurons, the previously observed increase in dopamine release following serotonergic stimulation in substantia nigra and VTA, and the serotonergic facilitation of dendritic calcium entry, and it could account for the generally inconsistent and weak effect of serotonergic agonists and dorsal raphé stimulation on dopaminergic neuron firing rate.

In addition, pars reticulata GABAergic neurons are excited by serotonin via both pre- and postsynaptic mechanisms (STANFORD and LACEY 1996). Given the feedforward inhibition of nigral dopaminergic neurons from pars reticulata (HAJOS and GREENFIELD 1994;TEPPER et al. 1995), the effects of serotonergic agonists and raphé input on dopaminergic neurons may also depend to an extent on the ratio of the opposing effects of direct activation of dopaminergic neurons and disynaptic input through pars reticulata, as well as on a balance between the action of serotonin on autoreceptors and different postsynaptic receptors.

Although not as well characterized nor as dense as the serotonergic input from the dorsal raphé, some retrograde tracing studies reveal a modest projection from the locus coeruleus to the VTA (PHILLIPSON 1979). Stimulation of the locus coeruleus produces excitatory responses in dopaminergic neurons recorded extracellularly in substantia nigra and VTA in vivo (GRENHOFF et al. 1993). Although α_1 adrenoceptor binding and message levels are extremely low or non-detectable in the midbrain (JONES et al. 1985; PIERIBONE et al. 1994), these responses were abolished by catecholamine depletion and were blocked by prazosin, indicating that they were mediated by an α_1 receptor. In vitro recordings provided largely consistent results, showing that about 60% of mesencephalic dopaminergic neurons respond to α_1 receptor stimulation with a depolarization due to a potassium conductance decrease (GRENHOFF et al. 1995). In addition, the α_2 agonist clonidine has been reported to promote a regularization of firing pattern in both substantia nigra (GRENHOFF and SVENSSON 1988) and VTA neurons (GRENHOFF and SVENSSON 1989), most likely by its presynaptic inhibitory effects on norepinephrine release.

E. Autoreceptor-Mediated Effects on Dopaminergic Neurons

I. Somatodendritic Autoreceptors

In 1973 BUNNEY and colleagues (BUNNEY et al. 1973a,b; BUNNEY and AGHAJANIAN 1973; AGHAJANIAN and BUNNEY 1973) published the first recordings from identified substantia nigra and VTA dopaminergic neurons. One of the key observations was that apomorphine, a direct-acting dopamine receptor agonist, potently inhibited dopaminergic neurons even when applied iontophoretically (AGHAJANIAN and BUNNEY 1977). This finding demonstrated that dopaminergic neurons possessed receptors for their own transmitter, dopamine, on their cell body and/or dendrites (somatodendritic region). These receptors were termed somatodendritic autoreceptors, to distinguish them from the axon terminal autoreceptors also expressed by dopaminergic neurons that play a role in the local regulation of dopamine release and synthesis (for review see STARKE et al. 1989).

The earliest pharmacological characterization of dopamine somatodendritic autoreceptors predated the current molecular biologically defined classification of dopamine receptors and indicated simply that they exhibited a pharmacological profile distinct from either α or β adrenoceptors, i.e., that they were a unique type of dopamine receptor (AGHAJANIAN and BUNNEY 1977). When dopamine neurons were classified into D1 or D2 subtypes (KEBABIAN and CALNE 1979), it became clear, based on the sensitivity of the receptor to haloperidol (GROVES et al. 1975), a moderately selective D_2 antagonist, that the dopamine autoreceptor was a D2 receptor. This was later confirmed with the use of highly selective D2 receptor agonists and antagonists in in vitro intracellular recordings (LACEY et al. 1987, 1988; LACEY 1993) and receptor binding (MORELLI et al. 1988). With the advent of the widespread use of molecular biological methods to isolate and identify neurotransmitter receptors in the last decade came the discovery that there are in fact two families of dopamine receptors, D1 and D2. Within each family exist subtypes, D_1 and D_5 for the D1 family and D_2 (both long and short isoforms), D_3 and D_4 for the D2 family (see for review, SIBLEY and MONSMA 1992). Although the most recent electrophysiological data confirm that the autoreceptor is a member of the D2 receptor family (DEVOTO et al. 1995), there remains some controversy as to whether the autoreceptor is exclusively a D_2 receptor, as suggested on the basis of experiments with transgenic D_2 (MERCURI et al. 1997) or D_3 (KOELTZOW et al. 1998) knockout mice, or instead comprises both D_2 and D_3 receptors, as suggested based on experiments localizing D_3 message and/or protein to midbrain dopaminergic neurons (TEPPER et al. 1997; SHAFER and LEVANT 1998; STANWOOD et al. 2000) or electrophysiological experiments in rats after antisense knockdown of dopamine D_2 and/or D_3 receptors (TEPPER et al. 1997). Using a very sensitive and specific polyclonal antibody raised against a synthetic peptide reflecting the amino acid sequence of the third cytoplasmic loop of the D_3 receptor, SOKOLOFF and associates have recently reported that all rat mesencephalic dopaminergic neurons express the D_3 receptor (DIAZ et al. 2000), which supports the notion that autoreceptors belong to both subclasses: D_2 and D_3.

In any event, somatodendritic autoreceptor stimulation leads to an hyperpolarization of dopaminergic neurons that is caused by an increase in conductance to potassium (LACEY et al. 1987, 1988). It is this hyperpolarization which can reach about 12mV in vitro in response to a maximal concentration of quinpirole (BOWERY et al. 1994) that is responsible for the inhibition of spontaneous activity seen after local or systemic administration of autoreceptor agonists. The potassium channel linked to the dopamine autoreceptor in situ appears to be the same one that is opened by activation of $GABA_B$ receptors since the autoreceptor-mediated potassium current is reversibly occluded by maximal stimulation of the $GABA_B$ receptor by baclofen (LACEY et al. 1988).

The D_2 somatodendritic autoreceptor is G-protein coupled and its function is disrupted by pertussis toxin (INNIS and AGHAJANIAN 1987; SHEPARD and CONNELLY 1999). Although the specifics of the G-protein coupling to D_2 or D_3 autoreceptors is unknown at present, it appears to be independent of protein kinase A or C pathways (CATHALA and PAUPARDIN-TRITSCH 1999). Transfection studies in MES-23.5, a dopaminergic neuroblastoma cell line in which D_2

receptor stimulation increases a potassium conductance, have revealed that the D_{2S} receptor is linked via a $G_{s\alpha}$ whereas the D_{2L} is linked via a $G_{o\alpha}$ (LIU et al. 1999).

Although commonly termed the somatodendritic autoreceptor, the D_2 autoreceptor may be preferentially located in the dendrites rather than the soma or pericellular region. Although electron microscopic immunocytochemistry revealed cellular D_2 receptor labeling in substantia nigra and VTA, the labeling of perikarya and large proximal dendrites was very weak compared to that of dendrites (SESACK et al. 1994). Almost exactly the same distribution of labeling was seen for the autoreceptor potassium channel subunit, Kir3.2 (IANOBE et al. 1999). Finally, in vivo extracellular recordings of dopaminergic neurons following local pressure injection of autoreceptor agonists showed that the neurons were more effectively inhibited when the drugs were applied several hundred micrometers distal to the recording site than when applied right at the recording site which was most often presumably at or near the soma (AKAOKA et al. 1992). Thus, the somatodendritic autoreceptor may be, in reality, principally expressed on the dendrites rather than the somata of dopaminergic neurons.

II. Axon Terminal Autoreceptors

As mentioned above, the first dopamine autoreceptors to be discovered were receptors located on the axon terminals of nigrostriatal fibers in slices of rat striatum (FARNEBO and HAMBERGER 1971; for review see STARKE et al. 1989). When rat striatal slices were incubated with ^{3}H-tyrosine and subjected to field electrical stimulation, radiolabeled dopamine was released. Addition of apomorphine to the bath significantly reduced the dopamine efflux. These data were correctly interpreted to mean that there existed a population of dopamine receptors on or near the release sites on dopaminergic axons in the dopamine terminal fields that served to inhibit the release of electrically evoked dopamine. Subsequent studies showed that release evoked by depolarization of the slices by high potassium was also subject to autoreceptor regulation but that release elicited by agents that interfered with the dopamine transporter, for example, amphetamine, was not subject to autoregulation (KAMAL et al. 1981). This turned out to be related to the calcium dependence of the releasing stimuli. Release that is calcium dependent, such as that evoked by electrical stimulation or high potassium, is subject to autoregulation, whereas calcium-independent release (e.g., by amphetamine) (ARNOLD et al. 1977; MEYERHOFF and KANT 1978) is not under autoreceptor control (KAMAL et al. 1981).

In addition to modulating the release of dopamine, dopamine terminal autoreceptors can also modulate the synthesis of dopamine by altering the rate of tyrosine hydroxylation (WALTERS and ROTH 1976; ROTH et al. 1978). A thorough discussion of autoreceptor effects on dopamine synthesis is beyond the scope of the present chapter and the reader is referred to WOLF and ROTH (1990) for a comprehensive review.

The terminal autoreceptor appears similar or identical in all respects to the somatodendritic autoreceptor. The axon terminal autoreceptor subtype is D_2 (BOYAR and ALTAR 1987; TEPPER et al. 1984a), and is a G-protein coupled receptor sensitive to pertussis toxin (BEAN et al. 1988). Stimulation of terminal autoreceptors in vivo produces an increase in the amount of current needed to evoke an antidromic action potential, indicating that autoreceptor activation is associated with a decrease in the excitability of the dopaminergic nerve terminals in the striatum (GROVES et al. 1981; TEPPER et al. 1984a,b, 1985), nucleus accumbens (MEREU et al. 1985), and cortex (GARIANO et al. 1989a). This is most likely due to an hyperpolarization of the terminal similar to that seen at the cell body, and can be reversed by local application of selective D_2 receptor antagonists including sulpiride (TEPPER et al. 1984a; TEPPER and GROVES 1990). In addition, application of D_2 antagonists by themselves results in an increase in the excitability of dopaminergic terminals indicating that the extracellular concentrations of dopamine in striatum, nucleus accumbens, and cortex are high enough to cause at least partial occupancy of the terminal autoreceptors in vivo (TEPPER et al. 1984a,b; MEREU et al. 1985; GARIANO et al. 1989a). In addition, there have been two reports of decreases in dopamine terminal excitability following D_1 receptor agonist SKF 38393 local administration that could be partially reversed by the D_1 selective antagonist SCH 23390 (DIANA et al. 1988, 1991a), But in view of the bulk of in vivo and in vitro electrophysiological, receptor binding, and in situ hybridization evidence it is unlikely that these effects reflect the presence of D_1 terminal autoreceptors.

III. Are D_2 Autoreceptors Different from Other D_2 Receptors?

It is often claimed that dopamine autoreceptors are "more sensitive" than other, postsynaptic D_2 receptors. One piece of evidence cited in support of this is the relatively low doses or concentrations of D2 agonists required to inhibit dopaminergic neuron firing (in the range of 4–8μg/kg, i.v. for apomorphine; CHIODO and ANTELMAN 1980; TEPPER et al. 1982), or to induce hyperpolarization of dopaminergic neurons in vitro (ED_{50} for quinpirole: 77nM; for apomorphine 205nm; BOWERY et al. 1994). The doses of D2 antagonists required to block the effects of dopamine or D2 agonists are similarly low; the selective D2 antagonist, sulpiride shows an apparent K_d of 13nM for antagonizing the effects of the selective D2 agonist, quinpirole (LACEY et al. 1987). This is indeed sensitive, but it is difficult to find something against which to compare this, since even though many other central neurons express postsynaptic D_2 and/or D_3 receptors, in most of them the receptor is not linked to the opening of a ligand-gated potassium channel as it is in substantia nigra (LACEY et al. 1988), but rather acts to modify the kinetics or gating of voltage gated channels (e.g., SURMEIER et al. 1992, 1996; SURMEIER and KITAI 1993). This difference creates problems when trying to compare the physiological effects of stimulating the dopamine autoreceptor with other populations of D_2 receptors.

For example, in one study that is widely cited as evidence that the dopamine autoreceptor is more sensitive than the postsynaptic D2 receptor, the ability of iontophoretically applied dopamine or intravenously administered apomorphine to inhibit the spontaneous activity of substantia nigra dopaminergic neurons or striatal neurons was compared (SKIRBOLL et al. 1979). In both cases the dopaminergic neurons were inhibited at much lower doses of agonist than the striatal neurons. However, since the dopamine receptors are linked to different effectors in the two neuronal populations (LACEY 1993; SURMEIER and KITA 1993; USIELLO et al. 2000), it is not valid to compare the ability of drugs to inhibit the spontaneous firing of striatal and dopaminergic neurons, nor to use differences in their ED_{50} as evidence that the autoreceptor is more sensitive than the postsynaptic D_2 receptor (SKIRBOLL et al. 1979). Studies which conclude that the autoreceptor is the same as the postsynaptic receptor from experiments comparing the ability of dopamine agonists to inhibit dopamine release with their ability to inhibit ACh release are similarly flawed (e.g., HELMREICH et al. 1982).

However, there is at least one place in which postsynaptic D_2 receptor signaling/linkage appears to be similar or identical to that in the dopaminergic neuron, and that is the lactotroph cells of the pituitary gland. Among these cells, dopamine acts through a D_2 receptor (VALLAR and MELDOLESI 1989) to open a potassium channel in concentrations as low as 100nM (ISRAEL et al. 1987), the same range as that required for activation of the autoreceptor (LACEY 1993). Based on these data, it seems likely that when coupled to a potassium conductance, the D_2 autoreceptor and the D_2 postsynaptic receptor exhibit similar or identical sensitivities.

IV. Are Autoreceptors Ubiquitous Among Dopaminergic Neurons?

Although the majority of the studies of dopamine autoreceptor pharmacology have been conducted in the nigrostriatal system, there have also been a large number of studies focusing on the mesoaccumbens and mesocortical dopaminergic projections. Although there is unanimous agreement about the existence of somatodendritic and axon terminal autoreceptors on dopaminergic neurons of the substantia nigra pars compacta, the situation has been more controversial with respect to the dopaminergic neurons of the VTA. The controversy arose when it was found that the turnover of dopamine was significantly faster in the frontal cortex than in the striatum and that the synthesis of dopamine in cortex appeared unaffected by apomorphine (BANNON et al. 1981, 1982). It was concluded that these neurons lacked "synthesis-modulating autoreceptors." Similar results and conclusions were reported for dopamine terminals in the amygdala, hypothalamus, and bed nucleus of the stria terminalis (KILTS et al. 1987). Furthermore, a subsequent study reported that iontophoretic application of dopamine failed to inhibit the spontaneous activity of dopaminergic neurons projecting to the prefrontal or cingulate cortices, whereas neurons projecting to the striatum or piriform cortices were readily inhibited (CHIODO et al. 1984). In addition, the mean spontaneous firing rates of the medial mesocortical dopaminergic neurons were reported to be relatively high (mesoprefrontal: 9.3 ± 0.6 Hz; mesocingulate: 5.9 ± 0.5 Hz), and the incidence of burst firing much higher than in nigrostriatal or mesopiriform neurons (CHIODO et al. 1984). Thus, it was concluded that these neurons were devoid of both "impulse-regulating somatodendritic and synthesis-modulating nerve terminal autoreceptors," although the possibility that these neurons might still possess terminal autoreceptors that modulate dopamine release was left open (CHIODO et al. 1984).

Subsequently, two groups reported that dopaminergic neurons that projected to prefrontal or cingulate cortex were inhibited by low "autoreceptor-specific" doses of apomorphine (5–6 μg/kg) to the same extent as nigrostriatal or meso-accumbens dopaminergic neurons (SHEPARD and GERMAN 1984; GARIANO et al. 1989a). Furthermore, these two studies reported that the mesocortical neurons also exhibited the same range of spontaneous firing rates as nigrostriatal neurons (SHEPARD and GERMAN 1984; GARIANO et al. 1989a), results that agreed well with earlier studies of the electrophysiological properties of VTA dopaminergic neurons in which the projection targets were not identified (e.g., WANG 1981a,b).

How can one resolve these discrepancies? It is possible that the electrophysiological results of CHIODO et al. (1984) derive from a small subpopulation of mesocortical dopaminergic neurons, located very close to the midline which were not sampled in the other studies. It should be noted that the cell bodies of origin of the nigrostriatal, mesolimbic, and mesocortical neurons reported in CHIODO et al. (1984) showed a much more restricted localization and projection topography with essentially no overlap than that reported by others (see for example, FALLON and LAUGHLIN 1995). Regardless, based on in situ hybridization studies and D_2 and/or D_3 receptor autoradiography, the dopaminergic neurons of origin of the nigrostriatal, mesolimbic, and mesocortical projections all express dopamine D_2 and/or D_3 mRNA and/or receptor protein (MORELLI et al. 1988; MEADOR-WOODRUFF et al. 1989; DIAZ et al. 2000), indicating the ubiquitous expression of the D_2 and/or D_3 autoreceptor on mesencephalic dopaminergic neurons. In vivo recording studies clearly show evidence for the existence of D2-family somatodendritic autoreceptors on VTA neurons projecting to prefrontal cortex (SHEPARD and GERMAN 1984; GARIANO et al. 1989b). Finally, retrograde tracing studies show clearly that a number of neurons in the substantia nigra and VTA collateralize to the striatum and cortical areas including prefrontal cortex (FALLON 1981). Although these results are in direct contradiction to those of CHIODO et al. (1984), the bulk of the evidence points strongly towards the idea that most or, more likely, all mesencephalic dopaminergic neurons express D_2 and/or D_3 somatodendritic autoreceptors.

What about nerve terminal autoreceptors? A large number of in vitro experiments have consistently shown that stimulus-evoked release of dopamine from all terminal regions, including prefrontal and cingulate cortices (PLANTJE et al. 1985, 1987) is modulated by D_2 and/or D_3 nerve terminal autoreceptors (for review see STARKE et al. 1989), although the sensitivity of release to autoreceptor agonists and antagonists in cortex is sometimes reported to be less than in striatum (e.g., CUBEDDU et al. 1990). In vivo electrophysiological experiments of changes in the excitability of dopamine nerve terminals in response to local infusion of D2 receptor agonists or antagonists or changes in impulse flow revealed that mesoprefrontal dopaminergic neurons responded exactly as did nigrostriatal neurons, reinforcing the idea that these mesoprefrontal dopaminergic neurons also possessed nerve terminal autoreceptors (TEPPER et al. 1984a,b; GARIANO et al. 1989a; TEPPER and GROVES 1990). It is still unclear why, if the cortical and mesolimbic dopaminergic terminals possess autoreceptors as they appear to, dopamine metabolism is different in the prefrontal cortex. One intriguing possibility is that the much lower levels of tissue dopamine (KILTS et al. 1987) and dopamine overflow (ABERCROMBIE et al. 1989), coupled with the far fewer functional reuptake sites in these structures (e.g., CASS and GERHARDT 1995; LETCHWORTH et al. 2000) interact to blunt autoinhibition. Interestingly, recent studies in a mouse mutant lacking the dopamine transporter show that interfering with the transporter severely attenuates autoreceptor function (JONES et al. 1999), although the mechanism for this is as yet unclear.

In any event, the bulk of the evidence now favors the conclusion that all mesencephalic dopaminergic neurons express D_2 and/or D_3 dopamine autoreceptors. Whether there are actually different "synthesis-modulating autoreceptors," "impulse-modulating autoreceptors," and "release-modulating autoreceptors" as proposed by some (see, for example, KILTS et al. 1987 or WOLF and ROTH 1990), or simply one autoreceptor (that may comprise both D_2 and D_3 receptors) that serves different functions depending on its subcellular location remains to be determined.

V. What Are the Physiological Roles of Autoreceptors?

The functional role of the axon terminal autoreceptor seems relatively clear. By making it possible to modulate dopamine release (and synthesis) locally, dopaminergic synaptic transmission can be fine-tuned to an extent simply not possible by modulating impulse activity along the main axon when each axon may give rise to several hundred thousand release sites (TEPPER et al. 1987a).

But what of the somatodendritic autoreceptor? Among the earliest ideas as to the physiological function of somatodendritic autoreceptors on dopaminergic neurons was the "self-inhibition" hypothesis of GROVES and associates (GROVES et al. 1975). According to this hypothesis, dopamine released from the dendrites of dopaminergic neurons activated somatodendritic autoreceptors thereby participating in a local negative feedback regulation of the electrophysiological and biochemical activity of the neurons. The self-inhibition hypothesis was consistent with the slow firing rate of dopaminergic neurons (BUNNEY et al. 1973a), the location of dopamine within dendrites of nigral dopaminergic neurons (e.g., BJORKLUND and LINDVALL 1975), and the inhibitory effects of dopamine or dopamine receptor agonists on the spontaneous activity of dopaminergic neurons (e.g., BUNNEY et al. 1973a,b). Furthermore, administration of dopamine receptor antagonists alone produced increases in the firing rate of dopaminergic neurons in vivo, suggesting that the neurons were under a tonic inhibition mediated by dopamine (BUNNEY and AGHAJANIAN 1973; BUNNEY et al. 1973a,b). Since there are no dopaminergic afferents to substantia nigra, and no local axon collaterals from the dopaminergic neurons (JURASKA et al. 1977; WASSEF et al. 1981; TEPPER et al. 1987b), the source of the endogenously released dopamine was most likely to be the dendrites of the dopamine neurons themselves. This hypothesis was borne out by subsequent demonstration that depolarizing stimuli such as high potassium (GEFFEN et al. 1976) as well as dopamine-releasing agents such as amphetamine (PADEN et al. 1976) elicited dopamine release from slices of substantia nigra.

From the earliest extracellular recordings in vivo, midbrain dopaminergic neurons were known to fire spontaneously at very low rates, rarely averaging more than eight spikes per second for prolonged periods, and it was natural to wonder if dopaminergic self-inhibition as originally proposed (GROVES et al. 1975) played a role in the slow firing and long post-spike refractoriness seen in autocorrelograms (WILSON et al. 1977). The earliest intracellular recordings from dopaminergic neurons revealed spontaneous action potentials that were followed by large, long-lasting afterhyperpolarizations (GRACE and BUNNEY 1980, 1983a,b) that seemed consistent with this idea, and administration of haloperidol was shown to alter the pattern of firing of these neurons in vivo, making the occurrence of shorter interspike intervals more common, a result that could sometimes be observed in the absence of a change in firing rate (WILSON et al. 1979). However, as described above, subsequent electrophysiological studies revealed that the prolonged spike afterhyperpolarization and long interspike intervals were due largely to a calcium activated potassium conductance (KITA et al. 1986; SHEPARD and BUNNEY 1988; PING and SHEPARD 1996), and not to dopamine. Interestingly enough, autoreceptor stimulation in dissociated dopaminergic neurons has been shown to reduce calcium entry through ω-conotoxin and w-AgaIVA-sensitive calcium channels which leads to a reduction in the calcium-activated potassium current (CARDOZO and BEAN 1995).

The dendritic tree of dopaminergic neurons is relatively sparse, but individual dendrites often extend for distances of a millimeter or more (JURASKA et al. 1977; TEPPER et al. 1987b; HAUSSER et al. 1995). One possible role for the autoreceptor-mediated hyperpolarization/conductance increase is to respond to dendritically released dopamine by attenuating or blocking the effects of afferent input or intrinsic voltage-dependent conductances (e.g., CARDOZO and BEAN 1995; WILSON and CALLAWAY 2000) of a dendrite or dendritic segment on which the autoreceptor is located. This type of action would be far more subtle than the more generally assumed classical function whereby autoreceptors function to limit or regulate the overall activity of dopaminergic neurons.

The classical idea of autoreceptor function derives from the many experiments in which autoreceptor agonists, administered either systemically or locally, have the effect of significantly hyperpolarizing the neuron and suppressing or completely inhibiting its spontaneous activity (BUNNEY et al. 1973a,b; GROVES et al. 1975; LACEY et al. 1987). In these experimental situations, exogenous application of autoreceptor agonists or dopamine releasing agents is likely to produce levels of autoreceptor occupancy that are significantly greater than those that obtain in vivo under normal physiological conditions. Evidence in support of a more subtle and localized physiological effect of somatodendritic autoreceptor activation comes from several lines of evidence.

The electrophysiological response of dopaminergic neurons to autoreceptor antagonists exhibits certain vagaries. Although early studies showed that systemic administration of chlorpromazine or haloperidol at low doses (1.25 mg/kg and 25–50 µg/kg, i.v., respectively) to unanesthetized, immobilized rats consistently produced large (approximately 100%) increases in the spontaneous firing rate (Bunney et al. 1973a,b; Wilson et al. 1979), this effect appeared to be mediated, at least in part, through the striatum since striatal lesions blunted or abolished the effect (KONDO and IWATSUBO 1980). In a recent re-examination of the effects of systemically administered haloperidol or sulpiride on dopaminergic neuron activity, PUCAK and GRACE (1994) did not find evidence of striatal involvement in the effects of autoreceptor antagonists, as there were no large difference between the effects of these drugs in hemitransected and intact rats. On the other hand, only about 50% of the dopaminergic neurons in their study were excited at all by haloperidol, even at 500 µg/kg, and in the excited cells the mean increase in firing rate was relatively modest, less than 20%. Although firing rate increases up to 56% were seen after administration of 4 mg/kg haloperidol, the significance of the response to such extremely high doses is unclear.

When administered locally in substantia nigra, autoreceptor antagonists (e.g. haloperidol) have been reported to be without effect (BUNNEY et al. 1973b; LACEY et al. 1990) or to cause large (GROVES et al. 1975) or modest (PUCAK and GRACE 1996) increases in firing of nigral dopaminergic neurons. Although it is clear that general anesthetics can interfere with the response of dopaminergic neurons to autoreceptor blockade (MEREU et al. 1984b), these inconsistent and surprisingly modest effects of D_2 receptor antagonists are hard to reconcile with the generally accepted idea that somatodendritic autoreceptors play a significant role in modulating the firing rate of dopaminergic neurons under physiological conditions.

Furthermore, when the autoreceptors are partially or completely inactivated by treatment with pertussis toxin or antisense knockdown, there are no significant changes in the spontaneous firing rate or pattern of substantia nigra dopaminergic neurons recorded in vivo (INNIS and AGHAJANIAN 1987; TEPPER et al. 1997; SHEPARD and CONNELLY 1999).

Experiments in which somatodendritic autoreceptors are stimulated by endogenous dopamine release by synaptic stimulation reveal changes in dendritic excitability with no significant alteration in mean firing rate (TRENT and TEPPER 1991). The absence of a gross change in neuronal activity is likely due to a more modest and localized activation of autoreceptors than is achieved by application of exogenous drugs, and is consistent with the functional compartmentalization of the dopaminergic neuron into different electroresponsive regions that may function independently (GRACE 1990). Thus, somatodendritic dopamine autoreceptors may serve as a mechanism for altering the excitability and/or response of specific dendritic segments of a neuron in a local manner in response to phasic afferent inputs, and in this way alter the way the neuron integrates its afferent inputs in a subtle and graded fashion.

F. Miscellaneous Neuropharmacology

I. Gamma-Hydroxybutyric Acid

Gamma-hydroxybutyric acid (GHBA) is a normal constituent of the mammalian brain and has been proposed as a putative neurotransmitter and/or neuromodulator (see MAITRE et al. 2000 for a recent review). GHBA administration has been shown to modify neuronal activity of dopaminergic neurons of the pars compacta in various ways. In chloral hydrate anesthetized rats, GHBA inhibits impulse flow and this inhibition is blocked by the selective GABA$_B$ antagonist, SCH 50911, but not by the selective GHBA-antagonist NCS-382, suggesting an action on GABA$_B$ receptors (ERHARDT et al. 1998). On the other hand when administered in low doses to unanesthetized rats, GHBA was found to increase the firing rate of pars compacta dopaminergic neurons (DIANA et al. 1991b) and to produce heterogeneous responses in non-dopaminergic pars reticulata cells (DIANA et al. 1993b). Unfortunately, no antagonism studies were performed, thus leaving open the possibility that GHBA in low doses may act through GHBA receptors (see MAITRE et al. 2000) to produce excitation of pars compacta neurons and GABA$_B$ receptors to produce inhibition and regularization of firing.

II. Glycine

Dopaminergic neurons respond to bath application of glycine in vitro with a chloride-dependent membrane hyperpolarization. This response is sensitive to strychnine and insensitive to bicuculline or picrotoxin, indicating that it is mediated by a glycine-specific receptor (MERCURI et al. 1990). The source of the glycinergic input is unknown, and could originate in as yet unidentified nigral interneurons and/or from the brainstem (McGEER et al. 1987).

III. Neuropeptides

Cholecystokinin-8 (CCK-8) is the carboxyterminal octapeptide of the peptide cholecystokinin, and is found in some dopaminergic neurons in rat VTA and substantia nigra (SKIRBOLL et al. 1981; KALIVAS 1993). CCK is co-released with dopamine from dopaminergic dendrites (FREEMAN et al. 1991), and when administered systemically in vivo or locally in vitro, CCK-8 excites dopaminergic neurons. In vivo, CCK-8 increases firing rate and burst firing (SKIRBOLL et al. 1981; FREEMAN and BUNNEY 1987). Thus, dopaminergic neurons may be considered to express a second class of autoreceptor, a CCK autoreceptor that acts to facilitate rather than depress the excitability of the neuron. In vitro studies in dissociated dopaminergic nigral neurons show that CCK-8 acts through CCK-A receptors to activate an inward G-protein coupled current. The current was insensitive to pertussis toxin but was abolished by intracellular heparin or calcium chelators, suggesting that it is mediated by IP_3-induced calcium release (WU and WANG 1994). However, in addition to its excitatory effects, CCK also appears to potentiate the inhibitory effects of dopamine autoreceptor stimulation through an unknown mechanism (HOMMER and SKIRBOLL 1983; FREEMAN and BUNNEY 1987; KALIVAS 1993), so the physiological significance of CCK release in substantia nigra remains to be determined.

Neurotensin and the related peptide, neuromedin N are also present in dopaminergic neurons in rat mesencephalic dopaminergic neurons, some of which also contain CCK. These neurons also express neurotensin receptors. In addition, neurotensin is contained in afferents to the substantia nigra and VTA. Similar to CCK, application of neurotensin in vivo or in vitro leads to increased firing rates of dopaminergic neurons (see KALIVAS 1993 for review). Part of this excitatory effect is due to the opening of a G-protein coupled non-selective inward cation conductance (CHIEN et al. 1996). However, neurotensin also affects autoreceptor responses, but in contrast to CCK, neurotensin attenuates the effects of dopamine autoreceptor agonists (WERKMAN et al. 2000) and does so by acting to close the same potassium conductance that is opened by dopamine autoreceptor and $GABA_B$ receptor agonists (LACEY et al. 1988; FARKAS et al. 1997).

Despite being contained in striatonigral neurons that synapse on dopaminergic neurons in substantia nigra (MAHALIK 1988), substance P has little or no effect when applied locally to substantia nigra dopaminergic neurons (COLLINGRIDGE and DAVIES 1982; PINNOCK and DRAY 1982), presumably because levels of substance P receptor binding are low or undetectable in substantia nigra (ROTHMAN et al. 1984). On the other hand, iontophoretic application of substance K or kassinin excites dopaminergic and non-dopaminergic nigral neurons in vivo (INNIS et al. 1985), and senktide, a selective selective neurokinin NK3 receptor agonist excites dopaminergic neurons in vitro (KEEGAN et al. 1992). The source and identity of the endogenous ligand is unclear, although nigral levels of both substance P and substance K decrease following excitotoxic lesions of striatum (ARAI et al. 1985). Since essentially all electrophysiological changes in nigral neurons following striatal stimulation appear to be due to GABA release, the physiological significance of these tachykinin effects is unclear at present.

G. Acute and Chronic Effects of Antipsychotics on Dopaminergic Neurons

I. Differences Between Effects of Typical and Atypical Antipsychotics

As discussed above, acute systemic administration of antipsychotics increases the activity of dopaminergic neurons in the different subdivisions of the midbrain. One potentially important difference that is apparent between A9 and A10 neurons is the response to "atypical" antipsychotics of which clozapine represents the prototype. These neuroleptics are distinguished from the "typical" antipsychotics because they have a much lower incidence of inducing extrapyramidal side effects (see MELTZER et al. 1999 for a recent review) and thus represent a pharmacological class with enormous clinical potential. One widely accepted hypothesis for the lack of extrapyramidal side effects from the atypical antipsychotics has been that the former have a preferential site of action in the mesolimbic and/or mesocortical dopaminergic system. Early in vivo recording studies following acute administration showed that these compounds increased the firing rate selectively in the A10 region without affecting neuronal activity in A9, whereas their chronic administration led to a reduction in the proportion of spontaneously active neurons as indexed by the cells per track ratio (see below) solely in A10 (CHIODO and BUNNEY 1983; WHITE and WANG 1983). Subsequent studies suggested a possible difference in interaction of the atypical antipsychotics with autoreceptors in A9 and A10 (e.g., STOCKTON and RASMUSSEN 1996). On the other hand, in vitro studies generally have not revealed a differential response of A9 and A10 neurons to typical and atypical antipsychotics (e.g., SUPPES and PINNOCK 1987; BOWERY et al. 1994) and a recent in vivo study showed that intravenous administration of clozapine increased the firing rate of nigrostriatal dopaminergic neurons to the same extent as seen in VTA neurons, but only in unanesthetized rats (MELIS et al. 1998). Thus, it is not yet clear that there is a preferential site of action of atypical antipsychotics for the mesolimbic versus nigrostriatal system, at least as far as autoreceptor blockade goes, nor what the pharmacological basis of such a preference might be. Alternative explanations include, for example, differences between the two classes of antipsychotics with respect to interaction with alpha$_2$ adrenergic receptors (HERTEL et al. 1999), a relatively more potent blockade of 5HT$_{2A}$ receptors coupled with a weak blockade of D$_2$ receptors (MELTZER et al. 1989, 1999), or a combination of properties (KINON and LIEBERMAN 1996), which may be the substrate for the differential incidence of extrapyramidal side effects resulting from chronic treatment with typical and atypical neuroleptics.

II. Effects of Chronic Antipsychotic Drug Administration – The Depolarization Block Hypothesis

While the acute administration of dopamine receptor antagonists leads to increased spontaneous firing of dopaminergic neurons (BUNNEY and AGHAJANIAN 1973; GROVES et al. 1975; WANG 1981b), chronic administration of antipsychotics has been suggested to reduce dopaminergic synaptic transmission not only by blocking postsynaptic dopamine receptors, but by a relatively novel mechanism in which a state of chronic depolarization of dopaminergic neurons is induced which, over time, renders a population of neurons unable to fire action potentials thereby reducing the population of spontaneously active dopaminergic neurons. This phenomenon was termed depolarization block (BUNNEY and GRACE 1978) and was measured experimentally by counting the number of neurons displaying the characteristics of dopaminergic neurons encountered while lowering an extracellular recording electrode through the region of the substantia nigra and/or VTA. Following chronic, but not acute antipsychotic treatment, the mean number of presumed dopaminergic neurons encountered per electrode track was found to be less than in controls. Iontophoresis of GABA or dopamine which would be expected to hyperpolarize the neurons reversed these effects. It was therefore proposed that the reduction in the number of cells encountered per track following chronic antipsychotic drug administration was a result of depolarization inactivation of the neurons (BUNNEY and GRACE 1978).

Considerable interest in this theory arose quickly as it provided the first compelling explanation of why the antipsychotic effects of neuroleptics usually take weeks to develop, despite the fact that the blockade of dopamine receptors occurs immediately upon drug administration. Subsequently, numerous reports consistent with the initial phenomenological description emerged (e.g., CHIODO and BUNNEY 1983; WHITE and WANG 1983; SKARSFELDT 1988, 1995). With additional evidence from intracellular and extracellular recordings consistent with the existence of depolarized dopaminergic neurons in animals chronically treated with neuroleptics (GRACE and BUNNEY 1986), the depolarization block theory gained widespread, although not universal (see MEREU et al. 1994, 1995), acceptance as the principal mechanism by which neuroleptics exert their clinically therapeutic antipsychotic action. The phenomenon appears to be fully reversible, as after withdrawal for 8–14 days after up to 14 months of chronic treatment with haloperidol there are no longer any changes in the number of cells per track or in any other measures of dopaminergic neuron activity compared to controls (CHIODO and BUNNEY 1987; GARIANO et al. 1990). The actual substrates of the depolarization inactivation are not known, although it appears that intact afferent input from the forebrain is essential for the development and maintenance of the phenomenon (see GRACE et al. 1997 for review).

There are actually two separate issues to consider with respect to the role of depolarization block in the clinical response to chronic administration of antipsychotic drugs. The first is whether depolarization block actually occurs in dopaminergic neurons in animals and/or humans chronically treated with neuroleptic drugs. The second is whether depolarization inactivation (assuming it occurs) accounts for the therapeutic action of antipsychotic drugs.

Much of the evidence for the existence of depolarization block relies on measurements of cells per track data described above. While drug-induced changes in the number of cells per track might well indicate changes in the proportion of spontaneously active neurons, alternative explanations have been proposed including changes in firing rate and/or changes in the extent to which the action potential invades the dendrites thereby altering the size of the extracellular field potential of the neuron. Both of these would alter the probability of encountering a neuron while lowering a microelectrode through a designated region of the brain (see discussions in DIANA et al. 1995a and DAI and TEPPER 1998). For example, a reduction in the number of dopaminergic cells per track was observed after chronic ethanol administration and subsequent withdrawal and attributed to a reduced number of spontaneously active neurons due to depolarization block (SHEN and CHIODO 1993). Subsequent experiments (DIANA et al. 1995a), however, revealed that during withdrawal, dopaminergic neurons exhibited reduced spontaneous activity (i.e. lower firing rates and burst firing) which could account for more difficult detection and hence a lower number of cells per track even though the neurons were not in depolarization block as evidenced by their slow spontaneous activity and the inability of apomorphine to increase the number of cells per track. Thus, although an interesting and potentially valuable tool, the interpretation of changes in the number of cells per track is complex and may be due to factors other than or in addition to a change in the number of spontaneously active neurons.

As to the second issue, although able to replicate the reduction in cells per track following chronic dopamine antagonists in anesthetized rats, MEREU et al. (1994, 1995) found no reduction in the number of cells per track in locally anesthetized, immobilized, and artificially respired rats. These authors argued that the appearance of depolarization block is an artifact of some type of interaction between general anesthetics and the neuroleptics, and hence is unlikely to account for the therapeutic effects of neuroleptics in (unanesthetized) humans. In addition, some predictions of the depolarization block hypothesis, for example the expected reduction in extracellular dopamine levels in striatal and/or cortical terminal fields following chronic neuroleptic treatment, have been difficult to demonstrate experimentally (e.g., HERNANDEZ and HOEBEL 1989; ZHANG et al. 1989; HOLLERMAN et al. 1992; MOGHADDAM and BUNNEY 1993 but see also MOORE et al. 1998). Furthermore, manipulations that increase dopaminergic neuron firing and dopamine release in normal animals also increase extracellular dopamine levels after chronic haloperidol treatment, although the hypothesis would seem to predict that dopaminergic neurons in depolarization block should be unable to respond to excitatory stimuli with an increase in firing rate and dopamine release (KLITENICK et al. 1996).

In conclusion, although there is electrophysiological evidence in support of the development of depolarization block in dopaminergic neurons following chronic neuroleptic treatment, some of these data, particularly the cells per track data, are open to alternative interpretations. In addition, the apparent dependency of the development of depolarization inactivation on anesthetic state or other aspects of the experimental preparation, coupled with the inability of a number of experiments to demonstrate the expected decrease in extracellular dopamine levels following chronic neuroleptic treatment, point toward the need for more research before a definitive conclusion about the role of depolarization inactivation in the therapeutic effects of neuroleptics can be reached.

H. Dopaminergic Neurons and Drugs of Abuse: Acute and Chronic Studies

I. Acute Effects of Drugs of Abuse on Dopaminergic Neurons

Dopaminergic systems of the mammalian brain are a major target of drugs of abuse and represent cellular systems which are considered crucial in conveying affect-related effects of various addicting drugs. Thus, dopaminergic neurons have been extensively studied in recent years and much is now known about their response to administration of drugs of abuse (WHITE 1996; DIANA 1998; PULVIRENTI and DIANA 2001).

In vivo, drugs as structurally and pharmacologically diverse as ethanol (GESSA et al. 1985), nicotine (LICHTENSTEIGER et al. 1982; GRENHOFF et al. 1986; MEREU et al. 1987), morphine (IWATSUBO and CLOUET 1977; GYSLING and WANG 1983; MATTHEWS and GERMAN 1984) and cannabinoids (FRENCH 1997; FRENCH et al. 1997; GESSA et al. 1998) increase the firing rate and bursting activity of mesencephalic dopaminergic neurons, resulting in augmented dopamine outflow in terminal areas when acutely administered (DI CHIARA and IMPERATO 1988). In contrast, psychostimulants such as amphetamine and cocaine decrease dopaminergic neuronal activity, principally through indirect actions at the somatodendritic autoreceptor (BUNNEY et al. 1973a,b; GROVES et al. 1975; EINHORN et al. 1988), although their effects on dopamine outflow in terminal regions are not dissimilar from other addicting compounds, i.e., they promote an increase in extracellular dopamine levels by blocking and/or reversing the dopamine uptake transporter (KUCZENKSI 1983).

In vitro recordings have provided useful insights into the cellular mechanisms which lead to the excitation of dopaminergic neurons after acute administration of drugs of abuse. Morphine does not act directly on dopaminergic neurons which lack μ-opioid receptors, but rather acts on μ-opioid receptors located on pars reticulata GABAergic neurons producing a potassium-mediated hyperpolarization, which in turn, leads to a depolarization and consequent excitation of dopaminergic neurons through disinhibition (LACEY et al. 1989; JOHNSON and NORTH 1992; KALIVAS 1993). Although the pars reticulata neuron mediating the disinhibitory effect of µ-opioids has not been conclusively identified and could be an interneuron (JOHNSON and NORTH 1992), other anatomical and electrophysiological studies have demonstrated that nigrothalamic and nigrotectal neurons exhibit the requisite synaptic arrangement to underlie the disinhibitory effect (HAJOS and GREENFIELD 1994; TEPPER et al. 1995, 2000).

A similar mechanism was proposed for the action of ethanol when it was demonstrated that the excitation of dopaminergic neurons induced by ethanol (MEREU et al. 1984a; GESSA et al. 1985) was accompanied by a reduction in pars reticulata non-dopaminergic neuronal activity (MEREU and GESSA 1985) of similar proportions. However, this is unlikely to be the sole mechanism of action of ethanol on dopaminergic neurons, since ethanol activates dopamine-containing cells even when these are mechanically dissociated or studied in slices (BRODIE et al. 1999a,b; BRODIE and APPEL 1998), and ethanol has been shown to have direct effects on the calcium-dependent potassium current in dopaminergic neurons. (BRODIE and APPEL 1998; BRODIE et al. 1999a,b).

Nicotine has been reported to activate dopaminergic neurons in vivo (LICHTENSTEIGER et al. 1982; GRENHOFF et al. 1986; MEREU et al. 1987) and in vitro (CALABRESI et al. 1989; PIDOPLICHKO et al. 1997), but in contrast to ethanol and opiates, its action is mediated by a direct action on nicotinic receptors located on dopaminergic neurons. Most of the nicotine-induced inward current in dopaminergic neurons is carried by β2-subunit-containing receptors with a minor component contributed by α7 subunit-containing receptors, and even when exposed to concentrations of nicotine found in the blood of smokers, exhibits rapid desensitization (PIDOPLICHKO et al. 1997; DANI et al. 2000).

Among various classes of drugs of abuse, cannabinoids rank high in the list especially in terms of spread of their use and recently have received much attention possibly owing to their social popularity. The actions of Δ^9-tetrahydrocannabinol (THC), the active principle of marijuana, and its synthetic analogues have been recently described in central dopaminergic systems. After acute administration, dopamine outflow is increased in the nucleus accumbens (GARDNER and LOWINSON 1991) and prefrontal cortex (CHEN et al. 1990) while dopaminergic neuronal activity in anesthetized rats is increased in the VTA and substantia nigra (FRENCH 1997; FRENCH et al. 1997) by an action on CB1 receptors. In unanesthetized rats, cannabinoids similarly activate mesolimbic (GESSA et al. 1998) and mesoprefrontal dopaminergic neurons (DIANA et al. 1998b) by a selective action on CB1 receptors. Although there is general agreement about the systems level effects of CB1 stimulation on dopaminergic systems (but see GIFFORD et al. 1997), their cellular site(s) of action remain controversial. Autoradiographic studies combined with 6-OHDA lesions of the ascending dopaminergic pathways have indicated that CB1 receptors are not expressed by dopaminergic neurons (HERKENHAM et al. 1991) while these receptors have been detected in high amounts on pars reticulata GABAergic neurons and on the terminals of striatonigral projection neurons in substantia nigra (HERKENHAM et al. 1991). The existence of CB1 receptors on pars reticulata GABAergic neurons coupled with the results of in vivo microdialysis studies in the shell of the nucleus accumbens has led to the suggestion that cannabinoids may increase dopaminergic transmission by acting through µ-opioid receptors in a disinhibitory fashion (TANDA et al. 1997) similar to that described above for opioids. However, such a mechanism seems incompatible with direct experimental evidence that shows that cannabinoid agonists increase rather than decrease pars reticulata neuronal activity (TERSIGNI and ROSENBERG 1996; MILLER and WALKER 1995; see MELIS et al. 2000 for discussion on this point) and that the cannabinoid-induced stimulation of firing rate of dopaminergic neurons is not antagonized by naloxone (FRENCH 1997; MELIS et al. 2000). Thus, at present, the cellular site of action for cannabinoid-induced increase of dopaminergic neuronal activity remains to be determined.

II. Chronic Effects of Drugs of Abuse on Dopaminergic Neurons

While studies of the acute effects of drug of abuse on dopaminergic neurons are extremely informative to identify primary sites of actions of addicting compounds, they are less helpful when trying to understand the general phenomenon of drug addiction. Drug addiction is induced by chronic administration of various substances and is now widely accepted as an example of drug-induced alterations in neuronal plasticity (NESTLER 1993; DIANA 1996, 1998; PULVIRENTI and DIANA 2001). Thus, the study of the activity of dopaminergic neurons after chronic administration of drugs of abuse is considered more pertinent and relevant in the context of drug dependence.

Chronic administration of psychostimulants such as cocaine and amphetamine have been shown to affect mesolimbic dopaminergic neurons at various levels (HENRY et al. 1989; ACKERMAN and WHITE 1990; WHITE et al. 1995; WHITE 1996). Firing rate appears to be higher in rats chronically treated with cocaine (ZHANG et al. 1992a), perhaps due to the reduced sensitivity of somatodendritic autoreceptors (ACKERMAN and WHITE 1990; ZHANG et al. 1992a), although administration regimen seems to be an important factor as it could affect differently A9 and A10 neurons (GAO et al. 1998). Chronic treatment with amphetamine leads to a reduction in the sensitivity of dopaminergic neurons to autoreceptor-mediated inhibition by apomorphine or amphetamine in a dose-dependent manner (KAMATA and REBEC 1983, 1984a,b). Further, an increased sensitivity to iontophoretically applied glutamate, which could push the cells to an apparent depolarization block (ZHANG et al. 1997), has been described after both cocaine and amphetamine, although it is unclear if these effects are related to the chronic regimen with cocaine and/or amphetamine or to their withdrawal, as investigations were carried out at variable lengths of time after last drug administration (for review see WHITE 1996). In addition, chronic amphetamine treatment affects dopaminergic neurons not only at the soma but also at the level of the synaptic endings. The ability of amphetamine to induce a decrease in striatal dopamine terminal excitability (TEPPER et al. 1984a) is blunted or eliminated in animals following 2 weeks of treatment with amphetamine (GARCIA-MUNOZ et al. 1996).

Morphine, when administered repeatedly, also produces a number of effects on the mesolimbic dopaminergic system. The firing rate of dopaminergic neurons is within control values 2h after the last morphine administration, but firing rate and burst firing are drastically reduced when the opiate antagonist, naloxone, is administered at this time (DIANA et al. 1995b). Further, the relative refractory period is consistently prolonged, supporting an increased refractoriness of the dopaminergic neuron in generating action potentials (DIANA et al. 1995b). In addition, dopaminergic cell bodies appear to "shrink" (SKLAIR-TAVRON et al. 1996) after chronic morphine administration, an effect consistent with the prolongation of refractory periods of these units (DIANA et al. 1995b,c; DIANA 1996) although is unclear if the reduction in cell body size is induced by chronic morphine or by its withdrawal. These effects, in any event, all point to a vulnerability of the mesolimbic dopaminergic system after chronic administration of morphine.

Ethanol, when chronically administered, has been shown to increase the basal activity of dopaminergic neurons projecting to the nucleus accumbens and no tolerance seems to develop (DIANA et al. 1992) to its stimulating properties on dopaminergic neurons (GESSA et al. 1985). Chronically administered nicotine, on the other hand, appears to affect dopaminergic neurons differently. In vitro studies have shown that the stimulating properties of nicotine upon dopaminergic neurons are rapidly lost after repeated exposure due to desensitization of nicotinic receptors present in the somatic region of dopaminergic neurons and helping in explaining acute tolerance to nicotine's rewarding effects (PIDOPLICHKO et al. 1997).

Another commonly abused drug is Δ^9-THC, the active principle of marijuana. Its actions on dopaminergic neurons have been recently elucidated and are similar from those reported above for other drugs, at least in terms of neuronal activity, in spite of the fact that cannabinoids are frequently considered only mildly addicting (GRINSPOON and BAKALAR 1997). Chronic administration of Δ^9-THC alters dopaminergic neuronal functioning in the limbic system in a way similar to that reported for morphine, and tolerance to the stimulating properties of Δ^9-THC seems to develop only in A9 but not in A10 neurons (WU and FRENCH 2000). Firing rate and burstiness are reduced after chronic exposure and are further reduced if the selective antagonist SR 1417116 A is administered (DIANA et al. 1998a). In contrast, overt behavioral signs of withdrawal are evident only in rats in which the selective antagonist, SR 141716 A, was administered, suggesting that the lack of withdrawal symptoms might be due to the presence of residual Δ^9-THC, which would counteract abstinence signs. This fact may also help in explaining why cannabinoids are traditionally considered devoid of withdrawal signs (GRINSPOON and BAKALAR 1997).

In conclusion, while acute administration of addicting drugs stimulates the activity of dopaminergic neurons and in particular the mesolimbic system, chronic administration alters neuronal functioning in various ways which indicate the mesolimbic dopaminergic pathway as a major target in the actions of chronic administration of addicting drugs, and provide the rationale for drug addiction viewed as an example of drug-induced alterations in neuronal plasticity (Koob and Bloom 1988; Nestler 1992, 1993, 2001; Diana 1996, 1998; Koob and Le Moal 1997; Pulvirenti and Diana 2001).

III. Withdrawal Following Chronic Administration

While repeated administration forms the basis of neurobiological changes induced by drugs of abuse, withdrawal is often a time-window which reveals enduring effects produced by the continued exposure. Indeed, drug-withdrawal offers the unique opportunity to study neurobiological alterations induced by chronic administration of addicting drugs in a drug-free condition, in which the abused substance may act as a potential confounding factor. It is often very difficult to discriminate between effects induced by the drug, when chronically administered, or by its absence after chronic administration. Thus, it is advisable to carefully discriminate between effects induced by drugs themselves and effects induced by their absence since interpretations are often opposite (Diana 1996; Sklair-Tavron et al. 1996; Diana et al. 1999).

The effect of withdrawal from various addicting drugs has recently been described in dopaminergic neurons. Ethanol withdrawal reduces the spontaneous activity (firing rate and burstiness) of dopaminergic neurons projecting to the nucleus accumbens, in rats in vivo (Diana et al. 1993a) and in mice in vitro (Bailey et al. 1998), and these effects are accompanied by an elongation of refractory periods and a reduction of dopamine dialysate in the nucleus accumbens (Fig. 6) (Diana et al. 1993a). The reduction in neuronal activity does not seem to be due to the depolarization block proposed for cocaine withdrawal (Ackerman and White 1990, 1992) as it persists in rats anesthetized with chloral hydrate which show the same sensitivity to apomorphine as unanesthetized rats (Diana et al. 1995a, but see Shen and Chiodo 1993). Further, hypofunctioning of dopaminergic neurons outlasts the behavioral manifestations of withdrawal, suggesting a role for dopaminergic neurons in subtle but reproducible and enduring modifications in cell physiology unrelated to somatic withdrawal but more closely linked to longer lasting changes occurring after ethanol withdrawal (Diana 1996, 1998).

Morphine withdrawal also produces a depression in firing rate and burst firing in dopaminergic neurons with no evidence of depolarization block (Diana et al. 1995b). These data are consistent with the hyperpolarization due to an increased GABA release seen in dopaminergic neurons in vitro during acute morphine withdrawal (Bonci and Williams 1997). In addition, morphine withdrawal produces a reduction in glutamatergic EPSCs in VTA dopaminergic neurons due to reduced glutamate release (Manzoni and Williams 1999). Furthermore, as in the case of ethanol, the reduction of dopaminergic activity after opiate withdrawal persists for 14 days, while behavioral measures

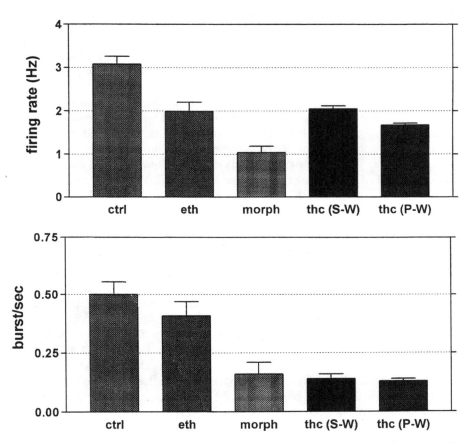

Fig. 6. Extracellular electrophysiological properties of mesolimbic dopaminergic neurons projecting to the nucleus accumbens in vivo after withdrawal from chronic administration of ethanol (*eth*), morphine (*morph*), and Δ^9-THC (*thc*) spontaneous (*S-W*) and pharmacologically precipitated (*P-W*). Note the parallel decline in firing rate (*top*) and bursting activity (*bottom*) irrespective of the substance administered. Due to the different baseline activity in treated and control rats, number of bursts is expressed as bursts per second. See details in DIANA et al. (1995c) and DIANA (1998)

of abstinence are within control values at 3 days (DIANA et al. 1999). Once again, these results would suggest that hypofunction of the mesolimbic dopaminergic system is related to the long-term consequences of chronic opiate abuse and not to behavioral signs of withdrawal (but see HARRIS and ASTON-JONES 1994). Furthermore, administration of morphine to rats with a history of morphine addiction results in an activation of dopaminergic firing rate far greater than that observed in saline-treated counterparts (DIANA et al. 1999). This suggests that although dopaminergic neurons have returned to apparent normality (extracellular electrophysiological indices are within control values), the mesolimbic dopamine system remains hyper-responsive (i.e., vulnerable) to opiates even longer, with profound implications for the phenomenon of relapse into opiate addiction in humans. Nicotine, the principal constituent of tobacco, seems to produce different effects upon discontinuation of chronic exposure (RASMUSSEN and CZACHURA 1995), at least in vivo. Indeed, chronic administration seems to produce a reduction of firing rate in the A10 region but not in the A9, whereas withdrawal restored control firing rates in A10 and increased above control in A9 (RASMUSSEN and CZACHURA 1995). Although stimulating, these results are flawed by the lack of antidromic identification of the neurons, which hampers firm conclusions on the regional selectivity of the effects observed, and thus we await confirmation in light of contrasting results obtained in vitro (PIDOPLICHKO et al. 1997) and in vivo with the microdialysis method (CARBONI et al. 2000).

Cannabis derivatives have long been seen as only mildly addicting and consequently as devoid of withdrawal manifestations. Recently, however, with the advent of appropriate pharmacological tools, it has been possible to demonstrate behavioral manifestations of cannabinoid withdrawal (ACETO et al. 1995, 1996; TSOU et al. 1995). On this basis we investigated the possibility that chronic treatment with Δ^9-THC affects the function of the mesolimbic dopamine system. We found that both withdrawal conditions (spontaneous and pharmacologically precipitated) reduced the firing rate of dopaminergic neurons projecting to the nucleus accumbens with behavioral manifestations of withdrawal evident only in the pharmacologically precipitated withdrawal group (DIANA et al. 1998a). These facts suggest that hypofunction of the dopaminergic mesolimbic system may participate in the neurobiological basis of long-term consequences of cannabinoid dependence, allowing us to extend this conclusion to the general phenomenon irrespective of the chemical class abused and further suggest that the failure to observe behavioral signs of cannabinoid withdrawal could be due to high lipophilicity of cannabinoids, which hampered observation of an abrupt somatic withdrawal (DIANA et al. 1998a).

I. Conclusions

In the last decade, electrophysiological studies have added significantly to our knowledge of the physiological activity and pharmacological responsiveness of dopaminergic neurons. Many of the intrinsic mechanisms that lead to action potential generation and the generation of different firing patterns, both under normal physiological conditions and after various pharmacological manipulations, have been described. Considerable advances have been made in understanding the pathways, neurotransmitters, and receptors that form the substrates for the afferent regulation of central dopaminergic systems.

These central dopaminergic systems have been demonstrated to be a major target for many psychotropic drugs including psychotherapeutic antipsychotics and drugs of abuse. Dopaminergic systems play a role in the response to drugs of abuse not only when administered acutely but, perhaps more importantly, following chronic administration and withdrawal. Under withdrawal, regardless of the specific drug, there is a depression in the spontaneous activity and burst firing of dopaminergic neurons projecting to the nucleus accumbens. This "hypodopaminergia" outlasts the behavioral signs of withdrawal and suggests that dopaminergic systems play an important role in the long-term consequences of prolonged drug intake and provides an example of drug-induced alterations in neuronal plasticity affecting the mesolimbic dopaminergic system. Identification of the etiological factors leading to the abnormal cellular physiology following chronic administration of, and withdrawal from, addictive drugs may pave the way for future pharmacological treatments of drug addiction.

Acknowledgements. The authors thank Fulva Shah and Stefano Aramo for expert technical assistance. Supported, in part, by the National Institute of Neurological Disease and Stroke (NS 34865 to JMT), the National Institute for Mental Health (MH58885 to JMT) and MURST (Cofinanziamento Progetti di Ricerca di rilevante interesse nazionale prot N¡ 9905043527 to M.D.).

References

Abercrombie ED, Keefe KA, DiFrischia DS, Zigmond MJ (1989) Differential effects of stress on in vivo dopamine release in striatum, nucleus accumbens and medial frontal cortex. J Neurochem 52:1655–1658

Aceto MD, Scates SM, Lowe JA, Martin BR (1995) Cannabinoid precipitated withdrawal by the selective cannabinoid receptor antagonist, SR 141716 A. Eur J Pharmacol 282:R1–2

Aceto MD, Scates SM, Lowe JA, Martin BR (1996) Dependence on delta 9-tetrahydrocannabinol: studies on precipitated and abrupt withdrawal. J Pharmacol Exp Ther 278:1290–1295

Ackerman JM, White FJ (1990) A10 somatodendritic dopamine autoreceptor sensitivity following withdrawal from repeated cocaine treatment. Neurosci Lett 117:181–187

Ackerman JM, White FJ (1992) Decreased activity of rat A10 dopamine neurons following withdrawal from repeated cocaine. Eur J Pharmacol 218:171–173

Aghajanian GK, Bunney BS (1973) Central dopaminergic neurons: Neurophysiological identification and responses to drugs. Snyder SH Usdin E (eds) Frontiers in Catecholamine Research, Pergamon Press, NY, pp 643–648

Aghajanian GK, Bunney BS (1977) Dopamine "autoreceptors": pharmacological characterization by microiontophoretic single cell recording studies. Naunyn-Schmiedeberg's Arch Pharmacol 297:1–7

Akaoka H, Charléty P, Saunier C-F, Buda M, Chouvet G (1992) Inhibition of nigral dopamine neurons by systemic and local apomorphine: Possible contribution of dendritic autoreceptors. Neuroscience 49:879–891

Anden NE, Carlsson A, Dahlstrom A, Fuxe K, Hillarp NA, Larsson K (1964) Demonstration and mapping out of nigro-neostriatal dopamine neurons. Life Sci 3:523–530

Anden NE, Dahlstrom A, Fuxe K, Larsson K (1965) Mapping out of catecholamine and 5-hydroxytryptamine neurons innervating the telencephalon and diencephalon. Life Sci 4:1275–1279

Arai H, Sirinathsinghiji DJS, Emson PC (1985) Depletion in substance P- and neurokinin A-like immunoreactivity in substantia nigra after ibotenate-induced lesions of striatum. Neurosci Res 5:167–171

Arnold EB, Molinoff PB, Rutledge CO (1977) The release of endogenous norepinephrine and dopamine from cerebral cortex by amphetamine. J Pharmacol Exp Ther 202:544–557

Bailey CP, Manley SJ, Watson WP, Wonnacott S, Molleman A, Little HJ (1998) Chronic ethanol administration alters activity in ventral tegmental area neurons after cessation of withdrawal hyperexcitability. Brain Res 803:144–152

Bannon MJ, Bunney EB, Roth RH (1981) Mesocortical dopamine neurons: Rapid transmitter turnover compraed to other brain catecholamine systems. Brain Res 218:376–382

Bannon MJ, Reinhard Jr. JF, Bunney EB, Roth RH (1982) Unique response to antipsychotic drugs is due to absence of terminal autoreceptors in mesocortical dopamine neurones. Nature 296:444–446

Bean AJ, Shepard PD, Bunney BS, Nestler EJ, Roth RH (1988) The effects of pertussis toxin on autoreceptor-mediated inhibition of dopamine synthesis in the rat striatum. Mol Pharmacol 34:715–718

Beart PM, McDonald D (1982) 5-Hydroxytryptamine and 5-hydroxytryptaminergic-dopaminergic interactions in the ventral tegmental area of rat brain. J Pharm Pharmacol 34:591–593

Beninato M, Spencer RF (1988) The cholinergic innervation of the rat substantia nigra: a light and electron microscopic immunohistochemical study. Exp Brain Res 72:178–184

Bertler A, Falck B, Gottfries CG, Lijunggren L, Rosengren E (1964) Some observations on adrenergic connections between mesencephalon and cerebral emispheres. Acta Pharmacol Toxicol 21:283–289

Bjorklund A, Lindvall O (1975) Dopamine in dendrites of substantia nigra neurons: suggestions for a role in dendritic terminals. Brain Res 83:531–537

Blaschko H (1939) The specific action of L-DOPA decarboxylase. J Physiol (Lond) 96:50–51

Bolam JP, Smith Y (1990) The GABA and substance P input to dopaminergic neurones in the substantia nigra of the rat. Brain Res 529:57–78

Bonci A, Williams JT (1997) Increased probability of GABA release during withdrawal from morphine. J Neurosci 17:796–803

Bonci A, Grillner P, Siniscalchi A, Mercuri NB, Bernardi G (1997) Glutamate metabotropic receptor agonists depress excitatory and inhibitory transmission on rat mesencephalic principal neurons. Eur J Neurosci 9:2359–2369

Bowery B, Rothwell LA, Seabrook GR (1994) Comparison between the pharmacology of dopamine receptors mediating the inhibition of cell firing in rat brain slices through the substantia nigra pars compacta and ventral tegmental area. Br J Pharmacol 112:873–880

Boyar WC, Altar CA (1987) Modulation of in vivo dopamine release by D_2 but not D_1 receptor agonists and antagonists. J Neurochem 48:824–831

Brodie MS, Appel SB (1998) The effects of ethanol on dopaminergic neurons of the ventral tegmental area studied with intracellular recording in brain slices. Alcohol Clin Exp Res 22:236–244

Brodie MS, McElvain MA, Bunney EB, Appel SB (1999a) Pharmacological reduction of small conductance calcium-activated potassium current (SK) potentiates the excitatory effect of ethanol on ventral tegmental area dopamine neurons. J Pharmacol Exp Ther 290:325–333

Brodie MS, Pesold C, Appel SB (1999b) Ethanol directly excites dopaminergic ventral tegmental area reward neurons. Alcohol Clin Exp Res 23:1848–1852

Bunney BS (1979) The electrophysiological pharmacology of midbrain dopaminergic systems. In: Horn AS, Korf J, Westerink BHC (ed) The Neurobiology of Dopamine. New York, Academic Press, pp 417–452

Bunney BS, Aghajanian GK (1973) Electrophysiological effects of amphetamine on dopaminergic neurons. In: Snyder SH, Usdin E (eds) Frontiers in Catecholamine Research. Pergamon Press, NY, pp 957–962

Bunney BS, Grace AA (1978) Acute and chronic haloperidol treatment: comparison of effects on nigral dopaminergic cell activity. Life Sci 23:1715–1727

Bunney BS, Aghajanian GK, Roth RH (1973a) Comparison of effects of L-dopa, amphetamine and apomorphine on firing rate of rat dopaminergic neurones. Nature New Biology 245:123–125

Bunney BS, Walters JR, Roth RH, Aghajanian GK (1973b) Dopaminergic neurons: effect of antipsychotic drugs and amphetamine on single cell activity. J Pharm Exp Ther 185:560–571

Calabresi P, Lacey MG, North RA (1989) Nicotinic excitation of rat ventral tegmental neurones in vitro studied by intracellular recording. Br J Pharmacol 98:135–140

Cameron DL, Williams JT (1993) Dopamine D1 receptors facilitate transmitter release. Nature 366:344–347

Canavier CC (1999) Sodium dynamics underlying burst firing and putative mechanisms for the regulation of the firing pattern in midbrain dopamine neurons: a computational approach. J Comput Neurosci 6:49–69

Carboni E, Bortone L, Giua C, Di Chiara G (2000) Dissociation of physical abstinence from changes in extracellular dopamine in the nucleus accumbens and in the prefrontal cortex of nicotine dependent rats. Drug and Alcohol Dependence 58: 93–102

Cardozo DL, Bean BP (1995) Voltage-dependent calcium channels in rat midbrain dopamine neurons: Modulations by dopamine and $GABA_B$ receptors. J Neurophysiol 74:1137–1148

Carlsson A, Lindqvist M (1963) Effects of chlorpromazine and haloperidol on formation of 3-methoxytyramine and normetanephrine in mouse brain. Acta Pharmacol Toxicol 20:140–144

Carlsson A, Lindqvist M, Magnusson T, Waldeck B (1958) On the presence of three-hydroxytyramine in brain. Science 27:471

Carlsson A, Kehr W, Lindqvist M, Magnusson T, Atack CV (1972) Regulation of monoamine metabolism in the central nervous system. Pharmacol Rev 24:371–384

Cass WA, Gerhardt GA (1995) In vivo assessment of dopamine uptake in rat medial prefrontal cortex: comparison with dorsal striatum and nucleus accumbens. J Neurochem 65:201–207

Cathala L, Paupardin-Tritsch D (1999) Effect of catecholamines on the hyperpolarization-activated cationic I_h and the inwardly rectifying potassium I_{Kir} currents in the rat substantia nigra pars compacta. Eur J Neurosci 11:398–406

Celada P, Paladini CA, Tepper JM (1999) GABAergic control of rat substantia nigra dopaminergic neurons: Role of globus pallidus and substantia nigra pars reticulata. Neuroscience 89:813–825

Chang HT, Kita H, Kitai ST (1984) The ultrastructural morphology of the subthalamic-nigral axon terminals intracellularly labeled with horseradish peroxidase. Brain Res 299:182–185

Charlety PJ, Grenhoff J, Chergui K, De La Chapelle Buda M, Svensson TH, Chouvet G (1991) Burst firing of mesencephalic dopamine neurons is inhibited by somatodendritic application of kynurenate. Acta Physiol Scand 142:105–112

Chen J, Paredes W, Lowinson JH, Gardner EL (1990) Delta 9-tetrahydrocannabinol enhances presynaptic dopamine efflux in medial prefrontal cortex. Eur J Pharmacol 190:259–262

Cheramy A, Leviel V, Glowinski J (1981) Dendritic release of dopamine in the substantia nigra. Nature 289:537–542

Chergui K, Charlety PJ, Akaoka H, Saunier CF, Brunet J-L, Buda M, Svensson TH, Chouvet G (1993) Tonic activation of NMDA receptors causes spontaneous burst discharge of rat midbrain dopamine neurons in vivo. Eur J Neurosci 5:137–144

Chergui K, Akaoka H, Charlety PJ, Saunier CF, Buda M, Chouvet G (1994) Subthalamic nucleus modulates burst firing of nigral dopamine neurones via NMDA receptors. NeuroReport 5:1185–1188

Chien P-Y, Farkas RH, Nakajima S, Nakajima Y (1996) Single-channel properties of the non-selective cation conductance induced by neurotensin in dopaminergic neurons. Proc Natl Acad Sci (USA) 93:14917–14921

Chiodo LA, Antelman SM (1980) Electroconvulsive shock: progressive dopamine autoreceptor subsensitivity independent of repeated treatment. Science 210:799–801

Chiodo LA, Bunney BS (1983) Typical and atypical neuroleptics: Differential effects of chronic administration on the activity of A9 and A10 midbrain dopaminergic neurons. J Neurosci 3:1607–1619

Chiodo LA, Bunney BS (1987) Population response of midbrain dopaminergic neurons to neuroleptics: further studies on time course and nondopaminergic neuronal influences. J Neurosci 7:629–33

Chiodo LA, Bannon MJ, Grace AA, Roth RH, Bunney BS (1984) Evidence for the absence of impulse-regulating somatodendritic and synthesis-modulating nerve terminal autoreceptors on subpopulations of mesocortical dopamine neurons. Neuroscience 12:1–16

Clark D, Chiodo LA (1988) Electrophysiological and pharmacological characterization of identified nigrostriatal and mesoaccumbens dopamine neurons in the rat. Synapse 2:474–485

Clarke PBS, Hommer DW, Pert A, Skirboll LR (1987) Innervation of substantia nigra neurons by cholinergic afferents from pedunculopontine nucleus in the rat: neuroanatomical and electrophysiological evidence. Neuroscience 23:1011–1019

Collingridge GL, Davies J (1982) Actions of substance P and opiates in the rat substantia nigra. Neuropharmacology 21:715–719

Collingridge GL, James TA, MacLeod NK (1980) Antidromic latency variations of nigral compacta neurones. Experientia 36:970–971

Connelly ST, Shepard PD (1997) Competitive NMDA receptor antagonists differentially affect dopamine cell firing pattern. Synapse 25:234–242

Coombs JS, Curtis DR, Eccles JC (1957) The interpretation of spike potentials of motoneurones. J Physiol (Lond) 139:198–231

Corvaja N, Doucet G, Bolam JP (1993) Ultrastructure and synaptic targets of the raphe-nigral projection in the rat. Neuroscience 55:417–427

Cubeddu, LX, Hoffmann IS, Talmaciu RK (1990) Is the release of dopamine from medial prefrontal cortex modulated by presynaptic autoreceptors? In: Kalsner S, Westfall TC (eds) Presynaptic Receptors and the Question of Autoregulation of Neurotransmitter Release, Ann NY Acad Sci 604:452–461

Dahlstrom A, Fuxe K (1964) Localization of monoamines in the lower brain stem. Experientia 15:398–399

Dai M, Tepper JM (1998) Do silent dopaminergic neurons exist in rat substantia nigra in vivo? Neuroscience 85:1089–1099

Damlama M (1994) Subthalamic and Pedunculopontine inputs to Substantia Nigra: A Light and Electron Microscopic Analysis. Ph.D. Thesis, Rutgers University, University Microfilms International, Ann Arbor

Damlama M, Tepper JM (1993) Subcortical excitatory inputs to nigral dopaminergic and non-dopaminergic neurons: A light and electron microscopic study. Proc. 51st Annual Meeting of the Microscopy Society of America. pp 94–95

Dani JA, Radcliffe KA, Pidoplichko VI (2000) Variations in desensitization of nicotinic acetylcholine receptors from hippocampus and midbrain dopamine areas. Eur J Pharmacol 393:31–38

De Blasi A, Conn PJ, Pin J, Nicoletti F (2001) Molecular determinants of metabotropic glutamate receptor signaling. Trends Pharmacol Sci 22:114–120

Deniau JM, Hammond C, Riszk A, Feger J (1978) Electrophysiological properties of identified output neurons of the rat substantia nigra (pars compacta and pars reticulata): evidences for the existence of branched neurons. Exp Brain Res 32:409–422

Deniau JM, Kitai ST, Donoghue JP, Grofova I (1982) Neuronal Interactions in the substantia nigra pars reticulata through axon collaterals of the projection neurons. Exp Brain Res 47:105–113

Devoto P, Collu M, Muntoni AL, Pistis M, Serra G, Gessa GL, Diana M (1995) Biochemical and electrophysiological effects of 7-OH-DPAT on the mesolimbic dopaminergic system. Synapse 20:153–155

Diana M (1996) Dopaminergic neurotransmission and drug withdrawal: relevance to drug craving. In: Ohye C, Kimura M, McKenzie J (eds) The Basal Ganglia V, Adv in Behav Biol 47: Plenum Press New York, pp 123–130

Diana M (1998) Drugs of abuse and dopamine cell activity. Adv Pharmacol 42:998–1001

Diana M, Young SJ, Groves PM (1988) Modulation of dopaminergic terminal excitability by D1 selective agents. Neuropharmacology 28:99–101

Diana M, Garcia-Munoz M, Richards J, Freed CR (1989) Electrophysiological analysis of dopamine cells from the substantia nigra pars compacta of circling rats. Exp Brain Res 74:625–630

Diana M, Young SJ, Groves PM (1991a). Modulation of dopaminergic terminal excitability by D_1 selective agents: Further characterization. Neuroscience 42:441–449

Diana M, Mereu G, Mura A, Fadda F, Passino N, Gessa GL (1991b) Low doses of gamma-hydroxybutyric acid stimulate the firing rate of dopaminergic neurons in unanesthetized rats. Brain Res 566:208–211

Diana M, Rossetti ZL, Gessa GL (1992) Lack of tolerance to ethanol-induced stimulation of mesolimbic dopamine system. Alcohol & Alcoholism 27:329–334

Diana M, Pistis M, Carboni S, Gessa GL, Rossetti ZL (1993a) Profound decrement of mesolimbic dopaminergic neuronal activity during ethanol withdrawal syndrome in rats: electrophysiological and biochemical evidence. Proc Natl Acad Sci (USA) 90:7966–7969

Diana M, Pistis M, Muntoni AL, Gessa GL (1993b) Heterogeneous responses of substantia nigra pars reticulata neurons to gamma-hydroxybutyric acid administration. Eur J Pharmacol 230:363–365

Diana M, Pistis M, Muntoni AL, Gessa GL (1995a) Ethanol withdrawal does not induce a reduction in the number of spontaneously active dopaminergic neurons in the mesolimbic system. Brain Res 682:29–34

Diana M, Pistis M, Muntoni AL, Gessa GL (1995b) Profound decrease of mesolimbic dopaminergic neuronal activity in morphine withdrawn rats. J Pharm Exp Ther 272:781–785

Diana M, Rossetti ZL, Gessa GL (1995c) Central dopaminergic mechanisms of alcohol and opiate withdrawal syndromes. In: Tagliamonte A, Maremmani I (eds) Drug Addiction and related clinical problems. Springer-Verlag Wien New York, pp 19–26

Diana M, Pistis M, Muntoni AL, Gessa GL (1996) Mesolimbic dopaminergic reduction outlasts ethanol withdrawal syndrome: evidence of protracted abstinence. Neuroscience 71:411–415

Diana M, Melis M, Muntoni AL, Gessa GL (1998a) Mesolimbic dopaminergic decline after cannabinoid withdrawal. Proc Natl Acad Sci (USA) 95:10269–10273

Diana M, Melis M, Gessa GL (1998b) Increase in meso-prefrontal dopaminergic activity after stimulation of CB1 receptors by cannabinoids. Eur J Neurosci 10:2825–2830

Diana M, Muntoni AL, Pistis M, Melis M, Gessa GL (1999) Lasting reduction in mesolimbic dopamine neuronal activity after morphine withdrawal. Eur J Neurosci 11:1037–1041

Di Chiara, G, Imperato A (1988) Drugs abused by humans preferentially increase synaptic dopamine concentrations in the mesolimbic system of freely moving rats. Proc Natl Acad Sci (USA) 85:5274–5278

Diaz J, Pilon C, Le Foll B, Triller A, Schwartz J-C, Sokoloff P (2000) Dopamine D_3 receptors expressed by all mesencephalic dopamine neurons. J Neurosci 20:8677–8684

Di Loreto S, Florio T, Scarnati E (1992) Evidence that non-NMDA receptors are involved in the excitatory pathway from the pedunculopontine region to nigrostriatal dopaminergic neurons. Exp Brain Res 89:79–86

Dray A, Gonye TJ, Oakley NR, Tanner T (1976) Evidence for the existence of a raphe projection to the substantia nigra in rat. Brain Res 113:45–57

Eccles JC (1957) The Physiology of Nerve Cells, Johns Hopkins Press, Baltimore, pp 47–56

Einhorn LC, Johansen PA, White FJ (1988) Electrophysiological effects of cocaine in the mesoaccumbens dopamine system: studies in the ventral tegmental area. J Neurosci 8:100–112

Engberg G, Kling-Petersen T, Nissbrandt H (1993) GABA$_B$-receptor activation alters the firing pattern of dopamine neurons in the rat substantia nigra. Synapse 15:229–238

Erhardt S, Andersson B, Nissbradt H, Engberg G (1998) Inhibition of firing rate and changes in the firing pattern on nigral dopaminergic neurons by γ-hydroxybutyric acid (GHBA) are specifically induced by activation of GABA$_B$ receptors. Naunyn-Schmiedeberg's Arch Pharmacol 357:611–619

Erhardt S, Nissbrandt H, Engberg G (1999) Activation of nigral dopamine neurons by the selective GABA(B)-receptor antagonist SCH 50911. J Neural Trans 106:383–394

Falck B, Hillarp NA, Thieme G, Torp A (1962) Fluorescence of catechol amines and related compounds condensed with formaldehyde. J Histochem Cytochem 10: 348–354

Fallon JH (1981) Collateralization of monoamine neurons: Mesotelencephalic dopamine projections to caudate, septum, and frontal cortex. J Neurosci 1:1361–1368

Fallon JH, Laughlin SE (1995) Substantia nigra. In: Paxinos G (ed) The Rat Nervous System, 2nd Edition. Academic Press, San Diego, pp 215–237

Farkas RH, Chien R-Y, Nakajima S, Nakajima Y (1997) Neurotensin and dopamine D2 activation oppositely regulate the same K+ conductance in rat midbrain dopaminergic neurons. Neurosci Lett 231:21–24

Farnebo LO, Hamberger B (1971) Drug-induced changes in the release of ^{3}H-monoamines from field stimulated rat brain slices. Acta Physiol Scand 371:35–44

Fibiger HC, Miller JJ (1977) An anatomical and electrophysiological investigation of the serotonergic projection from the dorsal raphe nucleus to the substantia nigra in the rat. Neuroscience 2:975–987

Fiorillo CD, Williams JT (1998) Glutamate mediates an inhibitory postsynaptic potential in dopamine neurons. Nature 394:78–82

Francois C, Percheron G, Yelnik J, Heyner S (1979) Demonstration of the existence of small local circuit neurons in the Golgi-stained primate substantia nigra. Brain Res 172:160–164

Freeman AS, Bunney BS (1987) Activity of A9 and A10 dopaminergic neurons in unrestrained rats: Further characterization and effects of apomorphine and cholecystokinin. Brain Res 405:46–55

Freeman AS, Meltzer LT, Bunney BS (1985) Firing properties of substantia nigra dopaminergic neurons in freely moving rats. Life Sci 36:1983–1994

Freeman AS, Chiodo LA, Lentz SI, Wade K, Bunney BS (1991) Release of cholecystokinin from rat midbrain slices and modulatory effect of D2 receptor stimulation. Brain Res 555:281–287

French ED (1997) Delta9-tetrahydrocannabinol excites rat VTA dopamine neurons through activation of cannabinoid CB1 but not opioid receptors. Neurosci Lett 226:159–162

French ED, Dillon K, Wu X (1997) Cannabinoids excite dopamine neurons in the ventral tegmentum and substantia nigra. Neuroreport 10:649–652

Freund TF, Powell JF, Smith AD (1984) Tyrosine hydroxylase-immunoreactive boutons in synaptic contact with identified striatonigral neurons, with particular reference to dendritic spines. Neuroscience 13:1189–1215

Futami T, Takakusaki K, Kitai ST (1995) Glutamatergic and cholinergic inputs from the pedunculopontine tegmental nucleus to dopamine neurons in the substantia nigra pars compacta. Neurosci Res 21:331–342

Galarraga E, Bargas J (1995) Firing patterns in substantia nigra compacta indentified neurons in vitro. Arch Med Res 26:191–199

Gao WY, Lee TH, King GR, Ellinwood EH (1998) Alterations in baseline activity and quinpirole sensitivity in putative dopamine neurons in the substantia nigra and ventral tegmental area after withdrawal from cocaine pretreatment. Neuropsychopharmacology 18:222–232

Garcia-Munoz M, Segal DS, Patino P, Young SJ, Kuczenski R, Groves PM (1996) Amphetamine-induced changes in nigrostriatal terminal excitability are modified following repeated amphetamine pretreatment. Brain Res 720:131–138

Gardner EL, Lowinson JH (1991) Marijuana's interaction with brain reward systems: update 1991. Pharmacol Biochem Behav 40:571–580

Gariano RF, Groves PM (1988) Burst firing in midbrain dopamine neurons by stimulation of the medial prefrontal and anterior cingulate cortices. Brain Res 462:194–198

Gariano RF, Sawyer SF, Tepper JM, Young SJ, Groves PM (1989a) Mesocortical dopaminergic neurons. 2. Electrophysiological consequences of terminal autoreceptor activation Brain Res Bull 22:517–523

Gariano RF, Tepper JM, Sawyer SF, Young SJ, Groves PM (1989b) Mesocortical dopaminergic neurons. 1. Electrophysiological properties and evidence for somadendritic autoreceptors. Brain Res Bull 22:511–516

Gariano RF, Young SJ, Jeste DV, Segal DS, Groves PM (1990) Effects of long-term administration of haloperidol on electrophysiolgic properties of rat mesencephalic neurons. J Pharmacol Exp Ther 255:108–113

Geffen LB, Jessell TM, Cuello AC, Iversen LL (1976) Release of dopamine from dendrites in rat substantia nigra. Nature 18:258–260

Gerfen CR, Herkenham M, Thibault J (1987) The neostriatal mosaic: II. Patch- and matrix-directed mesostriatal dopaminergic and non-dopaminergic systems. J Neurosci 7:3915–3934

Gessa GL, Muntoni F, Collu M, Vargiu L, Mereu G (1985) Low doses of ethanol activate dopaminergic neurons in the ventral tegmental area. Brain Res 348:201–203

Gessa GL, Melis M, Muntoni AL, Diana M (1998) Cannabinoids activate mesolimbic dopamine neurons by an action on cannabinoid CB1 receptors. Eur J Pharmacol 341:39–44

Gifford AN, Gardner EL, Ashby CR Jr (1997) The effect of intravenous administration of delta-9-tetrahydrocannabinol on the activity of A10 dopamine neurons recorded in vivo in anesthetized rats. Neuropsychobiology 36:96–99

Giralt MT, Bonanno G, Raiteri M (1990) GABA terminal autoreceptors in the pars compacta and in the pars reticulata of the rat substantia nigra are $GABA_B$. Eur J Pharmacol 175:137–144

Glowinski J, Cheramy A (1981) Dendritic release of dopamine: Its role in the substantia nigra. In: Stjarne L, Hedqvist P, Lagercrantz H, Wennmalm A (eds) Chemical Neurotransmission: 75 Years, Academic Press, New York, pp 285–299

Gonon FG (1988) Nonlinear Relationship Between Impulse Flow and Dopamine Released by Rat Midbrain Dopaminergic Neurons as Studied by In Vitro Electrochemistry. Neuroscience 24:19–28

Gonon F (1997) Prolonged and extrasynaptic excitatory action of dopamine mediated by D1 receptors in the rat striatum in vivo. J Neurosci 17:5972–5978

Gould E, Woolf NJ, Butcher LL (1989) Cholinergic projections to the substantia nigra from the pedunculopontine and laterodorsal tegmental nuclei. Neuroscience 28:611–623

Grace AA (1987) The regulation of dopamine neuron activity as determined by in vivo and in vitro intracellular recordings. In: Chiodo LA, Freeman AS (eds)

Neurophysiology of Dopaminergic Systems-Current Status and Clinical Perspectives. Grosse Pointe, Lakeshore Publishing Company, pp 1–66

Grace AA (1990) Evidence for the functional compartmentalization of spike generating regions of rat midbrain dopamine neurons recorded in vitro. Brain Res 524:31–41

Grace AA, Bunney BS (1979) Paradoxical GABA excitation of nigral dopaminergic cells: Indirect mediation through reticulata inhibitory neurons. Eur J Pharmacol 59:211–218

Grace AA, Bunney BS (1980) Nigral dopamine neurons: intracellular recording and identification with L-dopa injection and histofluorescence. Science 210:654–656

Grace AA, Bunney BS (1983a) Intracellular and extracellular electrophysiology of nigral dopaminergic neurons-1. Identification and characterization. Neuroscience 10:301–316

Grace AA, Bunney BS (1983b) Intracellular and extracellular electrophysiology of nigral dopaminergic neurons-2. Action potential generating mechanisms and morphological correlates. Neuroscience 10:317–331

Grace AA, Bunney BS (1984a) The control of firing pattern in nigral dopamine neurons: Single spike firing. J Neurosci 4:2866–2876

Grace AA, Bunney BS (1984b) The control of firing pattern in nigral dopamine neurons: Burst firing. J Neurosci 4:2877–2890

Grace AA, Bunney BS (1985) Opposing effects of striatonigral feedback pathways on midbrain dopamine cell activity. Brain Res 333:271–284

Grace AA, Bunney BS (1986) Induction of depolarization block in midbrain dopamine neurons by repeated administration of haloperidol: analysis using in vivo intracellular recording. J Pharmacol Exp Ther 238:1092–1100

Grace AA, Hommer DW, Bunney BS (1980) Peripheral and striatal influences on nigral dopamine cells: Mediations by reticula neurons. Brain Res Bull 5:105–109

Grace AA, Bunney BS, Moore H, Todd CL (1997) Dopamine-cell depolarization block as a model for the therapeutic actions of antipsychotic drugs. Trends Neurosci 20:31–37

Grenhoff J, Svensson TH (1988) Clonidine regularizes substantia nigra dopamine cell firing. Life Sci 42:2003–2009

Grenhoff J, Svensson TH (1989) Clonidine modulates dopamine cell firing in rat ventral tegmental area. Eur J Pharmacol 165:11–18

Grenhoff J, Aston-Jones G, Svensson TH (1986) Nicotinic effects on the firing pattern of midbrain dopamine neurons. Acta Physiol Scand 128:351–358

Grenhoff J, Ugedo L, Svensson TH (1988) Firing patterns of midbrain dopamine neurons differences between A9 and A10 cells. Acta Physiol Scand 134:127–132

Grenhoff J, Nisell M, Ferré S, Aston-Jones G, Svensson TH (1993) Noradrenergic modulation of midbrain dopamine cell firing elicited by stimulation of locus coeruleus in the rat. J Neural Trans 93:11–25

Grenhoff J, North RA, Johnson SW (1995) Alpha 1-adrenergic effects on dopamine neurons recorded intracellularly in the rat midbrain slice. Eur J Neurosci 7:1707–1713

Grillner P, Berretta N, Bernardi G, Svensson TH, Mercuri NB (2000) Muscarinic receptors depress GABAergic synaptic transmission in rat midbrain dopamine neurons. Neuroscience 96:299–307

Grinspoon L, Bakalar JB (1997) Marihuana. In: Substance abuse A comprehensive textbook. Third Edition. (Ed. Lowinson JH, Ruiz P, Millman RB, Langrod JG) Williams and Wilkins. pp 199–206

Grofova I, Rinvik E (1970) An experimental electron microscopic study on the striatonigral projection in the cat. Exp Brain Res 11:249–262

Grofova I, Deniau JM, Kitai ST (1982) Morphology of the substantia nigra pars reticulata projection neurons intracellularly labeled with HRP. J Comp Neurol 208:352–368

Groves PM, Wilson CJ, Young SJ, Rebec GV (1975) Self-inhibition by dopaminergic neurons. Science 190:522–529

Groves PM, Fenster GA, Tepper JM, Nakamura S, Young SJ (1981) Changes in dopaminergic terminal excitability induced by amphetamine and haloperidol. Brain Res 221:425–431

Guatteo E, Mercuri NB, Bernardi G, Knopfel T (1999) Group I metabotropic glutamate receptors mediate an inward current in rat substantia nigra dopamine neurons that is independent from calcium mobilization. J Neurophysiol 82: 1974–1981

Guyenet PG, Aghajanian GK (1978) Antidromic identification of dopaminergic and other output neurons of the rat substantia nigra. Brain Res 150:69–84

Gysling K, Wang RY (1983) Morphine-induced activation of A10 dopamine neurons in the rat. Brain Res 277:119–127

Hajos M, Greenfield SA (1994) Synaptic connections between pars compacta and pars reticulata neurones: Electrophysiological evidence for functional modules within the substantia nigra. Brain Res 660:216–224

Hammond C, Deniau JM, Rizk A, Feger J (1978) Electrophysiological demonstration of an excitatory subthalamonigral pathway in the rat. Brain Res 151:235–244

Harris GC, Aston-Jones G (1994) Involvement of D2 dopamine receptors in the nucleus accumbens in the opiate withdrawal syndrome. Nature 8371:155–157

Harris NC, Webb C, Greenfield SA (1989) A possible pacemaker mechanism in pars compacta neurons of the guinea-pig substantia nigra revealed by various ion channel blocking agents. Neuroscience 31:355–362

Harrison MB, Wiley RG, Wooten GF (1990) Selective localization of striatal D_1 receptors to striatonigral neurons. Brain Res 528:317–322

Hattori T, Fibiger HC, McGeer PL (1975) Demonstration of a pallido-nigral projection innervating dopaminergic neurons. J Comp Neurol 162:487–504

Hausser MA, Yung WH (1994) Inhibitory synaptic potentials in guinea-pig substantia nigra dopamine neurones in vitro. J Physiol (Lond) 479:401–422

Hausser M, Stuart G, Racca C, Sakmann B (1995) Axonal initiation and active dendritic propagation of action potentials in substantia nigra neurons. Neuron 15:637–647

Helmreich I, Reimann W, Hertting G, Starke K (1982) Are presynaptic dopamine autoreceptors and postsynaptic dopamine receptors in the rabbit caudate nucleus pharmacologically different? Neuroscience 7:1559–1566

Henry DJ, Greene MA, White FJ (1989) Electrophysiological effects of cocaine in the mesoaccumbens dopamine system: repeated administration. J Pharmacol Exp Ther 251:833–839

Herkenham M, Lynn AB, de Costa BR, Richfield EK (1991) Neuronal localization of cannabinoid receptors in the basal ganglia of the rat. Brain Res 547:267–274

Hernandez L, Hoebel BG (1989) Haloperidol given chronically decreases basal dopamine in the prefrontal cortex more than the striatum or nucleus accumbens as simultaneously measured by microdialysis. Brain Res Bull 22:763–769

Hertel P, Fagerquist MV, Svensson TH (1999) Enhanced cortical dopamine output and antipsychotic-like effects of raclopride by alpha2 adrenoceptor blockade. Science 286:105–107

Hervé D, Pickel VM, Joh TH, Beaudet A (1987) Serotonin axon terminals in the ventral tegmental area of the rat: fine structure and synaptic input to dopaminergic neurons. Brain Res 435:71–83

Inanobe A, Yoshimoto Y, Horio Y, Morishige K-I, Hibino H, Matsumoto S, Tokunaga Y, Maeda T, Hata Y, Takai Y, Kurachi Y (1999) Characterization of G-protein-gated K+ channels compsed of Kir3.2 subunits in dopaminergic neurons of the substantia nigra. J Neurosci 19:1006–1017

Hollerman JR, Abercrombie ED, Grace AA (1992) Electrophysiological, biochemical, and behavioral studies of acute haloperidol-induced depolarization block of nigral dopamine neurons. Neuroscience 47:589–601

Hommer DW, Skirboll LR (1983) Cholecystokinin-like peptides potentiate apomorphine-induced inhibition of dopamine neurons. Eur J Pharmacol 91:151–152

Ianobe A, Yoshimoto Y, Horio Y, Morishige KI, Hibino H, Matsumoto S, Tokunaga Y, Maeda T, Hata Y, Takai Y, Kurachi Y (1999) Characterization of G-protein-gated K+ channels composed of Kir3.2 subunits in dopaminergic neurons of the substantia nigra. J Neurosci 19:1006–10017

Innis RB, Aghajanian GK (1987) Pertussis toxin blocks autoreceptor-mediated inhibition of dopaminergic neurons in rat substantia nigra. Brain Res 411:139–143

Innis R, Andrade R, Aghajanian G (1985) Substance K excites dopaminergic and nondopaminergic neurons in rat substantia nigra. Brain Res 335:381–383

Iribe Y, Moore K, Pang KC, Tepper JM (1999) Subthalamic stimulation-induced synaptic responses in nigral dopaminergic neurons in vitro. J Neurophysiol 82:925–933

Israel JM, Kirk C, Vincent JD (1987) Electrophysiological responses to dopamine of rat hypophysial cells in lactotroph-enriched primary cultures. J Physiol 390:1–22

Iwatsubo K, Clouet DH (1977) Effects of morphine and haloperidol on the electrical activity of rat nigrostriatal neurons. J Pharmacol Exp Ther 202:429–436

Johnson SW, North RA (1992) Opioids excite dopamine neurons by hyperpolarization of local interneurons. J Neurosci 12:483–488

Johnson SW, Seutin V, North RA (1992) Burst firing in dopaminergic neurons induced by N-methyl-D aspartate: Role of electrogenic sodium pump. Science 258:665–667

Jones LS, Gauger L, Davis JN (1985) Anatomy of brain alpha-1-adrenergic receptors: In vitro autoradiography with [^{125}I]-HEAT. J Comp Neurol 231:190–208

Jones SR, Gainetdinov RR, Hu X-T, Cooper DC, Wightman RM, White FJ, aron MG (1999) Loss of autoreceptor functions in mice lacking the dopamine transporter. Nat Neurosci 2:649–655

Juraska JM, Wilson CJ, Groves PM (1977) The substantia nigra of the rat: A Golgi study. J Comp Neurol 4:585–599

Kalivas PW (1993) Neurotransmitter regulation of dopamine neurons in the ventral tegmental area. Brain Res Rev 18:75–113

Kamal LA, Arbilla S, anger SZ (1981) Presynaptic modulation of the release of dopamine from the rabbit caudate nucleus: Differences between electrical stimulation, amphetamine and tyramine. J Pharmacol Exp Ther 216:592–598

Kamata K, Rebec GV (1983) Dopaminergic and neostriatal neurons: dose-dependent changes in sensitivity to amphetamine following long-term treatment. Neuropharmacology 22:1377–1382

Kamata K, Rebec GV (1984a) Long-term amphetamine treatment attenuates or reverses the depression of neuronal activity produced by dopamine agonists in the ventral tegmental area. Life Sci 34:2419–2427

Kamata K, Rebec GV (1984b) Nigral dopaminergic neurons: Differential sensitivity to apomorphine following long-term treatment with low and high doses of amphetamine. Brain Res 321:147–150

Kang Y, Kitai ST (1993a) Calcium spike underlying rhythmic firing in the dopaminergic neurons of the rat substantia nigra. Neurosci Res 18:195–207

Kang Y, Kitai ST (1993b) A whole cell patch-clamp study on the pacemaker potential in dopaminergic neurons of rat substantia nigra compacta. Neurosci Res 18:209–221

Kang Y, Kubota Y, Kitai ST (1989) Synaptic action on substantia nigra compacta neurons by subthalamic inputs studied by a spike trigger averaging method. Soc Neurosci Abstr 15:900

Kebabian JW, Calne DB (1979) Multiple receptors for dopamine. Nature 277:93–96

Keegan KD, Woodruff GN, Pinnock RD (1992) The selective NK3 receptor agonist senktide excites a subpopulation of dopamine-sensitive neurones in the rat substantia nigra pars compacta in vitro. Br J Pharmacol 105:3–5

Kelland MD, Chiodo LA, Freeman AS (1990a) Anesthetic influences on the basal activity and pharmacological responsiveness of nigrostriatal dopamine neurons. Synapse 6:207–209

Kelland MD, Freeman AS, Chiodo LA (1990b) Serotonergic afferent regulation of the basic physiology and pharmacological responsiveness of nigrostriatal dopamine neurons. J Pharmacol Exp Ther 253:803–811

Kilts CD, Anderson CM, Ely TD, Nishita JK (1987) Absence of synthesis-modulating nerve terminal autoreceptors on mesoamygdaloid and other mesolimbic dopamine neuronal populations. J Neurosci 7:3961–3975

Kinon BJ, Lieberman JA (1996) Mechanisms of action of atypical antipsychotic drugs: a critical analysis. Psychopharmacology 124:2–34

Kita T, Kita H, Kitai ST (1986) Electrical membrane properties of rat substantia nigra compacta neurons in an in vitro slice preparation. Brain Res 372:21–30

Kita H, Kitai ST (1987) Efferent projections of the subthalamic nucleus in the rat: light and electron microscopic analysis with the PHA-L method. J Comp Neurol 260: 435–452

Kitai ST, Shepard PD, Callaway JC, Scroggs R (1999) Afferent modulation of dopamine neuron firing patterns. Curr Opin Neurobiol 9:690–697

Klitenick MA, Taber MT, Fibiger HC (1996) Effects of chronic haloperidol on stress- and stimulation-induced increases in dopamine release: Tests of the depolarization block hypothesis. Neuropsychopharmacology 15:424–428

Koeltzow TE, Xu M, Cooper DC, Hu XT, Tonegawa S, Wolf ME, White FJ (1998) Alterations in dopamine release but not dopamine autoreceptor function in dopamine D3 receptor mutant mice. J Neurosci 18:2231–2238

Kondo Y, Iwatsubo K (1980) Diminished responses of nigral dopaminergic neurons to haloperidol and morphine following lesions in the striatum. Brain Res 181:237–240

Konig JFR, Klippel RA (1963) The Rat Brain: A Stereotaxic Atlas of the Forebrain and Lower BrainStem, Williams and Wilkins, Baltimore, MD

Koob GF, Bloom FE (1988) Cellular and molecular mechanisms of drug dependence. Science 242:715–723

Koob GF, Le Moal M (1997) Drug abuse: hedonic homeostatic dysregulation. Science 278:52–58

Kuczenski R (1983) Biochemical actions of amphetamine and other stimulants In: Creese I (ed) Stimulants: Neurochemical Behavioral and Clinical Perspectives. New York, Raven Press, pp 31–62

Lacey MG (1993) Neurotransmitter receptors and ionic conductances regulating the activity of neurones in substantia nigra pars compacta and ventral tegmental area. In: Arbuthnott GW, Emson PC (ed) Progress in Brain Research. Vol. 99 Elsevier Science Publishers BV, pp 251–276

Lacey MG, Mercuri NB, North RA (1987) Dopamine acts on D2 receptors to increase potassium conductance in neurones of the rat substantia nigra zona compacta. J Physiol (Lond) 392:397–416

Lacey MG, Mercuri NB, North RA (1988) On the potassium conductance increase activated by $GABA_B$ and dopamine D_2 receptors in rat substantia nigra neurones. J Physiol (Lond) 401:437–453

Lacey MG, Mercuri NB, North RA (1989) Two cell types in rat substantia nigra zona compacta distinguished by membrane properties and the actions of dopamine and opioids. J Neurosci 9:1233–1241

Lacey MG, Mercuri NB, North RA (1990) Actions of cocaine on rat dopaminergic neurones in vitro. Br J Pharmacol 99:731–735

Letchworth SR, Smith HR, Porrino LJ, Bennett BA, Davies HM, Sexton T, Childers SR (2000) Characterization of a tropane radioligand, [(3)H]2beta-propanoyl-3beta-(4-tolyl) tropane ([(3)H]PTT), for dopamine transport sites in rat brain. J Pharmacol Exp Ther 293:686–696

Lichtensteiger W, Hefti F, Felix D, Huwyler T, Melamed E, Schlumpf M (1982) Stimulation of nigrostriatal dopamine neurones by nicotine. Neuropharmacology 21: 963–968

Liu L-X, Burgess LH, Gonzalez AM, Sibley DR, Chiodo LA (1999) D2 S, D2L, D3 and D4 dopamine receptors couple to a voltage-dependent potassium current in N18TG2 x mesencephalon hybrid cell (MES-23.5) via distinct G proteins. Synapse 31:108–118

Llinás R, Greenfield SA, Jahnsen HJ (1984) Electrophysiology of pars compacta cells in the in vitro substantia nigra – a possible mechanism for dendritic release. Brain Res 294:127–132

Lokwan SJ, Overton PG, Berry MS, Clark D (1999) Stimulation of the pedunculopontine tegmental nucleus in the rat produces burst firing in A9 dopaminergic neurons. Neuroscience 92:245–254

Mahalik TJ (1988) Direct demonstration of interactions between substance P immunoreactive terminals and tyrosine hydroxylase immunoreactive neurons in the substantia nigra of the rat: An ultrastructural study. Synapse 2:508–515

Maitre M, Andriamampandry C, Kemmel V, Schmidt C, Hode Y, Hechler V, Gobaille S (2000) Gamma-hydroxybutyric acid as a signaling molecule in brain. Alcohol 20:277–283

Manzoni OJ, Williams JT (1999) Presynaptic regulation of glutamate release in the ventral tegmental area during morphine withdrawal. J Neurosci 19:6629–6636

Matthews RT, German DC (1984) Electrophysiological evidence for excitation of rat ventral tegmental area dopamine neurons by morphine. Neuroscience 11:617–625

McGeer PL, Eccles JC, McGeer EL (1987) Molecular Neurobiology of the Mammalian Brain, 2nd Ed., New York, Plenum Press, pp 149–224; 553–594

Meador-Woodruff JH, Mansour A, Bunzow JR, Van Tol HH, Watson SJ Jr, Civelli O (1989) Distribution of D2 dopamine receptor mRNA in rat brain. Proc Natl Acad Sci (USA) 86:7625–7628

Melis M, Mereu GP, Lilliu V, Quartu M, Diana M, Gessa GL (1998) Haloperidol does not produce dopamine cell depolarization-block in immobilized rats. Brain Res 783:127–132

Melis M, Gessa GL, Diana M (2000) Different mechanisms for dopaminergic excitation induced by opiates and cannabinoids in the rat midbrain. Prog Neuropsychopharmacol Biol Psychiatry 24:993–1006

Meltzer HY, Matsubara S, Lee JC (1989) Classification of typical and atypical antipsychotic drugs on the basis of dopamine D-1, D-2 and serotonin2 pKi values. J Pharmacol Exp Ther 251:238–246

Meltzer HY, Park S, Kessler R (1999) Cognition, schizophrenia, and the atypical antipsychotic drugs. Proc Natl Acad Sci (USA) 96:13591–13593

Meltzer LT, Serpa KA, Christoffersen CL (1997) Metabotropic glutamate receptor-mediated inhibition and excitation of substantia nigra dopamine neurons. Synapse 26:184–193

Mercuri NB, Calabresi P, Bernardi G (1989) The mechanism of amphetamine-induced inhibition of rat substantia nigra compacta neurones investigated with intracellular recording in vitro. Br J Pharmacol 98:127–134

Mercuri NB, Calabresi P, Bernardi G (1990) Effects of glycine on neurons in the rat substantia nigra zona compacta: in vitro electrophysiological study. Synapse 5: 190–200

Mercuri NB, Stratta F, Calabresi P, Bernardi G (1992) Electrophysiological evidence for the presence of ionotropic and metabotropic excitatory amino acid receptors on dopaminergic neurons of the rat mesencephalon: an in vitro study. Funct Neurol 7:231–234

Mercuri NB, Stratta F, Calabresi P, Bonci A, Bernardi G (1993) Activation of metabotropic glutamate receptors induces an inward current in rat dopamine mesencephalic neurons. Neuroscience 56:399–407

Mercuri NB, Grillner P, Bernardi G (1996) N-Methyl-D-aspartate receptors mediate a slow excitatory postsynaptic potential in the rat midbrain dopaminergic neurons. Neuroscience 74:785–792

Mercuri NB, Saiardi A, Bonci A, Picetti R, Calabresi P, Bernardi G, Borrelli E (1997) Loss of autoreceptor function in dopaminergic neurons from dopamine D_2 receptor deficient mice. Neuroscience 79:323–327

Mereu G, Gessa GL (1985) Low doses of ethanol inhibit the firing of neurons in the substantia nigra pars reticulata: A GABAergic effect? Brain Res 360:325–330

Mereu G, Fadda F, Gessa GL (1984a) Ethanol stimulates the firing rate of nigral dopaminergic neurons in unanesthetized rats. Brain Res 292:63–69

Mereu G, Fanni B, Gessa GL (1984b) General anesthetics prevent dopaminergic neuron stimulation by neuroleptics. In: Catecholamines: Neuropharmacology and Central Nervous System-Theoretical Aspects. New York, Alan R Liss, pp 353–358

Mereu G, Westfall TC, Wang RY (1985) Modulation of terminal excitability of mesolimbic dopaminergic neurons by D-amphetamine and haloperidol. Brain Res 359:88–96

Mereu G, Yoon KW, Boi V, Gessa GL, Naes L, Westfall TC (1987) Preferential stimulation of ventral tegmental area dopaminergic neurons by nicotine. Eur J Pharmacol 141:395–399

Mereu G, Costa E, Armstrong DM, Vicini S (1991) Glutamate receptor subtypes mediate excitatory synaptic currents of dopamine neurons in midbrain slices. J Neurosci 11:1359–1366

Mereu G, Lilliu V, Vargiu P, Muntoni AL, Diana M, Gessa GL (1994) Failure of chronic haloperidol to induce depolarization inactivation of dopamine neurons in unanesthetized rats. Eur J Pharmacol 264:449–453

Mereu G, Lilliu V, Vargiu P, Muntoni AL, Diana M, Gessa GL (1995) Depolarization inactivation of dopamine neurons: an artifact? J Neurosci 15:1144–1149

Mereu G, Lilliu V, Casula A, Vargiu PF, Diana M, Musa A, Gessa GL (1997) Spontaneous bursting activity of dopaminergic neurons in midbrain slices from immature rats: Role of N-methyl-D-aspartate receptors. Neuroscience 77:1029–1036

Meyerhoff JL, Kant GJ (1978) Release of endogenous dopamine from corpus striatum. Life Sci 23:1481–1486

Miller AS, Walker JM (1995) Effects of a cannabinoid on spontaneous and evoked neuronal activity in the substantia nigra pars reticulata. Eur J Pharmacol 279:179–185

Moghaddam B, Bunney BS (1993) Depolarization inactivation of dopamine neurons: terminal release characteristics. Synapse 14:195–200

Moon-Edley S, Graybiel AM (1983) The afferent and efferent connections of the feline nucleus tegmenti pedunculopontinus, pars compacta. J Comp Neurol 217:187–215

Moore KE, Demarest KT, Lookingland KJ (1987) Stress, prolactin and hypothalamic dopaminergic neurons. Neuropharmacology 26:801–808

Moore H, Todd CL, Grace AA (1998) Striatal extracellular dopamine levels in rats with haloperidol-induced depolarization block of substantia nigra dopamine neurons. J Neurosci 18:5068–5077

Morelli M, Mennini T, DiChiara G (1988) Nigral dopamine autoreceptors are exclusively of the D_2 Type: Quantitative autoradiography of [^{125}I]iodosulpride and [^{125}I]SCH 23982 in adjacent brain sections. Neuroscience 27:865–870

Mori S, Matsuura T, Takino T, Sano Y (1987) Light and electron microscopic immunohistochemical studies of serotonin nerve fibers in the substantia nigra of the rat, cat and monkey. Anat Embryol (Berl) 176:13–18

Naito A, Kita H (1994) The cortico-nigral projection in the rat: an anterograde tracing study with biotinylated dextran amine. Brain Res 637:317–322

Nakamura S, Iwatsubo K, Tsai C-T, Iwama K (1979) Cortically evoked inhibition of neurons of rat substantia nigra (pars compacta). Jpn J Physiol 29:353–357

Nakanishi H, Kita H, Kitai ST (1987) Intracellular study of rat substantia nigra pars reticulata neurons in an in vitro slice preparation: electrical membrane properties and response characteristics to subthalamic stimulation. Brain Res 437:45–55

Nedergaard S, Greenfield SA (1992) Sub-populations of pars compacta neurons in the substantia nigra: The significance of qualitatively and quantitatively distinct conductances. Neuroscience 48:423–437

Nedergaard S, Bolam JP, Greenfield SA (1988) Facilitation of a dendritic calcium conductance by 5-hydroxytryptamine in the substantia nigra. Nature 333:174–177

Nedergaard S, Flatman JA, Engberg I (1991) Excitation of substantia nigra pars compacta neurones by 5-hydroxy-tryptamine in-$vitro$. NeuroReport 2:329–332

Nedergaard S, Flatman JA, Engberg I (1993) Nifedipine- and w-conotoxin-sensitive Ca^{2+} conductances in guinea-pig substantia nigra pars compacta neurones. J Physiol (Lond) 466:727–747

Nestler EJ (1992) Molecular mechanisms of drug addiction. J Neurosci 12:2439–2450

Nestler EJ (1993) Cellular responses to chronic treatment with drugs of abuse. Crit Rev Neurobiol 7:23–39

Nestler EJ (2001) Molecular basis of long-term plasticity underlying addiction. Nat Rev Neurosci 2:119–128

Nitsch C, Riesenberg R (1988) Immunocytochemical demonstration of GABAergic synaptic connections in rat substantia nigra after different lesions of the striatonigral projection. Brain Res 461:127–142

Overton P, Clark D (1992) Iontophoretically administered drugs acting and the N-methyl-D-aspartate receptor modulate burst firing in A9 dopamine neurons in the rat. Synapse 10:131–140

Overton PG, Clark D (1997) Burst firing in midbrain dopaminergic neurons. Brain Res Rev 25:312–334

Paden C, Wilson CJ, Groves PM (1976) Amphetamine-induced release of dopamine from the substantia nigra in vitro. Life Sci 19:1499–1506

Paladini CA, Tepper JM (1999) $GABA_A$ and $GABA_B$ antagonists differentially affect the firing pattern of substantia nigra dopaminergic neurons in vivo. Synapse 32:165–176

Paladini CA, Celada P, Tepper JM (1999a) Striatal, pallidal, and pars reticulata evoked inhibition of nigrostriatal dopaminergic neurons is mediated by $GABA_A$ receptors in vivo. Neuroscience 89:799–812

Paladini CA, Iribe Y, Tepper JM (1999b) $GABA_A$ receptor stimulation blocks NMDA-induced bursting of dopaminergic neurons in vitro by decreasing input resistance. Brain Res 832:145–151

Paladini CA, Fiorillo CD, Morikawa H, Williams JT (2001) Amphetamine selectively blocks inhibitory glutamate transmission in dopamine neurons. Nat Neurosci 4:275–281

Phillipson OT (1979) Afferent projections to the ventral tegmental area of Tsai and the interfascicular nucleus: A horseradish peroxidase study in the rat. J Comp Neurol 187:117–144

Pickel VM, Beckley SC, Joh TH, Reis DJ (1981) Ultrastructural immunocytochemical localization of tyrosine hydroxylase in the neostriatum. Brain Res 225:373–385

Pidoplichko VI, DeBiasi M, Williams JT, Dani JA (1997) Nicotine activates and desensitizes midbrain dopamine neurons. Nature 390:401–404

Pieribone VA, Nicholas AP, Dagerlind A, Hokfelt T (1994) Distribution of alpha-1 adrenoceptors in rat brain reveal;ed by in sitgu hybridization experiments utilizing subtype-specific probes. J Neurosci 14:4252–4268

Ping H, Shepard PD (1996) Apamine-sensitive Ca^{2+}-activated K^+ channels regulate pacemaker activity in nigral dopamine neurons. NeuroReport 73:809–814

Pinnock RD, Dray A (1982) Differential sensitivity of presumed dopaminergic and non-dopaminergic neurones in rat substantia nigra to electrophoretically applied substance P. Neurosci Lett 29:153–158

Plantje JF, Dijcks FA, Verheijden PFHM, Stoof JC (1985) Stimulation of D-2 dopamine receptors in rat mesocortical areas inhibits the release of [3H]dopamine. Eur J Pharmacol 114:401–402

Plantje JF, Steinbusch HWM, Schipper J, Dijcks FA, Verheijden PFHM, Stoof JC (1987) D-2 dopamine-receptors regulate the release of {3H}dopamine in rat cortical regions showing dopamine immunoreactive fibers. Neuroscience 20:157–168

Precht W, Yoshida M (1971) Monosynaptic inhibition of neurons of the substantia nigra by caudatonigral fibers. Brain Res 32:225–228

Pucak ML, Grace AA (1994) Evidence that systemically administered dopamine antagonists activate dopamine neuron firing primarily by blockade of somatodendritic autoreceptors. J Pharmacol Exp Ther 271:1181–1192

Pucak ML, Grace AA (1996) Effects of haloperidol on the activity and membrane physiology of substantia nigra dopamine neurons recorded in vitro. Brain Res 713:44–52

Pulvirenti L, Diana M (2001) Drug dependence as a disorder of neural plasticity: focus on dopamine and glutamate. Neurosci Rev 12:41–59

Rasmussen K, Czachura JF (1995) Nicotine withdrawal leads to increased firing rates of midbrain dopamine neurons. Neuroreport 7:329–332

Ribak CE, Vaughn JE, Saito K, Barber R, Roberts E (1976) Immunocytochemical localization of glutamate decarboxylase in rat substantia nigra. Brain Res 116:287–298

Ribak CE, Vaughn JE, Roberts E (1980) GABAergic nerve terminals decrease in the substantia nigra following hemitransections of the striatonigral and pallidonigral pathways. Brain Res 192:413–420

Robledo P, Féger J (1990) Excitatory influence of rat subthalamic nucleus to substantia nigra pars reticulata and the pallidal complex: electrophysiological data. Brain Res 518:47–54

Roth RH, Salzman PM, Nowycky MC (1978) Impulse flow and short-term regulation of transmitter biosynthesis in central catecholaminergic neurons. In: Lipton MA, DiMascio A, Killam KF (eds) Psychopharmacology: A Generation of Progress. New York, Raven Press, pp 185–198

Rothman RB, Herkenham M, Pert CB, Liang T, Cascieri MA (1984) Visualization of rat brain receptors for the neuropeptide, substance P. Brain Res 309:47–54

Rye DB, Saper CB, Lee RJ, Wainer BH (1987) Pedunculopontine tegmental nucleus of the rat: cytoarchitecture, cytochemistry, and some extrapyramidal connections of the mesopontine tegmentum. J Comp Neurol 252:483–528

Santiago M, Westerink BHC (1992) The role of GABA receptors in the control of nigrostriatal dopaminergic neurons: dual probe microdialysis study in awake rats. Eur J Pharmacol 219:175–181

Scarnati E, Pacitti C (1982) Neuronal responses to iontophoretically applied dopamine, glutamate, and GABA of identified dopaminergic cells in the rat substantia nigra after kainic acid-induced destruction of the striatum. Exp Brain Res 46:377–382

Scarnati E, Campana E, Pacitti C (1984) Pedunculopontine-evoked excitation of substantia nigra neurons in the rat. Brain Res 304:351–361

Schwyn RC, Fox CA (1974) The primate substantia nigra: A golgi and electron microscopic study. Journal Fur Hirnforschung 15:95–126

Sesack SR, Pickel VM (1992) Prefrontal cortical efferents in the rat synapse on unlabeled neuronal targets of catecholamine terminals in the nucleus accumbens septi and on dopamine neurons in the ventral tegmental area. J Comp Neurol 320:145–160

Sesack SR, Aoki C, Pickel VM (1994) Ultrastructural localization of D2 receptor-like immunoreactivity in midbrain dopamine neurons and their striatal targets. J Neurosci 14:88–106

Seutin V, Verbanck P, Massotte L, Dresse A (1990) Evidence for the presence of N-methyl-D-aspartate receptors in the ventral tegmental area of the rat: an electrophysiological in vitro study. Brain Res 514:147–150

Shafer RA, Levant B (1998) The D_3 dopamine receptor in cellular and organismal function. Psychopharmacology (Berl). 135:1–16

Shen RY, Chiodo LA (1993) Acute withdrawal after repeated ethanol treatment reduces the number of spontaneously active dopaminergic neurons in the ventral tegmental area. Brain Res 622:289–293

Shen KZ, Johnson SW (1997) Presynaptic $GABA_B$ and adenosine A_1 receptors regulate synaptic transmission to rat substantia nigra reticulata neurones. J Physiol (Lond) 505:153–163

Shepard PD, German DC (1984) A subpopulation of mesocortical dopamine neurons possesses autoreceptors. Eur J Pharmacol 98:455–456

Shepard PD, Bunney BS (1988) Effects of apamin on the discharge properties of putative dopamine-containing neurons in vitro. Brain Res 463:380–384

Shepard PD, Connelly ST (1999) Pertussis toxin lesions of the rat substantia nigra block the inhibitory effects of the γ-hydroxybutyrate agent, S(–)HA-966 without affecting the basal firing properties of dopamine neurons. Neuropsychopharmacology 21:650–661

Shim SS, Bunney BS, Shi W-X (1996) Effects of lesions in the medial prefrontal cortex on the activity of midbrain dopamine neurons. Neuropsychopharmacology 15: 437–441

Sibley DR, Monsma JFJ (1992) Molecular biology of dopamine receptors. Trends Pharmacol Sci 13:61–69

Silva NL, Pechura CM, Barker JL (1990) Postnatal rat nigrostriatal dopaminergic neurons exhibit five types of potassium conductances. J Neurophysiol 64:262–272

Skarsfeldt T (1988) Differential effects after repeated treatment with haloperidol, clozapine, thioridazine and tefludazine on SNC and VTA dopamine neurones in rats. Life Sci 42:1037–1044

Skarsfeldt T (1995) Differential effects of repeated administration of novel antipsychotic drugs on the activity of midbrain dopamine neurons in the rat. Eur J Pharmacol 281:289–294

Skirboll LR, Grace AA, Bunney BS (1979) Dopamine auto- and postsynaptic receptors: electrophysiological evidence for differential sensitivity to dopamine agonists. Science 206:80–82

Skirboll SR, Grace AA, Hommer DW, Goldstein M, Hokfelt T, Bunney BS (1981) Peptide-monoamine coexistence: Studies of the actions of cholecystokinin-like peptide on the electrical activity of midbrain dopamine neurons. Neuroscience 6:2111–2124

Sklair-Tavron L, Shi WX, Lane SB, Harris HW, Bunney BS, Nestler EJ (1996) Chronic morphine induces visible changes in the morphology of mesolimbic dopamine neurons. Proc Natl Acad Sci (USA) 93:11202–11207

Smith ID, Grace AA (1992) Role of the subthalamic nucleus in the regulation of nigral dopamine neuron activity. Synapse 12:287–303

Smith Y, Bolam JP (1989) Neurons of the substantia nigra reticulata receive a dense GABA-containing input from the globus pallidus in the rat. Brain Res 493:160–167

Smith Y, Charara A, Parent A (1996) Synaptic innervation of midbrain dopaminergic neurons by glutamate- enriched terminals in the squirrel monkey. J Comp Neurol 364:231–253

Somogyi P, Bolam JP, Totterdell S, Smith AD (1981) Monosynaptic input from the nucleus accumbens-ventral striatum region to retrogradely labelled nigrostriatal neurones. Brain Res 217:245–263

Sorenson EM, Shiroyama T, Kitai ST (1998) Postsynaptic nicotinic receptors on dopaminergic neurons in the substantia nigra pars compacta of the rat. Neuroscience 87:659–73

Stanford IM, Lacey MG (1996) Differential actions of serotonin, mediated by 5-HT_{1B} and 5-HT_{2C} receptors, on GABA-mediated synaptic input to rat substantia nigra pars reticulata neurons in vitro. J Neurosci 16:7566–7573

Stanwood GD, Artymyshyn RP, Kung M-P, Kung HF, Lucki I, McGonigle P (2000) Quantitative autoradiographic mapping of rat brain dopamine D_3 binding with [^{125}I]7-OH-PIPAT: Evidence for the presence of D_3 receptors on dopaminergic and nondopaminergic cell bodies and terminals. J Pharmacol Exp Ther 295:1223–1231

Starke K, Gothert M, Kilbinger H (1989) Modulation of neurotransmitter release by presynaptic autoreceptors. Physiol Rev 69:864–989

Stockton ME, Rasmussen K (1996) Olanzapine, a novel atypical antipsychotic, reverses d-amphetamine- induced inhibition of midbrain dopamine cells. Psychopharmacology 124:50–56

Sugimoto T, Hattori T (1984) Organization and efferent projections of nucleus tegmenti pedunculopontinus pars compacta with special reference to its cholinergic aspect. Neuroscience 11:931–946

Sugita S, Johnson SW, North RA (1992) Synaptic inputs to $GABA_A$ and $GABA_B$ receptors originate from discrete afferent neurons. Neurosci Lett 134:207–211

Suppes T, Pinnock RD (1987) Sensitivity of neuronal dopamine response in the substantia nigra and ventral tegmentum to clozapine, metoclopramide and SCH 23390. Neuropharmacology 26:331–337

Surmeier DJ, Kitai ST (1993) D_1 and D_2 dopamine receptor modulation of sodium and potassium currents in rat neostriatal neurons. In: Arbuthnott GW, Emson PC (ed) Progress in Brain Research. Vol. 99 Elsevier Science Publishers B.V., pp 309–324

Surmeier DJ, Eberwine J, Wilson CJ, Cao Y, Stefani A, Kitai ST (1992) Dopamine receptor subtypes colocalize in rat striatonigral neurons. Proc Natl Acad Sci (USA) 89:10178–10182

Surmeier DJ, Song W-J, Yan Z (1996) Coordinated expression of dopamine receptors in neostriatal medium spiny neurons. J Neurosci 16:6579–6591

Svensson TH, Tung C-S (1989) Local cooling of prefrontal cortex induced pacemaker-like firing of dopaminergic neurons in rat ventral tegmental area in vivo. Acta Physiol Scand 136:135–136

Takakusaki K, Shiroyama T, Yamamoto T, Kitai ST (1996) Cholinergic and noncholinergic tegmental pedunculopontine projection neurons in rats revealed by intracellular labeling. J Comp Neurol 371:345–361

Tanda G, Pontieri FE, Di Chiara G (1997) Cannabinoid and heroin activation of mesolimbic dopamine transmission by a common mu1 opioid receptor mechanism. Science 276:2048–2050

Tepper JM, Groves PM (1990) In vivo electrophysiology of central nervous system terminal autoreceptors. In: Kalsner S, Westfall TC (eds) *Presynaptic Autoreceptors and the Question of the Autoregulation of Neurotransmitter Release*. Ann New York Acad Sci 604:470–487

Tepper JM, Nakamura S, Spanis CW, Squire LR, Young SJ, Groves PM (1982) Subsensitivity of catecholaminergic neurons to direct acting agonists after single or repeated electroconvulsive shock. Biol Psych 17:1059–1079

Tepper JM, Nakamura S, Young SJ, Groves PM (1984a) Autoreceptor-mediated changes in dopaminergic terminal excitability: Effects of striatal drug infusions. Brain Res 309:317–333

Tepper JM, Young SJ, Groves PM (1984b) Autoreceptor-mediated changes in dopaminergic terminal excitability: Effects of increases in impulse flow. Brain Res 309:309–316

Tepper JM, Groves PM, Young SJ (1985) The neuropharmacology of the autoinhibition of monoamine release. Trends Pharmacol Sci 6:251–256

Tepper JM, Gariano RF, Groves PM (1987a) The neurophysiology of dopamine nerve terminal autoreceptors. In: Chiodo LA, Freeman AS (eds) Neurophysiology of Dopaminergic Systems – Current Status and Clinical Perspectives. Grosse Point, Lakeshore Publishing Co., pp 93–127

Tepper JM, Sawyer SF, Groves PM (1987b) Electrophysiologically identified nigral dopaminergic neurons intracellularly labeled with HRP: Light-microscopic analysis. J Neurosci 7:2794–2800

Tepper JM, Martin LP, Anderson DR (1995) GABA$_A$ receptor-mediated inhibition of nigrostriatal dopaminergic neurons by pars reticulata projection neurons. J Neurosci 15:3092–3103

Tepper JM, Sun B-C, Martin LP, Creese I (1997) Functional roles of dopamine D$_2$ and D$_3$ autoreceptors on nigrostriatal neurons analyzed by antisense knockdown in vivo. J Neurosci 17:2519–2530

Tepper JM, Celada P, Iribe Y, Paladini CA (2002) Afferent control of nigral dopaminergic neurons – The role of GABAergic inputs. In: Graybiel AM et al. (eds) The Basal Ganglia VI, Kluwer Academic Publishers, Norwell (in press)

Teranishi T, Negishi K, Kato S (1983) Dopamine modulates S-potential amplitude and dye-coupling between external horizontal cells in carp retina. Nature 301:243–246

Tersigni TJ, Rosenberg HC (1996) Local pressure application of cannabinoid agonists increases spontaneous activity of rat substantia nigra pars reticulata neurons without affecting response to iontophoretically-applied GABA. Brain Res 733:184–192

Tong Z-Y, Overton PG, Clark D (1996) Stimulation of the prefrontal cortex in the rat induces patterns of activity in midbrain dopaminergic neurons which resemble natural burst events. Synapse 22:195–208

Trent F, Tepper JM (1991) Dorsal raphé stimulation modifies striatal-evoked antidromic invasion of nigral dopaminergic neurons in vivo. Exp Brain Res 84:620–630

Tsou K, Patrick SL, Walker JM (1995) Physical withdrawal in rats tolerant to delta 9-tetrahydrocannabinol precipitated by a cannabinoid receptor antagonist. Eur J Pharmacol 280:R13–15

Ungerstedt U (1971) Stereotaxic mapping of the monoamine pathways in the rat brain. Acta Physiol Scand Suppl 367:1–48

Usiello A, Balk J-H, Rougé-Pont F, Picetti R, Dierich A, LeMeur M, Piazza PV, Borelli E (2000) Distinct functions of the two isoforms of dopamine D2 receptors. Nature 408:199–203

Usunoff KG (1984) Tegmentonigral projections in the cat: Electron microscopic observations. In: Hassler RG, Christ JF (eds) Advances in Neurology Vol. 40, Raven Press, New York, pp 55–61

Usunoff KG, Romansky KV, Malinov GB, Ivanov DP, Blagov ZA, Galabov GP (1982) Electron microscopic evidence for the existence of a corticonigral tract in the cat. J Hirnforsch 23:23–29

Vallar L, Meldolesi J (1989) Mechanisms of signal transduction at the dopamine D$_2$ receptor. Trends Pharmacol Sci 10:74–77

Walters JR, Roth RH (1976) Dopaminergic Neurons: An in vivo system for measuring drug interactions with presynaptic receptors. Naunyn Schmiedebergs Arch Pharmacol 296:5–14

Wang RY (1981a) Dopaminergic neurons in the rat ventral tegmental area. I. Identification and characterization. Brain Res Rev 3:123–140

Wang RY (1981b) Dopaminergic neurons in the rat ventral tegmental area. II. Evidence for autoregulation. Brain Res Rev 3:141–151

Wang RY (1981c) Dopaminergic neurons in the rat ventral tegmental area. III. Effects of D- and L-amphetamine. Brain Res Rev 3:152–165

Wang T, French ED (1993) L-Glutamate excitation of A10 dopamine neurons is preferentially mediated by activation of NMDA receptors: extra- and intracellular electrophysiological studies in brain slices. Brain Res 627:299–306

Wassef M, Berod A, Sotelo C (1981) Dopaminergic dendrites in the pars reticulata of the rat substantia nigra and their striatal input. Combined immunocytochemical localization of tyrosine hydroxylase and anterograde degeneration. Neuroscience 6:2125–2139

Werkman TR, Kruse CG, Nievelstein H, Long SK, Wadman WJ (2000) Neurotensin attenuates the quinpirole-induced inhibition of the firing rate of dopamine neurons in the rat substantia nigra pars compacta and ventral tegmental area. Neuroscience 95:417–423

White FJ (1996) Synaptic regulation of mesocorticolimbic dopamine neurons. Ann Rev Neurosci 19:405–436

White FJ, Wang RY (1983) Differential effects of classical and atypical antipsychotic drugs on A9 and A10 dopamine neurons. Science 221:1054–1057

Wilson CJ, Callaway JCV (2000) Coupled oscillator model of the dopaminergic neuron of the substantia nigra. J Neurophysiol 83:3084–3100

Wilson CJ, Young SJ, Groves PM (1977) Statistical properties of neuronal spike trains in the substantia nigra: cell types and their interactions. Brain Res 136:243–260

Wilson CJ, Fenster GA, Young SJ, Groves PM (1979) Haloperidol-induced alteration of post-firing inhibition in dopaminergic neurons of rat substantia nigra. Brain Res 179:165–170

Wolf ME, Roth RH (1990) Autoreceptor regulation of dopamine synthesis. In: Kalsner S, Westfall TC (eds) Presynaptic Receptors and the Question of Autoregulation of Neurotransmitter Release. Ann NY Acad Sci 604:232–343

Wu T, Wang H-L (1994) CCK-8 excites substantia nigra dopaminergic neurons by increasing a cationic conductance. Neurosci Lett 170:229–232

Wu X, French ED (2000) Effects of chronic delta9-tetrahydrocannabinol on rat midbrain dopamine neurons: an electrophysiological assessment. Neuropharmacology 39:391–398

Yoshida M, Precht W (1971) Monosynaptic inhibition of neurons of the substantia nigra by caudato-nigral fibers. Brain Res 32:225–228

Yung WH, Hausser MA, Jack JJB (1991) Electrophysiology of dopaminergic and nondopaminergic neurones of the guinea-pig substantia nigra pars compacta in vitro. J Physiol (Lond) 436:643–667

Zhang H, Lee TH, Ellinwood EH Jr (1992a) The progressive changes of neuronal activities of the nigral dopaminergic neurons upon withdrawal from continuous infusion of cocaine. Brain Res 594:315–318

Zhang J, Chiodo LA, Freeman AS (1992b) Electrophysiological effects of MK-801 on rat nigrostriatal and mesoaccumbal dopaminergic neurons. Brain Res 590:153–163

Zhang J, Chiodo LA, Freeman AS (1993) Effects of phencyclidine, MK-801 and 1,3-di(tolyl)guanidine on non-dopaminergic midbrain neurons. Eur J Pharmacol 230:371–374

Zhang J, Chiodo LA, Freeman AS (1994) Influence of excitatory amino acid receptor subtypes on the electrophysiological activity of dopaminergic and nondopaminergic neurons in rat substantia nigra. J Pharmacol Exp Ther 269:313–321

Zhang W, Tilson H, Stachowiak MK, Hong JS (1989) Repeated haloperidol administration changes basal release of striatal dopamine and subsequent response to haloperidol challenge. Brain Res 484:389–892

Zhang XF, Hu XT, White FJ, Wolf ME (1997) Increased responsiveness of ventral tegmental area dopamine neurons to glutamate after repeated administration of cocaine or amphetamine is transient and selectively involves AMPA receptors. J Pharmacol Exp Ther 281:699–706

CHAPTER 14
Presynaptic Regulation of Dopamine Release

J. Glowinski, A. Cheramy, and M.-L. Kemel

A. Introduction

The nigrostriatal dopaminergic pathway has generally been used as an experimental model for basic investigations into the release of dopamine (DA) from central dopaminergic neurons. The release of DA from striatal nerve endings is not only dependent on nerve impulse flow but also on regulation processes mediated by D_2 autoreceptors (Starke 1981; L'Hirondel et al. 1998). These autoreceptors are not only involved in the inhibitory control of DA release but also in its synthesis, and the efficacy of these presynaptic regulatory mechanisms depends on the state of depolarisation of the plasma membrane. While the DA autoreceptors involved in the regulation of the release process of DA are mainly coupled to potassium channels (Bowyer et al. 1989; Cass and Zahniser 1991), those which control the rate of the transmitter synthesis are negatively coupled to adenylyl cyclase (El Mestikawy et al. 1985, 1986; Onali et al. 1988). In addition, these D_2 autoreceptors regulate the state of excitability of nerve terminal arborisations (Romo and Schultz 1985; Tepper et al. 1986).

Besides DA autoreceptors, heteroreceptors participate in the presynaptic control of DA release in the striatum. The first indication of this type of heteroregulation was provided 30 years ago in our laboratory when acetylcholine and serotonin were shown to stimulate the release of newly synthesised DA from the isolated striatum of the rat (Besson et al. 1969). Since this early study, most transmitters and co-transmitters present in striatal afferent fibres, collaterals of efferent neurons and interneurons have been found to facilitate or reduce the spontaneous or evoked release of DA (see review in Chesselet 1984). These heteroregulation processes are either direct, mediated through receptors located on dopaminergic nerve endings, or indirect, involving local circuits. In most cases, direct presynaptic regulation has been demonstrated thanks to release studies performed on striatal slices in the presence of tetrodotoxin (a neurotoxin currently used to prevent most indirect effects by interrupting nerve impulse flow) or, more convincingly, on synaptosomes. Confirmation for the existence of the receptor subtypes involved in these forms of direct presynaptic regulation was obtained by identification of their mRNAs in dopaminergic cells ($D_{2,3}R$; $NMDAR_{1,2C,2D}$; $GLUR_{1,2,3,4C}$; $mGLUR_1$; M_5R; NK_3R; etc.) (VILARO et al. 1990; WEINER et al. 1990; SHIGEMOTO et al. 1992; FOTUHI et al. 1993; MARTIN et al. 1993; MEADOR-WOODRUFF et al. 1994; STANDAERT et al. 1994; STOESSL et al. 1994; TESTA et al. 1994; DIAZ et al. 1995; WHITTY et al. 1995). Due to the quasi-absence of heterologous synapses on dopaminergic nerve terminals, the physiological significance of these local heteroregulation processes of DA release has been challenged for several years. However, appositions of nerve terminals on dopaminergic nerve terminals have been observed and, in addition, the concept of volume transmission is now widely accepted.

The present review will be mainly dedicated to three main research developments from our laboratory on the presynaptic regulation of DA release: (1) the interactions between heteroreceptors located on dopaminergic nerve terminals, (2) the role of diffusible messengers and particularly of arachidonic acid and (3) the identity of local circuits contributing to the presynaptic regulation of DA release in striatal compartments. These developments largely derive from research on interactions between corticostriatal glutamatergic fibres and nerve terminals of the nigrostriatal dopaminergic neurons.

B. Interactions Between Heteroreceptors or Heteroreceptors and D_2 Autoreceptors Present on Dopaminergic Nerve Terminals

Studies performed in the cat implanted with push–pull cannulae have provided strong evidence for the occurrence of functional interactions between corticostriatal glutamatergic neurons and nerve terminals of the nigrostriatal dopaminergic neurons (CHÉRAMY et al. 1991). Indeed, the direct or indirect (through the thalamus, or even the substantia nigra pars reticulata) activation of the corticostriatal glutamatergic neurons that leads to the evoked release of glutamate in the caudate nucleus (BARBEITO et al. 1989) was shown to be associated with a marked and persistent stimulation of DA release (NIEOULLON et al. 1978; CHESSELET et al. 1983). Indicating the involvement of glutamate in the evoked release of DA, this latter response was prevented after the acute transection of the corticostriatal fibres (NIEOULLON et al. 1978; ROMO et al. 1984) and abolished by the application of riluzole (a compound which interrupts glutamatergic transmission) into the caudate nucleus (CHÉRAMY et al. 1986; ROMO et al. 1986a). Finally, demonstrating the presynaptic nature of this regulation, this stimulation of DA release resulting from the activation of the corticostriatal glutamatergic neurons persisted after the acute transection of the nigrostriatal dopaminergic pathway (ROMO et al. 1986b).

The involvement of glutamate in a presynaptic regulation of DA release was also demonstrated on striatal slices from rat by several groups. These authors indicated that the glutamate-evoked release of DA is concentration- and calcium-dependent and suggested that both α-amino-3-hydroxy-5-methyl-4-isoxazolepropionic acid (AMPA) and N-methyl-D-aspartate (NMDA) receptors are involved in the tetrodotoxin-resistant release of DA evoked by a high concentration of glutamate (ROBERTS and ANDERSON 1979; SNELL and JOHNSON 1986; CLOW and JHAMANDAS 1989; CAI et al. 1991; KREBS et al. 1989; JIN and FREDHOLM 1994). The presence of AMPA and NMDA receptors on dopaminergic nerve terminals was confirmed in studies performed on synaptosomes from rat and, more recently, mouse (DESCE et al. 1991, 1992; WANG 1991; CHÉRAMY et al. 1996a; KREBS et al. 1991a). These latter investigations allowed the occurrence of a co-operative effect between AMPA and NMDA receptors to be shown. Indeed, in the presence of magnesium, the NMDA-evoked release of DA could only be observed in the presence of AMPA which, by itself, stimulates also the release of DA and, in addition, eliminates the magnesium block of NMDA receptors by activating voltage-dependent calcium channels (DESCE et al. 1992). Experiments with appropriate antagonists also indicated that the prominent release of DA induced by a high concentration of glutamate results from the combined activation of both types of receptors.

Besides classical depolarising agents (potassium or veratridine), AMPA, or glutamate, others transmitters or receptor agonists that act on heteroreceptors located on dopaminergic nerve terminals may also suppress the magnesium block of NMDA receptors and thus allow the NMDA-evoked release of DA. This was particularly shown with acetylcholine and the agonists of muscarinic and nicotinic receptors, oxotremorine and nicotine, respectively. As expected, different molecular processes were found responsible for the suppression of the magnesium block of NMDA receptors evoked by either oxotremorine or nicotine (CHÉRAMY et al. 1996a).

One of the main problem which has still to be resolved is to understand the physiological significance of this type of co-operation between cholinergic and NMDA receptors, i.e. to determine in which circumstances the cholinergic interneurons facilitate the glutamatergic presynaptic control of DA release through NMDA receptors. Due to the well-known involvement of NMDA receptors in neuronal plasticity, this presynaptic co-operative process between cholinergic interneurons and corticostriatal glutamatergic neurons could decrease the amount of glutamate required for eventual long-term modifications in the reactivity of dopaminergic nerve terminals to incoming signals mediated by NMDA receptors (CALABRESI et al. 1992, 1997). Taking into consideration the hypothesis according to which cholinergic interneurons are involved in the transfer of information between striatal compartments (see below), such local co-operative processes could facilitate and amplify the necessary relationships between the sensory-motor and limbic networks.

As just indicated, specific chemical signals could facilitate, by synergistic processes, the presynaptic action of glutamate on DA transmission. We have also been interested to determine whether, reciprocally, glutamate itself could modify the efficacy of other presynaptic regulations of DA release and, more precisely, the potency of dopaminergic D_2 agonists to inhibit the release of DA through their effect on DA autoreceptors (Fig. 1). These experiments were performed on synaptosomes from mouse striatum.

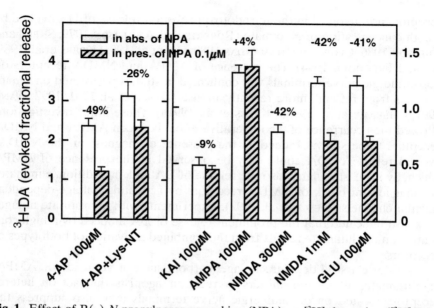

Fig. 1. Effect of R(–)-N-propylnorapomorphine (NPA) on [³H]-dopamine ([³H]-DA) release. Striatal synaptosomes from mouse, preloaded with [³H]-DA, were superfused with a normal or Mg^{++}-free (in NMDA experiments) CSF. 4-Aminopyridine (4-AP) and/or [lys^{8,9}]-neurotensin (8–13) (Lys-NT), kainate (KAI), AMPA (in the presence of cyclothiazide 10 μM), NMDA, L-glutamate and NPA were applied for 5 min, 40 min after the onset of superfusion. The average evoked fractional release of [³H]-DA during the 5-min treatment was calculated. Results are the mean ± SEM of data obtained with 12 superfusion chambers in six independent experiments. In all groups, the release of [³H]-DA was greater than in control groups. The inhibitory effect of NPA (indicated %) was always significant, except when kainate or AMPA were used. Cyclothiazide significantly increased the effect of AMPA alone (not shown). Lys-NT was without effect on basal [³H]-DA release when applied alone (not shown), but significantly reduced the inhibitory effect of NPA

Among different D$_2$ agonists, R(–)-propylnorapomorphine (NPA) was found to be the most potent in inhibiting the release of DA evoked by 4-aminopyridine, a potent blocker of potassium channels. As expected, the inhibitory effect of NPA was suppressed by sulpiride and not observed any longer on striatal synaptosomes from mice lacking D$_2$ receptors (L'HIRONDEL et al. 1998). In contrast to that observed under depolarisation with 4-aminopyridine, NPA did not inhibit the release of DA evoked by the stimulation of AMPA receptors with AMPA. This lack of inhibitory response also occurred under the combined application of AMPA and cyclothiazide, a compound which avoids the rapid desensitisation of AMPA receptors and thus markedly increases the AMPA-evoked release of DA (Fig. 1). Similarly, NPA was without inhibitory effect on the marked release of DA evoked by kainate, an agonist of presynaptic AMPA receptors which, in contrast to AMPA, is devoid of desensitising effect on AMPA receptors. In contrast and demonstrating the specificity of results obtained with AMPA or kainate, the D_2 autoreceptor-mediated inhibitory effect of NPA on the release of DA persisted with an amplitude similar to that observed with 4-aminopyridine under the NMDA-evoked release of DA (application of NMDA without magnesium or application of high concentration of glutamate allowing the combined stimulation of AMPA and NMDA receptors) (Fig. 1). Neurotensin receptors are also present on DA nerve terminals and binding as well as in vivo release studies have suggested that neurotensin reduces the sensitivity of D_2 autoreceptors (FUXE et al. 1992; TANGANELLI et al. 1989). Confirming these findings, we also observed that the inhibitory effect of NPA on the 4-aminopyridine-evoked release of DA was largely reduced in the presence of neurotensin or of its stable analogue, lys-neurotensin (Fig. 1).

These are a few examples of heteroregulations between heteroreceptors or heteroreceptors and D_2 autoreceptors located on DA nerve terminals. However, much has still to be learnt on these interactions in order to determine how functional units represented by the numerous varicosities of dopaminergic fibres integrate and react to simultaneous or successive incoming signals.

C. Role of Diffusible Messengers in the Presynaptic Control of Striatal Dopaminergic Transmission

As is well established, the stimulation of NMDA receptors can lead to several events involved in various processes such as protein synthesis regulation, cellular memory or cell death. Diffusible messengers such as nitric oxide (NO) or arachidonic acid can also be formed under the stimulation of NMDA receptors (DUMUIS et al. 1988; GARTHWAITE 1991; DAVIS and MURPHEY 1994; TENCÉ et al. 1995; RODRIGUEZ-ALVAREZ et al. 1997).

In the striatum, NMDA receptors located on the somatostatin-containing interneurons which possess the constitutive NO synthase are involved in the formation of NO (EMSON et al. 1993). However, in pathological states such as inflammation, NO synthase can also be expressed in glial cells. The facilitatory role of NO on the release of DA in the striatum was demonstrated by generating NO thanks to NO donors (ZHU and LUO 1992; LONART et al. 1993; GUEVARA-GUZMAN et al. 1994; BOWYER et al. 1995; STEWART et al. 1996) or by showing a reduction of the evoked release of DA following the stimulation of NMDA receptors in the presence of NO synthase inhibitors (HANBAUER et al. 1992; ISHIDA et al. 1994). This provided evidence for the involvement of a diffusible messenger in the presynaptic regulation of DA release, this effect requiring the presence of a guanylyl cyclase in dopaminergic nerve terminals. However, NO originating from somatostatin-containing interneurons may locally act, by several processes, on the release of DA. Indeed, contradictory results were obtained by several authors who investigated either in vitro or in vivo the effects of NO synthase inhibitors on either the glutamate- or the

NMDA-evoked release of DA (STRASSER et al. 1994; LIN et al. 1995; SANDOR et al. 1995; SHIBATA et al. 1996).

The NMDA-evoked formation of arachidonic acid has particularly been studied on striatal neuronal cultures (DUMUIS et al. 1988; TENCÉ et al. 1995; RODRIGUEZ-ALVAREZ et al. 1997). As generally assumed, this unsaturated fatty acid is mainly formed in the populations of efferent γ-aminobutyric acid (GABA)ergic neurons which represent more than 95% of the striatal neurons. However, its formation in interneurons cannot be excluded since these cells also possess NMDA receptors. Besides NMDA receptors, AMPA and metabotropic glutamatergic receptors are also involved in the glutamate-evoked formation of arachidonic acid (DUMUIS et al. 1990, 1993; PETITET et al. 1995; WILLIAMS and GLOWINSKI 1996) which depends on calcium influx and the activation of a phospholipase A_2. Interestingly, as shown by experiments from our laboratory performed on striatal neuronal cultures from mouse, marked synergistic effects in the formation of arachidonic acid occur under the combined application of glutamate and acetylcholine. Muscarinic receptors are involved in the effect of acetylcholine but the molecular processes responsible for this pronounced synergistic response are still unknown. Arachidonic acid can also originate from glial cells and particularly from astrocytes (MARIN et al. 1991; TENCÉ et al. 1992; STELLA et al. 1994a, 1997). Indeed, several transmitters alone or in association can lead to the production of arachidonic acid in striatal astrocytes (MARIN et al. 1991; EL-ETR et al. 1992). In particular, glutamate and ATP (the co-transmitter of acetylcholine in striatal cholinergic interneurons) stimulate the formation of arachidonic acid and their combined application leads to an important synergistic response in these cells (STELLA et al. 1994a,b).

These observations on the neuronal and astrocytic formation of arachidonic acid in the striatum led us to determine whether arachidonic acid, which its particularly known for its pleiotropic effects on ionic channels (ORDWAY et al. 1991; VOLTERRA et al. 1992a) and its ability to inhibit glutamate uptake in astrocytes (BARBOUR et al. 1989; VOLTERRA et al. 1992b) could also play a role in the presynaptic regulation of DA release. Synaptosomes or striatal slices from rat or mouse were used for this purpose. The first approach consisted in the investigation of the effect of arachidonic acid alone (L'HIRONDEL et al. 1995), and the second in the determination of the contribution of endogenously formed arachidonic acid in the release of DA evoked by the stimulation of NMDA and/or muscarinic receptors (L'HIRONDEL et al. 1999).

Arachidonic acid stimulates markedly in a concentration- and calcium-dependent manner the release of DA from striatal synaptosomes and a pronounced response can already be observed with a concentration as low as $2\mu M$ (L'HIRONDEL et al. 1999). This concentration is in the range of those evoking the various cellular effects of this unsaturated fatty acid (BARBOUR et al. 1989; CHAN et al. 1983; ORDWAY et al. 1991; VOLTERRA et al. 1992a,b). Arachidonic acid was also found to block the reuptake of DA. Nevertheless, it still markedly stimulates the release of the transmitter in the presence of classical blockers of the DA reuptake process such as nomifensine or mazindol (L'HIRONDEL et al. 1995). Thanks to a sensitive method ([^{3}H]-TPP$^+$), we also observed that arachidonic acid is a potent depolarising agent. However, its very potent stimulatory effect on the release of DA cannot be attributed to its depolarising action since changes in DA release of much lower amplitude are observed under large depolarisation induced by either potassium (25 mM) or 4 amino-pyridine (100 μM). In addition, while the potassium-evoked release of DA is not affected by the inhibition of protein kinase C, the arachidonic acid-evoked release of DA is completely inhibited by chelerythrine and RO 31-754, two potent inhibitors of protein kinase C (L'HIRONDEL et al. 1995). This latter observation is in agreement with the direct and potent stimulating action of the unsaturated fatty acid on protein kinase C activity (ASAOKA et al. 1992; ROBINSON 1992).

Several criteria of specificity were found in the arachidonic acid-evoked release of DA from striatal synaptosomes. First, the effect of arachidonic acid is still observed when the activity of either cytochrome P450 or cyclooxygenase and lipoxygenase is blocked with metyrapone (10 μM) or 5,8,11,14-eicosatetraynoic acid (ETYA, 100 μM), respectively (L'HIRONDEL et al. 1995). This indicates that in our experimental conditions, arachidonic acid alone and not one of its metabolites (which have the capacity to induce physiological responses) is responsible for the evoked release of DA. Secondly, several fatty acids, including oleic acid, the saturated fatty acid arachidic acid as well as their methyl ester derivatives are without effect on the release and the high-affinity uptake processes of DA (L'HIRONDEL et al. 1995). However, parallel experiments on the release and the reuptake processes of GABA performed on striatal synaptosomes from rat indicated that arachidonic acid is not only acting on dopaminergic nerve terminals. Indeed, arachidonic acid inhibits the reuptake and stimulates as well the release of GABA (CHÉRAMY et al. 1996b). However, slight differences can be observed since arachidonic acid is more potent and has a more rapid kinetic of action on GABA than on DA release. Moreover, the arachidonic acid-evoked release of GABA is reduced by 50% only by protein kinase C inhibitors, suggesting that different protein kinase C isoforms are present in the two types of nerve terminals. In this context, it should be recalled that arachidonic acid has also been shown to facilitate the release of glutamate from cortical nerve endings when co-applied with an agonist of metabotropic glutamatergic receptors (FREEMAN et al. 1990; LYNCH and VOSS 1990; HERRERO et al. 1992a,b; McGAHON and LYNCH 1996).

Experiments performed with several inhibitors of phospholipase A$_2$ [mepacrine, 4-bromophenacylbromide, 7,7-dimethyleicosadienoic acid (DEDA)] on microdiscs of tissues from mouse striatum have confirmed that endogenously formed arachidonic acid facilitates, indeed, the release of DA (L'HIRONDEL et al. 1999). For example, mepacrine (0.1 μM) reduces by about 40% the marked release of DA evoked by the combined stimulation of NMDA and muscarinic receptors with NMDA and carbachol (L'HIRONDEL et al. 1999), a treatment which, as already indicated, induced important synergistic effects on arachidonic acid formation in cultured striatal neurons from mouse (TENCÉ et al. 1995). Complementary data indicated that the effect of mepacrine (or other phospholipase A_2 inhibitors which induced similar reduction in DA release) results, indeed, from the inhibition of arachidonic acid formation and not from an unspecific action of the drug. For instance, in contrast, mepacrine (0.1 μM) modifies neither the potassium (25 mM)- nor the nicotine (1 mM)-evoked release of DA (L'HIRONDEL et al. 1999). Moreover, confirming that the stimulation of NMDA and muscarinic receptors are both involved in the endogenous formation of arachidonic acid in striatal microdiscs, mepacrine (0.1 μM) reduces as well, but with different kinetics, the NMDA (without magnesium)- or the oxotremorine-evoked release of DA. Finally, the amplitude and the pattern of the inhibitory effect of mepacrine depend on the concentration of NMDA (50 μM to 1 mM) (L'HIRONDEL et al. 1999). This is reminiscent of data obtained on neuronal cultures since the NMDA-evoked formation of arachidonic acid is concentration-dependent.

As already underlined, in striatal microdiscs from adult mouse, endogenously formed arachidonic acid originates for a large part from the populations of GABAergic efferent neurons. These cells possess both NMDA and muscarinic receptors and their spiny dendritic spines are the main targets of the nigrostriatal dopaminergic neurons. However, arachidonic acid could also be partially formed in dopaminergic nerve terminals. Indeed, phospholipase A_2 inhibitors were also shown to reduce the release of DA evoked by the combined application of NMDA and carbachol in striatal synaptosomes (L'HIRONDEL et al. 1999). As also demonstrated on synaptosomes, likely due to its depolarising effect, arachidonic acid can eliminate the magnesium block of NMDA receptors (Fig. 2). It was also found to reduce the inhibitory effect of NPA on the 4-aminopyridine- or the glutamate-evoked release of DA (reduced efficacy of D_2 autoreceptors) (Fig. 2). These latter effects could be partly responsible for the arachidonic acid-dependent release of DA evoked in synaptosomes by the combined stimulation of NMDA and muscarinic receptors.

D. Local Circuits Involved in the Control of DA Transmission in Striatal Compartments

As shown in several species including man, the striatum is an heterogeneous structure in which two main compartments can be distinguished, the striosomes and the matrix. These compartments appear at different stages during development and can be defined by specific biochemical markers but also by their afferent and efferent pathways (GRAYBIEL 1990; GERFEN and WILSON 1996). As generally assumed, the striosomes, which represent a three-dimensional labyrinthine network (DESBAN et al. 1989, 1993; GRAYBIEL 1990) are connected to the limbic system, while the matrix, which is mainly distributed in the dorsolateral part of the striatum, belongs to the sensory–motor

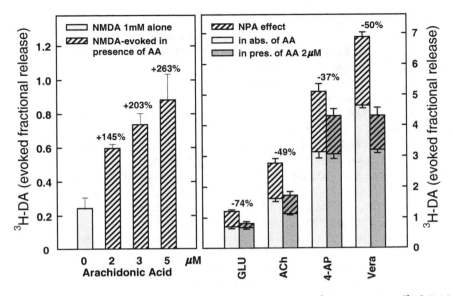

Fig. 2. Effects of arachidonic acid (AA) on the release of [³H]-dopamine ([³H]-DA). Striatal synaptosomes from rat, preloaded with [³H]-DA, were superfused with a normal CSF. *Left panel*: NMDA and/or AA, were applied for 5 min, 40 min after the onset of superfusion. The average NMDA-evoked fractional release of [³H]-DA during the 5-min treatment was calculated by subtracting the corresponding value obtained with AA alone. *Right panel*: Experiments were carried out as described in Fig. 1, but in the presence or absence of AA. L-Glutamate (GLU, 100 µM), acetylcholine (ACh, 100 µM), 4-aminopyridine (4-AP, 100 µM) or veratridine (Vera, 1 µM) were applied for 5 min, 40 min after the onset of superfusion. The release [³H]-DA evoked by each of these four drugs was calculated by subtracting the corresponding value obtained in absence of GLU, ACh, 4-AP or Vera. In all cases, results are the mean ± SEM of data obtained with 12 superfusion chambers in six independent experiments. In all groups, except NMDA alone, the release of [³H]-DA was greater than in control groups. The release of [³H]-DA evoked by NMDA in the presence of AA was significantly greater than when NMDA was applied alone. The inhibitory effect of NPA (indicated by a *dashed area*) was significantly reduced (indicated %) in the presence of AA

network. It has also been proposed that striatal interneurons and cholinergic interneurons, particularly, are involved in the transfer of information between these compartments (GRAYBIEL et al. 1986, 1994; KUBOTA and KAWAGUCHI 1993). In fact, the cholinergic interneurons which innervate all parts of the striatum are represented by two populations of cells. These cells are mainly located in the matrix either close to the striosomes or near a subcompartment of the matrix, the matrisomes (AOSAKI et al. 1995). While most striatal efferent neurons are silent in resting conditions, the cholinergic interneurons are tonically active (WILSON et al. 1990; AOSAKI et al. 1994; GRAYBIEL et al. 1994; KIMURA 1995; APICELLA et al. 1998). The dopaminergic innervation of the striatum is also heterogeneous since the striosomes are mainly innervated by a group of dopaminergic cells located in the densocellular zone of the pars compacta, while other nigral dopaminergic cells and those of the A8 group project to the matrix (GRAYBIEL 1990; GERFEN and WILSON 1996).

Several years ago, these anatomical observations led us to believe that the presynaptic regulations of DA release (either direct or indirect through local circuits) could differ from one striatal compartment to the other. Due to the small size of the striosomes and their complicated network, a new superfusion method in vitro was set up. This procedure allows the superfusion of discrete striatal areas enriched in either striosomes or matrix (KEMEL et al. 1989). Experiments were first carried out on coronal slices of cat brain and then on coronal or saggital slices of rat brain to study the direct and/or indirect effects of acetylcholine (cat) and glutamate (rat) on DA release in each compartment. In the latter case, for simplification, due to the diversity of glutamatergic receptors, the effects of NMDA (in the absence of magnesium) were particularly investigated. Since cholinergic and NMDA receptors are not only located on dopaminergic nerve terminals but mainly on most striatal neurons, indirect effects of either acetylcholine or NMDA on the release of DA were identified with appropriate antagonists. The role of GABA and of the peptidic cotransmitters contain in GABAergic efferent neurons (opioid peptides and tachykinins) were particularly investigated. In all cases, due to the very small volume of tissue superfused in our experimental conditions (less than $1\,mm^3$), radioactive DA continuously synthesised from tritiated tyrosine was estimated in superfusates.

I. Similarities and Differences in the Presynaptic Regulation of DA Release in Striatal Compartments

Although acetylcholine and NMDA experiments were performed in two distinct species, several general conclusions can already be drawn from these studies.

1. Direct (tetrodotoxin-insensitive) facilitatory presynaptic regulation of DA release occurs in both compartments under the local application of acetylcholine (muscarinic receptors) or NMDA (NMDA receptors) (KEMEL et al. 1989; KREBS et al. 1991a,b).
2. A direct facilitatory presynaptic regulation evoked by acetylcholine and involving nicotinic receptors is only observed in the matrix (KEMEL et al. 1989).
3. Important differences in the indirect presynaptic regulation of DA release triggered by either acetylcholine or NMDA are observed between striosomes and matrix. Therefore, different local circuits may contribute to the regulation of DA transmission in these compartments (KEMEL et al. 1989, 1992; GAUCHY et al. 1991; KREBS et al. 1991b, 1993, 1994).
4. Indirect inhibitory presynaptic regulation of DA transmission triggered by acetylcholine is only observed in the matrix while that evoked by NMDA

occurs in both compartments (KEMEL et al. 1989, 1992; GAUCHY et al. 1991; KREBS et al. 1993, 1994).
5. Indirectly, both acetylcholine (in the matrix) and NMDA (in both compartments) reduce DA transmission through a GABAergic link (KEMEL et al. 1992; KREBS et al. 1993).
6. Opioid peptides and/or tachykinins are also involved in the indirect presynaptic regulation of DA release (GAUCHY et al. 1991; KREBS et al. 1994).
7. In general, NMDA-sensitive local inhibitory circuits contributing to the control of DA transmission are more potent in striosomes than in matrix, but their complexity is much higher in matrix than in striosomes (KREBS et al. 1994).

II. The GABA- and Dynorphin-Dependent Inhibitions of DA Transmission Triggered by Acetylcholine Occur in Two Distinct Matrix Territories

As already indicated, in the cat experiments, the indirect cholinergic control of DA release was only observed in the matrix and, in addition, the identity of the transmitter involved in this indirect regulation was found to differ from one part of the matrix to another (KEMEL et al. 1992).

More precisely, thanks to experiments performed in the presence of biculline, acetylcholine was also shown to facilitate the release of GABA and, therefore, indirectly to exert an inhibitory effect on the direct cholinergic facilitation of DA release (GABA inhibits, indeed, the release of DA by acting through $GABA_A$ receptors located on dopaminergic nerve terminals). Both muscarinic and nicotinic receptors are involved in this inhibitory local circuit triggered by acetylcholine.

Similar experiments performed with naloxone indicated that, through its effect on muscarinic receptors, acetylcholine can also indirectly inhibit the evoked release of DA by stimulating the release of an opioid peptide. This indirect presynaptic inhibitory regulation of DA transmission results from the action of released dynorphin on kappa receptors located on DA nerve terminals. In agreement with the role of these opioid receptors in this regulation, dynorphin and another kappa agonist (U 50488) totally suppress the disinhibitory effect of naloxone on the acetylcholine-evoked release of DA.

Of particular interest, the inhibitory regulation triggered by acetylcholine which involves either GABA or dynorphin occur in distinct matrix territories. One of these territories is particularly rich in aggregated neurons projecting to the substantia nigra pars reticulata (GABA regulation), while the other contains non-aggregated cells projecting either to the substantia nigra pars reticulata and/or the internal globus pallidus (dynorphin regulation) (DESBAN et al. 1995; KEMEL et al.1992). According to GRAYBIEL et al. (1991), matrix territories enriched in aggregated neurons can be activated from somatosensory cortical areas and correspond to the matrisomes.

III. NMDA-Dependent Local Inhibitory Circuits of DA Transmission Occur in Both Striatal Compartments and Involve GABA and Dynorphin

In rat, bicuculline and naloxone were also shown to induce disinhibitory effects on the release of DA evoked by NMDA ($50\mu M$). However, these responses which result respectively from the blockade of the inhibitory effects of GABA and dynorphin on dopaminergic transmission, were observed in both striatal compartments. In addition, they were found to be much more potent in striosomes than in the matrix (KREBS et al. 1994).

The disinhibitory effects of bicuculline and naloxone on the NMDA-evoked release are not additive in the striosomes, but a complete additivity is observed in the matrix (Fig. 3). This latter observation, which is reminiscent of the results obtained in cat with acetylcholine, could reflect the heterogeneity of the matrix. Since NMDA stimulates as well the release of acetylcholine, a cholinergic link could be involved in the NMDA-sensitive inhibitory local circuits which contribute to the modulation of DA transmission. Supporting this statement, as observed with bicuculline and naloxone, the complete blockade of cholinergic transmission with atropine and pempidine resulted in a marked facilitation of the NMDA-evoked release of DA and this effect was only observed in the matrix (Fig. 3).

IV. Facilitation by DA of the NMDA-Sensitive Local Inhibitory Circuits Involved in the Presynaptic Regulation of DA Release in Striatal Compartments

Through its effects on D_2 and D_1 receptors, DA which is released under the local application of a small concentration of NMDA ($50\mu M$) regulates also some of the local circuits responsible for the presynaptic control of its own release process. Indeed, disinhibitory effects on NMDA-evoked responses were also observed in the presence of either sulpiride or SCH23390, the antagonists of D_2 and D_1 receptors, respectively. As observed with bicuculline and naloxone, the disinhibitory effects of the DA antagonists were of much larger amplitude in striosomes than in the matrix. These marked disinhibitory effects were suppressed in the presence of tetrodotoxin demonstrating that these responses result from the blockade of the action of DA on target cells of the striatum. From these results, it can be concluded that under the application of a moderate concentration of NMDA, through its effects on D_1 or D_2 receptors, released DA inhibits its own release process by facilitating NMDA-sensitive inhibitory local circuits involved in the control of DA transmission, and that these effects occur in both striatal compartments.

According to several groups (GERFEN et al. 1990; LE MOINE and BLOCH 1995; YUNG et al. 1995; INCE et al. 1997), in the matrix D1 receptors are mainly located on the GABAergic neurons which project to the substantia nigra pars reticulata and the entopeduncular nucleus, while D_2 receptors are mainly

Presynaptic Regulation of Dopamine Release

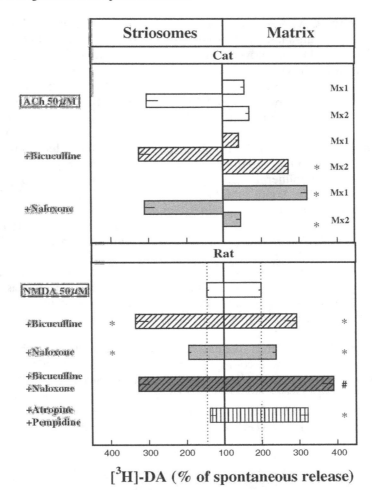

Fig. 3. Local inhibitory circuits of DA transmission triggered by acetylcholine or NMDA in striosomes and matrix. Selected areas of cat caudate nucleus (*upper part*) and of rat striatum (*lower part*) known to correspond to striosomes and matrix territories (Mx1 and Mx1, two distinct matrix areas in cat) were superfused using a microsuperfusion device and the release of [³H]-DA newly synthesized from [³H]-tyrosine was estimated in successive 5-min fractions. Acetylcholine (ACh) or NMDA (in a magnesium-free CSF) was applied during 25 min, 65 min after the onset of the superfusion. When used, bicuculline (5 μM) and/or naloxone (1 μM) or atropine (1 μM) and pempidine (10 μM) were present throughout the superfusion. Results correspond to the mean value of the evoked release of [³H]-DA (minus the spontaneous release) during the overall 25 min application of either ACh or NMDA. Due to the amplitude of the responses, NMDA data are expressed on a 5-min basis. Results are the mean ± SEM of data obtained in 8–17 experiments. *$p < 0.05$ effects of ACh (*upper part*) or of NMDA (*lower part*) in the presence of bicuculline, naloxone or atropine and pempidine when compared to the corresponding control response induced by ACh or NMDA alone; #$p < 0.05$ effect of NMDA in the presence of bicuculline and naloxone when compared to the effect of NMDA in the presence of either bicuculline or naloxone alone in the matrix compartment

found on the GABAergic neurons which project to the external globus pallidus. On this simplified basis, two distinct inhibitory circuits could be involved in the D_1 and D_2 receptor-mediated inhibitory control of DA transmission. Attempts were thus made to confirm this hypothesis by additivity experiments performed in the presence of sulpiride or SCH23390 with either bicuculline, naloxone or RP67580, a potent antagonist of NK1 tachykinin receptors (GARRET et al. 1991). RP67580 was also used in these experiments for several reasons: (1) substance P facilitates in a tetrodotoxin-sensitive manner the spontaneous release of DA in the matrix, and this effect which is blocked by RP67580 can also be partially blocked by cholinergic antagonists (TREMBLAY et al. 1992); (2) cholinergic interneurons possess NK1 receptors (GERFEN 1991; AUBRY et al. 1994; JAKAB ET GOLDMAN-RAKIC 1996), substance P stimulates the evoked release of acetylcholine (ARENAS et al. 1991; PETITET et al. 1991; ANDERSON et al. 1993; GUEVARA GUZMAN et al. 1993), and this effect is also blocked by RP67580; and (3) as with bicuculline and naloxone, RP67580 induced a disinhibitory effect on the NMDA ($50\mu M$)-evoked release of DA in the matrix (Fig. 4). This further indicates that substance P contributes also to the NMDA-dependent local control of DA transmission. This is not surprising since, as previously discussed, acetylcholine can be an intermediate link of the NMDA-sensitive inhibitory circuits involved in the presynaptic control of DA transmission.

Interestingly, additive disinhibitory effects were found when the D_2 antagonist sulpiride was co-applied with naloxone and RP67580, while no additivity occurred under the co-application of sulpiride and bicuculline. In contrast, additivity effects were found when the D1 antagonist was co-applied with either bicuculline or naloxone but not under the co-application of SCH23390 and RP67580 (Fig. 4). Several conclusions can be drawn from these experiments:

1. In agreement with our hypothesis, the disinhibitory effects of sulpiride and SCH23390 are mediated through distinct local circuits.
2. The prevention of the inhibitory effect of GABA on the evoked DA transmission could be the common link between the disinhibitory effects of sulpiride and bicuculline on the NMDA-evoked response.
3. The prevention of the inhibitory effect of substance P on the evoked DA transmission could be the common link between the disinhibitory effects of SCH23390 and RP67580. This could suggest that through its effects on D1 receptors, DA facilitates the NMDA-evoked release of substance P, which is in agreement with the co-localisation of substance P in the GABAergic neurons possessing D1 receptors.
4. The additivity of the disinhibitory effects of naloxone and sulpiride on one hand and of naloxone and SCH23390 on the other hand suggest that DA has little influence on the naloxone-sensitive inhibitory circuit triggered by NMDA and further underline the complexity of the matrix anatomical organisation.

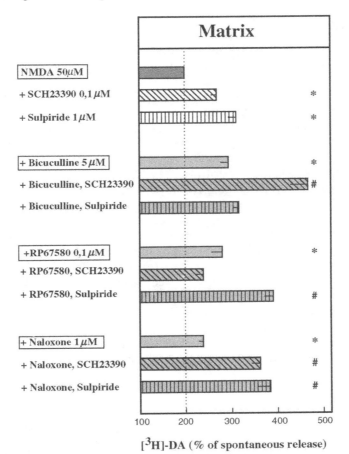

Fig. 4. Role of D$_1$ and D$_2$ receptors in the NMDA-sensitive inhibitory circuits involved in the presynaptic regulation of DA release in the matrix. Experiments and expression of data are as described in the legend of Fig. 3. NMDA (in a magnesium-free CSF) was applied for 25 min, 65 min after the onset of the superfusion. When used, SCH23390, sulpiride, bicuculline, RP67580 or naloxone were present throughout the superfusion. Results correspond to the mean value of the evoked release of [^{3}H]-DA (minus the spontaneous release) during the overall 25-min application of NMDA (expressed on a 5-min basis). Results are the mean ± SEM of data obtained in 10–17 experiments. *$p < 0.05$ effect of NMDA in the presence of either SCH23390, sulpiride, bicuculline, RP67580 or naloxone when compared to the effect of NMDA alone; #$p < 0.05$ effect of NMDA in the presence of the combined application of antagonists (bicuculline and SCH23390, RP68580 and sulpiride, naloxone and SCH23390 or naloxone and sulpiride) when compared to the effects of NMDA in the presence of either SCH23390, sulpiride, bicuculline, RP67580 or naloxone alone

E. Conclusions

Since the discovery that Parkinson's disease results from the degeneration of the nigrostriatal dopaminergic neurons, the crucial role of DA in the appropriate transfer of signals from the striatum to output structures from the basal ganglia has been well established. Several studies have been made to determine how released DA modulates signals delivered from various cortical areas or specific thalamic nuclei to different populations of striatal cells. Reciprocally, it seems important to precisely identify the mechanisms responsible for the regulation of DA transmission.

Due to the development of molecular biology, major efforts have been made during the last decade to increase our knowledge on DA receptors, their transduction processes and their effects on intracellular signalling cascades. Much has also been learnt about the processes of DA receptor expression and, due to the availability of specific antibodies, the cellular localisation of these receptors. Thanks to the development of the microdialysis technique, several release studies in vivo on unanaesthetised rats have allowed us to obtain some information on the relationships between changes in DA release and behavioural responses in pharmacological or physiological states. However, surprisingly, less attention has been made to explore more deeply the different types of presynaptic regulatory processes which contribute to the control of DA release in the striatum. In the present review, we have attempted to show that great progress can still be made in this particular field.

References

Anderson JJ, Chase TN, Engber TM (1993) Substance P increases release of acetylcholine in the dorsal striatum of freely moving rats. Brain Res 623:189–194

Aosaki T, Kimura M, Graybiel AM (1995) Temporal and spatial characteristics of tonically active neurons of the primate's striatum. J Neurophysiol 73:1234–1252

Aosaki T, Tsubokawa H, Ishida A, Watanabe K, Graybiel AM, Kimura M (1994) Responses of tonically active neurons in the primate's striatum undergo systematic changes during behavioral sensorimotor conditioning. J Neurosci 14: 3969–3984

Apicella P, Ravel S, Sardo P, Legallet E (1998) Influence of predictive information on reponses of tonically active neurons in the monkey striatum. J Neurophysiol 80: 3341–3344

Arenas E, Alberch J, Perez-Navarro E, Solsona C, Marsal J (1991) Neurokinin receptors differentially mediate endogenous acetylcholine release evoked by tachykinins in the neostriatum. J Neurosci 11:2332–2338

Asaoka Y, Nakamura S, Yoshida K, Nishizuka Y (1992) Protein kinase C, calcium and phospholipid degradation. TIBS 17:414–417

Aubry JM, Lundström K, Kawashima E, Ayala G, Schulz P, Bartanusz V, K iss JZ (1994) NK_1 receptor expression by cholinergic interneurons in human striatum. Neuroreport 5:1597–1600

Barbeito L, Girault JA, Godeheu G, Pittaluga A, Glowinski J, Chéramy A (1989) Activation of the bilateral cortico-striatal glutamatergic projection by infusion of GABA into thalamic motor nuclei in the cat: an in vivo release study. Neuroscience 28:365–374

Barbour B, Szatkowski M, Ingledew N, Attwell D (1989) Arachidonic acid induces a prolonged inhibition of glutamate uptake into glial cells. Nature 342:918–920

Besson MJ, Chéramy A, Feltz P, Glowinski J (1969) Release of newly synthesized dopamine from dopamine-containing terminals in the striatum of the rat. Proc Natl Acad Sci USA 62:741–748

Bowyer JF, Weiner N (1989) K^+ channel and adenylate cyclase involvement in regulation of Ca^{2+}-evoked release of 3H-dopamine from synaptosomes. J Pharmacol Exp Ther 248:514–520

Bowyer JF, Clausing P, Gough B, Slikker W, Jr, Holson RR (1995) Nitric oxide regulation of methamphetamine-induced dopamine release in caudate putamen. Brain Res 699:62–70

Cai NS, Kiss B, Erdo SL (1991) Heterogeneity of N-methyl-d-aspartate receptors regulating the release of dopamine and acetylcholine from striatal slices. J Neurochem 57:2148–2151

Calabresi P, Pisani A, Mercuri NB, Bernardi G (1992) Long-term potentiation in the striatum is unmasked by removing the voltage-dependent magnesium block of NMDA receptor channels. Eur J Neurosci 4:929–935

Calabresi P, Saiardi A, Pisani A, Baik JH, Centonze D, Mercuri NB, Bernardi G, Borrelli E (1997) Abnormal synaptic plasticity in the striatum of mice lacking dopamine D2 receptors. J Neurosci 17:4536–4544

Cass WA, Zahniser NR (1991) Potassium channel blockers inhibit D2 dopamine, but not A1 adenosine, receptor-mediated inhibition of striatal dopamine release. J Neurochem 57:147–152

Chan PH, Kerlan R, Fishman RA (1983) Reductions of gamma-aminobutyric acid and glutamate uptake and $(Na^+ + K^+)$-ATPase activity in brain slices and synaptosomes by arachidonic acid. J Neurochem 40:309–316

Chesselet MF, Chéramy A, Romo R, Desban M, Glowinski J (1983) GABA in the thalamic motor nuclei modulates dopamine release from the two dopaminergic nigrostriatal pathways in the cat. Exp Brain Res 51:275–282

Chesselet MF (1984) Presynaptic regulation of neurotransmitter release in the brain: facts and hypothesis. Neuroscience 12:347–375

Chéramy A, Romo R, Godeheu G, Baruch P, Glowinski J (1986) In vivo presynaptic control of dopamine release in the cat caudate nucleus II. Facilitatory or inhibitory influence of L-glutamate. Neuroscience 19:1081–1090

Chéramy A, Kemel ML, Gauchy C, Desce JM, Barbeito L, Glowinski J (1991) Role of excitatory amino acids in the direct and indirect presynaptic regulation of dopamine release from nerve terminals and dendrites of nigrostriatal dopaminergic neurons. Amino Acids 1:351–363

Chéramy A, Godeheu G, L'hirondel M, Glowinski J (1996a) Cooperative contributions of cholinergic and NMDA receptors in the presynaptic control of dopamine release from synaptosomes of the rat striatum. J Pharmacol Exp Ther 276:616–625

Chéramy A, Artaud F, Godeheu G, L'hirondel M, Glowinski J (1996b) Stimulatory effect of arachidonic acid on the release of GABA in matrix-enriched areas from the rat striatum. Brain Res 742:185–194

Clow DW, Jhamandas K (1989) Characterization of L-glutamate action on the release of endogenous dopamine from rat caudate-putamen. J Pharmacol Exp Ther 248:722–728

Davis GW, Murphey RK (1994) Long-term regulation of short-term transmitter release properties: Retrograde signaling and synaptic development. TINS 17:9–13

Desban M, Gauchy C, Glowinski J, Kemel ML (1995) Heterogeneous topographical distribution of the striatonigral and striatopallidal neurons in the matrix compartment of the cat caudate nucleus. J Comp Neurol 352:117–133

Desban M, Gauchy C, Kemel ML, Besson MJ, Glowinski J (1989) Three-dimensional organization of the striosomal compartment and patchy distribution of striatonigral projections in the matrix of the cat caudate nucleus. Neuroscience 29:551–566

Desban M, Kemel ML, Glowinski J, Gauchy C (1993) Spatial organization of patch and matrix compartments in the rat striatum. Neuroscience 57:661–671

Desce JM, Godeheu G, Galli T, Artaud F, Chéramy A, Glowinski J (1991) Presynaptic facilitation of dopamine release through AMPA receptors on synaptosomes from the rat striatum. J Pharmacol Exp Ther 259:692–698

Desce JM, Godeheu G, Galli T, Artaud F, Chéramy A, Glowinski J (1992) L-glutamate-evoked release of dopamine from synaptosomes of the rat striatum: involvement of AMPA and NMDA receptors. Neuroscience 47:333–339

Diaz J, Lévesque D, Lammers CH, Griffon N, Martres MP, Schwartz JC, Sokoloff P (1995) Phenotypical characterization of neurons expressing the dopamine D_3 receptor in the rat brain. Neuroscience 65:731–745

Dumuis A, Sebben M, Haynes L, Pin JP, Bockaert J (1988) NMDA receptors activate the arachidonic acid cascade system in striatal neurons. Nature 336:68–70

Dumuis A, Pin JP, Oomagari K, Sebben M, Bockaert J (1990) Arachidonic acid released from striatal neurons by joint stimulation of ionotropic and metabotropic quisqualate receptors. Nature 347:182–184

Dumuis A, Sebben M, Fagni L, Prézeau L, Manzoni OJ, Cragoe EJ, Bockaert J (1993) Stimulation by glutamate receptors of arachidonic acid release depends on the Na^+/Ca^{++} exchanger in neuronal cells. Mol Pharmacol 43:976–981

El-Etr M, Marin P, Tencé M, Delumeau JC, Cordier J, Glowinski J, Premont J (1992) 2-Chloroadenosine potentiates the a_1-adrenergic activation of phospholipase C through a mechanism involving arachidonic acid and glutamate in striatal astrocytes. J Neurosci 12:1363–1369

El Mestikawy S, Gozlan H, Glowinski J, Hamon M (1985) Characteristics of tyrosine hydroxylase activation by K^+ induced depolarization and/or forskolin in rat striatal slices. J Neurochem 45:173–184

El Mestikawy S, Glowinski J, Hamon M (1986) Presynaptic dopamine autoreceptors control tyrosine hydroxylase activation in depolarized striatal dopaminergic terminals. J Neurochem 46:12–22

Emson PC, Augood SJ, Senaris R, Guevara-Guzman R, Kishimoto J, Kadowaki K, Norris PJ, Kendrick KM (1993) Chemical signalling and striatal interneurons. Prog Brain Res 99:155–165

Fotuhi M, Sharp AH, Glatt CE, Hwang PM, von Krosigk M, Snyder SH, Dawson TM (1993) Differential localization of phosphoinositide-linked metabotropic glutamate receptor (mGluR1) and the inositol 1,4,5-triphosphate receptor in rat brain. J Neurosci 13:2001–2012

Freeman EJ, Terrian DM, Dorman RV (1990) Presynaptic facilitation of glutamate release from isolated hippocampal mossy fiber nerve endings by arachidonic acid. Neurochem Res 15:743–750

Fuxe K, O'Connor WT, Antonelli T, Osborne PG, Tanganelli S, Agnati LF, Ungerstedt U (1992) Evidence for a substrate of neuronal plasticity based on pre- and postsynaptic neurotensin-dopamine receptor interactions in the neostriatum. Proc Natl Acad Sci USA 89:5591–5595

Garthwaite J (1991) Glutamate, nitric oxide and cell-cell signalling in the nervous system. TINS 14:60–67

Garret C, Carruette A, Fardin V, Moussaoui S, Peyronel JF, Blanchart JC, Laduron P (1991) Pharmacological properties of a potent and selective nonpeptide substance P antagonist. Proc. Natl Acad. Sci. USA 88:10208–10212

Gauchy C, Desban M, Krebs MO, Glowinski J, Kemel ML (1991) Role of dynorphin-containing neurons in the presynaptic inhibitory control of the acetylcholine-evoked release of dopamine in the striosomes and the matrix of the cat caudate nucleus. Neuroscience 41:449–458

Gerfen CR (1991) Substance P (neurokinin-1) receptor mRNA is selectively expressed in cholinergic neurons in the striatum and basal forebrain. Brain Res 556:165–170

Gerfen CR, Engber TM, Mahan LC, Susel Z, Chase TN, Monsma FJ, Sibley DR (1990) D_1 and D_2 dopamine receptor-regulated gene expression of striatonigral and striatopallidal neurons. Science 250:1429–1432

Gerfen CR, Wilson CJ (1996) The basal ganglia. In: Swanson LW, Bjorklund A, Hokfelt T (eds) Handbook of chemical neuroanatomy, Integrated systems of the CNS, Part III. Elsevier Science vol 12:371–468

Graybiel AM (1990) Neurotransmitters and neuromodulators in the basal ganglia. Trends Neurosci 13:244–254

Graybiel AM, Aosaki T, Flaherty AW, Kimura M (1994) The basal ganglia and adaptative motor control. Science 265:1826–1831

Graybiel AM, Baughman RW, Eckenstein F (1986) Cholinergic neuropil of the striatum observes striosomal boundaries. Nature 323:625–627

Graybiel AM, Flaherty AW, Giménez-Amaya JM (1991) Striosomes and matrisomes, In: Bernardi G et al. (eds) The basal ganglia III. Plenum Press, New York 3–12

Guevara-Guzman R, Emson PC, Kendrick KM (1994) Modulation of in vivo striatal transmitter release by nitric oxide and cyclic GMP. J Neurochem 62:807–810

Guevara-Guzman R, Kendric KM, Emson PC (1993) Effect of substance P on acetylcholine and dopamine release in the rat striatum: a microdialysis study. Brain Res 622:147–154

Hanbauer I, Wink D, Osawa Y, Edelman GM, Gally JA (1992) Role of nitric oxide in NMDA-evoked release of [^{3}H]-dopamine from striatal slices. NeuroReport 3: 409–412

Herrero I, Miras-Portugal MT, Sanchez-Prieto J (1992a) Activation of protein kinase C by phorbol Esters and Arachidonic acid required for the optimal potentiation of glutamate exocytosis. J Neurochem 59:1574–1577

Herrero I, Miras-Portugal MT, Sanchez-Prieto JP (1992b) Positive feedback of glutamate exocytosis by metabotropic presynaptic receptor stimulation. Nature 360: 163–166

Ince E, Ciliax BJ, Levey AI (1997) Differential expression of D_1 and D_2 dopamine and m_4 muscarinic acetylcholine receptor proteins in identified striatonigral neurons. Synapse 27:357–366

Ishida Y, Yamamoto R, Mitsuyama Y (1994) Effects of L- and D-enantiomers of N omega-nitro-arginine on NMDA-evoked striatal dopamine overflow. Brain Res Bull 34:483–486

Jakab RL, Goldman-Rakic P (1996) Presynaptic and postsynaptic subcellular localization of substance P receptor immunoreactivity in the neostriatum of the rat and rhesus monkey. J Comp Neurol 369:125–136

Jin S, Fredholm BB (1994) Role of NMDA, AMPA and kainate receptors in mediating glutamate- and 4-AP-induced dopamine and acetylcholine release from rat striatal slices. Neuropharmacology 33:1039–1048

Kemel ML, Desban M, Glowinski J, Gauchy C (1989) Distinct presynaptic control of dopamine release in striosomal and matrix areas of the cat caudate nucleus. Proc. Natl. Acad. Sci. USA 86:9006–9010

Kemel ML, Desban M, Glowinski J, Gauchy C (1992) Functional heterogeneity of the matrix compartment in the cat caudate nucleus as demonstrated by the cholinergic presynaptic regulation of dopamine release. Neuroscience 50:597–610

Kimura M (1995) Role of basal ganglia in behavioral learning. Neurosci. Res 22: 353–358

Krebs MO, Desce JM, Kemel ML, Gauchy C, Godeheu G, Chéramy A, Glowinski J (1991a) Glutamatergic control of dopamine release in the rat striatum: evidence for presynaptic N-methyl-D-Aspartate receptors on dopaminergic terminals. J Neurochem 56:81–85

Krebs MO, Gauchy C, Desban M, Glowinski J, Kemel ML (1994) Role of dynorphin and GABA in the inhibitory regulation of NMDA-induced dopamine release in striosome- and matrix-enriched areas of the rat striatum. J Neurosci 14:2435–2443

Krebs MO, Kemel ML, Gauchy C, Desban M, Glowinski J (1989) Glycine potentiates the NMDA-induced release of dopamine through a strychnine-insensitive site in the rat striatum. Eur J Pharmacol 166:567–570

Krebs MO, Kemel ML, Gauchy C, Desban M, Glowinski J (1993) Local GABAergic regulation of the N-methyl-D-aspartate-evoked release of dopamine is more prominent in striosomes than in matrix of the rat striatum. Neuroscience 57: 249–260

Krebs MO, Trovéro F, Desban M, Gauchy C, Glowinski J, Kemel ML (1991b) Distinct presynaptic regulation of dopamine release through NMDA receptors in striosome- and matrix-enriched areas of the rat striatum. J Neurosci 11:1256–1262

Kubota Y, Kawaguchi Y (1993) Spatial distributions of chemically identified intrinsic neurons in relation to patch and matrix compartments of rat neostriatum. J Comp Neurol 332:499–513

Le Moine C, Bloch B (1995) D_1 and D_2 dopamine receptor gene expression in the rat striatum: sensitive cRNA probes demonstrate prominent segregation of D_1 and D_2 mRNAs in distinct neuronal populations of the dorsal and ventral striatum. J Comp Neurol 355:418–426

L'hirondel M, Chéramy A, Godeheu G, Glowinski J (1995) Effects of arachidonic acid on dopamine synthesis, spontaneous release and uptake in striatal synaptosomes from the rat. J Neurochem 64:1406–1409

L'hirondel M, Chéramy A, Godeheu G, Artaud F, Saiardi A, Borrelli E, Glowinski J (1998) Lack of autoreceptor-mediated inhibitory control of dopamine release in striatal synaptosomes of D2 receptor-deficient mice. Brain Res 792:253–262

L'hirondel M, Chéramy A, Artaud F, Godeheu G, Glowinski J (1999) Contribution of endogenously formed arachidonic acid in the presynaptic facilitatory effects of NMDA and carbachol on dopamine release in the mouse striatum. Eur J Neurosci 11:1292–1300

Lin A, Kao L, Chai C (1995) Involvement of nitric oxide in dopaminergic transmission in rat striatum: an in vivo electrochemical study. J Neurochem 65:2043–2049

Lonart G, Cassels KL, Johnson KM (1993) Nitric oxide induces calcium-dependent ³H-dopamine release from striatal slices. J Neurosci Res 35:192–198

Lynch MA, Voss KL (1990) Arachidonic acid increases inositol phospholipid metabolism and glutamate release in synaptosomes prepared from hippocampal tissue. J Neurochem 55:215–221

Marin P, Delumeau JC, Tencé M, Cordier J, Glowinski J, Prémont J (1991) Somatostatin potentiates the a_1-adrenergic activation of phospholipase C in striatal astrocytes through a mechanism involving arachidonic acid and glutamate. Proc Natl Acad Sci USA 88:9016–9020

Martin LJ, Blackstone CD, Levey AI, Huganir RL, Price DL (1993) AMPA glutamate receptor subunits are differentially distributed in rat brain. Neuroscience 53: 327–358

McGahon B, Lynch MA (1996) The synergism between ACPD and arachidonic acid on glutamate release in hippocampus is age-dependent. Eur J Pharmacol 309:323–326

Meador-Woodruff JH, Damask SP, Watson SJ (1994) Differential expression of autoreceptors in the ascending dopamine systems of the human brain. Proc Natl Acad Sci USA 91:8297–8301

Nieoullon A, Chéramy A, Glowinski J (1978) Release of dopamine evoked by electrical stimulation of the motor and visual areas of the cerebral cortex in both caudate nuclei and in the substantia nigra in the cat. Brain Res 145:69–83

Onali P, Olianas MC, Bunse B (1988) Evidence that adenosine A_2 and dopamine autoreceptors antagonistically regulate tyrosine hydroxylase activity in rat striatal synaptosomes. Brain Res 456:302–309

Ordway RW, Singer JJ, Walsh JV (1991) Direct regulation of ion channels by fatty acids. TINS 14:96–100

Petitet F, Blanchard JC, Doble A (1995) Effects of AMPA receptor modulators on the production of arachidonic acid from striatal neurons. Eur J Pharmacol 291:143–151

Petitet F, Glowinski J, Beaujouan JC (1991) Evoked release of acetylcholine in the rat striatum by stimulation of tachykinin NK$_1$ receptors. Eur J Pharmacol 192:203–204

Roberts PJ, Anderson SD (1979) Stimulatory effect of L-glutamate and related amino acids on 3H-dopamine release from rat striatum: an in vitro model for glutamate actions. J Neurochem 32:1539–1545

Robinson PJ (1992) Potencies of protein kinase C inhibitors are dependent on the activators used to stimulate the enzyme. Biochem Pharmacol 44:1325–1334

Rodriguez-Alvarez J, Lafon-Cazal M, Blanco I, Bockaert J (1997) Different routes of Ca^{2+} influx in NMDA-mediated generation of nitric oxide and arachidonic acid. Eur J Neurosci 9:867–870

Romo R, Chéramy A, Godeheu G, Glowinski J (1984) Distinct commissural pathways are involved in the enhanced release of dopamine induced in the contralateral caudate nucleus and substantia nigra by unilateral application of GABA in the cat thalamic motor nuclei. Brain Res 308:43–52

Romo R, Schultz W (1985) Prolonged changes in dopaminergic terminal excitability and short changes in dopaminergic neuron discharge rate after short peripheral stimulation in monkey. Neurosci Lett 62:335–340

Romo R, Chéramy A, Godeheu G, Glowinski J (1986a)In vivo presynaptic control of dopamine release in the cat caudate nucleus I. Opposite changes in neuronal activity and release evoked from thalamic motor nuclei. Neuroscience 19:1067–1079

Romo R, Chéramy A, Godeheu G, Glowinski J (1986b) In vivo presynaptic control of dopamine release in the cat caudate nucleus III. Further evidence for the implication of corticostriatal glutamatergic neurons. Neuroscience 19:1091–1099

Sandor NT, Brassai A, Puskas A, Lendvai B (1995) Role of nitric oxide in modulating neurotransmitter release from rat striatum. Brain Res Bull, 36:483–486

Shibata M, Araki N, Ohta K, Hamada J, Shimazu K, Fukuuchi Y (1996) Nitric oxide regulates NMDA-induced dopamine release in rat striatum. NeuroReport 7:605–608

Shigemoto R, Nakanishi S, Mizuno N (1992) Distribution of the mRNA for a metabotropic glutamate receptor (mGluR1) in the central nervous system: An in situ hybridization study in adult and developing rat. J Comp Neurol 322:121–135

Snell LD, Johnson KM (1986) Characterization of the inhibition of excitatory amino acid-induced neurotransmitter release in the rat striatum by phencyclidine-like drugs. J Pharmacol Exp Ther 238:938–946

Standaert DG, Testa CM, Young AB, Penney JB (1994) Organization of N-methyl-D-aspartate glutamate receptor gene expression in the basal ganglia of the rat. J Comp Neurol 343:1–16

Starke K (1981) Presynaptic receptors. Ann Rev Pharmacol Toxicol 21:7–30

Stella N, Tencé M, Glowinski J, Prémont J (1994a) Glutamate-evoked release of arachidonic acid from mouse brain astrocytes. J Neurosci 14:568–575

Stella N, Siciliano J, Piomelli D, El-Etr M, Glowinski J, Prémont J (1994b) Interleukin 1 enhances receptor-dependent and independent induced release of arachidonic acid from mouse striatal astrocytes. Soc Neurosci Abstr 20:1052–1052

Stella N, Estelles A, Siciliano J, Tencé M, Desagher S, Piomelli D, Glowinski J (1997) Interleukin-1 enhances the ATP-evoked release of arachidonic acid from mouse astrocytes. J Neurosci 17:2939–2946

Stewart TL, Michel AD, Black MD, Humphrey PPA (1996) Evidence that nitric oxide causes calcium-independent release of ^{3}H-dopamine from rat striatum in vitro. J Neurochem 66:131–137

Stoessl AJ (1994) Localization of striatal and nigral tachykinin receptors in the rat. Brain Res 646:13–18

Strasser A, McCarron RM, Ishii H, Stanimirovic D, Spatz M (1994) L-arginine induces dopamine release from the striatum in vivo. NeuroReport 5:2298–2300

Tanganelli S, Von Euler G, Fuxe K, Agnati LF, Ungerstedt U (1989) Neurotensin counteracts apomorphine-induced inhibition of dopamine release as studied by microdialysis in rat neostriatum. Brain Res 502:319–324

Tencé M, Cordier J, Glowinski J, Prémont J (1992) Endothelin-evoked release of arachidonic acid from mouse astrocytes in primary culture. Eur J Neurosci 4: 993–999

Tencé M, Murphy NP, Cordier J, Prémont J, Glowinski J (1995) Synergistic effects of acetylcholine and glutamate on the release of arachidonic acid from cultured striatal neurons. J Neurochem 64:1605–1613

Tepper JM, Sawyer SF, Young SJ, Groves PM (1986) Autoreceptor-mediated changes in dopaminergic terminal excitability: effects of potassium channel blockers. Brain Res 367:230–237

Testa CM, Standaert DG, Young AB, Penney JB (1994) Metabotropic glutamate receptor mRNA expression in the basal ganglia of the rat. J Neurosci 14:3005–3018

Tremblay L, Kemel ML, Desban M, Gauchy C, Glowinski J (1992) Distinct presynaptic control of dopamine release in striosomal- and matrix-enriched areas of the rat striatum by selective agonists of NK_1, NK_2, and NK_3 tachykinin receptors. Proc. Natl. Acad. Sci. USA 89:11214–11218

Vilaro MT, Palacios JM, Mengod G (1990) Localization of M5 muscarinic receptor mRNA in rat brain examined by in situ hybridization histochemistry. Neurosci Lett 114:154–159

Volterra A, Trotti D, Cassutti P, Tromba C, Galimberti R, Lecchi P, Racagni G (1992a) A role for the arachidonic acid cascade in fast synaptic modulation: ion channels and transmitter uptake systems as target proteins. Adv Exp Med Biol 318:147–158

Volterra A, Trotti D, Cassutti P, Tromba C, Salvaggio A, Melcangi RC, Racagni G (1992b) High sensitivity of glutamate uptake to extracellular free arachidonic acid levels in rat cortical synaptosomes and astrocytes. J Neurochem 59:600–606

Wang JK (1991) Presynaptic glutamate receptors modulate dopamine release from striatal synaptosomes. J Neurochem 57:819–822

Weiner DM, Levey AI, Brann MR (1990) Expression of muscarinic acetylcholine and dopamine receptor mRNAs in rat basal ganglia. Proc Natl Acad Sci USA 87: 7050–7054

Whitty CJ, Walker PD, Goebel DJ, Poosch MS, Bannon MJ (1995) Quantitation, cellular localization and regulation of neurokinin receptor gene expression within the rat substantia nigra. Neuroscience 64:419–425

Williams RJ, Glowinski J (1996) Cyclothiazide unmasks an AMPA-evoked release of arachidonic acid from cultured striatal neurones. J Neurochem 67:1551–1558

Wilson CJ, Chang HT, Kitai ST (1990) Firing patterns and synaptic potentials of identified giant aspiny interneurons in the rat neostriatum. J Neurosci 10:508–519

Yu AC, Chan PH, Fishman RA (1986) Effects of arachidonic acid on glutamate and gamma-aminobutyric acid uptake in primary cultures of rat cerebral cortical astrocytes and neurons. J Neurochem 47:1181–1189

Yung KKL, Bolam JP, Smith AD, Hersch SM, Ciliax BJ, Levey AI (1995) Immunocytochemical localization of D_1 and D_2 dopamine receptors in the basal ganglia of the rat: light and electron microscopy. Neuroscience 65:709–730

Zhu XZ, Luo LG (1992) Effect of nitroprusside (nitric oxide) on endogenous dopamine release from rat striatal slices. J Neurochem 59:932–935

CHAPTER 15
Dopamine – Acetylcholine Interactions

E. Acquas and G. Di Chiara

A. Introduction

Dopamine–acetylcholine interactions can take place within and outside the striatum. In the striatum, cholinergic neurons are large aspiny interneurons that comprise 1%–3% of the total neuronal population of the striatum in rats (Fibiger 1982; Phelps et al. 1985) and monkeys (Mesulam et al. 1984; Difiglia 1987), and by virtue of their dendritic arborization extend over large territories in the striatum (Woolf 1991). Striatal cholinergic neurons receive direct excitatory glutamatergic inputs from the cortex and in particular from the parafascicular thalamus (Lapper and Bolam 1992) and dopaminergic inputs from substantia nigra pars compacta (Kubota et al. 1987). Striatal cholinergic neurons receive inhibitory and modulatory influences from various interneurons and from γ-aminobutyric acid (GABA)ergic medium-size spiny neurons where they finally converge with DA neurons (Di Chiara et al. 1994a). Acetylcholine, on the other hand, modulates the function of dopamine mesencephalic neurons by an action on nicotinic receptors by virtue of cholinergic projections from pontomesencephalic cell groups (Garzon et al. 1999). Dopamine–acetylcholine interactions also take place outside the striatum; in fact, dopaminergic projections from the substantia nigra and ventral tegmental area (Zaborszky et al. 1991; Zilles et al. 1991) to Ch1–Ch4 cholinergic nuclei in the basal forebrain (Mesulam et al. 1994) or indirectly through interposed neurons in the nucleus accumbens, are responsible of the control exerted by dopamine over cortically and hippocampally projecting neurons.

B. Dopamine – Acetylcholine Interactions in the Basal Ganglia

Basal Ganglia are currently understood to gate executive cortical functions by parallel processing of neural information along somatotopically organized fast-transmitting cortico-striato-cortical modules (Chevalier and Deniau 1990; DeLong 1990); this hierarchical system is intersected at the level of the striatum by a network organized in a diffuse, non-somatotopic fashion and performing slow, synchronous modulatory operations (GRAYBIEL 1990; DI CHIARA et al. 1994a) . This striatal modulatory network is made up of two main components: an intrinsic component, made of striatal acetylcholine neurons (DI CHIARA and MORELLI 1994; GERFEN and WILSON 1996) and an extrinsic component, consisting of meso-striatal dopamine neurons (GERFEN and WILSON 1996). Dopamine and acetylcholine neurons might function as a co-ordinated modulatory device of the activity of striatal medium spiny neurons. Striatal acetylcholine neurons [which correspond to striatal tonically active neurons (TANs)] and dopamine neurons fire in a tonic, pacemaker-like mode interrupted by phasic changes in response to unexpected, motivationally salient stimuli (APICELLA et al. 1991; SCHULTZ et al. 1992). Phasic changes in firing activity are reciprocal (e.g. burst in dopamine neurons, pause in acetylcholine neurons) and largely synchronous. Dopamine and acetylcholine, in turn, exert reciprocal effects on striatal medium spiny neurons that encompass the segregation of dopamine and muscarinic receptor subtypes to different subpopulations of spiny neurons at the level of the transduction mechanisms (DI CHIARA et al. 1994a) (see Fig. 1). Thus, in striato-nigral neurons (direct pathway) stimulation of adenylate-cyclase and facilitation of N-methyl-D-aspartate (NMDA) transmission by D_1 receptors is associated to inhibition of adenylate cyclase by M_4 receptors (HULME et al. 1990); conversely, in striato-pallidal neurons (indirect pathway), inhibition of adenylate cyclase by D_2 receptors is associated with stimulation of phosphoinositol turnover and facilitation of NMDA transmission by M_1 receptors (HULME et al. 1990). Consistent with the co-ordinated nature of dopamine and acetylcholine striatal modulatory transmission is the direct control exerted by dopamine over acetylcholine transmission in the striatum.

I. Early Studies

The early understanding of the mechanism of the control by dopamine over acetylcholine transmission was based on studies of the effects of dopamine receptor agonists and antagonists on striatal acetylcholine levels assayed ex vivo and on acetylcholine release estimated in vitro in synaptosomal or slice preparations (LEHMANN and LANGER 1983; STOOF et al. 1992). Changes in turnover rates of acetylcholine, or in brain acetylcholine concentrations in post-mortem tissue, were utilised as an indirect index of in vivo acetylcholine release. These studies showed that non-selective dopamine receptor agonists decreased acetylcholine turnover rates (TRABUCCHI et al. 1975) and increased acetylcholine in brain tissue (MCGEER et al. 1974; WONG et al. 1983); conversely, dopamine D_2 receptor antagonists decreased acetylcholine concentrations in tissue (STADLER et al. 1973). In vitro studies, on the other hand, showed that dopamine, by acting onto D_2-like receptors inhibits K^+- or electrically evoked acetylcholine release from striatal slices (HERTTING et al. 1980; DRUKARCH et al. 1989; DRUKARCH et al. 1991). On this basis it was hypothesized that dopamine controls acetylcholine transmission in an inhibitory

Dopamine – Acetylcholine Interactions

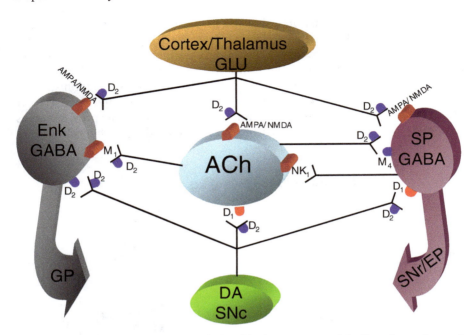

Fig. 1. Schematic diagram of the relationship between acetylcholine, dopamine and glutamate–*N*-methyl-D-aspartate (*NMDA*) transmission in the striatum. Dopamine (*DA*) input from substantia nigra pars compacta (*SNc*) and excitatory amino acid input from cerebral cortex and intralaminar thalamus impinges upon acetylcholine (*ACh*) interneurons, substance P/GABA projections to the substantia reticulata and to the entopeduncular nucleus and upon enkephalin (*Enk*)/GABA neurons to the globus pallidus (*GP*). *Red symbols* indicate receptors with excitatory actions, while *blue boxes* indicate receptors with inhibitory actions. Stimulation of post-synaptic D_1 receptors facilitates, while stimulation of D_2 receptors reduces the sensitivity of cholinergic medium-size spiny neurons to excitatory phasic input (α-amino-3-hydroxy-5-methyl-4-isoxazolepropionic acid) from the cerebral cortex and thalamus. Through the action of pre-synaptic D_2 receptors DA reduces ACh and glutamate (*GLU*) release. ACh would act on Enk neurons mainly through facilitatory M_1 receptors and on substance P (*SP*) neurons through inhibitory M_4 receptors. (Redrawn from Di Chiara et al. 1994)

fashion through pre-synaptic D_2 receptors (Lehmann and Langer 1983; Stoof et al. 1992).

While this hypothesis was established, drugs active with high selectivity on D_1-like receptors became available (Sethy and Van Woert 1974; Sethy 1979; Iorio et al. 1983). Initially it was shown that the D_1 receptor antagonist SCH 23390 increases striatal acetylcholine concentrations (Fage and Scatton 1986). However, in vitro studies failed to observe any effect on acetylcholine release (Scatton 1982a,b; Dolezal et al. 1992; Tedford et al. 1992) or obtained conflicting results (Gorell et al. 1986; Gorell and Czarnecki 1986). With the introduction of brain microdialysis for the estimation of the extracellular acetylcholine concentrations in vivo (Consolo et al. 1987a; Damsma et al. 1987) it was demonstrated that the D_1 receptor antagonist SCH 23390 decreases acetylcholine release (CONSOLO et al. 1987b) and blocks the increase of acetylcholine elicited by the D_1/D_2 agonist apomorphine (BERTORELLI and CONSOLO 1990). Subsequently the D_1 receptor agonist SKF 38393 was found to increase striatal acetylcholine release after systemic administration (CONSOLO et al. 1987b; DAMSMA et al. 1990; DAMSMA et al. 1991; IMPERATO et al. 1993) (see Fig. 2). These observations, together with the previous ones obtained with D_2 receptor agonists and antagonists, led to the hypothesis that

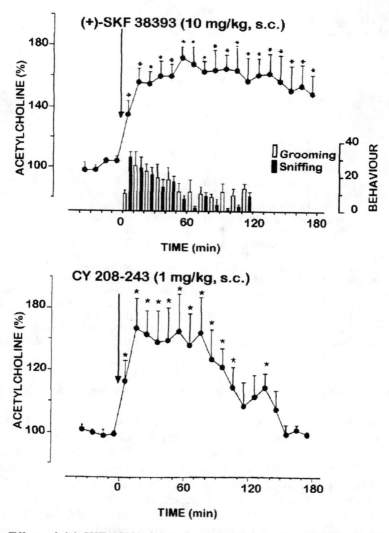

Fig. 2. Effect of (+)-SKF 38393 (10mg/kg s.c.) (*top*) or CY 208-243 (1 mg/kg s.c.) (*bottom*) on striatal acetylcholine output and (*top*) on grooming and sniffing behaviours. (Reproduced, modified, with permission from DAMSMA et al. 1990)

dopamine controls acetylcholine function in a reciprocal fashion, facilitating it by an action on D_1 receptors and inhibiting it by an action on D_2 receptors (BERTORELLI and CONSOLO 1990; DAMSMA et al. 1991; BERTORELLI et al. 1992; DI CHIARA et al. 1994a).

II. Direct D_1 Receptor-Mediated Facilitation of Striatal Acetylcholine Transmission

The neural mechanism by which the control of acetylcholine release takes place and in particular the location, intra- or extra-striatal, of the D_1 receptors facilitating striatal acetylcholine release has been the subject of much debate. Various observations point to a striatal location of D_1 receptors controlling striatal acetylcholine release. Thus, local striatal application of the D_1 antagonist SCH 23390 reduced striatal acetylcholine release (CONSOLO et al. 1992) while the D_1 agonist SKF 38393 stimulated it (AJIMA et al. 1990; ZOCCHI and PERT 1993; ANDERSON et al. 1994; SATO et al. 1994; STEINBERG et al. 1995). Consistent with an intra-striatal mechanism was also the observation that intra-striatal infusion of an antagonist of substance P receptors blocked the stimulant effects of D_1 receptor agonists on acetylcholine release (ANDERSON et al. 1994) and that intra-striatal dopamine receptor agonists affected, via a D_1-receptor dependent mechanism, the expression of genes for transcription factors and for peptides by specific subpopulations of striatal output neurons (e.g. preproenkephalin in striato-pallidal and preprodynorphin in striato-nigral neurons) (WANG and MCGINTY 1997).

Recent evidence has much strengthened the hypothesis of a striatal location of D_1 receptor-mediated influences on acetylcholine function and has definitely cleared some difficulties with that hypothesis. One such difficulty was the failure to demonstrate a D_1-mediated facilitation of acetylcholine release in synaptosomal preparations which consistently allowed the demonstration of a D_2-mediated inhibition (DOLEZAL et al. 1992; TEDFORD et al. 1992). However, it has been later reported that D_1 receptor agonists stimulate acetylcholine release in dissociated striatal cell preparations that maintain the integrity of acetylcholine somata and dendrites (LOGIN et al. 1995a,b). More recently, studies performed in striatal slices have shown that large aspiny neurons identified as cholinergic are slowly depolarized by dopamine and by SKF 38393 through a D_1 receptor-mediated mechanism related to the suppression of resting K^+ conductance and to opening of non-selective (mono and divalent) cation channels in a cyclic adenosine monophosphate (cAMP)-dependent fashion (AOSAKI et al. 1998) (see Fig. 3).

Another difficulty with the hypothesis of a striatal location of D_1 receptors controlling acetylcholine release was the low prevalence of D_1 receptor expression (30%) on striatal acetylcholine neurons reported by early in situ hybridization studies (LE MOINE et al. 1991).

However, application of more sensitive techniques of detection of the dopamine receptor message in striatal acetylcholine neurons has resulted in increase of the proportion of cells expressing D_1 receptors from the initial 30%

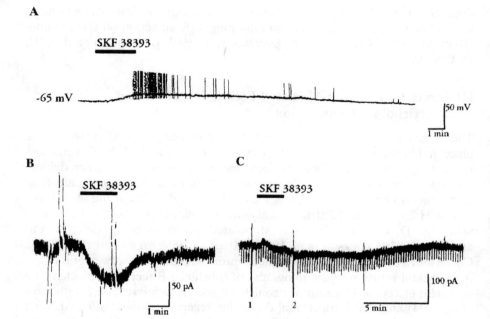

Fig. 3. Effects of the D_1-like agonist, SKF 38393, on striatal large aspiny neurons. **A** A whole-cell current-clamp recording with a resting membrane potential of −65 mV illustrates a slowly rising, prolonged and reversible membrane depolarization with actions potentials occurring during the peak of the response. **B, C** Voltage-clamp traces (holding potential −60 mV) recorded from a large aspiny neuron in saline containing tetrodotoxin illustrate a slow inward current induced by SKF 38393. (Reproduced, modified, with permission from AOSAKI et al. 1998)

(LE MOINE et al. 1991) to 70% (JONGEN-RELO et al. 1995) and to 95% (YAN et al. 1997). Specifically, 88% of striatal acetylcholine neurons express the D_5/D_{1B} subtype and 17% the D_{1A} subtype (YAN et al. 1997).

Difficulties with the hypothesis of a striatal location of D_1 receptors controlling acetylcholine release have also arisen from the failure of some Authors to observe in vivo changes in acetylcholine release following intrastriatal infusions of D_1 receptor antagonists (DAMSMA et al. 1991; DE BOER et al. 1992; ACQUAS et al. 1997). However, evidence has been provided that this failure could be due to an interaction between the rat strain (Wistar) and the anaesthetic (pentobarbital) utilized for probe implantation (CONSOLO et al. 1996a). Further support of an intra-striatal location of D_1 receptors controlling striatal acetylcholine release has been provided by the report that local infusion of the D_1 receptor antagonist SCH 39166, at concentrations of 5 and 10 μM, reduces in vivo acetylcholine release in a concentration-dependent and reversible manner (ACQUAS and DI CHIARA 1999a) (see Fig. 4).

Finally, a somato-dendritic localization of D_1 receptors and a pre-synaptic localization of D_2 receptors on acetylcholine neurons can explain the finding

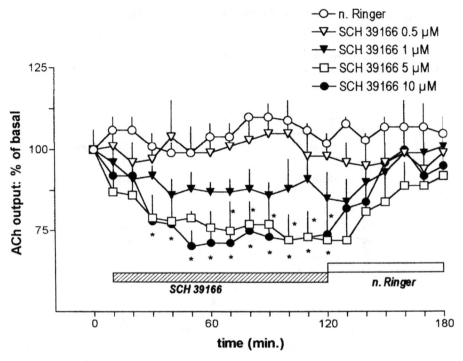

Fig. 4. Effect of SCH 39166 (0.5–10 μM) in presence of 0.01 μM neostigmine and the reversal of the effect during perfusion with SCH 39166-free Ringer on in vivo striatal acetylcholine release. Values are expressed as percentage baseline. *Vertical bars* represent standard error of mean (SEM). (Reproduced with permission from Acquas and Di Chiara 1999a)

that intrastriatal infusion of amphetamine reduces striatal acetylcholine release (De Boer et al. 1992; Abercrombie and DeBoer 1997) and that subsequent systemic administration of amphetamine increases it (Abercrombie and DeBoer 1997). Thus, after local amphetamine, a preferential release of dopamine onto pre-synaptic D_2 receptors located on acetylcholine terminals in the immediate vicinity of the dialytic membrane would take place; after systemic administration, instead, amphetamine, by distributing to the whole striatum, would reach a sufficient number of dopamine terminals to affect the firing activity of acetylcholine neurons and stimulate acetylcholine release by a D_1 receptor-mediated mechanism. According to this hypothesis, local intrastriatal amphetamine reduces acetylcholine release by acting mainly on pre-synaptic D_2 receptors, while systemic amphetamine stimulates acetylcholine transmission by releasing dopamine on somato-dendritic D_1 receptors.

In conclusion, the available evidence strongly suggests that D_1-mediated influences on striatal acetylcholine release arise from an action on dopamine receptors located on striatal acetylcholine neurons.

III. Separate Transduction Pathways for D_1 and D_2 Receptor-Mediated Influences on Acetylcholine Transmission

No matter what is the final effect of stimulation of D_1 and D_2 receptors on the activity of acetylcholine neurons is, they act post-synaptically by different transduction mechanisms that might operate independently (DRUKARCH et al. 1989; STOOF et al. 1992). For example, it has been reported that under conditions in which D_1 receptor stimulation depolarizes acetylcholine neurons (by suppressing a resting K^+ conductance and/or by opening a non-selective cation channel) (AOSAKI et al. 1998), D_2 receptor stimulation fails to elicit consistent changes in membrane conductance (YAN et al. 1997). On the other hand, while D_1 actions on acetylcholine neurons are reportedly mediated by cAMP, D_2 receptor activation in acetylcholine neurons inhibits N-type Ca^{++} channels through a cAMP-independent mechanism (DRUKARCH et al. 1989; YAN et al. 1997). Therefore, at the somato-dendritic level the functional pathway activated by D_1 receptor stimulation might carry its neural computations without interference from the pathway activated by D_2 receptors even in the instance in which both pathways are activated concurrently.

It should also be pointed out that, in contrast to D_1 receptors, D_2 receptors are present not only on somata and dendrites but even more so on the terminals of acetylcholine neurons (JOYCE and MARSHALL 1987) where, by inhibiting N-type Ca^{++} channels (YAN et al. 1997), they can modulate acetylcholine release. This, coupled to the different affinity for dopamine, might result in different outcomes in relation to different levels of activity of the dopamine input. Therefore, the contemporary activation of D_1 and D_2 receptors on acetylcholine neurons, rather than cancelling each other, might affect the reactivity of the acetylcholine neuron in a concerted, functionally meaningful manner.

IV. Independent Gating of Input to Striatal Acetylcholine Neurons by Dopamine Receptor Subtypes

Modulatory influences exert their effects mainly by affecting the sensitivity of the neuron to fast synaptic input mediated by ionotropic receptors (DI CHIARA et al. 1994b). This principle might be particularly valid for dopamine input whose main function might be that of gating fast synaptic input on striatal spiny neurons and acetylcholine neurons (KITAI and SURMEIER 1993; DI CHIARA et al. 1994b). Therefore, a fundamental role in the concerted action of dopamine receptor subtypes on acetylcholine function might be played by the input that drives the striatal acetylcholine neuron.

Striatal acetylcholine neurons are under at least two excitatory inputs, both mediated by glutamate: a major one originating from the intralaminar thalamus (LAPPER and BOLAM 1992) and a minor one originating from the cerebral cortex (DIVAC et al. 1977; MCGEER et al. 1977; FONNUM et al. 1981). In vivo microdialysis studies indicate that the two inputs control acetylcholine release

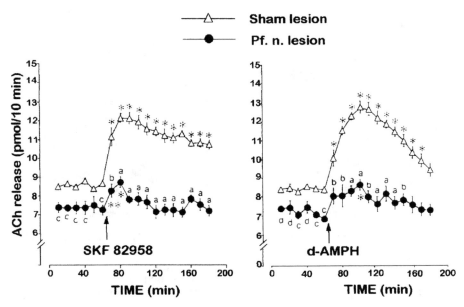

Fig. 5. Effect of SKF 82958 (3mg/kg) (*left*) of *d*-amphetamine (*d*-AMPH; 2mg/kg) (*right*) on acetylcholine output from rat striatum after bilateral electrolytic lesion of the nucleus parafascicularis of the thalamus (*Pf*). (Reproduced, modified, with permission from CONSOLO et al. 1996)

by different glutamate receptor subtypes (GIOVANNINI et al. 1995; STARR 1995). Thus, striatal acetylcholine release can be induced by focal electrical stimulation of the intralaminar thalamus (BALDI et al. 1995; CONSOLO et al. 1996b) or of the cerebral cortex (TABER and FIBIGER 1994); however, while the first seems dependent upon NMDA receptors, being blocked by MK-801, the second is dependent upon α-amino-3-hydroxy-5-methyl-4-isoxazolepropionic acid (AMPA)/kainate receptors, being blocked by L-glutamate diethyl ester (SPENCER 1976; CONSOLO et al. 1996b). D_1 receptor-induced stimulation of striatal acetylcholine release is in turn dependent upon an intact intralaminar thalamus and upon the availability of NMDA receptors in the striatum (BALDI et al. 1995; CONSOLO et al. 1996c) (see Figs. 5 and 6). These observations have been interpreted to indicate that D_1 receptor stimulation amplifies the effect of glutamate on striatal acetylcholine neurons released from thalamic afferents at NMDA receptors. These observations raise the possibility that the two excitatory inputs to striatal acetylcholine neurons are gated by different dopamine receptor subtypes. Thus, similarly to what has been reported for striatal medium-size spiny neurons (KITAI and SURMEIER 1993), while D_1 receptor activation might specifically facilitate thalamic excitation mediated by NMDA receptors, D_2 receptor activation might selectively reduce cortical excitation mediated by AMPA/kainate receptors (GLUR). In this manner, activation of

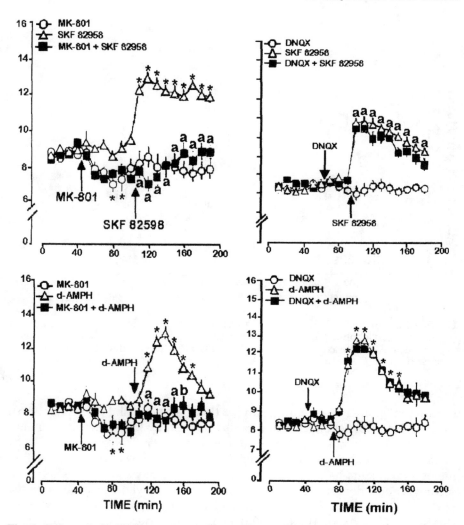

Fig. 6. Effect of the NMDA receptor antagonist MK-801 (0.1 mg/kg) (*left panels*) or the non-NMDA receptor antagonist, DNQX (3 µg/i.c.v. each side) (*right panels*) on SKF 82958 (3 mg/kg) (*top*) of d-AMPH (*bottom*) on acetylcholine output from rat striatum. (Reproduced, modified, with permission from CONSOLO et al. 1996)

D_1 receptors by dopamine would shift the excitatory input to the acetylcholine neuron in favour of the thalamic one.

Acetylcholine neurons also receive two distinct inhibitory inputs, a sparse one provided by $GABA_A$ receptors and activated by recurring collaterals of medium-size spiny neurons (BOLAM et al. 1986; BOLAM and Izzo 1988) and, probably, by GABA interneurons, and a more robust one, provided by muscarinic M_2 receptors. Both these inputs are able to generate inhibitory postsynaptic potentials (IPSPs): a rapid one, related to influx of Cl⁻ ions, and a slow one, due to a G protein-mediated facilitation of K⁺-conductance. It has been reported that D_{1B} (D_5) receptors enhance a Zn-sensitive component of GABA currents through a protein kinase A/protein phosphatase 1 pathway (YAN et al. 1997).

Finally, another way by which dopamine can gate neural information onto acetylcholine neurons is by presynaptic inhibition of transmitter release. This influence is mediated by D_2 receptors through a reduction of N-type calcium currents (YAN et al. 1997) and seems to affect mostly $GABA_A$ and acetylcholine inputs (i.e. the inhibitory inputs) and to a lesser extent excitatory inputs (PISANI et al. 2000). As dopamine is present extracellularly in the striatum in concentrations sufficient to activate high-affinity D_2 receptors, it is likely that dopamine exerts a tonic inhibitory barrage on inhibitory inputs over acetylcholine neurons.

V. Relative Role of D_1 and D_2 Receptors in the Control of Striatal Acetylcholine Function

Once established that acetylcholine neurons possess both D_1-like and D_2-like receptors capable of directly modulating the function of acetylcholine neurons, the problem has arisen as to the physiological role of these receptors in the control of acetylcholine function by endogenous dopamine. A complicating circumstance for the appraisal of the function of each dopamine receptor subtype independently from the other is the fact that, due to the existence of a feedback control of endogenous dopamine release by both D_1-like and D_2-like receptors, any manipulation of each receptor subtype by agonists or antagonists invariably affects the release of endogenous dopamine and therefore the input on the other receptor subtype. This circumstance has generated three hypotheses: one hypothesis is that dopamine controls acetylcholine function primarily by D_1 receptors (DAMSMA et al. 1991; IMPERATO et al. 1993; IMPERATO et al. 1994a). This hypothesis is based on the observation that even the D_2 antagonist-induced increase of acetylcholine release is reversed by a D_1 antagonist and that combined administration of a D_1 antagonist and of a D_2 agonist is no more effective in reducing acetylcholine release than each drug given alone (IMPERATO et al. 1994a). Accordingly, D_2 antagonists would increase acetylcholine release indirectly by stimulating dopamine release onto D_1 receptors (DAMSMA et al. 1991). This hypothesis predicts that any change in the absolute levels of dopamine would result in a correspondent change in D_1-mediated stimulation of acetylcholine function and release. Accordingly, dopamine depletion should reduce acetylcholine release; this, however, is not the case (BERTORELLI et al. 1992; IMPERATO et al. 1994b). This hypothesis is also unable to explain the observation that by reducing the Ringer concentration of the acetylcholine-esterase inhibitor, neostigmine (from $100 nM$ to $10 nM$), D_2-mediated inhibition of acetylcholine release increases independently from an action on D_1 receptors (DEBOER and ABERCROMBIE 1996). Another hypothesis, opposite to the above one, posits that dopamine controls acetylcholine function primarily through inhibitory D_2 receptors (DeBoer et al. 1996). Accordingly, changes in acetylcholine release elicited by D_1 receptor antagonists would be the result of changes in the release of endogenous dopamine onto D_2 receptors. This hypothesis rests on the observation that large doses of amphetamine (10mg/kg) reduce striatal acetylcholine release (DeBoer and Abercrombie 1996; Acquas et al. 1998) in dialysates and increase post-mortem brain levels of acetylcholine while very low doses of the D_2/D_3 agonist quinpirole (3µg/kg s.c.) that reduce dopamine release, increase acetylcholine release (DeBoer et al. 1996). However, the physiological relevance of the first observations is doubtful, given the non-physiologic increase of extracellular dopamine (20 times or more) by such doses of amphetamine. As to the second observation, closer examination of the results obtained shows that the increase of acetylcholine is small and biphasic (at 15 and 60min but not at 30 and 45min) and dissociated from the reduction of dopamine (peak acetylcholine effect: 15min; peak dopamine effect: 45–60min). On the other hand this hypothesis, like the first one, does not account for the observation that drugs which reduce extracellular dopamine (e.g. reserpine and α-methyl tyrosine) or increase it (e.g. 2mg/kg of amphetamine) fail to modify striatal acetylcholine release. A third hypothesis, integrative of the previous two, posits that endogenous dopamine controls acetylcholine transmission in a reciprocal manner through both facilitatory D_1 and inhibitory D_2 receptors (Bertorelli et al. 1992; Di Chiara and Morelli 1994). This hypothesis is confirmed by the recent observation that low doses of the D_2/D_3 agonist quinpirole and of the preferential D_3 agonist PD 128,907, while prevent the feedback stimulation of dopamine release by the D_1 antagonist SCH 39166, potentiate the reduction of acetylcholine release induced by SCH 39166 (Acquas and Di Chiara 1999b) (see Fig. 7).

Electrophysiological studies performed on striatal acetylcholine neurons isolated in vitro, have revealed the possibility that D_1 receptor stimulation reduces the activity of acetylcholine neurons. D_1 receptors would exert this effect by at least two mechanisms: by facilitating of after-hyperpolarization with prolongation of interspike interval (Bennett and Wilson 1998) and by potentiating GABA-mediated inhibition (Yan and Surmeier 1997). Although this possibility apparently contrasts with the observation that D_1 receptor stimulation depolarizes acetylcholine neurons (Aosaki et al. 1998), it is not unlikely that, given the relativistic nature of modulatory influences, stimulation of D_1 receptors exerts, depending on the state of the acetylcholine neuron, a facilitatory or inhibitory influence an acetylcholine neurons. Recently the above mechanisms have been implicated in the pause of TANs in response to conditional stimuli (Bennett and Wilson 1998). Thus, it has been suggested that firing of dopamine neurons in response to stimuli results in activation of D_1 receptors on striatal TANs with secondary prolongation of after hyperpolarization and potentiation of GABA-mediated inhibition (Bennett and Wilson 1998). However, the possibility that phasic changes in TAN activity

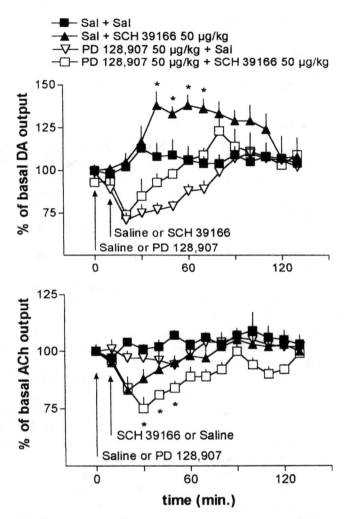

Fig. 7. Effects of saline followed by a second administration of saline or SCH 39166 (50µg/kg) and effect of the administration of PD 128,907 (50µg/kg), followed by the administration of saline or SCH 39166 (50µg/kg) thereafter, on striatal dopamine release on striatal dopamine (*top*) or acetylcholine (*bottom*) release. Values are expressed as percentage baseline. *Vertical bars* represent SEM. *Arrows* indicate the last pretreatment sample. *$p < 0.05$ with respect to the correspondent point of the PD 128,907 (µg/kg)+SCH 39166 (50µg/kg) group. (Reproduced with permission from Acquas and Di Chiara 1999b)

are secondary to phasic changes in dopamine neuron activity is made unlikely by the fact that the latency of phasic events in dopamine neurons (~100ms) (SCHULTZ et al. 1993) and in TANs (67–150ms) (AOSAKI et al. 1995) is superimposable, suggesting that these events are synchronous rather than sequential. Therefore, if dopamine is essential for TAN responses, its action should be regarded as tonic rather than phasic. Consistent with this possibility is the observation that administration of a dopamine receptor agonist, apomorphine, reinstates phasic TAN responses in animals lesioned with the dopaminergic neurotoxin MPTP (AOSAKI et al. 1994).

VI. Nicotinic Receptors and Dopamine Neurons

On the basis of various criteria (ligand binding affinity estimated by autoradiography, desensitization kinetics estimated in electrophysiological experiments, presence of a β_2 subunit and sensitivity to blockade by α-bungarotoxin and methyllycaconitine), four different nicotinic receptor subtypes have been distinguished: type 1, containing α_7 subunits; type 2, containing β_2 subunits either with α_4 (most abundant), α_2, α_5 or with α_6, and β_3 subunits; type 3, containing β_4 subunits with α_3 or α_5 ($\alpha_3\beta_4$ or $\alpha_5\beta_4$), and type 4, containing β_4 subunits with α_2 or α_4 ($\alpha_2\beta_4$, $\alpha_4\beta_4$) (similar to type 3 but rapidly desensitizing) (ZOLI et al. 1998).

Cholinergic projections from pontomesencephalic cell groups (Ch5 and Ch6) to mesencephalic dopamine cell bodies in the substantia nigra (SN) pars compacta and ventral tegmental area (VTA) have been demonstrated (HENDERSON and SHERRIFF 1991; OAKMAN et al. 1995). Studies of mRNA expression have shown that the dopamine neurons of the substantia nigra, VTA and retrorubral field express mRNA for α_2, α_3, α_4, α_5, α_6, β_2 and β_3 (DENERIS et al. 1989; WADA et al. 1989). α_7-like immunoreactivity has also been demonstrated in the mesencephalic tegmentum (SEGUELA et al. 1993; DOMINGUEZ et al. 1994; SCHILSTROM et al. 1998a). Single cell reverse transcriptase polymerase chain reaction (RT-PCR) studies revealed the presence of mRNA for α_7 subunit in ~40% of dopamine neurons in the SN and VTA (KLINK et al. 2001) and the strict correspondence between detection of α_7-mRNA and electrophysiological response to choline (KLINK et al. 2001) suggests the existence of functional α_7-containing nicotinic receptors in SN and VTA neurons. A minority (~10%) of dopamine neurons also express the β_4 subunit (KLINK et al. 2001); double labelling studies for the assessment of α_4 subunit-like immunoreactivity and tyrosine hydroxylase unequivocally demonstrated that α_4 subunit immunoreactivity is present in mesencephalic dopaminergic cells (ARROYO-JIMENEZ et al. 1999).

VII. Actions of Nicotine on Dopamine Function

Nicotine stimulates the synthesis, metabolism and release of dopamine and the functional activity of dopamine neurons both in vitro and in vivo. Early in vitro studies showed that nicotine stimulates the release of [^{3}H] dopamine from striatal slices, minced tissue and synaptosomes (GOODMAN 1974; WESTFALL 1974; ARQUEROS et al. 1978; CONNELLY and LITTLETON 1983; GIORGUIEFF-CHESSELET et al. 1979; SAKURAI et al. 1982; MARIEN et al. 1983; TAKANO et al. 1983; WESTFALL et al. 1983). Nicotine also reportedly stimulates [³H]dopamine release from minced nucleus accumbens tissue in a range of concentrations ($4\cdot10^{-7}M$ in nucleus accumbens [NAc] tissue [ROWELL et al. 1987] and $3\cdot10^{-7}M$ in mouse striatal synaptosomes [GRADY et al. 1994]), in good agreement with the concentration of nicotine found in the blood of smokers (ARMITAGE et al. 1975; RUSSELL et al. 1980; KOGAN et al. 1981).

Intracellular recording studies from ventral tegmental dopamine neurons in vitro have shed light on the cellular mechanism of nicotine actions on dopamine neurons. CALABRESI et al. (1989) showed that nicotine (10–100μM) depolarizes dopamine neurons in a tetrodotoxin (TTX) and cobalt-resistant manner thus excluding a role of voltage dependent Na^+ and Ca^{++} channels. The reversal potential for these actions of nicotine was –4mV, consistent with that estimated on the basis of the current flow through nicotinic receptor channels in various tissues. Notably, nicotinic current was voltage-dependent, a feature also observed in autonomic ganglia (RANG et al. 1982). K-bungarotoxin, but not α-bungarotoxin, blocked the current activated by nicotine, consistent with a role of nAChRs containing α_3/α_4 submits but not by α_7 subunits (CALABRESI et al. 1989).

The stimulant action of nicotine on dopamine neurons of the VTA was described by GRENHOFF et al. (1986) as an increase in burst firing rather than in total firing activity. Doses of 50–500μg/kg i.v. of nicotine increase the frequency of firing of extracellularly recorded dopamine neurons in the A_9 and in the A_{10} region of the mesencephalon, in paralysed, unanaesthetized rats (MEREU et al. 1987) and, as shown more recently, also in awake, un-paralysed animals (FA et al. 2000). In agreement with early observations by (CLARKE et al. 1985) after systemic nicotine in chloral hydrate anaesthetized rats and by (LICHTENSTEIGER et al. 1982) after iontophoretic application of nicotine, comparative dose-response studies showed that A_{10} neurons are more sensitive than A_9 neurons to the stimulant action of nicotine (MEREU et al. 1987).

Systemic administration of nicotine increases in vivo dopamine function. Thus, nicotine stimulates the synthesis, metabolism, turnover and release of dopamine in specific brain areas. Early studies showed that nicotine, either injected or inhaled from tobacco smoke, increases the rate of disappearance of dopamine fluorescence after blockade of dopamine synthesis in terminal dopamine areas, in particular in areas innervated by the mesolimbic dopamine system such as the ventral striatum (NAc/olfactory tubercle); on this basis it was concluded that nicotine increases the impulse flow and the release of dopamine from mesolimbic dopamine neurons (ANDERSSON et al. 1981; FUXE et al. 1986). A study of the effect of acute nicotine on DOPAC/dopamine ratio in different terminal dopamine areas showed that nicotine (0.4–0.9mg/kg s.c.) increases dopamine metabolism to a larger extent in the NAc, followed by the antero-medial caudate-putamen but fails to do so in the prefrontal cortex and in the latero-dorsal caudate-putamen (VEZINA et al. 1992). According to

GEORGE et al. (1998), however, nicotine stimulates dopamine metabolism in the prefrontal cortex at low doses (0.15 mg/kg s.c.), but this effect is lost at higher doses of the drug (0.4 mg/kg s.c.).

In vivo monitoring of extracellular dopamine by microdialysis demonstrated that nicotine acutely increases extracellular dopamine in terminal dopaminergic areas and the preferential stimulant effects of nicotine on A_{10} dopamine neurons (MEREU et al. 1987) is consistent with the preferential stimulant effects of dopamine release by nicotine on the NAc shown in microdialysis studies (IMPERATO et al. 1986; PONTIERI et al. 1996). Within the NAc, a preferential stimulatory effect of nicotine (25–50 µg/kg i.v.) on dopamine release in the shell compartment compared to the core has been observed by two different groups (PONTIERI et al. 1996; NISELL et al. 1997). These two subdivisions of the NAc have been attributed different functions consistent with their different connections (the extended amygdala for the NAc shell and the striato-pallidal system for the NAc core) (HEIMER et al. 1991). Prefrontal cortex dopamine is released by acute nicotine in naïve rats only at doses higher than those that are fully active in releasing dopamine in the NAc shell (BASSAREO et al. 1996). Finally, nicotine stimulates dopamine release also in the bed nucleus of stria terminalis (CARBONI et al. 2000) which is at least as sensitive to nicotine as the NAc shell, in agreement with its assignment to the extended amygdala and with the suggestion that the NAc shell is an area of transition from the ventral striatum to the extended amygdala (HEIMER et al. 1991). Dopamine release in the NAc by nicotine is blocked by intra-VTA but not by intra-accumbens mecamylamine (NISELL et al. 1994a) (see Fig. 8) and is mimicked by intra-VTA but not by intra-NAc nicotine; thus, while intra-VTA nicotine elicits a sustained release of dopamine in the NAc, intra-NAc nicotine elicits a transient effect (NISELL et al. 1994b). These observations are consistent with a proximal action of nicotine on the mechanism of spike generation in the cell body region of dopamine neurons.

VIII. Mechanism of Nicotine Actions on Dopamine Function

The mechanism by which nicotine increases dopamine transmission in the nucleus accumbens is likely to be a complex one. The principal mechanism might be a proximal being related to stimulation of the frequency of spike generation (firing) in dopamine neurons and to an increase in the proportion of burst firing. This mode is most efficient for transmitter release and synaptic transmission; dopamine neurons, in contrast to other monoaminergic neurons, possess this modality, indicative of the ability of dopamine transmission to respond not only tonically but also phasically to stimuli.

Nicotine elicits these changes both in vivo, as shown by extracellular single-unit recording, as well as in vitro, as shown by intracellular recording in mesencephalic slices.

In vitro studies have provided evidence on the receptor mechanism by which the effects of nicotine on dopamine neurons could take place. Thus,

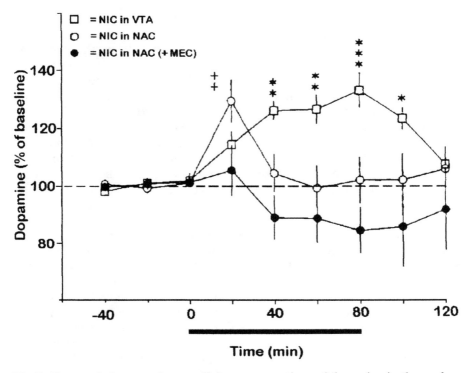

Fig. 8. Temporal changes of extracellular concentrations of dopamine in the nucleus accumbens (*NAC*) after local infusion of nicotine (*NIC*) 1,000 μ*M* alone in the ventral tegmental area (*VTA*) and in the NAC, or in the NAC after injection of mecamylamine (*MEC*) 1 mg/kg s.c. The *horizontal bar* indicates the duration (80 min) of NIC infusion. (Reproduced with permission from NISELL et al. 1994)

pressure injection of acetylcholine on VTA neurons in vitro showed two components, a fast one, peaking at about 30 ms, and a slower one, peaking at about 50 ms. These two components had different pharmacological properties and resistance to desensitization. Thus, the fast component was sensitive to α-bungarotoxin and methyllycaconitine blockade but not to mecamylamine blockade and more prone to desensitization than the slower, mecamylamine-sensitive component. These properties have lead to the assignment of the fast component to α_7-containing nicotinic acetylcholine receptors and of the slow component to an $\alpha_3/\alpha_4 \beta_2$ nicotinic acetylcholine receptor (PIDOPLICHKO et al. 1997).

Nicotinic receptors might influence the activity of dopamine neurons also in an indirect manner, by promoting release of an excitatory transmitter (glutamate) onto dopamine neurons through an action on pre-synaptic α_7-containing nicotinic receptors. The evidence for this mechanism is indirect. Thus, the ability of local intra-tegmental infusion of methyllycaconitine to reduce nicotine-induced release of dopamine in the nucleus accumbens implicates an α_7-containing receptor (SCHILSTROM et al. 1998a), not necessarily a presynaptic receptor; indeed, the α_7 receptors demonstrated to date in relation to dopamine neurons are localized post-synaptically on the dopamine neurons themselves rather than pre-synaptically on terminals impinging on them [see above and PIDOPLICHKO et al. (1997)]. Similarly, the ability of glutamate antagonists infused in the ventral tegmentum to impair nicotine-induced release of dopamine in the nucleus accumbens is not necessarily indicative of a presynaptic mechanism (SCHILSTROM et al. 1998b; SVENSSON et al. 1998).

Distal mechanisms, related to an action of nicotine in terminal dopamine areas, have also been implicated in the mechanism of the stimulant action of nicotine on dopamine transmission. Two possibilities have been envisioned, a direct pre-synaptic action of nicotine on dopamine terminals or an indirect action via nicotinic receptors located on terminals impinging on dopamine neurons.

Although nicotinic acetylcholine receptors controlling dopamine release have been demonstrated also in synaptosomes from the nucleus accumbens, most studies, for obvious practical reasons, have been performed in whole striatal preparations. Instead, in vivo studies have been performed mainly in the nucleus accumbens (if not in its shell subdivision), given the relative insensitivity of neo-striatal dopamine transmission to systemic nicotine. Because of this, the relationship between the studies made in striatal in vitro preparations and the in vivo effects of nicotine is obscure; this, in turn, makes difficult to utilize in vitro dopamine release studies as a basis for explaining the mechanism of the in vivo effects of nicotine on dopamine transmission.

An indirect test of the role of distal mechanisms, however, is offered by studies on the effect of local infusion of nicotinic antagonists on the release of dopamine in the nucleus accumbens after systemic administration of nicotine. In these studies, nicotine effects were impaired by intra-tegmental but not intra-accumbens mecamylamine (NISELL et al. 1994b). However, it has been reported that intra-accumbens α-bungarotoxin, a selective blocker of α_7 containing nicotinic acetylcholine receptors, reduces the release of dopamine stimulated by systemic nicotine in this area (FU et al. 1999). This issue, therefore, awaits clarification.

Among other mechanisms that might contribute to the effects of nicotine on dopamine transmission in vivo, the possibility of an impairment of dopamine-reuptake by nicotine, reported by (IZENWASSER et al. 1991) in vitro, is unlikely, given the observation that the clearance of dopamine in the nucleus accumbens in vivo is increased rather than decreased by nicotine (KSIR et al. 1995).

Thus, in light of the results of studies directly estimating dopamine transmission in vivo by microdialysis, earlier reports of stimulation by nicotine of the synthesis, metabolism and turnover of dopamine in terminal areas of the mesolimbic system (see above) can be explained as secondary to stimulation of its exocytotic release from the terminals of mesolimbic dopamine neurons.

In conclusion, nicotine acutely stimulates the release of dopamine, estimated by brain microdialysis, specifically in the NAc shell/extended amygdala at doses that are well in the range of those self-administered i.v. by rats (around 0.05 mg/kg). At higher doses, dopamine release is increased also in the dorsolateral caudate-putamen and in the prefrontal cortex.

The main mechanism of these acute effects appears to be the activation of non-α_7- as well as α_7-containing nicotinic acetylcholine receptors with resulting depolarization of dopamine neurons and firing of action potentials. This primary action might be modulated at the somato-dendritic region by an NMDA input on dopamine neurons, eventually facilitated by a pre-synaptic action of nicotine on glutamate terminals (see above), which promotes burst firing. An action of nicotine on pre-synaptic receptors in the terminal regions of dopamine neurons might further modulate dopamine transmission by affecting the efficiency of stimulus-secretion coupling rather than by directly releasing dopamine. The notion that the primary action of nicotine on dopamine transmission is mediated by non-α_7 nicotinic acetylcholine receptors is indirectly confirmed by the observation of (PICCIOTTO et al. 1998) that mutant mice not expressing the β_2 subunit of the nicotinic acetylcholine receptor (which is not known to associate with α_7 subunits) also do not show a stimulatory dopamine response to nicotine both in vivo, estimated by microdialysis, as well as in vitro, by electrophysiology.

C. Dopamine – Acetylcholine Interactions Outside the Basal Ganglia

The organization of the central cholinergic systems, besides striatal interneurons, has been described by MESULAM and co-workers (1983) as organized into Ch_1 to Ch_4, Ch_5 and Ch_6 nuclei. Ch_1–Ch_4 nuclei constitute the so-called basal forebrain cholinergic nuclear complex (SCHWABER et al. 1987) that includes along a rostro-caudal axis the medial septum, the horizontal and vertical limb of the diagonal band of Broca and the nucleus basalis magnocellularis and innervates the entire cortical mantle and the hippocampal formation (FIBIGER 1982; MESULAM et al. 1983). Ch_5 and Ch_6 nuclei correspond, respectively, to the pedunculopontine tegmental nucleus and the laterodorsal tegmental nucleus, and heavily project to all thalamic nuclei (WOOLF et al. 1990; WOOLF 1991) to the SN/VTA (MESULAM et al. 1983; CLARKE et al. 1987; WOOLF et al. 1990, 1991; BOLAM et al. 1991; MESULAM et al. 1992) and to the basal forebrain cholinergic nuclear complex (Ch_1–Ch_4) (BOLAM et al. 1991). Dopamine can modulate cortical and hippocampal cholinergic function through direct projections from the SN and from the VTA to the basal forebrain (ZABORSZKY et al. 1991; ZILLES et al. 1991) or indirectly trough projections from the nucleus accumbens (YANG and MOGENSON 1989; ZABORSZKY and CULLINAN 1992) and lateral septum (SWANSON and COWAN 1979; WOOLF 1991) to the basal forebrain nuclei (see Fig. 9). Therefore, there are anatomical grounds for direct interactions between dopamine and acetylcholine in cortical areas.

Experimental evidence points to a role of cholinergic projections to the neocortex and the hippocampus in arousal, attention, learning and memory

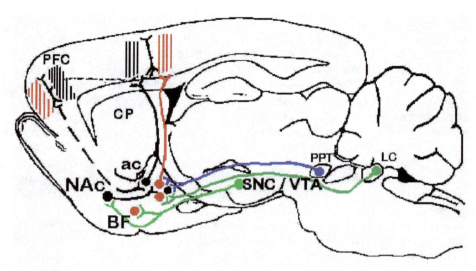

Fig. 9. Hypothetical circuitry between prefrontal cortex (*PFC*) and specific somatosensory cortical areas via GABAergic local and projections neurons of the basal forebrain. Cholinergic (*red*) and noncholinergic (*black*) neurons in the basal forebrain receive identified synaptic input from the nucleus accumbens (*NAc*), the locus coeruleus (*LC*), the substantia nigra (*SN*), and the mesopontine tegmentum (*PPT*). (Reproduced, modified, with permission from ZABORSZKY et al. 1999)

(FIBIGER 1991; ROBBINS and EVERITT 1994; WILLIAMS et al. 1994; MCCORMICK and BAL 1997; SARTER and BRUNO 2000; SARTER et al. 2001). Moreover, behavioural (INGLIS and WINN 1995; OLMSTEAD et al. 1998), pharmacological and lesion studies (BLAHA and WINN 1993; KLITENICK and KALIVAS 1994; BLAHA et al. 1996; GRONIER and RASMUSSEN 1998; OLMSTEAD et al. 1998; GRONIER et al. 2000) indicate the existence of a functional relationship between cholinergic neurons of the Ch_5–Ch_6 nuclei (MESULAM et al. 1983) and dopaminergic ones in the SN and VTA. In relation to this, it has been speculated that cholinergic nuclei of the mesencephalic tegmentum and of the brainstem, via their projections to mesolimbic DA neurons in the VTA (BLAHA and WINN 1993; BLAHA et al. 1996), modulate the expression of positive symptoms of schizophrenia (SARTER 1994; SARTER and BRUNO 2000), schizophrenic hallucinations (GRAY et al. 1991; YEOMANS 1995) and latency of rapid eye movement (REM) sleep in schizophrenics (SILBERSWEIG et al. 1995; YEOMANS 1995).

I. Dopaminergic Regulation of Cortical and Hippocampal Acetylcholine Transmission

Early in vivo studies, performed with the cortical cup technique showed that *d*-amphetamine and other non-specific dopaminergic drugs could positively modulate acetylcholine neurotransmission in vivo, thus indicating the existence of a dopaminergic regulation of cortical and hippocampal cholinergic transmission (PEPEU and BARTOLINI 1968; PEPEU and MANTOVANI 1978).

Brain microdialysis studies subsequently showed that dopamine facilitates acetylcholine release in the frontal cortex and hippocampus by acting on D_1-like receptors (DAY and FIBIGER 1992; DAY and FIBIGER 1993, 1994; ACQUAS et al. 1994; HERSI et al. 1995; ACQUAS and FIBIGER 1996). D_2-like receptors also facilitate cortical and hippocampal acetylcholine release (IMPERATO et al. 1993; IMPERATO et al. 1996); moreover, the stimulant effect of d-amphetamine on acetylcholine release is prevented by 6-OHDA lesions of dopamine but not noradrenaline neurons (DAY et al. 1994).

Cortical and hippocampal acetylcholine neurotransmission, estimated by in vivo brain microdialysis, has recently been proposed as a neurochemical index of arousal and attention. In fact, cortical and hippocampal acetylcholine release is activated by unexpected, salient or motivationally relevant stimuli and their effect is attenuated by habituation (MOORE et al. 1992; ACQUAS et al. 1996). Simultaneous blockade of D_1-like and D_2-like dopamine receptors significantly reduces the increase of acetylcholine release in the rat frontal cortex evoked by unconditioned sensory stimuli (ACQUAS et al. 1998); the dopamine receptors responsible for these actions might be located either onto cholinergic neurons of the basal forebrain, (ZABORSZKY et al. 1991; ZILLES et al. 1991) or on GABAergic neurons in the NAc (YANG and MOGENSON 1989; ZABORSZKY and CULLINAN 1992).

Consistent with this possibility is the finding that the stimulatory effects of d-amphetamine on cortical acetylcholine release are inhibited by electrical stimulation of the nucleus accumbens (CASAMENTI et al. 1986). In further agreement with a role of the NAc, it has been shown that the increases of cortical acetylcholine release evoked by the partial inverse agonist of benzodiazepines receptors, FG 7142, are blocked by local injections of D_2-like antagonists into the shell of the accumbens (MOORE et al. 1999). However, there are instances in which an increase of cortical acetylcholine release escapes dopaminergic control: thus, the local application of dopamine antagonists into the accumbens fails to prevent the effects on acetylcholine release of systemic d-amphetamine given in combination with sensory stimuli known to activate acetylcholine output in the cortex (MOORE et al. 1999; ARNOLD et al. 2000).

In this regard, the increases in cortical acetylcholine might also be related to cortical and behavioural arousal and to the complex interplay between the classically recognized arousal systems (dopaminergic, cholinergic, noradrenergic and serotonergic) (ROBBINS and EVERITT 1994; SARTER and BRUNO 2000; SARTER et al. 2001).

References

Abercrombie ED, DeBoer P (1997) Substantia nigra D1 receptors and stimulation of striatal cholinergic interneurons by dopamine: a proposed circuit mechanism. J Neurosci 17:8498–8505

Acquas E, Fibiger HC (1996) Chronic lithium attenuates dopamine D1-receptor mediated increases in acetylcholine release in rat frontal cortex. Psychopharmacology (Berl) 125:162–167

Acquas E, Fibiger HC (1998) Dopaminergic regulation of striatal acetylcholine release: the critical role of acetylcholinesterase inhibition. J Neurochem 70(3):1088–1093

Acquas E, Di Chiara G (1999a) Local application of SCH 39166 reversibly and dose-dependently decreases acetylcholine release in the rat striatum. Eur J Pharmacol 383:275–279

Acquas E, Di Chiara G (1999b) Dopamine D(1) receptor-mediated control of striatal acetylcholine release by endogenous dopamine. Eur J Pharmacol 383:121–127

Acquas E, Day JC, Fibiger HC (1994) The potent and selective dopamine D1 receptor agonist A-77636 increases cortical and hippocampal acetylcholine release in the rat. Eur J Pharmacol 260:85–87

Acquas E, Wilson C, Fibiger HC (1996) Conditioned and unconditioned stimuli increase frontal cortical and hippocampal acetylcholine release: effects of novelty, habituation, and fear. J Neurosci 16:3089–3096

Acquas E, Wilson C, Fibiger HC (1997) Nonstriatal dopamine D1 receptors regulate striatal acetylcholine release in vivo. J Pharmacol Exp Ther 281:360–368

Acquas E, Wilson C, Fibiger HC (1998) Pharmacology of sensory stimulation-evoked increases in frontal cortical acetylcholine release. Neuroscience 85:73–83

Ajima A, Yamaguchi T, Kato T (1990) Modulation of acetylcholine release by D1, D2 dopamine receptors in rat striatum under freely moving conditions. Brain Res 518:193–198

Anderson JJ, Kuo S, Chase TN, Engber TM (1994) Dopamine D1 receptor-stimulated release of acetylcholine in rat striatum is mediated indirectly by activation of striatal neurokinin1 receptors. J Pharmacol Exp Ther 269:1144–1151

Andersson K, Fuxe K, Agnati LF (1981) Effects of single injections of nicotine on the ascending dopamine pathways in the rat. Evidence for increases of dopamine turnover in the mesostriatal and mesolimbic dopamine neurons. Acta Physiol Scand 112:345–347

Aosaki T, Graybiel AM, Kimura M (1994) Effect of the nigrostriatal dopamine system on acquired neural responses in the striatum of behaving monkeys. Science 265:412–415

Aosaki T, Kimura M, Graybiel AM (1995) Temporal and spatial characteristics of tonically active neurons of the primate's striatum. J Neurophysiol 73:1234–1252

Aosaki T, Kiuchi K, Kawaguchi Y (1998) Dopamine D1-like receptor activation excites rat striatal large aspiny neurons in vitro. J Neurosci 18:5180–5190

Apicella P, Scarnati E, Schultz W (1991) Tonically discharging neurons of monkey striatum respond to preparatory and rewarding stimuli. Exp Brain Res 84:672–675

Armitage AK, Dollery CT, George CF, Houseman TH, Lewis PJ, Turner DM (1975) Absorption and metabolism of nicotine from cigarettes. Br Med J 4:313–316

Arnold HM, Nelson CL, Neigh GN, Sarter M, Bruno JP (2000) Systemic and intra-accumbens administration of amphetamine differentially affects cortical acetylcholine release. Neuroscience 96:675–685

Arqueros L, Naquira D, Zunino E (1978) Nicotine-induced release of catecholamines from rat hippocampus and striatum. Biochem Pharmacol 27:2667–2674

Arroyo-Jimenes MM, Bourgeois J-P, Marubio LM, Le Sourd A-M, Ottersen OP, Rinvik E, Fairen A, Changeux J-P (1999) Ultrastructural localization if the α4-subunit of the neuronal acetylcholine nicotinic receptor in the rat substantia nigra. J Neurosci 19(15):6475–6487

Baldi G, Russi G, Nannini L, Vezzani A, Consolo S (1995) Trans-synaptic modulation of striatal ACh release in vivo by the parafascicular thalamic nucleus. Eur J Neurosci 7:1117–1120

Bassareo V, Tanda G, Petromilli P, Giua C, Di Chiara G (1996) Non-psychostimulant drugs of abuse and anxiogenic drugs activate with differential selectivity dopamine transmission in the nucleus accumbens and in the medial prefrontal cortex of the rat [published erratum appears in Psychopharmacology (Berl) 1996 Oct;127(3):289–90]. Psychopharmacology (Berl) 124:293–299

Bennett BD, Wilson CJ (1998) Synaptic regulation of action potential timing in neostriatal cholinergic interneurons. J Neurosci 18:8539–8549

Bertorelli R, Consolo S (1990) D1 and D2 dopaminergic regulation of acetylcholine release from striata of freely moving rats. J Neurochem 54:2145–2148

Bertorelli R, Zambelli M, Di Chiara G, Consolo S (1992) Dopamine depletion preferentially impairs D1- over D2-receptor regulation of striatal in vivo acetylcholine release. J Neurochem 59:353–357

Blaha CD, Winn P (1993) Modulation of dopamine efflux in the striatum following cholinergic stimulation of the substantia nigra in intact and pedunculopontine tegmental nucleus-lesioned rats. J Neurosci 13:1035–1044

Blaha CD, Allen LF, Das S, Inglis WL, Latimer MP, Vincent SR, Winn P (1996) Modulation of dopamine efflux in the nucleus accumbens after cholinergic stimulation of the ventral tegmental area in intact, pedunculopontine tegmental nucleus-lesioned, and laterodorsal tegmental nucleus-lesioned rats. J Neurosci 16:714–722

Bolam JP, Ingham CA, Izzo PN, Levey AI, Rye DB, Smith AD, Wainer BH (1986) Substance P-containing terminals in synaptic contact with cholinergic neurons in the neostriatum and basal forebrain: a double immunocytochemical study in the rat. Brain Res 397:279–289

Bolam JP, Izzo PN (1988) The postsynaptic targets of substance P-immunoreactive terminals in the rat neostriatum with particular reference to identified spiny striatonigral neurons. Exp Brain Res 70:361–377

Bolam JP, Francis CM, Henderson Z (1991) Cholinergic input to dopaminergic neurons in the substantia nigra: a double immunocytochemical study. Neuroscience 41:483–494

Calabresi P, Lacey MG, North RA (1989) Nicotinic excitation of rat ventral tegmental neurones in vitro studied by intracellular recording. Br J Pharmacol 98:135–140

Carboni E, Silvagni A, Rolando MT, Di Chiara G (2000) Stimulation of in vivo dopamine transmission in the bed nucleus of stria terminalis by reinforcing drugs. J Neurosci 20:RC102

Casamenti F, Deffenu G, Abbamondi AL, Pepeu G (1986) Changes in cortical acetylcholine output induced by modulation of the nucleus basalis. Brain Res Bull 16:689–695

Chevalier G, Deniau JM (1990) Disinhibition as a basic process in the expression of striatal functions. Trends Neurosci 13:277–280

Clarke PB, Schwartz RD, Paul SM, Pert CB, Pert A (1985) Nicotinic binding in rat brain: autoradiographic comparison of [3H]acetylcholine, [3H]nicotine, and [125I]-alpha-bungarotoxin. J Neurosci 5:1307–1315

Clarke PB, Hommer DW, Pert A, Skirboll LR (1987) Innervation of substantia nigra neurons by cholinergic afferents from pedunculopontine nucleus in the rat: neuroanatomical and electrophysiological evidence. Neuroscience 23:1011–1019

Connelly MS, Littleton JM (1983) Lack of stereoselectivity in ability of nicotine to release dopamine from rat synaptosomal preparations. J Neurochem 41:1297–1302

Consolo S, Wu CF, Fiorentini F, Ladinsky H, Vezzani A (1987a) Determination of endogenous acetylcholine release in freely moving rats by transstriatal dialysis coupled to a radioenzymatic assay: effect of drugs. J Neurochem 48:1459–1465

Consolo S, Wu CF, Fusi R (1987b) D-1 receptor-linked mechanism modulates cholinergic neurotransmission in rat striatum. J Pharmacol Exp Ther 242:300–305

Consolo S, Girotti P, Russi G, Di Chiara G (1992) Endogenous dopamine facilitates striatal in vivo acetylcholine release by acting on D1 receptors localized in the striatum. J Neurochem 59:1555–1557

Consolo S, Colli E, Caltavuturo C, Di Chiara G (1996a) Surgical anaesthesia with pentobarbital prevents the effect of local SCH 23390 on rat striatal acetylcholine release in a strain-dependent manner. Behavioural Pharmacology 7:663–668

Consolo S, Baldi G, Giorgi S, Nannini L (1996b) The cerebral cortex and parafascicular thalamic nucleus facilitate in vivo acetylcholine release in the rat striatum through distinct glutamate receptor subtypes. Eur J Neurosci 8:2702–2710

Consolo S, Baronio P, Guidi G, Di Chiara G (1996c) Role of the parafascicular thalamic nucleus and N-methyl-D-aspartate transmission in the D1-dependent control of in vivo acetylcholine release in rat striatum. Neuroscience 71:157–165

Damsma G, Westerink BH, de Vries JB, Van den Berg CJ, Horn AS (1987) Measurement of acetylcholine release in freely moving rats by means of automated intracerebral dialysis. J Neurochem 48:1523–1528

Damsma G, Tham CS, Robertson GS, Fibiger HC (1990) Dopamine D1 receptor stimulation increases striatal acetylcholine release in the rat. Eur J Pharmacol 186:335–338

Damsma G, Robertson GS, Tham CS, Fibiger HC (1991) Dopaminergic regulation of striatal acetylcholine release: importance of D1 and N-methyl-D-aspartate receptors. J Pharmacol Exp Ther 259:1064–1072

Day J, Fibiger HC (1992) Dopaminergic regulation of cortical acetylcholine release. Synapse 12:281–286

Day J, Fibiger HC (1993) Dopaminergic regulation of cortical acetylcholine release: effects of dopamine receptor agonists. Neuroscience 54:643–648

Day JC, Fibiger HC (1994) Dopaminergic regulation of septohippocampal cholinergic neurons. J Neurochem 63:2086–2092

Day JC, Tham CS, Fibiger HC (1994) Dopamine depletion attenuates amphetamine-induced increases of cortical acetylcholine release. Eur J Pharmacol 263:285–292

DeBoer P, Abercrombie ED (1996) Physiological release of striatal acetylcholine in vivo: modulation by D1 and D2 dopamine receptor subtypes. J Pharmacol Exp Ther 277:775–783

DeBoer P, Damsma G, Schram Q, Stoof JC, Zaagsma J, Westerink BH (1992) The effect of intrastriatal application of directly and indirectly acting dopamine agonists and antagonists on the in vivo release of acetylcholine measured by brain microdialysis. The importance of the post-surgery interval. Naunyn Schmiedebergs Arch Pharmacol 345:144–152

DeBoer P, Heeringa MJ, Abercrombie ED (1996) Spontaneous release of acetylcholine in striatum is preferentially regulated by inhibitory dopamine D2 receptors. Eur J Pharmacol 317:257–262

DeLong MR (1990) Primate models of movement disorders of basal ganglia origin. Trends Neurosci 13:281–285

Deneris ES, Boulter J, Swanson LW, Patrick J, Heinemann S (1989) Beta 3: a new member of nicotinic acetylcholine receptor gene family is expressed in brain. J Biol Chem 264:6268–6272

Di Chiara G, Morelli M (1994) Acetylcholine, Dopamine and NMDA transmission in the caudate-putamen: their interaction and functions as a striatal modulatory system. In: Percheron G (ed) The Basal ganglia IV. Plenum Press, New York, pp 491–505

Di Chiara G, Morelli M, Consolo S (1994) Modulatory functions of neurotransmitters in the striatum: ACh/dopamine/NMDA interactions. Trends Neurosci 17:228–233

Difiglia M (1987) Synaptic organization of cholinergic neurons in the monkey neostriatum. J Comp Neurol 255:245–258

Divac I, Fonnum F, Storm-Mathisen J (1977) High affinity uptake of glutamate in terminals of corticostriatal axons. Nature 266:377–378

Dolezal V, Jackisch R, Hertting G, Allgaier C (1992) Activation of dopamine D1 receptors does not affect D2 receptor-mediated inhibition of acetylcholine release in rabbit striatum. Naunyn Schmiedebergs Arch Pharmacol 345:16–20

Dominguez DT, Juiz JM, Peng X, Lindstrom J, Criado M (1994) Immunocytochemical localization of the alpha 7 subunit of the nicotinic acetylcholine receptor in the rat central nervous system. J Comp Neurol 349:325–342

Drukarch B, Schepens E, Schoffelmeer AN, Stoof JC (1989) Stimulation of D-2 dopamine receptors decreases the evoked in vitro release of [3H]acetylcholine from rat neostriatum: role of K+ and Ca2+. J Neurochem 52:1680–1685

Drukarch B, Schepens E, Stoof JC (1991) Sustained activation does not desensitize the dopamine D2 receptor-mediated control of evoked in vitro release of radiolabeled acetylcholine from rat striatum. Eur J Pharmacol 196:209–212

Fa M, Carcangiu G, Passino N, Ghiglieri V, Gessa GL, Mereu G (2000) Cigarette smoke inhalation stimulates dopaminergic neurons in rats. Neuroreport 11:3637–3639

Fage D, Scatton B (1986) Opposing effects of D-1 and D-2 receptor antagonists on acetylcholine levels in the rat striatum. Eur J Pharmacol 129:359–362

Fibiger HC (1982) The organization and some projections of cholinergic neurons of the mammalian forebrain. Brain Res 257:327–388

Fibiger HC (1991) Cholinergic mechanisms in learning, memory and dementia: a review of recent evidence [see comments]. Trends Neurosci 14:220–223

Fonnum F, Storm-Mathisen J, Divac I (1981) Biochemical evidence for glutamate as neurotransmitter in corticostriatal and corticothalamic fibres in rat brain. Neuroscience 6:863–873

Fu Y, Matta SG, Sharp BM (1999) Local alpha-bungarotoxin-sensitive nicotinic receptors modulate hippocampal norepinephrine release by systemic nicotine. J Pharmacol Exp Ther 289:133–139

Fuxe K, Andersson K, Harfstrand A, Agnati LF (1986) Increases in dopamine utilization in certain limbic dopamine terminal populations after a short period of intermittent exposure of male rats to cigarette smoke. J Neural Transm 67:15–29

Garzon M, Vaughan RA, Uhl GR, kuhar MJ, Pickel VM (1999) Cholinergic axon terminals in the ventral tegmental area target a subpopulation of neurons expressing low levels of the dopamine transporter. J Comp Neurol 410(2):197–210

George TP, Verrico CD, Roth RH (1998) Effects of repeated nicotine pre-treatment on mesoprefrontal dopaminergic and behavioral responses to acute footshock stress. Brain Res 801:36–49

Gerfen CR, Wilson CJ (1996) The Basal Ganglia. In: Swanson LW, Bjorklund A, Hockfelt T, Elsevier SV (eds) Handbook of Chemical Neuroanatomy, Vol. 12: Integrates Systems of the CNS, Part III. (Amsterdam), pp 371–468

Giorguieff-Chesselet MF, Kemel ML, Wandscheer D, Glowinski J (1979) Regulation of dopamine release by presynaptic nicotinic receptors in rat striatal slices: effect of nicotine in a low concentration. Life Sci 25:1257–1262

Giovannini MG, Camilli F, Mundula A, Bianchi L, Colivicchi MA, Pepeu G (1995) Differential regulation by *N*-methyl-D-aspartate and non-*N*-methyl-D-aspartate receptors of acetylcholine release from the rat striatum in vivo. Neuroscience 65:409–415

Goodman FR (1974) Effects of nicotine on distribution and release of 14C-norepinephrine and 14C-dopamine in rat brain striatum and hypothalamus slices. Neuropharmacology 13:1025–1032

Gorell JM, Czarnecki B (1986) Pharmacologic evidence for direct dopaminergic regulation of striatal acetylcholine release. Life Sci 38:2239–2246

Gorell JM, Czarnecki B, Hubbell S (1986) Functional antagonism of D-1 and D-2 dopaminergic mechanisms affecting striatal acetylcholine release. Life Sci 38:2247–2254

Grady SR, Marks MJ, Collins AC (1994) Desensitization of nicotine-stimulated [3H]dopamine release from mouse striatal synaptosomes. J Neurochem 62:1390–1398

Gray JA, Feldon J, Rawlins JN, Hemsley DR, Smith AD (1991) The neuropsychology of schizophrenia. Behav Brain Sci 14:1–81

Graybiel AM (1990) Neurotransmitters and neuromodulators in the basal ganglia. Trends Neurosci 13:244–254

Grenhoff J, Aston-Jones G, Svensson TH (1986) Nicotinic effects on the firing pattern of midbrain dopamine neurons. Acta Physiol Scand 128(3):351–358

Gronier B, Rasmussen K (1998) Activation of midbrain presumed dopaminergic neurones by muscarinic cholinergic receptors: an in vivo electrophysiological study in the rat. Br J Pharmacol 124:455–464

Gronier B, Perry KW, Rasmussen K (2000) Activation of the mesocorticolimbic dopaminergic system by stimulation of muscarinic cholinergic receptors in the ventral tegmental area. Psychopharmacology (Berl) 147:347–355

Heimer L, Zahm DS, Churchill L, Kalivas PW, Wohltmann C (1991) Specificity in the projection patterns of accumbal core and shell in the rat. Neuroscience 41:89–125

Henderson Z, Sherriff FE (1991) Distribution of choline acetyltransferase immunoreactive axons and terminals in the rat and ferret brainstem. J Comp Neurol 314:147–163

Hersi AI, Richard JW, Gaudreau P, Quirion R (1995) Local modulation of hippocampal acetylcholine release by dopamine D1 receptors: a combined receptor autoradiography and in vivo dialysis study. J Neurosci 15:7150–7157

Hertting G, Zumstein A, Jackisch R, Hoffmann I, Starke K (1980) Modulation by endogenous dopamine of the release of acetylcholine in the caudate nucleus of the rabbit. Naunyn Schmiedebergs Arch Pharmacol 315:111–117

Hulme EC, Birdsall NJ, Buckley NJ (1990) Muscarinic receptor subtypes. Annu Rev Pharmacol Toxicol 30:633–673

Imperato A, Mulas A, Di Chiara G (1986) Nicotine preferentially stimulates dopamine release in the limbic system of freely moving rats. Eur J Pharmacol 132:337–338

Imperato A, Obinu MC, Casu MA, Mascia MS, Dazzi L, Gessa GL (1993) Evidence that neuroleptics increase striatal acetylcholine release through stimulation of dopamine D1 receptors. J Pharmacol Exp Ther 266:557–562

Imperato A, Obinu MC, Dazzi L, Gessa GL (1994b) Does dopamine exert a tonic inhibitory control on the release of striatal acetylcholine in vivo? Eur J Pharmacol 251:271–279

Imperato A, Obinu MC, Carta G, Mascia MS, Casu MA, Dazzi L, Gessa GL (1994a) Neuroleptics cause stimulation of dopamine D1 receptors and their desensitization after chronic treatment. Eur J Pharmacol 264:55–60

Imperato A, Obinu MC, Mascia MS, Casu MA, Zocchi A, Cabib S, Puglisi-Allegra S (1996) Strain-dependent effects of dopamine agonists on acetylcholine release in the hippocampus: an in vivo study in mice. Neuroscience 70:653–660

Inglis WL, Winn P (1995) The pedunculopontine tegmental nucleus: where the striatum meets the reticular formation. Prog Neurobiol 47:1–29

Iorio LC, Barnett A, Leitz FH, Houser VP, Korduba CA (1983) SCH 23390, a potential benzazepine antipsychotic with unique interactions on dopaminergic systems. J Pharmacol Exp Ther 226:462–468

Izenwasser S, Jacocks HM, Rosenberger JG, Cox BM (1991) Nicotine indirectly inhibits [3H]dopamine uptake at concentrations that do not directly promote [3H]dopamine release in rat striatum. J Neurochem 56:603–610

Jongen-Relo AL, Docter GJ, Jonker AJ, Voorn P (1995) Differential localization of mRNAs encoding dopamine D1 or D2 receptors in cholinergic neurons in the core and shell of the rat nucleus accumbens. Brain Res Mol Brain Res 28:169–174

Joyce JN, Marshall JF (1987) Quantitative autoradiography of dopamine D2 sites in rat caudate-putamen: localization to intrinsic neurons and not to neocortical afferents. Neuroscience 20:773–795

Kitai ST, Surmeier DJ (1993) Cholinergic and dopaminergic modulation of potassium conductances in neostriatal neurons. Adv Neurol 60:40–52

Klink R, de Kerkove d'Exaerde A, Zoli M, Changeaux J-P (2001) Molecular and physiological diversity of nicotinic acetylcholine receptors in the midbrain dopaminergic nuclei. J Neurosci 21(5):1452–1463

Klitenick MA, Kalivas PW (1994) Behavioral and neurochemical studies of opioid effects in the pedunculopontine nucleus and mediodorsal thalamus. J Pharmacol Exp Ther 269:437–448

Kogan MJ, Verebey K, Jaffee JH, Mule SJ (1981) Simultaneous determination of nicotine and cotinine in human plasma by nitrogen detection gas-liquid chromatography. J Forensic Sci 26:6–11

Ksir C, Mellor G, Hart C, Gerhardt GA (1995) Nicotine enhances dopamine clearance in rat nucleus accumbens. Prog Neuropsychopharmacol Biol Psychiatry 19: 151–156

Kubota Y, Inagaki S, Shimada S, Kito S, Eckenstein F, Tohyama M (1987) Neostriatal cholinergic neurons receive direct synaptic inputs from dopaminergic axons. Brain Res 413:179–184

Lapper SR, Bolam JP (1992) Input from the frontal cortex and the parafascicular nucleus to cholinergic interneurons in the dorsal striatum of the rat. Neuroscience 51:533–545

Le Moine C, Normand E, Bloch B (1991) Phenotypical characterization of the rat striatal neurons expressing the D1 dopamine receptor gene. Proc Natl Acad Sci USA 88:4205–4209

Lehmann J, Langer SZ (1983) The striatal cholinergic interneuron: synaptic target of dopaminergic terminals? Neuroscience 10:1105–1120

Lichtensteiger W, Hefti F, Felix D, Huwyler T, Melamed E, Schlumpf M (1982) Stimulation of nigrostriatal dopamine neurones by nicotine. Neuropharmacology 21: 963–968

Login IS, Borland K, Harrison MB (1995a) Acute dopamine depletion potentiates independent stimulatory and inhibitory D1 DA receptor-mediated control of striatal acetylcholine release in vitro. Brain Res 681:209–212

Login IS, Borland K, Harrison MB, Ragozzino ME, Gold PE (1995b) Acetylcholine release from dissociated striatal cells. Brain Res 697:271–275

Marien M, Brien J, Jhamandas K (1983) Regional release of [3H]dopamine from rat brain in vitro: effects of opioids on release induced by potassium, nicotine, and L-glutamic acid. Can J Physiol Pharmacol 61:43–60

McCormick DA, Bal T (1997) Sleep and arousal: thalamocortical mechanisms. Annu Rev Neurosci 20:185–215

McGeer PL, Grewaal DS, McGeer EG (1974) Influence of noncholinergic drugs on rat striatal acetylcholine levels. Brain Res 80:211–217

McGeer PL, McGeer EG, Scherer U, Singh K (1977) A glutamatergic corticostriatal path? Brain Res 128:369–373

Mereu G, Yoon KW, Boi V, Gessa GL, Naes L, Westfall TC (1987) Preferential stimulation of ventral tegmental area dopaminergic neurons by nicotine. Eur J Pharmacol 141:395–399

Mesulam MM, Mufson EJ, Wainer BH, Levey AI (1983) Central cholinergic pathways in the rat: an overview based on an alternative nomenclature (Ch1–Ch6). Neuroscience 10:1185–1201

Mesulam MM, Mufson EJ, Levey AI, Wainer BH (1984) Atlas of cholinergic neurons in the forebrain and upper brainstem of the macaque based on monoclonal choline acetyltransferase immunohistochemistry and acetylcholinesterase histochemistry. Neuroscience 12:669–686

Mesulam MM, Mash D, Hersh L, Bothwell M, Geula C (1992) Cholinergic innervation of the human striatum, globus pallidus, subthalamic nucleus, substantia nigra, and red nucleus. J Comp Neurol 323:252–268

Moore H, Sarter M, Bruno JP (1992) Age-dependent modulation of in vivo cortical acetylcholine release by benzodiazepine receptor ligands. Brain Res 596:17–29

Moore H, Fadel J, Sarter M, Bruno JP (1999) Role of accumbens and cortical dopamine receptors in the regulation of cortical acetylcholine release. Neuroscience 88: 811–822

Nisell M, Nomikos GG, Svensson TH (1994a) Infusion of nicotine in the ventral tegmental area or the nucleus accumbens of the rat differentially affects accumbal dopamine release. Pharmacol Toxicol 75:348–352

Nisell M, Nomikos GG, Svensson TH (1994b) Systemic nicotine-induced dopamine release in the rat nucleus accumbens is regulated by nicotinic receptors in the ventral tegmental area. Synapse 16:36–44

Nisell M, Marcus M, Nomikos GG, Svensson TH (1997) Differential effects of acute and chronic nicotine on dopamine output in the core and shell of the rat nucleus accumbens. J Neural Transm 104:1–10

Oakman SA, Faris PL, Kerr PE, Cozzari C, Hartman BK (1995) Distribution of pontomesencephalic cholinergic neurons projecting to substantia nigra differs significantly from those projecting to ventral tegmental area. J Neurosci 15:5859–5869

Olmstead MC, Munn EM, Franklin KB, Wise RA (1998) Effects of pedunculopontine tegmental nucleus lesions on responding for intravenous heroin under different schedules of reinforcement. J Neurosci 18:5035–5044

Pepeu G, Bartolini A (1968) Effect of psychoactive drugs on the output of acetylcholine from the cerebral cortex of the cat. Eur J Pharmacol 4:254–263

Pepeu G, Mantovani P (1978) Effect of bromocriptine on acetylcholine output from the cerebral cortex. Pharmacology 16 Suppl 1:204–206

Phelps PE, Houser CR, Vaughn JE (1985) Immunocytochemical localization of choline acetyltransferase within the rat neostriatum: a correlated light and electron microscopic study of cholinergic neurons and synapses. J Comp Neurol 238:286–307

Picciotto MR, Zoli M, Rimondini R, Lena C, Marubio LM, Pich EM, Fuxe K, Changeux JP (1998) Acetylcholine receptors containing the beta2 subunit are involved in the reinforcing properties of nicotine. Nature 391:173–177

Pidoplichko VI, DeBiasi M, Williams JT, Dani JA (1997) Nicotine activates and desensitizes midbrain dopamine neurons. Nature 390:401–404

Pisani A, Bonsi P, Centonze D, Calabresi P, Bernardi G (2000) Activation of D2-like dopamine receptors reduces synaptic inputs to striatal cholinergic interneurons. J Neurosci (Online.) 20:RC69

Pontieri FE, Tanda G, Orzi F, Di Chiara G (1996) Effects of nicotine on the nucleus accumbens and similarity to those of addictive drugs [see comments]. Nature 382:255–257

Rang HP, Colquhoun D, Rang HP (1982) The action of ganglionic blocking drugs on the synaptic responses of rat submandibular ganglion cells. Br J Pharmacol 75:151–168

Robbins TW, Everitt BJ (1994) Arousal systems and attention. In: Gazzaniga MS (ed). The Cognitive Neurosciences, MIT (Cambridge), pp 703–720

Rowell PP, Carr LA, Garner AC (1987) Stimulation of [3H]dopamine release by nicotine in rat nucleus accumbens. J Neurochem 49:1449–1454

Russell MA, Jarvis M, Iyer R, Feyerabend C (1980) Relation of nicotine yield of cigarettes to blood nicotine concentrations in smokers. Br Med J 280:972–976

Sakurai Y, Takano Y, Kohjimoto Y, Honda K, Kamiya HO (1982) Enhancement of [3H]dopamine release and its [3H]metabolites in rat striatum by nicotinic drugs. Brain Res 242:99–106

Sarter M (1994) Neuronal mechanisms of the attentional dysfunctions in senile dementia and schizophrenia: two sides of the same coin? Psychopharmacology (Berl) 114:539–550

Sarter M, Bruno JP (2000) Cortical cholinergic inputs mediating arousal, attentional processing and dreaming: differential afferent regulation of the basal forebrain by telencephalic and brainstem afferents. Neuroscience 95:933–952

Sarter M, Givens B, Bruno JP (2001) The cognitive neuroscience of sustained attention: where top-down meets bottom-up. Brain Res Brain Res Rev 35:146–160

Sato K, Ueda H, Okumura F, Misu Y (1994) Supersensitization of intrastriatal dopamine receptors involved in opposite regulation of acetylcholine release in Parkinson's model rats. Neurosci Lett 173:59–62

Scatton B (1982a) Further evidence for the involvement of D2, but not D1 dopamine receptors in dopaminergic control of striatal cholinergic transmission. Life Sci 31:2883–2890

Scatton B (1982b) Effect of dopamine agonists and neuroleptic agents on striatal acetylcholine transmission in the rat: evidence against dopamine receptor multiplicity. J Pharmacol Exp Ther 220:197–202

Schilstrom B, Svensson HM, Svensson TH, Nomikos GG (1998a) Nicotine and food induced dopamine release in the nucleus accumbens of the rat: putative role of alpha7 nicotinic receptors in the ventral tegmental area. Neuroscience 85: 1005–1009

Schilstrom B, Nomikos GG, Nisell M, Hertel P, Svensson TH (1998b) N-methyl-D-aspartate receptor antagonism in the ventral tegmental area diminishes the systemic nicotine-induced dopamine release in the nucleus accumbens. Neuroscience 82:781–789

Schultz W, Apicella P, Scarnati E, Ljungberg T (1992) Neuronal activity in monkey ventral striatum related to the expectation of reward. J Neurosci 12:4595–4610

Schultz W, Apicella P, Ljungberg T (1993) Responses of monkey dopamine neurons to reward and conditioned stimuli during successive steps of learning a delayed response task. J Neurosci 13:900–913

Schwaber JS, Rogers WT, Satoh K, Fibiger HC (1987) Distribution and organization of cholinergic neurons in the rat forebrain demonstrated by computer-aided data acquisition and three- dimensional reconstruction. J Comp Neurol 263:309–325

Seguela P, Wadiche J, Dineley-Miller K, Dani JA, Patrick JW (1993) Molecular cloning, functional properties, and distribution of rat brain alpha 7: a nicotinic cation channel highly permeable to calcium. J Neurosci 13:596–604

Sethy VH (1979) Regulation of striatal acetylcholine concentration by D2-dopamine receptors. Eur J Pharmacol 60:397–398

Sethy VH, Van Woert MH (1974) Modification of striatal acetylcholine concentration by dopamine receptor agonists and antagonists. Res Commun Chem Pathol Pharmacol 8:13–28

Silbersweig DA, Stern E, Frith C, Cahill C, Holmes A, Grootoonk S, Seaward J, McKenna P, Chua SE, Schnorr L (1995) A functional neuroanatomy of hallucinations in schizophrenia. Nature 378:176–179

Spencer HJ (1976) Antagonism of cortical excitation of striatal neurons by glutamic acid diethyl ester: evidence for glutamic acid as an excitatory transmitter in the rat striatum. Brain Res 102:91–101

Stadler H, Lloyd KG, Gadea-Ciria M, Bartholini G (1973) Enhanced striatal acetylcholine release by chlorpromazine and its reversal by apomorphine. Brain Res 55: 476–480

Starr MS (1995) Glutamate/dopamine D1/D2 balance in the basal ganglia and its relevance to Parkinson's disease. Synapse 19:264–293

Steinberg R, Rodier D, Souiclhac J, Bougault I, Emonds-Alt X, Soubrie P, Le Fur G (1995) Pharmacological characterization of tachykinin receptors controlling acetylcholine release from rat striatum: an in vivo microdialysis study. J Neurochem 65:2543–2548

Stoof JC, Drukarch B, De Boer P, Westerink BH, Groenewegen HJ (1992) Regulation of the activity of striatal cholinergic neurons by dopamine. Neuroscience 47: 755–770

Svensson TH, Mathe JM, Nomikos GG, Schilstrom B (1998) Role of excitatory amino acids in the ventral tegmental area for central actions of non-competitive NMDA-receptor antagonists and nicotine. Amino Acids 14:51–56

Swanson LW, Cowan WM (1979) The connections of the septal region in the rat. J Comp Neurol 186:621–656

Taber MT, Fibiger HC (1994) Cortical regulation of acetylcholine release in rat striatum. Brain Res 639:354–356

Takano Y, Sakurai Y, Kohjimoto Y, Honda K, Kamiya HO (1983) Presynaptic modulation of the release of dopamine from striatal synaptosomes: differences in the effects of high K+ stimulation, methamphetamine and nicotinic drugs. Brain Res 279:330–334

Tedford CE, Crosby G, Jr, Iorio LC, Chipkin RE (1992) Effect of SCH 39166, a novel dopamine D1 receptor antagonist, on [3H]acetylcholine release in rat striatal slices. Eur J Pharmacol 211:169–176

Trabucchi M, Cheney DL, Racagni G, Costa E (1975) In vivo inhibition of striatal acetycholine turnover by L-DOPA, apomophine and (plus)-amphetamine. Brain Res 85:130–134

Vezina P, Blanc G, Glowinski J, Tassin JP (1992) Nicotine and morphine differentially activate brain dopamine in prefrontocortical and subcortical terminal fields: effects of acute and repeated injections. J Pharmacol Exp Ther 261:484–490

Wada E, Wada K, Boulter J, Deneris E, Heinemann S, Patrick J, Swanson LW (1989) Distribution of alpha 2, alpha 3, alpha 4, and beta 2 neuronal nicotinic receptor subunit mRNAs in the central nervous system: a hybridization histochemical study in the rat. J Comp Neurol 284:314–335

Wang JQ, McGinty JF (1997) The full D1 dopamine receptor agonist SKF-82958 induces neuropeptide mRNA in the normosensitive striatum of rats: regulation of D1/D2 interactions by muscarinic receptors. J Pharmacol Exp Ther 281:972–982

Westfall TC (1974) Effect of nicotine and other drugs on the release of 3H-norepinephrine and 3H-dopamine from rat brain slices. Neuropharmacology 13: 693–700

Westfall TC, Grant H, Perry H (1983) Release of dopamine and 5-hydroxytryptamine from rat striatal slices following activation of nicotinic cholinergic receptors. Gen Pharmacol 14:321–325

Williams JA, Comisarow J, Day J, Fibiger HC, Reiner PB (1994) State-dependent release of acetylcholine in rat thalamus measured by in vivo microdialysis. J Neurosci 14:5236–5242

Wong DT, Bymaster FP, Reid LR, Fuller RW, Perry KW, Kornfeld EC (1983) Effect of a stereospecific D2-dopamine agonist on acetylcholine concentration in corpus striatum of rat brain. J Neural Transm 58:55–67

Woolf NJ (1991) Cholinergic systems in mammalian brain and spinal cord. Prog Neurobiol 37:475–524

Woolf NJ, Harrison JB, Buchwald JS (1990) Cholinergic neurons of the feline pontomesencephalon. II. Ascending anatomical projections. Brain Res 520:55–72

Yan Z, Surmeier DJ (1997) D5 dopamine receptors enhance Zn2+-sensitive GABA(A) currents in striatal cholinergic interneurons through a PKA/PP1 cascade. Neuron 19:1115–1126

Yan Z, Song WJ, Surmeier J (1997) D2 dopamine receptors reduce N-type Ca2+ currents in rat neostriatal cholinergic interneurons through a membrane-delimited, protein-kinase-C- insensitive pathway. J Neurophysiol 77:1003–1015

Yang CR, Mogenson GJ (1989) Ventral pallidal neuronal responses to dopamine receptor stimulation in the nucleus accumbens. Brain Res 489:237–246

Yeomans JS (1995) Role of tegmental cholinergic neurons in dopaminergic activation, antimuscarinic psychosis and schizophrenia. Neuropsychopharmacol 12:3–16

Zaborszky L, Cullinan WE, Braun A (1991) Afferents to basal forebrain cholinergic projection neurons: an update. Adv Exp Med Biol 295:43–100

Zaborszky L, Cullinan WE (1992) Projections from the nucleus accumbens to cholinergic neurons of the ventral pallidum: a correlated light and electron microscopic double-immunolabeling study in rat. Brain Res 570:92–101

Zaborszky L, Pang K, Somogyi J, Nadasdy Z, Kallo I (1999) The basal forebrain corticopetal system revisited. Ann NY Acad Sci 877:339–367

Zilles K, Werner L, Qu M, Schleicher A, Gross G (1991) Quantitative autoradiography of 11 different transmitter binding sites in the basal forebrain region of the rat – evidence of heterogeneity in distribution patterns. Neuroscience 42:473–481

Zocchi A, Pert A (1993) Increases in striatal acetylcholine by SKF-38393 are mediated through D1 dopamine receptors in striatum and not the frontal cortex. Brain Res 627:186–192

Zoli M, Lena C, Picciotto MR, Changeux JP (1998) Identification of four classes of brain nicotinic receptors using beta2 mutant mice. J Neurosci 18:4461–4472

CHAPTER 16
Dopamine – Glutamate Interactions

C. KONRADI, C. CEPEDA, and M.S. LEVINE

A. Introduction

Dopamine (DA) and glutamate interact in the brain on a number of different levels. In this chapter we will illustrate both interneuronal and intraneuronal interactions of the DA and glutamate neurotransmitter systems, ranging from reciprocal release regulation of neurotransmitters to an interactive control of membrane depolarization and gene expression. Although many of these interactions are reciprocal, our approach in this review will be to consider the glutamate system to function as the prime mover, while the DA system provides a strong modulatory influence on responses mediated by glutamate release or activation of glutamate receptors. The characteristics of the modulation by the DA system depend on a number of factors. These include, but certainly are not limited to, the DA and glutamate receptor subtypes involved, the baseline activity-state of the neuron, the location of the receptors on pre- and/or postsynaptic elements, and endogenous concentrations of glutamate and DA. In our view a very important factor is receptor subtype. The combinations of DA and glutamate receptor subtypes activated determines, to a large extent, the outcome of the interaction. Thus, depending on the subtypes of DA and glutamate receptors involved, the interactions can be cooperative or opposing. This chapter will review the present knowledge of the different levels and types of interaction between both neurotransmitter systems. Because an exhaustive analysis of DA-glutamate interactions in different regions of the brain is beyond the scope of this review, we are going to limit our discussion to only a few regions. One area that is particularly well suited to illustrate the complexity of DA-glutamate interactions in the brain is the dorsal striatum, which will be the major focus of this chapter. Also not covered is an exhaustive account of glutamate-DA interactions from a historical perspective. These have been summarized in a previous publication (CEPEDA and LEVINE 1998).

B. Neuropharmacological Interactions
I. Dopamine and Glutamate Act Within the Same Neuronal Circuits

DA neurons which project to forebrain structures emanate predominantly from two areas in the midbrain, the substantia nigra and the ventral tegmental area (see Vol. I, Chap. 3). Neurons from each area are involved in elaborate circuits that rely heavily on glutamate neurotransmission. Neurons from the substantia nigra project primarily to the dorsal striatum and are part of a circuit that includes the thalamus and cortex. Glutamate-containing projections can be found in many places within the circuit. Most directly, neurons in the substantia nigra receive glutamate-containing inputs and these cells express *N*-methyl-D-aspartate (NMDA) receptors (COUNIHAN et al. 1998; GAUCHY et al. 1994; SMITH et al. 1996). Moreover, the corticostriatal projection puts the glutamate system in axo-axonal contact with DA terminals in the striatum.

Neurons from the ventral tegmental area project predominantly to the nucleus accumbens, to the glutamate-containing neurons of the medial prefrontal cortex, and to other cortical areas. In addition, the nucleus accumbens is innervated by glutamate-containing axon terminals emanating from the medial prefrontal cortex (see also Vol. I, Chap. 3).

The interaction between the DA and glutamate neurotransmitter systems takes place on so many different levels that an accurate assessment of the role of distinct glutamate pathways in the regulation of the DA system and vice versa can be very difficult. To gain a better understanding of the interactions between both neurotransmitters in the various parts of the circuitry, information from studies using brain slices, or cultured or isolated neurons have been combined with information from studies of the intact brain.

II. Dopamine and Glutamate Receptors in the Striatum

DA receptors are distinguished pharmacologically into the D1 family of receptors (D_1, D_5) and the D2 family of receptors (D_2, D_3 and D_4) (KEBABIAN and CALNE 1979; SEEMAN and VAN TOL 1994), and are described in detail in Chaps. 5–7 of Vol. I. In this chapter, we will use D1 and D2 to refer to receptor families and subscript notation to refer to family members (i.e., D_1 and D_5). In the striatum and the nucleus accumbens, of the five DA receptor subtypes known, D_1, D_2, and D_3 are abundant, while D_4 and D_5 are sparse (BUNZOW et al. 1988; MONSMA et al. 1990; SOKOLOFF et al. 1990; SUNAHARA et al. 1991; TIBERI et al. 1991; VAN TOL et al. 1991).

Glutamate receptors are classified into ionotropic receptors, which gate ion channels, and metabotropic receptors, which are linked to G proteins (HOLLMANN and HEINEMANN 1994). Ionotropic glutamate receptors are further subdivided into NMDA receptors, and into α-amino-3-hydroxy-5-methylisoxazole-4-propionic acid (AMPA)/kainate receptors depending on their affinities for specific agonists (HOLLMANN and HEINEMANN 1994). We will refer to this latter group as non-NMDA ionotropic receptors. NMDA receptors are blocked by physiologic levels of Mg^{2+} and need depolarization in addition to ligand binding (glutamate and glycine) to open (MAYER et al. 1984). The NMDA receptor is assembled from NR1 subunits with various combinations of NR2 (A-D) subunits (HOLLMANN and HEINEMANN 1994). AMPA and kainate receptors are assembled from various combinations of subunits (GluR1–4 for AMPA and GluR5–7 and KA1–2 for kainate). The neurons of the striatum express NMDA, AMPA, and kainate receptors, as well as metabotropic glutamate receptors (BAHN et al. 1994; HOLLMANN and HEINEMANN 1994; TESTA et al. 1994).

III. Reciprocal Release Regulation of Dopamine and Glutamate by Dopamine Receptors and Ionotropic Glutamate Receptors

Glutamate receptors on DA neurons and DA receptors on glutamate neurons play a role in the reciprocal regulation of neurotransmitter release. While neurotransmitter release of any neuron can be manipulated via activation of pre- and postsynaptic receptors, receptors that are located presynaptically on axons are particularly interesting in striatal neurotransmitter release-regulation since DA and glutamate axons converge in the striatum. The most abundant presynaptic DA receptor in the striatum is the D_2 receptor (SESACK et al. 1994; HERSCH et al. 1995; MERCURI et al. 1997). Of the glutamate receptors, presynaptic location of metabotropic glutamate receptors is generally accepted (PETRALIA et al. 1996), while low levels of presynaptic ionotropic glutamate receptors were demonstrated in the nucleus accumbens, cortex, and hippocampus (GRACY and PICKEL 1996; CHARTON et al. 1999).

Functional assays that examine neurotransmitter release point to a regulation of DA release by axonal glutamate receptors, and a regulation of glutamate release by axonal DA receptors (Table 1). It has been shown that glutamate facilitates basal DA release in the striatum (SHIMIZU et al. 1990; DESCE et al. 1992). NMDA receptors mediate DA release in the absence of Mg^{2+} or during depolarization (DESCE et al. 1992, 1994; MARTINEZ-FONG et al.

Table 1. Neuropharmacological interactions of DA and glutamate

Reciprocal release regulation of DA and glutamate by DA receptors and ionotropic glutamate receptors

By:	DA release		
Glutamate	Up		
NMDA	Up		
AMPA/kainate	Up		
By:	Glutamate release		
DA	Down		
D_1	Up or no change		
D_2	Down		

Reciprocal regulation of receptor synthesis

By:	D_1 receptor synthesis	D_2 receptor synthesis	
NMDA	Down	Down	
By:	NR1 subunit synthesis	NR2A subunit synthesis	AMPA/kainate receptor synthesis
DA	No change	Down	No change
D_1	Up	–	–
D_2	Down	–	–

1992), while AMPA/kainate receptors mediate DA release independent of Mg^{2+} or depolarization (IMPERATO et al. 1990; DESCE et al. 1992). AMPA/kainate receptors also contribute to the depolarization needed to remove the Mg^{2+} block of NMDA receptors. Ionotropic glutamate receptors are not the only glutamate receptors involved in DA release, as metabotropic glutamate receptors have been implicated in the regulation of striatal DA release as well (VERMA and MOGHADDAM 1998). Finally, DA levels are attenuated after decortication, a procedure that greatly reduces glutamate input to the striatum, confirming the facilitatory role of glutamate on DA release (SMOLDERS et al. 1996).

The role of DA in the regulation of glutamate release is less transparent (Table 1). In KCl-depolarized neurons, stimulation of D_2 receptors decreases glutamate transmission (MAURA et al. 1989; YAMAMOTO and DAVY 1992). Activation of D_1 receptors has been reported to increase glutamate transmission or to have no measurable effect (YAMAMOTO and DAVY 1992). Lesions of the substantia nigra, which diminish DA levels, result in increased glutamate release in the striatum (LINDEFORS and UNGERSTEDT 1990), affirming an overall inhibitory role of the DA system on glutamate release.

IV. Reciprocal Regulation of Receptor Synthesis

In rats treated chronically with the NMDA antagonist MK801, a significant increase in D_1- and D_2-receptor mRNA is observed in the striatum (MICHELETTI et al. 1992; HEALY and MEADOR-WOODRUFF 1996). In the reversed experimental paradigm, chronic treatment with D_1 antagonists decreases the expression of the NR1 subtype of the NMDA receptor in the striatum, while chronic treatment with D_2 antagonists increases the expression of NR1 in the striatum (FITZGERALD et al. 1995). This reciprocal regulation may be responsible for the lack of net change of NR1 levels after DA denervation (FITZGERALD et al. 1995; ULAS and COTMAN 1996). Of the NR2 subunits of the NMDA receptor, the NR2A subunit is increased in the striatum after DA depletion (ULAS and COTMAN 1996). DA denervation has little effect on expression of striatal AMPA receptor subtypes (FITZGERALD et al. 1995; BERNARD et al. 1996).

Taken together, NMDA receptor activation inhibits the synthesis of D_1 and D_2 receptors, while DA receptor activation decreases expression of the NR2A subunit of the NMDA receptor. Synthesis of the NR1 subunit is facilitated by activation of D_1 receptors and inhibited by activation of D_2 receptors, leading to a net effect of no change in the presence of DA. There appear to be no effects of DA on striatal AMPA receptor expression and the effects of DA on kainate receptor expression have not yet been evaluated.

V. Glutamate Regulates the Synthesis of Dopamine in Striatal Synaptosomes

When synaptosomal preparations of the striatum are treated with glutamate, a decrease in the synthesis of DA is observed (DESCE et al. 1994). In concurrence, inhibition of glutamate receptors causes an increase in striatal DA levels (RICHARD and BENNETT 1995). The action of glutamate on DA concentrations in the striatum is antagonistic; it increases release and decreases synthesis (DESCE et al. 1994). However, these data were collected after acute treatment, and to our knowledge, the effect of chronic glutamate receptor inhibition on DA synthesis has not been investigated.

VI. Glutamate and Dopamine Are Co-released from Dopamine Neurons

In a recent study in monkey and rat, DA neurons co-immunostained for glutamate (SULZER et al. 1998). Moreover, stimulation of DA neurons in single cell microcultures evoked rapid synaptic actions via glutamate synapses and slower, modulatory actions via DA synapses (SULZER et al. 1998). These data suggest that glutamate co-transmission may occur in central monoaminergic neurons. In many instances, DA and glutamate could be simultaneously released upon stimulation of midbrain DA neurons. This observation, if corroborated in physiological conditions, can have important implications. For example, simultaneous release of glutamate and DA can provide the depolarization necessary to remove the Mg^{2+} block of NMDA receptors in striatal neurons.

C. Intraneuronal Interactions

I. Dopamine Receptors and NMDA Receptors Cooperatively Modulate Gene Expression

DA receptors are linked to a signal transduction cascade that regulates the expression of various genes. Functional glutamate receptors, in particular NMDA receptors, are a requirement for DA receptor-mediated gene regulation. It has been proposed that an intraneuronal interaction between DA and glutamate signal transduction pathways leads to the cooperative regulation of gene expression (KONRADI 1998).

D1 and D2 receptors are oppositely linked to second messenger pathways, which determines their interaction with glutamate receptors. D1 receptors are coupled to G_s proteins (see Vol. I, Chap. 5). Stimulation of D1 receptors activates adenylate cyclase and increases the levels of cyclic AMP. D2 receptors are coupled to G_i proteins, inhibit adenylate cyclase, and cause a decrease in the levels of cyclic AMP (see Vol. I, Chap. 6). Therefore, stimulation of D1 receptors or inhibition of D2 receptors leads to the activation of the cyclic AMP signal transduction pathway. Genes that are stimulated by the cyclic AMP pathway and that are colocalized with either receptor, such as the immediate early gene c-*fos*, are induced after D_1 receptor activation or D_2 receptor inhibition. Other genes that are colocalized predominantly with one receptor subtype, such as prodynorphin (colocalized with D_1 receptors) or proenkephalin (colocalized with D_2 receptors), respond to manipulation of the respective receptor only (TANG et al. 1983; ENGBER et al. 1992; COLE et al. 1995). When NMDA receptors are blocked, induction of gene expression after manipulation of D_1 or D_2 receptors is prevented (ZIOLKOWSKA and HOLLT 1993; KONRADI et al. 1996). A closer examination reveals that the cyclic AMP-mediated second messenger pathway that is activated by DA receptors on the dendrites modulates NMDA receptor activity (KONRADI et al. 1996, 1998), a modulation that is required for DA receptor-mediated gene expression. There is an indication that this modulation is accomplished by protein kinase A (PKA), which is activated by cyclic AMP and which phosphorylates the NMDA receptor (RAJADHYAKSHA et al. 1998). Phosphorylation of the NMDA receptor seems to increase its responsiveness to ambient glutamate, possibly by removing the Mg^{2+} block (KONRADI 1998) (Fig. 1). Thus, manipulation of DA receptor activity changes the response threshold of NMDA receptors to ambient glutamate and activates an NMDA receptor-mediated intraneuronal signal transduction pathway. The resultant change in gene expression is initiated by activation of DA receptors, but mediated via NMDA receptors and, therefore, is sensitive to NMDA antagonists. On the other hand, if gene expression is initiated by NMDA receptors, DA can act as a modulator of NMDA receptor-mediated gene expression.

D. Electrophysiological Interactions

I. Striatal Organization

The predominant neuron in the striatum is the γ-aminobutyric acid (GABA)-containing projection neuron. These striatal neurons receive DA-containing inputs from the substantia nigra and glutamate-containing inputs from the cortex and the thalamus that are capable of activating both NMDA and non-NMDA ionotropic receptors (CHERUBINI et al. 1988; SMITH and BOLAM 1990; SMITH et al.1994; LEVINE et al. 1996). The GABA-containing projection neurons in the striatum have been roughly divided into two subpopulations, those expressing primarily D_1 receptors, projecting to the substantia nigra or internal pallidal segment and colocalizing substance P or dynorphin and those expressing primarily D_2 receptors, projecting to the external segment of the globus pallidus and colocalizing enkephalin (GERFEN et al. 1990; LEMOINE et al. 1990, 1991). This initial dichotomy has been questioned by studies demonstrating colocalization of D_1 and D_2 receptors to striatal output neurons (SURMEIER et al. 1992, 1993, 1996). Part of the controversy is due to differences in experimental approaches and methods of analysis of expression patterns (SURMEIER et al. 1993, 1998). In addition, electrophysiological analyses that support colocalization of D_1 and D_2 family receptors to striatal output neurons tend to use pharmacological tools that do not differentiate among the subtypes of receptors in each family. A recent study has attempted to account for the differences by demonstrating that although D_1 and D_2 receptors do not

Dopamine – Glutamate Interactions

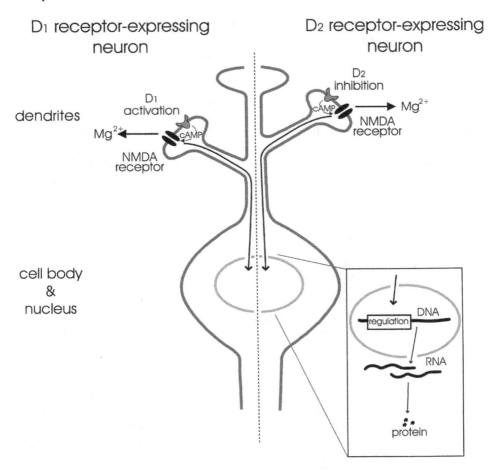

Fig. 1. NMDA receptors play a crucial role in DA receptor-mediated gene expression. The *left side* of the model depicts a D_1 receptor-expressing neuron, the *right side* depicts a D_2 receptor-expressing neuron. Activation of D_1 receptors or inhibition of D_2 receptors initiates a second messenger pathway that stimulates NMDA receptor function, e.g., by removing the Mg^{2+}-block. Activation of NMDA receptors initiates a signal transduction pathway that translocates to the nucleus and activates the expression of specific genes (*insert*). Newly synthesized mRNA is transported out of the nucleus and translated into protein (*insert*). *cAMP*, cyclic AMP-mediated second messenger pathway; *regulation*, regulation of gene expression

appear to colocalize frequently, other members of each family do colocalize to substance P- and enkephalin-containing neurons (Surmeier et al. 1996, 1998).

In addition to the medium-sized spiny projection neurons, the striatum contains interneurons. These neurons are not as prevalent but have important consequences for striatal function. There are multiple classes of interneurons and they have been identified both by their electrophysiological properties and their neurochemical signatures (KAWAGUCHI 1993). The most frequently studied interneuron is the large cholinergic neuron. This cell type expresses both D1 and D2 family receptors. However, recent evidence points to an abundance of the D_5 receptor mRNA versus D_1 mRNA (YAN and SURMEIER 1997). These neurons also express both NMDA and non-NMDA ionotropic glutamate receptors (STANDAERT et al. 1999). In recent experiments, we have demonstrated that NMDA receptor-mediated current density is smaller in these large interneurons than in the medium-sized neurons (CEPEDA et al. 2001a) while current density in response to activation of kainate receptors is similar in the large and medium-sized cells. Unfortunately, little is currently known about DA–glutamate interactions in the large cholinergic interneurons or in the other subpopulations of interneurons in the striatum. We have observed that DA receptor activation increases NMDA responses in these interneurons (LEVINE et al. 1998); however, the subtype of DA receptor mediating this effect remains to be determined.

II. Dopamine Modulates Glutamate Inputs

The strategic location of DA terminals on the neck of dendritic spines of the striatal medium-sized spiny output neurons allows a tight regulation of responsiveness of glutamate terminals located on the head of the same spines (SMITH and BOLAM 1990). The outcome of this regulation depends in great measure on the glutamate and DA receptor subtypes activated. A working hypothesis of glutamate–DA interactions in the striatum has been proposed recently (CEPEDA and LEVINE 1998; LEVINE and CEPEDA 1998) (Fig. 2). Accordingly, D_1 receptor activation enhances responses due to activation of glutamate receptors, particularly those mediated by activation of NMDA receptors. In contrast, D_2 receptor activation reduces responses due to activation of glutamate receptors, particularly those mediated by activation of non-NMDA receptors. The enhancing effects of D_1 receptor activation appear to involve postsynaptic actions (at least in striatum), whereas the attenuating effects mediated by D_2 receptors may involve both postsynaptic as well as presynaptic actions on corticostriatal terminals (CEPEDA et al. 1993; LEVINE et al. 1996; CEPEDA et al. 2001b).

D_1 receptor enhancement of NMDA-evoked responses appears to be mediated by multiple mechanisms, involving DA's effects on intrinsic, voltage-gated currents as well as more direct actions in which activation of transduction pathways changes the phosphorylation state of the NMDA receptor. Agonists of L-type calcium channels potentiate NMDA responses and blockade of L-type calcium channels reduce the modulation of NMDA responses by D_1 receptor activation (CEPEDA et al. 1998a). Stimulation of the cyclic AMP-PKA cascade with forskolin also enhances NMDA responses (COLWELL and LEVINE 1995; BLANK et al. 1997) and blockade of this pathway reduces the modulation (BLANK et al. 1997). D_1 receptor activation phosphorylates the NMDA NR1 subunit (SNYDER et al. 1998). The same effect is observed after

Prediction of Direction of DA Modulation in Neostriatum

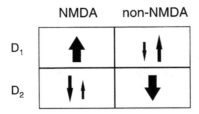

Fig. 2. Schematic representation of DA modulation of responses mediated by activation of glutamate receptor subtypes. *Arrows* indicate direction of modulation. When D_1 receptors are activated, almost all responses mediated by activation of NMDA receptors are potentiated (*left upper box*), while responses mediated by activation of non-NMDA receptors can either be potentiated or attenuated, although proportionately more appear to be potentiated (*right upper box*). When D_2 receptors are activated, virtually all responses mediated by activation of non-NMDA receptors are attenuated (*right lower box*), while responses mediated by activation of NMDA receptors can either be potentiated or attenuated, although proportionately more appear to be attenuated (*left lower box*)

forskolin (RAJADHYAKSHA et al. 1998). As described above, phosphorylation of the NMDA receptor could increase its responsiveness to ambient glutamate. DARPP-32, a substrate for PKA that selectively inhibits protein phosphatase-1, is also involved in this modulation (BLANK et al. 1997; FLORES-HERNÁNDEZ et al. 1999). Non-NMDA receptors can also be phosphorylated by similar mechanisms. There is very recent evidence that protein phosphatase-1 modulates striatal AMPA channels by regulation of DARPP-32 and spinophilin (YAN et al. 1999).

The effects of DA on responses mediated by activation of glutamate receptors are activity-state dependent. Studies in behaving animals have demonstrated that DA may exert potentiating or attenuating effects depending on the level of cortical glutamate input onto striatal neurons (REBEC 1998). Intracellular studies have revealed that the actions of DA are also related to the level of membrane polarization. When the membrane is hyperpolarized (more negative than −60mV), glutamate activates non-NMDA receptors preferentially, and the predominant effects of DA are inhibitory. In contrast, when the membrane is more depolarized (less than −60mV) glutamate can activate NMDA receptors and the effects of DA become facilitatory (HERNÁNDEZ-LOPEZ et al. 1997; CEPEDA and LEVINE 1998). The inhibitory effects may involve a cooperative interaction of D_2 and D_1 receptors acting pre- and postsynaptically (CEPEDA et al. 1993; CEPEDA et al. 2001b). The facilitatory effects involve principally postsynaptic activation of D_1 receptors (CEPEDA et al. 1993; LEVINE et al. 1996; FLORES-HERNÁNDEZ et al. 1999). Activation of postsynaptic D_2 receptors may prevent excessive facilitation, thus counterbalancing D_1 effects.

At the cellular level, there are physiological mechanisms that allow these differential interactions to occur. For example, the resting membrane potential of medium-spiny neurons oscillates between a depolarized and a hyperpolarized state (WILSON and KAWAGUCHI 1996). The depolarization is produced by a barrage of glutamate inputs from cortex. Such depolarization removes the Mg^{2+} block and permits NMDA receptor activation (KITA 1996). Under these conditions the effects of DA are facilitatory. During the hyperpolarized state DA's actions are inhibitory. This means that if a DA signal coincides with the depolarized state of the membrane, the glutamate signal will be potentiated. In contrast, if the DA signal coincides with the hyperpolarized state, the glutamate signal will be decreased. In addition, it has been shown that in vivo, DA itself can produce membrane depolarizations (BERNARDI et al. 1978; HERRLING and HULL 1980). This depolarization can also remove the Mg^{2+} block and allow NMDA receptor activation. How DA produces a direct membrane depolarization is still unknown; however, enhancement of L-type calcium currents is one possibility. Another is inhibition of K^+ conductances or decreases in GABA-mediated responses (FLORES-HERNÁNDEZ et al. 2000; SURMEIER and KITAI 1993). Finally, if co-release of DA and glutamate (SULTZER et al. 1998) after substantia nigra stimulation occurs under physiological conditions, another potential mechanism for membrane depolarization in striatal cells would be available.

Although many electrophysiological studies have shown that DA modulates glutamate transmission in the dorsal striatum, others have not found such regulation (CALABRESI et al.1995; NICOLA and MALENKA 1998). The reason for these differences is not known and we have discussed potential possibilities for lack of modulation elsewhere (see CEPEDA and LEVINE 1998).

E. Interaction of the Dopamine and Glutamate Neurotransmitter Systems in Other Brain Areas

As shown in Chap. 3 (Vol. I), DA neurons innervate many brain areas in addition to the striatum. In the majority of these brain areas, DA neurons are in close contact with the glutamate-containing inputs. A cursory review of the literature indicates that the effects of DA modulation appear to be variable in the different brain regions. As we have pointed out in the dorsal striatum, modulation by DA depends on the type of DA receptor, the pre- and postsynaptic topography, the co-expression with non-DA receptors, and the membrane potential of the neurons involved. As these parameters vary in different brain areas, an impression of great diversity and unpredictability is created. However, if experimental conditions are tightly controlled, the DA system modulates the glutamate system in a consistent fashion.

In the nucleus accumbens, D_1 receptor activation has also been shown to enhance NMDA responses (HARVEY and LACEY 1997). Protein kinase C activation plays an important role in this potentiation (CHERGUI and LACEY 1999).

The cerebral cortex is another region where DA–glutamate interactions occur. In human cortex, a differential modulation of excitatory inputs by DA has been demonstrated (CEPEDA et al. 1992). Here again, DA can potentiate or reduce responses mediated by activation of NMDA or non-NMDA receptors depending on which DA receptor subtype is activated (CEPEDA et al. 1999; ZHENG et al. 1999). D_1 receptor activation potentiates whereas D_2 receptor activation attenuates responses.

F. Functional Consequences of the Interaction of the Glutamate and Dopamine Systems

The interaction between DA and glutamate plays a prominent role in the pathophysiology of drug addiction, movement disorders, schizophrenia, as well as in the physiological processes underlying learning and memory formation. We have suggested that differential modulation of glutamate inputs by D_1 and D_2 receptors plays an important role as a filtering device that can effectively alter the signal-to-noise ratio (CEPEDA et al. 1992, 1993). Thus, the DA signal is extremely important for extracting relevant information, and could be used as a global reinforcement signal for adapting behavior according to the motivational value of environmental stimuli (SCHULTZ 1998). Alternatively, it could promote the switching of attentional and behavioral resources towards significant stimuli (REDGRAVE et al. 1999).

Diverse forms of synaptic plasticity occur in the striatum depending on which glutamate receptor subtypes are preferentially activated. If non-NMDA receptors are activated long-term depression is produced (CALABRESI et al. 1992; LOVINGER et al. 1993). If NMDA receptors are unmasked, short- or long-term potentiation is observed (CALABRESI et al. 1992; WALSH and DUNIA 1993). Activation of DA receptors modulates these forms of synaptic plasticity (CALABRESI et al. 1992a). In the absence of D_2 receptors, long-term potentiation occurs (CALABRESI et al. 1997), and activation of D_1 receptors facilitates the induction of long-term potentiation (KERR and WICKENS 2001). It has also been proposed that long-term depression in the striatum may reflect extinction and long-term potentiation the reinforcement of specific behaviors (ARBUTHNOTT and WICKENS 1996). In consequence, the enhancement of NMDA responses by activation of D_1 receptors and the attenuation of non-NMDA responses by activation of D_2 receptors becomes particularly relevant at the behavioral level. It is tempting to speculate that one consequence of the potentiation of NMDA responses by D_1 receptors in the striatum could be the consolidation of motor programs.

DA–glutamate interactions in the nucleus accumbens are critically involved in the regulation of sensorimotor gating (WAN et al. 1995). Prepulse inhibition, a measure of sensorimotor gating, is obliterated by manipulations affecting glutamate and DA inputs (WAN and SWERDLOW 1996). Deficits in prepulse inhibition have been found in schizophrenia (SWERDLOW and GEYER

1998) and Huntington's disease (SWERDLOW et al. 1995). Recently, prepulse inhibition has also been found to be disrupted in mice that lack expression of D_2 receptors (RALPH et al. 1999). In the cerebral cortex, an interaction between D_1 receptor activation and glutamate inputs appears critical for the formation of memory traces (GOLDMAN-RAKIC 1998).

On the other side of the spectrum, many DA-mediated processes are facilitated by activation of the glutamate system. NMDA antagonists are effective tools for intervention in animal models of DA-mediated behavior (KARLER et al. 1989; SCHENK et al. 1993; WOLF et al. 1994; KIM and JANG 1997). Delayed consequences of reduced DA neurotransmission, an important factor in movement disorders, can be prevented with NMDA antagonists (BOLDRY et al. 1995). If either system malfunctions, the interdependence can make it difficult to expose the primary problem. However, the close interaction of both systems also opens additional therapeutic avenues in that each system can be pharmacologically adjusted to counterbalance problems of the other.

Acknowledgements. C.K. is supported by USPHS Grant DA07134 and the National Alliance for Research on Schizophrenia and Depression. C.C. and M.S.L. are supported by USPHS Grants NS33538, NS35649, and the National Alliance for Research on Schizophrenia and Depression.

References

Arbuthnott GW, Wickens JR (1996) Dopamine cells are neurones too! Trends Neurosci 19:279–280

Bahn S, Volk B, Wisden W (1994) Kainate receptor gene expression in the developing rat brain. J Neurosci 14:5525–5547

Bernard V, Gardiol A, Faucheux B, Bloch B, Agid Y, Hirsch EC (1996) Expression of glutamate receptors in the human and rat basal ganglia: effect of the dopaminergic denervation on AMPA receptor gene expression in the striatopallidal complex in Parkinson's disease and rat with 6-OHDA lesion. J Comp Neurol 368: 553–568

Bernardi G, Marciani MG, Morocutti C, Pavone F, Stanzione P (1978) The action of dopamine on rat caudate neurones intracellularly recorded. Neurosci Lett 8: 235–240

Blank T, Nijholt I, Teichert U, Kugler H, Behrsing H, Fienberg A, Greengard P, Spiess J (1997) The phosphoprotein DARPP-32 mediates cAMP-dependent potentiation of striatal N-methyl-D-aspartate responses. Proc Natl Acad Sci (USA) 94: 14859–14864

Boldry RC, Papa SM, Kask AM, Chase TN (1995) MK-801 reverses effects of chronic levodopa on D1 and D2 dopamine agonist-induced rotational behavior. Brain Res 692:259–264

Bunzow JR, Van Tol HH, Grandy DK, Albert P, Salon J, Christie M, Machida CA, Neve KA, Civelli O (1988) Cloning and expression of a rat D2 dopamine receptor cDNA. Nature 336:783–787

Calabresi P, Maj R, Pisani A, Mercuri NB, Bernardi G (1992) Long-term synaptic depression in the striatum: physiological and pharmacological characterization. J Neurosci 12:4224–4233

Calabresi P, Maj R, Mercuri NB, Bernardi G (1992a) Coactivation of D1 and D2 dopamine receptors is required for long-term synaptic depression in the striatum. Neurosci Lett 142:95–99

Calabresi P, De Murtas M, Pisani A, Stefani A, Sancesario G, Mercuri NB, Bernardi G (1995) Vulnerability of medium spiny striatal neurons to glutamate: role of Na+/K+ ATPase. Eur J Neurosci 7:1674–1683

Calabresi P, Saiardi A, Pisani A, Baik JH, Centonze D, Mercuri NB, Bernardi G, Borrelli E (1997) Abnormal synaptic plasticity in the striatum of mice lacking dopamine D2 receptors. J Neurosci 17:4536–4544

Cepeda C, Radisavljevic Z, Peacock W, Levine MS, Buchwald NA (1992) Differential modulation by dopamine of responses evoked by excitatory amino acids in human cortex. Synapse 11:330–341

Cepeda C, Buchwald NA, Levine MS (1993) Neuromodulatory actions of dopamine in the neostriatum are dependent upon the excitatory amino acid receptor subtypes activated. Proc Natl Acad Sci (USA) 90:9576–9580

Cepeda C, Levine MS (1998) Dopamine and N-methyl-D-aspartate receptor interactions in the neostriatum. Dev Neurosci 20:1–18

Cepeda C, Colwell CS, Itri JN, Chandler SH, Levine MS (1998a) Dopaminergic modulation of NMDA-induced whole cell currents in neostriatal neurons in slices: contribution of calcium conductances. J Neurophysiol 79:82–94

Cepeda C, Li Z, Cromwell HC, Altemus KL, Crawford CA, Nansen EA, Ariano MA, Sibley DR, Peacock WJ, Mathern GW, Levine MS (1999) Electrophysiological and morphological analyses of cortical neurons obtained from children with catastrophic epilepsy: dopamine receptor modulation of glutamatergic responses. Dev Neurosci 21:223–235

Cepeda C, Itri JN, Flores-Hernández J, Hurst RS, Calvert CR, Levine MS (2001a) Differential sensitivity of medium- and large-sized striatal neurons to NMDA but not kainate receptor activation in the rat. Eur J Neurosci 14:1577–1589

Cepeda C, Hurst RS, Altemus KL, Flores-Hernández J, Calvert CR, Jokel ES, Grandy DK, Low MJ, Rubinstein M, Ariano MA, Levine MS (2001b) Facilitated glutamatergic transmission in the striatum of D2 dopamine receptor-deficient mice. J Neurophysiol 85:659–670

Charton JP, Herkert M, Becker CM, Schroder H (1999) Cellular and subcellular localization of the 2B-subunit of the NMDA receptor in the adult rat telencephalon. Brain Res 816:609–617

Chergui K, Lacey MG (1999) Modulation by dopamine D1-like receptors of synaptic transmission and NMDA receptors in rat nucleus accumbens is attenuated by the protein kinase C inhibitor Ro 32–0432. Neuropharm 38:223–231

Cherubini E, Herrling PL, Lanfumey L, Stanzione P (1988) Excitatory amino acids in synaptic excitation of rat striatal neurones in vitro. J Physiol (Lond) 400:677–690

Cole RL, Konradi C, Douglass J, Hyman SE (1995) Neuronal adaptation to amphetamine and dopamine: molecular mechanisms of prodynorphin gene regulation in rat striatum. Neuron 14:813–823

Colwell CS, Levine MS (1995) Excitatory synaptic transmission in neostriatal neurons: regulation by cyclic AMP-dependent mechanisms. J Neurosci 15:1704–1713

Counihan TJ, Landwehrmeyer GB, Standaert DG, Kosinski CM, Scherzer CR, Daggett LP, Velicelebi G, Young AB, Penney JB, Jr (1998) Expression of N-methyl-D-aspartate receptor subunit mRNA in the human brain: mesencephalic dopaminergic neurons. J Comp Neurol 390:91–101

Desce JM, Godeheu G, Galli T, Artaud F, Cheramy A, Glowinski J (1992) L-glutamate-evoked release of dopamine from synaptosomes of the rat striatum: involvement of AMPA and N-methyl-D-aspartate receptors. Neuroscience 47:333–339

Desce JM, Godeheu G, Galli T, Glowinski J, Cheramy A (1994) Opposite presynaptic regulations by glutamate through NMDA receptors of dopamine synthesis and release in rat striatal synaptosomes. Brain Res 640:205–214

Engber TM, Boldry RC, Kuo S, Chase TN (1992) Dopaminergic modulation of striatal neuropeptides: differential effects of D1 and D2 receptor stimulation on somatostatin, neuropeptide Y, neurotensin, dynorphin and enkephalin. Brain Res 581: 261–268

Fitzgerald LW, Deutch AY, Gasic G, Heinemann SF, Nestler EJ (1995) Regulation of cortical and subcortical glutamate receptor subunit expression by antipsychotic drugs. J Neurosci 15:2453–2461

Flores-Hernandez J, Cepeda C, Fienberg AA, Greengard P, Levine MS (1999) Multiple pathways are involved in the enhancement of NMDA responses by activation of dopamine D1 receptors in neostriatal neurons. Soc Neurosci Abstr

Flores-Hernández J, Hernández S, Snyder GL, Yan Z, Fienberg AA, Moss SJ, Greengard P, Surmeier DJ (2000) D(1) dopamine receptor activation reduces GABA(A) receptor currents in neostriatal neurons through a PKA/DARPP-32/PP1 signaling cascade. J Neurophysiol 83:2996–3004

Gauchy C, Desban M, Glowinski J, Kemel ML (1994) NMDA regulation of dopamine release from proximal and distal dendrites in the cat substantia nigra. Brain Res 635:249–256

Gerfen CR, Engber TM, Mahan LC, Susel Z, Chase TN, Monsma FJ, Jr, Sibley DR (1990) D1 and D2 dopamine receptor-regulated gene expression of striatonigral and striatopallidal neurons. Science 250:1429–1432

Goldman-Rakic PS (1998) The cortical dopamine system: role in memory and cognition. Adv Pharmacol 42:707–711

Gracy KN, Pickel VM (1996) Ultrastructural immunocytochemical localization of the N-methyl-D-aspartate receptor and tyrosine hydroxylase in the shell of the rat nucleus accumbens. Brain Res 739:169–181

Harvey J, Lacey MG (1997) A postsynaptic interaction between dopamine D1 and NMDA receptors promotes presynaptic inhibition in the rat nucleus accumbens via adenosine release. J Neurosci 17:5271–5280

Healy DJ, Meador-Woodruff JH (1996) Differential regulation, by MK-801, of dopamine receptor gene expression in rat nigrostriatal and mesocorticolimbic systems. Brain Res 708:38–44

Hernandez-Lopez S, Bargas J, Surmeier DJ, Reyes A, Galarraga E (1997) D1 receptor activation enhances evoked discharge in neostriatal medium spiny neurons by modulating an L-type Ca2+ conductance. J Neurosci 17:3334–3342

Herrling PL, Hull CD (1980) Iontophoretically applied dopamine depolarizes and hyperpolarizes the membrane of cat caudate neurons. Brain Res 192:441–462

Hersch SM, Ciliax BJ, Gutekunst CA, Rees HD, Heilman CJ, Yung KK, Bolam JP, Ince E, Yi H, Levey AI (1995) Electron microscopic analysis of D1 and D2 dopamine receptor proteins in the dorsal striatum and their synaptic relationships with motor corticostriatal afferents. J Neurosci 15:5222–5237

Hollmann M, Heinemann S (1994) Cloned glutamate receptors. Annu Rev Neurosci 17:31–108

Imperato A, Honore T, Jensen LH (1990) Dopamine release in the nucleus caudatus and in the nucleus accumbens is under glutamatergic control through non-NMDA receptors: a study in freely-moving rats. Brain Res 530:223–228

Karler R, Calder LD, Chaudhry IA, Turkanis SA (1989) Blockade of "reverse tolerance"to cocaine and amphetamine by MK-801. Life Sci 45:599–606

Kawaguchi Y (1993) Physiological, morphological, and histochemical characterization of three classes of interneurons in rat neostriatum. J Neurosci 13:4908–4923

Kebabian JW, Calne DB (1979) Multiple receptors for dopamine. Nature 277:93–96

Kerr JN, Wickens JR (2001) Dopamine D-1/D-5 receptor activation is required for long-term potentiation in the rat neostriatum in vitro. J Neurophysiol 85:117–124

Kim HS, Jang CG (1997) MK-801 inhibits methamphetamine-induced conditioned place preference and behavioral sensitization to apomorphine in mice. Brain Res Bull 44:221–227

Kita H (1996) Glutamatergic and GABAergic postsynaptic responses of striatal spiny neurons to intrastriatal and cortical stimulation recorded in slice preparations. Neuroscience 70:925–940

Konradi C, Leveque JC, Hyman SE (1996) Amphetamine and dopamine-induced immediate early gene expression in striatal neurons depends on postsynaptic NMDA receptors and calcium. J Neurosci 16:4231–4239

Konradi C (1998) The molecular basis of dopamine and glutamate interactions in the striatum. Adv Pharmacol 42:729–733

Le Moine C, Normand E, Guitteny AF, Fouque B, Teoule R, Bloch B (1990) Dopamine receptor gene expression by enkephalin neurons in rat forebrain. Proc Natl Acad Sci (USA) 87:230–234

Le Moine C, Normand E, Bloch B (1991) Phenotypical characterization of the rat striatal neurons expressing the D1 dopamine receptor gene. Proc Natl Acad Sci (USA) 88:4205–4209

Levine MS, Li Z, Cepeda C, Cromwell HC, Altemus KL (1996) Neuromodulatory actions of dopamine on synaptically-evoked neostriatal responses in slices. Synapse 24:65–78

Levine MS, Cepeda C (1998) Dopamine modulation of responses mediated by excitatory amino acids in the neostriatum. Adv Pharmacol 42:724–729

Levine MS, Cepeda C, Colwell CS, Yu Q, Chandler SH (1998) Infrared video microscopy: Visualization and manipulation of neurons in neostriatal slices. In: Ariano MA (ed) Receptor Localization: Laboratory Method Procedure New York: John Wiley, pp 128–139

Lindefors N, Ungerstedt U (1990) Bilateral regulation of glutamate tissue and extracellular levels in caudate-putamen by midbrain dopamine neurons. Neurosci Lett 115:248–252

Lovinger DM, Tyler EC, Merritt A (1993) Short- and long-term synaptic depression in rat neostriatum. J Neurophysiol 70:1937–1949

Martinez-Fong D, Rosales MG, Gongora-Alfaro JL, Hernandez S, Aceves J (1992) NMDA receptor mediates dopamine release in the striatum of unanesthetized rats as measured by brain microdialysis. Brain Res 595:309–315

Maura G, Carbone R, Raiteri M (1989) Aspartate-releasing nerve terminals in rat striatum possess D-2 dopamine receptors mediating inhibition of release. J Pharmacol Exp Ther 251:1142–1146

Mayer ML, Westbrook GL, Guthrie PB (1984) Voltage-dependent block by Mg^{2+} of NMDA responses in spinal cord neurones. Nature 309:261–263

Mercuri NB, Saiardi A, Bonci A, Picetti R, Calabresi P, Bernardi G, Borrelli E (1997) Loss of autoreceptor function in dopaminergic neurons from dopamine D2 receptor deficient mice. Neuroscience 79:323–327

Micheletti G, Lannes B, Haby C, Borrelli E, Kempf E, Warter JM, Zwiller J (1992) Chronic administration of NMDA antagonists induces D2 receptor synthesis in rat striatum. Brain Res Mol Brain Res 14:363–368

Monsma FJ, Jr, Mahan LC, McVittie LD, Gerfen CR, Sibley DR (1990) Molecular cloning and expression of a D1 dopamine receptor linked to adenylyl cyclase activation. Proc Natl Acad Sci (USA) 87:6723–6727

Nicola SM, Malenka RC (1998) Modulation of synaptic transmission by dopamine and norepinephrine in ventral but not dorsal striatum. J Neurophysiol 79:1768–1776

Petralia RS, Wang YX, Niedzielski AS, Wenthold RJ (1996) The metabotropic glutamate receptors, mGluR2 and mGluR3, show unique postsynaptic, presynaptic and glial localizations. Neuroscience 71:949–976

Rajadhyaksha A, Leveque J, Macias W, Barczak A, Konradi C (1998) Molecular components of striatal plasticity: the various routes of cyclic AMP pathways. Dev Neurosci 20:204–215

Ralph RJ, Varty GB, Kelly MA, Wang YM, Caron MG, Rubinstein M, Grandy DK, Low MJ, Geyer MA (1999) The dopamine D2, but not D3 or D4, receptor subtype is essential for the disruption of prepulse inhibition produced by amphetamine in mice. J Neurosci 19:4627–4633

Rebec GV (1998) Dopamine, glutamate, and behavioral correlates of striatal neuronal activity. Adv Pharmacol 42:737–740

Redgrave P, Prescott TJ, Gurney K (1999) Is the short-latency dopamine response too short to signal reward error? Trends Neurosci 22:146–151

Richard MG, Bennett JP, Jr (1995) NMDA receptor blockade increases in vivo striatal dopamine synthesis and release in rats and mice with incomplete, dopamine-depleting, nigrostriatal lesions. J Neurochem 64:2080–2086

Schenk S, Valadez A, McNamara C, House DT, Higley D, Bankson MG, Gibbs S, Horger BA (1993) Development and expression of sensitization to cocaine's reinforcing properties: role of NMDA receptors. Psychopharmacology (Berl) 111:332–338

Schultz W (1998) The phasic reward signal of primate dopamine neurons. Adv Pharmacol 42:686–690

Seeman P, Van Tol HH (1994) Dopamine receptor pharmacology. Trends Pharmacol Sci 15:264–270

Sesack SR, Aoki C, Pickel VM (1994) Ultrastructural localization of D2 receptor-like immunoreactivity in midbrain dopamine neurons and their striatal targets. J Neurosci 14:88–106

Shimizu N, Duan SM, Hori T, Oomura Y (1990) Glutamate modulates dopamine release in the striatum as measured by brain microdialysis. Brain Res Bull 25:99–102

Smith AD, Bolam JP (1990) The neural network of the basal ganglia as revealed by the study of synaptic connections of identified neurones. Trends Neurosci 13:259–265

Smith Y, Bennett BD, Bolam JP, Parent A, Sadikot AF (1994) Synaptic relationships between dopaminergic afferents and cortical or thalamic input in the sensorimotor territory of the striatum in monkey. J Comp Neurol 344:1–19

Smith Y, Charara A, Parent A (1996) Synaptic innervation of midbrain dopaminergic neurons by glutamate-enriched terminals in the squirrel monkey. J Comp Neurol 364:231–253

Smolders I, Sarre S, Vanhaesendonck C, Ebinger G, Michotte Y (1996) Extracellular striatal dopamine and glutamate after decortication and kainate receptor stimulation, as measured by microdialysis. J Neurochem 66:2373–2380

Snyder GL, Fienberg AA, Huganir RL, Greengard P (1998) A dopamine/D1 receptor/protein kinase A/dopamine- and cAMP-regulated phosphoprotein (Mr 32 kDa)/protein phosphatase-1 pathway regulates dephosphorylation of the NMDA receptor. J Neurosci 18:10297–10303

Sokoloff P, Giros B, Martres MP, Bouthenet ML, Schwartz JC (1990) Molecular cloning and characterization of a novel dopamine receptor (D3) as a target for neuroleptics. Nature 347:146–151

Standaert DG, Friberg IK, Landwehrmeyer GB, Young AB, Penney JB, Jr (1999) Expression of NMDA glutamate receptor subunit mRNAs in neurochemically identified projection and interneurons in the striatum of the rat. Brain Res Mol Brain Res 64:11–23

Sulzer D, Joyce MP, Lin L, Geldwert D, Haber SN, Hattori T, Rayport S (1998) Dopamine neurons make glutamatergic synapses in vitro. J Neurosci 18:4588–4602

Sunahara RK, Guan HC, BF OD, Seeman P, Laurier LG, Ng G, George SR, Torchia J, Van Tol HH, Niznik HB (1991) Cloning of the gene for a human dopamine D5 receptor with higher affinity for dopamine than D1. Nature 350:614–619

Surmeier DJ, Eberwine J, Wilson CJ, Cao Y, Stefani A, Kitai ST (1992) Dopamine receptor subtypes colocalize in rat striatonigral neurons. Proc Natl Acad Sci (USA) 89:10178–10182

Surmeier DJ, Kitai ST (1993) D1 and D2 dopamine receptor modulation of sodium and potassium currents in rat neostriatal neurons. Prog Brain Res 99:309–324

Surmeier DJ, Reiner AJ, Levine MS, Ariano MA (1993) Are neostriatal dopamine receptors co-localized? Trends Neurosci 16:299–305

Surmeier DJ, Song WJ, Yan Z (1996) Coordinated expression of dopamine receptors in neostriatal medium spiny neurons. J Neurosci 16:6579–6591

Surmeier DJ, Yan Z, Song WJ (1998) Coordinated expression of dopamine receptors in neostriatal medium spiny neurons. Adv Pharmacol 42:1020–1023

Swerdlow NR, Paulsen J, Braff DL, Butters N, Geyer MA, Swenson MR (1995) Impaired prepulse inhibition of acoustic and tactile startle response in patients with Huntington's disease. J Neurol Neurosurg Psychiatry 58:192–200

Swerdlow NR, Geyer MA (1998) Using an animal model of deficient sensorimotor gating to study the pathophysiology and new treatments of schizophrenia. Schizophr Bull 24:285–301

Tang F, Costa E, Schwartz JP (1983) Increase of proenkephalin mRNA and enkephalin content of rat striatum after daily injection of haloperidol for 2 to 3 weeks. Proc Natl Acad Sci (USA) 80:3841–3844

Testa CM, Standaert DG, Young AB, Penney JB, Jr (1994) Metabotropic glutamate receptor mRNA expression in the basal ganglia of the rat. J Neurosci 14:3005–3018

Tiberi M, Jarvie KR, Silvia C, Falardeau P, Gingrich JA, Godinot N, Bertrand L, Yang-Feng TL, Fremeau RT, Jr, Caron MG (1991) Cloning, molecular characterization, and chromosomal assignment of a gene encoding a second D1 dopamine receptor subtype: differential expression pattern in rat brain compared with the D1A receptor. Proc Natl Acad Sci (USA) 88:7491–7495

Ulas J, Cotman CW (1996) Dopaminergic denervation of striatum results in elevated expression of NR2A subunit. Neuroreport 7:1789–1793

Van Tol HH, Bunzow JR, Guan HC, Sunahara RK, Seeman P, Niznik HB, Civelli O (1991) Cloning of the gene for a human dopamine D4 receptor with high affinity for the antipsychotic clozapine. Nature 350:610–614

Verma A, Moghaddam B (1998) Regulation of striatal dopamine release by metabotropic glutamate receptors. Synapse 28:220–226

Walsh JP, Dunia R (1993) Synaptic activation of N-methyl-D-aspartate receptors induces short-term potentiation at excitatory synapses in the striatum of the rat. Neuroscience 57:241–248

Wan FJ, Geyer MA, Swerdlow NR (1995) Presynaptic dopamine-glutamate interactions in the nucleus accumbens regulate sensorimotor gating. Psychopharmacology (Berl) 120:433–441

Wan FJ, Swerdlow NR (1996) Sensorimotor gating in rats is regulated by different dopamine-glutamate interactions in the nucleus accumbens core and shell subregions. Brain Res 722:168–176

Wilson CJ, Kawaguchi Y (1996) The origins of two-state spontaneous membrane potential fluctuations of neostriatal spiny neurons. J Neurosci 16:2397–2410

Wolf ME, White FJ, Hu XT (1994) MK-801 prevents alterations in the mesoaccumbens dopamine system associated with behavioral sensitization to amphetamine. J Neurosci 14:1735–1745

Yamamoto BK, Davy S (1992) Dopaminergic modulation of glutamate release in striatum as measured by microdialysis. J Neurochem 58:1736–1742

Yan Z, Surmeier DJ (1997) D5 dopamine receptors enhance Zn2+-sensitive GABA(A) currents in striatal cholinergic interneurons through a PKA/PP1 cascade. Neuron 19:1115–1126

Yan Z, Hsieh-Wilson L, Feng J, Tomizawa K, Allen PB, Fienberg AA, Nairn AC, Greengard P (1999) Protein phosphatase 1 modulation of neostriatal AMPA channels: regulation by DARPP-32 and spinophilin. Nat Neurosci 2:13–17

Zheng P, Zhang XX, Bunney BS, Shi WX (1999) Opposite modulation of cortical N-methyl-D-aspartate receptor-mediated responses by low and high concentrations of dopamine. Neurosci 91:527–535

Ziolkowska B, Hollt V (1993) The NMDA receptor antagonist MK-801 markedly reduces the induction of c-fos gene by haloperidol in the mouse striatum. Neurosci Lett 156:39–42

CHAPTER 17
Dopamine – Adenosine Interactions

M. Morelli, E. Acquas, and E. Ongini

A. Adenosine in the CNS

Adenosine, which is formed by the purine base adenine and the ribose moiety, is present in all tissues in the mammalian organism, where it has a variety of important physiological functions. Linked to phosphate groups to form ATP, adenosine is an integral part of the cellular energy system. At synapses, adenosine is a mediator in many biological systems.

Adenosine originates within the cells from the hydrolysis of AMP through the action of the enzyme ecto-5' nucleotidase. Therefore, adenosine formation is dependent upon ATP breakdown and synthesis. Another pathway contributing to intracellular adenosine formation is from S-adenosylhomocysteine. In the extracellular compartment, the levels of adenosine also depend upon the rate of hydrolysis of ATP that is released from either neurons or glial cells. Extracellularly, adenosine concentrations are kept in equilibrium by specific reuptake mechanisms occurring through the action of specialized transporter proteins. It is estimated that the levels of adenosine in the CNS range between 30 and 300 nM. Adenosine is then catabolized by the action of enzymes such as adenosine kinases and adenosine deaminase.

The action of adenosine as neuromodulator occurs through the stimulation of specific receptors, the adenosine receptors, located on cell membranes which belong to the family of G protein-coupled receptors. Currently, four adenosine receptors have been cloned and characterized, A_1, A_{2A}, A_{2B}, and A_3. The main intracellular signaling pathways are through the formation of cAMP, with A_1 and A_3 causing inhibition of adenylate cyclase, and A_{2A} and A_{2B} activating it. Other transduction mechanisms are also involved for each of the adenosine receptor, e.g., voltage-sensitive Ca^{2+} channels. The molecular characteristics of the receptors and intracellular signaling are described in detail elsewhere (Fredholm et al. 1998; Olah and Stiles 2000). Their profile is summarized in Table 1.

I. Receptor Distribution

Adenosine receptors are located on membranes of several cell types. There are adenosine receptors on circulating blood elements such as platelets,

Table 1. Adenosine receptors in the brain

Receptor subtypes	Major transduction mechanism	Receptor distribution	Selective agonists	Selective antagonists
A_1	G_i and G_o, inhibition adenylate cyclase	Widely distributed in cortex, hippocampus, cerebellum	CPA	DPCPX
A_{2A}	G_s and G_{olf}, stimulation adenylate cyclase	High density in caudate putamen, nucleus accumbens, olfactory tubercle	CGS 21680	SCH 58261, KF 17837, KW 6002
A_{2B}	G_s, stimulation adenylate cyclase	Low density, glial cells	–	–
A_3	G_{i-1} and G_q, inhibition adenylate cyclase	Low density, widely distributed	2-Cl-IB-MECA	MRS 1220, MRE 3008F20

2-HE-NECA, 2-hexyl-5'-N-ethylcarboxamidoadenosine; CGS 21680, 2-[4-(2-carbonylethyl)-phenylethylamino]-5'-N-ethylcarboxamidoadenosine; Cl-IB-MECA, chloro-N^6-(3-iodobenzyl)-5'-(N-methylcarbamoyl)adenosine; CPA, N^6-cyclopentyladenosine; DPCPX, 1,3 dipropyl-8-cyclopentylxanthyne; MRE 3008F20, 5N-(4-methoxyphenylcarbamoyl)amino-8-propyl-2-(2-furyl)pyrazolo[4,3-e]-1,2,4-triazolo[1,5-c]pyrimidine; MRS 1220, 9-chloro-2-(2-furyl)-5-phenylacetylamino[1,2,4]triazolo[1,5-c]quinazoline; SCH 58261, 5-amino-7-(2-phenylethyl-2-(2-furyl)-pyrazolo[4,3-e]-1,2,4-triazolo[1,5-c]pyrimidine; ZM 241385, 4-(2-[7-amino-2-(2-furyl)1,2,4-triazolo[2,3-a][1,3,5]triazin-5-ylamino]ethylphenol.

neutrophils, and lymphocytes, on smooth muscle cells, cardiac myocytes, mast cells, and, within the CNS, on neurons and glial cells. The wide distribution of these receptors has important implications in pharmacology, since most drugs producing their action through receptors located in the CNS can also interact with receptors in the periphery, which may also contribute to the overall biological activity. In this review we will examine the receptors whose function is relevant in the CNS and, when important, we will mention any other contributing action deriving from effects in periphery.

A variety of studies based on autoradiography using radiolabeled ligands, in situ hybridization, and reverse transcription-polymerase chain reaction have shown that A_1 receptors are widely distributed in the brain, whereas A_{2A} receptors are abundant in discrete brain regions such as the striatum. Distribution and density of A_{2B} and A_3 receptors are less clear.

The higher density of A_1 receptors is found in hippocampus, cerebral cortex, cerebellum, and thalamic nuclei. A_1 receptors are also present, although to a lower level, in the rat basal ganglia (striatum and nucleus accumbens) (JARVIS and WILLIAMS 1989); however, receptor number was found to be high in basal ganglia structures in the human brain (SVENNINGSSON et al. 1997a).

There is evidence for both presynaptic and postsynaptic localization of A_1 receptors. Their presence at the presynaptic level in several neuronal pathways appears to be responsible for A_1 receptor-mediated inhibition of a variety of neurotransmitters release including glutamate, GABA, noradrenaline, acetylcholine (ACh), and dopamine (DUNWIDDE and FREDHOLM 1997).

A_{2A} receptors are predominant in several basal ganglia structures such as the striatum, globus pallidus, nucleus accumbens, and tuberculum olfactorium (JARVIS and WILLIAMS 1989; ROSIN et al. 1998). There are A_{2A} receptors in other brain areas, e.g., hippocampus, cerebral cortex, and thalamic nuclei, with some differences found between human brain and that of other animal species (SVENNINGSSON et al. 1997a). It remains, however, that using different methodological approaches, all studies are consistent in describing high levels of A_{2A} receptors in the striatum (ONGINI and FREDHOLM 1996). With regard to specific neuronal populations, A_{2A} receptors are present in striatopallidal enkephalin-expressing neurons (SCHIFFMAN et al. 1991; FINK et al. 1992). The same cells also express dopamine D_2 receptors; therefore, both A_{2A} and D_2 receptors are segregated on the same neuronal pathway. In contrast, there are no A_{2A} receptors in neurons expressing D_1 receptors, substance P, and dynorphin, which project from striatum to substantia nigra (SCHIFFMAN et al. 1991; FINK et al. 1992). It is worth noting that A_{2A} receptors are also present on glial cells.

A_{2B} and A_3 receptors appear to be important in the CNS, although the lack of selective ligands has hampered the characterization of these receptors. A_{2B} receptors are widely distributed with low density in the brain, and they require a high concentration of adenosine to be activated above the range available under physiological conditions. The A_3 receptors are localized in astrocytes and widespread in neurons; the function in the brain of A_{2B} and A_3 receptors, however, remains largely unknown.

II. Adenosine in CNS Pathology

The most impressive changes of adenosine metabolism occur under conditions leading to states of hypoxia/hypoglycemia. Thus, adenosine levels rise rapidly in the cortical area after sudden interruption of cerebral blood flow in a variety of animal species (ONGINI and SCHUBERT 1998 for review). Elevated levels of adenosine influence biochemical processes, e.g., excitatory amino acid release or Ca^{2+} influx, which ultimately result in neuroprotective actions. Through inhibition of adenosine transport, attempts have been made to create drugs that could reduce neuronal damage. Both A_1 and A_{2A} receptors appear to be involved in neuroprotective mechanisms and several data have been generated showing that the same net result can be achieved by either stimulating A_1 receptors or blocking A_{2A} receptors (ONGINI and SCHUBERT 1998). Another area of pathology involving adenosine is that of epilepsy. It is known that adenosine levels in the brain rapidly increase immediately after seizure onset. A_1 receptor agonists or inhibitors of adenosine kinase have been shown to possess anticonvulsant properties in a variety of animal models (WIESNER et al. 1999). A_{2A} receptors appear also to be involved, but their role in mechanisms underlying seizures is less defined (ADAMI et al. 1995).

Through the modulation of either A_1 or A_{2A} receptors, adenosine participates in the regulation of key processes under normal or altered conditions. For example, whereas A_1 receptors appear to be relevant for pain modulation (SAWYNOK 1998), A_{2A} receptors located in discrete brain areas are involved in mediating sleep mechanisms (SATOH et al. 1999).

A critical area of CNS pathology where adenosine-related drugs may have interesting perspectives is that of motor disorders, specifically Parkinson's disease (ONGINI and FREDHOLM 1996; RICHARDSON et al. 1997). There is evidence in animal models of Parkinson's disease that motor dysfunction is significantly reduced by blocking A_{2A} receptors (see Sect. D, this chapter). However, despite the great interest, currently there is no clear-cut data showing specific changes of adenosine levels or receptors in discrete brain regions in patients suffering from Parkinson's disease (MARTINEZ-MIR et al. 1991).

The recent development of genetically manipulated mice makes it possible to take a step forward in understanding the role of specific receptors. So far, knock-out mice for A_{2A} and A_3 receptors have been generated (LEDENT et al. 1997; CHEN et al. 1999; SALVATORE et al. 2000). Interestingly, mice bearing deletion of A_{2A} receptor gene show hypoalgesia and enhanced levels of anxiety (LEDENT et al. 1997). Most recently, it has been found that A_{2A} receptor knock-out mice display low susceptibility to cerebral ischemia showing that adenosine and A_{2A} receptors are important in controlling the response to hypoxia/hypoglycemia (CHEN et al. 1999).

B. Pharmacology of Adenosine Receptors

Of the four adenosine receptors, two of them, namely A_1 and A_{2A} receptors, have gained importance for their role in the modulation of CNS functions. Stimulation of A_1 receptors by specific agonists (see Table 1) leads to a variety of behavioral effects, ranging from sedation, anticonvulsant activity, decreased locomotor activity, analgesia, and neuroprotection. In an opposite manner, blockade of A_1 receptors through xanthine derivatives tends to produce stimulatory effects such as increased locomotor activity and enhanced susceptibility to seizures. Regulation of transmitters release by A_1 receptors located on neuronal terminals is considered to be the critical mechanism underlying the various CNS effects following administration of A_1 receptor agonists or antagonists (DUNWIDDE and FREDHOLM 1997).

The A_{2A} receptors are strongly involved in mediating effects related to the central control of motor activity. A_{2A} agonists reduce locomotor activity whereas selective A_{2A} antagonists enhance it. A_{2A} receptors appear to be also involved in the modulation of the sleep–waking continuum. Agonists reduce sleep patterns in a variety of experimental conditions and attenuate seizures in models of chemically induced convulsions (ONGINI and FREDHOLM 1996). Blockade of A_{2A} receptors tends to increase the level of wakefulness and also produce neuroprotection in animal models of cerebral ischemia (ONGINI and SCHUBERT 1998). A role of A_{2A} receptors in mediating transmitter release has also been reported (see Sect. C.III., this chapter). A_{2A} agonists stimulate the release of excitatory amino acids and ACh in the rat striatum (SEBASTIAO and RIBEIRO 1996); however, other studies have not confirmed such an interaction between ACh and A_{2A} receptors (DUNWIDDE and FREDHOLM 1997).

Currently there are no drugs used in therapy showing selectivity either as agonists or antagonists for adenosine receptors. While clinical trials are ongoing to develop adenosine-related therapeutics, some compounds available show interactions with adenosine receptors in the CNS. The most known of such drugs are the xanthines: caffeine and theophylline. These compounds block both A_1 and A_{2A} receptors in the micromolar range, and part of their action is believed to be mediated by adenosine receptors. The recent review by FREDHOLM et al. (1999) provides a thorough analysis of CNS pharmacology of caffeine whose action is attributed mainly to blockade of A_{2A} receptors in the brain.

C. Adenosine – Dopamine Interactions

Adenosine plays a role opposite to dopamine in the mediation of psychomotor behaviors originated in the dorsal and ventral striatum. Like dopamine receptor antagonists, adenosine receptor agonists induce sedation and catalepsy in a dose-dependent manner and inhibit the motor activating effects of dopamine receptor agonists (VELLUCCI et al. 1993; MORELLI et al. 1994; FERRE 1997; RIMONDINI et al. 1997). The depressant effects of adenosine receptor agonists better correlate with their affinity for A_{2A} than A_1 receptors. Such effects are obtained by N-ethylcarboxamide adenosine (NECA) which preferentially, but not selectively, acts on A_{2A} receptors, or by the most selective A_{2A} receptor agonist CGS 21680, following either parenteral administration or local infusion in the dorsal and ventral striatum (DURCAN and MORGAN 1989; HEFFNER et al. 1989; BARRACO et al. 1993).

In contrast, generic adenosine receptor antagonists, including caffeine and related methylxanthines, produce psychomotor stimulant effects by enhancing locomotor activity and schedule-controlled behavior (GARRET and GRIFFITHS 1997; FREDHOLM et al. 1999) whereas, as shown in Fig. 1, selective A_{2A} antagonists as SCH 58261 potentiate the turning behavior induced by L-dopa in 6-hydroxydopamine (6-OHDA)-lesioned rats (FENU et al. 1997). The expression of caffeine-induced motor behaviors largely depends upon dopamine transmission as shown by the sensitization of caffeine effects obtained after dopamine agonist administration (FENU and MORELLI 1998; FENU et al. 2000) or by the counteraction of locomotor and turning behavior by either reserpine

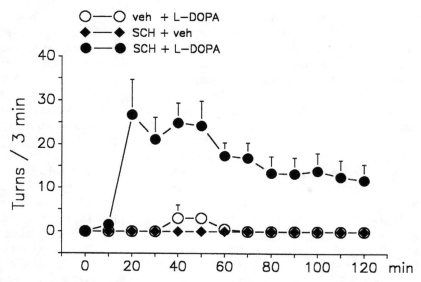

Fig. 1. Contralateral turning behavior in 6-OHDA-lesioned rats after vehicle (*veh*) + L-dopa (2 mg/kg i.p.), SCH 58261 (5 mg/kg i.p.) + vehicle (*veh*) or SCH 58261 (5 mg/kg i.p.) + L-dopa (2 mg/kg i.p.). *Ordinate* represents the rate of contralateral rotations, *abscissa* indicates the time after L-dopa administration

and α-methyltyrosine or by dopamine receptor antagonists (HERRERA-MARSCHITZ et al. 1988; JOSSELYN and BENINGER 1991; GARRETT and HOLTZMANN 1994a). In line with the dopamine dependence of xanthine-mediated motor effects, caffeine potentiates cocaine and amphetamine discriminative effects and produces a partial generalization to them (HARLAND et al. 1989; GAUVIN et al. 1990), whereas in self-administration tests, caffeine increases and reinstates cocaine self-administration (SCHENK et al. 1994; WORLEY et al. 1994). Moreover, rats rendered tolerant to caffeine exhibit cross-tolerance to dopamine receptor agonists (GARRETT and HOLTZMAN 1994b). Similarly to what is observed with adenosine agonists, the motor stimulant effects of caffeine appear to be related to an action on A_{2A} rather than A_1 receptors, since drugs blocking A_{2A} receptors such as CGS 15943 and SCH 58261 induce motor stimulant effects, whereas the A_1 antagonist DPCPX does not (GRIEBEL et al. 1991; SVENNINGSSON et al. 1997b).

I. Dopamine D_1 and Adenosine Receptors

Specific interactions between D_1 receptors and A_1 and A_2 receptors have been described either in reserpinized mice or in 6-OHDA-lesioned rats.

A preferential antagonism of D_1-mediated motor behavior by A_1 rather than A_{2A} receptor agonists has been reported in reserpinized mice, whereas in the presence of dopamine receptor supersensitivity, such as in 6-OHDA-lesioned rats, A_{2A} receptor agonists, more efficiently than A_1 receptor agonists, counteract turning behavior induced by D_1 receptor agonists (FERRE et al. 1994b; MORELLI et al. 1994). Similarly, A_{2A} receptor antagonists, more powerfully than A_1 receptor antagonists, potentiate D_1-mediated turning behavior in 6-OHDA-lesioned rats and increase striatal c-*fos* expression (JANG et al.1993; PINNA et al. 1996; POLLACK and FINK 1996; POPOLI et al. 1996; LE MOINE et al. 1997).

A_1 receptors are colocalized in striatal efferent neurons containing either D_1 or D_2 receptors whereas A_{2A} receptors are segregated in striatal neurons which do not contain D_1 receptors (SCHIFFMAN et al. 1991; FINK et al. 1992). Therefore, whereas an interaction between D_1 and A_1 receptors at the second messenger level or directly at the receptor level may explain the behavioral effects described above, an interaction at different basal ganglia levels and not only in the striatum is clearly important for the D_1/A_{2A} interaction. As shown in Fig. 2, the direct striatonigral efferent pathway containing D_1 receptors and the indirect striato-pallido-nigral efferent pathway, mainly containing D_2 receptors, control in an inhibitory and excitatory way, respectively, the activity of the substantia nigra, which is the most important basal ganglia output structure. Manipulation of the indirect pathway by A_{2A} receptor agonists or antagonists has, therefore, similar to D_2 receptors, the ability of influencing D_1-mediated responses extensively (FERRÉ et al. 1997).

II. Dopamine D_2 and Adenosine Receptors

Differently from D_1 receptors which interact with either A_1 and A_{2A} receptors, dopamine D_2 receptors exclusively interact with A_{2A} receptors in the mediation of motor behavior. Although A_1 receptors are partially colocalized with D_2 receptors in striatal efferent neurons, no evidence of an interaction at either receptor or behavioral level has been evidenced (FERRE et al. 1994b; PINNA et al. 1996; POPOLI et al. 1996) showing that A_1 receptors play a marginal role in the modulation of D_2-mediated motor responses.

By contrast, A_{2A} receptor agonists effectively counteract the effects of D_2 agonists on motor activity in reserpinized mice and turning behavior in 6-OHDA-lesioned rats (FERRE 1994b; MORELLI et al. 1994). In agreement, caffeine or selective A_{2A} receptor antagonists potentiate bromocriptine-induced motor activity in reserpinized mice (FERRE et al. 1991a) and quinpirole-induced turning behavior in 6-OHDA-lesioned rats (FENU et al. 1997).

At least three mechanisms might be responsible for the interaction between A_{2A} and D_2 receptors. A negative direct interaction at the receptor level, as shown by the decrease in the affinity of the D_2 receptor for the agonist in brain homogenates or in fibroblast cell lines, cotransfected with A_{2A} and D_2 receptors, after stimulation of A_{2A} receptors (FERRE et al. 1991b; DASGUPTA et al. 1996). An interaction at the second messenger level is also likely to

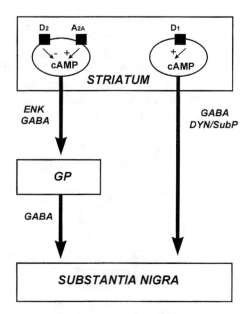

Fig. 2. Schematic representation of dopamine D_1, D_2, and adenosine A_{2A} receptor interaction in the nigrostriatal system. A_{2A} and D_2 receptors are colocalized in the indirect striato-pallido-nigral γ-aminobutyric acid (GABA)ergic pathway, whereas D_1 receptors are localized in the direct striatonigral pathway. D_1 and A_{2A} receptors stimulate cAMP formation whereas D_2 receptors inhibit cAMP formation. Stimulation of the direct pathway (D_1) or inhibition of the indirect pathway (D_2) inhibits substantia nigra activity. Stimulation of the indirect pathway (A_{2A}) disinhibits substantia nigra activity. Blockade of A_{2A} receptors can, therefore, potentiate dopamine-mediated inhibition of substantia nigra activity by either a direct interaction with D_2 receptors at the level of the striato-pallido-nigral pathway or by an indirect interaction with D_1 receptors at the substantia nigra level where the responses mediated by the striato-pallido-nigral and striatonigral pathway are integrated. *GP*, globus pallidus; *ENK*, enkephalin; *DYN*, dynorphin; *SubP*, substance P

contribute to this interaction since the two receptors affect adenylate cyclase in an opposite direction (Table 1). An indirect mechanism involving cholinergic transmission has also been reported to play an important role since either atropine administered intrastriatally, or scopolamine administered parenterally, reduce the inhibitory effects of CGS 21680 on dopamine receptor agonist-induced turning behavior and c-*fos* induction (VELLUCCI et al. 1993; MORELLI et al. 1995). These results are in line with the negative role played by ACh release on dopamine-mediated responses and with the increase of ACh release induced in the striatum and in motor nerve terminals by A_{2A} receptor stimulation (KUROKAWA et al. 1994; CORREIRA-DE-SA and RIBEIRO 1996). It is worth noting, however, that in striatal slices A_{2A} receptor agonists do not induce ACh release although they counteract the D_2 receptor-mediated decrease in ACh release (JIN et al. 1993).

III. Modulation of Dopamine Release

Early in vitro studies on the regulation of dopamine synthesis and release by adenosine provided the original demonstration that adenosine, acting on A_{2A} receptors, stimulates tyrosine hydroxylase in striatal synaptosomal preparations in antagonistic manner with dopamine D_2 receptors (ONALI et al. 1988; BOOTH and BALDESSARINI 1990). Similarly, in vitro analysis of the regulation of [^{3}H] dopamine release from striatal slices showed that adenosine (CASS and ZAHNISER 1991) and both A_1 and A_{2A} selective agonists inhibit electrically evoked [^{3}H] dopamine release (JIN et al. 1993).

In vivo brain microdialysis studies on the role of adenosine receptor subtypes in the control of dopamine neurotransmission are mostly restricted to the striatum, and the role of A_1 and A_{2A} adenosine receptors has been studied after either systemic administration or local application, by reverse dialysis, of adenosinergic compounds. The results so far obtained, however, appear conflicting. Whereas some groups reported a decrease of dopamine release after A_1 receptor stimulation by adenosine or 2-chloro-N6-cyclopentyladenosine (CCPAs) (OKADA et al. 1996; OKADA and KANEKO 1998), and by 2-chloroadenosine (2-CADO) (BALLARIN et al. 1995), others reported no effects on dopamine release after local application of the A_1 agonist CPA (GOLEMBIOWSKA and ZYLEWSKA 1997).

Similar conflicting results were obtained on the effects of adenosine A_2 agonists and antagonists on dopamine release. The A_{2A} agonist, CGS 21680, increases dopamine release in a concentration-dependent fashion (GOLEMBIOWSKA and ZYLEWSKA 1997) and stimulation of dopamine release was also described after local application of another A_{2A} agonist, CPCA, by ZETTERSTOM and FILLENZ (1990). In contrast, OKADA and coworkers reported that A_{2A} agonists and antagonists do not affect striatal dopamine release (OKADA et al. 1996; OKADA and KANEKO 1998), unless agonists and antagonists for A_{2A} receptors are given when adenosine A_1 receptors are previously blocked, in which case they could, respectively, increase and decrease striatal dopamine release (OKADA and KANEKO 1998).

Among the few in vivo microdialysis studies that investigated on dopamine-adenosine interaction in other brain areas, a recent one showed that the intravenous administration of caffeine (0.0625–5 mg/kg) fails to affect dopamine release in the core subdivision of the nucleus accumbens (a brain area anatomically related to the dorsal striatum) (TANDA et al. 1998). These data are in agreement with the finding that systemic administration of caffeine (in a range of behaviorally relevant doses, 5–10 mg/kg) fails to stimulate dopamine release in the dorsal striatum (G. Di Chiara, unpublished observations). It is interesting to observe that, whereas intravenous administration of caffeine and the selective antagonists for A_1 and A_{2A} receptors, DPCPX and SCH 58261, fails to affect dopamine release in the core and the shell of the nucleus accumbens (ACQUAS et al. 1999), they stimulate, dose-dependently, dopamine release in the medial prefrontal cortex (TANDA et al. 1998; ACQUAS et al. 1999). Thus, the failure of caffeine to stimulate dopamine release in the mesolimbic system might be related to its lack of addictive properties (DI CHIARA 1999). On the other hand, the ability of caffeine, DPCPX, and SCH 58261 to stimulate dopamine release in prefrontal cortex might account for its reinforcing psychostimulant properties and also suggests that caffeine's actions on prefrontal dopamine arise from blockade of both A_1 and A_{2A} adenosine receptors.

D. Therapeutic Implications

The critical role played by the dopamine system in pathological conditions such as schizophrenia and Parkinson's disease has suggested that its modulation by adenosine receptor agonists and antagonists may be beneficial in the treatment of these diseases (FERRE et al. 1997; RICHARDSON et al. 1997). The demonstration that dopamine-innervated areas have abundant adenosine A_{2A} receptors, whereas these receptors are rarely expressed outside these areas, supports this hypothesis and underlines the importance of A_{2A} receptors in the interaction with the dopamine system. The possible utilization of adenosine agonists and antagonists in pathologies correlated to the dopamine system has been highlighted only recently with the introduction of selective A_{2A} receptor agonists and antagonists.

After an initial suggestion that the adenosine A_1/A_{2A} receptor antagonist caffeine could be useful in the treatment of Parkinson's disease (MALLY and STONE 1996), a recent clinical survey has shown that heavy caffeine drinkers have a low risk to develop Parkinson's disease (Ross et al. 2000). Experimental studies, using antagonists with high affinity and selectivity for the A_{2A} receptor have shown an improvement in motor disabilities in Parkinson's disease rodent and primate models (PINNA et al. 1996; POLLACK and FINK 1996; FENU et al. 1997; KANDA et al. 1998; GRONDIN et al. 1999). The A_{2A} antagonist SCH 58261 potentiates the contralateral turning behavior induced by threshold dose of L-dopa (Fig. 1) or dopamine receptor agonists in unilaterally 6-OHDA-lesioned rats, an effect accompanied by an increase in Fos-like-immunoreactivity in neurons of the lesioned striatum (PINNA et al. 1996; FENU et al. 1997). Likewise, another A_{2A} receptor antagonist, 3,7-dimethyl-1-propargylxanthine (DMPX), antagonizes catalepsy induced by haloperidol in the rat (MANDHANE et al. 1997), whereas in non-human primate models of Parkinson's disease, the xanthine derivative KW 6002 reduces rigidity and improves the disability score of 1-methyl-4-phenyl-1,2,3,6-tetrahydropyridine (MPTP)-treated marmoset (KANDA et al. 1998) and cynomolgus monkeys (GRONDIN et al. 1999). Chronic administration of L-dopa in parkinsonian patients is accompanied by severe side effects such as dyskinesia and motor fluctuations. A_{2A} antagonists, in contrast to L-dopa, revert motor disability score and are less likely to reproduce dyskinesia (KANDA et al. 1998). At this stage, A_{2A} antagonists are one of the most promising pharmacological treatments for Parkinson's disease.

In an opposite manner, A_{2A} receptor agonists were shown to reduce the psychomotor stimulant effect of dopamine agonists like amphetamine at doses not inducing catalepsy (FERRE 1997; RIMONDINI et al. 1997). These effects and those on experimental models of schizophrenia such as pre-pulse-inhibition (HAUBER and KOCK 1997), conditioned avoidance responding (MARTIN et al. 1993) and climbing assay (KAFKA and CORBETT 1996) were also shared by A_1 receptor agonists; however, whereas A_{2A} agonists display a clear separation between doses inducing sedation and motor incoordination, A_1 agonists induce ataxia and sedation at similar doses. The motor depressant effects of A_{2A} agonists are therefore qualitatively similar to those induced by dopamine antagonists. Further indications of the close relationship between drugs which block the dopamine receptors and A_{2A} receptor agonists can be found in the modifications at the level of adenosine receptors after chronic neuroleptic administration and in the interaction between haloperidol and A_{2A} receptor antagonists on c-*fos* induction (PARSON et al. 1995; BOEGMAN and VINCENT 1997; PINNA et al. 1999).

Of relevance in respect to these effects is the preferential activity of CGS 21680 in the ventral striatum, since the nucleus accumbens has been shown to play a fundamental role in the therapeutic effects of antipsychotics. Studies examining the pattern of induction of Fos-like-immunoreactivity after CGS 21680 have shown a preferential induction in the nucleus accumbens rather than the dorsal striatum (PINNA et al. 1997), whereas dialysis studies on GABA release have reported a preferential effect of CGS 21680 in ventral rather than dorsal striatum (FERRÉ et al. 1994a). These studies, together with receptor binding studies showing a stronger interaction between A_{2A} and D_2 receptor in the ventral than in the dorsal striatum (FERRÉ et al. 1997), have suggested that the postulated antipsychotic activity of CGS 21680 closely resembles that of atypical antipsychotics. Atypical neuroleptics, in fact, preferentially affect the activity of the nucleus accumbens shell, whereas classical neuroleptics also influence the dorsal striatum. In line with these results is the antagonism of clozapine-induced Fos-like-immunoreactivity by blockade of A_{2A} receptors with SCH 58261 (PINNA et al. 1999). However, despite these promising results in experimental animals, A_{2A} receptor agonists such as CGS 21680 are known to induce marked hypotension, an effect which has limited the clinical development of these compounds (CASATI et al. 1995).

Thanks to the interaction with the dopaminergic system, compounds which either stimulate or block A_{2A} receptors have the potential to become new drug candidates for the treatment of CNS disorders such as schizophrenia and Parkinson's disease.

References

Acquas E, Tanda G, Loddo P, Di Chiara G (1999) Is caffeine additive? Effect of intravenous caffeine on dopamine and acetylcholine transmission. Society for Neuroscience, 29th Annual Meeting, 226.4, October 23–28, Miami Beach, FL

Adami M, Bertorelli R, Ferri N, Foddi MC, Ongini E (1995) Effects of repeated administration of selective adenosine A_1 and A_{2A} receptor agonists on pentylenetetrazole-induced convulsions in the rat. Eur J Pharmacol 294:383–389

Ballarin M, Reiriz J, Ambrosio S, Mahy N (1995) Effect of locally infused 2-chloroadenosine, an A_1 receptor agonist, on spontaneous and evoked dopamine release in rat neostriatum. Neurosci Lett 185:29–32

Barraco RA, Martens KA, Parizon M, Normile HJ (1993) Adenosine A_{2a} receptors in the nucleus accumbens mediate locomotor depression. Brain Res Bull 31:397–404

Boegman RJ, Vincent SR (1996) Involvement of adenosine and glutamate receptors in the induction of c-fos in the striatum by haloperidol. Synapse 22:70–77

Booth RG, Baldessarini RJ (1990) Adenosine A2 stimulation of tyrosine hydroxylase in rat striatal minces is reversed by dopamine D2 autoreceptor activation. Eur J Pharmacol 185:217–221

Casati C, Monopoli A, Forlani A, Bonizzoni E, Ongini E (1995) Telemetry monitoring of hemodynamic changes induced over time by adenosine agonists in spontaneously hypertensive rats J Pharmacol Exp Ther 275:914–919

Cass WA, Zahniser NR (1991) Potassium channel blockers inhibit D2 dopamine, but not A1 adenosine, receptor inhibition of striatal dopamine release. J Neurochem 57:147–152

Chen JF, Huang Z, Ma J, Zhu JM, Moratalla R, Standaert D, Moskowitz MA, Fink JS, Schwarzschild MA (1999) A_{2A} adenosine receptors deficiency attenuate brain injury induced by transient focal ischemia in mice. J Neurosci 19:9192–9200

Correia-de-Sá P, Ribeiro JA (1996) Adenosine uptake and deamination regulate tonic A_{2a} receptor facilitation of evoked [^{3}H]acetylcholine release from the rat motor nerve terminals. Neuroscience 73:85–92

Dasgupta S, Ferré S, Kull B, Hedlund PB, Finnman UB, Ahlberg S, Arenas E, Fredholm BB, Fuxe K (1996) Adenosine A_{2A} receptors modulate the binding characteristics of dopamine D_2 receptors in stably cotransfected cells. Eur J Pharmacol 316: 325–331

Di Chiara G (1999) Drug addiction as dopamine-dependent associative learning disorder. Eur J Pharmacol 375:13–30

Dunwiddie TV, Fredholm BB (1997) Adenosine neuromodulation. In: Jacobson KA, Jarvis MF (eds) Purinergic approaches in experimental therapeutics, Wiley Liss, New York, pp 359–382

Durcan MJ, Morgan PF (1989) Evidence for adenosine A_2 receptor involvement in the hypomobility effects of adenosine analogues in mice. Eur J Pharmacol 168:285–290

Fenu S, Morelli M (1998) Motor stimulant effects of caffeine in 6-hydroxydopamine-lesioned rats are dependent on previous stimulation of dopamine receptors: a different role of D_1 and D_2 receptors. Eur J Neurosci 10:1878–1884

Fenu S, Pinna A, Ongini E, Morelli M (1997) Adenosine A_{2A} receptor antagonism potentiates L-DOPA-induced turning behaviour and c-fos expression in 6-hydroxydopamine-lesioned rats. Eur J Pharmacol 321:143–147

Fenu S, Cauli O, Morelli M (2000) Cross-sensitization between the motor activating effects of bromocriptine and caffeine: role of adenosine A_{2A} receptors. Behav Brain Res 114:97–105

Ferré S (1997) Adenosine-dopamine interactions in the ventral striatum. Implications for the treatment of schizophrenia. Psychopharmacology 133:107–120

Ferré S, Herrera-Marschitz M, Grabowska-Andén M, Casas M, Ungerstedt U, Andén NE (1991a) Postsynaptic dopamine/adenosine interaction: II. Postsynaptic dopamine agonism and adenosine antagonism of methylxanthines in short-term reserpinized mice. Eur J Pharmacol 192:31–37

Ferré S, Von Euler G, Johansson B, Fredholm BB, Fuxe K (1991b) Stimulation of high affinity adenosine A-2 receptors decreases the affinity of dopamine D-2 receptors in rat striatal membranes. Proc Natl Acad Sci USA 88:7238–7241

Ferré S, O'Connor WT, Snaprud P, Ungerstedt U, Fuxe K (1994a) Antagonistic interaction between adenosine A_{2A} receptors and dopamine D_2 receptors in the ventral striopallidal system implications for the treatment of schizophrenia. Neuroscience 63:765–773

Ferré S, Popoli P, Giménez-Llort L, Finnman UB, Martinez E, Scotti de Carolis A, Fuxe K (1994b) Postsynaptic antagonistic interaction between adenosine A_1 and dopamine D_1 receptors. NeuroReport 6:73–76

Ferré S, Fredholm BB, Morelli M, Popoli P, Fuxe K (1997) Adenosine-dopamine receptor-receptor interactions as an integrative mechanism in the basal ganglia. Trend Neurosci 20:482–486

Fink JS, Weaver DR, Rivkees SA, Peterfreund RA, Pollack AE, Adler EM, Reppert SM (1992) Molecular cloning of the rat A_2 adenosine receptor: selective co-expression with D_2 dopamine receptors in rat striatum. Mol Brain Res 14:186–190

Fredholm BB, Ijzerman AP, Jacobson KA, Linden J, Stiles GL (1998) Adenosine receptors. In: IUPHAR compendium of receptor characterization and classification. IUPHAR Media, London, pp 48–57

Fredholm BB, Battig K, Holmen J, Nehlig A, Zvartau EE (1999) Actions of caffeine in the brain with special reference to factors that contribute to its widespread use. Pharmacol Rev 51:83–133

Garrett BE, Griffiths RR (1997) The role of dopamine in the behavioral effects of caffeine in animals and humans. Pharmacol Biochem Behav 57:533–541

Garrett BE, Holtzman SG (1994a) D_1 and D_2 dopamine receptor antagonists block caffeine-induced stimulation of locomotor activity in rats. Pharmacol Biochem Behav 47:89–94

Garrett BE, Holtzman SG (1994b) Caffeine cross-tolerance to selective dopamine D_1 and D_2 receptor agonists but not to their synergistic interaction. Eur J Pharmacol 262:65–75

Gauvin DV, Criado JR, Moore KR, Holloway FA (1990) Potentiation of cocaine's discriminative effects by caffeine: a time-effect analysis. Pharmacol Biochem Behav 36:195–197

Golembiowska K, Zylewska A (1997) Adenosine receptors – the role in modulation of dopamine and glutamate release in the rat striatum. Pol J Pharmacol 49(5):317–322

Griebel G, Saffroy-Spittler M, Misslin R, Remmy D, Vogel E, Bourguignon JJ (1991) Comparison of the behavioural effects of an adenosine A1/A2-receptor antagonist, CGS 15943A, and an A1-selective antagonist, DPCPX. Psychopharmacology 103:541–544

Grondin R, Bédard PJ, Hadj Tahar A, Grégoire L, Mori A, Kase H (1999) Antiparkinsonian effect of a new selective adenosine A_{2A} receptor antagonist in MPTP-treated monkeys. Neurology 52:1673–1677

Harland RD, Gauvin DV, Michaelis RC, Carney JM, Seale TW, Holloway FA (1989) Behavioral interaction between cocaine and caffeine: a drug discrimination analysis in rats. Pharmacol Biochem Behav 32:1017–1023

Hauber W, Koch M (1997) Adenosine A_{2a} receptors in the nucleus accumbens modulate prepulse inhibition of the startle response. NeuroReport 8:1515–1518

Heffner TG, Wiley JN, Williams AE, Bruns RF, Coughenour LL, Downs DA (1989) Comparison of the behavioral effects of adenosine agonists and dopamine antagonists in mice. Psychopharmacology 98:31–37

Herrera-Marschitz M, Casas M, Ungerstedt U (1988) Caffeine produces contralateral rotation in rats with unilateral dopamine denervation: comparisons with apomorphine-induced responses. Psychopharmacology 94:38–45

Jarvis MF, Williams M (1989) Direct autoradiographic localization of adenosine A_2 receptors in the rat brain using the A_2-selective agonist, [^{3}H]CGS 21680. Eur J Pharmacol 168:243–246

Jiang H, Jackson-Lewis V, Muthane U, Dollison A, Ferreira M, Espinosa A, Parsons B, Przedborski S (1993) Adenosine receptor antagonists potentiate dopamine receptor agonist-induced rotational behavior in 6-hydroxydopamine-lesioned rats. Brain Res 613:347–351

Jin S, Johansson B, Fredholm BB (1993) Effects of adenosine A_1 and A_2 receptor activation on electrically evoked dopamine and acetylcholine release from rat striatal slices. J Pharmacol Exp Therap 267:801–808

Josselyn SA, Beninger RJ (1991) Behavioral effects of intrastriatal caffeine mediated by adenosinergic modulation of dopamine. Pharmacol Biochem Behav 39:97–103

Kafka SH, Corbett R (1996) Selective adenosine A_{2A} receptor/dopamine D_2 receptor interactions in animal models of schizophrenia. Eur J Pharmacol 295:147–154

Kanda T, Jackson MJ, Smith LA, Pearce RKB, Nakamura J, Kase H, Kuwana Y, Jenner P (1998) Adenosine A_{2A} antagonist: A novel antiparkinsonian agent that does not provoke dyskinesia in parkinsonian monkeys. Ann Neurol 43:507–513

Kurokawa M, Kirk IP, Kirkpatric KA, Kase H, Richardson PJ (1994) Inhibition by FK17837 of adenosine receptor-mediated modulation of striatal GABA and ACh release. Br J Pharmacol 113:43–48

Ledent C, Vaugeois JM, Schiffman SN, Pedrazzini T, El Yacoubi M, Vanderhaeghen JJ, Costantin J, Heath JK, Vassart G, Parmentier M (1997) Aggressiveness, hypoalgesia and high blood pressure in mice lacking the adenosine A_{2A} receptor. Nature: 388:674–678

Le Moine C, Svenningsson P, Fredholm BB, Bloch B (1997) Dopamine-adenosine interactions in the striatum and globus pallidum: inhibition of striatopallidal neurons through either D_2 or A_{2A} receptors enhances D_1 receptor-mediated effects on c-foc expression. J Neurosci 17:8038–8048

Mally J, Stone TW (1996) Potential role of adenosine antagonist therapy in pathological tremor disorders. Pharmacol Ther 72:243–250

Mandhane SN, Chopde CT, Ghosh AK (1997) Adenosine A2 receptors modulate haloperidol-induced catalepsy in rats. Eur J Pharmacol 328:135–141

Martin GE, Rossi DJ, Jarvis MF (1993) Adenosine agonists reduce conditioned avoidance responding in the rat. Pharmacol Biochem Behav 45:951–958

Martinez-Mir MI, Probst A, Palacios JM (1991) Adenosine A_2 receptors: selective localization in the human basal ganglia and alterations with disease. Neuroscience 42:697–706

Morelli M, Fenu S, Pinna A, Di Chiara G (1994) Adenosine A_2 receptors interact negatively with dopamine D_1 and D_2 receptors in unilaterally 6-hydroxydopamine-lesioned rats. Eur J Pharmacol 251:21–25

Morelli M, Pinna A, Wardas J, Di Chiara G (1995) Adenosine A_2 receptors stimulate c-fos expression in striatal neurons in 6-hydroxydopamine-lesioned rats. Neuroscience 67:49–55

Okada M, Kaneko S (1998) Pharmacological interactions between magnesium ion and adenosine on monoaminergic system in the central nervous system. Magnes Res 11(4):289–305

Okada M, Mizuno K, Kaneko S (1996) Adenosine A1 and A2 receptors modulate extracellular dopamine levels in rat striatum. Neurosci Lett 212:53–56

Olah M, Stiles GL (2000) The role of receptor structure in determining adenosine receptor activity Pharmacol & Therap 85:55–75

Onali P, Olianas MC, Bunse B (1988) Evidence that adenosine and dopamine autoreceptors antagonistically regulate tyrosine hydroxylase activity in rat striatal synaptosomes. Brain Res 456:302–309

Ongini E, Fredholm BB (1996) Pharmacology of adenosine A_{2A} receptors. Trends Pharmacol Sci 17:364–372

Ongini E, Schubert P (1998) Neuroprotection induced by stimulating A_1 or blocking A_{2A} adenosine receptors: an apparent paradox. Drug Dev Res 45:387–393

Ongini E, Dionisotti S, Morelli M, Ferré S, Svenningsson P, Fuxe K, Fredholm BB (1996) Neuropharmacology of the adenosine A_{2A} receptors. Drug Dev Res 39: 450–460

Parsons B, Togasaki DM, Kassir S, Przedborski S (1995) Neuroleptics up-regulate adenosine A_{2a} receptors in rat striatum: implications for the mechanism and the treatment of tardive dyskinesia. J Neurochem 65:2057–2064

Pinna A, Di Chiara G, Wardas J, Morelli M (1996) Blockade of A_{2a} adenosine receptors positively modulates turning behaviour and c-Fos expression induced by D_1 agonists in dopamine-denervated rats. Eur J Neurosci 8:1176–1181

Pinna A, Wardas J, Cristalli G, Morelli M (1997) Adenosine A_{2A} receptor agonists increase fos-like immunoreactivity in mesolimbic areas. Brain Res 759:41–49

Pinna A, Wardas J, Cozzolino A, Morelli M (1999) Involvement of adenosine A_{2A} receptors in the induction of c-Fos expression by clozapine and haloperidol. Neuropsychopharmacology 20:44–51

Pollack AE, Fink JS (1996) Synergistic interaction between an adenosine antagonist and a D_1 dopamine agonist on rotational behavior and striatal c-Fos induction in 6-hydroxydopamine-lesioned rats. Brain Res 743:124–130

Popoli P, Gimenez-Llort L, Pezzola A, Reggio R, Martinez E, Fuxe K, Ferré S (1996) Adenosine A1 receptor blockade selectively potentiates the motor effects induced by dopamine D1 receptor stimulation in rodents. Neurosci Lett 218:209–213

Richardson PJ, Kase H, Jenner PG (1997) Adenosine A_{2A} receptor antagonists as new agents for the treatment of Parkinson's disease. Trends Pharmacol Sci 18:338–344

Rimondini R, Ferré S, Ögren SO, Fuxe K (1997) Adenosine A_{2A} agonists: A potential new type of atypical antipsychotic. Neuropsychopharmacology 17:81–91

Rosin DL, Robeva A, Woodard RL, Guyenet PG, Linden J (1998) Immunohistochemical localization of adenosine A_{2A} receptors in the rat central nervous system. J Comp Neurol 401:163–186

Ross GW, Abbott RD, Petrovitch H, Morens DM, Grandinetti A, Tung K-H, Tanner CM, Masaki KH, Blanchette PL, Curb JD, Popper JS, White LR (2000) Association of coffee and caffeine intake with the risk of Parkinson disease. J Med American Ass 283:2674–2679

Salvatore CA, Tilley SL, Latour AM, Fletcher DS, Koller BH, Jacobson MA (2000) Disruption of the A_3 adenosine receptor gene in mice and its effect on stimulated inflammatory cells. J Biol Chem 275:4429–4434

Satoh S, Matsumura H, Koike N, Tokunaga Y, Maeda T, Hayaishi O (1999) Region-dependent difference in the sleep-promoting potency of an adenosine A_{2A} receptor agonist. Eur J Neurosci 11:1587–1597

Sawynok J (1998) Adenosine receptor activation and nociception. Eur J Pharmacol 317:1–11

Schenk S, Valadez A, Horger BA, Snow S, Wellman PJ (1994) Interactions between caffeine and cocaine in tests of self-administration. Behav Pharmacol 5:153–158

Schiffmann SN, Jacobs O, Vanderhaeghen JJ (1991) The striatal restricted adenosine A2 receptor (RDC8) is expressed by enkephalin but not by substance P neurons. An in situ hybridization histochemistry study. J Neurochem 57:1062–1067

Sebastiao AM, Ribeiro JA (1996) Adenosine A_{2A} receptor-mediated excitatory actions on the nervous system. Prog in Neurobiol 48:167–189

Svenningsson P, Hall H, Sedvall G, Fredholm BB (1997a) Distribution of adenosine receptors in the post-mortem human brain: an extended autoradiographic study. Synapse 27:322–335

Svenningsson P, Nomikos GG, Ongini E, Fredholm BB (1997b) Antagonism of adenosine A_{2A} receptors underlies the behavioural activating effect of caffeine and is associated with reduced expression of messenger RNA for NGFI-A and NGFI-B in caudate-putamen and nucleus accumbens. Neuroscience 79:753–764

Tanda G, Loddo P, Frau R, Acquas E, Di Chiara G (1998) Effect of intravenous caffeine on limbic and cortical transmission in the rat: a microdialysis study. 6th International Symposium on Adenosine and Adenine Nucleotides, May 19–24, Ferrara (Italy)

Vellucci SV, Sirinathsinghji DJS, Richardson PJ (1993) Adenosine A_2 receptor regulation of apomorphine-induced turning in rats with unilateral striatal dopamine denervation. Psychopharmacology 111:383–388

Wiesner JB, Ugarkar BG, Castellino AJ, Barankiewicz J, Dumas DP, Gruber HE, Foster AC, Eiron MD (1999) Adenosine kinase inhibitors as a novel approach to anticonvulsant therapy. J Pharmacol Exp Ther 289:1669–1677

Worley CM, Valadez A, Schenk S (1994) Reinstatement of extinguished cocaine-taking behavior by cocaine and caffeine. Pharmacol Biochem Behav 48:217–221

Zetterstrom T, Fillenz M (1990) Adenosine agonists can both inhibit and enhance in vivo striatal dopamine release. Eur J Pharmacol 180(1):137–143

CHAPTER 18
Dopamine – GABA Interactions

M.-F. CHESSELET

A. Introduction

In contrast to γ-aminobutyric acid (GABA)ergic neurons which are ubiquitous in the brain, dopaminergic systems are restricted to a few well-characterized pathways. Dopaminergic cell bodies are for the most part concentrated in the mesencephalon and give rise to three main pathways: the nigrostriatal (or mesostriatal) system, innervating the caudate putamen (or striatum) and other regions of the basal ganglia, the mesolimbic system, innervating the nucleus accumbens and other parts of the limbic system, and the mesocortical pathway, innervating the prefrontal cortex (see CHESSELET 1999). All these dopaminergic systems interact with GABAergic neurons both at the level of their cell bodies and in their terminal regions. However, the mesostriatal system has been more extensively studied because the loss of these dopaminergic neurons leads to Parkinson's disease, and GABAergic neurons normally controlled by nigrostriatal dopamine are thought to play a critical role in the symptoms of the disease (see CHESSELET and DELFS 1996).

This review will highlight some aspects of GABA–dopamine interactions in brain with a particular focus on the nigrostriatal system.

B. The Anatomical Relationship Between Nigrostriatal Dopamine and GABAergic Neurons

I. Substantia Nigra

The cell bodies of the nigrostriatal neurons are concentrated in the substantia nigra pars compacta. This region is immediately adjacent to the substantia nigra pars reticulata containing the cell bodies of GABAergic neurons projecting to the thalamus, superior colliculus, and reticular formation (DENIAU and CHEVALIER 1992). Both regions of the substantia nigra interact directly by way of the dendrites of dopaminergic neurons, which extend deeply into the pars reticulata where they release dopamine (CHERAMY et al. 1981). Conversely, collaterals of GABAergic neurons of the pars reticulata synapse onto dopaminergic neurons (TEPPER et al. 1995). In addition, the dopaminergic neurons receive inputs from GABAergic neurons originating in the striatum and forming the striatonigral pathway (SMITH et al. 1998).

II. Striatum

The main targets of nigrostriatal neurons are the GABAergic efferent neurons of the striatum. These GABAergic neurons represent approximately 95% of striatal neurons and form two distinct populations based on their terminal field and the neuropeptides co-localized with GABA (see CHESSELET 1999). Striatal neurons containing enkephalin project exclusively to the external pallidum (often referred to as the globus pallidus in rats). In contrast, striatal neurons that contain substance P and dynorphin, although they project mainly to the internal pallidum and substantia nigra pars reticulata, also send axon collaterals to the external pallidum, at least in rats (KAWAGUCHI et al. 1990). Therefore, the two GABAergic output pathways of the striatum are not as strictly separated as previously thought (ALBIN et al. 1989).

From the point of view of dopamine–GABA interactions, a main difference between the two systems is that enkephalin-containing GABAergic neurons express primarily dopamine D_2 receptors, whereas substance P-containing neurons express primarily D_1-like receptors (GERFEN et al. 1990), although more recent evidence suggests that the segregation of these receptor subtypes is not absolute (SURMEIER et al. 1996).

In addition to these numerous GABAergic efferent neurons, the striatum contains several classes of GABAergic interneurons. Among these, a population of neurons characterized by the presence of the calcium-binding protein parvalbumin is particularly remarkable because it expresses very high levels of GABA and of the GABA-synthesizing enzyme glutamic acid decarboxylase (GAD) M_r 67,000 (GAD67) (SOGHOMONIAN et al. 1992; KAWAGUCHI et al. 1995). These neurons also express parvalbumin (SOGHOMONIAN et al. 1992; KAWAGUCHI et al. 1995) and the Shaw-like potassium channel Kv3.1 (LENZ et al. 1994). These neurons have rapid firing rates (KAWAGUCHI et al. 1995), suggesting that they may contribute to a significant amount of the GABA released in the striatum.

III. Other Basal Ganglia Regions

GABAergic neurons also form the main output pathways of both the external (globus pallidus in rats) and internal (entopeduncular nucleus) pallidum. It has long been thought that dopaminergic control of these GABAergic neurons was mostly indirect, by way of the striatal output neurons. However, it is now clear that both pallidal segments contain dopaminergic receptors, and recent evidence suggests that they receive collaterals from the nigrostriatal dopaminergic system, suggesting the possibility of direct dopamine–GABA interactions in these regions (see CHESSELET 1999).

The subthalamic nucleus, which is anatomically and functionally part of the basal ganglia, does not contain intrinsic GABAergic neurons but it receives both GABAergic inputs from the globus pallidus and dopaminergic collaterals from the nigrostriatal pathway (PARENT and HAZRATI 1995). Therefore, GABA and dopamine are likely to interact in the control of the output neurons of the subthalamic nucleus, which are glutamatergic.

C. Functional Interactions Between GABA and Nigrostriatal Dopaminergic Neurons

I. Striatum

1. Effects of GABA on Dopaminergic Neurons

The N-methyl-D-aspartate (NMDA)-induced release of dopamine is increased by GABA antagonists, particularly in the striosomal compartment of the striatum (KREBS et al. 1993). A tonic inhibition of dopamine release by endogenous GABA in the striatum is also suggested by evidence that GABA antagonists increase extracellular levels of endogenous dopamine in vivo (GRUEN et al. 1992). An inhibitory effect of GABA on endogenous dopamine release in vivo has been further suggested by positron emission tomography (PET) studies in humans (DEWEY et al. 1992).

It has been proposed that GABA influences the spontaneous release of endogenous dopamine by a direct action on presynaptic GABA B receptors, whereas GABA A receptors may be primarily post-synaptic and their effects on dopamine release mostly indirect (SMOLDERS et al. 1995). However, GABA and GABA A agonists inhibit the evoked release of preloaded ^{3}H-dopamine in striatal synaptosomes, suggesting a direct presynaptic effect (RONKEN et al. 1993).

2. Effects of Dopamine on GABAergic Output Neurons

Measuring GABAergic function in the basal ganglia is particularly difficult because most regions of the basal ganglia contain both intrinsic GABAergic neurons and GABAergic afferents. Biochemical measurements of levels of GABA or GAD activity are not very informative because they do not distinguish between different GABAergic systems (CHESSELET and DELFS 1996; CALON et al. 1999). For this reason, investigators turned to the measurement of receptor binding sites, which are usually regulated in the opposite direction as that of the neurotransmitter input, and of GAD mRNA, which is expressed in neuronal cell bodies and not axon terminals.

Lesions of the nigrostriatal dopaminergic pathway cause an increase in GAD activity in the striatum, and elevate the level of expression of the mRNA encoding GAD67 (but not GAD65) in rats (SOGHOMONIAN et al. 1992; CONSOLO et al. 1999), and of both GAD67 and GAD M_r 65,000 (GAD65) in primates (SOGHOMONIAN et al. 1994; PEDNEAULT and SOGHOMONIAN 1994). This effect does not require stimulation of NMDA receptors (HAJJI et al. 1996). Increases in GAD mRNA after nigrostriatal dopamine lesions are paralleled by increases in GAD immunoreactivity (SOGHOMONIAN et al. 1992; SEGOVIA et al. 1990) and GAD activity. This suggests that dopamine normally exerts an inhibitory effect on striatal GABAergic efferent neurons. Supporting this hypothesis, dopaminergic lesions increase GABA release in the striatum and in the globus pallidus, the brain region that contains the axon terminals of a subpopulation of striatal GABAergic efferent neurons (Tossman et al. 1986; Lindefors et al. 1989). An increase in GABA release from striatopallidal neurons after dopaminergic lesions is also supported by binding studies. Indeed, GABAergic binding sites decrease in the globus pallidus of both rats (Pan et al. 1985) and primates (Robertson et al. 1990) with lesions of nigrostriatal dopaminergic neurons.

Studies of peptides, however, suggest that dopamine may differentially affect striatal efferent neurons projecting to the globus pallidus and entopeduncular nucleus/substantia nigra. Specifically, measurement of peptides and their mRNA indicate that dopamine inhibits efferents to the globus pallidus by acting on D_2 dopaminergic receptors whereas it stimulates neurons projecting to the entopeduncular nucleus and substantia nigra through a D_1-mediated mechanism (see Chesselet 1999). In agreement with a dual effect of dopamine in the regulation of GAD mRNA, indirect evidence discussed in Soghomonian et al. 1992, suggested that the increase in GAD67 mRNA observed after 6-hydroxydopamine lesions was limited to a subset of striatal neurons. This hypothesis has been supported by recent double-label in situ hybridization studies showing that the blockade of dopamine D_2 receptors increased GAD67 mRNA selectively in those striatal neurons projecting to the globus pallidus (Laprade and Soghomonian 1995). When dopaminergic lesions were performed in the neonate, GAD67 was also increased in enkephalinergic neurons. However, in this case, GAD65 was increased in enkephalin-negative neurons, presumably neurons projecting to the internal pallidum/substantia nigra (Laprade and Soghomonian 1999).

Supporting a role for D_2 receptors in the increase in GAD67 mRNA observed after dopaminergic lesions, the neuroleptic haloperidol, administered at a dose that preferentially blocks the D_2 receptor, also increased GAD67 mRNA in rat striatum (Delfs et al. 1995a). Conversely, D_2 agonists decrease both GAD67 and enkephalin mRNA in rat striatum (Caboche et al. 1991). In contrast, administration of D_1 agonists to adult rats selectively increased GAD65 mRNA in striatal neurons projecting to the internal pallidum and substantia nigra pars reticulata (Laprade and Soghomonian 1997) whereas D_1 antagonists decrease both GAD67 and enkephalin mRNA in rat striatum (Caboche et al. 1991).

The striatum is the main target of nigrostriatal dopaminergic neurons, and alterations in striatal output neurons are usually thought to be critical for the symptoms of Parkinson's disease, a neurodegenerative illness characterized by the progressive loss of nigrostriatal neurons. Transplants of dopamine-producing cells in the striatum, which improve motor behavior after dopaminergic lesions, also reverse the increase in GAD activity induced by the lesion (Segovia et al. 1989). However, a direct link between changes in striatal

GABAergic transmission (as evidenced by changes in GAD activity and GAD mRNA) and motor symptoms remains unclear. Indeed, changes in GAD mRNA in the striatum do not parallel the time course of haloperidol-induced catalepsy, a motor symptom similar to the akinesia of patients with Parkinson's disease (OSBORNE et al. 1994; DELFS et al. 1995a). Furthermore, L-dopa, which improves motor deficits secondary to nigrostriatal dopamine lesions, not only does not reverse but rather potentiates the increase in GAD67 mRNA in rat striatum (CONSOLO et al. 1999). Similarly, administration of D_1 agonists further increases GAD67 mRNA in the striatum after neonatal lesions (LAPRADE and SOGHOMONIAN 1999). These data suggest that although changes in striatal GABAergic transmission are likely to contribute to the symptoms of Parkinson's disease, the effects of dopaminergic lesions on other neuronal systems may also be critical. Identifying these effects and their mechanisms will be crucial in developing better treatments for Parkinson's disease.

3. Dopaminergic Regulation of Striatal GABAergic Interneurons

An intriguing consequence of unilateral lesions of the nigrostriatal pathway in rats is that, while it increases GAD67 mRNA in striatal efferent neurons, it decreases GAD67 mRNA in parvalbumin-containing striatal GABAergic interneurons (SOGHOMONIAN et al. 1992). These interneurons are likely to perform critical functions in the striatum. Indeed, evidence suggests that the fast-firing GABAergic interneurons mediate the feed-forward inhibition of striatal efferent neurons by the cerebral cortex (PLENZ and KITAI 1998). Therefore, a decreased GABA production in these interneurons, as suggested by the decrease in GAD mRNA, could contribute to the increased activity of GABAergic efferent neurons after dopaminergic lesions described earlier. The effects of selective dopaminergic antagonists and of L-dopa on these interneurons have yet to be elucidated.

4. Dopaminergic Regulation of Striatal GABA Release

As expected from alterations in striatal GABAergic neurons after dopaminergic lesions, dopamine affects striatal GABA release. This effect is complex, however. Recent studies showed that a dopamine uptake inhibitor, nomifensine, increased GABA release, an effect attenuated by either D_1 or D_2 antagonists (EXPÓSITO et al. 1999), suggesting that endogenous dopamine may stimulate GABA release. This effect is at odds with observations made after dopaminergic lesions (LINDEFORS et al. 1989). Extensive studies of dopaminergic modulation of GABA release in vitro have indeed shown that multiple dopaminergic receptors are involved in this complex regulation. Specifically, stimulation of D_1 receptors increases GABA release in striatal slices; however, D_1 antagonists have no effect, arguing against a role of D_1 receptor in a tonic regulation of GABA release by dopamine in the striatum (WANG and JOHNSON 1995; HARSING and ZIGMOND 1997). In contrast, D_2 agonist decrease while D_2 antagonists increase GABA release, suggesting a tonic regulation of GABA release by D_2-mediated mechanisms (MAYFIELD et al. 1996; HARSING and ZIGMOND 1997).

Interactions between dopaminergic and GABAergic mechanisms in the striatum could be of particular importance for the management of cocaine abuse. Increased dopamine release by cocaine is thought to be critical for drug addiction. Repeated cocaine use decreases the function of striatal GABA A receptors (PERIS 1996) and cocaine-abusing subjects show an increased sensitivity to benzodiazepines (VOLKOW et al. 1998). Interestingly, increasing GABA levels by inhibiting its catabolism attenuates the ability of cocaine to induce dopamine release. This led to the hypothesis that increasing GABAergic function in the striatum could be beneficial in the treatment of cocaine addiction (DEWEY et al. 1997).

II. Dopamine – GABA Interactions in the Globus Pallidus

The basal ganglia comprise a succession of GABAergic neurons contributing to the regulation of motor output by way of the thalamus, superior colliculus, and reticular formation. Therefore, although the major target of the nigrostriatal pathway is the striatum, direct and indirect effects of dopamine on non-striatal GABAergic neurons are likely to be functionally important as well.

As indicated previously, dopamine lesions increase GABA release in the globus pallidus (TOSSMAN et al. 1986), which is likely due to an increased GABA outflow from striatal output neurons. Indeed, it is difficult to distinguish between GABA originating from afferents and from collaterals of intrinsic neurons in release studies. Measurements of GAD mRNA with high cellular resolution in intrinsic GABAergic neurons of the pallidum strongly suggest that alterations in dopaminergic transmission also affect intrinsic pallidal GABAergic neurons. Indeed, unilateral lesions of the nigrostriatal pathway increase GAD mRNA in neurons of the external pallidum in rats and non-human primates (KINCAID et al. 1992; SOGHOMONIAN and CHESSELET 1992; SOGHOMONIAN et al. 1994). This effect was blocked by L-dopa administration in monkeys, probably explaining why it was difficult to detect in post-mortem brains of patients with Parkinson's disease (HERRERO et al. 1996). This suggests that increased GAD mRNA in the external pallidal neurons may reflect changes in GABAergic transmission that are directly related to the motor symptoms of Parkinson's disease.

In support of this hypothesis, changes in GAD mRNA in the globus pallidus in rats parallel the motor symptoms induced by short- and long-term administration of neuroleptics (DELFS et al. 1995a,c). Indeed, short-term administration of haloperidol at a dose that preferentially blocks dopamine D_2 receptors increased GAD67 mRNA in the globus pallidus in rats (DELFS et al. 1995a). In contrast, long-term treatments, which induce orofacial dyskinesia instead of akinesia, decreased GAD67 mRNA in the same region (DELFS et al. 1995c). Furthermore, blockade of haloperidol-induced catalepsy by the cholinergic muscarinic antagonist scopolamine, did not block GAD67 mRNA increases in the striatum, but abolished changes in GAD mRNA in the globus pallidus (DELFS et al. 1995a). A decrease in GAD mRNA in the globus pallidus has been recently reported after higher doses of haloperidol (MAVRIDIS and BESSON 1999). However, as indicated earlier, increases in GAD67 mRNA correlate well with catalepsy. Furthermore, increases in GAD67 mRNA were also observed in the globus pallidus after dopaminergic lesions both in rats and in primates. Taken together, these observations suggest that the increase in GAD67 mRNA observed after low doses of haloperidol in neurons of the external pallidum is more relevant to Parkinson's disease.

Changes in activity of GABAergic neurons in the globus pallidus, as evidenced by changes in GAD mRNA levels, may contribute to the functional consequences of dopamine depletion or dopaminergic receptor blockade. The mechanisms by which decreased dopamine transmission induces these effects is not fully elucidated. A major input to the globus pallidus is formed by the GABAergic, enkephalin-containing striatal ouput neurons. As indicated earlier, all evidence points to an increased activity of this pathway after nigrostriatal lesion or short-term blockade of dopaminergic D_2 receptors (see Sect. C.II., this chapter). This has led to the hypothesis that one of the main consequences of nigrostriatal dopaminergic lesions is a decreased activity of neurons in the globus pallidus (ALBIN et al. 1989; DELONG 1990). Electrophysiological recordings have confirmed a decrease in spontaneous firing in neurons of the globus pallidus after dopaminergic lesions (PAN and WALTERS 1988; FILLION et al. 1991). However, the firing pattern of these neurons is also altered, with an increase in bursting activity (PAN and WALTERS 1989; FILLION et al. 1991). Changes in neuronal patterns are increasingly recognized as critical for synaptic transmission, and bursting patterns are associated with an increased neurotransmitter release (SUAUD-CHAGNY et al. 1992). Therefore, changes in firing patterns in neurons of the globus pallidus could account for the increase in GAD67 mRNA observed in the globus pallidus after dopaminergic lesions.

Changes in the firing pattern of these neurons after dopaminergic lesions could result from the combined effect of several inputs to the globus pallidus. Indeed, in addition to GABAergic inputs from the striatum, the globus pallidus receives glutamatergic inputs from the subthalamic nucleus (PARENT and HAZRATI 1995). It is known that a major consequence of dopaminergic lesions is an increase in the firing rate and bursting pattern of subthalamic neurons, suggesting that this input could play a critical role in the regulation of pallidal neurons after dopamine depletion (BERGMAN et al. 1990). Supporting this hypothesis, lesions of the subthalamic nucleus abolish the increased GAD67 mRNA expression in the globus pallidus in rats with lesions of the dopaminergic nigrostriatal pathway (DELFS et al. 1995b).

Long-term administration of L-dopa, the main treatment for Parkinson's disease, can induce invalidating dyskinesia. This severe motor side effect can also occur after long-term neuroleptic treatment. The mechanism of dyskinesia remains poorly understood. However, recent evidence suggests an involvement of pallidal output neurons. Indeed, dyskinesia resulting from long-term treatment with classical antipsychotic agents, such as haloperidol, are accompanied by a decrease in GAD mRNA in the globus pallidus (DELFS et al. 1995a) and an increase in GABA A binding in the substantia nigra pars reticulata (SEE et al. 1990; SHIRAKAWA and TAMMINGA 1994). These two effects are compatible with a decreased GABAergic output from pallidonigral neurons in dyskinetic rats.

III. GABA – DA Interactions in the Internal Pallidum

The internal pallidal segment (entopeduncular nucleus in rats) constitutes, with the substantia nigra pars reticulata, the main output pathway of the basal ganglia (see CHESSELET 1999). The output neurons of the internal pallidum are GABAergic and project mainly to the thalamic motor nuclei (KULTAS-ILINSKY et al. 1983). Classically, it has been considered that the main GABAergic input to the entopeduncular nucleus originates in the striatum (see SOGHOMONIAN and CHESSELET 1999). However, the internal pallidum also receives GABAergic afferents from the external pallidum (SMITH et al. 1998). The anatomical organization of these inputs suggests that they exert a powerful effect on pallidal neurons. Indeed, they primarily synapse onto the initial segments of dendrites and neuronal soma, whereas GABAergic inputs from the striatum and glutamatergic inputs form the subthalamic nucleus terminate on distal dendrites (SMITH et al. 1998).

The mechanisms by which dopamine influences GABAergic neurons of the internal pallidum are not completely understood. Clearly, dopamine regulates GABA output from the internal pallidum through its effects on D_1-bearing striatal neurons that project directly to the internal pallidum (GERFEN et al. 1990). This projection has received the name "direct pathway" in basal ganglia circuitry (ALBIN et al. 1989). However, this is not the only way dopamine interacts with GABAergic neurons in this region. Dopamine also influences D_2-bearing neurons that project directly to the external pallidum (or globus pallidus) as described above. By regulating the GABAergic neurons of the external pallidum, dopamine indirectly affects the internal pallidum by way (1) of the direct GABAergic connection between external and internal pallidum and (2) of the indirect connection between the two pallidal segments that involve a relay in the subthalamic nucleus, and glutamatergic subthalamic inputs to the internal pallidum. In addition, the internal pallidum contains a high concentration of dopaminergic D_1 receptors and also some D_2-binding sites (see CHESSELET 1999).

The majority of D_1 receptors are located presynaptically on striatopallidal inputs and are likely to control the release of GABA from these neurons. Although the direct dopaminergic innervation of the internal pallidum may be minimal, dopamine released from collaterals of the nigrostriatal pathway terminating into the internal pallidum is likely to be the endogenous ligand at these receptors (LAVOIE et al. 1989). The net effect of a loss of nigrostriatal dopaminergic neurons on GABAergic neurons of the internal pallidum is an increased activity, which is thought to play a critical role in the resulting akinesia (ALBIN et al. 1989; DELONG 1990).

IV. DA – GABA Interactions in the Substantia Nigra

The substantia nigra comprises two adjacent and closely interrelated regions, the pars compacta and the pars reticulata. The pars compacta contains the cell bodies of dopaminergic nigrostriatal neurons. Electrophysiological studies have described different properties of "principal neurons," probably dopaminergic, and "secondary neurons" in the substantia nigra pars compacta (LACEY et al. 1989). Although the neurotransmitter of the "secondary neurons" has not been identified in these studies, the data clearly indicate that the substantia nigra pars compacta does not exclusively contain dopaminergic neurons. Indeed, in the rat, GABAergic neurons are intermingled with the dopaminergic neurons in the pars compacta (EBERLE-WANG et al. 1997; RODR'GUEZ and GONZALES-HERNANDEZ 1999). It is possible that some of these GABAergic neurons project to the striatum because a GABAergic nigrostriatal pathway has been described (RODR'GUEZ and GONZALES-HERNANDEZ 1999). However, some evidence suggests that they may also project to the thalamus, as do GABAergic neurons of the pars reticulata (HERKENHAM and NAUTA 1979).

Whether a direct relationship exists between these two populations of neurons (dopaminergic and GABAergic) in the substantia nigra pars compacta is not known. However, powerful GABAergic influences on nigrostriatal neurons at the level of the substantia nigra are well documented. Dopaminergic neurons receive GABAergic inputs from the striatum and from the globus pallidus (BOLAM and SMITH 1990). Furthermore, they receive inputs from collaterals of GABAergic pars reticulata neurons (TEPPER et al. 1995). GABAergic neurons from the striatum and globus pallidus exert a direct inhibitory influence on nigrostriatal dopaminergic neurons (PALADINI et al. 1999). However, by way of their input on GABAergic pars reticulata neurons, they also modulate dopaminergic neurons indirectly.

Dopaminergic neurons express GABA A receptors, mostly including the alpha3/4beta2/3gamma3 subunits (GUYON et al. 1999). Stimulation of GABA A and GABA B receptors differentially modulate the firing of dopaminergic neurons. In vivo experiments showed that GABA A antagonists induce burst firing in dopaminergic neurons, suggesting that dopaminergic neurons are tonically inhibited by GABAergic afferents through an action on GABA A receptors (PALADINI and TEPPER 1999). GABA B agonists decrease firing rate, regularize firing rhythm, and decrease burst activity in dopaminergic neurons in vivo (ERHARDT et al. 1998). These effects on firing pattern of dopaminergic neurons were observed at much lower doses than firing inhibition (ENGBERG and NISSBRANDT 1993). However, GABA B antagonists only produce a modest shift to a more regular pattern of firing in half the neurons (PALADINI and

TEPPER 1999), suggesting that control of dopaminergic neurons by GABA B mechanisms is not tonic in vivo. This may be due to the fact that GABA B receptors exert their major role in the substantia nigra pars reticulata as pre- rather than postsynaptic receptors (CHAN et al. 1998). Accordingly, GABA B receptors may primarily regulate the firing pattern of dopaminergic nigral neurons by mediating the effects of pallidonigral inputs, because their effects can be mimicked by lesions of the pallidum (TEPPER et al. 1995). The complexity of interactions between GABA inputs and dopaminergic neurons in the substantia nigra may explain why GABA A agonists applied into the substantia nigra can exert disinhibitory effects on dopaminergic neurons and actually increase dopamine release in the striatum (SPERBER et al. 1989; SANTIAGO and WESTERINK 1992). Yet, GABA A antagonists also increase dopamine release, and GABA B agonists decrease it (SANTIAGO and WESTERINK 1992), as does intranigral injection of GABA (REID et al. 1990).

Interactions between GABA and dopamine in the substantia nigra are reciprocal. Dopamine D_1 receptors are present on GABAergic afferents from the striatum and dopamine alters GABA release form these neurons. Although some studies have suggested an inhibitory effect of dopamine on nigral GABA release (MARTIN and WASZCZAK 1994), most evidence points to a facilitatory effect of dopamine on GABA release in substantia nigra through a D_1-mediated action (FLORAN et al. 1990; CAMERON and WILLIAMS 1993; TIMMERMAN and WESTERINK 1995; BYRNES et al. 1997; GARCIA et al. 1997; RADNIKOW and MISGELD 1998; MATUSZEWICH and YAMAMOTO 1999). Regulation of GABA release by stimulation of D_1 receptors in the substantia nigra indirectly alters the release of acetylcholine in the striatum (ABERCROMBIE and DEBOER 1997) and may be involved in the circling behavior induced by intranigral administration of D_1 agonists (STARR and STARR 1989).

D_2 agonists applied into the substantia nigra attenuate the inhibitory response of pars reticulata neurons to GABA (MARTIN and WASZCZAK 1996), and decrease the stimulated release of GABA (MATUSZEWICH and YAMAMOTO 1999) suggesting an opposite effect of dopamine by way of D_1 and D_2 receptors in the substantia nigra, as demonstrated in the striatum. D_2 agonists applied into the substantia nigra increase GABA release in the superior colliculus, which receives GABAergic inputs from the substantia nigra (LANTIN LE BOULCH et al. 1991) consistent with an inhibitory effect of D_2 agonists on GABA release within the substantia nigra. It should be noted that D_4 receptors have been detected by immunohistochemistry in the primate substantia nigra (MRZLJAK et al. 1996). Therefore, some of the effects of D_2/D_4 agonists and antagonists could be due to an action at the D_4 receptors.

V. Functional Implications of GABA – DA Interactions Within the Basal Ganglia

In conclusion, dopamine and GABA interact directly and indirectly at all levels of the basal ganglia. The complexity of the anatomical substrates for these interactions and the multiplicity of receptors involved explain that contradictory results have sometimes been obtained in experimental studies. This complexity offers the possibility of subtle interactions between these neuronal systems for the control of movement. In pathological conditions, this equilibrium is compromised. The best-documented case is that of Parkinson's disease, characterized by the loss of dopaminergic nigrostriatal neurons. In this situation, the dual control of GABAergic output neurons of the striatum is altered and dopaminergic regulation of pallidal and nigral GABAergic neurons, both directly and by way of the subthalamic nucleus, is lost (ALBIN et al. 1989; DELONG 1990). This results in an increased GABAergic output from the basal ganglia, causing akinesia, and an alteration in the firing pattern of these neurons, which may be critical for tremor and dystonia. By elevating the level of remaining dopamine, L-dopa therapy corrects these defects to a certain point, but sensitization of dopaminergic responses and/or the lack of appropriate timing of dopaminergic control may eventually result in invalidating dyskinesia.

A situation analogous to that of Parkinson's disease is observed after treatment with classical antipsychotic drugs that block dopamine receptors, such as haloperidol. Repeated low doses of haloperidol produce the same effects on GABAergic neurons of the basal ganglia as dopaminergic lesions. Chronic treatments, however, induce dyskinesia. Despite these opposite behavioral effects, short- and long-term haloperidol treatments produce similar effects on GABAergic neurons of the striatum but they have opposite effects in the globus pallidus, suggesting that dopaminergic regulation of GABAergic neurons in this region may be critical for the generation of dyskinesia.

D. DA – GABA Interactions in the Mesolimbic Pathway

Dopaminergic neurons of the mesolimbic system originate in the ventral tegmental area and project to the nucleus accumbens. As in the pars compacta, these neurons are intermingled with GABAergic neurons that also project to the nucleus accumbens (KALIVAS et al. 1990). Regulation of dopaminergic neurons of the mesolimbic system has been examined in great detail because evidence suggests that dopaminergic mechanisms in this pathway play a critical role in reinforcement mechanisms (MCBRIDE et al. 1999), drug addiction (HENRY and WHITE 1995), and the rewarding properties of ethanol (DIANA et al. 1993).

The anatomical organization of the mesolimbic system presents many similarities with that of the basal ganglia. Notably, the ventral tegmental area receives GABAergic inputs from the nucleus accumbens and output neurons of the nucleus accumbens shell are GABAergic. These project onto GABAergic neurons of the ventral pallidum, which in turn influence both the ventral tegmental area and the cerebral cortex by way of the thalamus. In this case, however, the main thalamic relay nucleus is the mediodorsal nucleus (BERGER et al. 1991).

GABA agonists in the ventral tegmental area inhibit mesolimbic dopaminergic neurons (SAUD-CHAGNY et al. 1992; WESTERINK et al. 1996) whereas GABA antagonists disinhibit these neurons, leading to an increase in dopamine release in the shell of the nucleus accumbens (WESTERINK et al. 1996; IKEMOTO et al. 1997). In agreement with evidence for a role of dopamine release in nucleus accumbens in addiction, GABA antagonists in the ventral tegmental area have reinforcing properties (IKEMOTO et al. 1997). These data suggest that mesolimbic dopaminergic neurons are tonically inhibited by GABAergic neurons at the level of the ventral tegmental area. Other studies, however, have shown that GABA A agonists injected into the ventral tegmental area increase both dopamine release in the nucleus accumbens and locomotor activity, an effect mediated by increased dopamine in the nucleus accumbens (KALIVAS et al. 1990). In this study, pharmacological characterization of the effects suggested a direct inhibitory effect of GABA by way of GABA B receptor stimulation but an indirect disinhibition of mesolimbic dopamine by stimulation of GABA A receptors in the ventral tegmental area (KALIVAS et al. 1990). GABAergic regulation of mesolimbic dopaminergic neurons may also mediate the complex regulation of these neurons by mu-opioid agonists in the ventral tegmental area (KALIVAS et al. 1990; DEVINE et al. 1993).

Increased dopaminergic transmission in the nucleus accumbens induces locomotor activity (ESSMAN et al. 1993). Evidence suggests that this effect is mediated by an inhibitory action of dopamine on GABAergic output neurons from the nucleus accumbens to the ventral pallidum (YANG and MOGENSON 1989; BOURDELAIS and KALIVAS 1992). Dopaminergic inhibition of GABAergic projections from the nucleus accumbens to the ventral pallidum may also mediate dopamine-induced sensorimotor gating deficits of acoustic startle (SWERDLOW et al. 1990).

A tonic inhibitory effect of dopaminergic neurons on GABAergic output neurons of the nucleus accumbens core is supported by the increase in GAD67 mRNA observed in the core and the anterior part of the nucleus accumbens, but not the shell, after lesions of the mesolimbic dopaminergic pathway in adult rats (RÉTAUX et al. 1994). In contrast, acute cocaine increased GAD mRNA in nucleus accumbens shell, suggesting a differential interaction between dopamine and GABAergic neurons in nucleus accumbens core and shell (SORG et al. 1995).

Dopamine – GABA interactions in the nucleus accumbens are likely to be mediated in part by dopamine D_2 receptors which are located in the dendrites and perikarya of both GABA-immunoreactive spiny neurons with the morphology of output neurons, and interneurons (DELLE DONNE et al. 1997). Interestingly, a greater abundance of D_2 receptors was found on GABA-immunoreactive terminals in the nucleus accumbens shell than in the dorsal striatum, suggesting a significant presynaptic effect of dopamine on GABAergic transmission in this region (DELLE DONNE et al. 1997). In addition to D_2 receptors, GABAergic neurons in the nucleus accumbens respond to stimulation of D_1 dopaminergic receptors, an effect that displays sensitization after repeated administration of drug of abuse such as cocaine (HENRY and WHITE 1995) or morphine (SCHOFFELMEER et al. 1995). In slices of the nucleus accumbens, stimulation of D_1 dopamine receptors increases GABA release, an effect attenuated by concurrent activation of adenosine A1 receptors (MAYFIELD et al. 1999).

A presynaptic effect of GABA on dopamine release in the nucleus accumbens has been less well documented than in the striatum. However, evidence exists in favor of either a stimulatory or an inhibitory effect of GABA on dopamine release, probably because either direct or indirect mechanisms predominate depending on the conditions of the experiment. However, modafinil decreased GABA release in the nucleus accumbens and induced an increase in dopamine release that is blocked by phaclofen, an antagonist of GABA B receptors (FERRARO et al. 1996). These data suggest that GABA may inhibit dopamine release in the accumbens. In support of this, local applications of the GABA agonist muscimol inhibits dopamine release in the nucleus accumbens (YOSHIDA et al. 1997). Furthermore, GABA A antagonists in the nucleus accumbens increase locomotor activity through a mechanism that requires dopamine, also suggesting a tonic inhibition of dopamine in the nucleus accumbens by GABA (WONG et al. 1991). However, the increase in dopamine release induced by local applications of the neuropeptide neurotensin in the nucleus accumbens (which also increases GABA release) is blocked by the GABA antagonist bicuculline suggesting that GABA can increase dopamine release in the nucleus accumbens through a presynaptic mechanism (TANGANELLI et al. 1994).

These data suggest an interaction between GABA and dopamine within the nucleus accumbens, with a net effect depending on the experimental conditions and the systems involved. For example, it has been proposed that the disinhibition of mesolimbic dopaminergic activity by infusion of muscimol into the nucleus accumbens may involve an indirect effect at the level of the cell bodies of these neurons because this treatment increases the immediate early gene c-*fos* in tyrosine-hydroxylase positive neurons of the ventral tegmental area (YOSHIDA et al. 1997).

E. DA – GABA Interactions in the Mesocortical System

Dopaminergic projections to the prefrontal cortex also originate in the ventral tegmental area (BERGER et al. 1991). Although restricted to discrete cortical regions in rats, this mesocortical projection is more extensive in primates. Dopaminergic neurons of the ventral tegmental area are intermingled with GABAergic neurons that project to the same area of the prefrontal cortex that is innervated by dopaminergic mesocortical neurons (STEFFENSEN et al. 1998; PIROT et al. 1992). The prefrontal dopaminergic system is regulated by GABAergic mechanisms in the mediodorsal nucleus of the thalamus. Indeed,

GABA antagonists in the mediodorsal nucleus activate dopaminergic transmission in cortical regions that receive inputs form this thalamic nucleus (JONES et al. 1988), whereas GABA agonists have the opposite effect (CHURCHILL et al. 1996).

The mesocortical dopaminergic pathway is selectively activated by stress (DEUTCH et al. 1991) and has been shown to play a role in working memory (GOLDMAN-RAKIC 1999; ROMANIDES et al. 1999). It can also control locomotor activity and movement, and the subcortical response to stress, by controlling dopaminergic mechanisms in subcortical structures such as the striatum and nucleus accumbens (CHURCHILL et al. 1996; KARLER et al. 1998; DOHERTY and GRATTON 1999). It has been proposed that GABAergic mechanisms play a role in the activation of the mesocortical dopaminergic system by stress. Indeed, benzodiazepines, which have GABA antagonistic effects, block stress-induced increases in cortical dopamine release (DEUTCH et al. 1991; FINLAY et al. 1995). However, this effect is due to a decrease in basal concentration of dopamine and the net outflow of the amine is not reduced (FINLAY et al. 1995). Handling stress also increases dopamine release in the prefrontal cortex, an effect attenuated by infusion of the GABA B agonist baclofen in the ventral tegmental area (ENRICO et al. 1998).

In the prefrontal cortex, dopamine can regulate pyramidal neurons not only directly, but also by way of its effects on GABAergic interneurons (KEVERNE 1999). Expression of dopaminergic receptors has been detected in subpopulations of cortical GABAergic interneurons. In particular, D_1 dopaminergic receptors are not only present in pyramidal neurons, but also in parvalbumin-positive GABAergic interneurons (LE MOINE and GASPAR 1998). They are present on the plasma membrane of distal dendrites of these interneurons (MULY et al. 1998). In addition, they are present presynaptically on axon terminals forming asymmetric synapses, thus presumably inhibitory. D_2 dopamine receptors are also primarily found in the parvalbumin-positive GABAergic neurons, whereas only a small subpopulation of calbindin-containing GABAergic interneurons express D_1 dopaminergic receptor (MULY et al. 1998). There is also ultrastructural evidence for direct synaptic contacts between dopaminergic nerve terminals and parvalbumin-positive GABAergic interneurons in the prefrontal cortex (SESACK et al. 1998). These GABAergic interneurons correspond to the wide arbor and chandelier neurons that target pyramidal cell soma and axon initial segments, respectively.

Lesions of dopaminergic afferents to the prefrontal cortex decrease GAD67 mRNA in GABAergic interneurons located in deep cortical layers, suggesting a tonic excitatory effect of dopamine on these interneurons (RETAUX et al. 1994). It is interesting to note that lesions of dopaminergic output neurons from the ventral tegmental area (VTA) increases GAD67 mRNA in efferent GABAergic neurons of the nucleus accumbens but decreases GAD67 mRNA in GABAergic interneurons of the cerebral cortex (RETAUX et al. 1994). This dual effect of dopamine on output versus interneurons is reminiscent of observations in the striatum after lesions of the nigrostriatal system, which leads to increased GAD mRNA in efferent neurons and decrease of the mRNA in interneurons of the striatum, respectively (SOGHOMONIAN et al. 1992).

Local application of dopamine into the prefrontal cortex of rats inhibits most cortical efferent neurons. This effect is blocked by D_2 and GABA A antagonists, suggesting it is mediated by stimulation of cortical GABAergic interneurons through an action of dopamine on D_2 dopaminergic receptors (PIROT et al. 1992). Although electrically evoked release of GABA in vitro was decreased by D_2 agonists (RETAUX et al. 1991a) through a synergistic stimulation of D_1 and D_2 receptor activation (RETAUX et al. 1991b), D_2 agonists increase spontaneous GABA release in slices of rat prefrontal cortex (RETAUX et al. 1991) and in vivo (GROBIN and DEUTCH 1998). In agreement with a stimulatory effect of dopamine on GABAergic interneurons in prefrontal cortex, dopamine enhances inhibitory neurons excitability through depolarization and increased frequency and amplitude of spontaneous inhibitory postsynaptic currents in both interneurons and pyramidal cells in the absence of tetrodotoxin (ZHOU and HABLITZ 1999).

By stimulating GABA interneurons in the cerebral cortex, dopamine may inhibit cortical output neurons and influence subcortical structures. A recent study suggests that, by acting on GABA B receptors, GABA released by dopamine in prefrontal cortex inhibits dopamine release induced by stress in the nucleus accumbens (DOHERTY and GRATTON 1999). Similarly, increased dopamine in the prefrontal cortex in response to administration of amphetamine or cocaine decreases dopaminergic and glutamatergic activity in the striatum by way of the activation of GABAergic neurons in the cerebral cortex (KARLER et al. 1998).

In conclusion, a general theme that emerges from the numerous studies that have examined GABA–DA interactions is that these neurotransmitters are engaged in complex mutual regulations involving a variety of receptors, often leading to opposite effects.

References

Abercrombie ED, DeBoer P (1997) Substantia nigra D_1 receptors and stimulation of striatal cholinergic interneurons by dopamine: a proposed circuit mechanism. J Neurosci 17:8498–8505

Albin RL, Young AB, Penney JB (1989) The functional anatomy of basal ganglia disorders [see comments]. Trends Neurosci 12:366–375

Berger B, Gaspar P, Verney C (1991) Dopaminergic innervation of the cerebral cortex: unexpected differences between rodents and primates [published erratum appears in Trends Neurosci 1991 Mar;14(3):119]. Trends Neurosci 14:21–27

Bergman H, Wichmann T, DeLong MR (1990) Reversal of experimental parkinsonism by lesions of the subthalamic nucleus. Science 249:1436–1438

Bolam JP, Smith Y (1990) The GABA and substance P input to dopaminergic neurones in the substantia nigra of the rat. Brain Res 529:57–78

Bourdelais AJ, Kalivas PW (1992) Modulation of extracellular gamma-aminobutyric acid in the ventral pallidum using in vivo microdialysis. J Neurochem 58:2311–2320

Byrnes EM, Reilly A, Bruno JP (1997) Effects of AMPA and D1 receptor activation on striatal and nigral GABA efflux. Synapse 26:254–268

Caboche J, Vernier P, Julien JF, Rogard M, Mallet J, Besson MJ (1991) Parallel decrease of glutamic acid decarboxylase and preproenkephalin mRNA in the rat striatum following chronic treatment with a dopaminergic D1 antagonist and D2 agonist. J Neurochem 56:428–435

Calon F, Morissette M, Goulet M, Grondin R, Blanchet PJ, Bédard PJ, D i Paolo T (1999) Chronic D1 and D2 dopaminomimetic treatment of MPTP-denervated monkeys: effects on basal ganglia GABA(A)/benzodiazepine receptor complex and GABA content. Neurochem Intl 35:81–91

Cameron DL, Williams JT (1993) Dopamine D1 receptors facilitate transmitter release. Nature 366:344–347

Chan PK, Leung CK, Yung WH (1998) Differential expression of pre- and postsynaptic GABA(B) receptors in rat substantia nigra pars reticulata neurones. Eur J Pharm 349:187–197

Cheramy A, Leviel V, Glowinski J (1981) Dendritic release of dopamine in the substantia nigra. Nature 289:537–542

Chesselet MF, Delfs JM (1996) Basal ganglia and movement disorders: an update [see comments]. Trends Neurosci 19:417–422

Chesselet M-F (1999) Mapping the basal ganglia. In: Toga AW, Mazziotta JC (eds) "Brain Mapping: The Applications," Academic Press. In press

Churchill L, Zahm DS, Duffy P, Kalivas PW (1996) The mediodorsal nucleus of the thalamus in rats – II. Behavioral and neurochemical effects of GABA agonists. Neurosci 70:103–112

Consolo S, Morelli M, Rimoldi M, Giorgi S, Di Chiara G (1999) Increased striatal expression of glutamate decarboxylase 67 after priming of 6-hydroxydopamine-lesioned rats. Neurosci 89:1183–1187

Delfs JM, Anegawa NJ, Chesselet MF (1995a) Glutamate decarboxylase messenger RNA in rat pallidum: comparison of the effects of haloperidol, clozapine and combined haloperidol-scopolamine treatments. Neurosci 66:67–80

Delfs JM, Ciaramitaro VM, Parry TJ, Chesselet MF (1995b) Subthalamic nucleus lesions: widespread effects on changes in gene expression induced by nigrostriatal dopamine depletion in rats. J Neurosci 15:6562–6575

Delfs JM, Ellison GD, Mercugliano M, Chesselet MF (1995c) Expression of glutamic acid decarboxylase mRNA in striatum and pallidum in an animal model of tardive dyskinesia. Exp Neurol 133:175–188

Delle Donne KT, Sesack SR, Pickel VM (1997) Ultrastructural immunocytochemical localization of the dopamine D2 receptor within GABAergic neurons of the rat striatum. Brain Res 746:239–255

DeLong MR (1990) Primate models of movement disorders of basal ganglia origin. Trends Neurosci 13:281–285

Deniau JM, Chevalier G (1992) The lamellar organization of the rat substantia nigra pars reticulata: distribution of projection neurons. Neurosci 46:361–377

Deutch AY, Lee MC, Gillham MH, Cameron DA, Goldstein M, Iadarola MJ (1991) Stress selectively increases fos protein in dopamine neurons innervating the prefrontal cortex. Cereb Ctx 1:273–292

Devine DP, Leone P, Wise RA (1993) Mesolimbic dopamine neurotransmission is increased by administration of mu-opioid receptor antagonists. Eur J Pharm 243:55–64

Dewey SL, Smith GS, Logan J, Brodie JD, Yu DW, Ferrieri RA, King PT, MacGregor RR, Martin TP, Wolf AP, et al. (1992) GABAergic inhibition of endogenous dopamine release measured in vivo with 11C-raclopride and positron emission tomography. J Neurosci 12:3773–3780

Dewey SL, Chaurasia CS, Chen CE, Volkow ND, Clarkson FA, Porter SP, Straughter-Moore RM, Alexoff DL, Tedeschi D, Russo NB, Fowler JS, Brodie JD (1997) GABAergic attenuation of cocaine-induced dopamine release and locomotor activity. Synapse 25:393–398

Diana M, Rossetti ZL, Gessa G (1993) Rewarding and aversive effects of ethanol: interplay of GABA, glutamate and dopamine. Alcohol Alcoholism Supp 2:315–319

Doherty MD, Gratton A (1999) Effects of medial prefrontal cortical injections of GABA receptor agonists and antagonists on the local and nucleus accumbens dopamine responses to stress. Synapse 32:288–300

Eberle-Wang K, Mikeladze Z, Uryu K, Chesselet MF (1997) Pattern of expression of the serotonin2C receptor messenger RNA in the basal ganglia of adult rats. J Comp Neurol 384:233–247

Engberg G, Nissbrandt H (1993) gamma-Hydroxybutyric acid (GHBA) induces pacemaker activity and inhibition of substantia nigra dopamine neurons by activating GABAB-receptors. Naunyn-Schmiedebergs Archives Of Pharmacology 348:491–497

Enrico P, Bouma M, de Vries JB, Westerink BH (1998) The role of afferents to the ventral tegmental area in the handling stress-induced increase in the release of dopamine in the medial prefrontal cortex: a dual-probe microdialysis study in the rat brain. Brain Res 779:205–213

Erhardt S, Andersson B, Nissbrandt H, Engberg G (1998) Inhibition of firing rate and changes in the firing pattern of nigral dopamine neurons by gamma-hydroxybutyric acid (GHBA) are specifically induced by activation of GABA(B) receptors. Naunyn-Schmiedebergs Archives Of Pharmacology 357:611–619

Essman WD, McGonigle P, Lucki I (1993) Anatomical differentiation within the nucleus accumbens of the locomotor stimulatory actions of selective dopamine agonists and d-amphetamine. Psychopharmacol 112:233–241

Expóto I, Del Arco A, Segovia G, Mora F (1999) Endogenous dopami ne increases extracellular concentrations of glutamate and GABA in striatum of the freely moving rat: involvement of D1 and D2 dopamine receptors. Neurochem Res 24:849–856

Ferraro L, Tanganelli S, O'Connor WT, Antonelli T, Rambert F, Fuxe K (1996) The vigilance promoting drug modafinil increases dopamine release in the rat nucleus accumbens via the involvement of a local GABAergic mechanism. Eur J Pharm 306:33–39

Filion M, Tremblay L (1991) Abnormal spontaneous activity of globus pallidus neurons in monkeys with MPTP-induced parkinsonism. Brain Res 547:142–151

Finlay JM, Zigmond MJ, Abercrombie ED (1995) Increased dopamine and norepinephrine release in medial prefrontal cortex induced by acute and chronic stress: effects of diazepam. Neurosci 64:619–628

Floran B, Aceves J, Sierra A, Martinez-Fong D (1990) Activation of D1 dopamine receptors stimulates the release of GABA in the basal ganglia of the rat. Neurosci Let 116:136–140

Garcia M, Floran B, Arias-Montañ JA, Young JM, Aceves J (1997) His tamine H3 receptor activation selectively inhibits dopamine D1 receptor-dependent [3H]GABA release from depolarization-stimulated slices of rat substantia nigra pars reticulata. Neurosci 80:241–249

Gerfen CR, Engber TM, Mahan LC, Susel Z, Chase TN, Monsma FJ Jr, Sibley DR (1990) D1 and D2 dopamine receptor-regulated gene expression of striatonigral and striatopallidal neurons [see comments]. Science 250:1429–1432

Goldman-Rakic PS (1999) The psychic neuron of the cerebral cortex. Ann NY Acad Sci 868:13–26

Grobin AC, Deutch AY (1998) Dopaminergic regulation of extracellular gamma-aminobutyric acid levels in the prefrontal cortex of the rat. J Pharm & Exp Ther 285:350–357

Gruen RJ, Friedhoff AJ, Coale A, Moghaddam B (1992) Tonic inhibition of striatal dopamine transmission: effects of benzodiazepine and GABAA receptor antagonists on extracellular dopamine levels. Brain Res 599:51–56

Guyon A, Laurent S, Paupardin-Tritsch D, Rossier J, Eugène D (1999) Incremental conductance levels of GABAA receptors in dopaminergic neurones of the rat substantia nigra pars compacta. J Phys 516(Pt 3):719–737

Hajji MD, Salin P, Kerkerian-Le Goff L (1996) Repeated injections of dizocilpine maleate (MK-801) do not suppress the effects of nigrostriatal dopamine deafferentation on glutamate decarboxylase (GAD67) mRNA expression in the adult rat striatum. Brain Res Molec Brain Res 43:219–224

Harsing LGJ, Zigmond MJ (1997) Influence of dopamine on GABA release in striatum: evidence for D1-D2 interactions and non-synaptic influences. Neurosci 77: 419–429

Henry DJ, White FJ (1995) The persistence of behavioral sensitization to cocaine parallels enhanced inhibition of nucleus accumbens neurons. J Neurosci 15:6287–6299

Herkenham M, Nauta WJ (1979) Efferent connections of the habenular nuclei in the rat. J Comp Neurol 187:19–47

Herrero MT, Levy R, Ruberg M, Javoy-Agid F, Luquin MR, Agid Y, Hirsch EC, Obeso JA (1996) Glutamic acid decarboxylase mRNA expression in medial and lateral pallidal neurons in the MPTP-treated monkey and patients with Parkinson's disease. Adv Neurol 69:209–216

Ikemoto S, Murphy JM, McBride WJ (1997) Self-infusion of GABA(A) antagonists directly into the ventral tegmental area and adjacent regions. Behav Neurosci 111: 369–380

Jones MW, Kilpatrick IC, Phillipson OT (1988) Dopamine function in the prefrontal cortex of the rat is sensitive to a reduction of tonic GABA-mediated inhibition in the thalamic mediodorsal nucleus. Exp Brain Res 69:623–634

Kalivas PW, Duffy P, Eberhardt H (1990) Modulation of A10 dopamine neurons by gamma-aminobutyric acid agonists. J Pharm & Exp Therap 253:858–866

Karler R, Calder LD, Thai DK, Bedingfield JB (1998) The role of dopamine and GABA in the frontal cortex of mice in modulating a motor-stimulant effect of amphetamine and cocaine. Pharm, Biochem & Behav 60:237–244

Kawaguchi Y, Wilson CJ, Emson PC (1990) Projection subtypes of rat neostriatal matrix cells revealed by intracellular injection of biocytin. J Neurosci 10:3421–3438

Kawaguchi Y, Wilson CJ, Augood SJ, Emson PC (1995) Striatal interneurones: chemical, physiological and morphological characterization [published erratum appears in Trends Neurosci 1996 Apr;19(4):143]. Trends Neurosci 18:527–535

Keverne EB (1999) GABA-ergic neurons and the neurobiology of schizophrenia and other psychoses. Brain Res Bull 48:467–473

Kincaid AE, Albin RL, Newman SW, Penney JB, Young AB (1992) 6-Hydroxydopamine lesions of the nigrostriatal pathway alter the expression of glutamate decarboxylase messenger RNA in rat globus pallidus projection neurons. Neurosci 51:705–718

Krebs MO, Kemel ML, Gauchy C, Desban M, Glowinski J (1993) Local GABAergic regulation of the N-methyl-D-aspartate-evoked release of dopamine is more prominent in striosomes than in matrix of the rat striatum. Neurosci 57:249–260

Kultas-Ilinsky K, Ilinsky I, Warton S, Smith KR (1983) Fine structure of nigral and pallidal afferents in the thalamus: an EM autoradiography study in the cat. J Comp Neurol 216:390–405

Lacey MG, Mercuri NB, North RA (1989) Two cell types in rat substantia nigra zona compacta distinguished by membrane properties and the actions of dopamine and opioids. J Neurosci 9:1233–1241

Lantin Le Boulch N, Truong-Ngoc NA, Gauchy C (1991) Role of dendritic dopamine of the substantia nigra in the modulation of nigrocollicular gamma-aminobutyric acid release: in vivo studies in the rat. J Neurochem 57:1080–1083

Laprade N, Soghomonian JJ (1995) Differential regulation of mRNA levels encoding for the two isoforms of glutamate decarboxylase (GAD65 and GAD67) by dopamine receptors in the rat striatum. Brain Res Molec Brain Res 34:65–74

Laprade N, Soghomonian JJ (1997) Glutamate decarboxylase (GAD65) gene expression is increased by dopamine receptor agonists in a subpopulation of rat striatal neurons. Brain Res Molec Brain Res 48:333–345

Laprade N, Soghomonian JJ (1999) Gene expression of the GAD67 and GAD65 isoforms of glutamate decarboxylase is differentially altered in subpopulations of striatal neurons in adult rats lesioned with 6-OHDA as neonates. Synapse 33:36–48

Lavoie B, Smith Y, Parent A (1989) Dopaminergic innervation of the basal ganglia in the squirrel monkey as revealed by tyrosine hydroxylase immunohistochemistry. J Comp Neurol 289:36–52

Le Moine C, Gaspar P (1998) Subpopulations of cortical GABAergic interneurons differ by their expression of D1 and D2 dopamine receptor subtypes. Brain Res Molec Brain Res 58:231–236

Lenz S, Perney TM, Qin Y, Robbins E, Chesselet MF (1994) GABA-ergic interneurons of the striatum express the Shaw-like potassium channel Kv3.1. Synapse 18:55–66

Lindefors N, Brodin E, Tossman U, Segovia J, Ungerstedt U (1989) Tissue levels and in vivo release of tachykinins and GABA in striatum and substantia nigra of rat brain after unilateral striatal dopamine denervation. Exp Brain Res 74:527–534

Martin LP, Waszczak BL (1994) D1 agonist-induced excitation of substantia nigra pars reticulata neurons: mediation by D1 receptors on striatonigral terminals via a pertussis toxin-sensitive coupling pathway. J Neurosci 14:4494–4506

Martin LP, Waszczak BL (1996) Dopamine D2, receptor-mediated modulation of the GABAergic inhibition of substantia nigra pars reticulata neurons. Brain Res 729:156–169

Matuszewich L, Yamamoto BK (1999) Modulation of GABA release by dopamine in the substantia nigra. Synapse 32:29–36

Mavridis M, Besson MJ (1999) Dopamine-opiate interaction in the regulation of neostriatal and pallidal neuronal activity as assessed by opioid precursor peptides and glutamate decarboxylase messenger RNA expression. Neurosci 92:945–966

Mayfield RD, Larson G, Orona RA, Zahniser NR (1996) Opposing actions of adenosine A2a and dopamine D2 receptor activation on GABA release in the basal ganglia: evidence for an A2a/D2 receptor interaction in globus pallidus. Synapse 22:132–138

Mayfield RD, Jones BA, Miller HA, Simosky JK, Larson GA, Zahniser NR (1999) Modulation of endogenous GABA release by an antagonistic adenosine A1/dopamineD1 receptor interaction in rat brain limbic regions but not basal ganglia. Synapse 33:274–281

McBride WJ, Murphy JM, Ikemoto S (1999) Localization of brain reinforcement mechanisms: intracranial self-administration and intracranial place-conditioning studies. Behav Brain Res 101:129–152

Mrzljak L, Bergson C, Pappy M, Huff R, Levenson R, Goldman-Rakic PS (1996) Localization of dopamine D4 receptors in GABAergic neurons of the primate brain. Nature 381:245–248

Muly ECR, Szigeti K, Goldman-Rakic PS (1998) D1 receptor in interneurons of macaque prefrontal cortex: distribution and subcellular localization. J Neurosci 18:10553–10565

Osborne PG, O'Connor WT, Beck O, Ungerstedt U (1994) Acute versus chronic haloperidol: relationship between tolerance to catalepsy and striatal and accumbens dopamine, GABA and acetylcholine release. Brain Res 634:20–30

Paladini CA, Celada P, Tepper JM (1999) Striatal, pallidal, and pars reticulata evoked inhibition of nigrostriatal dopaminergic neurons is mediated by GABA(A) receptors in vivo. Neurosci 89:799–812

Paladini CA, Tepper JM (1999) GABA(A) and GABA(B) antagonists differentially affect the firing pattern of substantia nigra dopaminergic neurons in vivo. Synapse 32:165–176

Pan HS, Penney JB, Young AB (1985) Gamma-aminobutyric acid and benzodiazepine receptor changes induced by unilateral 6-hydroxydopamine lesions of the medial forebrain bundle. J Neurochem 45:1396–1404

Pan HS, Walters JR (1988) Unilateral lesion of the nigrostriatal pathway decreases the firing rate and alters the firing pattern of globus pallidus neurons in the rat. Synapse 2:650–656

Parent A, Hazrati LN (1995) Functional anatomy of the basal ganglia. II. The place of subthalamic nucleus and external pallidum in basal ganglia circuitry. Brain Res Brain Res Rev 20:128–154

Pedneault S, Soghomonian JJ (1994) Glutamate decarboxylase (GAD65) mRNA levels in the striatum and pallidum of MPTP-treated monkeys. Brain Res Molec Brain Res 25:351–354

Peris J (1996) Repeated cocaine injections decrease the function of striatal gamma-aminobutyric acid(A) receptors. J Pharmacology & Exp Therap 276:1002–1008

Pirot S, Godbout R, Mantz J, Tassin JP, Glowinski J, Thierry AM (1992) Inhibitory effects of ventral tegmental area stimulation on the activity of prefrontal cortical neurons: evidence for the involvement of both dopaminergic and GABAergic components. Neurosci 49:857–865

Plenz D, Kitai S (1998) Up and down States in Striatal Medium Spiny Neurons Simultaneously Recorded with Spontaneous Activity in Fast-Spiking Interneurons Studied in Cortex-Striatum-Substantia Nigra Organotypic Cultures. J Neurosci 18:266–283

Radnikow G, Misgeld U (1998) Dopamine D1 receptors facilitate GABAA synaptic currents in the rat substantia nigra pars reticulata. J Neurosci 18:2009–2016

Reid MS, O'Connor WT, Herrera-Marschitz M, Ungerstedt U (1990) The effects of intranigral GABA and dynorphin A injections on striatal dopamine and GABA release: evidence that dopamine provides inhibitory regulation of striatal GABA neurons via D2 receptors. Brain Res 519:255–260

Rétaux S, Besson MJ, Penit-Soria J (1991b) Synergism between D1 a nd D2 dopamine receptors in the inhibition of the evoked release of [3H]GABA in the rat prefrontal cortex. Neurosci 43:323–329

Rétaux S, Besson MJ, Penit-Soria J (1991a) Opposing effects of do pamine D2 receptor stimulation on the spontaneous and the electrically evoked release of [3H]GABA on rat prefrontal cortex slices. Neurosci 42:61–71

Rétaux S, Trovero F, Besson MJ (1994) Role of dopamine in the pla sticity of glutamic acid decarboxylase messenger RNA in the rat frontal cortex and the nucleus accumbens. Eur J Neurosci 6:1782–1791

Robertson RG, Clarke CA, Boyce S, Sambrook MA, Crossman AR (1990) The role of striatopallidal neurones utilizing gamma-aminobutyric acid in the pathophysiology of MPTP-induced parkinsonism in the primate: evidence from [3H]flunitrazepam autoradiography. Brain Res 531:95–104

Rodrǵuez M, Gonzlèz-Hernńdez T (1999) Electrophysiological an d morphological evidence for a GABAergic nigrostriatal pathway. J Neurosci 19:4682–4694

Romanides AJ, Duffy P, Kalivas PW (1999) Glutamatergic and dopaminergic afferents to the prefrontal cortex regulate spatial working memory in rats. Neurosci 92:97–106

Ronken E, Mulder AH, Schoffelmeer AN (1993) Interacting presynaptic kappa-opioid and GABAA receptors modulate dopamine release from rat striatal synaptosomes. J Neurochem 61:1634–1639

Santiago M, Westerink BH (1992) The role of GABA receptors in the control of nigrostriatal dopaminergic neurons: dual-probe microdialysis study in awake rats. Eur J Pharm 219:175–181

Schoffelmeer AN, De Vries TJ, Vanderschuren LJ, Tjon GH, Nestby P, Wardeh G, Mulder AH (1995) Glucocorticoid receptor activation potentiates the morphine-induced adaptive increase in dopamine D-1 receptor efficacy in gamma-aminobutyric acid neurons of rat striatum/nucleus accumbens. J Pharmacology & Exp Therap 274:1154–1160

See RE, Toga AW, Ellison G (1990) Autoradiographic analysis of regional alterations in brain receptors following chronic administration and withdrawal of typical and atypical neuroleptics in rats. J Neural Trans General Section 82:93–109

Segovia J, Meloni R, Gale K (1989) Effect of dopaminergic denervation and transplant-derived reinnervation on a marker of striatal GABAergic function. Brain Res 493:185–189

Segovia J, Tillakaratne NJ, Whelan K, Tobin AJ, Gale K (1990) Parallel increases in striatal glutamic acid decarboxylase activity and mRNA levels in rats with lesions of the nigrostriatal pathway. Brain Res 529:345–348

Sesack SR, Hawrylak VA, Melchitzky DS, Lewis DA (1998) Dopamine innervation of a subclass of local circuit neurons in monkey prefrontal cortex: ultrastructural analysis of tyrosine hydroxylase and parvalbumin immunoreactive structures. Cereb Ctx 8:614–622

Shirakawa O, Tamminga CA (1994) Basal ganglia GABAA and dopamine D1 binding site correlates of haloperidol-induced oral dyskinesias in rat. Exp Neurol 127: 62–69

Smith Y, Bolam JP (1990) The output neurones and the dopaminergic neurones of the substantia nigra receive a GABA-containing input from the globus pallidus in the rat. J Comp Neurol 296:47–64

Smith Y, Bevan MD, Shink E, Bolam JP (1998) Microcircuitry of the direct and indirect pathways of the basal ganglia. Neurosci 86:353–387

Smolders I, De Klippel N, Sarre S, Ebinger G, Michotte Y (1995) Tonic GABA-ergic modulation of striatal dopamine release studied by in vivo microdialysis in the freely moving rat. Eur J Pharm 284:83–91

Soghomonian JJ, Chesselet MF (1992) Effects of nigrostriatal lesions on the levels of messenger RNAs encoding two isoforms of glutamate decarboxylase in the globus pallidus and entopeduncular nucleus of the rat. Synapse 11:124–133

Soghomonian JJ, Gonzales C, Chesselet MF (1992) Messenger RNAs encoding glutamate-decarboxylases are differentially affected by nigrostriatal lesions in subpopulations of striatal neurons. Brain Res 576:68–79

Soghomonian JJ, Pedneault S, Audet G, Parent A (1994) Increased glutamate decarboxylase mRNA levels in the striatum and pallidum of MPTP-treated primates. J Neurosci 14:6256–6265

Soghomonian JJ, Chesselet M-F (in press) GABA in the Basal Ganglia. In: DLM, RWO (eds) "GABA in the Nervous System," Lippincott Williams & Willkins, Philadelphia

Sorg BA, Guminski BJ, Hooks MS, Kalivas PW (1995) Cocaine alters glutamic acid decarboxylase differentially in the nucleus accumbens core and shell. Brain Res Molec Brain Res 29:381–386

Sperber EF, Wurpel JN, Sharpless NS, MoshéSL (1989) Intranigral G ABAergic drug effects on striatal dopamine activity. Pharm, Biochem & Behav 32:1067–1070

Starr MS, Starr BS (1989) Circling evoked by intranigral SKF 38393: a GABA-mediated D-1 response? Pharm, Biochem & Behav 32:849–851

Steffensen SC, Svingos AL, Pickel VM, Henriksen SJ (1998) Electrophysiological characterization of GABAergic neurons in the ventral tegmental area. J Neurosci 18: 8003–8015

Suaud-Chagny MF, Chergui K, Chouvet G, Gonon F (1992) Relationship between dopamine release in the rat nucleus accumbens and the discharge activity of dopaminergic neurons during local in vivo application of amino acids in the ventral tegmental area. Neurosci 49:63–72

Surmeier DJ, Song WJ, Yan Z (1996) Coordinated expression of dopamine receptors in neostriatal medium spiny neurons. J Neurosci 16:6579–6591

Swerdlow NR, Braff DL, Geyer MA (1990) GABAergic projection from nucleus accumbens to ventral pallidum mediates dopamine-induced sensorimotor gating deficits of acoustic startle in rats. Brain Res 532:146–150

Tanganelli S, O'Connor WT, Ferraro L, Bianchi C, Beani L, Ungerstedt U, Fuxe K (1994) Facilitation of GABA release by neurotensin is associated with a reduction of dopamine release in rat nucleus accumbens. Neurosci 60:649–657

Tepper JM, Martin LP, Anderson DR (1995) GABAA receptor-mediated inhibition of rat substantia nigra dopaminergic neurons by pars reticulata projection neurons. J Neurosci 15:3092–3103

Timmerman W, Westerink BH (1995) Extracellular gamma-aminobutyric acid in the substantia nigra reticulata measured by microdialysis in awake rats: effects of various stimulants. Neurosci Lett 197:21–24

Tossman U, Segovia J, Ungerstedt U (1986) Extracellular levels of amino acids in striatum and globus pallidus of 6-hydroxydopamine-lesioned rats measured with microdialysis. Acta Physiologica Scandinavica 127:547–551

Volkow ND, Wang GJ, Fowler JS, Hitzemann R, Gatley SJ, Dewey SS, Pappas N (1998) Enhanced sensitivity to benzodiazepines in active cocaine-abusing subjects: a PET study. Amer J Psychia 155:200–206

Wang J, Johnson KM (1995) Regulation of striatal cyclic-3',5'-adenosine monophosphate accumulation and GABA release by glutamate metabotropic and dopamine D1 receptors. J Pharm & Exper Therap 275:877–884

Westerink BH, Kwint HF, deVries JB (1996) The pharmacology of mesolimbic dopamine neurons: a dual-probe microdialysis study in the ventral tegmental area and nucleus accumbens of the rat brain. J Neurosci 16:2605–2611

Wong LS, Eshel G, Dreher J, Ong J, Jackson DM (1991) Role of dopamine and GABA in the control of motor activity elicited from the rat nucleus accumbens. Pharm, Biochem & Behav 38:829–835

Yang CR, Mogenson GJ (1989) Ventral pallidal neuronal responses to dopamine receptor stimulation in the nucleus accumbens. Brain Res 489:237–246

Yoshida M, Yokoo H, Nakahara K, Tomita M, Hamada N, Ishikawa M, Hatakeyama J, Tanaka M, Nagatsu I (1997) Local muscimol disinhibits mesolimbic dopaminergic activity as examined by brain microdialysis and Fos immunohistochemistry. Brain Res 767:356–360

Zhou FM, Hablitz JJ (1999) Dopamine modulation of membrane and synaptic properties of interneurons in rat cerebral cortex. J Neurophys 81:967–976

CHAPTER 19
Dopamine – Its Role in Behaviour and Cognition in Experimental Animals and Humans

T.W. ROBBINS and B.J. EVERITT

A. Introduction

The importance of dopaminergic transmission for normal behaviour has been evident since the initial characterization of the organization and functioning of the dopamine (DA) pathways, as well as the subsequent discovery and mapping of the DA receptor systems, comprising the D_1-like receptors (i.e. D_1, D_4 and D_5 receptor subtypes) and the $D_{2/3}$-like receptors. The search for functional correlates of DA function has been given great impetus by its undoubted involvement in Parkinson's disease, in the mediation of reinforcing effects of drugs of abuse such as the amphetamine-like psychomotor stimulants and in the anti-psychotic effects of neuroleptic drugs. The purpose of this chapter is to build upon the syntheses provided by several previous reviews and to reach conclusions about the nature of the contribution of DA neurotransmission to behaviour, with particular emphasis on its possible role in cognition.

The preponderance of the dopaminergic innervation of the basal ganglia (i.e. striatum) in the mammalian brain has highlighted its neuromodulatory effects on motor function. However, it is becoming increasingly clear that the striatum subserves important functions in sensorimotor integration. Such integration can occur at many functional levels, from the organization of reflexes to that of complex response sequences which entail the formation and integration of plans, motor programmes and goal-directed behaviour. Moreover, the fact that DA-containing neurons also innervate the prefrontal cortex (PFC), as well as other structures with potential roles in learning, (e.g. the amygdala) has further raised the possibility of additional functions in higher cognitive functions for DA which intervene between the processing of sensory input and motor output such as learning, working memory and aspects of attentional functioning.

Previous reviews have focused on the importance of DA in gain-amplification processes that contribute to the preparatory phases of responding to external incentive or reinforcing stimuli and also to the production of efficient responses to specific stimuli (as occurs, for example, in reaction time situations) (BLACKBURN et al. 1992; ROBBINS and EVERITT 1992). These processes can be considered as parallel modulations of functions dependent, respectively, on the ventral (including the nucleus accumbens) and dorsal (including the caudate-putamen) striatum (ROBBINS and EVERITT 1992).

Both the aforementioned reviews considered the possible role of striatal DA in plastic processes, including associative learning. For example, the enhancement of response set, that process by which predispositions to respond to a particular stimulus with a particular response are increased, could potentially facilitate stimulus-response learning. Moreover, the evident role of DA in reinforcement mechanisms within the nucleus accumbens suggests that stimulus-reward learning may occur at this site, as also posited by other authors (WHITE 1989). More recently, there has been increasing interest in the possible contribution of DA to reinforcement learning. This has been given impetus by a combination of electrophysiological findings and computational modelling approaches (HOUK et al. 1995; MONTAGUE et al. 1996; SCHULTZ et al. 1997). These investigations have made clear that the critical issue is not that DA contributes to reinforcement per se, but in specifying its exact role in associative learning.

B. Electrophysiological and Neurocomputational Approaches

Precise data concerning the possible coding of reinforcement by DA neurons have been obtained from experiments in which their activity is recorded in alert monkeys while they perform in situations where their behaviour earns food rewards (SCHULTZ 1992). In such experiments DA neurons in the midbrain ventral tegmental region respond with short, phasic activity when monkeys are presented with appetitive stimuli. DA neurons are also transiently activated by novel stimuli that elicit behavioural orienting over the first few presentations. However, it has been shown that at least some aversive stimuli, such as air puffs to the hand, or drops of saline to the tongue, do not generally elicit firing. When repeated presentations of food reward are reliably predicted by other cues such as lights or noises, the activity of the DA neurons is advanced temporally to the time of onset of these conditioned stimuli (CS) and responding to the reward stimulus itself is no longer present. However, if the reward is omitted then the activity of the DA neuron is depressed at exactly the point in time at which it would normally have occurred, suggesting that it contributes to an internal representation of the reward.

These results are consistent with theories of associative conditioning such as that of RESCORLA and WAGNER (1972), which place emphasis on the importance of the predictability of the unconditioned reinforcer. Learning occurs as a consequence of reducing error feedback signals, such that when reward is completely predictable no further learning occurs. The activity of the DA neurons appears to provide a "teaching signal" that provides information about the expected time and magnitude of reinforcement (MONTAGUE et al.

1996; SCHULTZ et al. 1997). These teaching signals potentially can alter the synaptic weights of neural networks within terminal structures such as the striatum. Theorists have also speculated how the delay of reinforcement could be mediated by biochemical changes in striatal neurons initiated by the binding of DA to its receptors (HOUK et al. 1995). It is important to note, however, that there is no evidence that dopaminergic activity represents sensory properties of the reinforcer (e.g. its precise visual or olfactory nature), which are presumably encoded by other neural networks in non-striatal structures (for example, the orbitofrontal cortex).

Whilst it appears that there is ample circumstantial evidence for a role for central dopaminergic mechanisms in neural plasticity and mechanisms of appetitive learning, this hypothesis must be considered in the light of existing evidence that DA has other, more immediate, functions that directly affect processing in its terminal regions, producing, for example, general changes in locomotor activity (e.g. in the rat, KELLY et al. 1975). This consideration is also relevant to the locus of the inferred changes in learning, plausibly in the dorsal and ventral striatum, or in the prefrontal cortex. The second major issue that has to be raised is the extent to which the electrophysiological and neuro-computational findings are supported by direct evidence that DA plays a causal role in learning and memory processes. It is possible that the changes in activity in DA neurons reflect plastic changes occurring elsewhere, being consequences rather than causes of learning. As such, the changes in DA activity would still play a crucial role in behaviour, but may not, for example, be necessary for learning to occur. For example, REDGRAVE et al. (1999) have recently suggested that the changes in DA activity might function as a signal to switch from one form of behaviour to another (e.g. from lever pressing to food consumption), consistent with established roles for striatal DA in the control of behavioural orienting, (see below). At the neurobiological level of analysis, there is also no convincing evidence that dopaminergic activity in the ventral striatum is necessary for, or augments, processes of neuronal plasticity, as exemplified for example, by long-term potentiation (LTP) (PENNARTZ et al. 1994, 1995). It is of interest that other, possibly DA-dependent, forms of neuronal plasticity have been demonstrated within the dorsal striatum (including long-term depression, CALABRESI et al. 1995) but their possible relevance to behavioural learning is only just beginning to be explored (GRAYBIEL 1995). This issue could be resolved by evidence that pharmacological interventions that reduce or enhance DA activity should produce predictable changes in learning – impairment or facilitation, respectively. In fact, as will be seen, the apparently important roles that DA has in behavioural performance means that it is more difficult to provide this decisive evidence of its role in learning than might at first be thought. Converging lines of evidence are required to resolve these issues, including tests which isolate causal relationships between DA and behaviour.

C. Neuropharmacological Evidence for a Role for Dopamine in Learning

There are two major neuropharmacological approaches for investigating the functions of DA. The first approach monitors the fluctuations in extracellular DA that occur in behavioural situations using in vivo dialysis or voltammetry. The levels change as a consequence of altered release and re-uptake mechanisms and of course do not only reflect synaptic concentrations, but also gradients of local concentrations distal from the synapses themselves. These techniques thus potentially provide important converging evidence for the role of DA neurons in associative processes, although over a longer time scale (minutes in the case of in vivo dialysis) than in the case of electrophysiological recording from identified DA neurons. Microdialysis offers the considerable advantage over voltammetry of chemical specificity, but the disadvantage of poor temporal resolution. This means that the capacity to establish temporal precedence of the effect of one event over another is diminished, and thus compromises the use of the technique for establishing causal relationships between behavioural contingencies and chemical events. Moreover, the lack of temporal resolution also means that the sign of any change might reflect "rebound" or compensatory processes that overwhelm the immediate effect of the discrete event.

The second, classical approach for demonstrating a selective role for a neural structure or neurotransmitter pathway in associative learning is to show a specific effect of a given manipulation on acquisition, but not pre-established performance. This pattern of results would normally indicate that the manipulation has probably interfered with processes of associative learning rather than other non-associative processes inevitably confounded with learning, including perception, attention, motivation and motor function. As a facilitation of learning is always a more impressive demonstration than its impairment, this provides the gold standard for interpretations of specific effects on learning. Of course, if a given manipulation affects performance as well as learning, then parsimony dictates that a non-associative effect can account for both sets of findings. Alternatively, it is plausible that the manipulation separately interferes with both associative and non-associative factors; however, for that interpretation to hold, it might be expected that the effect on learning would be quantitatively greater than any effect on performance. With these general points in mind, it is evident in reviewing the experimental literature that it is still quite difficult to find consistent evidence for a specific role in learning for brain DA systems that matches predictions from the electrophysiological evidence.

I. Overview of Results from In Vivo Monitoring Studies

DA neurons appear to be responsive to a variety of stimuli and states, as well as pharmacological challenges. These responses cannot be reviewed completely or in detail because of limitations of space. However, consistent with Schultz' electrophysiological data in monkeys, presentation of food or water to rats can lead to increases in extracellular DA, sometimes in the dorsal as well as the ventral striatum (see review by Blackburn et al. 1992). Moreover, the responses are greater if food is presented in an intermittent, periodic manner than all at once (Salamone et al. 1994). However, there is evidence that foot shock can also increase extracellular striatal DA, and that some stimuli (e.g. loud noises) leading to startle responses (Humby et al. 1996) and aversively conditioned taste stimuli (Mark et al. 1991) produce reductions in ventral striatal DA. Also consistent with some of Schultz' evidence is that DA neurons show changes in activity to previously neutral environmental stimuli (e.g. lights and auditory tones) which are conditioned to important events such as food delivery (Blackburn et al. 1992). These latter effects can be elusive, however, and may depend on precise conditions of food deprivation (Wilson et al. 1995).

A particularly revealing study (Bassereo and Di Chiara 1999) has shown that the medial, so-called shell region of the nucleus accumbens responds to novel presentations of novel palatable food with increased concentrations of extracellular DA, a response which habituates even though the rat may be consuming more food with repeated presentation. This is strong evidence that the response may be related to the salience of the food, and possibly, at a behavioural level, to the motivational excitement likely to occur in the presence of a highly appetitive reinforcer. DA levels in the medial prefrontal cortex also increase, but fail to show such clear-cut habituation (Bassareo and Di Chiara 1997). Conditioned stimuli (largely olfactory) predicting the presentation of the food also increased DA in the medial prefrontal cortex, a response not initially seen in the nucleus accumbens (Bassareo and Di Chiara 1997). However, the later study (Bassareo and Di Chiara 1999) clarified the situation by showing how the conditioned stimuli led to increases in DA concentrations in the core region of the nucleus accumbens, but inhibited the response to food itself in the shell. These experiments suggest that the mesolimbic-cortical DA system is modulating different aspects of appetitive behaviour; possibly aspects of the representation of the unconditioned reinforcer in the shell, and those of the conditioned stimulus or reinforcer in the core regions of the nucleus accumbens. The latter results are broadly consistent with the electrophysiological data of Schultz, in showing some connection between associative mechanisms and striatal DA transmission, even though the methods employed are probably monitoring different temporal modes of dopaminergic transmission, in terms of tonic (steady-state) extracellular levels and phasic release, associated with burst firing patterns (Moore and Grace 1999).

However, we re-iterate that it is difficult to resolve the question of whether such changes are causally involved in the associative process itself, as they could reflect the expression of some behavioural correlate of learning. An alternative way of addressing this issue is to utilize preparations of Pavlovian aversive conditioning, which lead to behavioural suppression rather than locomotor activation. Several studies have been able to show increases in DA concentrations within the ventral striatum as a consequence of such conditioning (YOUNG et al. 1993; BESSON and LOUILOT 1995; SAULSKAYA and MARSDEN 1995) although so far none have addressed whether the changes are related to specific accumbens sub-regions. A related study by WILKINSON et al. (1998) has investigated parallel changes during acquisition and extinction of aversive conditioning in rats of DA in the nucleus accumbens and medial prefrontal cortex. This study showed greater changes initially during acquisition in the medial prefrontal cortex, but then subsequently greater responses in the nucleus accumbens that appeared to map onto the changes in behavioural freezing seen as a consequence of such conditioning and extinction in these rats.

Of particular interest is the study by YOUNG et al. (1998), which utilized sensory preconditioning. Initially, dialysis showed increased overflow of ventral striatal DA in response to a pairing of motivationally neutral visual and auditory stimuli. Then, one of the stimuli (e.g. tone) was paired with an aversive foot-shock, after which the response to tone and light was measured separately, in the absence of the shock. The impressive finding was that accumbens DA was elevated in response to the light when it had been previously paired with the tone, but not when it had been unpaired. This suggests that associative conditioning does lead to an increase in accumbens DA in a situation in which it is far from clear that the effect can be explained simply in terms of an orientational behavioural response to the light (although that cannot be entirely excluded). An earlier study had shown, in fact, that the latent inhibition of aversive conditioning to a tone by its previous non-reinforced exposure to the animals produced parallel reductions in extracellular accumbens DA (YOUNG et al. 1993). In general, it appears that the work from the in vivo monitoring of DA by dialysis and other neurochemical techniques is supportive of a role for DA in aversive, as well as appetitive behaviour. This is in line with other evidence from a variety of sources, indicating that DA turnover is increased tonically during stress, particularly in the medial prefrontal cortex, but also in striatal regions, such as the nucleus accumbens shell (KALIVAS and DUFFY 1995).

II. Psychopharmacological Evidence of Specific Actions of Dopamine on Learning and Memory

The fact that the release of DA can function as a reinforcing event, as inferred, for example, from studies on the self-administration of dopaminergic drugs (reviewed by other contributors), suggests that it has some role in learning, if only by contributing to the affective representation of the unconditioned reinforcer. In general, psychopharmacological evidence showing a specific role for DA in learning is rather limited because drugs have generally been administered to animals exhibiting steady-state performance. There is no doubt that drugs such as amphetamine, as well as more specific DA agonists and antagonists, have profound effects on performance in a variety of appetitive and aversive situations. However, such effects potentially confound an analysis of their possible effects on learning. It is clear, for example, that the acquisition of responding with conditioned reinforcers is potentiated by amphetamine-like drugs via DA-dependent mechanisms of the nucleus accumbens that include the shell region (ROBBINS et al. 1989; PARKINSON et al. 1999; review by SUTTON and BENINGER 1999). However, it is more dubious that this potentiation reflects a facilitation of associative learning rather than a potentiation of instrumental responding produced by exaggerations of the efficacy of the conditioned reinforcer. Thus, neither mesolimbic DA depletion achieved via 6-hydroxydopamine (6-OHDA) lesions of the nucleus accumbens (TAYLOR and ROBBINS 1986) nor intra-accumbens infusions of selective DA D_1 or D_2 receptor antagonists (WOLTERINCK et al. 1993) in themselves appear to impair the acquisition of a new instrumental response for a conditioned reinforcer, as distinct from blocking the potentiative effects of d-amphetamine.

A similar analysis can be applied to the symmetrical issue of selective impairments in the acquisition of active avoidance behaviour produced by neuroleptic drugs and the role of negative conditioned reinforcers (see BLACKBURN et al. 1992). In one early experiment (BENINGER et al. 1980), systemic pimozide was shown to have its normal disruptive effect on signalled avoidance behaviour. However, the additional, revealing finding was that when the capacity of the signal to act as a fear signal was assessed independently, animals receiving pimozide during the Pavlovian conditioning phase nevertheless exhibited normal levels of conditioned suppression to the CS on a food-reinforced baseline, thus demonstrating intact associative fear conditioning.

Another early experiment by BENINGER and PHILLIPS (1980) focused on appetitive associative learning by showing that systemic injections of the DA-receptor antagonist pimozide may have impaired the acquisition by rats of an association between a specific CS and food presentation. When the rats were subsequently tested in the undrugged state in a situation requiring the new learning of a response to produce the CS as a conditioned reinforcer, this effect was attenuated in the rats previously treated with pimozide. However, it is difficult to be sure that some unmeasured effects of the drug (e.g. to change eating rate) actually did not interfere with the associative process indirectly. As with effects on active avoidance acquisition (see BLACKBURN et al. 1992) it is unclear that the drug effect does not simply reflect an effect on motor performance (see also SALAMONE 1994).

On the other hand, in the investigation of systemic effects of a low and a high dose of the D_2/D_3 receptor agonist quinpirole, NADER and LEDOUX (1999) have recently employed an inactive response (defensive freezing) and a sophisticated design which separates basic effects on associative learning and sensory processing via a comparison of groups of rats subjected to second order fear learning or sensory preconditioning. They found that when quinpirole was administered prior to the CS1–CS2 pairing stage, there was a subsequent block of aversively-motivated freezing behaviour in the quinpirole-treated rats, suggesting an attenuation of the retrieval of the fear associated with the CS that is hypothetically mediated via a reduction of DA neurotransmission through D_1-like post-synaptic mechanisms in unspecified anatomical structures. The lack of effect on sensory preconditioning is somewhat surprising in view of the demonstration by YOUNG et al. (1999) of an elevation of nucleus accumbens DA during CS1–CS2 sensory preconditioning, employing stimuli of neutral motivational salience.

Experiments using the neurotoxin 6-OHDA to produce selective and profound depletions of DA in certain regions, such as the nucleus accumbens or caudate-putamen, have also been shown to impair instrumental visual discrimination learning (EVENDEN et al. 1989; ROBBINS et al. 1990). However, in most of these experiments, impairments produced by such DA-depleting lesions are also seen in control experiments, using previously trained rats. A case might then be made for some effects of DA loss on memory retrieval, but there are also ancillary actions, for example on attentional function, to take into account. In fact, as will be seen below, distinguishing effects on attention from those on associative factors is a particularly difficult problem.

We will therefore focus on studies in which these potentially confounding effects on learning are minimised by post-training administration. There is, in fact, a considerable literature showing that post-trial administration of amphetamine under certain conditions can subsequently enhance memory when retention is tested several days later, for both appetitively and aversively motivated tasks (e.g. KRIVANEK and MCGAUGH 1969 – see Table 1). For a while, it was thought that such actions were largely mediated peripherally, as the "memory-enhancing" effects could be blocked by adrenalectomy (MARTINEZ et al. 1980). However, experiments by CARR and WHITE (1984) and others have shown that a central, probably, caudate site could at least contribute. In further extensions of the work, intracaudate administration of the D_2/D_3 agonist quinpirole produced enhanced retention of a conditioned suppression task. In theory, such effects could still be explained if, for example, the drug directly strengthened the unconditioned stimulus (US), i.e. increased the subjective sensation of the shock for the animal. However, ingenious experiments seem to have excluded this possible interpretation. For example, WHITE and VIAUD (1991) also varied not only the site of infusion within the caudate but also the sensory modality of the CS. When the dopaminergic agent was infused into that anatomical region of the rat caudate-putamen known to receive input from visual areas, it only subsequently enhanced learning of the visually cued learning; the same was true of the enhancement of olfactory cued aversive learning. Therefore, the enhancement only occurred when the DA agonist interacted with the that region of the stratum processing the CS, and also only affected the response to this stimulus if it had been contingently related to the shock US – suggesting some specific modulation of post-trial associative processing.

Table 1. Dopamine and memory consolidation: landmark studies in rodents

Study	Paradigm	Post-training manipulation	Conclusions
KRIVANEK and McGAUGH 1969	Y-maze appetitive discrimination	Amphetamine (systemic)	Improved retention
CARR and WHITE 1984	Conditioned suppression (aversive)	Intra-caudate d-amphetamine	Improved retention
WHITE and VIAUD 1991	Conditioned suppression (visual or olfactory)	Intra-caudate d-amphetamine, D_1,D_2 agonists posteroventral or ventrolateral sites	Modality/region enhanced retention D_2 agonists
PACKARD and WHITE 1991	Radial maze (win-stay and win-shift food-foraging tasks)	Intra-caudate or -hippocampal infusions of d-amphetamine or DA agonists	Enhanced win-stay (intra-caudate), enhanced win-shift (hippocampus)
SETLOW and McGAUGH 1999	Morris water-maze spatial learning	Intra-accumbens sulpiride (D_2 antagonist)	Deficit

This technique of post-trial manipulation of the modulation by DA of memory consolidation processes has now been extended to forms of memory mediated by other terminal domains. PACKARD and WHITE (1991) showed that post-trial administration of d-amphetamine, or the $D_{2/3}$ agonist quinpirole, or the D_1 receptor agonist SKF-38393 to the caudate (but not the hippocampus) all enhanced subsequent retention of an appetitive "win-stay" task carried out in a radial maze, whereas similar administrations to the hippocampus (but not the caudate) enhanced learning of a "win-shift" procedure in the same apparatus. These effects seem very difficult to explain simply in terms of general performance-altering effects of the drug.

A possible role for DA in modulating longer-term spatial memories known to depend on hippocampal functions has been extended to the nucleus accumbens. PLOEGER et al. (1994) were initially able to show that intra-accumbens haloperidol impaired acquisition of the Morris water maze escape task, but a yet more significant demonstration is that of SETLOW and McGAUGH (1998) with immediate post-trial administration of the DA D_2 receptor antagonist sulpiride, leading to a retention deficit 2 days later. Delayed infusions or immediate post-trial infusions of sulpiride, using an externally cued version of the task, failed to affect retention, suggesting a specific effect on the consolidation of long-term spatial memory. These authors speculate on the basis of other results that these DA-dependent processes of the nucleus accumbens are only implicated in consolidation of the memory and not in its storage. The consolidation of long-term spatial memory, however, is unlikely only to involve the ventral and not the dorsal striatum. In a follow-up experiment, the same authors (SETLOW and McGAUGH 1999) reported on results obtained following post-trial sulpiride infusions into the posteroventral caudate-putamen, which they interpreted to reflect memory for procedural aspects of the task. Specifically, sulpiride-treated rats spent less time swimming in the vicinity of the previously trained platform, while reaching the platform location with a normal latency. Thus dopaminergic processes appear to modulate several aspects of memory associated with this task in different regions of the striatum that are in receipt of different limbic-cortical afferents. The dopaminergic influences may also include projections within such limbic structures themselves. Thus, the above results have been extended by the demonstration that post-trial infusions of amphetamine into the amygdala modulate retention of both a cued and a spatial version of the Morris water maze (PACKARD et al. 1994), potentially via dopaminergic mechanisms.

A parallel set of experiments has now been completed that analyse the effects of specific manipulations of dopaminergic transmission on the consolidation of stimulus-reward learning or "emotional memory". HITCHCOTT et al. (1997a) first found that intra-amygdaloid, post-trial amphetamine enhanced the acquisition of a discriminative approach response to sucrose solution. To follow this, HITCHCOTT et al. (1997b) examined effects of the DA receptor agonists SKF-398393 (D_1), quinpirole and 7-OH-DPAT (both D_2/D_3). Significant enhancement of discriminative approach was found at certain doses of 7-OH-DPAT. However, the precise locus of this effect within the amygdala (e.g. central nucleus or basolateral amygdala) is somewhat unclear, although presumably the greater density of D_2/D_3 receptors in the central nucleus implicates that structure, possibly through its known involvement in Pavlovian appetitive conditioning (PARKINSON et al. 2000).

III. The Possible Complication of a Role for Dopamine in Attentional Function

Unilateral striatal DA depletion in the rat was originally reported to produce behavioural symptoms in addition to the well-known effects on rotational behaviour that were interpreted as forms of attentional or "sensori-motor" neglect (UNGERSTEDT 1971; MARSHALL and TEITELBAUM 1977). Studies utilizing primates (Schneider 1990; Annett et al. 1992) have found analogous symptoms. Detailed analysis in rats of the "neglect" syndrome has shown that it is mainly attributable to DA depletion from the dorsal striatum (caudate-putamen) and that it may result from impairments in such processes as the preparatory readiness of orienting responses (see review by ROBBINS and BROWN 1990; WARD and BROWN 1996).

Three other main paradigms have been utilized that also bear on possible attentional dysfunction following manipulations of dopaminergic function: latent inhibition, prepulse inhibition (PPI) and continuous performance (the

5-choice serial reaction time task) – all notable for their correspondence to parallel tests for human subjects. Curiously, for each of these paradigms, the main emphasis of investigations has been on mesolimbic rather than mesostriatal systems.

Latent inhibition (LI) refers to the retardation of conditioning that occurs following non-reinforced pre-exposures of the CS (MACKINTOSH 1983). This behaviour is impaired following systemic doses of d-amphetamine, so that learning is actually facilitated in the pre-exposed condition. These effects, however, are apparently restricted to the learning rather than the pre-exposure stages of the test, to the use of low and intermediate doses of the drug, and are more readily obtained following chronic administration (WEINER et al. 1984, 1987; WEINER 1990). Similar effects are also much more difficult to obtain following treatment with DA receptor agonists such as apomorphine. Thus, from the perspective of dopaminergic function, more impressive evidence derives from effects of systemically administered DA receptor antagonists, which consistently facilitate LI in rats. The position in humans is a little more equivocal. One study (WILLIAMS et al. 1997) has reported enhancement of LI using a visual task following low i.v. doses of haloperidol. However, the same group have also now reported the opposite result in young volunteers with an auditory paradigm – namely impaired LI (WILLIAMS et al. 1998). This is a particularly important result, as schizophrenics naïve to neuroleptic medication were shown not to have the usual deficits in LI associated with chronic (and medicated) schizophrenia. The implications appear clear. DA receptor antagonism may impair LI, possibly via attentional factors. But the deficits in LI in schizophrenia may arise, at least in part, as side-effects of such medication.

Original theorizing focused on the likely role of the nucleus accumbens in mediating effects of dopaminergic drugs on LI, but this conclusion remains controversial. Specifically, KILLCROSS and ROBBINS (1993) found that intra-accumbens infusions of d-amphetamine, while impairing aversive conditioning per se, did not differentially affect pre-exposed versus non pre-exposed stimuli, in a within-subject design. Systemic treatments with either d-amphetamine or a neuroleptic drug (alpha-flupenthixol) did produce the commonly found effects. However, these were later shown to depend on apparent drug-reinforcer interactions. Amphetamine appeared to enhance conditioning by enhancing the impact of the reinforcers (electric shock or sucrose). By contrast, the neuroleptic had the opposite type of effect on the reinforcers, possibly accounting for its contrasting effect on LI. Consistent with the findings of KILLCROSS and ROBBINS (1993), ELLENBROEK et al. (1997) found impaired LI following dorsal rather than ventral striatal infusions of amphetamine, but they employed a taste aversion procedure for assessing LI.

In the original study, SOLOMON and STATON (1982) demonstrated impaired LI following chronic ventral rather than dorsal striatal infusions of amphetamine, though using an active avoidance rather than a conditioned suppression procedure. Other authors have found that mesolimbic DA depletion appears to facilitate LI, apparently consistent with the results of microdialysis studies and the effects of DA receptor antagonists, described above (GRAY et al. 1995). Perhaps it is safest to conclude at this juncture that effects of intra-accumbens manipulations on LI may depend on the chronicity of treatment, the precise nature of the behavioural paradigm employed for measuring LI, and possible side-effects of the drug on the impact of the reinforcer. An over-riding consideration is that effects on LI may not arise directly from actions on attentional processes but instead reflect effects on the unconditioned reinforcer, or as has been argued previously (KILLCROSS et al. 1994a,b), memory retrieval processes based on contextual processing. Specifically, drugs such as amphetamine, which enhance the effectiveness of the reinforcer, might increase the difference in context between the pre-exposure and testing stages of the LI paradigm, which would of itself attenuate LI. DA receptor antagonists could be expected to have the opposite effect.

A probably distinct form of attention is likely exemplified by the phenomenon of PPI, in which a less-intense surrogate stimulus reduces the magnitude of the acoustic startle response to an intense loud noise (BRAFF and GEYER 1992) – paralleling its apparent action to protect against the reduction in extracellular DA levels produced by such a startle stimulus (HUMBY et al. 1996). DA-dependent mechanisms of the nucleus accumbens are certainly implicated in this response, although deficits in this "sensori-motor gating" process are produced by both DA D_2 receptor agonists and antagonists (SWERDLOW et al. 1994). Recent studies with transgenically modified mice have confirmed a possibly key role for the DA D_2, rather than the D_3 or D_4, receptor (RALPH et al. 1999). However, there are evidently considerable strain differences in the role of D_2 receptors within the nucleus accumbens for the PPI response in the rat (KINNEY et al. 1999). WAN and SWERDLOW (1998) have further provided evidence that this form of "sensorimotor gating" is mediated by DA-glutamate interactions within both the core and shell sub-regions of the nucleus accumbens.

To date there have been relatively few direct comparisons of PPI and LI, but one such was made in a study that investigated the responses of rats reared in social isolation, which have elevated levels of extracellular striatal DA (WILKINSON et al. 1994). The main finding of interest is that social isolation impaired PPI, but not LI. The PPI deficit is of considerable interest, not least because of possible relevance in schizophrenia, and may illustrate how descending forebrain influences, including the nucleus accumbens, modulate the tone of a set of reflexes organized in the brain stem. This alteration of "tone" may be but one consequence of reinforcing events that produce changes in dopaminergic function.

Possible effects of DA on attentional functions have also been investigated using a number of tasks which require animals to detect signals over a protracted period of stable performance. For reasons of space these cannot be reviewed in detail here. One such paradigm, the 5-choice serial reaction time task, was developed by analogy from human studies (see ROBBINS 1998 for review). Rats are required to detect brief visual stimuli that are presented randomly in one of five locations in a specially designed apparatus. The temporal predictability of the stimuli can also be varied, as well as their detectability via manipulations of stimulus illuminance and duration. Initial experiments focused on neuropharmacological probes of mesolimbic DA function. Depletion of mesolimbic DA using 6-OHDA had little effect on the accuracy of stimulus detection under any experimental conditions. However, the latency of responding was lengthened, errors of omission were increased and premature responses reduced (COLE and ROBBINS 1989). This pattern of effects is consistent with effects of mesolimbic DA on the invigoration of behaviour, perhaps via motivational influences, rather than a disruption of attention. Complementary effects were obtained when d-amphetamine was infused into the nucleus accumbens; again there were no effects on choice accuracy, but premature responses were greatly increased in frequency (COLE and ROBBINS 1987).

These early results have now been augmented by parallel studies of 6-OHDA-induced lesions of the mesostriatal and mesocortical DA systems (ROBBINS et al. 1998; BAUNEZ and ROBBINS 1999). Both studies produced results that were different from those of mesolimbic DA loss, in that there were impairments in choice accuracy when the visual stimuli were presented in a temporally unpredictable manner. Following mesocortical DA loss, there were few other impairments in this task, but the specific deficit in accuracy might just have been attributable to the almost unavoidable depletion of noradrenaline from the prefrontal cortex following such 6-OHDA lesions. Further specific evidence for a role of DA receptors in attentional accuracy is provided by recent results following infusion of specific DA receptor agonist and antagonists into the prefrontal cortex. Intra-cortical infusions of the D_1 DA receptor antagonist SCH-23390, but not the DA D_2 receptor antagonist sulpiride, produced selective impairments in the accuracy of responding, whereas similar infusions of the partial D_1 receptor agonist SKF-38393 actually improved choice accuracy under some conditions (GRANON et al. 2000).

The impaired choice accuracy resulting from mesostriatal DA depletion was found in the context of many other behavioural deficits, including slowed responding and large increases in response latency (similar to those seen following mesolimbic DA loss, see above). However, despite these effects, no deficits in accuracy were observed under baseline conditions. The selective disruption produced by the variable inter-trial intervals may be related to the basic impairments in the readiness to respond described in earlier studies on simple and choice reaction time (BROWN and ROBBINS 1991).

IV. Models of ADHD

The phenomenon of attention deficit hyperactivity disorder (ADHD) and the ameliorative effects of methylphenidate (Ritalin) and amphetamine have led some investigators to attempt to produce animal models of this syndrome and the apparently paradoxical effects of psychomotor stimulants in reducing high levels of locomotor activity (see ROBBINS and SAHAKIAN 1979; SEIDEN et al. 1989). This has proved to be an elusive problem which has recently, however, capitalized on genetic technology. The DA transporter knockout (DAT) mouse has elevated dopaminergic tone, is hyperactive and also exhibits deficits in tests of spatial memory (GAINETDINOV et al. 1999). Methylphenidate antagonized this hyperactivity, although possible beneficial actions on spatial or other forms of cognition were not apparently investigated (in common with most of the studies in this field). The mechanisms of action of methylphenidate in this model, and indeed in ADHD itself, are unclear. They could include an action on another neurotransmitter such as serotonin (SEIDEN et al. 1989; GAINETDINOV et al. 1999). Hyperactivity in DAT knockout mice could also be treated with chronic fluoxetine, a selective serotonin reuptake blocker, but this by itself does not establish how methylphenidate itself works. The reader is referred to a more detailed discussion in a book devoted to this topic (SOLANTO et al. 2001).

An overall evaluation of the role of DA in attentional function in experimental animals may be premature. It seems difficult to maintain that DA, within subcortical regions at least, has a direct role in selective attentional functions. Rather it appears that manipulations of DA may affect, perhaps phasically, the salience or impact of intense stimuli or reinforcers and, on a more tonic basis, states of activation that modulate basic behavioural reflexes, including the orienting response. Further research in this area is important because it bears on processes related to attention that have been linked especially to prefrontal cortical DA function, namely working memory, in which stimuli are maintained "on-line" for further processing after their initial detection and selection.

D. Working Memory

When used in the animal literature, this construct generally refers to the capacity to hold information "on-line" in a period during which the eliciting stimulus is no longer present. According to GOLDMAN-RAKIC (e.g. 1987), therefore, this form of working memory thus has a crucial role in the intermediate stages of stimulus processing, to provide input to brain structures that form representations of the world. A related perspective is that of OLTON (e.g. OLTON et al. 1979) based on his distinction of performance by rats in radial mazes between behavioural contingencies based on recently acquired information and those based on permanent, long-lasting "response rules". Thus, within a single set of trials, perhaps with interpolated delays, rats will learn systematically not to return to recently baited arms within the maze, this "win-shift" tendency exhibiting what he denotes as "working memory". On the other hand, they will consistently avoid arms never baited with food over repeated test sessions ("reference memory").

These concepts, therefore, have something in common with the more extended concept of working memory in human cognition introduced by BADDELEY (1986), which includes two distinct short-term memory stores (the "articulatory loop", a form of sub-vocal rehearsal mechanism, and a "visuospatial sketchpad", a short-term memory buffer for visuospatial imagery). Both of these stores, in a sense, hold stimuli "on-line" for further processing. The additional, and most controversial, element of BADDELEY's scheme is the positing of a "central executive" system which co-ordinates processing between the various dedicated satellite systems. This is commonly related to the functioning of the prefrontal cortex, although a simple mapping of psychological processes onto anatomical structures is, of course, not viable. In fact, the "central executive" system of BADDELEY (1986) has much in common with another possible model of frontal lobe functioning termed the "supervisory attentional system", in which control over instrumental choice behaviour is exerted through "attention to action" (SHALLICE 1982). This concept is particularly relevant to paradigms such as the spatial delayed response task in which there are other cognitive requirements besides "holding stimuli on-line". For example, the animal has to inhibit making repeated responses to prepotent stimuli (DIAMOND 1996), and this potentially is also under dopaminergic modulation regardless of whether one considers the inhibitory function to be dependent on working memory or alternatively to be a relatively independent form of executive function.

The partial correspondence of concepts of working memory in animal research, with those from the domain of human cognitive psychology, therefore, provides many opportunities for misunderstanding, especially in the context of the functioning of the prefrontal cortex (ROBERTS et al. 1998). As we have seen, the debate centres around the interpretation of behavioural processes required for tests of "working memory" function in experimental animals such as the delayed response task, used mainly for primates, but also the delayed alternation task, which has analogies with the radial arm maze paradigm of OLTON described above and is more often used when testing rodents.

There is little doubt that the pharmacological manipulation of DA, probably within mesostriatal as well as mesofrontal domains, has profound effects on performance in these situations in both rodents and monkeys (see Table 2). For example, early work (reviewed by LEMOAL and SIMON 1991) demonstrated that 6-OHDA-induced lesions of the mesoaccumbens or mesostriatal, as well as the mesocortical DA projections, led to impaired delayed alternation performance in rats. However, there is a question of whether the capacity to hold "on-line" the location of the previous goal or choice response has been impaired or whether other behavioural capacities, such as the inhibition that is normally required for the spontaneous alternation of choices is disrupted.

In monkeys, a landmark study on the role of DA in working memory function was that of BROZOSKI et al. (1979). These investigators used a delayed-

Table 2. Dopamine and working memory: landmark cross-species studies

Study	Paradigm/species	Manipulation	Conclusions
Brozoski et al. 1979	Spatial delayed response/monkeys	DA depletion from PFC by 6-OHDA	Impaired
Sahakian et al. 1985	Delayed alternation/rats	DOPAC/DA measures	Behavioural/ neurochemical relationship
Sawaguchi and Goldman-Rakic 1991	Delayed saccade/monkeys	Iontophoresis of selective D_1/D_2 antagonists	Selective D_1 impairment
Arnsten et al. 1995	Delayed response/monkeys	Systemic D_2 agonist	Low-dose deficit, high-dose benefit
Luciana et al. 1992	Delayed saccade/normal humans	Oral D_2 agonist bromocriptine	Improved
Williams and Goldman-Rakic 1995	Delayed saccade/monkeys	Iontophoresis-selective DA antagonists	Enhanced firing with D_1 antagonist
Zahrt et al. 1997	Delayed alternation/rats	Intra-PFC D_1 agonist	Impaired
Seamans et al. 1998	Win-shift task/rats	Intra-PFC DA antagonist	Impaired
Muller et al. 1998	Spatial working memory/normal humans	DA agonists (mixed D_1/D_2)	Improved

response-type procedure to show that 6-OHDA-induced depletion of DA in the vicinity of the principal sulcus of the dorsolateral prefrontal cortex in macaques produced an impairment every bit as profound as ablation of the region itself. Depletion of either noradrenaline or 5-hydroxytryptamine (5-HT) in the prefrontal cortex had little effect. Further evidence for a specific role of DA came from additional evidence that the deficits could be remediated by systemic treatment with drugs such as apomorphine and L-dopa. In a follow-up study, Arnsten et al. (1994) have shown beneficial effects of systemically administered DA D_1 receptor agonists in aged macaques and catecholamine-depleted younger animals.

Mindful of the possible behavioural interpretation that these effects might reflect some possible action of dopaminergic manipulations on the performance of "mediating responses" which obviate the necessity to hold specific information "on-line", Goldman-Rakic and collaborators have more latterly employed a "delayed saccade" procedure in which monkeys have to hold fixation of a central spot before shifting making an eye-movement to the location of a brief visual stimulus presented a few seconds previously. Selective disruptions in the accuracy of the "memory saccades" were produced by iontophoretic application to the PFC of doses of DA D_1, but not D_2, receptor antagonists (e.g. SAWAGUCHI and GOLDMAN-RAKIC 1991). These findings have been supported by experiments with a delayed response procedure in marmosets which removed the possibility of mediating responses by distracting the animal to the back of the testing chamber during the delay period (ROBERTS et al. 1994). Once again, DA depletion from the PFC was found to impair the acquisition of a spatial delayed response task, though not to quite the same extent as an excitotoxic lesion of most of the PFC itself. However, the key finding from a further study (COLLINS et al. 1998) was the sparing, following mesocortical DA depletion, of the capacity to self-order responses without perseveration, which was markedly impaired by excitotoxic lesions. Thus, it appeared from this study that DA normally modulates mnemonic functions associated with the working memory task rather than the "executive" operations of producing the optimal response sequence.

In monkeys, investigators have been rather slow to test the hypothesis of possible striatal involvement in working memory function, as measured by delayed response performance. ARNSTEN et al. (1995) found some significant benefit in delayed response performance in young macaques, following systemic treatment with high doses of quinpirole, but impairment at low doses (probably acting at pre-synaptic autoreceptors) consistent with a possible striatal role, in view of the much greater density of D_2-like receptors in this region as compared with the prefrontal cortex. These effects were blunted in aged monkeys, possibly due to a loss of D_2 receptors. SCHNEIDER (1990) has tested spatial delayed response in monkeys following treatment with the neurotoxin 1-methyl-4-phenyl-1,2,3,6-tetrahydropyridine (MPTP), a notable model of Parkinson's disease, and found significant deficits. A recent paper (FERNANDEZ-RUIZ et al. 1999) has shown beneficial effects of L-dopa treatment on MPTP-treated monkeys in the spatial delayed response task. However, MPTP produces DA lesions, which are not restricted to the striatum nor indeed to DA itself. COLLINS et al. (2000) have nonetheless recently produced selective lesions of the caudate DA system, using infusions of 6-OHDA in the terminal fields, and also found evidence for a delayed response deficit.

However, the precise nature of these deficits remains unclear. Demonstration of a role for DA in performance of an object retrieval task is particularly germane, as this paradigm emphasizes to a much greater extent the role of response inhibitory rather than working memory functions. TAYLOR et al. (1990) showed clearly that treatment of monkeys with MPTP, leading to profound central DA loss, also impaired the ability of these animals to inhibit reaching through a transparent barrier rather than making a more effective "detour reach" – although it is unclear to what extent this deficit depends on striatal or cortical DA loss. A future challenge will be to delineate the relative contributions of prefrontal and striatal DA to spatial delayed response performance in monkeys.

One way forward in this endeavour has been indicated by the recent series of elegant studies by SEAMANS et al. (e.g. 1998) on the role of DA D_1 receptors in mediating foraging performance by rats in a number of radial 8-armed maze tests. Microinjections of the D_1 receptor antagonist SCH-23390 (but not the D_2 receptor antagonist, sulpiride) into the prelimbic region of the PFC disrupted performance of a delayed version of the task (similar to that used by PACKARD and WHITE 1991) in which spatial information acquired during a training phase was used prospectively 30 min later to guide responses, but had no effect on choice performance in the maze without the delay. These effects were further shown probably to depend on the modulation of hippocampal inputs to the PFC. The authors' hypothesis was that the information may be held within the hippocampus until required for formulating a subsequent plan to guide action. Thus, DA hypothetically modulates a circuitry including the hippocampus at the level of the PFC that affects spatial working memory functioning, including its "executive aspects".

The issue of the possible contribution of the striatum to working memory was also examined by these authors following intra-accumbens infusions of haloperidol (FLORESCO et al. 1996). This treatment did not affect performance on the delayed task described above, but did impair performance on the non-delayed, random foraging task in which rats have to retrieve within a single session four pellets from four different arms of the 8-armed maze. Haloperidol increased errors to both previously baited and non-baited arms. They attributed the deficits to the processing of information from hippocampus to the nucleus accumbens normally implicated in the organization of foraging behaviour. These data should also be interpreted in the light of evidence of a role for DA in the consolidation of long-term spatial memories, for example in the Morris water maze escape task (SETLOW and McGAUGH, 1998) considered above.

I. Problems of Interpretation of the Role of the PFC in Working Memory

The effects on working memory processing shown following PFC infusions of a D_1 receptor antagonist (SEAMANS et al. 1998) can usefully be compared to other findings in rats produced by similar infusions, but using a delayed matching-to-position operant procedure (BROERSEN et al. 1995). The effects of the DA receptor antagonist were not clearly delay-dependent in this latter study, unlike those of the muscarinic receptor antagonist scopolamine. The very different nature of the tasks and concepts of working memory, compared to those used by SEAMANS et al. (1998) may have contributed to this apparent discrepancy. Whereas the SEAMANS et al. study looked at how DA modulation modulated choice on the basis of retrieval of a memory occurring some 30 min previously, BROERSEN et al. attempted more faithfully to reproduce the repeated short-term spatial memory requirements, in terms of seconds rather than minutes, of the delayed response or delayed alternation task. Further work is required to resolve this issue, particularly as there is additional evidence for a form of spatial attentional deficit following intra-PFC SCH-23390 (ROBBINS et al. 1998a; GRANON et al. 2000), which might be related to the results found by BROERSEN et al. (1995).

A further complication for the hypothesis of an enabling role for PFC DA in working memory comes from findings that increments in DA function can lead to decrements in working memory performance. This has come from a variety of sources. Elevated PFC turnover produced by environmental or pharmacological stressors can disrupt working memory performance in rats in the delayed alternation paradigm, effects that can be remediated by treatment with D_1 receptor antagonists (MURPHY et al. 1996). Moreover, intra-PFC infusion of a full DA agonist can also impair delayed alternation performance, accompanied by perseverative responding, an effect which is also blocked by the D_1 receptor antagonist (ZAHRT et al. 1997). Finally, it has been reported that performance of a group of normal rats in this task is inversely related to DOPAC/DA indices of DA utilization or turnover within the cortex, but not the nucleus accumbens or dorsal striatum (ROBBINS 1985; SAHAKIAN et al. 1985). Thus, variations in DA turnover produced by stress in the normal population hypothetically modulate working memory performance. These findings have been related to a hypothetical inverted U-shaped function relating performance to level of D_1 receptor stimulation and the concomitant modulation of pyramidal cell functioning within the PFC (ARNSTEN 1997; ZAHRT et al. 1997).

Significantly, a similar complication has arisen in work with primates, as SCH-23390 and other D_1 receptor antagonists have been shown to enhance, rather than degrade, processing of single units in delayed saccade paradigms when administered iontophoretically (WILLIAMS and GOLDMAN-RAKIC 1995). Presumably, the earlier apparent discrepancy with the work of SAWAGUCHI and GOLDMAN-RAKIC (1991) arose because of the larger doses employed in that study. Nevertheless, WILLIAMS and GOLDMAN-RAKIC (1995) clearly conclude that, under many conditions, blockade of D_1 receptors can potentially enhance spatial working memory performance. Overall, as with experiments on the effects of intra-PFC infusions of D_1 agonists in rats, it does seem as though the effects of DA manipulations will depend on the underlying state of the animal and its baseline level of performance, rather than simply the dose of agent administered. There is thus the potential for DA D_1 receptor agonists and antagonists alike to exert opposite effects on performance, i.e. facilitation as well as impairment, depending on such conditions.

Recent evidence comparing the effects of prefrontal cortical DA depletion in monkeys on different aspects of cognition all known to be dependent on intact prefrontal functioning has extended the notion of a Yerkes-Dodson type inverted U-shaped curve by showing that the effects are task-dependent. Thus, mesofrontal DA loss does indeed impair delayed response performance, but it is also associated with an enhancement of extra-dimensional shift performance, a paradigm tapping a form of selective attention in which responding has to be switched from one perceptual dimension to another (ROBERTS et al. 1994). Additionally, such DA loss has no effect on the actual sequencing of spatial responses in a working memory paradigm in a task on which frontal lesions profoundly disrupt performance by inducing perseverative responding (COLLINS et al. 1998). This has led to the notion that fluctuations in mesofrontal DA activity, possibly representing a central correlate of enhanced stress or activation, impact upon behaviour in ways depending on environmental demands and the nature of the task at hand. An important related issue to be resolved is the exact relationship of cortical to subcortical DA function, as levels of frontal and striatal DA activity quite often appear to be inversely related, at least in functional terms (e.g. see ROBERTS et al. 1994).

These considerations are important when considering complex behaviour or higher cognitive functioning in which a variety of different capacities have to be co-ordinated effectively, as originally envisaged in the BADDELEY (1986) "working memory" model. So, for example, the effective planning of goal-directed behaviour requires identification and attention to one of several goals, the capacity to compute the optimal route to the goal (involving working memory) and the selection and the execution of the appropriate response sequence leading to that goal. Each of these processes may be best performed in different forebrain regions under different optimal levels of dopaminergic modulation. Thus pharmacological modification of DA is likely to affect performance in different ways. Even a relatively simple procedure such as the spatial delayed response test is known to be subject to demands of attention and response inhibition, as well as "holding stimuli online". Consequently, it is unsurprising that other components of performance can potentially be affected by prefrontal DA loss, and for example, the attentional lability of the animal with prefrontal DA loss must be taken into account. Such lability is often deleterious to good performance; however, it may aid performance of tasks requiring attentional disengagement, such as the extra-dimensional shifting task.

E. Evidence for a Role for Dopamine in Cognition in Humans

Not surprisingly, the analysis of the role of DA in human cognition has been somewhat dominated by the history of the extensive research in experimental animals of the functions of cortical DA in working memory, although there are now signs of more broadly based analyses. The critical evidence derives from two main sources: studies of patients with disorders that implicate the DA system; and studies on the effects of dopaminergic drugs in normal subjects. Such work is beginning to be augmented by the use of functional neuroimaging, generally employing positron emission tomography (PET) but most recently functional magnetic resonance imaging (fMRI), to measure interactions between task and drug effects on regional cerebral blood flow.

I. Dopamine and Cognition in Clinical Disorders: Parkinson's Disease, Schizophrenia, Acute Brain Injury and ADHD

Restorations of underactive (or alternatively, reductions in overactive) dopaminergic transmission are generally assumed to be beneficial for cognitive function and motivate attempts to treat such diverse disorders as Parkinson's disease, schizophrenia, ADHD, and more recently, acute brain injury.

There is little doubt that there is a cognitive deficit syndrome in idiopathic Parkinson's disease, even early in its course (TAYLOR et al. 1986; OWEN et al. 1992) and also following MPTP-induced parkinsonism (STERN and LANGSTON 1985). Many of these deficits are similar to those seen after PFC dysfunction, and include impairments in working memory, planning and set-shifting (ROBBINS et al. 1998b) in the relatively early stages of the disease, although a range of other memory and learning impairments are also evident (e.g. KNOWLTON et al. 1996). However, it is more difficult to be sure which, if any, of these deficits are linked specifically to the loss of central DA function, because of the multivariate nature of the neurochemical pathology of this neurodegenerative disease.

A certain amount can be inferred from a cross-sectional comparison of patients that are initially unmedicated and then treated with L-dopa or related dopaminergic preparations, including DA receptor agonists such as apomorphine, bromocriptine or pergolide. The cognitive deficits seen in Parkinson's disease patients medicated with mild clinical disability may even be less than those seen in patients earlier in the course of the disease and yet to receive medication (DOWNES et al. 1988; OWEN et al. 1995). And inferences can also be made on the basis of longitudinal studies, in which the effects of medication are assessed prior to and following medication, as long as one assumes that the disease itself pursues an unremitting course of further deficit. In one such large-scale study, GROWDON et al. (1998) reported that L-dopa improves motor function without impairing cognition in mild, nondemented Parkinson's disease patients; in fact, performance in tests of executive function, supposed to be sensitive to frontal lobe dysfunction, showed some benefit of medication. However, potentially the most informative evidence is that in which Parkinson's disease patients have their medication removed in a controlled manner. In one study of this type, LANGE et al. (1992) showed that L-dopa withdrawal from a small group ($n = 10$) of Parkinson's disease patients selectively impaired their performance in tests from the Cambridge Neuropsychological Test Automated Battery (CANTAB) of spatial working memory, planning and varieties of visual discrimination learning. However, it was not possible to assess performance in this relatively severely affected group of patients on tests of extra-dimensional set-shifting because of the low number of patients attempting this task. Of interest was that the latency and accuracy of thinking on the planning task were both affected in this group, seemingly paralleling the beneficial effects of medication on bradykinesia in Parkinson's disease.

Dopaminergic medication does not always have beneficial effects on cognition in Parkinson's disease. There is now quite extensive evidence of psychosis-inducing effects of dopaminergic medication including hallucinations (VERHOEVEN and TUINIER 1993), presumably related to the extensive older literature on psychotic effects of amphetamine and related drugs. Moreover, GOTHAM et al. (1988) provided evidence that certain aspects of cognitive performance in Parkinson's disease could actually be worsened by L-dopa. They proposed a hypothesis that related the effects of L-dopa to the pattern and course of DA loss within the striatum in Parkinson's disease. Those regions suffering extensive DA depletion, such as the putamen, would have their functions optimally titrated by DA medication. By contrast, those regions that were relatively spared in the early stages, such as the caudate and ventral striatum, would potentially be disrupted by medication, as the level of DA function would presumably be set supra-optimally by the drug. This hypothesis thus invokes the same inverted U-shaped function as used above to explain the deleterious effects of excessive DA activity in the PFC. Deleterious, as well as beneficial, effects of L-dopa treatment have also been reported in a subset of Parkinson's disease patients in which the motor response to therapy is showing signs of "wearing-off" (KULISEVSKY et al. 1996). Further evidence to support the GOTHAM et al. (1988) hypothesis comes from a recent study by SWAINSON et al. (2000) which showed mild medicated Parkinson's disease patients to perform poorly in tests of probability reversal learning probably associated with ventral striatal and orbitofrontal function – whilst the same Parkinson's disease patients were relatively improved on tests of spatial memory function. A potentially related study by CHARBONNEAU et al. (1996) demonstrated that medicated Parkinson's disease patients were impaired in stimulus-reward but not stimulus–stimulus learning; they hypothesized that the precise timing of DA release necessary for learning would be disrupted in Parkinson's disease by the disease itself, despite, or possibly because of, the medication.

The use of dopaminergic medication in other forms of neurological disturbance is more limited, but case study reports and experimental studies (e.g. MCDOWELL et al. 1998) are suggesting possible applications for brain-injured patients. MCDOWELL et al. (1998) examined the effects of a low dose of the DA D_2 receptor agonist bromocriptine on working memory and other executive forms of cognitive function in individuals with traumatic brain injury in a double-blind cross-over trial with placebo. Consistent with the findings for Parkinson's disease, bromocriptine improved performance on some but not all tasks thought to be subserved by the PFC. Also consistent with the Parkinson's disease literature, no effects were observed for control tasks not thought to be subserved by the PFC. More controversially, and seemingly at odds with both the animal literature and that on normal individuals to be reviewed below, bromocriptine exerted no effects on working memory tasks with minimal additional demands on executive function.

Making inferences about the functions of DA in cognition is less promising in the case of schizophrenia, as anti-psychotic medication may produce indirect effects on performance by the remediation of disruptive positive symptoms. Additionally, neuroleptic drugs, as we have seen above (e.g. WILLIAMS et al. 1998) can impair cognitive functioning (KING 1990). In a comprehensive review, MORTIMER (1997) concluded that much remained unclear about whether neuroleptic treatment affected the cognitive deficit syndrome present in schizophrenia. The effects of conventional neuroleptics are quite small, often being beneficial and related to the remission of psychosis. The possibility that the so-called atypical neuroleptics such as clozapine exert "cognitive facilitatory" as well as "cognitive sparing" effects needs to be resolved using more sophisticated neuropsychological methods and study designs.

The potential complexity of this area can be gauged from a functional neuroimaging study using positron emission tomography (PET) to measure regional cerebral blood flow (rCBF) in normal and unmedicated schizophrenic subjects following challenge with apomorphine or placebo (DOLAN et al. 1995) – extending an analogously-motivated study of the effects of d-amphetamine in schizophrenia (DANIEL et al. 1991). DOLAN et al. found that rCBF was enhanced in the anterior cingulate cortex in the schizophrenic patients under the conditions of a verbal fluency task. However, one problem of interpretation with these is assessing whether the effects of apomorphine depended on an enhancement of DA neurotransmission, or alternatively on reductions, via its pre-synaptic action at D_2 receptors. Another problem of interpretation is posed by the lack of reported data on verbal fluency performance in that study; so although the therapeutic implications may be evident, the actual impact on cognition of cortical actions of apomorphine in the schizophrenic or normal individuals, is a little unclear.

Similar uncertainties about whether treatment is "damping down" unwanted activity or boosting deficient functioning also hinder our understanding of the basis of the apparently effective strategy of treating ADHD with methylphenidate and amphetamine-like compounds (MEHTA et al. 2000; SOLANTO et al. 2000). Converging evidence implicates the dopaminergic system and the prefrontal and nigrostriatal regions in the pathophysiology of childhood ADHD and prefrontal dopaminergic dysfunction in adult ADHD (ERNST et al. 1998), but it remains unclear to what extent the beneficial effects of drugs such as methylphenidate (Ritalin) depend on modulation of dopaminergic or noradrenergic neurotransmission, or both. The neural site of such effects is also unclear. VAIDYA et al. (1998) have recently employed fMRI in a "Go/No Go" functional imaging paradigm to show that methylphenidate attenuated blood flow in the basal ganglia of normal children, but increased blood flow in children with ADHD. On the other hand, equivalent degrees of frontal activation were seen in both groups. Improvements in behavioural performance were also seen in both groups following the drug, but it is difficult to be sure at which neural loci the stimulant is acting to produce these effects. Studies by MATTAY et al. (1996) and MEHTA et al. (2000b) on the effects respectively of d-amphetamine and methylphenidate in normal volunteers, implicate cortical networks that include the dorsolateral PFC. These latter experiments also utilized tasks that normally require PFC functioning (respectively, performance on the Wisconsin Card Sorting Test [WCST] and self-ordered spatial working memory tasks, respectively), and so the identity of the neural networks upon which stimulant drugs exert their effects on performance – for both normal and clinical populations – may hinge on the nature of the task under study.

II. Effects of Dopaminergic Drugs on Cognition in Normal Human Volunteers

The early literature showing that amphetamine-like drugs had beneficial effects on vigilance functions has generally been supported by more recent work (KOELEGA 1993). Despite its use in ADHD, the effects of methylphenidate on other aspects of cognition until recently have not been widely investigated. CLARK et al. (1986) showed that methylphenidate (0.65 mg/kg po) reversed impairments in a dichotic auditory attention task produced by the neuroleptic droperidol. By itself, however, methylphenidate had little effect except to enhance subjective increases in elation, energy and alertness. It was not possible to attribute significant improvements of a similar oral dose in CANTAB tests of self-ordered spatial working memory and planning function (ELLIOTT et al. 1997), which were limited mainly to the first test session. Indeed, when taken on a second session, the drug sometimes increased the speed of responding on certain tests at the expense of reduced accuracy. Also evident were effects to enhance retrieval of certain aspects of performance, consistent with other data (EVANS et al. 1986). A more recent study (ROGERS et al. 1999) has shown that methylphenidate (at the same dose to that employed by ELLIOTT et al. 1997), can improve performance on an extra-dimensional set-shift task, similar to that employed in monkeys by ROBERTS et al. (1994), but at the cost of slowing performance and increasing errors in the control test of intra-dimensional set-shifting. These results are important in showing that it is possible to demonstrate improvements in normal individuals treated with methylphenidate, as well as patients with ADHD. However, consistent with the animal and clinical data reviewed above, other functions may also show impairment. Thus, drugs such as methylphenidate (and presumably also amphetamine) seem to place the subject into an altered mode of functioning that is optimal for certain forms of performance, such as working memory, memory retrieval functions and responding to previously irrelevant stimulus dimensions, though at the cost of other capacities. The challenge now is to determine the contribution of DA itself to these effects and also to identify the neural loci of the drug–task interactions in the intact brain.

The most direct means of addressing this challenge is to study the effects of specific dopaminergic agonists and antagonists on human cognition, ideally also incorporating a functional imaging approach where feasible. Unfortunately, the lack of suitably selective compounds that are also suitable for administering to normal human volunteers (e.g. without emetic and dyskinetic side-effects) has somewhat retarded progress. DA D_2 receptor antagonists generally impair cognitive function in normals. However, the impairments are not simply linked to sedative actions, as for example, sulpiride produces relatively little effect on tests of sustained attention and associative learning that are sensitive to benzodiazepines such as diazepam (MEHTA et al. 1999b). In the same study, however, sulpiride (400mg po) did produce a pattern of impairments that is qualitatively similar to that seen in Parkinson's disease, including deficits in spatial but not visual pattern recognition memory, planning performance and attentional set-shifting – again reflecting capacities mediated by fronto-striatal systems.

The preponderance of D_2 receptor binding in striatal as distinct from cortical regions implicates the striatum as a probable site of action of many of these effects. This is consistent with evidence of correlations between DA D_2 receptor binding in both normal volunteers and patients. For example, VOLKOW et al. (1998) found several significant correlations between performance measures (on tasks administered outside the scanner) and indices of D_2 receptor binding using [^{11}C]-raclopride. Although these were greatest for motor tasks such as finger tapping, significant correlations were also found for measures of cognitive function, including performance on Raven's Matrices, and the Stroop and WCST (categories attained measure) tests, even after correcting for the considerable decline in D_2 receptor binding that occurs with normal ageing. Additionally, LAWRENCE et al. (1998) found that several aspects of performance on spatial working memory and planning tasks exhibited significant correlations with indices of striatal D_2 receptor binding in patients at various stages of Huntington's disease. An exciting prospect would be to attempt to confirm such findings using functional imaging paradigms to effect DA receptor displacement – in other words, directly to relate DA release to cognitive performance in conscious human subjects. Some progress in attaining this goal has been made in what promises to be a seminal study by KOEPP et al. (1998). They were able to show that performance in a motivating video game could be used to reduce binding of raclopride to DA receptors in the region of the ventral striatum, presumably because of striatal DA release engendered by the task. Whilst the nature of the cognitive operations engaged by this task within the striatum could not be identified from this study alone, it nevertheless offers considerable promise for making future advances, particularly if used in combination with the other approaches we have surveyed.

Most impressive of all would be the demonstration of significant facilitation in aspects of cognitive function following specific DA receptor agonists. For the most part, it has only proven feasible to assess performance-altering effects of DA D_2 receptor agents such as bromocriptine, or alternatively, of mixed D_1-D_2 agents such as apomorphine and pergolide. Even though only a handful of studies have emerged so far, significant improvements in certain aspects of cognitive performance have been seen in most of these. The main reported exception used a rather different cognitive task: GRASBY et al. (1992)

showed that the effects of apomorphine (5 and 10 μg s.c.) to impair learning of an auditory-verbal word list in a PET-scanning paradigm were related to its effects to reduce prefrontal cortical regional cerebral blood flow.

The improvements in cognitive function have mainly been observed in visuospatial working memory tasks. Luciana et al. (1992) were the first to demonstrate that bromocriptine (2.5 mg p.o.) enhanced the accuracy of performance in a delayed saccade task. Luciana et al. (1997) extended the result to show improvement of memory for spatial but not object cues at a lower dose of bromocriptine (1.25 mg), and they further demonstrated pharmacological specificity by demonstrating opposed effects of a serotoninergic drug (fenfluramine) (Luciana and Collins 1998). By contrast, Muller et al. (1998), using a rather different delayed matching, working memory task in which subjects had to match the location of a complex visual pattern within a spatial frame of reference, failed to find significant improvement with bromocriptine (2.5 mg). They were able, however, to demonstrate significant benefits of the mixed DA agonist, pergolide, which they attributed potentially to its D_1 receptor agonist properties. Further light has been thrown on the variables controlling these effects from the findings of Kimberg et al. (1997) that the effect of bromocriptine (2.5 mg) in normal young adults depended on their baseline working memory capacity. High-capacity subjects performed more poorly on a range of executive and working memory tasks whereas low-capacity subjects performed better after this dose. This is reminiscent of the inverted U-shaped Yerkes-Dodson-like functions already shown above to be important for determining the effects of dopaminergic manipulations. although Kimberg et al. (1997) invoke more computationally rigorous applications of the sigmoid activation function (Servan-Schreiber et al. 1990). Kimberg et al. thus failed to replicate Luciana's (1992) effects with a task that was slightly different from that used by her, in its inclusion of a central distractor condition. While Kimberg et al. suggest that the discrepancy between their results and those of Luciana might reflect differences in the baseline working memory capacities of their subject samples, another plausible explanation is that the less-complex visuospatial form of the memory task, requiring memory for only the location of a simple stimulus at a single spatial location, may be more sensitive to improvement than the more complex forms of this task. Mehta et al. (2001) have shown that a lower dose of bromocriptine (1.25 mg) improves performance of the CANTAB spatial span task but not its self-ordered spatial working memory equivalent.

Evidently, the effects of dopaminergic agents such as bromocriptine are quite weak and subtle, depending on both the nature of the task under study as well as on baseline capacities of normal individuals. One issue to be resolved is whether a direct agonist is the most effective way of enhancing normal function, as compared to a drug that modulates neurotransmitter release. Nevertheless, the data are exciting in helping to remove the prospect of "cognitive-enhancing" drugs for normal individuals from the realms of science fiction. However, it already seems quite clear that enhancement is only likely to be achieved in certain situations and only at the possible cost of inefficiency in other domains, The apparent susceptibility of individuals low in baseline working memory capacity to cognition-enhancing effects of bromocriptine may be a useful portent for the use of D_2 agonists in clinical applications.

The study of effects of dopaminergic drugs on other aspects of learning and memory, including, for example, acquisition and retrieval in procedural and semantic memory, has been somewhat neglected. This is surprising, given the considerable interest in the roles of the basal ganglia themselves in procedural memory and the promising animal research in this area reviewed above (WHITE 1989; GRAYBIEL 1995). An interesting recent example of dopaminergic effects on associative thought processes, based partly on semantic network theory, concerns the effects of oral L-dopa in normal volunteers tested in a lexical decision paradigm in which direct semantic priming (e.g. by a word such as "black" for the response "white") and indirect semantic priming (e.g. "summer" and "snow", mediated indirectly by associations with the word "winter") were directly compared KISCHKA et al. (1996). The significance of these two types of priming is that direct versus indirect priming represents the spread of activation within semantic networks that encode these associative semantic relationships. Thus, higher signal-to-noise ratios, hypothetically produced by dopaminergic activity within the cortex, are equivalent to more focused activation and a greater degree of direct versus indirect priming. By contrast, low signal-to-noise ratios represent the opposite type of profile. In fact, L-dopa produced evidence of more selective reductions in indirect priming. These results obviously have possible relevance for understanding how associative thought processes might be influenced by DA, and how, for example, schizophrenic thought disorder might implicate more indirect forms of semantic priming, associated with possible reductions in prefrontal dopamine function. However, as the effects of L-dopa were quite subtle, the specificity of these results should be substantiated using a more detailed pharmacological analysis that includes dose-response functions and comparisons with DA receptor blockers.

F. Conclusions and Future Directions

Now it is apparent that brain DA has important roles in many aspects of cognition as well as overt motor behaviour, it is timely to bring into focus future research priorities. These priorities include understanding the relative contributions of the striatal and cortical (mainly frontal) DA systems to behavioural and cognitive functioning, and the extent to which sometimes they appear to be co-ordinated in enabling such functions, but also sometimes opposed, through counter-balancing influences (see Fig. 1). One way of formulating this question is to consider the possibility that the behavioural activation produced by tonically enhanced sub-cortical DA activity normally functionally opposes the modulation of mnemonic and selection functions mediated by neocortical structures. Two of the principal aspects of such behavioural activation in the

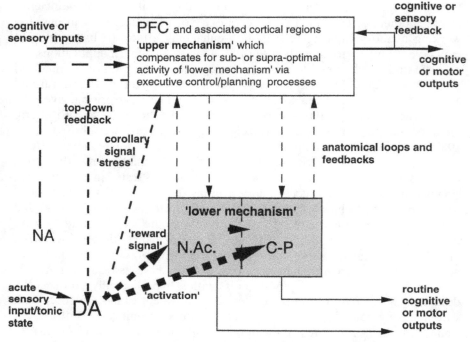

Fig. 1. Schematic diagram to indicate functional relationships between mesocortical and subcortical dopamine systems based on the "two arousal scheme" of BROADBENT (1970), as adapted by ROBBINS (1984). In this diagram, the activation of mesofrontal dopamine systems is seen as a form of "corollary discharge" of the general state of activity in the subcortical systems that helps to modulate the descending influence of the prefrontal cortex (*PFC*) on behavioural regulation, which provides executive control over action selection and performance. Thus, the reinforcement learning systems that are informed and modulated by signals from the subcortical dopamine systems are subject to "top-down" influences of the prefrontal cortex, engaged, for example, during times of stress, or when novel contingencies arise NA = noradenaline

rat include effects on general locomotor activity and also on reinforcing functions, possibly including long term memory consolidation. Such functions, commonly associated with behavioural responses to potent reinforcers, may be suboptimal for "on-line" processing in working memory and the formulation and selection of optimal response sequences or "plans" and better suited for the efficient performance of well-learned or routine actions or habits. Improved understanding of this question will come from a greater knowledge of the relative roles of the PFC and striatum themselves, within the same corticostriatal circuitry. An interesting possibility is that the mesofrontal projection represents in part a sort of "corollary discharge" of the level of activity of the subcortical dopaminergic systems (Fig. 1) Potent reinforcing events (including rewards, novelty and also aversive stimuli), probably mediated via diffuse inputs to activate the midbrain DA cells, thus impact on those regions of the ventral and dorsal striatum implicated in the preparation and initiation of goal-directed behaviour. However, this behaviour has to be performed in an optimal manner, for example, without perseveration and not so rapid as to lead to inaccurate performance. Such regulation requires "executive" or "top-down" adjustments from regions such as the prefrontal cortex – especially when, for example, new routines have to be developed and old ones inhibited, in the manner suggested by SHALLICE and NORMAN's theoretical scheme of the supervisory control over the "contention scheduling" of actions (see SHALLICE 1987).

The scheme shown in Fig. 1 is based on a previous conceptualization that the different chemically identified neurotransmitters of the reticular core of the brain, including the catecholamines DA and noradrenaline, as well as their separate cortical and subcortical projections, have roles in optimizing different forms of processing occurring in the various domains they innervate (see ROBBINS 1984). This scheme incorporates the view that there are "upper" and "lower" arousal mechanisms (BROADBENT 1970) that control different aspects of processing, but which are also interactive. Thus, in a previous scheme, cortical noradrenaline was seen as contributing powerfully to the "upper" mechanism by preserving attentional selectivity under conditions of high levels of behavioural arousal of the "lower" mechanism. The scheme presented in Fig. 1 suggests that prefrontal DA, as also pointed out by ARNSTEN (1998), might fulfill an analogous role. However, we must now attempt to differentiate more clearly the specific roles of these distinct catecholamine projections to the prefrontal cortex (see also ARNSTEN 1998).

The relationship between subcortical and cortical DA function needs to be understood in the context of human cognition, as well as animal behaviour. To some extent the study of the role of DA in reinforcement processes, and in cortically mediated functions such as working memory, has proceeded independently and in separate species. However, this may well change as a consequence of the extensive effort now being devoted to drug abuse, and the realization that chronic drug treatment can impact on cognitive function (e.g. JENTSCH et al. 1997; JENTSCH and TAYLOR 1999; ROGERS et al. 1999b). Another promising direction is that of relating DA-mediated reinforcement learning to more complex decision-making processes in humans (e.g. EGELMAN et al. 1998).

We have identified the possibility of identifiable "states" or "modes" of function associated with low or high levels of DA release (e.g. those associated with elevated "stress" or states associated with the expectation and processing of reinforcers), which optimize different patterns of cognitive as well as behavioural outputs. Here, an improved specification of the genetic, developmental and (perhaps above all) the environmental influences that normally drive DA activity would undoubtedly be useful in understanding why certain cognitive functions (e.g. spatial working memory) appear to be more susceptible to modulation by fronto-striatal DA systems than others. On the other hand, it should be evident from this review that DA modulates a vast range of different aspects of behaviour and cognition, possibly by virtue of its functions within subcortical as well as cortical regions.

We are aware that this chapter has several limitations. While pointing in the Introduction to the diversity of DA receptor types, we have been able to pay only a little attention to their respective roles, mainly because of the lack of suitably selective agents, especially in humans. In similar vein, we have mentioned as a possible approach the use of computational modelling, (e.g. via models of reinforcement learning or constrained neural networks) but we have not been able to invoke it to tackle these residual problems directly. Given the consistency of certain types of finding (e.g. seemingly ubiquitous inverted U-shaped dose-response functions and baseline-dependent effects), nevertheless, it seems likely that such modelling will eventually help to clarify our ideas about the underlying processes, especially when further data have been collected. Nor has there yet been sufficient exploitation of transgenic animals to provide unambiguous extensions of existing knowledge about DA and cognition: this also depends on the development of sensitive tests for such functions in mice, if they can be convincingly demonstrated.

Above all comes the suspicion that it is ultimately simplistic to consider the functions of DA in a single circumscribed area of cognition or behaviour. The role of DA systems in a wide range of behavioural functions, from simple movements through reinforcement mechanisms to advanced planning cognition, suggests that we need to know more about how the "building-blocks" of behaviour are integrated to produce complex behavioural or cognitive output. And the importance of DA in processes that enable rapid responding in the current context has also to be weighed against its possible role in feedback mechanisms leading to information storage – therefore invoking mechanisms of neuronal plasticity. The various modes of functioning of the DA systems (e.g. phasic versus tonic) may best be understood in this context simply in terms of the homeostatic regulation of the activity of this system within narrowly defined limits. In considering both the role of DA in humans as well as for other animals, evolutionary factors come to the fore. We do not consider it appropriate, for example, to consider that DA modulates only simple forms of motor expression in rodents and exerts influences on cognition only in human subjects. Either of these stances would render impossible, for example, the development of animal models of mental illness, and would hinder our understanding of basic cognitive mechanisms. From the commentary we have provided here, however, we hope that the comparative approach, including the identification of behavioural homologies and the utilization of cognitive theory derived from human experimental psychology to neuroscientific endeavours, will continue to be stimulating and productive.

Acknowledgements. We acknowledge helpful discussions and collaborations with our colleagues, as well as financial support from the Wellcome Trust and the MRC.

References

Annett LE, Rogers DC, Hernandez TD, Dunnett SB (1992) Behavioural analysis of unilateral monoamine depletion in the marmoset. Brain 115:825–856

Arnsten AFT (1997) Catecholaminergic regulation of the prefrontal cortex. J Psychopharmacology 11:151–162

Arnsten AFT (1998) Catecholaminergic modulation of prefrontal cortical cognitive function. Trends Cog Sci 2:436–447

Arnsten AFT, Cai JX, Murphy BL, Goldman-Rakic PS (1994) Dopamine D-1 receptor mechanisms in the cognitive performance of young adult and aged monkeys. Psychopharmacology 116:143–151

Arnsten AFT, Cai JK, Steere JC, Goldman-Rakic PS (1995) D2 receptor mechanisms contribute to age-related decline-the effects of quinpirole om memory and motor performance in monkeys. J Neurosci 15:3429–3439

Baddeley AD (1986) Working Memory. Oxford University Press

Bassareo V, Di Chiara G (1997) Differential influence of associative and nonassociative learning mechanisms on the responsiveness of prefrontal and accumbal dopamine transmission to food stimuli in rats fed ad libitum. J Neurosci 17:851–861

Bassareo V, Di Chiara G (1999) Differential responsiveness of dopamine transmission to food-stimuli in nucleus accumbens shell/core compartments. Neuroscience 89:637–641

Baunez C, Robbins TW (1999) Effects of dopamine depletion of the dorsal striatum and further interaction with subthalamic nucleus lesions in an attentional task in the rat. Neuroscience 92:1343–1356

Beninger RJ, Phillips AG (1980) The effect of pimozide on the establishment of conditioned reinforcement. Psychopharmacology 68:147–153

Beninger R, Mason ST, Phillips AG, Fibiger HC (1980) The use of conditioned suppression to evaluate the nature of neuroleptic induced avoidance deficits. J Pharm Exp Ther 213:623–627

Besson C, Louilot A (1995) Asymmetrical involvement of mesolimbic dopaminergic neurons in affective perception. Neuroscience 68:963–968

Blackburn JR, Pfaus JG, Phillips AG (1992) Dopamine functions in appetitive and defensive behaviours. Progress in Neurobiology 39:247–279

Braff DL, Geyer MA (1992) Sensorimotor gating and schizophrenia: Human and animal model studies. Archives of General Psychiatry 47:181–188

Broadbent DE (1970) Decision and Stress. Academic Press, London

Broersen LM, Heinsbroek RPW, Debruin JPC, Uylings HBM, Olivier B (1995) The role of the medial prefrontal cortex of rats in short-term-memory functioning – further support for involvement of cholinergic, rather than dopaminergic mechanisms. Brain Research 674(2):221–229

Brown VJ, Robbins TW (1991) Simple and choice reaction time performance following unilateral striatal dopamine depletion in the rat. Brain 114:513–525

Brozoski T, Brown RM, Rosvold HE, Goldman PS (1979) Cognitive deficit caused by regional depletion of dopamine in prefrontal cortex of rhesus monkey. Science 205:929–931

Calabresi P, Pisani A, Mercuri NB, Gattoni G, Tolu M, Bernardi G (1995) Long-term changes of corticostriatal synaptic transmission: Possible implication for motor memory. In: Kimura M, Graybiel AM (eds) Functions of the cortico-basal ganglia loop. pp 89–103

Carr GD, White NM (1984) The relationship between stereotypy and memory improvement produced by amphetamine. Psychopharmacology 82:203–209

Charbonneau D, Riopelle RJ, Beninger RJ (1996) Impaired incentive learning in treated Parkinson's disease. Canadian Journal of Neurological Science 23:271–278

Clark CR, Geffen GM, Geffen LB (1986) Role of monoamine pathways in the control of attention: Effects of droperidol and methylphenidate in normal adult humans. Psychopharmacology 90:28–34

Cole BJ, Robbins TW (1987) Amphetamine impairs the discriminative performance of rats with dorsal noradrenergic bundle lesions on a 5-choice serial reaction-time-task: New evidence for central dopaminergic-noradrenergic interactions. Psychopharmacology 91:4, 458–466

Cole BJ, Robbins TW (1989) Effects of 6-hydroxydopamine lesions of the nucleus accumbens septi on performance of a 5-choice serial reaction-time task in rats: Implications for theories of selective attention and arousal. Behavioural Brain Research 33:165–179

Collins P, Roberts AC, Dias R, Everitt BJ, Robbins TW (1998) Perseveration and strategy in a novel spatial self-ordered sequencing task for nonhuman primates: Effects of excitotoxic lesions and dopamine depletions of the prefrontal cortex. Journal of Cognitive Neuroscience 10:332–354

Collins P, Wilkinson LS, Everitt BJ, Robbins TW, Roberts AC (2000) The effect of dopamine depletion from the caudate nucleus of the common marmoset on tests of prefrontal cognitive function. Behavioral Neuroscience 114:3–17

Daniel DG, Weinberger DR, Jones DW, Zigun JR, Coppola R, Handel S, Goldberg TE, Berman KF, Kleinman JE (1991) The effect of amphetamine on regional cerebral blood flow during cognitive activation in schizophrenia. Journal of Neuroscience 11:1907–1917

Diamond A (1996) Evidence for the importance of dopamine for prefrontal cortex functions early in life. Phil Trans R Soc Lond 351:1483–1494

Dolan RJ, Fletcher P, Frith CD, Friston KJ, Frackowiak RSJ, Grasby PM (1995) Dopaminergic modulation of impaired cognitive activation in the anterior cingulate cortex in schizophrenia. Nature 378:180–182

Downes JJ, Roberts AC, Sahakian BJ, Evenden JL, Robbins TW (1989) Impaired extra-dimensional shift performance in medicated and unmedicated Parkinson's disease: Evidence for a specific attentional dysfunction. Neuropsychologia 27:1329–1344

Egelman DM, Person C, Montague PR (1998) A computational role for dopamine delivery in human decision-making. Journal of Cognitive Neuroscience 10:623–630

Ellenbroek BA, Knobbout DA, Cools AR (1997) The role for mesolimbic and nigrostriatal dopamine in latent inhibition as measured with the conditioned taste aversion paradigm. Psychopharmacology 129:112–120

Elliott R, Sahakian BJ, Matthews K, Bannerjea A, Rimmer J, Robbins TW (1997) Effects of methylphenidate on spatial working memory and planning in healthy young adults. Psychopharmacology 131:196–206

Ernst M, Zametkin AJ, Matochik JA, Jons PH, Cohen RM (1998) DOPA decarboxylase activity in attention deficit hyperactivity disorder adults: A (fluorine-18) fluorodopa positron emission tomographic study. Journal of Neuroscience 18:5901–5907

Evans RW, Gualtieri T, Amara I (1986) Methyphenidate and memory: dissociated effects in hyperactive children. Psychopharmacology 90:211–216

Evenden JL, Marston HM, Jones GH, Giardini V, Lenard L, Everitt BJ, Robbins TW (1989) Effects of excitotoxic lesions of the substantia innominata, ventral and dorsal globus pallidus on visual discrimination, acquisition, performance and reversal in the rat. Behavioural Brain Research 32:129–149

Fernadez-Ruiz J, Doudet D, Aigner TC (1999) Spatial memory improved by levodopa in parkinsonian MPTP-treated monkeys. Psychopharmacology 147:104–107

Floresco SB, Seamans JK, Phillips AG (1996) A selective role for dopamine in the nucleus accumbens of the rat in random foraging but not delayed spatial win-shift based foraging. Behavioural Brain Research 80:161–168

Gainetdinov RR, Wetsel WC, Jones SR, Levin ED, Jaber M, Caron MG (1999) Role of serotonin in the paradoxical calming effect of psychostimulants on hyperactivity. Science 283:397–401

Goldman-Rakic PS (1987) Circuitry of primate prefrontal cortex and the regulation of behaviour by representational memory. In: Plum F, Mountcastle V (eds) Handbook of Physiology, vol 5. American Physiological Society: Bethesda, MD pp 373–417

Gotham AM, Brown RG, Marsden CD (1988) "Frontal" cognitive function in patients with Parkinson's disease "on" and "off" levodopa. Brain 111:299–321

Granon S, Passetti F, Thomas KL, Dalley JW, Everitt BJ, Robbins TW (2000) Enhanced and impaired attentional performance following infusion of dopamine receptor agents into rat prefrontal cortex. Journal of Neuroscience 20:1208–1215

Grasby PM, Friston KJ, Bench CJ, Frith CD, Paulesu E, Cowen PJ, Liddle PF, Frackowiak RSJ, Dolan R (1992) The effect of apomorphine and buspirone on regional cerebral blood flow during the performance of a cognitive task – measuring neuromodulatory effects of psychotropic drugs in man. Eur J Neuroscience 4:1203–1212

Gray JA, Joseph MH, Hemsley DR, Young AMJ, Warburton EC, Bouleguez P, Grigoryan G, Peter SL, Rawlings JNP, Taib C-T, Yee BK, Cassaday H, Weiner I, Gal G, Gusak O, Joel D, Shadach E, Shalev U, Tarrasch R, Feldon J (1995) The role of mesolimbic dopaminergic and retrohippocampal afferents to the nucleus accumbens in latent inhibition: implications for schizophrenia. Behavioural Brain Research 71:19–31

Graybiel AM (1995) Building action repertoires: memory and learning functions of the basal ganglia. Current Opinion in Neurobiology 5:733–741

Growdon JH, Kieburtz K, McDermott MP, Panisset M, Friedman JH, Shoulson I et al. (1998) Levodopa improves motor function without impairing cognition in mild nondemented Parkinson's disease patients. Neurology 50:1327–1331

Hitchcott PK, Harmer CJ, Phillips GD (1997a) Enhanced acquisition of discriminative approach following intra-amygdala d-amphetamine. Psychopharmacology 132:237–246

Hitchcott PK, Bonardi CMT, Phillips GD (1997b) Enhanced stimulus-reward learning by intra-amygdala administration of a D3 dopamine receptor agonist. Psychopharmacology 133:240–248

Houk JC, Adams JL, Barto AG (1995) A model of how the basal ganglia generate and use neural signals that predict reinforcement. In: Houk JC, Davis JL, Beiser DG (eds) Models of Information Processing in the Basal Ganglia. MIT Press, Cambridge Massachusetts pp 249–270

Humby T, Wilkinson LS, Robbins TW, Geyer M (1996) Prepulses inhibit startle-induced reductions of extracellular dopamine in the nucleus accumbens of rat. Journal of Neuroscience 16:2149–2156

Jentsch JD, Redmond DE, Elsworth JD, Taylor JR, Youngren KD, Roth RH (1997) Enduring cognitive deficits and cortical dopamine dysfunction in monkeys after long-term administration of phencyclidine. Science 277:953–955

Jentsch JD, Taylor JR (1999) Impulsivity resulting from frontostriatal dysfunction in drug abuse: implications for the control of behavior by reward-related stimuli. Psychopharmacology 146:373–390

Kalivas PW, Duffy P (1995) Selective activation of dopamine transmission in the shell of the nucleus accumbens by stress, Brain Res 675:325–328

Kelly PH, Seviour PW, Iversen SD (1975) Amphetamine and apomorphine responses in the rat following 6-OHDA lesions of the nucleus accumbens septia and corpus striatum. Brain Research 94:507–522

Killcross AS, Robbins TW (1993) Differential-effects of intraaccumbens and systemic amphetamine on latent inhibition using an on-base-line, within-subject conditioned suppression paradigm. Psychopharmacology 110:479–489

Killcross AS, Dickinson A, Robbins TW (1994a) Amphetamine-induced abolition of latent inhibition are reinforcer-mediated: Implications for animal models of schizophrenic attentional dysfunction. Psychopharmacology 115:185–195

Killcross AS, Dickinson A, Robbins TW (1994b) Effects of the neuroleptic alpha-flupentixol on latent inhibition in aversively-motivated and appetitively-motivated paradigms: Evidence for dopamine-reinforcer interactions. Psychopharmacology 115:196–205

Kimberg DY, D'Esposito M, Farah MJ (1997) Effects of bromocriptine on human subjects depend on working memory capacity. Neuroreport 8:3581–3585

King DJ (1990) The effect of neuroleptics on cognitive and psychomotor function. Br J Psychiatry 157:799–811

Kinney GG, Wilkinson LO, Saywell KL, Tricklebank MD (1999) Rat strain differences in the ability to disrupt sensorimotor gating are limited to the dopaminergic system, specific to prepulse inhibition, and unrelated to changes in startle amplitude or nucleus accumbens dopamine receptor sensitivity. Journal of Neuroscience 19:5644–5653

Kischka U, Kammer TH, Maier S, Weisbrod M, Thimm M, Spitzer M (1996) Dopaminergic modulation of semantic network activation. Neuropsychologia 34:1107–1113

Knowlton BJ, Mangels JA, Squire LR (1996) A neostriatal habit learning system in humans. Science 273:1399–1402

Koelega HS (1993) Stimulant drugs and vigilance performance. Psychopharmacology 111:1–16

Koepp MJ, Funn RN, Lawrence AD, Cunningham VJ, Dagher A, Jones T, Brooks DJ, Bench CJ, Grasby PM (1998) Evidence for striatal dopamine release during a video game. Nature 393:266–268

Krivanek JA, McGaugh JL (1969) Facilitating effects of pre- and posttrial amphetamine administration on discrimination learning in mice. Agents and Actions 1:36–42

Kulisevsky J, Avila A, Barbaboj M, Antonijoan RLBM, Gironell A (1996) Acute effects of levodopa on neuropsychological performance in stable and fluctuating Parkinson's disease patients at different levodopa plasma levels. Brain 119:2121–2132

Lange KW, Robbins TW, Marsden CD, James M, Owen AM, Paul GM (1992) L-dopa withdrawal selectively impairs performance in tests of frontal lobe function in Parkinson's disease. Psychopharmacology 107:394–404

Lawrence AD, Weeks RA, Brooks DJ, Andrews TC, Watkins LHA, Harding AE, Robbins TW, Sahakian BJ (1998) The relationship between striatal dopamine receptor binding and cognitive performance in Huntington's disease. Brain 121:1343–1355

LeMoal M, Simon H (1991) Mesocorticolimbic dopaminergic network: Functional and regulatory roles. Physiological Reviews 71:155–234

Luciana M, Collins PF (1998) Dopaminergic modulation of working memory for spatial but not object cues in normal volunteers. Journal of Cognitive Neuroscience 9:330–347

Luciana M, Collins PF, Depue RA (1998) Opposing roles for dopamine and serotonin in the modulation of human spatial working memory functions. Cerebral Cortex 8:218–226

Luciana M, Depue RA, Arbisi P, Leon A (1992) Facilitation of working memory in humans by a D2 dopamine receptor agonist. Journal of Cognitive Neuroscience 4:58–68

Mackintosh NJ (1983) Conditioning and associative learning. The Clarendon Press, Oxford

Mark GP, Blander DS, Hoebel BG (1991) A conditioned stimulus decreases extracellular dopamine in the nucleus accumbens after the development of a learned taste aversion, Brain Res 551:308–310

Marshall JF, Teitelbaum P (1977) New considerations in the neuropsychology of motivated behaviors. In: Iversen LL, Iversen SD, Snyder SH (eds) Handbook of Psychopharmacology, vol 7 Plenum Press, New York pp 201–229

Martinez JL, Jensen RA, Messing RB, Vasquez BJ, Soumireu-Mourat B, Gedden D, Liang KC, McGaugh JL (1980) Central and peripheral actions of amphetamine on memory storage. Brain Research 182:157–176

Mattay VS, Berman KP, Ostrem JL, Esposito G, Van Horn JD, Bigelow LB, Weinberger DR (1996) Dextroamphetamine enhances "neural network-specific" physiological

signals: A positron-emission tomography rCBF study. Journal of Neuroscience 16:4816–4822
McDowell S, Whyte J, D'Esposito M (1998) Differential effect of a dopaminergic agonist on prefrontal function in traumatic brain injury patients. Brain 121:1155–1164
Mehta MA, Sahakian BJ, McKenna PJ, Robbins TW (1999) Systemic sulpiride in young adult volunteers simulates the profile of cognitive deficits in Parkinson's disease. Psychopharmacology 146:162–174
Mehta MA, Owen AM, Sahakian BJ, Mavaddat N, Pickard JD, Robbins TW (2000) Methylphenidate enhances working memory by modulating discrete frontal and parietal lobe regions in the human brain. Journal of Neuroscience 20:RC65, 1–6
Mehta M, Swainson R, Ogilvie AD, Sahakian BJ, Robbins TW (2001) Improved short-term spatial memory but impaired reversal learning following the dopamine D2 agonist bromocriptine in human volunteers. Psychopharmacology 159:10–20
Montague PR, Dayan P, Sejnowski TJ (1996) A framework for mesencephalic dopamine systems based on predictive hebbian learning. Journal of Neuroscience 16:1936–1947
Moore H, West AR, Grace AA (1999) The regulation of forebrain dopamine transmission: Relevance to the pathophysiology and psychopathology of schizophrenia. Biological Psychiatry 46:40–55
Mortimer A (1997) Cognitive function in schizophrenia – do neuroleptics make a difference? Pharmacology Biochemistry and Behavior 56:789–795
Murphy BL, Arnsten AFT, Goldman-Rakic PS, Roth RH (1996) Increased dopamine turnover in the prefrontal cortex impairs spatial working memory performance in rats and monkeys. Proc Natl Acad Sci USA 93:1325–1329
Mller U, von Cramon DY, Pollmann S (1998) D1- versus D2-receptor modulation of visuospatial working memory in humans. Journal of Neuroscience 18:2720–2728
Nader K, LeDoux J (1999) The dopaminergic modulation of fear: Quinpirole impairs the recall of emotional memories in rats. Behavioural Neuroscience 113:152–165
Olton DS, Becker JT, Handelmans GE (1979) Hippocampus, space and memory. Behavior and Brain Sciences 2:315–365
Owen AM, James M, Leigh PH, Summers BA, Marsden CD, Quinn NP, Lange KW, Robbins TW (1992) Fronto-striatal cognitive deficits at different stages of Parkinson's disease. Brain 115:1727–1751
Owen AM, Sahakian BJ, Hodges JR, Summers BA, Polkey CE, Robbins TW (1995) Dopamine-dependent fronto-striatal planning deficits in early Parkinson's disease. Neuropsychology 9:126–140
Packard MG, Cahill L, McGaugh JL (1994) Amygdala modulation of hippocampal-dependent and caudate nucleus-dependent memory processes. Proceedings of the National Academy of Sciences 91:8477–8481
Packard MG, White NM (1991) Dissociation of hippocampal and caudate nucleus memory systems by post-training intracerebral injection of dopamine agonists. Behavioral Neuroscience 105:295–306
Parkinson JA, Olmstead MC, Burns LH, Robbins TW, Everitt BJ (1999) Dissociation in effects of lesions of the nucleus accumbens core and shell on appetitive Pavlovian approach behavior and the potentiation of conditioned reinforcement and locomotor activity by D-Amphetamine Journal of Neuroscience 19:6, 2401–2411
Parkinson JA, Robbins TW, Everitt BJ (2000) Dissociable roles of the central and basolateral amygdala in appetitive emotional conditioning. European Journal of Neuroscience 12:405–413
Pennartz CMA (1995) The ascending neuromodulatory systems in learning by reinforcement: Comparing computational conjectures with experimental findings. Brain Research Reviews 21:219–245
Pennartz CMA, Groenewegen HJ, Lopes Da Silva FJ (1994) The nucleus accumbens as a complex of functionally distinct neuronal ensembles: An integration of behav-

ioral, electrophysiological and anatomical data. Progress in Neurobiology 42: 719–761

Ploeger GE, Spruijt BM, Cools AR (1994) Spatial localization in the Morris watermaze task in rats: Acquisition is affected by intra-accumbens injections of the dopaminergic antagonist haloperidol. Behavioral Neuroscience 108:927–934

Ralph RJ, Varty GB, Kelly MA, Wang YM, Caron MG, Rubinstein M, Grandy DK, Low MJ, Geyer MA (1999) The dopamine D-2, but not D-3 or D-4, receptor subtype is essential for the disruption of prepulse inhibition produced by amphetamine in mice. Journal of Neuroscience 19:11, 4627–4633

Redgrave P, Prescott TJ, Gurney K (1999) Is the short-latency dopamine response too short to signal reward error? Trends in Neurosciences 22:146–151

Rescorla RA, Wagner AR (1972) A Theory of Pavlovian conditioning: Variations in the effectiveness of reinforcement and non-reinforcement. In: Black AH, Prokasy WF (eds) Classical Conditioning II: Current research and theory. Appleton-Century-Crofts, New York, pp 64–99

Robbins TW (1984) Cortical noradrenaline, attention and arousal. Psychological Medicine 14:13–21

Robbins TW (1985) Neuropsychological evaluation of higher cortical function in animals and man: Can neuropsychology contribute to psychopharmacology? In: Iversen SD (ed) Psychopharmacology: Recent advances and future prospects, Oxford University Press, Oxford, pp 155–169

Robbins TW (1998) The psychopharmacology and neuropsychology of attention in experimental animals. In: Parasuraman R (ed) The Attentive Brain. MIT Press, Cambridge MA pp 189–221

Robbins TW, Brown VJ (1990) The role of the striatum in the mental chronometry of action: a theoretical review. Reviews in the Neurosciences 2:181–213

Robbins TW, Everitt BJ (1992) Functions of dopamine in the dorsal and ventral striatum. In: Robbins TW (ed) Seminars in the Neurosciences Vol 4. Saunders, London pp 119–127

Robbins TW, Owen AM, Sahakian BJ (1998b) The neuropsychology of basal ganglia disorders: An integrative cognitive and comparative approach. In: Ron MA, David AS (eds) Disorders of Brain and Mind Cambridge University Press pp 57–84

Robbins TW, Sahakian BJ (1979) "Paradoxical" effects of psychomotor stimulant drugs in hyperactive children from the standpoint of behavioural pharmacology. Neuropharmacology 18:931–950

Robbins TW, Taylor JR, Cador M, Everitt BJ (1989) Limbic-striatal interactions and reward-related processes. Neruoscience and Biobehavioural Reviews 13:155–162

Robbins TW, Giardini V, Jones GH, Reading PE, Sahakian BJ (1990) Effects of dopamine depletion from the caudate-putamen and nucleus accumbens septi on the acquisition and performance of a conditional discrimination task. Behavioural Brain Research 38:243–262

Robbins TW, Granon S, Muir JL, Durantou F, Harrison A, Everitt BJ (1999) Neural systems underlying arousal and attention. In: Harvey JA, Kosofsky BE (eds) Cocaine: Effects on the developing brain Vol 846. Annals of the New York Academy of Sciences, pp 222–237

Roberts AC, De Salvia MA, Wilkinson LS, Collins P, Muir JL, Everitt BJ, Robbins TW (1994) 6-hydroxydopamine lesions of the prefrontal cortex in monkeys enhance performance on an analogue of the Wisconsin Card Sorting Test: Possible interactions with subcortical dopamine. Journal of Neuroscience 14:2531–2544

Roberts AC, Robbins TW, Weiskrantz L (1998) Discussion and conclusions. In: Roberts AC, Robbins TW, Weiskrantz L (eds) The Prefrontal Cortex: Executive and Cognitive Functions Oxford University Press, pp 221–242

Rogers RD, Blackshaw AJ, Middleton HC, Matthews K, Hawtin K, Crowley C, Hopwood A, Wallace C, Deakin JFW, Sahakian BJ, Robbins TW (1999a) Tryptophan depletion impairs stimulus-reward learning while methylphenidate disrupts attentional control in healthy young adults: implications for the monoaminergic basis of impulsive behaviour. Psychopharmacology 146:482–492

Sahakian BJ, Sarna GS, Kantamaneni BD, Jackson A, Hutson PH, Curzon G (1985) Association between learning and cortical catecholamines in non-drug-treated rats. Psychopharmacology 86:339–343

Salamone JD (1994) The involvement of nucleus accumbens dopamine in appetitive and aversive motivation. Behavioural Brain Research 61:117–133

Salamone JD, Cousins MS, McCullough LD, Carriero DL, Berkowitz RJ (1994) Nucleus accumbens dopamine release increases during instrumental lever pressing for food but not free food consumption. Pharmacological Biochemical Behaviour 49:25–31

Saulskaya N, Marsden CA (1995) Conditioned dopamine release: Dependence upon N-methyl-D-aspartate receptors. Neuroscience 67:57–63

Sawaguchi T, Goldman-Rakic PS (1991) D1 dopamine receptors in prefrontal cortex: Involvement in working memory. Science 251:947–950

Schneider JS (1990) Chronic exposure to low doses of MPTP. II Neurochemical and pathological consequences in cognitively impaired, motor asymptomatic monkeys. Brain Research 534:25–36

Schultz W (1992) Activity of dopamine neurons in the behaving primate. Seminars in Neuroscience 4:129–138

Schultz W, Dayan P, Montague PR (1997) A neural substrate of prediction and reward. Science 275:1593–1599

Seamans JK, Floresco SB, Phillips AG (1998) D1 receptor modulation of hippocampal-prefrontal cortical circuits integrating spatial memory with executive control functions in the rat. Journal of Neuroscience 18:1613–1621

Seiden LS, Miller FE, Heffner TG (1989) Neurotransmitters in Attention Deficit Disorder. In: Sagvolden T, Archer T (eds) Attention Deficit Disorder. Erlbaum, London, pp 223–253

Servan-Schneider D, Printz H, Cohen JD (1990) A network model of catecholamine effects: gain, signal-to-noise ratio, and behaviour. Science 249:892–895

Setlow B, McGaugh JL (1998) Sulpiride infused into the nucleus accumbens post-training impairs memory of spatial water maze training. Behavioral Neuroscience 112:603–610

Setlow B, McGaugh JL (1999) Involvement of the posteroventral caudate-putamen in memory consolidation in the Morris water-maze. Neurobiology of Memory and Learning 71:240–247

Shallice T (1982) Specific impairments of planning. Phil. Trans. Royal Society London B 298:199–209

Solanto M, Arnsten AFT, Castellanos FX (Eds) Stimulant drugs and ADHD: Basic and Clinical Neuroscience. Oxford University Press, New York 2001

Solomon PR, Staton DM (1982) Differential effects of microinjections of d-amphetamine into the nucleus accumbens or the caudate-putamen on the rat's ability to ignore an irrelevant stimulus. Biological Psychiatry 17:743–756

Stern Y, Langston W (1985) Intellectual changes in patients with MPTP-induced parkinsonism. Neurology 35:1506–1509

Sutton MA, Beninger RJ (1999) Psychopharmacology of conditioned reward: evidence for a rewarding signal at D-1-like dopamine receptors. Psychopharmacology 144(2):95–110

Swainson R, Rogers RD, Sahakian BJ, Summers BA, Polkey CE, Robbins TW (2000) Probabilistic learning and reversal deficits in patients with Parkinson's disease or frontal or temporal lobe lesions: possible adverse effects of dopaminergic medication. Neuropsychologia 38:596–612

Swerdlow NR, Braff DL, Taiad N, Geyer MA (1994) Assessing the validity of an animal model of deficient sensori-motor gating in schizophrenic patients. Arch Gen Psychiat 51:139–154

Taylor AE, Saint-Cyr JA, Lang AE (1986) Frontal lobe dysfunction in Parkinson's disease. Brain 109:845–843

Taylor JR, Robbins TW (1986) 6-hydroxydopamine lesions of the nucleus accumbens, but not of the caudate nucleus, attenuate enhanced responding with reward-related stimuli produced by intra-accumbens d-amphetamine. Psychopharmacology 90:390–397

Taylor JR, Roth RH, Sladek JR Jr, Redmond DE Jr (1990) Cognitive and motor deficits in the performance of an object retrieval task with a barrier-detour in monkeys (Cercopithecus aethiops sabaeus) treated with MPTP: Long-term performance and effect of transparency of the barrier. Behavioral Neuroscience 104:564–576

Ungerstedt U (1971) Striatal dopamine release after amphetamine or nerve regeneration revealed by rotational behaviour. Acta Physiol Scand 367:49–68

Vaidya CJ, Austin G, Kirkorian G, Ridlehuber HW, Desmond JE, Glover GH, Gabrieli JDE (1998) Selective effects of methylphenidate in attention deficit hyperactivity disorder: A functional magnetic resonance study. Proc Natl Acad Sci 95:14494–14499

Verhoeven WMA, Tuinier S (1993) Dopaminometic psychosis: Thoughts on etiology. In: Wolters ECH, Scheltens P (eds) Mental Dysfunction in Parkinson's Disease, ICG Printing, Dordrecht

Volkow ND, Gur RC, Wange C-J, Fowler JS, Moberg PJ, Ding Y-S, Hitzemann R, Smith G, Logan J (1998) Association between decline in brain dopamine activity with age and cognitive and motor impairment in healthy individuals. Am J Psychiatry 155:344–349

Wan FJ, Swerdlow NR (1996) Sensori-motor gating is regulated by different dopamine-glutamate interactions in the nucleus accumbens core and shell sub-regions. Brain Research 722:168–176

Ward NM, Brown VJ (1996) Covert orienting of attention in the rat and the role of striatal dopamine. Journal of Neuroscience 16:3082–3088

Weiner I (1990) Neural substrates of latent inhibition: The switching model. Psychological Bulletin 108:442–461

Weiner I, Feldon J, Katz J (1987) Facilitation of the expression but not the acquisition of latent inhibition by haloperidol in rats. Pharmacol Biochem Behav 26:241–246

Weiner I, Lubow TE, Feldon J (1984) Abolition of the expression but not the acquisition of latent inhibition by chronic amphetamine in rats. Psychopharmacology 83:194–199

White NM (1989) A functional hypothesis concerning the striatal matrix and patches: mediation of S-R memory and reward. Life Science 45:1943–1957

White NM, Viaud MD (1991) Localized intracaudate dopamine D2 receptor activation during the post-training period improves memory for visual or olfactory conditioned emotional responses in the rat. Behavioural and Neural Biology 55:255–269

Wilkinson LS, Killcross AS, Humby T, Hall FS, Geyer MA, Robbins TW (1994) Social isolation produces developmentally-specific deficits in pre-pulse inhibition of the acoustic startle response but does not disrupt latent inhibition. Neuropsychopharmacology 10:61–72

Wilkinson LS, Humby T, Killcross AS, Torres EM, Everitt BJ, Robbins TW (1998) Dissociations in dopamine release in medial prefrontal cortex and ventral striatum during the acquisition and extinction of classical aversive conditioning in the rat. European Journal of Neuroscience 10:1019–1026

Williams GV, Goldman-Rakic PS (1995) Modulation of memory fields by dopamine D1 receptors in prefrontal cortex. Nature 376:572–575

Williams JH, Wellman NA, Geaney DP, Feldon J, Cowen PJ, Rawlins JNP (1997) Haloperidol enhances latent inhibition in visual tasks in healthy people. Psychopharmacology 133:262–268

Williams JH, Wellman NA, Geaney DP, Cowen PJ, Feldon J, Rawlins JWP (1998) Residual latent inhibition in people with schizophrenia: an affect of psychosis or of its treatment. British Journal of Psychiatry 172:243–249

Wilson C, Nomikos GC, Collu M, Fibiger HC (1995) Dopaminergic correlates of motivated behavior: importance of drive. J Neurosci 15:5169–5178

Wolterink G, Phillips G, Cador M, Donselaar-Wolterink I, Robbins TW, Everitt BJ (1993) Relative roles of ventral striatal D1 and D2 dopamine receptors in responding with conditioned reinforcement. Psychopharmacology 110:355–364

Young AMJ, Joseph MH, Gray JA (1993) Latent inhibition of conditioned dopamine release in rat nucleus accumbens. Neuroscience 54:5–9

Young AMJ, Ahier RG, Upton RL, Joseph MH, Gray JA (1998) Increased extracellular dopamine in the nucleus accumbens of the rat during associative learning of neutral stimuli. Neuroscience 83:1175–1183

Zahrt J, Taylor JR, Mathew RG, Arnsten AFT (1997) Supranormal stimulation of D1 dopamine receptors in the rodent prefrontal cortex impairs working memory performance. Journal of Neuroscience 17:8528–8535

CHAPTER 20
Molecular Knockout Approach to the Study of Brain Dopamine Function

G.F. KOOB, S.B. CAINE, and L.H. GOLD

A. Introduction

Excellent pharmacological tools are available for manipulation of various neuropharmacological components of the dopamine system. However, a major drawback of the pharmacological approach is that almost all drug antagonists or agonists have multiple sites of action, certainly at higher doses. A molecular biological approach provides a means of selectively manipulating the genes that encode the proteins responsible for a given neuropharmacological site with little concern of crosstalk or pharmacological interaction. The cloning of the genes responsible for encoding dopamine receptors and transporter proteins, as well as the proteins responsible for the synthesis of dopamine, has provided the molecular information necessary to decrease or eliminate these proteins and assess function. Such a knockout approach has a number of advantages over traditional pharmacological approaches but also a number of disadvantages.

The present chapter will briefly describe what constitutes the molecular pharmacological approach, define knockouts, and review the results obtained to date with this approach. Evidence exists for phenotypes produced by knockout of the D_1, D_2, D_3, D_4, and D_5 dopamine receptor subtypes, knockout of DARPP-32 (dopamine and cAMP-regulated phosphoprotein of molecular weight 32,000), knockout of the dopamine transporter, and knockout of tyrosine hydroxylase. A "knockout" or knockout mouse will refer in this review to mice carrying a specific mutation through the process of gene targeting by homologous recombination.

In the knockout approach, specified changes are introduced into the nucleotide sequence of a chosen gene, and through either insertions or deletions the gene becomes inactivated resulting in a consequent absence of the gene product. To produce such an inactivation for a gene of interest, a DNA-targeting vector is employed to generate a chromosome with the targeted gene mutation through the process of homologous recombination. Here, DNA molecules with identical sequences line up next to each other, are cut, and subsequently spliced at the cut ends. This results in homologous regions of the genomic DNA in the vector replacing the original gene in the chromosome and then transferring the modified genetic material (responsible for inactivation of the normal protein) into the genome of a living cell (see Fig. 1). Two types of vectors can be used either where the endogenous sequence is replaced by an exogenous sequence or where the entire vector DNA sequence is inserted. A neomycin resistance gene also is inserted to serve as a positive marker to identify which chromosomes have received the vector. A negative

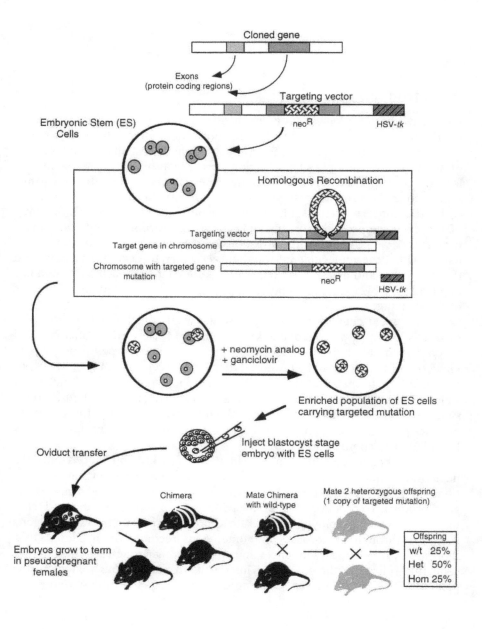

selection marker such as the thymidine kinase gene is attached to one end of the vector to identify cells that have incorporated the targeting vector at a random location (see Fig. 1). The vector is transfected, or passively introduced into embryonic stem cells that are maintained in culture. Cells possessing the random insertions then are removed by exposing the culture to an agent such as ganciclovir that kills the cells bearing the negative selectable marker. Cells in which the cloned gene has replaced the targeted sequences in the chromosome are selected by exposing the culture to neomycin, since the neomycin resistance gene is incorporated into the vector with the desired knockout sequence.

The embryonic stem cells often are derived from the 129 strain mouse (a brown mouse) and then are microinjected into embryos usually derived from the C57BL/6 mouse (a black mouse) at the blastocyst stage. These embryos are then implanted into surrogate mothers. The offspring can be sorted by coat color and the chimeric males are crossed with females from the C57BL/6 strain. Progeny with brown coats then are screened for inheritance of the targeted mutation (some brown coats will not be mutants). Offspring with the targeted mutation are identified by analysis of genomic tail DNA. Subsequent mating of the males and females carrying the mutation will result in some mice (25%) that possess two copies of the mutated gene (for reviews of the theoretical and methodological aspects of knockout technology see CAMPBELL and GOLD 1996; PHILLIPS 1996; TECOTT and BARONDES 1996; PICCIOTTO and WICKMAN 1998; MULLER 1999).

Fig. 1. Generation of knockout mouse mutants by targeted gene replacement. A targeting vector is constructed in which a neomycin resistance (*neo-R*) gene is inserted into a protein-coding region of the gene of interest. Attached to one end of the targeting vector is a thymidine kinase (*tk*) gene from a herpes simplex virus that will serve as a negative selectable marker. The vectors are introduced into embryonic stem (*ES*) cells by exposing the cells to a brief pulse of electrical current, called electroporation. This transiently opens pores in the cell membranes, permitting passage of the vectors into the cells. The vectors then line up with the chromosomes allowing exchange of identical regions of the chromosomes and the targeting vector, termed homologous recombination. Homologous recombination results in chromosomes that possess the targeted insertion, and thus are resistant to neomycin analogues. Also associated with homologous recombination is a loss of the terminal *tk* gene, thus eliminating the susceptibility to ganciclovir. Random insertion events may take place in non-target genes and result in a retention of the *tk* gene, and likewise ganciclovir sensitivity. In situations where the vector does not become integrated at all, the chromosome will lack both the *neo-R* and *tk* genes. Taking advantage of these positive and negative selectable markers by treatment of the ES cells with a neomycin analogue and ganciclovir eliminates those cells that do not possess a *neo-R* gene and those that retain the *tk* gene. The ES cell population is thus enriched for cells carrying the targeted mutation. These cells are injected into a blastocyst stage embryo derived from a second mouse strain. Chimeric male mice bearing cells from both mouse strains are bred with wild-type mice. The heterozygous offspring, possessing one copy of the mutated gene, then are mated. The genotype of the progeny is determined by analysis of tail DNA to be either wild-type (*w/t*), heterozygous (*Het*) or homozygous (*Hom*) for the targeted mutation. (With permission from GOLD 1996)

B. Limitations of the Knockout Approach: Compensation and Epistasis

The promise of the knockout approach is to reveal the in vivo function of a gene of interest because the mutant organism resulting from gene targeting completely lacks the gene product in question, in this chapter a dopaminergic function. One of the major advantages of the knockout approach is that the targeted protein is completely removed and in effect represents a complete lesion. Thus, the hypothesis that a given phenotype is only manifested by a specific genotype can be readily tested. Also, the complete removal of a given receptor conveys a means of assessing the selectivity of allegedly selective neuropharmacological agonists and antagonists.

However, there are a number of potential problems that arise from using the traditional knockout approach and they center on two major issues: (1) there may be compensatory changes in response to the primary effects of the mutation that produce the phenotype, and (2) the phenotype may be masked or exaggerated by the nature of the genetic background. First, it is clear that mutations can lead to an "avalanche of compensatory processes" (up- or downregulation of gene products; GERLAI 1996). Thus, phenotypical changes might not be directly related to the mutation at a functional level but may reflect secondary changes. Clearly, the use of selective pharmacological agents in combination with gene targeting can help resolve some of the discrepancies (see Sect. E.). As with the lesion approaches of earlier decades, a confluence of information using various techniques will be required to confirm any hypotheses regarding function from the knockout approach. Compensatory changes are presumably more likely to occur for mutations that are critical for survival of the species, and the absence of effects of a knockout on any particular dependent variable may reflect redundancy in the system in question or possible compensatory changes in other systems. Perhaps most consequential for interpreting the effects of knockouts is the variability in genetic background that contributes to the mutant mouse. Phenotypic variations in the parental strains also may mask the expression of an underlying mutation, a phenomenon known as epistasis.

Gene targeting has been carried out largely by using embryonic stem cells from the mouse 129 strain. Chimeras are mated to wild-type mice, which are invariably a different strain C57BL/6 (see below). The offspring are not only heterozygous for the null mutant but also have one set of genes from the 129 strain and one set from C57BL/6. The F-2 generation then provides the null mutants in 25% of the animals according to Mendelian genetics. However, there are three problems with this approach that derive from the segregation of a population from two parental mouse strains with recombinant genotypes (GERLAI 1996). First, the recombination pattern (that is, which gene locus contains 129 alleles and which gene locus contains C57BL/6 alleles) may be different between littermates, thus suggesting that even wild-type littermates may not be a good control population. Second, the genetic variation resulting from the hybrid background may mask significant effects. Finally, there may be a problem of gene linkage where the alleles that are near the targeted locus will be 129-type in the null mutant mice and C57BL/6 in the wild-type mice. Because the probability of genetic recombination is inversely related to the distance from two genes, the 129-type alleles of the embryonic stem cells that are close to the locus of the mutated gene will remain with the mutated allele of the knockout gene (GERLAI 1996). These problems cannot be readily dismissed, because a number of recent studies have shown that knockout effects depend on the background phenotype (SIBILIA and WAGNER 1995; OLSON et al. 1996). Certainly, the differences in the phenotypes of D_1- and D_2-null mutants generated from different laboratories may be explained by such factors (see below).

Solutions to these background problems could include classical genetic approaches and some additional molecular physiology (BANBURY CONFERENCE 1997). For example, backcrossing the mutant hybrid animals to a strain of choice would effectively eliminate many of the above concerns. However, complete elimination of the genes associated with the embryonic stem cell 129-strain could take more than 12 generations (2 years of breeding). Alternatively, rescue experiments could be conducted where the missing protein product is delivered chronically to the animal or by introducing a transgene that expresses the protein in question. Other alternatives would be knockin mice in addition to knockout mice, where homologous recombination is used to insert a small DNA marker flanking the gene of interest without altering the gene of interest's function. These knockin mice would have the full function of the gene in question but the same linkage associations of the knockout mice, providing an excellent control (GERLAI 1996; CRAWLEY 1996). Other solutions might be the generation of null mutant mice with a pure genetic background. Nevertheless, the significant genetic differences between inbred strains that are well established in the literature will have to be considered carefully in choosing an appropriate strain for the gene target in question (PLOMIN et al. 1990). At the very least, steps can be taken to ensure that knockouts do not genetically drift away from their wild-type controls (PHILLIPS et al. 1999). For example, periodic interbreeding of mutant and control populations to yield heterozygotes that serve as a renewed source of knockout and wild-type offspring restores the commonality of the genetic backgrounds. Exclusive heterozygote breeding, though inefficient, is an even more rigorous approach to maintaining similar genetic backgrounds of knockout and control populations. Future manipulations that can complement the standard knockout approach will be attempts to rescue the phenotype of a knockout by gene transfer, the use of conditional knockouts, the use of tissue-specific knockouts, and the use of multiple knockouts and pharmacological probes to assess potential compensatory changes. A convergence of evidence using multiple approaches will provide a powerful means of utilizing the knockout technique to its fullest potential, a conceptual position not unlike that considered for lesion studies.

C. Overview of the Midbrain Dopamine System in Motor Behavior and Reward

Dopamine neurons that project to the forebrain long have been associated with initiation of behavior, reward, and motivational processes. The cell bodies of origin of the forebrain dopamine projections can be found in the ventral part of the midbrain, and they project to the forebrain in two major functional systems. The nigrostriatal dopamine system projects from the substantia nigra to the corpus striatum, and degeneration of this system is the primary basis for many of the motor dysfunctions associated with Parkinson's disease (MOORE and BLOOM 1978). The nigrostriatal dopamine system also is implicated in the focused repetitive behavior, called stereotyped behavior, associated with high doses of stimulants (CREESE and IVERSEN 1974). The mesocorticolimbic dopamine system, in contrast, projects from the ventral tegmental area to the limbic forebrain (nucleus accumbens, olfactory tubercle, amygdala, and frontal cortex; MOORE and BLOOM 1978). The mesocorticolimbic dopamine system has been implicated in activation and locomotor behavior, psychostimulant-induced locomotor behavior, drug reward, and non-drug motivational attributes (KELLY et al. 1975; LE MOAL and SIMON 1991; KOOB 1996).

Pharmacological manipulations which increased or decreased dopaminergic function provided some of the early evidence for a role of the midbrain dopamine systems in reward. Pharmacological activation of dopamine synaptic activity produced behavioral activation, facilitated responding for many reinforcers, and decreased reward thresholds (LE MOAL and SIMON 1991; KOOB 1992; ROBBINS and EVERITT 1992). Blockade of dopamine function produced decreases in responding for both positive and negative reinforcers (WISE 1978, 1980, 1982). In addition, electrophysiological studies have shown that unpredictable appetitive stimuli and conditioned reward-predicting stimuli activate the actual physiological firing of midbrain dopamine neurons. In studies of responses to stimuli of specific motivational valence, only appetitive events and not aversive events activated dopamine neurons in the mesocorticolimbic dopamine system of monkeys (MIRENOWICZ and SCHULTZ 1996). Such hedonic selectivity of the activation of these neurons also provides an intriguing insight into the conceptualization of what constitutes positive rewards or incentives. One interpretation of these results is that midbrain dopamine neurons may be part of the process by which rewards motivate or guide behavior (incentive motivation). Changes in positive incentives would, through an activation of the mesocorticolimbic dopamine system, allow or actually release species-specific approach responses or changes in direction toward these larger incentives. The mechanism for this enabling function could be hypothesized to be through additional activation of the central motive state (in addition to primary drives) or by feeding directly to motor routines in the extrapyramidal motor system or both (KOOB 1996).

D. Overview of the Dopamine Receptor Subtypes in Motor Behavior and Reward

Five different dopamine receptors, D_1 through D_5, have been identified through which dopamine may act to produce its functional effects (SOKOLOFF and SCHWARTZ 1995). Most pharmacological studies have been performed using agonists and antagonists for the D_1 and D_2 receptors because selective agents for these receptors have been available. D_1 and D_2 receptors are widely distributed throughout the terminal areas of both the mesocorticolimbic and nigrostriatal dopamine systems, but D_3 receptors are localized to specific subregions of the mesocorticolimbic dopamine system of the rat, namely the shell subdivision of the nucleus accumbens and the Islands of Calleja (SOKOLOFF et al. 1990). Interestingly, few D_2 receptors are found in these subregions, but these subregions are rich in D_1 receptors (SOKOLOFF et al. 1997).

Dopamine D_1 and D_2 antagonists in general block motor activity and block the locomotor activation associated with psychostimulant drugs that are direct or indirect dopamine agonists (ARNT 1985; AMALRIC et al. 1986), and in general, agonists for the D_1 and D_2 dopamine receptors produce locomotor activation and arousal (MOLLOY and WADDINGTON 1984; WADDINGTON et al. 1994). However, dose-effect functions and more selective compounds for the dopamine receptors have revealed some functional distinctions. Low doses of D_1 antagonists can block the locomotor activation produced by d-amphetamine without producing motor effects such as catalepsy or increases in reaction time in a sensitive reaction time task (AMALRIC and KOOB 1993; AMALRIC et al. 1993; SMITH et al. 2000). In contrast, D_2 antagonists at very low doses effectively block reaction time performance, whereas D_1 and D_3 selective antagonists are ineffective (AMALRIC et al. 1993; SMITH et al. 2000). D_3 receptor antagonists actually produce increases in locomotor activity at low doses, an effect attributed to a subset of postsynaptic receptors mediating tonic behavioral inhibition (WATERS et al. 1993; SAUTEL et al. 1995). However, D_3 receptors also may act synergistically with D_1 receptors to produce locomotor sensitization (BORDET et al. 1997).

All three major dopamine receptor subtypes have been implicated in psychostimulant drug reward (KOOB et al. 1996). Antagonists of D_1, D_2, and D_3 receptors dose-dependently decrease the interinjection interval for intravenous cocaine self-administration in rats (MORETON 1991; CAINE and KOOB 1994; HUBNER and KOOB et al. 1996; CAINE et al. 1997). Both D_1 and D_2 antagonists have been shown to shift dose-effect functions for cocaine to the right (BERGMAN et al. 1990; CAINE and KOOB 1995). D_2 and D_3 agonists potentiate or supplement the reinforcing and discriminative stimulus effects of cocaine, and these drugs also maintain self-administration behavior when substituted for cocaine (WOOLVERTON et al. 1984; CAINE and KOOB 1993, 1995; LAMAS et al. 1996; NADER and MACH 1996; SPEALMAN 1996). D_1 agonists appear to have a more complex profile – these drugs are self-administered under some conditions (SELF and STEIN 1992; WEED and WOOLVERTON 1995; GRECH et al. 1996) but not others (GRECH et al. 1996; CAINE et al. 1999). Moreover, unlike D_2 and D_3 agonists, D_1 agonists do not "prime" reinstatement of cocaine self-administration (SELF et al. 1996; BARRETT-LARIMORE and SPEALMAN 1997), nor do they shift the dose-effect function for cocaine self-administration leftward (CAINE et al. 1999, 2000).

E. D_1 Receptor Knockouts

Studies with knockouts of dopamine receptor subtypes have provided two major and important sources of information on the functioning of the midbrain dopamine systems. First, they have provided new insights into the functional role of these dopamine effector systems, and second, they provide a means of evaluating the pharmacological specificity and selectivity of purported selective ligands.

Two groups have generated D_1 receptor knockouts (D_{1A} receptor knockouts) (DRAGO et al. 1994; XU et al. 1994a,b). Targeted gene deletion was constructed from 129 embryonic stem cells and male chimeras were mated with C57BL/6 females to produce heterozygous mutants. Constructs for the targeting vector were made from the 129/Sv (DRAGO et al. 1994) or mouse 129 genomic library (XU et al. 1994a,b).

Mice that lack the dopamine D_1 receptor have a phenotype that confirms hypotheses regarding the functional role of the dopamine system. D_1 knockouts show no locomotor activity response to D_1 agonists and antagonists (XU et al. 1994b; Fig. 2) and a blunted locomotor stimulation to cocaine (XU et al. 1994a) and amphetamine (CRAWFORD et al. 1997). These mice also show spontaneous hyperactivity to vehicle injections (MINER et al. 1995), increases in grooming (CLIFFORD et al. 1998), but decreases in rearing (DRAGO et al. 1994; CLIFFORD et al. 1998). Others have observed decreases in novelty- and neuropeptide-induced grooming (DRAGO et al. 1999) and decreases in exploration (initiation of movement and reactivity to external stimuli) in an open field test (SMITH et al. 1998). Such knockout mice also were impaired in the visual-orienting response (SMITH et al. 1998) and in learning a water maze task (place training or cue training; SMITH et al. 1998) yet showed no deficit in acquisition

Fig. 2. Effects of D_1 agonist SKF 81297 and D_1 antagonist SCH 23390 on locomotor activity and catalepsy. **A** Following a 2-h habituation period, mice were injected with saline or increasing doses of SKF 81297. Values represent mean+standard error of mean (SEM) total photocell beam interruptions (horizontal and vertical activity combined) for 2-h test sessions tested during the light phase of the light-dark cycle. **B** Following a 2-h habituation period, mice were injected with vehicle or increasing doses of SCH 23390. Values represent mean+SEM total photocell beam interruptions (horizontal and vertical activity combined) for 2-h test sessions tested during the dark phase of the light-dark cycle. **C** Catalepsy testing was conducted 15min following injection with vehicle or SCH 23390. Values represent mean+SEM time (in seconds) immobile during a 5-min test. In all cases, *WT* represents the wild-type mice and *D1KO* represents the mutants. (With permission from XU et al. 1994b)

Molecular Knockout Approach to the Study of Brain Dopamine Function 221

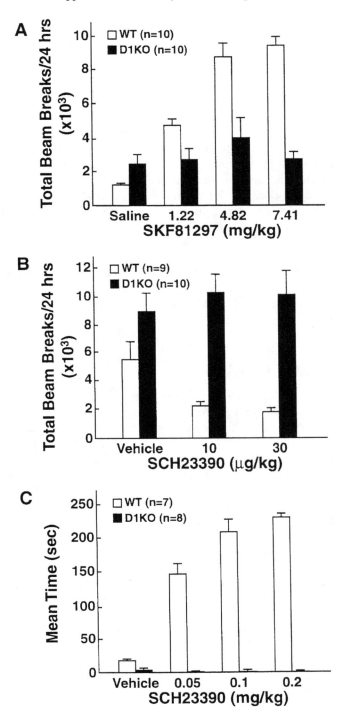

of a conditioned place preference for cocaine (MINER et al. 1995) nor any deficit in acquisition of an odor discrimination (SMITH et al. 1998). D_1 knockout mice were impaired in their acquisition of cocaine self-administration compared to wild-type mice, but dose-effect functions for cocaine self-administration were similar between mutant and control mice (CAINE et al. 1995).

These results are consistent with data showing that D_1 antagonists can block the psychostimulant effects of cocaine and impair learning in certain situations (BENINGER and MILLER 1998). Studies in these mice also support an essential role for the D_1 receptor in dopamine-mediated inhibitory effects within the nucleus accumbens, measured electrophysiologically. In particular, reduced efficacy of cocaine, dopamine, and D_1 and D_2 receptor agonists was found in $D_1^{-/-}$ mice compared to wild-type controls (XU et al. 1994a). In summary, the data to date on mice bearing dopamine D_1 receptor deletions demonstrate an important contribution of the D_1 receptor to spontaneous activity and activation and suggest that such a function may extend to more complex behaviors. How deficits in learning and orientation relate to the perennial question of motor versus motivational behavior will require further testing and further development of neuropsychological tests in mice.

F. D_2 Receptor Knockouts

Knockouts of the dopamine D_2 receptor have produced phenotypes that support the hypothesis that D_2 receptors have a role in motor behavior and have focused largely on behavior mediated by the striatum. An initial report described animals that were dramatically akinetic and showed major decreases in locomotor activity, and the authors speculated that this gene deletion might be a model of Parkinson's disease (BAIK et al. 1995). However, a subsequent study showed that this phenotype has epistatic qualities in that the severe behavioral deficits are largely manifested only in the mouse with C57BL/6 background because the 129 background produces a "floor" effect (KELLY et al. 1998; Fig. 3). In the rotorod test of motor coordination, the knockout $F2^{-/-}$ mice and the wild-type 129 mice showed severe deficits, but the wild-type C57BL/6 and congenic $B6^{-/-}$ mice successfully learned the task, although the $B6^{-/-}$ mice were slower (KELLY et al. 1998). A third more recent study reports $D_2^{-/-}$ mice exhibiting locomotor activity reductions that are exacerbated during the dark periods (JUNG et al. 1999). D_2 homozygous mutants also were found to have a 50% increase in dopamine metabolites in the striatum and compensatory increases in the D_3 receptor protein measured using immunoprecipitation (JUNG et al. 1999). Studies in mutant mice also have revealed an essential role for the D_2, but not the D_3 or D_4, receptor subtype in the disruption of prepulse inhibition produced by amphetamine in mice (RALPH et al. 1999).

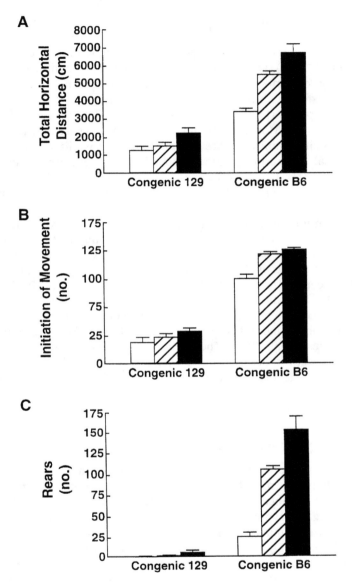

Fig. 3. Locomotor activity in congenic 129 and B6 strains of D_2 receptor mutant mice. **A** Total horizontal distance traveled. **B** Initiation of movement. **C** Vertical rears in 30 min by drug-naïve mice in an open field. Data are mean±SEM. $129^{-/-}$, $n = 9$ (*open bars*); $129^{+/-}$, $n = 20$ (*striped bars*); $129^{+/+}$, $n = 16$ (*black bars*); $B6^{-/-}$, $n = 16$ (*open bars*); $B6^{+/-}$, $n = 36$ (*striped bars*); $B6^{+/+}$, $n = 19$ (*black bars*). Statistical analyses revealed significantly lower scores for the B6 congenic $-/-$ mice compared to $+/+$ siblings in total horizontal distance, initiation of movement, and rears. The B6 congenic $+/-$ mice also showed significantly reduced total horizontal distance and rears compared to $+/+$ siblings. In the congenic $129^{-/-}$ mice compared to 129 congenic $+/+$ siblings there were significant deficits in total horizontal distance and initiation of movement, whereas the $+/-$ mice were only different in total horizontal distance. $p < 0.05$, ANOVA (analysis of variance), Tukey post-hoc tests. (With permission from KELLY et al. 1998)

$D_2^{-/-}$ mice exhibited a marked aversion to ethanol in a two-bottle choice procedure and reduced sensitivity to ethanol-induced locomotor impairments, pointing to a role for the D_2 receptor in the behavioral effects of alcohol (PHILLIPS et al. 1998). D_2 receptor knockout mice also exhibited a deficit in the acquisition of a morphine-conditioned place preference (MALDONADO et al. 1997) consistent with a role for D_2 receptors in some of the motivational effects of mu opioids (DI CHIARA and IMPERATO 1988; KOOB 1992; HARRIS and ASTON-JONES 1994). Moreover, preliminary results suggest that mice lacking D_2 receptors self-administer more cocaine than their wild-type littermates, an effect identical to pharmacological blockade of D_2 receptors in intact mice (CAINE et al. 2002). Collectively, these results suggest a role for D_2 receptors in the behavioral effects of a variety of drugs that are abused by humans.

The null mutant mice for the D_2 receptor also have provided insight into the role of the D_2 receptors in the intrinsic functioning of the basal ganglia. D_2 mutant mice failed to show autoreceptor-mediated inhibition of dopaminergic cell firing or the evoked release of dopamine, suggesting an important role for D_2 receptors in autoreceptor function (MERCURI et al. 1997; L'HIRONDEL et al. 1998). In addition, corticostriatal slices of D_2 mutant mice show long-term potentiation instead of long-term depression to tetanic stimulation of the corticostriatal fibers, and this effect was reversed by an NMDA receptor antagonist (CALABRESI et al. 1997). The authors hypothesized that an imbalance between D_2 receptor activity and NMDA receptor activity may produce changes in synaptic organization that lead to some of the symptoms of Parkinson's disease.

G. D_3 Receptor Knockouts

In contrast to the decreases in locomotor activity and motor behavior associated with null mutant mice for D_1 or D_2 receptors, D_3 knockout mice express a phenotype of enhanced locomotor activity (ACCILI et al. 1996). Mice were generated using embryonic stem cells from the 129/Sv strain and the chimeras were mated with female C57BL/6 strain mice. These mice showed no D_3 binding and normal D_2 receptor binding, and they showed hyperactivity and increased rearing in an open field relative to wild-type controls of the F-2 generation (ACCILI et al. 1996). Subsequent testing of D_3-null mutant mice showed similar hyperactivity in a novel environment (XU et al. 1997) and in an open field and elevated plus-maze (XU et al. 1997; STEINER et al. 1998), suggesting an anxiolytic-like effect or enhanced responsiveness to novelty. D_3-null mutants also showed a hyperresponsiveness to dopamine agonists when both D_1 and D_2 receptors were activated simultaneously, suggesting that D_3 receptors likely dampen normal responses to combined D_1 and D_2 stimulation postsynaptically through a post-synaptic mechanism (XU et al. 1997).

A more recent investigation with a third D_3 mutant mouse found no differences in locomotor activity in D_3 mutant mice during the light or the dark period when a longer test session was implemented (JUNG et al. 1999). These results suggest that the hyperactivity of the D_3 mutants habituates rapidly. Interestingly, in this same study, creation of a D_2/D_3 double mutant produced a motor phenotype more severe than the D_2 single mutants. Double mutants also exhibited increased levels of dopamine metabolites in the striatum compared to single mutants. These authors postulate that the D_3 receptor may compensate for the lack of D_2 receptor function, but this compensation remains masked in the presence of abundant D_2 receptors.

The pharmacology of the D_3 receptor system also has been evaluated in D_3 mutant mice. Putative selective D_3 receptor agonists and antagonists were found to produce similar responses in mutant and wild-type mice for locomotor activity and hypothermia effects purportedly mediated by the D_3 receptor (BOULAY et al. 1999; XU et al. 1999). These results call into question the selectivity of the currently available pharmacological agents, and future studies will be necessary to fully characterize the functional role of the D_3 receptor subtype.

H. Knockout of the Dopamine Transporter

Another protein target for molecular neuropharmacological manipulation of the dopamine system using the knockout technique is the dopamine transporter, but in this case molecular loss of function conveys a neuropharmacological increase in dopamine activity. The dopamine transporter controls the quantity and temporal characteristics of dopamine released into the presynaptic terminal, and pharmacological blockade of the dopamine transporter pharmacologically with drugs such as cocaine and amphetamine results in an increase in extracellular dopamine. Disruption of the dopamine transporter by homologous recombination using embryonic stem cells from 129Sv/J mice and the mating of chimeric males with C57BL/6J females produced $DAT^{-/-}$ mice, $DAT^{+/-}$, and $DAT^{+/+}$ mice (GIROS et al. 1996). These mice were spontaneously hyperactive and, neuropharmacologically, dopamine persisted over 100-times longer in the extracellular space. Psychostimulant drugs, including d-amphetamine, had no effect on dopamine release or on locomotor activity in the $DAT^{-/-}$ mice, suggesting that in fact the dopamine-releasing effects of d-amphetamine involve a neuropharmacological action to actually reverse the dopamine transporter (GIROS et al. 1996; Fig. 4).

Given that cocaine is a major drug of abuse and long has been hypothesized to produce its neuropharmacological effects by increasing the availability of dopamine in the terminal areas of the mesocorticolimbic dopamine system, an important question was whether the reinforcing effects of cocaine would be blocked in $DAT^{-/-}$ mice. Two independent studies using two differ-

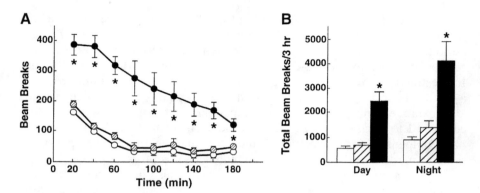

Fig. 4. A Spontaneous locomotor activity and habituation of naïve wild-type DAT$^{+/+}$ (*open circles*), heterozygote DAT$^{+/-}$ (*striped circles*) and homozygote DAT$^{-/-}$ (*black circles*) mice. Locomotor activity was recorded every 20 min for a period of 3h, $n = 12$ mice per group. *$p < 0.01$ compared to DAT$^{+/+}$ using the student's *t*-test. SEM is less than 5% of the mean if not stated otherwise. **B** Spontaneous locomotor activity of naïve DAT$^{+/+}$ (*open bars*), DAT$^{+/-}$ (*striped bars*), and DAT$^{-/-}$ (*black bars*) mice. Accumulated locomotor activity was recorded for 3h during the light (1100–1400 hours) or dark (2300–0200 hours) phase of the light-dark cycle. *$p < 0.001$ compared to DAT$^{+/+}$, $n = 10$–12 mice per group. The spontaneous locomotor activity of the homozygote animals was significantly higher during the dark cycle compared to the light cycle ($p < 0.05$). The heterozygotes are consistently more active than the wild-type mice, but this increase is of marginal significance ($p < 0.06$) during the dark phase of the cycle. (With permission from GIROS et al. 1996)

ent measures of cocaine reinforcement and two separate DAT$^{-/-}$ constructs have shown that cocaine reinforcement persists in DAT$^{-/-}$ mice (ROCHA et al. 1998; SORA et al. 1998). DAT$^{-/-}$ mice, carrying the same construct as described in the above locomotor studies, were implanted with intravenous catheters and successfully learned to self-administer cocaine, although twice as many sessions were required to meet acquisition criteria compared to wild-type mice (ROCHA et al. 1998). DAT$^{-/-}$ mice prepared using embryonic stem cells from 129/Sv mice and mated with C57BL/6J mice showed spontaneous hyperactivity and a blunted locomotor response to cocaine but a significant dose-dependent place preference for cocaine (SORA et al. 1998; Fig. 5). A simple explanation that serotonin may be the site for the reinforcing actions of cocaine was not supported by the observation that mice with knockout of the serotonin transporter also showed a robust place preference for cocaine (SORA et al. 1998). Clearly, other neuropharmacological mechanisms such as activation of norepinephrine and even opioid peptides may have to be considered for conveying redundancy in mediation of the reinforcing effects of psychostimulants in the absence of DAT. In addition, although a single intraperitoneal injection of cocaine did not apparently increase extracellular dopamine levels in DAT knockout mice, evidence that cocaine produces reinforcing effects in these mice independently of changes in dopamine transmission should be held up to more rigorous experimental scrutiny (CAINE 1998; CARBONI et al. 2001).

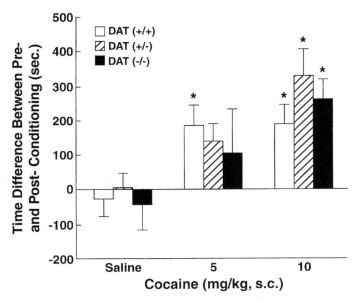

Fig. 5. Cocaine conditioned place preferences in DAT knockout mice. Conditioned place preference induced by cocaine in wild-type (+/+, *open bars*), heterozygous (+/−, *striped bars*), and homozygous (−/−, *black bars*) DAT knockout mice. Time scores shown represent differences between post-conditioning (*Post*) and pre-conditioning (*Pre*) time spent in the cocaine-paired environment. Wild-type mice displayed significant place preference associated with 5 and 10 mg/kg cocaine, whereas heterozygous and homozygous animals showed significant place preferences associated with 10 mg/kg cocaine. *$p < 0.05$ compared to saline-injected group by ANOVA, $n = 8$–23 mice per genotype. (With permission from SORA et al. 1998)

I. Knockout of Tyrosine Hydroxylase Gene

In an elegant demonstration of the power of the molecular pharmacological approach, a double construct was used to produce selective dopamine-deficient mice. The gene encoding tyrosine hydroxylase (TH) was inactivated and selectively restored in noradrenergic neurons (ZHOU and PALMITER 1995). Disruption of the TH gene results in both a dopamine and norepinephrine deficiency, which is lethal (ZHOU et al. 1995). To restore expression of the TH gene in noradrenergic neurons, the TH coding sequence was linked to the noradrenergic-specific dopamine beta-hydroxylase (DBH) promotor in embryonic stem cells by homologous recombination (ZHOU and PALMITER 1995; Fig. 6). Transgenic DBH-TH mice were mated to produce offspring with a selective dopamine or norepinephrine deficiency by intercrossing DBH-TH$^{+/-}$ mice. The homozygous DBH-TH$^{+/+}$ mice were deficient in DBH function and presented a phenotype like DBH$^{-/-}$ mice (THOMAS et al. 1995). When TH$^{+/-}$ mice were crossed with DBH$^{+/-}$, six different genotypes were formed, one in which TH$^{-/-}$ DBH-TH$^{+/-}$ has a selective dopamine deficiency (ZHOU and

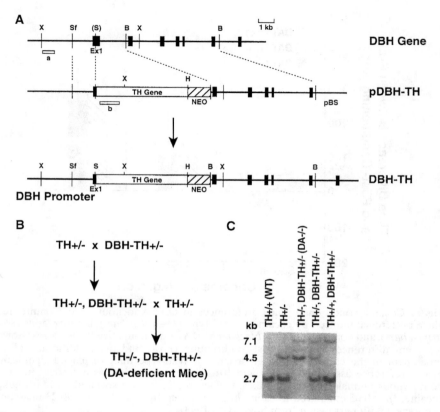

Fig. 6. Gene targeting, mating strategy, and genetic diagnosis of DA$^{-/-}$ mice. **A** The murine DBH gene and targeting vector pDBH-TH. A region of the DBH gene that includes the proximal promoter is shown. The entire TH coding region, including the 1-kb sequence after the polyadenylation site and a neo cassette, was inserted between exons 1 and 2 of the DBH gene. Locations of probes from DBH gene (*Box a*) and the TH gene (*Box b*) that were used for screening ES cell clones and mice are indicated. Abbreviations: *B*, *Bam*HI; *H*, *Hin*dIII; *Sf*, *Sfi*; *X*, *Sba*; *S*, artificial *Sa*II; *pBS*, *p*Bluescript (Strategene). **B** Breeding strategy for generating DA$^{-/-}$ mice. **C** Southern blot analysis of representative tail DNA samples. DNA was digested with XbaI and hybridized with probe b from the TH gene. The 2.7-kb band is the wild-type (*WT*) TH allele; the 4.5-kb band is the disrupted TH allele; the 7.1-kb band represents the DBH-TH allele. The faint 5.5-kb band is due to partial digestion. (With permission from ZHOU and PALMITER 1995)

PALMITER 1995; Fig. 6). These mice, considered DA$^{-/-}$ by the authors, were severely impaired in motor behavior, feeding, and drinking. The animals would die from aphagia and adipsia unless treated chronically with L-dopa (ZHOU and PALMITER 1995). Interestingly, as with animals bearing 6-hydroxydopamine lesions of the mesocorticolimbic dopamine system, the mice showed increased locomotion in response to a selective D$_1$ and D$_2$ agonist, suggesting parallel pathways for D$_1$ and D$_2$ activation.

An important molecular target for the actions of dopamine is dopamine and adenosine 3',5'-monophosphate-regulated phosphoprotein (32 kDa, DARPP-32), which is converted in response to dopamine into a potent protein phosphatase inhibitor and thus regulates the physiological activity of a wide array of neuronal phosphoproteins. DARPP-32 mutant mice have been created by disruption of the targeted gene in a 129/Ola-derived embryonic stem cell line (FIENBERG et al. 1998). Homologous recombination at the endogenous locus was designed to result in the replacement of a 400-base-pair genomic DNA fragment containing the start of translation with a neomycin resistance gene. After C57BL/6J blastocyst injection and embryo transfer, chimeric offspring were crossed to C57BL/6J females, and those mice carrying the mutation were crossed to generate heterozygous and homozygous mutants. Mice generated to contain the targeted mutation exhibited deficits in their molecular, electrophysiological, and behavioral responses to dopamine, drugs of abuse, and antipsychotic medication. In mutant mice there was a loss of D_1 agonist-induced inhibition of glutamate-evoked activity in the nucleus accumbens. Similarly, cocaine and amphetamine-stimulated locomotion were attenuated in DARPP-32$^{-/-}$ mice, and raclopride produced catalepsy with reduced efficacy in mutant mice. Interestingly, DARPP-32$^{-/-}$ mice demonstrate a higher rate of sensitization to cocaine compared to wild-type mice, but no increases in delta Fos-b expression in the striatum following repeated cocaine administration, suggesting DARPP-32 is involved in regulating biochemical and behavioral plasticity associated with repeated administration of cocaine (HIROI et al. 1999). Mutant mice also exhibit a significant impairment in reversal learning, providing evidence for a functional role for DARPP-32 in the processes underlying learning and memory (HEYSER et al. 2000).

J. Other Knockouts

The dopamine D_4 receptor is expressed in high amounts in terminal areas of the mesocorticolimbic dopamine system such as the frontal cortex, and in low amounts in the nigrostriatal dopamine system such as the striatum and globus pallidus (ARIANO et al. 1997) and has received considerable interest because it shows the highest affinity for the atypical antipsychotic clozapine (SEEMAN and VAN TOL 1994). However, this receptor is of low abundance, and a role for this receptor in the therapeutic or other effects of antipsychotic drugs remains controversial (BRISTOW et al. 1997; MANSBACH et al. 1998; MILLAN et al. 1998). Targeted removal of the dopamine D_4 receptor was produced by use of homologous recombination in embryonic stem cells using a 129/SvEv mouse genomic phase library screened with a human D_4R cDNA. These embryonic stem cells were injected into C57BL/6J blastocysts, and the chimeras were mated with C57BL/6J females. The F-1 heterozygotes were mated to produce $D_4R^{-/-}$ mice. These $D_4^{-/-}$ mice grew and reproduced normally but were less sensitive to the blockade of apomorphine-induced locomotor activity produced by clozapine (RUBINSTEIN et al. 1997). $D_4^{-/-}$ mice were also less active in locomotor activity and rearing. However, reductions in startle amplitude and prepulse inhibition produced by amphetamine were measured in both $D_4^{-/-}$ and $D4^{+/+}$ mice (RALPH et al. 1999). These mice performed better on a rotarod test, remaining on the rod 2.5 times longer than wild-type littermates, and were also more responsive to the locomotor-activating effects of ethanol, cocaine, and methamphetamine. One possible explanation for this complex phenotype is that there was increased synthesis and turnover of dopamine in the $D_4^{-/-}$ mice, and the enhanced turnover of dopamine may be acting via D_1, D_2, or D_3 receptors in the striatum. Another potential explanation is that the loss of D_4 receptors in the cortex produces a loss of inhibitory tone that would normally be present. Paradoxically, in rats a selective D_4 antagonist blocked sensitization to the locomotor and accumbens dopamine-enhancing effects of amphetamine (FELDPAUSCH et al. 1998). Collectively, these results suggest a role for D_4 dopamine receptors in sensitization to the behavioral and neurochemical effects of psychomotor stimulants. They also underscore the paradoxical effects that sometimes are observed in comparisons of acute pharmacological treatments and chronic targeted genetic mutations.

D_5 dopamine-deficient mice recently have been generated using homologous recombination techniques and mating the chimeras with C57BL/6 mice (HOLMES et al. 1998). The resulting mutant D_5 receptor mice developed normally and showed loss of D_5 receptor staining in the central nervous system. Preliminary behavioral tests revealed hyperactivity in an open field test and increased latency to fall from an accelerating rotarod compared to wild-type controls (HOLMES et al. 1998).

Another target for potential disruption of dopaminergic function at the molecular level is the vesicular monoaminergic transporter that transports monoamines from the cytoplasm into secretory vesicles. Using homologous recombination, mutant mice lacking the vesicular monoamine transporter 2 (VMAT-2) have been generated (WANG et al. 1997). A polymerase chain reaction-generated probe from the rat cDNA was used to isolate the VMAT-2 gene from a 129/SvJ genomic library, and transfection using an embryonic stem cell line isogenic with the 129/SvJ substrain was performed. Clones were injected into C57BL/6 blastocysts, and the chimeric offspring were mated to produce F-1 and F-2 offspring, both of which were used. The mice homozygous for VMAT-2 were not viable, but heterozygous adults showed decreased basal extracellular dopamine and decreased K^+ and amphetamine-evoked release. These mice showed a pronounced increase in sensitivity to the locomotor stimulant effects of the dopamine agonist apomorphine, cocaine, amphetamine, and ethanol (TAKAHASHI et al. 1997; WANG et al. 1997). These VMAT-2 mice failed to show further increases in activity after repeated cocaine administration. Diminished amphetamine reinforcement measured by conditioned place preference also was displayed in VMAT-$2^{+/-}$ mice (TAKAHASHI et al. 1997).

K. Summary and Conclusions: What We Know That We Did Not Know Before Knockouts

The major contribution to date of the molecular pharmacological approach to the study of dopaminergic function can be summarized in three domains. Knockout studies have (1) confirmed many pre-existing hypotheses regarding the role of specific elements of dopamine neuropharmacology, (2) confirmed or cast doubt on the selectivity of action of a number of neuropharmacological agents, and (3) uncovered novel functional effects within the dopamine system.

The pre-existing hypotheses confirmed by knockout studies range from the importance of the dopamine transporter and vesicular transporter in maintaining extracellular dopaminergic tone to a role for dopamine in certain types of learning. Clearly, as has been known for some time, animals without dopamine (tyrosine-hydroxylase knockouts) do not do well and are severely hypoactive, aphagic, and adipsic. Mice without D_2 receptors on certain background strains also are hypoactive and show motor deficits associated with striatal dysfunction. Mice without D_1 or D_2 receptors show blunted responses to psychostimulant drugs further confirming an important role for the D_1 and D_2 receptor subtypes in psychostimulant activation. In contrast, both D_3 and D_4 knockouts show enhanced responsiveness to the activating effects of psychostimulant drugs.

Finally, the knockout approach provides an excellent validation of the selectivity of a given agonist or antagonist in vivo. For example, if a D_3 agonist produces a functional effect in a D_3 knockout mouse, one has reason to suspect a lack of selectivity to the D_3 receptor or other neuropharmacological actions. This has been shown for a variety of D_3 receptor agonists and antagonists (BOULAY et al. 1999; XU et al. 1999). Knockout mice also will provide a means of evaluating crosstalk or lack of interaction between dopamine receptors. The unknown effects revealed by knockout studies include the discovery of novel functional effects within the dopamine system and outside the dopamine system. Apparently the effects of amphetamine to release dopamine require an intact dopamine transporter, suggesting that amphetamine actually produces monoamine release by reversing transporter function. One surprise revealed by knockout studies is that both self-administration and place preference for cocaine remain intact in dopamine transporter knockout mice, suggesting that neurotransmitter systems other than dopamine may contribute to the reinforcing effects of cocaine, or that these transmitter systems are capable of compensating rapidly for the loss of dopamine activity.

Interesting challenges remain for the study of the brain dopamine systems using the knockout approach. Clearly, procedures will be needed to isolate confounds due to epistasis and background strains. The use of site-directed knockouts will allow a means of evaluating not only contributions of specific brain regions to the function of specific dopaminergic neuropharmacological agents but also compensatory responses to dysfunction of one or more elements. Ultimately, one could imagine that such an approach might model early stages of the pathogenesis of disorders such as Parkinson's disease and perhaps elements of affective disorders and schizophrenia. The use of conditional knockouts will eliminate the compensatory responses observed during development but produce new challenges to understand compensatory changes possible in adult animals.

Acknowledgements. This is publication number 12667-NP from The Scripps Research Institute. Research was supported by National Institutes of Health grants DA04398 (GFK) and DA12142 (SBC) from the National Institute on Drug Abuse. The authors would like to thank Mike Arends for his help with the preparation of this manuscript.

References

Accili D, Fishburn CS, Drago J, Steiner H, Lachowicz JE, Park B-H, Gauda EB, Lee EJ, Cool MH, Sibley DR, Gerfen CR, Westphal H, Fuchs S (1996) A targeted mutation of the D-3 dopamine receptor gene is associated with hyperactivity in mice. Proc Natl Acad Sci USA 93:1945–1949

Amalric M, Berhow M, Polis I, Koob GF (1993) Selective effects of low-dose D2 dopamine receptor antagonism in a reaction-time task in rats. Neuropsychopharmacology 8:195–200

Amalric M, Koob GF (1993) Functionally selective neurochemical afferents and efferents of the mesocorticolimbic and nigrostriatal dopamine system. Prog Brain Res 99:209–226

Amalric M, Koob GF, Creese I, Swerdlow NR (1986) "Selective" D-1 and D-2 receptor antagonists fail to differentially alter supersensitive locomotor behavior in the rat. Life Sci 39:1985–1993

Ariano MA, Wang J, Noblett KL, Larson ER, Sibley DR (1997) Cellular distribution of the rat D4 dopamine receptor protein in the CNS using anti-receptor antisera. Brain Res 752:26–34

Arnt J (1985) Hyperactivity induced by stimulation of separate dopamine D1 and D2 receptors in rats with bilateral 6-OHDA lesions. Life Sci 37:717–723

Baik JH, Picetti R, Saiardi A, Thiriet G, Dierich A, Depaulis A, Le Meur M, Borrelli E (1995) Parkinsonian-like locomotor impairment in mice lacking dopamine D2 receptors. Nature 377:424–428

Banbury Conference on Genetic Background in Mice (1997) Mutant mice and neuroscience: recommendations concerning genetic background. Neuron 19:755–759

Barrett-Larimore RL, Spealman RD (1997) Reinstatement of cocaine-seeking behavior in a nonhuman primate model of relapse: effects of preferential D1 and D2 agonists. NIDA Res Monogr 178:283

Beninger RJ, Miller R (1998) Dopamine D1-like receptors and reward-related incentive learning. Neurosci Biobehav Rev 22:335–345

Bergman J, Kamien JB, Spealman RD (1990) Antagonism of cocaine self-administration by selective dopamine D1 and D2 antagonists. Behav Pharmacol 1:355–363

Bordet R, Ridray S, Carboni S, Diaz J, Sokoloff P, Schwartz JC (1997) Induction of dopamine D3 receptor expression as a mechanism of behavioral sensitization to levodopa. Proc Natl Acad Sci USA 94:3363–3367

Boulay D, Depoortere R, Rostene W, Perrault Gh, Sanger DJ (1999) Dopamine D3 receptor agonists produce similar decreases in body temperature and locomotor activity in D3 knock-out and wild-type mice. Neuropharmacology 38:555–565

Bristow LJ, Collinson N, Cook GP, Curtis N, Freedman SB, Kulagowski JJ, Leeson PD, Patel S, Ragan CI, Ridgill M, Saywell KL, Tricklebank MD (1997) L-745,870, a subtype selective dopamine D4 receptor antagonist, does not exhibit a neuroleptic-like profile in rodent behavioral tests. J Pharmacol Exp Ther 283:1256–1263

Caine SB (1998) Cocaine abuse: hard knocks for the dopamine hypothesis? Nature Neurosci 1:90–92

Caine SB, Gold LH, Koob GF, Deroche V, Heyser C, Polis I, Roberts A, Xu M, Tonegawa S (1995) Intravenous cocaine self-administration in mice: Strain differences and effects of D-1 dopamine receptor targeted gene mutation. Society Neurosci Abstr 21:719

Caine SB, Koob GF (1993) Modulation of cocaine self-administration in the rat through D-3 dopamine receptors. Science 260:1814–1816

Caine SB, Koob GF (1994) Effects of dopamine D1 and D2 antagonists on cocaine self-administration under different schedules of reinforcement in the rat. J Pharmacol Exp Ther 270:209–218

Caine SB, Koob GF (1995) Pretreatment with the dopamine agonist 7-OH-DPAT shifts the cocaine self-administration dose-effect function to the left under different schedules in the rat. Behav Pharmacol 6:333–347

Caine SB, Koob GF, Parsons LH, Everitt BJ, Schwartz J-C, Sokoloff P (1997) D3 receptor test in vitro predicts decreased cocaine self-administration in rats. Neuroreport 8:2373–2377

Caine SB, Negus SS, Mello NK (2000) Effects of dopamine D1-like and D2-like agonists on cocaine self-administration in rhesus monkeys: Rapid assessment of cocaine dose-effect functions. Psychopharmacology 148:41–51

Caine SB, Negus SS, Mello NK, Bergman J (1999b) Effects of D1-like and D2-like agonists in rats that self-administer cocaine. J Pharmacol Exp Ther 291:353–360

Caine SB, Negus SS, Mello NK, Patel S, Bristow L, Kulagowski J, Vallone D, Saiardi A, Borrelli E (2002) Role of dopamine D2-like receptors in cocaine self-administration: Studies with D2 receptor mutant mice and novel D2 receptor antagonists. J Neurosci 22:in press

Calabresi P, Saiardi A, Pisani A, Baik J-H, Centonze D, Mercuri NB, Bernardi G, Borrelli E (1997) Abnormal synaptic plasticity in the striatum of mice lacking dopamine D2 receptors. J Neurosci 17:4536–4544

Campbell IL, Gold LH (1996) Transgenic modeling of neuropsychiatric disorders. Mol Psychiatry 1:105–120

Carboni E, Spielewoy C, Vacca C, Nosten-Bertrand M, Giros B, Di Chiara G (2001) Cocaine and amphetamine increase extracellular dopamine in the nucleus accumbens of mice lacking the dopamine transporter gene. J Neurosci 21:RC141–RC144

Clifford JJ, Tighe O, Croke DT, Sibley DR, Drago J, Waddington JL (1998) Topographical evaluation of the phenotype of spontaneous behaviour in mice with targeted gene deletion of the D-1 A dopamine receptor: paradoxical elevation of grooming syntax. Neuropharmacology 37:1595–1602

Crawford CA, Drago J, Watson JB, Levine MS (1997) Effects of repeated amphetamine treatment on the locomotor activity of the dopamine D-1A-deficient mouse. Neuroreport 8:2523–2527

Crawley JN (1996) Unusual behavioral phenotypes of inbred mouse strains. Trends Neurosci 19:181–182

Creese I, Iversen SD (1974) The role of forebrain dopamine systems in amphetamine-induced stereotyped behavior in the rat. Psychopharmacologia 39:345–357

Di Chiara G, Imperato A (1988) Drugs abused by humans preferentially increase synaptic dopamine concentrations in the mesolimbic system of freely moving rats. Proc Natl Acad Sci USA 85:5274–5278

Drago F, Contarino A, Busa L (1999) The expression of neuropeptide-induced excessive grooming behavior in dopamine D-1 and D-2 receptor-deficient mice. Eur J Pharmacol 365:125–131

Drago J, Gerfen CR, Lachowicz JE, Steiner H, Hollon TR, Love PE, Ooi GT, Grinberg A, Lee EJ, Huang SP, Bartlett PF, Jose PA, Sibley DR, Westphal H (1994) Altered striatal function in a mutant mouse lacking D-1 A dopamine receptors. Proc Natl Acad Sci USA 91:12564–12568

Fienberg AA, Hiroi N, Mermelstein PG, Song W-J, Snyder GL, Nishi A, Cheramy A, O'Callaghan JP, Miller DB, Cole DG, Corbett R, Haile CN, Cooper DC, Onn SP, Grace AA, Ouimet CC, White FJ, Hyman SE, Surmeier DJ, Girault J-A, Nestler EJ, Greengard P (1998) DARPP-32: regulator of the efficacy of dopaminergic neurotransmission. Science 281:838–842

Feldpausch DL, Needham LM, Stone MP, Althaus JS, Yamamoto BK, Svensson KA, Merchant KM (1998) The role of dopamine D4 receptor in the induction of behavioral sensitization to amphetamine and accompanying biochemical and molecular adaptations. J Pharmacol Exp Ther 286:497–508

Gerlai R (1996) Gene-targeting studies of mammalian behavior: is it the mutation or the background genotype? Trends Neurosci 19:177–181

Giros B, Jaber M, Jones SR, Wightman RM, Caron MG (1996) Hyperlocomotion and indifference to cocaine and amphetamine in mice lacking the dopamine transporter. Nature 379:606–612

Gold LH (1996) Integration of molecular biological techniques and behavioural pharmacology. Behav Pharmacol 7:589–615

Grech DM, Spealman RD, Bergman J (1996) Self-administration of D1 receptor agonists by squirrel monkeys. Psychopharmacology 125:97–104

Harris GC, Aston-Jones G (1994) Involvement of D2 dopamine receptors in the nucleus accumbens in the opiate withdrawal syndrome. Nature 371:155–157

Heyser C, Fienberg AA, Greengard P, Gold LH (2000) DARPP-32 knockout mice exhibit impaired reversal learning in a discriminated operant task. Brain Res 867:122–130

Hiroi N, Fienberg AA, Haile CN, Alburges M, Hanson GR, Greengard P, Nestler EJ (1999) Neuronal and behavioural abnormalities in striatal function in DARPP-32 mutant mice. Eur J Neurosci 11:1114–1118

Holmes A, Hollon TR, Gleason TC, Liu Z, Drieling J, Sibley DR, Crawley JN (2001) Behavioral characterization of dopamine D5 receptor null mutant mice. Behav Neurosci 115:1129–1144

Hubner CB, Moreton JE (1991) Effects of selective D1 and D2 dopamine antagonists on cocaine self-administration in the rat. Psychopharmacology 105:151–156

Jung M-Y, Skryabin BV, Arai M, Abbondanzo S, Fu D, Brosius J, Robakis NK, Polites HG, Pintar JE, Schmauss C (1999) Potentiation of the D-2 mutant motor phenotype in mice lacking dopamine D-2 and D-3 receptors. Neuroscience 91:911–924

Kelly MA, Rubinstein M, Phillips TJ, Lessov CN, Burkhart-Kasch S, Zhang G, Bunzow JR, Fang Y, Gerhardt GA, Grandy DK, Low MJ (1998) Locomotor activity in D2 dopamine receptor-deficient mice is determined by gene dosage, genetic background, and developmental adaptations. J Neurosci 18:3470–3479

Kelly PH, Seviour PW, Iversen SD (1975) Amphetamine and apomorphine responses in the rat following 6-OHDA lesions of the nucleus accumbens septi and corpus striatum. Brain Res 94:507–522

Koob GF (1992) Drugs of abuse: anatomy, pharmacology, and function of reward pathways. Trends Pharmacol Sci 13:177–184

Koob GF (1996) Hedonic valence, dopamine and motivation. Mol Psychiatry 1:186–189

Koob GF, Parsons LH, Caine SB, Weiss F, Sokoloff P, Schwartz J-C (1996) Dopamine receptor subtype profiles in cocaine reward. In: Beninger RJ, Palomo T, Archer T (eds) Dopamine disease states. Editorial CYM, Madrid, pp 433–445

Lamas X, Negus SS, Nader MA, Mello NK (1996) Effects of the putative dopamine D3 receptor agonist 7-OH-DPAT in rhesus monkeys trained to discriminate cocaine from saline. Psychopharmacology 124:306–314

Le Moal M, Simon H (1991) Mesocorticolimbic dopaminergic network: functional and regulatory roles. Physiol Rev 71:155–234

L'hirondel M, Cheramy A, Godeheu G, Artaud F, Saiardi A, Borrelli E, Glowinski J (1998) Lack of autoreceptor-mediated inhibitory control of dopamine release in striatal synaptosomes of D2 receptor-deficient mice. Brain Res 792:253–262

Maldonado R, Saiardi A, Valverde O, Samad TA, Roques BP, Borrelli E (1997) Absence of opiate rewarding effects in mice lacking dopamine D2 receptors. Nature 388:586–589

Mansbach RS, Brooks EW, Sanner MA, Zorn SH (1998) Selective dopamine D4 receptor antagonists reverse apomorphine-induced blockade of prepulse inhibition. Psychopharmacology 135:194–200

Mercuri NB, Saiardi A, Bonci A, Picetti R, Calabresi P, Bernardi G, Borrelli E (1997) Loss of autoreceptor function in dopaminergic neurons from dopamine D2 receptor deficient mice. Neuroscience 79:323–327

Millan MJ, Newman-Tancredi A, Brocco M, Gobert A, Lejeune F, Audinot V, Rivet J-M, Schreiber R, Dekeyne A, Spedding M, Nicolas J-P, Peglion J-L (1998) S18126 ([2-[4-(2,3-dihydrobenzo[1,4]dioxin-6-yl)piperazin-1-yl-methyl]indan-2-yl]), a potent, selective and competitive antagonist at dopamine D4 receptors: an in vitro and in vivo comparison with L-745,870 (3-(4-[4-chlorophenyl]piperazin-1-yl) methyl-1H-pyrrolol[2,3b]pyridine) and raclopride. J Pharmacol Exp Ther 287: 167–186

Miner LL, Drago J, Chamberlain PM, Donovan D, Uhl GR (1995) Retained cocaine conditioned place preference in D1 receptor deficient mice. Neuroreport 6:2314–2316

Mirenowicz J, Schultz W (1996) Preferential activation of midbrain dopamine neurons by appetitive rather than aversive stimuli. Nature 379:449–451

Molloy AG, Waddington JL (1984) Dopaminergic behaviour stereospecifically promoted by the D1 agonist R-SK & F 38393 and selectively blocked by the D1 antagonist SCH 23390. Psychopharmacology 82:409–410

Moore RY, Bloom FE (1978) Central catecholamine neuron systems: anatomy and physiology of the dopamine systems. Annu Rev Neurosci 1:129–169

Muller U (1999) Ten years of gene targeting: targeted mouse mutants, from vector design to phenotype analysis. Mech Dev 82:3–21

Nader MA, Mach RH (1996) Self-administration of the dopamine D3 agonist 7-OH-DPAT in rhesus monkeys is modified by previous cocaine exposure. Psychopharmacology 125:13–22

Olson EN, Arnold H-H, Rigby PWJ, Wold BJ (1996) Know your neighbors: three phenotypes in null mutants of the myogenic bHLH gene MRF4. Cell 85:1–4

Phillips H (1996) Use of transgenics and knockouts. Seminars Neurosci 8:115–116

Phillips TJ, Brown KJ, Burkhart-Kasch S, Wenger CD, Kelly MA, Rubinstein M, Grandy DK, Low MJ (1998) Alcohol preference and sensitivity are markedly reduced in mice lacking dopamine D-2 receptors. Nature Neurosci 1:610–615

Phillips TJ, Hen R, Crabbe JC (1999) Complications associated with genetic background effects in research using knockout mice. Psychopharmacology 147:5–7

Picciotto MR, Wickman K (1998) Using knockout and transgenic mice to study neurophysiology and behavior. Physiol Rev 78:1131–1163

Plomin R, DeFries JC, McClearn GE (1990) Behavioral genetics: a primer, 2nd edn. Freeman, New York, pp 262–295

Ralph RJ, Varty GB, Kelly MA, Wang Y-M, Caron MG, Rubinstein M, Grandy DK, Low MJ, Geyer MA (1999) The dopamine D-2, but not D-3 or D-4, receptor subtype is essential for the disruption of prepulse inhibition produced by amphetamine in mice. J Neurosci 19:4627–4633

Robbins TW, Everitt BJ (1992) Functions of dopamine in the dorsal and ventral striatum. Seminars Neurosci 4:119–127

Rocha BA, Fumagalli F, Gainetdinov RR, Jones SR, Ator R, Giros B, Miller GW, Caron MG (1998) Cocaine self-administration in dopamine-transporter knockout mice. Nature Neurosci 1:132–137

Rubinstein M, Phillips TJ, Bunzow JR, Falzone TL, Dziewczapolski G, Zhang G, Fang Y, Larson JL, McDougall JA, Chester JA, Saez C, Pugsley TA, Gershanik O, Low MJ, Grandy DK (1997) Mice lacking dopamine D4 receptors are supersensitive to ethanol, cocaine, and methamphetamine. Cell 90:991–1001

Sautel F, Griffon N, Levesque D, Pilon C, Schwartz JC, Sokoloff P (1995) A functional test identifies dopamine agonists selective for D3 versus D2 receptors. Neuroreport 6:329–332

Seeman P, Van Tol HH (1994) Dopamine receptor pharmacology. Trends Pharmacol Sci 15:264–270

Self DW, Barnhart WJ, Lehman DA, Nestler EJ (1996) Opposite modulation of cocaine-seeking behavior by D1- and D2-like dopamine receptor agonists. Science 271:1586–1589

Self DW, Stein L (1992) The D1 agonists SKF 82958 and SKF 77434 are self-administered by rats. Brain Res 582:349–352

Sibilia M, Wagner EF (1995) Strain-dependent epithelial defects in mice lacking the EGF receptor. Science 269:234–238

Smith AD, Smith DL, Zigmond MJ, Amalric M, Koob GF (2000) Differential effects of dopamine receptor subtype blockade on motor performance of rats in a reaction time paradigm. Psychopharmacology 148:355–360

Smith DR, Striplin CD, Geller AM, Mailman RB, Drago J, Lawler CP, Gallagher M (1998) Behavioural assessment of mice lacking D-1 A dopamine receptors. Neuroscience 86:135–146

Sokoloff P, Giros B, Martres MP, Bouthenet ML, Schwartz JC (1990) Molecular cloning and characterization of a novel dopamine receptor (D3) as a target for neuroleptics. Nature 347:146–151

Sokoloff P, Griffon N, Sautel F, Levesque D, Pilon C, Schwartz J-C, Ridray S, Diaz J, Simon P, Costentin J, Mann A, Wermuth CG, Caine SB, Parsons LH, Koob GF (1997) The dopamine D3 receptor: from cloning to function. In: Jenner P, Demirdamar R (eds) Dopamine receptors: from basic science to clinic. IOS Press, Amsterdam, pp 1–13

Sokoloff P, Schwartz JC (1995) Novel dopamine receptors half a decade later. Trends Pharmacol Sci 16:270–275

Sora I, Wichems C, Takahashi N, Li X-F, Zeng Z, Revay R, Lesch K-P, Murphy DL, Uhl GR (1998) Cocaine reward models: conditioned place preference can be established in dopamine- and in serotonin-transporter knockout mice. Proc Natl Acad Sci USA 95:7699–7704

Spealman RD (1996) Dopamine D3 receptor agonists partially reproduce the discriminative stimulus effects of cocaine in squirrel monkeys. J Pharmacol Exp Ther 278:1128–1137

Steiner H, Fuchs S, Accili D (1998) D-3 dopamine receptor-deficient mouse: evidence for reduced anxiety. Physiol Behav 63:137–141

Takahashi N, Miner LL, Sora I, Ujike H, Revay RS, Kostic V, Jackson-Lewis V, Przedborski S, Uhl GR (1997) VMAT2 knockout mice: heterozygotes display reduced amphetamine-conditioned reward, enhanced amphetamine locomotion, and enhanced MPTP toxicity. Proc Natl Acad Sci USA 94:9938–9943

Tecott LH, Barondes SH (1996) Genes and aggressiveness: behavioral genetics. Curr Biol 6:238–240

Thomas SA, Matsumoto AM, Palmiter RD (1995) Norepinephrine is essential for mouse fetal development. Nature 374:643–646

Waddington JL, Daly SA, McCauley PG, O'Boyle KM (1994) Levels of functional interaction between D-1-like and D-2-like dopamine receptor systems. In: Niznik HB (ed) Dopamine receptors and transporters: pharmacology, structure, and function. Marcel Dekker, New York, pp 513–538

Wang YM, Gainetdinov RR, Fumagalli F, Xu F, Jones SR, Bock CB, Miller GW, Wightman RM, Caron MG (1997) Knockout of the vesicular monoamine transporter 2 gene results in neonatal death and supersensitivity to cocaine and amphetamine. Neuron 19:1285–1296

Waters N, Svensson K, Haadsma-Svensson SR, Smith MW, Carlsson A (1993) The dopamine D3-receptor: a postsynaptic receptor inhibitory on rat locomotor activity. J Neural Transm 94:11–19

Weed MR, Woolverton WL (1995) The reinforcing effects of dopamine D1 receptor agonists in rhesus monkeys. J Pharmacol Exp Ther 275:1367–1374

Wise RA (1978) Catecholamine theories of reward: a critical review. Brain Res 152:215–247

Wise RA (1980) The dopamine synapse and the notion of "pleasure centers" in the brain. Trends Neurosci 3:91–94

Wise RA (1982) Neuroleptics and operant behavior: the anhedonia hypothesis. Behav Brain Sci 5:39–88

Woolverton WL, Goldberg LI, Ginos JZ (1984) Intravenous self-administration of dopamine receptor agonists by rhesus monkeys. J Pharmacol Exp Ther 230:678–683

Xu M, Hu X-T, Cooper DC, Moratalla R, Graybiel AM, White FJ, Tonegawa S (1994a) Elimination of cocaine-induced hyperactivity and dopamine-mediated neurophysiological effects in dopamine D1 receptor mutant mice. Cell 79:945–955

Xu M, Koeltzow TE, Cooper DC, Tonegawa S, White FJ (1999) Dopamine D3 receptor mutant and wild-type mice exhibit identical responses to putative D3 receptor-selective agonists and antagonists. Synapse 31:210–215

Xu M, Koeltzow TE, Santiago GT, Moratalla R, Cooper DC, Hu X-T, White NM, Graybiel AM, White FJ, Tonegawa S (1997) Dopamine D3 receptor mutant mice exhibit increased behavioral sensitivity to concurrent stimulation of D1 and D2 receptors. Neuron 19:837–848

Xu M, Moratalla R, Gold LH, Hiroi N, Koob GF, Graybiel AM, Tonegawa S (1994b) Dopamine D1 receptor mutant mice are deficient in striatal expression of dynorphin and in dopamine-mediated behavioral responses. Cell 79:729–742

Zhou QY, Palmiter RD (1995) Dopamine-deficient mice are severely hypoactive, adipsic, and aphagic. Cell 83:1197–1209

Zhou QY, Quaife CJ, Palmiter RD (1995) Targeted disruption of the tyrosine hydroxylase gene reveals that catecholamines are required for mouse fetal development. Nature 374:640–643

CHAPTER 21

Behavioural Pharmacology of Dopamine D$_2$ and D$_3$ Receptors: Use of the Knock-out Mice Approach

R. Depoortere, D. Boulay, G. Perrault, and D.J. Sanger

A. Introduction

Since the first evidence that dopamine (DA) serves as a central neurotransmitter became available (Carlsson et al. 1958), this catecholamine has generated enormous interest among neuroscientists, and would probably qualify as the most studied central neurotransmitter. Its pivotal role in numerous physiological processes (Jaber et al. 1996) and in major pathological conditions, in particular psychoses (Snyder 1976), has certainly contributed greatly to this privileged status. It appeared fairly early that the effects of DA are mediated by at least two types of receptors, named the D1 and the D2 receptors (Kebabian and Calne 1979). This conclusion was based on the dissociated effects that stimulation of each type had on the activity of adenylate cyclase, the enzyme responsible for the production of cyclic adenosine monophosphate (c-AMP). Levels of c-AMP are increased by activation of D1 receptors (Kebabian and Calne 1979) and decreased by activation of D2 receptors (De camilli et al. 1979). This opposite role of the two types of DA receptors is not ubiquitous, as they have also been shown to act in a cooperative manner in several models (see Waddington 1989 for review). For example, the two subtypes act synergistically to promote locomotor activity when activated (Molloy et al. 1986) or to produce catalepsy when blocked (Klemm and Block 1988).

The advent of molecular biology has expanded the field of DA receptor research with the cloning of five subtypes during the last 10 years or so. DA D1 and D2 receptors have given way to the DA D1-like family, that comprises the D$_1$ (cloned by Dearry et al. 1990; Monsma et al. 1990; Sunahara et al. 1990; Zhou et al. 1990) and the D$_5$ (cloned by Grandy et al. 1991; Sunahara et al. 1991; Tiberi et al. 1991; Weinshank et al. 1991) subtypes, and the D2-like family, that encompasses the D$_2$, D$_3$ and D$_4$ subtypes (cloned respectively by Bunzow et al. 1988; Sokoloff et al. 1990; Van Tol et al. 1991). The DA D2-like family has been postulated to play a central role in the therapeutic action and certain side effects of antipsychotic drugs (Seeman 1992; Wilson et al. 1998). Following the discovery that the newly cloned D$_3$ subtype was preferentially localised in limbic areas (where blockade of D$_2$-like receptors by antipsychotic drugs is believed to mediate therapeutic effects) but was absent from the striatal and tuberoinfundibular systems (where blockade of D_2-like receptors is believed to mediate extrapyramidal symptoms and hyperprolactinaemia, respectively), it was hypothesised that a selective DA D_3 receptor antagonist would possess marked advantages compared to drugs currently used in the treatment of schizophrenia (SNYDER 1990; SOKOLOFF et al. 1990). This concentration of DA D_3 receptors in limbic structures (which are believed to be heavily involved in the mediation of reward mechanisms: FIBIGER and PHILLIPS 1988) has also prompted some authors to propose using D_3 receptor agonists as substitution strategies for drug abuse therapy (CAINE and KOOB 1993).

B. Behavioural Pharmacology of DA D_2/D_3 Receptor Agonists

In the original paper that described the cloning of the rat DA D_3 receptor (SOKOLOFF et al. 1990), it was reported that some DA receptor agonists presented a preferential D_3 over D_2 affinity (i.e. lower K_i for inhibition of [^{125}I]iodosulpride binding in CHO cells transfected with D_3 than in cells transfected with D_2 receptors). Follow-up studies (LEVESQUE et al. 1992; GACKENHEIMER et al. 1995; MIERAU et al. 1995; PUGSLEY et al. 1995) have extended the list, and one of these DA D_2/D_3 receptor agonists, 7-hydroxy-2-(di-N-propylamino)-tetralin (7-OH-DPAT) rapidly gained the status of prototypical DA D_3 receptor agonist. This compound (in the context of a selective D_3 agonist) has been assessed on rat spontaneous locomotor activity by DALY and WADDINGTON (1993) who reported a biphasic effect: low doses reduced, whereas higher doses increased spontaneous locomotor activity in rats. This initial finding of 7-OH-DPAT affecting locomotor activity was confirmed by several other laboratories (AHLENIUS and SALMI 1994; SVENSSON et al. 1994; STARR and STARR 1995) and extended to other DA D_2/D_3 receptor agonists, such as quinpirole and PD 128,907 (PUGSLEY et al. 1995; DEPOORTERE et al. 1996).

On the basis that 7-OH-DPAT decreased locomotor activity in rats at doses that did not affect DA release or synthesis (and that the putative DA D_3 receptor antagonist U 99194A – see below – increased locomotor activity without concomitant changes in DA neurochemical parameters), CARLSSON and colleagues hypothesised the existence of a post-synaptic D_3 receptor with a motor-inhibitory function (WATERS et al. 1993; SVENSSON et al. 1994). However, a subsequent study showed that 7-OH-DPAT was at least as potent in decreasing locomotor activity when microinjected into the ventral tegmental area than when microinjected into the nucleus accumbens (KLING-PETERSEN et al. 1995b), which does not seem to fit with the above-mentioned hypothesis. Also, a dissociation between the effects of 7-OH-DPAT on locomotor activity and its effects on DA metabolism has not been observed by others (GAITNETDINOV et al. 1996).

Numerous behaviours have been described following administration of putative DA D_3 receptor agonists. The following list, far from being exhaustive, is more of a representative sample to exemplify the diversity of behaviours elicited by treatment with D_3 selective agonists. These compounds can, depending on the dose, produce a conditioned place preference or conditioned place aversion (MALLET and BENINGER 1994; KHROYAN et al. 1995, 1997; KLING-PETERSEN et al. 1995a; CHAPERON and THIEBOT 1996). They also induce yawning and penile erection (KOSTRZEWA and BRUS 1991; DAMSMA et al. 1993; FERRARI and GIULIANI 1995; KURASHIMA et al. 1995), induce sniffing-gnawing (DALY and WADDINGTON 1993; DAMSMA et al. 1993; MCELROY et al. 1993) increase duration of sleep (LAGOS et al. 1998), facilitate ejaculatory behaviour (ALHENIUS and LARSON 1995), affect intracranial self-stimulation (NAKAJIMA et al. 1993; GILBERT et al. 1995; KLING-PETERSEN et al. 1995a; DEPOORTERE et al. 1996, 1999; HATCHER and HAGAN 1998), reduce oral ethanol (RUSSEL et al. 1996; SILVESTRE et al. 1996) or i.v. cocaine intake (CAINE and KOOB 1993), produce conditioned taste aversion (BEVINS et al. 1996), increase or decrease (depending on the dose) immobility time in the tail-suspension test (FERRARI and GIULIANI 1997), substitute for a cocaine discriminative cue (ACRI et al. 1995), prevent the acquisition or expression of morphine-induced conditioned place preference (DE FONSECA et al. 1995), and attenuate the discriminative cue (COOK and PICKER 1998) or antinociceptive effects (COOK et al. 1999) of mu opioids.

It should be emphasised that most of the studies listed above used 7-OH-DPAT, or a very limited range of DA D_3 receptor agonists, so that definitive conclusions regarding the implication of the D_3 subtype in these behaviours might have been premature. The gradually increasing availability of other DA D_3 receptor agonists, some of them claimed to possess a greater D_3 selectivity than 7-OH-DPAT (e.g. PD 128,907: SAUTEL et al. 1995a), coupled with the development of functional in vitro tests (see below), opened new avenues for the exploration of the functions of the D_3 receptor subtype.

The use of compounds with a clear preference for D_2 versus D_3 receptors might have offered a complementary approach for a better understanding of the pharmacology of the DA D_2/D_3 system. Unfortunately, the search for such compounds has not been very successful so far. The first agonist that could qualify as being selective for the D_2 receptor appears to be bromocriptine, with a D_2 versus D_3 selectivity ratio of about 5 (see Table 1 in LEVANT 1997). More recently, two compounds, U-91356 A (ratio of D_2 versus D_3: 23) and U-95666 A, were described as being selective for the D_2 subtype, but pharmacological data on these compounds are rather scarce (CAMACHO-OCHOA et al. 1995; SCHREUR and NICHOLS 1995; PIERCEY et al. 1996; CALON et al. 1995).

C. Correlational Studies Using DA D_2/D_3 Receptor Agonists

The availability of functional in vitro tests such as mitogenesis assessed by [^{3}H]thymidine uptake (CHIO et al. 1994; SAUTEL et al. 1995a) has allowed behavioural pharmacologists to investigate the correlations between the potencies of

DA D_2/D_3 receptor agonists to produce a given in vivo effect, and their in vitro potency. This comparative in vivo/in vitro approach has yielded a series of informative data from which it has been possible to establish that there was a significant correlation between the potency of DA D_2/D_3 receptor agonists to produce a given in vivo effect (see list below) and their potency to induce mitogenesis (Sautel et al. 1995a) in DA D_3, but not D_2, receptor transfected CHO cells. These in vivo effects were: decrease of body temperature (Perrault et al. 1996; Varty and Higgins 1998), reduction of spontaneous locomotor activity (Sautel et al. 1995a), decrease of operant responding (Sanger et al. 1996), disruption of the prepulse inhibition of the startle reflex (Caine et al. 1995; Varty and Higgins 1998), reduced intake of i.v. cocaine (Caine et al. 1997) or oral ethanol (Cohen et al. 1998), and substitution for the discriminative stimuli produced by 7-OH-DPAT, apomorphine or d-amphetamine (Sanger et al. 1997, 1999; Varty and Higgins 1997). Two other studies reported similar correlations between the potency of DA D_2/D_3 receptor agonists to substitute for the discriminative stimuli produced by cocaine (Spealman 1996) or to produce hypothermia (Millan et al. 1995) and their in vitro affinity for the D_3 receptor. Of note was the finding that the potency of DA D_2/D_3 receptor agonists to produce eye blinking in monkeys correlated better with affinity for the D_2 than for the D_3 subtype (Kleven and Koek 1996).

These correlational studies appeared to confirm the implication of the D_3 subtype in functions that had been suggested by studies that used a single or a limited number of DA D_3 receptor agonists (see previous paragraph).

D. Behavioural Pharmacology of DA D_2/D_3 Receptor Antagonists

In the initial paper that described the D_3 versus D_2 selectivity ratio of dopaminergic compounds, it was clear that, in contrast to agonists, most DA receptor antagonists tested showed a preference for the D_2 over the D_3 subtype (Sokoloff et al. 1990). The substituted benzamide amisulpride, which is selective for DA D_2/D_3 receptors and has no known affinity for any of the other major types of receptors (Scatton et al. 1997), appeared to have among the lowest D_2 versus D_3 selectivity ratio, and was found in a subsequent study to have similar affinity for both subtypes (Schoemaker et al. 1997). Such a binding profile might explain the atypical neuropharmacological profile of this compound, characterised by selectivity for presynaptic DA receptors and for DA receptors localised in limbic structures (Perrault et al. 1997). These two properties might underlie the demonstrated atypical nature of this antipsychotic in clinical practice, with therapeutic efficacy against both negative and positive symptoms, associated with a low propensity to produce extrapyramidal side effects (Boyer et al. 1995; Moller et al. 1997).

AJ 76 and UH 232 were the only two antagonists showing very modest preferences for the D_3 subtype in the study of Sokoloff and colleagues (1990).

Both compounds had been considered as preferential presynaptic DA receptor antagonists (SVENSSON et al. 1986), but their very limited preference for D_3 receptors (D_3 versus D_2 selectivity ratio of 3–4) would make them poor pharmacological tools for the in vivo probing of D_3 receptor function. Nafadotride is another example of a claimed D_3 preferential antagonist, though its D_3 versus D_2 selectivity ratio (5–9: SAUTEL et al. 1995b; AUDINOT et al. 1998) does not seem to make it much more valuable than AJ 76 or UH 232 as a compound for probing D_3 function. Nafadotride was found to have activating properties in rodents (SAUTEL et al. 1995b). Similar results were obtained by XU and collaborators (1999), but not by other laboratories (CLIFFORD and WADDINGTON 1998; unpublished data from our laboratory). Also, due to the marked affinity of this compound for the D_2 subtype ($K_i = 4.5\,nM$: SAUTEL et al. 1995b), one cannot exclude the possibility that some of the effects observed, even at fairly low doses (i.e. less than 1 mg/kg) could be partly or totally due to activity at the D_2 subtype (see Sect. E).

PNU-99194A (previously referred to as U 99194) appears to have been the first antagonist with a D_3 versus D_2 selectivity ratio greater than 20. WATERS and colleagues (1993) found that this compound enhanced rat locomotor activity in the absence of increased release or turnover of DA. This led the authors to speculate about the existence of a post-synaptic D_3 receptor with an inhibitory role on locomotor activity. However, recent data obtained in D_3 knock-out (KO) mice (see relevant section below) do not argue in favour of a D_3 receptor-mediated effect for the increase of locomotor activity produced by PNU-99194A. Further, the recent findings that this compound does not antagonise, but instead potentiates, the discriminative cue produced by 7-OH-DPAT (DEPOORTERE et al. 2000), along with the observation that morphine-induced hyperlocomotion can be antagonised both by PNU-99194A (MANZANEDO et al. 1999) and 7-OH-DPAT (SUZUKI et al. 1995), are also inconsistent with a claimed D_3 receptor antagonist property of this compound.

More recently, several compounds with an even greater selectivity than PNU-99194A have been described. S 14297 (D_3 versus D_2 selectivity ratio of 23–60: AUDINOT et al. 1998) was shown to reverse hypothermia produced by 7-OH-DPAT and PD 128,907, to be devoid of cataleptogenic activity and to antagonise haloperidol-induced catalepsy (MILLAN et al. 1995, 1997; AUDINOT et al. 1998). PD 152255 (D_3 versus D_2 selectivity ratio of about 40) reduced locomotor activity in mice, as well as spontaneous and amphetamine-stimulated locomotion in non-habituated rats (CORBIN et al. 1998). L-745,829, an antagonist with a 40-fold selectivity for D_3 receptors, failed to antagonise the discriminative cue produced by PD 128,907 (BRISTOW et al. 1998). In the same study, the 100-fold D_3 selective antagonist GR 103,691 was also found to be inactive to block the PD 128,907 discriminative cue. GR 103,691 has been found to be basically devoid of in vivo activity in two other studies (AUDINOT et al. 1998; CLIFFORD and WADDINGTON 1998), a finding that was attributed to limited bioavailability in the former paper.

We have studied GR 231,218 – another putative D_3 receptor antagonist (MURRAY et al. 1996) – but have not found any consistent effects of this compound in various rat or mice behavioural tests (all unpublished data). Administered i.p. either acutely or chronically (five daily treatments), this compound did not reverse hypothermia produced by the DA D_2/D_3 receptor agonist quinelorane. It inconsistently produced small increases in locomotor activity in mice habituated to the activity chamber, did not reverse catalepsy induced by haloperidol in rats, and failed to consistently and dose-dependently reverse d-amphetamine-induced hyperactivity, or hypoactivity produced by DA D_2/D_3 receptor agonists. This compound possesses a wide preferential affinity for D_3 ($IC_{50} < 5\,nM$) versus D_2 ($IC_{50} > 1\,\mu M$) receptors. Furthermore, its very low affinity for the D_2 receptor (contrary to compounds such as nafadotride) makes it more suitable for investigating the function of D_3 receptors. It was also found to behave as an antagonist in a mitogenesis test in CHO cells transfected with D_3 receptors. Finally, it has good brain penetration, so that its lack of in vivo activity is probably not due to an inability to obtain meaningful brain concentrations following i.p. treatment (unpublished data from the Neurochemistry and Pharmacokinetic Depts. of Sanofi-Synthelabo).

To summarise, it is at present difficult to determine a consensual profile of the behavioural effects of D_3 preferring antagonists. Antagonists of the first generation (AJ76, UH 232 and nafadotride) are probably not selective enough to be used in vivo for the elucidation of the function of the D_3 receptor. PNU-99194A has seen its claimed preference for the D_3 subtype challenged by recent data in D_3 KO mice (see below). The remaining candidates, S 14297, PD 152255, GR 103,691 and GR 231,218 appear to share too little in common to draw firm conclusions regarding the behavioural profile of D_3 preferring antagonists (see also CLIFFORD and WADDINGTON 1998).

One of these compounds, GR 231,218 – which would appear to be the most selective D_3 antagonist – was, in our hands, basically devoid of activity on behaviour (see above). There seem to be no published data on the effects on this compound on behaviour, so that comparison with findings from other laboratories is not possible. Whether or not this absence of effects reflects a particularity of the compound, or indicates that in vivo blockade of D_3 receptors is mostly inconsequential at the behavioural level, cannot be ascertained at present.

The case for selective D_2 antagonists is hardly more rosy, with the best candidates showing D_2 over D_3 selectivity ratios in the order of 10 (Table 1 in LEVANT 1997). To the best of our knowledge, there is a single compound – L741,626 – that has been reported to possess a 40-fold preferential affinity for D_2 ($K_i = 2.4\,nM$) over D_3 ($K_i = 100\,nM$) receptors (KULAGOWSKI et al. 1996). L741,626 was found to block the discriminative cue induced by the DA D_2/D_3 receptor agonist PD 128,907 (whereas the preferential D_3 receptor antagonists L745,829 and GR 103,691 were without effect: BRISTOW et al. 1998). The authors concluded that the D_2 receptor might be more likely than the D_3 subtype to mediate the discriminative effects of this compound. It is worth noting that

KLEVEN and KOEK (1997) reached a similar conclusion concerning the role of the D_2 receptor in the discriminative stimulus effects of PD 128,907.

Despite the rather impressive number of studies that have investigated the behavioural pharmacology of D_3 preferring agonists (and to a lesser extent antagonists), and in spite of the correlational studies that established a link between in vitro potency to produce mitogenesis in D_3 receptor transfected cells and several in vivo effects, definite conclusions regarding the role of the D_3 (or D_2) receptors, based on these classical pharmacological approaches, are missing. This is partly due to doubt over the selectivity of compounds for the D_3 subtype (CHIO et al. 1994; LARGE and STUBB 1994; BURRIS et al. 1995), but also because of a lack of convincing demonstrations that effects produced by apparently selective DA D_3 receptor agonists can be dose-dependently reversed by several putative DA D_3 receptor antagonists.

The use of pharmacological tools to explore the function of a receptor subtype is not without drawbacks. In particular, the observation of a given behaviour following treatment with a compound showing a certain level of selectivity for a receptor, does not necessarily mean that activity at this receptor is primarily responsible for the elicited behaviour. The observed effect could result from activity of metabolites with affinity for other types of receptors, or from activity of the parent compound at sites additional to the site(s) for which it shows selectivity. Likewise, failure to observe in vivo activity might critically depend on several pharmacokinetic parameters of the compound, such as half-life, bioavailability, or central penetration.

Recent progress in molecular biology has permitted the generation of mice in which a gene (or several genes) can be selectively deleted, giving rise to mutant individuals lacking the protein that was coded for by the deleted gene. Availability of these KO mice has spurred considerable interest in the community of behavioural pharmacologists as they offer the advantage – at least theoretically – of providing a model which should give an insight into the function of the protein coded for by the deleted gene. Despite the possible caveats that have been linked to the use of these KO mice in behavioural studies (the reader is referred to CRUSIO 1996; GERLAI 1996; GOLD 1996 or KOOB et al. in Chap. 20 of this volume for in-depth discussion), these KO mice offer an elegant complementary approach to classical pharmacology (or to lesion or antisense approaches, both of which, for reasons of space, will not be reviewed in this chapter). In recent years, DA D_3 and DA D_2 receptor KO mice have generated interesting data that have started to show some promise in shedding light on the function of DA D_2 and D_3 receptors.

E. Dopamine D_3 Receptor Knock-out Mice

Dopamine D_3 receptor KO mice were first engineered by ACCILI and collaborators (1996), followed by XU and colleagues (1997) and more recently by JUNG and co-workers (1999).

I. Analysis of the Phenotype of D$_3$ Receptor Knock-out Mice

As a general rule, and in distinction to the literature on DA D$_2$ KO mice (see below), all studies published so far have noted a lack of overt phenotypical differences between DA D$_3$ KO mice and their wild-type counterparts (ACCILI et al. 1996; XU et al. 1997; BOULAY et al. 1999a; JUNG et al. 1999). They grow and breed normally, and present no obvious neurological defect or abnormal reactivity to handling.

In both of the early studies, mice homozygous for the deletion (D$_3^{-/-}$ mice) were reported to show increased levels of spontaneous activity (ACCILI et al. 1996; XU et al. 1997). Jung and colleagues, however, found no hyperactivity in their D$_3^{-/-}$ mice: according to these authors, this apparent difference might have been due to variations in the time for which locomotor activity was recorded between their experiment and the two studies mentioned above. In our first study (BOULAY et al. 1999a) on DA D$_3$ mutant mice (colony generated from an individual issued from the laboratory of ACCILI and colleagues), analysis of the level of spontaneous activity did not reveal any consistent differences in levels of spontaneous locomotor activity amongst the three genotypes (wild-types, heterozygotes and homozygotes: D$_3^{+/+}$, D$_3^{+/-}$ and D$_3^{-/-}$). In the five batches of mice that we have tested so far (three of them used in the published study), D$_3^{-/-}$ mice were not systematically hyperactive, and were occasionally less active than their controls. In these conditions, it was not possible to conclude whether or not, under our experimental conditions, deletion of the gene coding for the DA D$_3$ receptor is associated with an enhanced level of spontaneous locomotor activity. This uncertainty concerning the association between the deletion of the gene coding for the DA D$_3$ receptor and enhanced levels of spontaneous locomotor activity is reminiscent of that concerning the locomotor-enhancing effects of DA D$_3$ receptor antagonists (see discussion above).

D$_3^{-/-}$ mice (issued from the colony bred by ACCILI and colleagues) were studied by STEINER and co-workers (1998) in animal models of anxiety. These mutants were found to spend more time on the open arm of an elevated plus maze, and in the centre of an open quadrant, behaviours interpreted as reflecting a reduced level of anxiety, leading the authors to conclude that the D$_3$ receptor was involved in anxiety-related behaviours. Such evidence for reduced levels of anxiety was not observed by others in the plus maze (XU et al. 1997; BOULAY et al. 1998) or in the "light-dark box" test, another test of anxiety (BOULAY et al. 1998).

II. Effects of DA Receptor Ligands in D$_3$ Receptor Knock-out Mice

Given the uncertainty about the in vivo selectivity of compounds used in behavioural studies to characterise the function of the DA D$_3$ receptor (see discussion above), testing the effects of DA D$_2$/D$_3$ receptor ligands in D$_3$ KO mice appeared to be warranted.

1. DA D_2/D_3 Receptor Agonists

Xu and colleagues (KOELTZOW et al. 1995; Xu et al. 1995) were the first to report in abstracts that the DA D_2/D_3 receptor agonist PD 128,907 – which has been claimed to be the compound with the highest D_3 versus D_2 selectivity ratio (SAUTEL et al. 1995a) – produced identical effects on locomotor activity in $D_3^{-/-}$ and $D_3^{+/+}$ mice. This was the first indication that an agonist shown to have selectivity for D_3 receptors in vitro was still able to induce a behavioural effect in mice lacking D_3 receptors. Subsequent studies confirmed this initial finding, with PD 128,907 as well as the other putative DA D_3 receptor selective agonists quinelorane and 7-OH-DPAT (BOULAY et al. 1999a; Xu et al. 1999). It was also reported in these two papers that the temperature-decreasing effects of PD 128,907, quinelorane or 7-OH-DPAT were still observed in $D_3^{-/-}$ mice. These results led both teams to conclude that the D_3 receptor does not mediate those pharmacologically induced decreases of locomotor activity and core temperature, and that these compounds may lack in vivo selectivity for this receptor subtype.

2. DA D_2/D_3 Receptor Antagonists

The two studies referred to above also tested the effects of putative DA D_3 receptor antagonists. Xu and colleagues (1999) showed that the locomotor-enhancing effects of nafadotride and PNU-99194A were still present in $D_3^{-/-}$ mice; we obtained similar results with PNU-99194A, which was found to enhance locomotor activity and reverse quinelorane-induced hypothermia in $D_3^{-/-}$ mice (BOULAY et al. 1999a). As was the case for the DA D_2/D_3 receptor agonists mentioned above, both studies cast doubt on the claimed D_3 selectivity of these two antagonists, and indicate that these pharmacological effects are not mediated by activity at D_3 receptors.

3. Psychostimulants

$D_3^{-/-}$ mice have been shown to present increased sensitivity to a low dose, but not to high doses, of cocaine and to concurrent activation of DA D_1 (using the agonist SKF 81297) and DA D_2 (using the agonist PD 128,907) receptors, both in normal and in reserpinised mice (Xu et al. 1997). $D_3^{-/-}$ mice were also shown to be more sensitive than controls to the effects of d-amphetamine in a conditioned place preference test (Xu et al. 1997). These two results prompted the authors to propose that the D_3 subtype modulates behaviour by interfering with the synergistic interaction between D_1-like and the other members of the D_2-like family of DA receptors.

F. Dopamine D_2 Receptor Knock-out Mice

At the same time as DA D_3 receptor KO mice were engineered, BORRELLI and colleagues published their first study on mice lacking the gene coding for the

DA D_2 receptor (BAIK et al. 1995). These investigators were followed shortly after by other teams (YAMAGUCHI et al. 1996; KELLY et al. 1997). More recently, a fourth line of these mutant mice has been generated by JUNG and collaborators (1999).

I. Analysis of the Phenotype of D_2 Receptor Knock-out Mice

In the original paper (BAIK et al. 1995), $D_2^{-/-}$ mice were described as being "parkinsonian-like", that is bradykinetic, ataxic, spontaneously cataleptic in a "ring test", showing severe motor incoordination, and presenting deficits in growth and fecundity. DA D_2 receptor KO mice studied by YAMAGUCHI and colleagues (1996) were succinctly described as being hypoactive with a slow and creeping movement. However, most of these motor characteristics were not seen in $D_2^{-/-}$ mice from the colony generated by KELLY and colleagues (KELLY et al. 1997, 1998). These mice did not present ataxia and abnormal stance, were described as looking healthy, having good muscle tone and alert behaviour, and growing and breeding normally. Similarly, analysis of the spontaneous behaviour of the colony of mice used in our laboratory (generated from $D_2^{+/-}$ mice from the colony used in the study of BAIK and colleagues, 1995) did not show evidence for the severe neurological defects that were originally described. Visual inspection of this colony did not allow us to distinguish between the three genotypes. More specifically, these mice were not ataxic, presented a normal posture and did not present abnormalities of forelimb muscle strength. When tested in a "bar test" (in which mice are positioned so that both front paws rest on a 0.4cm diameter steel rod 3.5cm above the surface of the bench), neither $D_2^{-/-}$ nor $D_2^{+/-}$ mice were spontaneously cataleptic (BOULAY et al. 1999b). In addition, we found no evidence for a deficit of rotarod performance in either type of mutant (unpublished results). The phenotype of the third line (JUNG et al. 1999), on the whole, was reminiscent of the one described for mice generated by BAIK and colleagues (1995). However, the severity of motor abnormalities of these mice was reported to vary with their age. Motor dysfunction appeared two weeks postnatal, and significantly improved from day 45 onwards. Nonetheless, all reports agree on the findings that $D_2^{-/-}$ mice show reduced levels of spontaneous locomotor activity (BAIK et al. 1995; KELLY et al. 1997, 1998; MALDONADO et al. 1997; BOULAY et al. 1999b; JUNG et al. 1999).

II. Effects of DA Receptor Ligands in D_2 Receptor Knock-out Mice

The search for compounds, in particular antagonists, showing a preference for the D_2 subtype was not justified on theoretical grounds (see the introductory section for the purported advantages of selective D_3 compounds, and the disadvantages thought to be linked to activity at D_2 receptors). This might partially explain the relative lack of interest of testing dopaminergic compounds in DA D_2 KO mice.

1. DA D_2/D_3 Receptor Agonists

To the best of our knowledge, the first study that investigated the effects of a DA receptor agonist in D_2 KO mice was done by KELLY and colleagues in 1998. Akinesia induced by reserpine was reversed by a combination of a subthreshold dose of the DA D_1 receptor agonist SKF 38393 and the DA D_2/D_3 receptor agonist quinpirole in $D_2^{+/+}$ and $D_2^{+/-}$, but not in $D_2^{-/-}$ mice. This indirectly suggested that $D_2^{-/-}$ mice were unresponsive to the effects of quinpirole.

Following the findings indicating that the D_3 subtype was not necessary for the expression of DA D_2/D_3 receptor agonist-induced decreases of locomotor activity and core temperature (BOULAY et al. 1999a; XU et al. 1999), the obvious next experimental step was to verify whether or not deletion of the DA D_2 receptor would interfere with these pharmacologically-induced effects. It was found that $D_2^{-/-}$ and $D_2^{+/-}$ mice were respectively unresponsive and markedly less responsive to the hypolocomotor and hypothermic effects of PD 128,907 and quinelorane (BOULAY et al. 1999b). These data, along with those obtained in DA D_3 KO mice, provide unambiguous experimental arguments that the DA D_2, but not the D_3, receptor subtype mediates these two in vivo effects.

2. DA D_2/D_3 Receptor Antagonists

KELLY and collaborators (1998) were also the first to study the behavioural effects of DA receptor antagonists in D_2 KO mice. Haloperidol, at the dose of 0.6mg/kg i.p., was found to be without effects on locomotor activity in these $D_2^{-/-}$ mice.

3. Psychotropic Agents

$D_2^{-/-}$ mice have also been shown to present reduced sensitivity to the reinforcing effects of morphine as assessed by the conditioned place preference test, but to respond normally to the locomotor-enhancing effects of this drug, and to show conditioned place preference produced by food reward (MALDONADO et al. 1997). Likewise, DA D_2 KO mice were markedly less sensitive to several effects of alcohol: these mice showed reduced oral intake of ethanol, and were less sensitive both to its depressant effects on locomotor activity and to its ataxic effects (PHILLIPS et al. 1998). However, these mutant mice responded normally to a saccharin and quinine consumption preference test.

We had previously shown that in rats trained to discriminate apomorphine, 7-OH-DPAT or d-amphetamine from saline in a two-lever, food-reinforced discriminative task (SANGER et al. 1997, 1999), the potency of DA D_2/D_3 receptor agonists to substitute for these discriminative cues correlated with their in vitro potency in a mitogenesis test for the D_3, but not the D_2, receptor. This led us to speculate that these cues were mediated by the DA D_3

receptor. An ideal experiment to test this hypothesis would have consisted in subjecting $D_2^{+/+}$, $D_2^{+/-}$, $D_2^{-/-}$, $D_3^{+/+}$, $D_3^{+/-}$ and $D_3^{-/-}$ mice to a protocol of drug discrimination adapted from the one used with rats, to verify if the deletion of the gene coding for the D_3 or the D_2 receptor could prevent acquisition of, say, a 7-OH-DPAT discrimination. Unfortunately, we found in a pilot study that C57BL.6 J mice did not readily discriminate 0.1 mg/kg i.p. of 7-OH-DPAT from saline, using a protocol adapted from the one used to train rats to discriminate a similar dose (SANGER et al. 1997). However, under the same conditions, another batch of C57BL.6J mice rapidly discriminated 1 mg/kg i.p. d-amphetamine from saline. Given that DA D_2/D_3 receptor agonists substitute for d-amphetamine (BEVINS et al. 1997; SANGER et al. 1999; but see VARTY and HIGGINS 1997), and due to the fact that all effects of DA D_2/D_3 receptor agonists tested so far appear to be mediated through activity at the D_2 subtype, it was decided to train $D_2^{+/+}$, $D_2^{+/-}$ and $D_2^{-/-}$ mice to discriminate 1 mg/kg i.p. of d-amphetamine from saline.

$D_2^{-/-}$ mice acquired the discrimination almost as rapidly as $D_2^{+/+}$ mice. The mean ± standard error of mean (SEM) numbers of training sessions to reach an arbitrary discrimination criterion were: 41.2 ± 6.6 versus 38.2 ± 6.0, for $D_2^{-/-}$ and $D_2^{+/+}$ mice, respectively. $D_2^{+/-}$ mice were notably faster than the other two phenotypes (27.1 ± 2.9). The DA D_2/D_3 receptor agonists quinelorane and 7-OH-DPAT engendered dose-dependent substitution in $D_2^{+/+}$ mice and, to a variable extent, in $D_2^{+/-}$ mice, but failed to substitute at all in $D_2^{-/-}$ mice (Fig. 1).

These results indicate that the common element between the discriminative cues of d-amphetamine and DA D_2/D_3 receptor agonists is not likely to be the D_3 subtype, but rather the D_2 subtype. These findings indirectly suggest that the D_3 subtype is probably not implicated in the d-amphetamine discriminative cue. One possibility is that the discriminative cue is mediated by D_1-like receptors. Alternatively, it might be that in $D_2^{-/-}$ mice the d-amphetamine cue is not preferentially mediated by enhancement of dopaminergic transmission, but rather by noradrenergic and/or serotonergic mechanisms. This could also explain the lack of substitution with these two DA D_2/D_3 receptor agonists in $D_2^{-/-}$ mice. However, the lesser potency of these agonists to substitute in $D_2^{+/-}$ mice, which have been shown to have D_2 receptor expression reduced by about 50% (BAIK et al. 1995; KELLY et al. 1997; Boulay et al. 2000), would indirectly argue in favour of a central role of the D_2 subtype in the discriminative cue of d-amphetamine in wild-type animals.

G. Direct Comparison Between D_2 and D_3 Receptor Knock-out Mice

The identification of the key role played by the D_2 subtype, and the lack of involvement of the D_3 subtype, in mediating the hypolocomotor and hypothermic effects of DA D_2/D_3 receptor agonists (BOULAY et al. 1999a,b; XU et al.

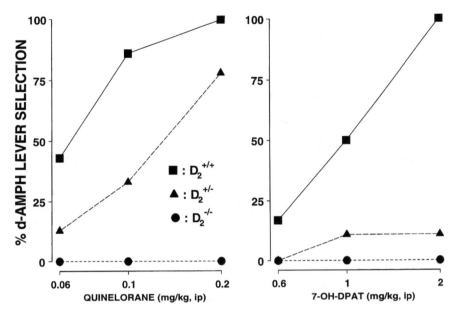

Fig. 1. Generalisation curves for quinelorane and 7-OH-DPAT in DA D_2 KO mice trained to discriminate d-amphetamine from saline. Each symbol represents the percentage of mice selecting the drug-associated lever during a session when the training drug (d-amphetamine, 1 mg/kg i.p.) was replaced by quinelorane or 7-OH-DPAT. Doses of substitution drugs where administered i.p., 30 min pre-test and in a counterbalanced order. $n = 10$ for each genotype

1999) were rather unexpected in the light of the results of the correlational studies (see above). For that reason, we consider that hypotheses implicating the D_3 subtype in other in vivo functions (suggested to be under the control of the D_3 subtype by these correlational studies) should be tested using both D_3 KO and D_2 KO mice. Furthermore, the two types of KO mice should preferably be studies in parallel, so as to minimise extraneous variables that might differ from one experiment to the other.

I. Comparison of Avoidance Behaviour of D_2 and D_3 Receptor Knock-out Mice

Despite the correlation between the clinical efficacy of antipsychotics and their affinity for DA D2-like receptors, the behavioural profile of D_2 or D_3 (or D_4) KO mice, in tests known to be sensitive to the effects of antipsychotics, seems not to have been explored yet. The active avoidance test has been claimed to show selective sensitivity to antipsychotic compounds (NIEMEGEERS et al. 1969; DAVIDSON and WEIDLEY 1976; ARNT 1982), as they interfere both with the acquisition of this behavioural task and with performance in trained animals by reducing the number of shock avoidance responses without increasing escape failures. In the two-way shuttle box version of this procedure, mice are required, in order to avoid an electric shock, to move from one compartment of a box to the opposite one, following the onset of a stimulus (a light for example) that precedes the delivery of the shock. We have recently started to analyse the behaviour of D_2 and D_3 KO mice in such a procedure (BOULAY et al. 1999c). $D_2^{-/-}$ mice were markedly impaired in their ability to acquire this task, whilst $D_2^{+/-}$ siblings were between $D_2^{-/-}$ and $D_2^{+/+}$ mice in their speed of acquisition (Fig. 2, top panel). At the fourth training session, $D_2^{-/-}$ mice avoided shocks on about 30% of trials (while controls avoided almost all shocks), and took twice as long (10 versus 4 sessions) as controls to reach their asymptotic level of performance (about 80% of avoidance trials). DA $D_3^{+/+}$, $D_3^{+/-}$ and $D_3^{-/-}$ mice did not differ appreciably in their speed of acquisition of this avoidance task (Fig. 2, bottom panel). To the extent that this preclinical model has been claimed to detect antipsychotic activity, these results show that the deletion of the D_2, but not the D_3 subtype, mimics the effects of antipsychotics (i.e. interferes with the acquisition of this task).

II. Comparison of Effects of DA Receptor Ligands in D_2 and D_3 Receptor Knock-out Mice

1. Psychotropic Agents

Considering the preferential localisation of DA D_3 receptors in the limbic system – which is thought to be of major importance in the control of reward mechanisms (FIBIGER and PHILLIPS 1988) – we investigated in DA D_2 and DA D_3 KO mice, the effects of two psychotropic drugs well known for their ability to enhance spontaneous locomotor activity. d-Amphetamine and phencyclidine enhanced locomotor activity in the colony of DA D_3 KO mice to similar extents, irrespective of the genotype (Fig. 3, right panels). Contrary to what has been reported by XU and collaborators (1997), we found no evidence for an increased sensitivity of our $D_3^{-/-}$ mice to a low (5 mg/kg) or higher (10 and 20 mg/kg) doses of cocaine (data not shown). If one accepts the proposal that the increase of locomotor activity is predictive of the positive reinforcing value of a drug (WISE and BOZARTH 1987) then the present data, at first sight, do not argue in favour of an implication of the D_3 subtype in reward mechanisms. However, testing these mice in other behavioural tests, such as conditioned place preference, intracranial self-stimulation or drug auto-administration, will be necessary to further explore this matter.

$D_2^{-/-}$ and $D_2^{+/-}$ mice responded quite normally to the locomotor-enhancing effects of d-amphetamine; this mirrors our findings that DA D_2 KO mice can be trained to discriminate the cue produce by this compound (see above). The nature of the compensatory mechanisms that counteract the effects of the absence of D_2 receptors for these psychostimulant-induced locomotor enhancing effects is unknown. By contrast, these mutant mice were markedly less hyperactive than controls when injected with phencyclidine. This finding

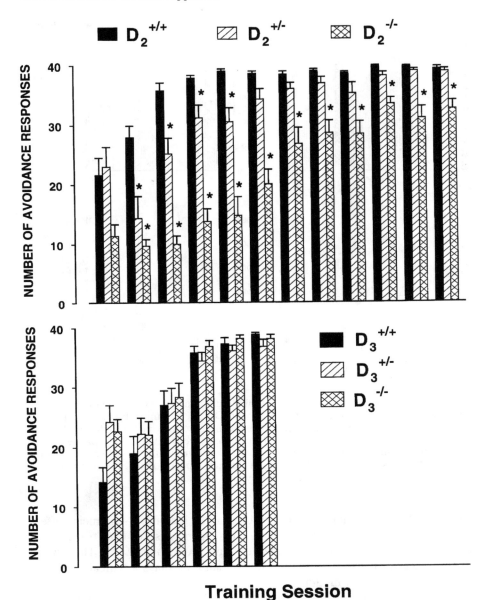

Fig. 2. Acquisition of an active avoidance task by DA D_2 and DA D_3 KO mice. Each bar represents the mean (+ the SEM) number of avoidance responses (out of a maximum of 40) as a function of the training session number. *$p < 0.05$, **$p < 0.01$: significantly different from control (+/+) mice, at the considered session number [Dunnett's post-hoc test following one-way analyses of variance (ANOVAs)]. $n = 10–12$ mice for each genotype

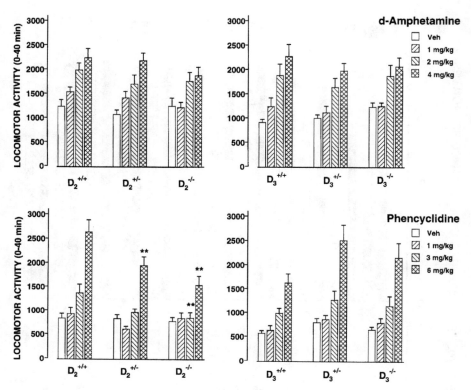

Fig. 3. Effects of pretreatment with *d*-amphetamine and phencyclidine on the level of spontaneous locomotor activity in DA D$_2$ and DA D$_3$ KO mice. Each bar represents the mean (+SEM) number of infrared beam interruptions in the activity chamber. Treatments [i.p. injection of compound or vehicle (*Veh*) immediately pre-test] were applied in a counterbalanced order. **p < 0.01: significantly different from control (+/+) mice, at the considered dose of the drug (Dunnett's post-hoc test following one-way ANOVAs). *n* = 10–12 mice for each genotype

would tend to strengthen the supposition that the locomotor-enhancing effects of this compound are partly mediated by an increase of DA transmission (HERNANDEZ et al. 1988; MCCULLOUGH and SALAMONE 1992). They would further suggest that the D$_2$, but not the D$_3$, subtype plays a critical role in the expression of this DA-mediated effect.

An article by RALPH and collaborators (1999) compared mice in which the gene coding for either the D$_2$, D$_3$ or D$_4$ subtype had been knocked out, for their sensitivity to the disrupting effects of *d*-amphetamine on the prepulse inhibition (PPI) of the startle reflex. The authors found that only DA D$_2$ KO mice were resistant to the PPI-disrupting effects of *d*-amphetamine. To the extent that PPI-disrupting effects produced by drugs that increase DA neurotransmission has been proposed as a preclinical test to detect antipsychotic activity, the authors concluded that the D_2, but not the D_3 or D_4, subtype is relevant to antipsychotic drug action.

2. DA Receptor Antagonists

We have analysed the sensitivity of colonies of DA D_2 or D_3 receptor KO mice to the cataleptogenic effects of the DA D2-like receptor antagonist haloperidol (BOULAY et al. 2000). $D_2^{-/-}$ mice were totally unresponsive to the cataleptogenic effects of haloperidol, while $D_2^{+/-}$ mice, at the highest doses tested, showed a level of catalepsy about half that of wild-type controls. However, $D_2^{-/-}$ and $D_2^{+/-}$ mice were as sensitive as their wild-type counterparts to the cataleptogenic effects of the DA D1-like receptor antagonist SCH 23390. The ability of SCH 23390 to induce catalepsy in $D_2^{-/-}$ mice suggests that their resistance to haloperidol-induced catalepsy is consecutive to the absence of DA D_2 receptors, and not to the abnormal striatal synaptic plasticity that has been shown to occur in these mice (CALABRESI et al. 1997). DA $D_3^{-/-}$ and $D_3^{+/-}$ mice, on the whole, did not differ from their controls in the time spent in a cataleptic position following administration of either haloperidol or SCH 23390. Also, D_3 mutant mice were no more responsive than wild-type controls when co-administered subthreshold doses of haloperidol and SCH 23390, suggesting that DA D_3 KO mice are not more sensitive than wild-types to the synergistic effects of concurrent blockade of DA D_2 and D_1 receptors in this model of catalepsy. Together, these results suggest that the DA D_2 subtype is necessary for haloperidol to produce catalepsy, and that – contrary to what could be expected from pharmacological studies (MILLAN et al. 1997) – the DA D_3 subtype appears to exert no observable control over the catalepsy produced by D2-like, in addition to catalepsy produced by D1-like and combination of D1-like and D2-like, receptor antagonists.

We have mentioned above that mice lacking the gene coding for the D_2, but not the D_3, subtype showed a deficit of acquisition of a two-way active avoidance response, indicating that the absence of the D_2 receptor mimics the effects of antipsychotic drugs in this test. Antipsychotic drugs have also been shown to have deleterious effects on the performance phase of this task, i.e. once the animals have learned the task. Indeed, we have also observed that once trained, $D_2^{-/-}$ mice were totally unresponsive to the performance-disrupting effects of the prototypical antipsychotic haloperidol, while $D_2^{+/-}$ mice were only mildly affected (Fig. 4, left panel). $D_3^{+/+}$, $D_3^{+/-}$ and $D_3^{-/-}$ mice responded with similar sensitivities to the disrupting effects of haloperidol, i.e. all three phenotypes showed dose-dependent decreases in the number of avoidance responses (Fig. 4, right panel). These results suggest that the D_2 subtype is responsible for the effects of the antipsychotic haloperidol in this preclinical test claimed to have predictive validity to detect potential antipsychotic activity. Further experiments will need to be carried out to assess if this is also the case with other antipsychotics.

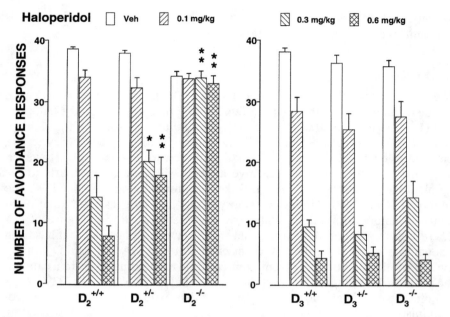

Fig. 4. Effects of pretreatment with haloperidol on the performance phase of an active avoidance task in DA D_2 and DA D_3 KO mice. Each bar represents the mean (+SEM) number of avoidance responses (out of a maximum of 40) recorded during a single session. Treatments [i.p. injection of haloperidol or vehicle (*Veh*) 30 min pre-test] were applied in a counterbalanced order. $*p < 0.05$, $**p < 0.01$: significantly different from control (+/+) mice, at the considered dose of haloperidol (Dunnett's post-hoc test following one-way ANOVAs). $n = 10$–12 mice for each genotype

H. Conclusions

Respectively 13 and 11 years after their cloning, the D_2 and D_3 subtypes of DA receptors still offer a fantastic challenge to behavioural pharmacologists. Despite the early availability of agonists that showed a certain in vitro selectivity for the D_3 subtype, the advance in defining the in vivo role of this subtype has been slow, punctuated with some early warnings suggesting that the extent of the selectivity of some agonists for the D_3 subtype might have been exaggerated (CHIO et al. 1994; LARGE and STUBB 1994; BURRIS et al. 1995). To make things worse, the rather late development of selective D_3 receptor antagonists slowed the progress of antagonist/agonist interaction studies that could have helped to refine our knowledge of the behavioural pharmacology of this subtype. Also, the D_2 subtype has been the object of relatively little research, which may be explained by the relative paucity of D_2-preferring compounds, and an obvious limited interest for such compounds (see in the introductory section the rationale for the presumed advantages offered by DA D_3 receptor selective agents). The use of D_2 KO and D_3 KO mice in the field of DA research has been punctuated both by rather deceptive findings when used alone (lack of consistency in phenotypic differences between lines, or within lines but tested in different laboratories, etc.), and by key findings when associated with pharmacological studies. Recent data collected with KO mice strongly suggest that selectivity profiles of DA D_2/D_3 receptor ligands inferred from in vitro studies might not necessarily apply in vivo, or that, contrary to what was concluded from correlational and other studies, the D_2, but not the D_3, receptor is implicated in some (or perhaps all) in vivo effects produced by these agents. Although exploration of the pharmacological reactivity of these KO mice is at an early stage, results gathered so far are mostly inconsistent with those obtained with classical pharmacology. Consequently, those functions previously ascribed to the D_3 receptor based on the in vivo effects of these DA D_2/D_3 receptor ligands might have to be reappraised.

In spite of the considerable – though still only theoretical – advantages that DA D_3 selective agents might present as pharmacotherapy for schizophrenia, drug abuse and possibly depression (see SCHWARTZ et al. 1995; WILLNER 1995), some in vivo data (especially those collected recently in KO mice) should prompt one to be cautious in drawing conclusions about the clinical application of these compounds (see RALPH et al. 1999 for antipsychotics).

Acknowledgements. We wish to express our gratitude to S. Ho Van, M. Lacave and M.T. Lucas for their participation and assistance in setting up or conducting some of the experiments, and to B. Kleinberg for his expertise in electronics.

References

Accili D, Fishburn CS, Drago J, Steiner H, Lachowicz JE, Park B-H, Gauda EB, Lee EJ, Cool MH, Sibley DR, Gerfen CR, Westphal H, Fuchs S (1996) A targeted mutation of the D_3 dopamine receptor gene is associated with hyperactivity in mice. Proc Natl Acad Sci USA 93:1945–1949

Acri JB, Carter SR, Alling K, Geter-Douglass B, Dijkstra D, Wikström H, Katz JL, Witkin JM (1995) Assessment of cocaine-like discriminative stimulus effects of dopamine D_3 receptor ligands. Eur J Pharmacol R7–R9

Ahlenius S, Salmi P (1994) Behavioral and biochemical effects of the dopamine D3 receptor-selective ligand, 7-OH-DPAT, in the normal and the reserpine-treated rat. Eur J Pharmacol 260:177–181

Ahlenius S, Larsson K (1995) Effects of the dopamine D_3 receptor ligand 7-OH-DPAT on male rat ejaculatory behavior. Pharmacol Biochem Behav 51:545–547

Arnt J (1982) Pharmacological specificity of conditioned avoidance response inhibition in rats: inhibition by neuroleptics and correlation to dopamine receptor blockade. Acta Pharmacol Toxicol 51:321–329

Audinot V, Newman-Tancredi A, Gobert A, Rivet JM, Brocco M, Lejeune F, Gluck L, Despote I, Bervoets K, Dekeyne A, Millan MJ (1998) A comparative in vitro and in vivo pharmacological characterization of the novel dopamine D_3 receptor antagonists (+)-S 14297, nafadotride, GR 103,691 and U 99194. J Pharmacol Exp Ther 287:187–197

Baik JH, Picetti R, Saiardi A, Thiriet G, Dierich A, Depaulis A, Le Meur M, Borrelli E (1995) Parkinsonian-like locomotor impairment in mice lacking dopamine D2 receptors. Nature 377:424–428

Bevins RA, Delzer TA, Bardo MT (1996) Characterisation of the conditioned taste aversion produced by 7-OH-DPAT in rats. Pharmacol Biochem Behav 53:695–699

Bevins RA, Klebaur JE, Bardo MT (1997) 7-OH-DPAT has *d*-amphetamine-like discriminative stimulus properties. Pharmacol Biochem Behav 58:485–490

Boulay D, Depoortere R, Perrault Gh, Sanger DJ (1998) No evidence for differences between dopamine D_3 receptor knock-out and wild-type mice. J Psychopharmacology 12, suppl. A: 26

Boulay D, Depoortere R, Rostene W, Perrault Gh, Sanger DJ (1999a) Dopamine D_3 receptor agonists produce similar decreases in body temperature and locomotor activity in D_3 knock-out and wild-type mice. Neuropharmacology 38:555–565

Boulay D, Depoortere R, Perrault Gh, Borrelli E, Sanger DJ (1999b) Dopamine D_2 receptor knock-out mice are insensitive to the hypolocomotor and hypothermic effects of DA D_2/D_3 receptor agonists. Neuropharmacology 38:1389–1396

Boulay D, Depoortere R, Sanger DJ, Perrault Gh (1999c) Knocking out the D_2, but not the D_3, dopamine receptor produces an acquisition deficit, and an insensitivity to the deleterious effects of haloperidol in an active avoidance task. Society for Neuroscience 25:30.19

Boulay D, Depoortere R, Oblin A, Claustre Y, Sanger DJ, Schoemaker H, Perrault Gh (2000) Cataleptogenic effects of haloperidol are absent in dopamine D_2 receptor knock-out mice but maintained in D_3 receptor knock-out mice. Eur J Pharmacol 391:63–73

Boyer P, Lecrubier Y, Puech AJ, Dewailly J, Aubin F (1995) Treatment of negative symptoms in schizophrenia with amisulpride. Br J Psychiatry 166:68–72

Bristow LJ, Cook GP, Patel S, Curtis N, Mawer I, Kulagowski JJ (1998) Discriminative stimulus properties of the putative dopamine D_3 receptor agonist, (+)-PD 128907: Role of presynaptic dopamine D_2 autoreceptors. Neuropharmacology 37:793–802

Bunzow JR, Van Tol HHM, Grandy DK, Albert P, Salon J, Christie M, Machida CA, Neve KA, Civelli O (1988) Cloning and expression of a rat D_2 dopamine receptor cDNA. Nature 336:783–787

Burris KD, Pacheco MA, Filtz TM, Kung MP, Kung HF, Molinoff PB (1995) Lack of discrimination by agonists for D_2 and D_3 dopamine receptors. Neuropsychopharmacology 12:335–345

Caine SB, Koob GF (1993) Modulation of cocaine self-administration in the rat through D-3 dopamine receptors. Science 260:1814–1816

Caine SB, Geyer MA, Swerdlow NR (1995) Effects of D_3/D_2 dopamine receptor agonists and antagonists on prepulse inhibition of acoustic startle in the rat. Neuropsychopharmacology 12:139–145

Caine SB, Koob GF, Parsons LH, Everitt BJ, Schwartz JC, Sokoloff P (1997) D_3 receptor test in vitro predicts decreased cocaine self-administration in rats. NeuroReport 8:2373–2377

Calabresi P, Saiardi A, Pisani A, Baik JH, Centonze D, Mercuri NB, Bernardi G, Borrelli E (1997) Abnormal synaptic plasticity in the striatum of mice lacking dopamine D2 receptors. J Neurosci 17:4536–4544

Calon F, Goulet M, Blanchet PJ, Martel JC, Piercey MF, Bedard PJ, Di paolo T (1995) Levodopa or D2 agonist induced dyskinesia in MPTP monkeys: correlation with changes in dopamine and $GABA_A$ receptors in the striatopallidal complex. Brain Res 680:43–52

Camacho-Ochoa M, Hoffmann WE, Moon MW, Figur LM, Tang AH, Himes CS, Nichols NF, Piercey MF (1995) Presynaptic and postsynaptic pharmacology of U-95666 A, a dopamine agonist selective for the D_2 receptor subtype: increase in postsynaptic response in Parkinson's disease (PD) model. Society for Neuroscience 21:340.9

Carlsson A, Lindqvist M, Magnusson T, Waldeck B (1958) On the presence of 3-hydroxytyramine in brain. Science 127:471

Chaperon F, Thiébot, MH (1996) Effects of dopaminergic D3-receptor-preferring ligands on the acquisition of place conditioning in rats. Behav Pharmacol 7:105–109

Chio CL, Lajiness ME, Huff RM (1994) Activation of heterologously expressed D3 dopamine receptors: comparison with D2 dopamine receptors. Mol Pharmacol 45:51–60

Clifford JJ, Waddington JL (1998) Heterogeneity of behavioural profile between three new putative selective D_3 dopamine receptor antagonists using an ethologically based approach. Psychopharmacology 136:284–290

Cohen C, Perrault GH, Sanger DJ (1998) Preferential involvement of D_3 versus D_2 dopamine receptors in the effects of dopamine receptor ligands on oral ethanol self-administration in rats. Psychopharmacology 140:478–485

Cook CD, Picker MJ (1998) Dopaminergic activity and the discriminative stimulus effects of mu opioids in pigeons: importance of training dose and attenuation by the D_3 agonist (±)-7-OH-DPAT. Psychopharmacology 136:59–69

Cook CD, Rodefer JS, Picker MJ (1999) Selective attenuation of the antinociceptive effects of μ opioids by the putative dopamine D_3 agonist 7-OH-DPAT. Psychopharmacology 144:239–247

Corbin AE, Pugsley TA, Akunne HC, Whetzel SZ, Zoski KT, Georgic LM, Nelson CB, Wright JL, Wise LD, Heffner TG (1998) Pharmacological characterization of PD152255, a novel dimeric benzimidazole dopamine D_3 antagonist. Pharmacol Biochem Behav 59:487–493

Crusio WE (1996) Gene-targeting studies: New methods, old problems – Commentary. Trends Neurosci 19:186–187

Daly SA, Waddington JL (1993) Behavioural effects of the putative D-3 dopamine receptor agonist 7-OH-DPAT in relation to other "D-2-like" agonists. Neuropharmacology 32:509–510

Damsma G, Bottema T, Westerink BHC, Tepper PG, Dijkstra D, Pugsley TA, MacKenzie RG, Heffner TG, Wikström H (1993) Pharmacological aspects of R-(+)-7-OH-DPAT, a putative dopamine D_3 receptor ligand. Eur J Pharmacol 249: R9–R10

Davidson AB, Weidley E (1976) Differential effects of neuroleptic and other psychotropic agents on acquisition of avoidance in rats. Life Sci 18:1279–1284

Dearry A, Gingrich JA, Falardeau P, Fremeau RT Jr, Bates MD, Caron MG (1990) Molecular cloning and expression of the gene for a human D_1 dopamine receptor. Nature 347:72–76

De Camilli P, Macconi D, Spada A (1979) Dopamine inhibits adenylate cyclase in human prolactin-secreting pituitary adenoma. Nature 278:252–254

De Fonseca FR, Rubio P, Martin-Calderon JL, Caine SB, Koob GF, Navarro M (1995) The dopamine receptor agonist 7-OH-DPAT modulates the acquisition and expression of morphine-induced place preference. Eur J Pharmacol 274:47–55

Depoortere R, Perrault GH, Sanger, DJ (1996) Behavioural effects in the rat of the putative dopamine D_3 receptor agonist 7-OH-DPAT: Comparison with quinpirole and apomorphine. Psychopharmacology 124:231–240

Depoortere R, Perrault GH, Sanger, DJ (1999) Intracranial self-stimulation under a progressive-ratio schedule in rats: Effects of strength of stimulation, d-amphetamine, 7-OH-DPAT and haloperidol. Psychopharmacology 142:221–229

Depoortere R, Perrault GH, Sanger, DJ (2000) The D_3 antagonist PNU-99194A potentiates the discriminative cue produced by the D_3 agonist 7-OH-DPAT. Pharmacol Biochem Behav 65:31–34

Ferrari F, Giuliani D (1995) Behavioural effects of the dopamine D3 receptor agonist 7-OH-DPAT in rats. Pharmacol Res 32:63–68

Ferrari F, Giuliani D (1997) Effects of (−)eticlopride and 7-OH-DPAT on the tail suspension test in mice. J Psychopharmacol 11:339–344

Fibiger H, Phillips A (1988) Mesocorticolimbic dopamine systems and reward. Ann NY Acad Sci 537:206–215

Gackenheimer SL, Schaus JM, Gehlert DR (1995) [^{3}H]-quinelorane binds to D_2 and D_3 dopamine receptors in the rat brain. J Pharmacol Exp Ther 274:1558–1565

Gaitnetdinov RR, Sotnikova TD, Grekhova TV, Rayevsky KS (1996) In vivo evidence for preferential role of dopamine D_3 receptor in the presynaptic regulation of dopamine release but not synthesis. Eur J Pharmacol 308:261–269

Gerlai R (1996) Gene-targeting studies of mammalian behavior: Is it the mutation or the background genotype? Trends Neurosci 19:177–182

Gilbert DB, Millar J, Cooper SJ (1995) The putative dopamine D_3 agonist, 7-OH-DPAT, reduces dopamine release in the nucleus accumbens and electrical self-stimulation to the ventral tegmentum. Brain Res 681:1–7

Gold LH (1996) Integration of molecular biological techniques and behavioural pharmacology. Behav Pharmacol 7:589–615

Grandy DK, Zhang YA, Bouvier C, Zhou QY, Johnson RA, Allen L, Buck K, Bunzow JR, Salon J, Civelli O (1991) Multiple human D5 dopamine receptor genes: a functional receptor and two pseudogenes. Proc Natl Acad Sci USA 88:9175–9179

Hatcher JP, Hagan JJ (1998) The effects of dopamine D_3/D_2 receptor agonists on intracranial self stimulation in the rat. Psychopharmacology 140:405–410

Hernandez L, Auerbach S, Hoebel BG (1988) Phencyclidine (PCP) injected in the nucleus accumbens increases extracellular dopamine and serotonin as measured by microdialysis. Life Sci 42:1713–1723

Jaber M, Robinson SW, Missale C, Caron MG (1996) Dopamine receptors and brain function. Neuropharmacology 35:1503–1519

Jung MY, Skryabin BV, Arai M, Abbondanzo S, Fu D, Brosius J, Robakis NK, Polites HG, Pintar JE, Schmauss C (1999) Potentiation of the D_2 mutant motor phenotype in mice lacking dopamine D_2 and D_3 receptors. Neuroscience 91:911–924

Kebabian JW, Calne DB (1979) Multiple receptors for dopamine. Nature 277:93–96

Kelly MA, Rubinstein M, Asa SL, Zhang G, Saez C, Bunzow JR, Allen RG, Hnasko R, Ben-Jonathan N, Grandy DK, Low MJ (1997) Pituitary lactotroph hyperplasia and chronic hyperprolactinemia in dopamine D2 receptor-deficient mice. Neuron 19:103–113

Kelly MA, Rubinstein M, Phillips TJ, Lessov CN, Burkhart-Kasch S, Zhang G, Bunzow JR, Fang Y, Gerhardt GA, Grandy DK, Low MJ (1998) Locomotor activity in D2 dopamine receptor-deficient mice is determined by gene dosage, genetic background, and developmental adaptations. J Neurosci 18:3470–3479

Khroyan TV, Baker DA, Neisewander JL (1995) Dose dependent effects of the D3-preferring agonist 7-OH-DPAT on motor behaviors and place conditioning. Psychopharmacology 122:351–357

Khroyan TV, Fuchs RA, Baker DA, Neisewander JL (1997) Effects of D3-preferring agonists 7-OH-PIPAT and PD-128,907 on motor behaviors and place conditioning. Behav Pharmacol 8:65–74

Klemm WR, Block H (1988) D-1 and D-2 receptor blockade have additive cataleptic effects in mice, but receptor effects may interact in opposite ways. Pharmacol Biochem Behav 29:223–229

Kleven MS, Koek W (1996) Differential effects of direct and indirect dopamine agonists on eye blink rate in cynomolgus monkeys. J Pharmacol Exp Ther 279:1211–1219

Kleven MS, Koek W (1997) Dopamine D_2 receptors play a role in the (−)-apomorphine-like discriminative stimulus effects of (+)-PD 128907. Eur J Pharmacol 321:1–4

Kling-Petersen T, Ljung E, Wollter L, Svensson K (1995a) Effects of dopamine D_3 preferring compounds on conditioned place preference and intracranial self-stimulation in the rat. J Neural Transm [Gen Sect] 101:27–39

Kling-Petersen T, Ljung E, Svensson K (1995b) Effects on locomotor activity after local application of D3 preferring compounds in discrete areas of the rat brain. J Neural Transm [Gen Sect] 102:209–220

Koeltzow TE, Cooper DC, Hu X-T, Xu M, Tonegawa S, White, FJ (1995) *In vivo* effects of dopaminergic ligands in dopamine D3 receptor deficient mice. Society for Neuroscience 21:149.3

Kostrzewa RM, Brus R (1991) Is dopamine-agonist induced yawning behavior a D3 mediated event? Life Sci 48:129

Kulagowski JJ, Broughton HB, Curtis NR, Mawer IM, Ridgill MP, Baker R, Emms F, Freedman SB, Marwood R, Patel S, Patel S, Ragan CI, Leeson PD (1996) 3[[4-(4-

Chlorophenyl)piperazin-1-yl]-methyl]-1H-pyrrolo[2,3-b]pyridine: an antagonist with high affinity and selectivity for the human dopamine D4 receptor. J Med Chem 39:1941–1942

Kurashima M, Yamada K, Nagashima M, Shirakawa K, Furukawa T (1995) Effects of putative dopamine D_3 receptor agonists, 7-OH-DPAT, and quinpirole, on yawning, stereotypy, and body temperature in rats. Pharmacol Biochem Behav 52:503–508

Lagos P, Scorza C, Monti JM, Jantos H, Reyes-Parada M, Silveira R, Ponzoni A (1998) Effects of the D_3 preferring dopamine agonist pramipexole on sleep and waking, locomotor activity and striatal dopamine release in rats. Eur Neuropsychopharmacol 8:113–120

Large CH, Stubbs CM (1994) The dopamine D_3 receptor: Chinese hamsters or Chinese whispers? Trends Pharmacol Sci 15:46–47

Levant B (1997) The D_3 dopamine receptor: Neurobiology and potential clinical relevance. Pharmacol Rev 49:231–252

Lévesque D, Diaz J, Pilon C, Martres MP, Giros B, Souil E, Schott D, Morgat JL, Schwartz JC, Sokoloff P (1992) Identification, characterization, and localization of the dopamine D_3 receptor in rat brain using 7-[^{3}H]hydroxy-N,N-di-n-propyl-2-aminotetralin. Proc Natl Acad Sci USA 89:8155–8159

Maldonado R, Saiardi A, Valverde O, Samad TA, Roques BP, Borrelli E (1997) Absence of opiate rewarding effects in mice lacking dopamine D2 receptors. Nature 388: 586–589

Mallet PE, Beninger RJ (1994) 7-OH-DPAT produces place conditioning in rats. Eur J Pharmacol 261:R5–R6

Manzanedo C, Aguilar MA, Minarro J (1999) The effects of dopamine D_2 and D_3 antagonists on spontaneous motor activity and morphine-induced hyperactivity in male mice. Psychopharmacology 143:82–88

McCullough LD, Salamone JD (1992) Increases in extracellular dopamine levels and locomotor activity after direct infusion of phencyclidine into the nucleus accumbens. Brain Res 577:1–9

McElroy J, Zeller KL, Amy KA, Ward KA, Cawley JF, Mazzola AL, Keim W, Rohrbach K (1993) In vivo agonist properties of 7-hydroxy-N,N-Di-N-propyl-2-aminotetralin, a dopamine D_3-selective receptor ligand. Drug Dev Res 30:257–259

McElroy JF (1994) Discriminative stimulus properties of 7-OH-DPAT, a dopamine D3-selective receptor ligand. Pharmacol Biochem Behav 48:531–533

Mierau J, Schneider FJ, Ensinger HA, Chio CL, Lajiness ME, Huff RM (1995) Pramipexole binding and activation of cloned and expressed dopamine D2, D3 and D4 receptors. Eur J Pharmacol 290:29–36

Millan MJ, Peglion JL, Vian J, Rivet JM, Brocco M, Gobert A, Newman-Tancredi A, Dacquet C, Bervoets K, Girardon S, Jacques V, Chaput C, Audinot V (1995) Functional correlates of dopamine D_3 receptor activation in the rat in vivo and their modulation by the selective antagonist, (+)-S 14297: I. Activation of postsynaptic D_3 receptors mediates hypothermia, whereas blockade of D_2 receptors elicits prolactin secretion and catalepsy. J Pharmacol Exp Ther 275:885–898

Millan MJ, Gressier H, Brocco M (1997) The dopamine D_3 receptor antagonist, (+)-S 14297, blocks the cataleptic properties of haloperidol in rats. Eur J Pharmacol 321:R7–R9

Moller HJ, Boyer P, Fleurot O, Rein W (1997) Improvement of acute exacerbations of schizophrenia with amisulpride: a comparison with haloperidol. Psychopharmacology 132:396–401

Molloy AG, O'Boyle KM, Pugh MT, Waddington JL (1986) Locomotor behaviors in response to new selective D-1 and D-2 dopamine receptor agonists, and the influence of selective antagonists. Pharmacol Biochem Behav 25:249–253

Monsma FJ Jr, Mahan LC, McVittie LD, Gerfen CR, Sibley DR (1990) Molecular cloning and expression of a D1 dopamine receptor linked to adenylyl cyclase activation. Proc Natl Acad Sci USA 87:6723–6727

Murray PJ, Helden RM, Johnson MR, Robertson GM, Scopes DIC, Stokes M, Wadman S, Whitehead JWF, Hayes AG, Kilpatrick GJ, Large C, Stubbs CM, Turpin MP (1996) Novel 6-substituted 2-aminotetralins with potent and selective affinity for the dopamine D_3 receptor. Bioorg Med Chem Lett 6:403–408

Nakajima S, Liu X, Lau CL (1993) Synergistic interaction of D1 and D2 dopamine receptors in the modulation of the reinforcing effect of brain stimulation. Behav Neurosci 107:161–165

Niemegeers CJE, Verbrugen FJ, Janssen PAJ (1969) The influence of various neuroleptic drugs on shock avoidance responding in rats. Psychopharmacologia 16:161–174

Perrault Gh, Depoortere R, Sanger DJ (1996) Hypothermia and climbing behaviour induced by D2/D3 dopamine agonists in mice. Behav Pharmacol 7, sup 1:83

Perrault Gh, Depoortere R, Morel E, Sanger DJ, Scatton B (1997) Psychopharmacological profile of amisulpride: an antipsychotic drug with presynaptic D_2/D_3 dopamine receptor antagonist activity and limbic selectivity. J Pharmacol Exp Ther 280:73–82

Piercey MF, Moon MW, Sethy VH, Schreur PJKD, Smith MW, Tang AH, VonVoigtlander PF (1996) Pharmacology of U-91356 A, an agonist for the dopamine D_2 receptor subtype. Eur J Pharmacol 317:29–38

Phillips TJ, Brown KJ, Burkhart-Kasch S, Wenger CD, Kelly MA, Rubinstein M, Grandy DK, Low MJ (1998) Alcohol preference and sensitivity are markedly reduced in mice lacking dopamine D_2 receptors. Nature Neurosci 7:610–615

Pugsley TA, Davis MD, Akunne HC, Mackenzie RG, Shih YH, Damsma G, Wikstrom H, Whetzel SZ, Georgic LM, Cooke LW, Demattos SB, Corbin AE, Glase SA, Wise LD, Dijkstra D, Heffner TG (1995) Neurochemical and functional characterization of the preferentially selective dopamine D3 agonist PD 128907. J Pharmacol Exp Ther 275:1355–1366

Ralph RJ, Varty GB, Kelly MA, Wang YM, Caron MG, Rubinstein M, Grandy DK, Low MJ, Geyer MA (1999) The dopamine D_2, but not D_3 or D_4, receptor subtype is essential for the disruption of prepulse inhibition produced by amphetamine in mice. J Neurosci 19:4627–4633

Russel RN, Mc Bride WJ, Lumeng L, Li TK, Murphy JM (1996) Apomorphine and 7-OH-DPAT reduce ethanol intake of P and HAD rats. Alcohol 13:515–519

Sanger DJ, Depoortere R, Perrault Gh (1996) Evidence for a role for dopamine D3 receptors in the effects of dopamine agonists on operant behaviour in rats. Behav Pharmacol 7:477–482

Sanger DJ, Depoortere R, Perrault Gh (1997) Discriminative stimulus effects of apomorphine and 7-OH-DPAT: a potential role for dopamine D_3 receptors. Psychopharmacology 130:387–395

Sanger DJ, Perrault Gh, Depoortere R, Cohen C (1999) Using drug discrimination in rats to investigate the pharmacology of dopamine D_3 receptors. In: Palomo T, Beninger RJ, Archer T (eds) Interactive monoaminergic disorders. Editorial Sintesis, Madrid, pp 481–498

Sautel F, Griffon N, Levesque D, Pilon C, Schwartz JC, Sokoloff P (1995a) A functional test identifies dopamine agonists selective for D3 versus D2 receptors. Neuroreport 6:329–332

Sautel F, Griffon N, Sokoloff P, Schwartz JC, Launay C, Simon P, Costentin J, Schoenfelder A, Garrido F, Mann A (1995b) Nafadotride, a potent preferential dopamine D_3 receptor antagonist, activates locomotion in rodents. J Pharmacol Exp Ther 275:1239–1246

Scatton B, Claustre Y, Cudennec A, Oblin A, Perrault GH, Sanger DJ, Schoemaker H (1997) Amisulpride: from animal pharmacology to therapeutic action. Int Clin Psychopharmacol 12:S29–S36

Schoemaker H, Claustre Y, Fage D, Rouquier L, Chergui K, Curet O, Oblin A, Gonon F, Carter C, Benavides J, Scatton B (1997) Neurochemical characteristics of amisul-

pride, an atypical dopamine D_2/D_3 receptor antagonist with both presynaptic and limbic selectivity. J Pharmacol Exp Ther 280:83–97
Schreur PJKD, Nichols NF (1995) U-95666 A, a dopamine D2 agonist: Behavioral studies in rats. Society for Neuroscience 21:340.7
Schwartz J-C, Griffon N, Diaz J, Levesque D, Sautel F, Sokoloff P, Simon P, Costentin J, Garrido F, Mann A, Wermuth C (1995) The D_3 receptor and its relevance in psychiatry. Int Clin Psychopharmacol 10:15–20
Seeman P (1992) Dopamine receptor sequences. Therapeutic levels of neuroleptics occupy D_2 receptors, clozapine occupies D_4. Neuropsychopharmacology 7:261–284
Silvestre JS, O'Neill MF, Fernandez AG, Palacios JM (1996) Effects of a range of dopamine receptor agonists and antagonists on ethanol intake in the rat. Eur J Pharmacol 318:257–265
Snyder SH (1976) The dopamine hypothesis of schizophrenia: Focus on the dopamine receptor. Am J Psychiatry 133:197–202
Snyder SH (1990) The dopamine connection. Nature 347:121–122
Sokoloff P, Giros B, Martres MP, Bouthenet ML, Schwartz JC (1990) Molecular cloning and characterization of a novel dopamine receptor (D_3) as a target for neuroleptics. Nature 347:146–151
Spealman RD (1996) Dopamine D_3 receptor agonists partially reproduce the discriminative stimulus effects of cocaine in Squirrel monkeys. J Pharm Exp Ther 278:1128–1137
Starr MS, Starr BS (1995) Motor actions of 7-OH-DPAT in normal and reserpine-treated mice suggest involvement of both dopamine D_2 and D_3 receptors. Eur J Pharmacol 277:151–158
Steiner H, Fuchs S, Accili D (1998) D_3 dopamine receptor-deficient mouse: evidence for reduced anxiety. Physiol Behav 63:137–141
Sunahara RK, Niznik HB, Weiner DM, Stormann TM, Brann MR, Kennedy JL, Gelernter JE, Rozmahel R, Yang YL, Israel Y, Seeman P, O'Dowd BF (1990) Human dopamine D1 receptor encoded by an intronless gene on chromosome 5. Nature 347:80–83
Sunahara RK, Guan HC, O'Dowd BF, Seeman P, Laurier LG, Ng G, George SR, Torchia J, Van Tol HHM, Niznik HB (1991) Cloning of the gene for a human dopamine D_5 receptor with higher affinity for dopamine than D_1. Nature 350:614–619
Suzuki T, Maeda J, Funada M, Misawa M (1995) The D_3-receptor agonist (±)-7-hydroxy-N,N-di-n-propyl-2-aminotetralin (7-OH-DPAT) attenuates morphine-induced hyperlocomotion in mice. Neurosci Lett 187:45–48
Svensson K, Johansson AM, Magnusson T, Carlsson A (1986) (+)-AJ 76 and (+)-UH 232: central stimulants acting as preferential dopamine autoreceptor antagonists. Naunyn Schmiedebergs Arch Pharmacol 334:234–245
Svensson K, Carlsson A, Waters N (1994) Locomotor inhibition by the D_3 ligand R-(+)-7-OH-DPAT is independent of changes in dopamine release. J Neural Transm [Gen Sect] 95:71–74
Tiberi M, Jarvie KR, Silvia C, Falardeau P, Gingrich JA, Godinot N, Bertrand L, Yang-Feng TL, Fremeau RT Jr, Caron MG (1991) Cloning, molecular characterization, and chromosomal assignment of a gene encoding a second D1 dopamine receptor subtype: differential expression pattern in rat brain compared with the D1A receptor. Proc Natl Acad Sci USA 88:7491–7495
Van Tol HHM, Bunzow JR, Guan HC, Sunahara RK, Seeman P, Niznik HB, Civelli O (1991) Cloning of the gene for a human dopamine D_4 receptor with high affinity for the antipsychotic clozapine. Nature 350:610–614
Varty GB, Higgins GA (1997) Investigations into the nature of a 7-OH-DPAT discriminative cue: Comparison with D-amphetamine. Eur J Pharmacol 339:101–107
Varty GB, Higgins GA (1998) Dopamine agonist-induced hypothermia and disruption of prepulse inhibition: Evidence for a role of D3 receptors? Behav Pharmacol 9:445–455

Waddington JL (1989) Functional interactions between D1 and D2 dopamine receptor systems: Their role in the regulation of psychomotor behaviour, putative mechanisms, and clinical relevance. J Psychopharmacol 3:54–63

Waters N, Svensson K, Haadsma-Svensson SR, Smith MW, Carlsson A (1993) The dopamine D3-receptor: a postsynaptic receptor inhibitory on rat locomotor activity. J Neural Transm [Gen Sect] 94:11–19

Waters N, Lofberg L, Haadsma-Svensson S, Svensson K, Sonesson C, Carlsson A (1994) Differential effects of dopamine D2 and D3 receptor antagonists in regard to dopamine release, in vivo receptor displacement and behaviour. J Neural Transm [Gen Sect] 98:39–55

Weinshank RL, Adham N, Macchi M, Olsen MA, Branchek TA, Hartig PR (1991) Molecular cloning and characterization of a high affinity dopamine receptor (D1 beta) and its pseudogene. J Biol Chem 266:22427–22435

Willner P (1995) Dopaminergic mechanisms in depression and mania. In: Bloom FE, Kupfer DJ (eds) Psychopharmacology, the fourth generation of progress. Raven Press, New York, pp 921–931

Wilson JM, Sanyal S, Van Tol HHM (1998) Dopamine D2 and D4 receptor ligands: Relation to antipsychotic action. Eur J Pharmacol 351:273–286

Wise RA, Bozarth MA (1987) A psychomotor stimulant theory of addiction. Psychol Rev 94:469–492

Xu M, Caine SB, Cooper DC, Gold LH, Graybiel AM, Hu XT, Koeltzow TE, Koob GF, Moratalla R, White FJ, Tonegawa S (1995) Analyses of dopamine D3 and D1 receptor mutant mice. Society for Neuroscience 21:149.2

Xu M, Koeltzow TE, Santiago GT, Moratalla R, Cooper DC, Hu X-T, White NM, Graybiel AM, White FJ, Tonegawa S (1997) Dopamine D3 receptor mutant mice exhibit increased behavioral sensitivity to concurrent stimulation of D1 and D2 receptors. Neuron 19:837–848

Xu M, Koeltzow TE, Cooper DC, Tonegawa S, White FJ (1999) Dopamine D3 receptor mutant and wild-type mice exhibit identical responses to putative D3 receptor-selective agonists and antagonists. Synapse 31:210–215

Yamaguchi H, Aiba A, Nakamura K, Nakao K, Sakagami H, Goto K, Kondo H, Katsuki M (1996) Dopamine D2 receptor plays a critical role in cell proliferation and proopiomelanocortin expression in the pituitary. Genes Cells 1:253–268

Zhou QY, Grandy DK, Thambi L, Kushner JA, Van Tol HHM, Cone R, Pribnow D, Salon J, Bunzow JR, Civelli O (1990) Cloning and expression of human and rat D_1 dopamine receptors. Nature 347:76–80

CHAPTER 22
Dopamine and Reward

G. Di Chiara

A. Introduction

This chapter is specifically devoted to the role of dopamine (DA) in conventional, non-drug reward [food, water, sex, intracranial self-stimulation (ICSS)]. This topic is among the most controversial of the whole neurobiology of motivation (FIBIGER 1978; WISE 1978, 1982; BENINGER 1983; FIBIGER and PHILLIPS 1986; ETTENBERG 1989; MILLER et al. 1990; LE MOAL and SIMON 1991; BLACKBURN et al. 1992; SALAMONE 1992; DI CHIARA 1995; SALAMONE et al. 1997; BERRIDGE and ROBINSON 1998; SCHULTZ 1998). There are many reasons for this: first, the uncertain definition and limited knowledge of the basic processes into which DA is thought to be involved; second, the multiplicity of the role of DA in behaviour; third, the lack of specificity of the measures taken as an expression of that role and of the paradigms utilized to investigate it. In addition to these reasons, that are general and apply to DA as to any other neurobiological substrate of behaviour, there is a difficulty intrinsic to the status of DA: its prominent role in drug-reward. In fact, drug-induced stimulation of DA transmission is itself rewarding (WISE and BOZARTH 1987; DI CHIARA 1995). This circumstance, while acting as a powerful incentive of studies on the role of DA in drug-reinforcement and drug-addiction, has been also utilized as the basis for interpreting the role of DA in conventional, non-drug reward. If the role of DA in psychostimulant reward is a case of a general role of DA in reward than the effect of drugs that impair DA transmission on conventional reward should be reciprocally symmetric to the effect of drugs that stimulate it. Thus, since pharmacological stimulation of DA transmission is rewarding, blockade of DA transmission should impair non-drug reward. A vivid expression of such reasoning has been provided by WISE (1982, p 39): "Because direct activation of dopaminergic synaptic activity by amphetamine, cocaine, and apomorphine is reinforcing in its own right ... and because selective dopaminergic lesions or receptor blockade attenuates the reinforcing actions of these agents ..., we have come to suspect that neuroleptic drugs ... interfere with operant behavior in a more subtle ... and important way than simply reducing performance capacity." This reasoning has been the basis for at least two influential accounts of the role of DA in reward, the anhedonia hypothesis and the incentive-motivational hypothesis. It is now increasingly clear, however, that the reciprocal symmetry between the effect of DA-stimulant and DA-blocking drugs on reward is an oversimplification. Thus, not only is psychostimulant reward not a model of conventional reward but DA-receptor blockers can block psychostimulant reward without affecting conventional reward. Since this review specifically addresses the issue of the role of DA in conventional reward, we will limit our analysis to studies of the effect of pharmacological manipulations and lesions on non-drug reward; studies on drug reward will be considered only to mark the differences with non-drug reward.

B. Terminology

Part of the difficulty in the understanding of the role of DA in motivated behaviour arises from problems related to the definition of the terms utilized to describe that role. Scientific terms are defined operationally by the specific criteria that allow one to measure, assay or estimate the object to which the term is referred (BORING 1945). Operational definition, in turn, is at the core of the scientific process, being essential for the development of testable hypotheses and valid experimental models (FEIGL 1945).

For the purpose of the present discussion, a case can be made for the use of two terms taken from common parlance, namely "wanting" and "liking". Each of these terms has been referred to by proponents as states of will (wanting) and of pleasure (liking) that are regarded as main factors of motivated responding (BERRIDGE 1996; BERRIDGE and ROBINSON 1998). By the same principle, "craving" is defined as abnormal "wanting", i.e. compulsive will (ROBINSON and BERRIDGE 1993). These terms have apparently been selected on the basis of the meaning they have in their non-scientific use and, therefore, have the apparent advantage of having self-evident definitions. This property, however, turns out to be counterproductive for the scientific use of these terms because it leads one to regard their operational definition as redundant or superfluous, if not eventually confusing, being in contrast with their meaning in common parlance. Whatever the reason, an operational definition of these terms has not been provided yet. The fact that these states are not necessarily conscious makes them impervious to operational definition in humans, since they cannot be estimated as self-reported measures. On the other hand, their operational definition for the study of animal behaviour suffers the additional difficulties encountered in any comparative definition of human psychological states.

An example of how terminology can affect testing of a scientific hypothesis is provided by the term "hedonia" as referred to in the anhedonia hypothesis (WISE 1982). Here, depending on the fact that hedonia is intended as a stimulus-bound cue (e.g. taste hedonia) or a state (e.g. euphoria, euthymia), quite different experiments would provide testing of the hypothesis. In the first case, given the assumption that taste reactivity reflects hedonic value and valence of taste stimuli (BERRIDGE 1996, 2000), testing could be performed in animals by means of the taste-reactivity paradigm; in the second case, testing of the hypothesis can be only performed in humans since it would require the use of verbally reported self-estimation of affective states (ECKMAN 1967; MCNAIR et al. 1971). This dichotomy can lead to misleading conclusions over the validity of the anhedonia hypothesis if each type of hedonia is differentially affected by a given experimental manipulation. Recently, evidence has been provided that amphetamine-induced euphoria is correlated to DA release in the ventral striatum (DREVETS et al. 2001); moreover, risperidone (NEWTON et al. 2001), an atypical antipsychotic, and SCH39166 (ROMACH et al. 1999), a D_1 receptor antagonist, have been reported to reduce cocaine-induced euphoria. Previous studies by BRAUER and DE WIT (1996) did not observe any effect of pimozide on amphetamine euphoria; however, the maximal doses of pimozide administered in this study were more than ten times lower than those utilized in rats (WISE 1982). Thus, it could be indeed the case that DA-receptor blockers, while not affecting taste-hedonia, effectively blunt state-hedonia. Therefore, rejection or confirmation of the anhedonia hypothesis seems to critically depend on the definition of the term hedonia.

I. Reward, Reinforcer, Incentive

An often used term in accounts of the role of DA in behaviour is that of "reward". This term has been utilized in so many different senses that some investigators prefer to stay respectfully away from it, substituting it with other terms like incentive and reinforcer. "Reward" refers to objects (e.g. food, water) and organisms (e.g. sexual partners) provided with intrinsic biological value (metabolic, genetic, etc.) and capable of promoting consummatory behaviours intended to utilize their biological resources for the survival of the self and of its species. Rewards are essentially a special class of motivational stimuli (i.e. stimuli that provide a motivation for behaviour). A fundamental property of rewards is that of transferring their motivational properties to stimuli that predict their occurrence and to strengthen responses upon which they are contingent. For this reason, rewards are reinforcers. Specifically, rewards are primary, unconditioned stimuli that utilize proximal sensory modalities (e.g. gustatory, tactile, thermic) involving close contact and interaction between the organism and the object stimulus and promoting specific and almost stereotyped patterns of consummatory responses.

Terms like "reinforcer" and "incentive" have a meaning that is linked to the theoretical context in which these terms were introduced. "Reinforcer" is commonly defined, following SKINNER (1938), as a stimulus that strengthens responses upon which it is contingent (i.e. to which it reliably follows). This definition, that reflects the behaviourist sense of "reinforcer", refers specifically to response reinforcement. However, the term "reinforcer" was also utilized by PAVLOV (1927) and later by incentive theorists [e.g. BINDRA's "reinforcing stimulation" (BINDRA 1974)] in the sense of strengthening of the response-eliciting (incentive) properties of a stimulus by another stimulus contingent upon the first (stimulus reinforcement). In this case, responding would not be directly strengthened as a result of its contingency with the reward but indirectly as a result of contingency of the reward upon the stimulus that elicits responding.

The term "incentive" was originally referred to as the innate or learned ability of certain stimuli of eliciting species-specific response patterns related to their biologically relevant value (e.g. orienting, approaching, exploring, etc.) (BOLLES 1972; BINDRA 1974; TOATES 1986). This term implies the assumption that responding is not a function of its consequences (as stated by response-reinforcement theories) but is itself the consequence of stimuli (incentives). Accordingly, while reinforcers act as consequences of responding, incentives act as premises. In agreement with its derivation from incentive theory, the term "incentive" should be reserved to indicate stimuli that elicit responding on the basis of their contingency with other stimuli (Pavlovian principle); similarly, the term "reinforcer" should be used to specifically indicate stimuli that elicit responding on the basis of a contingency upon a response (instrumental principle). Incentive theorists, however, have offered explanations of instrumental responding based on Pavlovian stimulus-contingency learning (see, for example, BINDRA 1974). For the above reasons the terms "reinforcer" and "incentive" have lost much of their theoretical roots and are often used interchangeably, eventually with the further connotation of "primary" and "secondary", depending on whether they refer to "unconditioned" or "conditioned" stimuli. This practice is sometimes extended to rewards (primary and secondary reward).

Here we will clearly distinguish between "reward" and "incentive" by reserving the term "reward" for primary unconditioned stimuli that elicit consummatory responses (e.g. the taste of food) and the term "incentive" to primary unconditioned or secondary conditioned stimuli that elicit appetitive responses of orienting, searching and approaching through distal sensory modalities (visual, acoustic, olfactory). This distinction between "incentive" and "reward" takes into account the different biological significance of appetitive-preparatory as opposed to consummatory patterns of response (KONORSKI 1967) and their relation to distinct classes of stimuli (distal/incentive versus proximal/reward) as well as the primary versus secondary nature of the stimulus. The term "reinforcer" will be used here in the sense of response-reinforcement to refer to primary and secondary stimuli either positive or negative.

II. Motivation and Instrumental Responding

Motivation is the process by which organisms emit responses to stimuli in relation to their predicted consequences in terms of survival of the self and of the species. Motivation consists, therefore, in learning of predictive relationships (contingencies) between neutral stimuli and biologically meaningful ones and between responses and their outcome. Learning of these contingencies enables the subject to actively promote by its actions the occurrence of biologically valuable events (instrumental action). The understanding of the mechanism by which goal-directed action is learned and maintained is a central issue in the study of behaviour and the subject of much debate and speculation. It is recognized that both cognitive, conscious (explicit/declarative), as well as associative, unconscious (implicit/procedural) mechanisms contribute to purposeful behaviour (DICKINSON and BALLEINE 1994; TOATES 1998).

1. Incentive-Motivational Responding

The basis for incentive-motivational responding is hard-wired by evolution in the brain of organisms, including man. Organisms are provided with the innate ability of coding the intrinsic biological value of objects and organisms on the basis of the signals they emit and to respond to these signals in a manner consistent with that code (GLICKMAN and SCHIFF 1967). Thus, certain stimuli such as the taste of sweet, the smell of a female, the cry of a predator, evoke behaviours that, depending on the stimulus, consist in approach or avoidance of the object or organism from which they originate. These responses are not the result of learning by experience of the consequences of the stimuli or of sheer imitation of the behaviour of conspecifics. The biological code of a primary stimulus (reward or punisher) can be extended to other stimuli by learning of stimulus contingencies (Pavlovian learning). By this mechanism, novel salient stimuli (CSs) that reliably predict the occurrence of the unconditioned stimulus (US) acquire conditioned response (CR)-eliciting properties consistent with the valence of the US. By this simple associative process, otherwise-neutral stimuli acquire the ability of eliciting US-non-specific preparatory responses of search, approach and general arousal as well as US-specific consummatory responses (KONORSKI 1967). Pavlovian incentive learning can be explained in the framework of the theory developed by KONORSKI for explaining the effects of Pavlovian CSs on behaviour. According to KONORSKI (1967), depending on its nature, the CS can elicit conditioned consummatory or preparatory responses. Conditioned consummatory responses are phenomenologically similar to the correspondent unconditioned response (UR) and can be understood as the result of the excitation by the CS of a representation of the US. Conditioned preparatory responses, instead, are not specific to a given US since, irrespective of a specific US, they consist of flexible patterns of orienting, approaching and exploring the CS. These typically incentive responses, in contrast to consummatory CRs, are quite different from the response to the US (UR). For this reason, it is difficult to explain their emission as the result of the excitation by the CS of the representation of the US. According to KONORSKI (1967), these preparatory (incentive) responses to the CS can be explained as due to excitation of a motivational system common to different USs. Thus, as a result of the association with the US, the representation of the CS establishes a connection with this motivational system (Pavlovian incentive learning; DICKINSON et al. 2000) thus acquiring the ability of inducing preparatory/incentive responses (KONORSKI 1967).

Incentives show two properties: (1) a directional property that promotes responses directed towards the incentive itself and, through it, towards the reward to which the incentive has been conditioned; (2) an activational property consisting of a state of motivational arousal (incentive arousal) that increases in a non-specific manner the incentive properties of other stimuli present in the environment but not necessarily related to the reward to which the triggering incentive has been conditioned. This arousing property of incentives can explain their ability to trigger, under appropriate conditions, the repetitive and excessive emission of behaviours that are part of the species repertoire (adjunctive or displacement behaviour) (FALK 1977; KILLEEN et al. 1978a). Another property of incentives is that of increasing the emission of responses instrumental to the presentation of the reward to which the incentive has been conditioned and, eventually, of other rewards. Thus, incentives acquired through Pavlovian (stimulus–reward) associations are capable of energizing primary reinforcement (ESTES 1943; BOWER and GRUSEC 1964; TRAPOLD et al. 1968; MELLGREN and OST 1969; LOVIBOND 1983). This property has been termed "transfer from Pavlovian to instrumental responding" (DICKINSON 1994). Another property of incentives acquired through Pavlovian contingency learning is that of acting as secondary reinforcers, i.e. of promoting responding instrumental to their own presentation in the absence (extinction) of response reinforcement by the reward (secondary reinforcement).

2. Instrumental Responding

Pavlovian incentive learning provides organisms with the ability of preparing themselves for the occurrence of biologically significant events by making them responsive to stimuli that are predictive of those events. Pavlovian incentive learning, apart from its ability, through its activational effects, of increasing the probability of encountering a reward present in the environment, does not provide, per se, organisms with the ability of controlling by their actions the occurrence of biologically significant events. That is instead what instrumental learning does and what instrumental responding is about.

In the past, instrumental responding was entirely explained in terms of strengthening of the tendency to emit a response to a situational stimulus by its satisfying consequences (THORNDIKE 1898) or as strengthening of the association between an arbitrary stimulus (S) and a response (R) by its consequences (e.g. feeding) (WATSON 1913; MOSS and THORNDIKE 1934; HULL 1943). HULL (1943) further refined this model by introducing the drive term (D) as a factor that multiplies the strength of the S–R association (habit, H) to increase response strength (E), therefore, $E = H \times D$. According to HULL (1943), however, the consequence and the motive of responding is drive reduction; therefore, as response is reinforced by its consequences, drive reduction is reinforcing. Drive reduction, however, is not the immediate consequence of response; thus, in feeding, drive reduction takes place long after consumption and only during the post-consummatory phase, the immediate consequence of responding being instead rewarding stimulation by food taste. Thus, contrary to HULL's assumption, the strengthening factor of the S–R association is not drive reduction but reward by consummatory stimuli (primary reinforcers).

The current understanding of the role of response reinforcement in motivated behaviour regards it as a special modality of instrumental responding, namely, habit responding. In this modality, response is relatively independent from its outcome, being controlled by stimuli that precede it (DICKINSON 1994). As a result of this, devaluation of response outcome or degradation of the instrumental act-outcome contingency fails to impair habit responding. Habit responding takes place as a result of exhaustive training on high-ratio schedules or under variable interval schedules where reinforcement is loosely related to response (DICKINSON 1994). Under continuous reinforcement schedules, where every response is reinforced, responding is tightly controlled by an act-outcome contingency and by the value of the outcome. The dependence of responding from a tight response-reward contingency would indicate that this form of instrumental responding (incentive instrumental responding) is controlled by the establishment of a declarative (conscious) representation of the cause-effect relationship between each act and its outcome (TOLMAN 1959; DICKINSON and BALLEINE 1994) or by a procedural (unconscious) response-outcome association (COLWILL and RESCORLA 1990); on the other hand, the circumstance that responding is controlled by the current value of the reward would indicate that incentive instrumental responding is truly goal-directed in nature (DICKINSON and BALLEINE 1994). Pavlovian stimulus-contingency and instrumental response-contingency mechanisms are not mutually exclusive; rather, they contribute to the overall pattern of motivated responding (RESCORLA and SOLOMON 1967). Pavlovian processes are likely to be operative in tasks utilizing mazes and alleyways that are based on a natural response such as orienting and approach toward the stimulus. In contrast, the mechanisms based on response-contingency are operative in tasks involving a manipulandum such as a lever pressing, chain pulling and the like. Even in this case, however, stimulus reinforcement can be demonstrated in the form of consummatory responses directed at the manipulandum (licking, gnawing, etc.). By arranging the Pavlovian stimulus and the manipulandum in two spatially distinct locations, first-order Pavlovian and instrumental conditioning can be distinguished and differentially impaired by excitotoxic lesions of central and basolateral amygdala, respectively (GALLAGHER et al. 1990; KILLCROSS et al. 1997; PARKINSON et al. 2000).

With practice, incentive (act-outcome) instrumental responding is transformed into automatic habit responding based on S–R associations (DICKINSON 1994); this modality ensures response to stimuli at a speed that would be unattainable by incentive instrumental responding, due to its dependence on outcome. Habit responding, although automatic, is not impervious to adaptive control by its outcome. Thus, repeated failure to meet the requirements of a situational change results in switching back from the habit modality to the incentive instrumental modality and then, after stabilization and practice, in the acquisition of a new habit. In this manner, intentional act-outcome modalities (incentive instrumental responding) alternate with automatic habit modalities of responding in relation to the changing needs of the external world.

From this necessarily synthetic overview it is clear that the various theories about the mechanism of motivated responding [incentive-motivational, habit (S–R) strengthening, incentive instrumental], rather than mutually excluding each other, are actually descriptive of specific processes and modalities of response along a continuum that corresponds to the complexity of purposive behaviour in a naturalistic setting.

C. Early Studies: The Original and the Revised Anhedonia Hypothesis

Early studies on the behavioural effects of systemic DA-receptor blockers showed that these drugs disrupt instrumental responding (DEWS and MORSE 1961). This effect was initially interpreted as an impairment of motor performance (ROLLS et al. 1974; FIBIGER et al. 1976), a suggestion consistent with the dominant views of the time about DA and its role in extrapyramidal functions. Still, a purely motor explanation of the behavioural effects of DA-receptor blockers could not account for their antipsychotic properties. Moreover, that DA could play a more complex role in behaviour than that related to extrapyramidal functions was suggested by mapping and lesion studies on ICSS, showing a consistent association between the known origin, course and termination of DA neurons and the sites from which ICSS could be evoked or disrupted (CROW 1972; FIBIGER 1978; WISE 1978). At the same time, anatomical studies had shown that DA was not restricted to extrapyramidal motor areas but extended to "limbic" areas (UNGERSTEDT 1971; LINDVALL and BJORKLUND 1974) and even to specific areas of the cerebral cortex (THIERRY et al. 1973; BERGER et al. 1974; LINDVALL et al. 1974). It was at this time that the hypothesis of a role of DA in reward independent from motor function grew.

The first indication that there was more than an impairment of motor function in the disruption of instrumental responding by neuroleptics came from the observation that these drugs typically induce a delayed, within-session decrement of the rate of lever pressing in continuous reinforcement schedules (WISE 1982). This peculiarity was a general one as it applied to responding for ICSS as well as for conventional (water, food) and psychostimulant reward (see WISE 1982 and SALAMONE 1987 for an account and a discussion of these studies). The delayed character of the action of neuroleptics on responding, while ruling out a performance effect, also made it similar to the effect of non-reinforcement, i.e. of extinction (WISE et al. 1978a,b), forming the basis for its interpretation as a reduction of the impact of rewards. Specifically, it was hypothesized that neuroleptics impair responding by blunting the hedonic impact of rewards (original anhedonia hypothesis). Soon after, however, GRAY and WISE (1980) and PHILLIPS and FIBIGER (1979) observed that DA antagonists impaired responding on variable interval schedules even before the first reward had been earned, i.e. before reinforcement had taken place. Superimposed to this effect was a progressive reduction of responding (GRAY and WISE 1980). On this basis, GRAY and WISE (1980) hypothesized that DA mediates the incentive-motivational properties of both primary (rewards) and secondary reinforcers and accordingly revised the original anhedonia hypothesis. As explicitly stated by WISE (1982), "Dopaminergic impairment disrupts first and most strongly the motivational arousal function of external rather than internal stimuli" and, even more explicitly, "I am suggesting that reinforcers and their associated environmental cues lose their sensory impact in terms of arousal function but not in terms of cue function." The idea that DA mediates the quantitative rather than the directional aspects of response to reinforcers has been later included in many accounts of DA function (SALAMONE 1987; ROBBINS et al. 1989a). This revised anhedonia hypothesis derived from the incentive-motivational theories of BINDRA (1974 and 1978) the notion that incentives acquire not only the response-eliciting but also the hedonic properties of the reward to which they have been conditioned. The observation made by PHILLIPS and FIBIGER (1979) that haloperidol given during extinction further reduces responding compared to extinction alone was apparently consistent with blunting by neuroleptics of the influence of conditioned incentives in addition to that of the reward. These authors argued, however, that, if neuroleptics act by blunting the incentive properties of primary and secondary stimuli, as proposed by WISE, it remained unexplained why, given during reinforcement sessions, haloperidol reduced responding to the same extent as extinction, a condition in which secondary reinforcers are still present (PHILLIPS and FIBIGER 1979). On this basis, PHILLIPS and FIBIGER (1979) concluded that neuroleptics exert multiple effects on instrumental behaviour that can be explained by a combination of blockade of primary reinforcement and of impairment of performance. Thus, these early studies set the theoretical stage for later studies on the role of DA in behaviour. Three main hypotheses were considered to explain the effect of DA-receptor blockers on instrumental behaviour: (1) DA is the substrate of the rewarding properties of primary reinforcers (original anhedonia hypothesis); (2) DA mediates the incentive-motivational and arousal properties of primary reinforcers (rewards) and of stimuli conditioned to them (secondary reinforcers, conditioned incentives) (revised anhedonia hypothesis); (3) DA is essential for performance and sensory-motor functions.

D. Testing the Original Anhedonia Hypothesis

The original anhedonia hypothesis has been and still is extremely influential for the interpretation of the role of DA in behaviour and in the effects of drugs that act on DA transmission (antipsychotics, antidepressants, psychostimulants, etc.). Unfortunately, the fact that the theory relies on a term like "hedonia" which refers to a psychological state rather than to an operationally defined measure, inevitably makes difficult its testing in animals and weakens its working character. Although the anhedonia hypothesis was essentially based on the similarity between the effect of neuroleptics and that of nonreinforcement (extinction) in paradigms of instrumental responding, the assumption that blunting of hedonia was the basis of the effect of neuroleptics involved a major leap from reinforcement theories of the time. In fact, the concept of hedonia is traditionally not intrinsic to the definition of reinforcement, at least in the Skinnerian sense (SKINNER 1938). Even from a behaviourist point of view (WATSON 1913), an impairment of reinforcement would be explained as an impairment of S–R association rather than as blunting of the "satisfying" properties of the reward. Therefore, specific testing of the anhedonia hypothesis should consist of testing the role of DA specifically in the hedonic properties of reinforcers rather than generically in response reinforcement.

I. Sweet Reward

The study of the effect of DA-receptor antagonists on consumption of and on taste reactivity to sweet reward should provide the simplest and more direct means to test the anhedonia hypothesis. Indeed, the innate and relatively stereotyped nature of sweet reward makes this behaviour a more direct expression of the impact of a reward than instrumental responding, being relatively independent from an impairment of performance that instead is more likely to affect instrumental responses, particularly such unnatural ones as bar pressing.

In one of the first studies addressing this issue, XENAKIS and SCLAFANI (1981) showed that pimozide dose-dependently reduces the 30-min intake as well as the lick rate and lick efficiency of saccharin-glucose solutions early in the session. These effects were less pronounced when water rather than a saccharin-sucrose solution was the drinking fluid. A similar pattern of changes was elicited by quinine-adulterated solutions. Therefore, reduction of the hedonic properties of the solution by quinine adulteration elicited effects similar to those of DA-receptor blockade. On the other hand, DA-receptor blockade was more effective in reducing the consumption of the hedonically stronger sweet reward than of water (XENAKIS and SCLAFANI 1981).

Subsequently, the same authors performed a parametric study on the consumption of a saccharin-glucose solution in control rats and in rats lesioned in the ventromedial (VM) hypothalamus (XENAKIS and SCLAFANI 1982). After these lesions, rats became hyperphagic and more "finicky", i.e. more sensitive to the hedonic quality of food. Independent variables were the dose of pimozide, the concentration of quinine and the dilution of the saccharin-glucose solution. Dependent variables were 30-min fluid intake and 3-min (initial) cumulative licks. The effects of pimozide on control and lesioned rats were more similar to those of dilution of the sweet solution than to quinine adulteration, consistently with an anhedonic interpretation. In particular, quinine reduced intake more effectively in lesioned than in control rats, while pimozide and dilution affected it to a similar extent in the same groups. Initial lick rate was similarly suppressed by quinine in control and lesioned rats while was more effectively reduced by pimozide and by dilution in lesioned as compared to control rats.

Further studies by WILLNER and colleagues in intact rats are consistent with those of XENAKIS and SCLAFANI (1981, 1982). Thus, pimozide (TOWELL et al. 1987), sulpiride (MUSCAT and WILLNER 1989), and raclopride (PHILLIPS et al. 1991b), reduced the intake of low concentrations of sucrose (0.7%) on single-bottle tests or in two-bottle tests against water but increased water intake. At high concentrations of sucrose, however (32%–34%), both pimozide (TOWELL et al. 1987) and sulpiride (MUSCAT and WILLNER 1989) did not modify intake in one-bottle test. In two-bottle tests against water, both sulpiride (MUSCAT and WILLNER 1989) and raclopride (PHILLIPS et al. 1991b) actually increased intake of concentrated sucrose solutions. It is notable that while the decrease in consumption of 0.7% solutions by raclopride was obtained immediately on the first exposure trial to sucrose and in the first 5 min of a 15-min session, the increase of 34% sucrose consumption was time-dependent, taking place on the last 5 min of the first trial and progressively anticipating its appearance on each trial so that by the third trial the effect was observed on the first 5-min bin (PHILLIPS et al. 1991b).

The authors interpreted these findings as the result of the ability of neuroleptics to shift to the right of the overall bell-shaped concentration/intake curve of sucrose and, therefore, as evidence of their property of blunting sweet reward. This interpretation, although parsimonious, does not fully account for the fact that in the conditions of these studies different factors are likely to be operative in controlling intake at low- and at high-sucrose concentrations. Thus, at high-sucrose concentrations, post-ingestive inhibitory mechanisms are recruited that reduce intake and might result in the bell-shaped concentration/intake curve of sucrose. In fact, if post-ingestive effects are prevented by sham-feeding, the bell-shaped concentration/intake curve is converted into a monotonic one (SMITH and GIBBS 1979; WEINGARTEN and WATSON 1982). Therefore, the different outcome of DA-receptor blockade on sucrose intake at low and high sucrose concentrations might be related to an effect on post-ingestive mechanisms that limits sucrose intake at high concentrations. Indeed, even the studies by XENAKIS and SCLAFANI (1981 and 1982), although consistent with the original anhedonia hypothesis, could not exclude the possibility that the effect of neuroleptics on sweet reward was the result of a specific reduction of the motivational impact of satiety. This possibility however, contrasts with the observation that reduction in the level of motivation for food by reduction of the degree of food deprivation, reduces drinking of 7% and 34% sucrose solutions, an effect just opposite to that of neuroleptics (MUSCAT and WILLNER 1989).

In order to directly exclude the possibility of an interaction between neuroleptics and post-ingestive effects of sweet solutions, the effect of pimozide on sucrose sham-feeding was investigated in rats bearing a gastric cannula that drained the whole fluid ingested (GEARY and SMITH 1985). Under these conditions pimozide, at the low dose of 0.25 mg/kg, reduced the rate of sucrose intake without altering its time-course. In the same sham-fed/food-deprived model, SCHNEIDER et al. (1986) estimated the ID_{50} (dose for half maximal inhibition of intake in mg/kg) of four neuroleptics specific for D_2 receptors to be 0.15 for haloperidol, 0.3 for pimozide, 20 for sultopride and 100 for (–) sulpiride. In another experiment of the same study low doses of pimozide (0.25 mg/kg) and haloperidol (0.1 mg/kg) reduced sham-intake of 1.0% sucrose solution by about 30% in food-deprived rats but only marginally reduced (–5.6/6.0%) the intake of water in water-deprived rats.

These results are consistent with the hypothesis that neuroleptics impair sweet reward by an action different from simple motor impairment, either by blunting the hedonic properties of sweet reward (original anhedonia hypothesis) or by impairing its incentive properties (revised anhedonia hypothesis). In further agreement with the above hypothesis is the observation by WEATHERFORD et al. (1990) that the D_1 antagonist SCH23390 and the D_2 antagonist raclopride reduce drinking of 100% corn oil and 5% and 10% sucrose solutions in a manner inversely proportional to their rewarding value as deduced from preference tests (100% corn oil > 10% sucrose > 6% sucrose).

The anatomical substrate of the effect of neuroleptics on sweet reward has been studied by PHILLIPS et al. (1991c) after infusion of sulpiride in various terminal DA areas in two-bottle tests with sucrose (0.7%, 7.0%, 34%) against water. Intra-accumbens sulpiride best reproduced the systemic effect of the drug on intake of low concentrations of sucrose (0.7%) as characterized by an early in the session (first 5 min) decrease and by a parallel increase in water intake. Intra-caudate sulpiride elicited a late in the session (16–60 min) decrease of sucrose intake and an early in the session (first 5 min) increase in water intake without change in sucrose intake. After 1 h pretreatment, the differences between intra-accumbens and intra-caudate sulpiride were even clearer than when the drug was given immediately before the session. Under these conditions, a low dose of sulpiride (0.625 µg) strongly reduced 0.7% sucrose intake and conversely increased water intake specifically in the first 5 min of the session. Intra-caudate infusion of the same dose of sulpiride, instead, decreased sucrose intake only in the last part of the session (16–60 min) and did not increase water intake. In contrast to the site-specificity of the decrease of the intake of 0.7% sucrose, the increase of 34% sucrose by intracerebral sulpiride was not site-specific, since it could be elicited by infusion in the accumbens, in the dorsal caudate and even in the basolateral amygdala. The existence of these topographic differences in the effect of neuroleptics on the intake of low and high concentrations of sucrose after intracerebral infusion confirm the suspicion that these effects involve different mechanisms.

Studies on the microstructure of licking (lick rate and duration, interlick interval and distribution, size of lick cluster and intercluster interval) have specifically addressed the issue of an impairment of motor performance in the effect of neuroleptics on sweet reward (GRAMLING et al. 1984; GRAMLING and FOWLER 1986; SCHNEIDER et al. 1990). While the studies by GRAMLING et al. (1984) and GRAMLING and FOWLER (1986) have been performed in nondeprived intact rats, that of SCHNEIDER et al. (1990) utilized sham-fed, food-deprived rats. Further differences are related to the different concentrations of sucrose utilized. GRAMLING et al. (1984) compared the effect of pimozide (0.5 and 1.0 mg/kg) to that of extinction in daily 10-min sessions for 8 days. Pimozide reduced the lick rate, but this effect was relatively minor when compared with that on extinction that resulted in a dramatic drop of licking. Therefore the effect of extinction was quite different from that of pimozide. The marked reduction of licking under extinction makes difficult its comparison with the effect of pimozide. A more instructive comparison between neuroleptic and extinction was performed by GRAMLING and FOWLER (1986) by studying the differences between pimozide and upshifts and downshifts in sucrose concentration. Downshifts from 32% to 4% reduced the lick rate and increased the proportion of pauses (20.5-s interlick interval). Pimozide (1.0 mg/kg) also reduced lick rate and increased the proportion of pauses, but in addition increased interlick interval. Curiously, however, upshifts from 4% to 32% sucrose did not affect any measure except for the incidence of pauses that, however, was increased, just like after downshifts in sucrose concentrations.

Apart from the generic conclusion that pimozide-induced changes in licking topography are quantitatively different from extinction, the differences between more graded changes in reward density and pimozide in their effects on licking topography do not enable one to make a firm conclusion about the mechanism of neuroleptic action on sweet reward.

The design of the study by SCHNEIDER et al. (1990) enables a more direct test of the anhedonia hypothesis. In this study, the dose of raclopride was individually adjusted to provide a reduction of the intake of 10% sucrose correspondent to that induced by a reduction of sucrose concentration from 10% to 5%. None of these manipulations modified the interlick interval, but both raclopride and dilution reduced the size of lick clusters and increased the interval between clusters to a similar extent. The same doses of raclopride, while reduced water intake and increased cluster size, decreased (rather than increasing as in the case of sucrose) the intercluster interval. The similarity between reward dilution and pimozide for their effects on licking topography and the differences between their effects in relation to the nature of the reward (sucrose versus water) speaks in favour of an interference of raclopride with the reinforcing properties of the reward rather than with motor performance. A purely motor interpretation is also excluded by the fact that raclopride, rather than reducing the efficiency of licking, actually increased it in the case of 10% sucrose.

Although these studies would exclude a role of a simple motor impairment in the effects of DA-receptor blockers on sweet reward, it is unclear if neuroleptics blunt the value of the reward or its impact on response instrumental to its consumption.

The studies by TYRKA and SMITH (1991 and 1993) and by TYRKA et al. (1992) provide evidence to select among these possibilities. Thus, raclopride, while reducing the sucrose intake from drinking tubes in adult rats (SCHNEIDER et al. 1990) and from tissue on the bottom of a beaker in rat pups (TYRKA et al. 1992), failed to decrease the intake of sucrose infused intraorally through cannulas both in rat pups (TYRKA et al. 1992) and in adult rats (TYRKA and SMITH 1993). Similar observations were made with SCH23390 in rat pups of 7 and 14 days (TYRKA and SMITH 1991). In adults, SCH23390 reduced intake of intraoral sucrose only at doses much higher (~10 times) than those that inhibit intake by drinking from tubes. A role of a generalized motor deficit in the inhibitory effect of raclopride on sucrose intake was excluded on the basis of the observation that raclopride did not affect latency to initiate ingestion, motor activity scores or latency to withdraw one hindlimb from a horizontal bar (TYRKA et al. 1992). SCH23390 given to rat pups of postnatal age 21 days at doses that inhibit intake from intraoral cannulas also increased the latency to initiate ingestion and reduced motor activity scores (TYRKA and SMITH 1991). This suggests that SCH23390 did not reduce sucrose intake from intraoral cannulas by blunting the reward value of sucrose but rather by impairing motor performance.

These observations can be explained if one assumes that sucrose intake in independent consumption tests such as licking from the floor in pups or drinking from tubes involves two phases, (1) an appetitive/preparatory phase which consists of the approach by the subject to the source of sucrose thus leading to contact of the mouth with the sweet source, tongue protrusion and licking, with consequent stimulation of gustatory receptors by the sweet taste, and (2) a consummatory phase, characterized by a rigid, almost stereotyped sequence of licking and swallowing, which is initiated and carried on until it is progressively reduced and terminated by satiety. Blockade of DA receptors would impair the first appetitive/preparatory phase but not the second, purely consummatory one. As these effects cannot be explained by a generalized motor impairment (except at high doses of SCH23390), it is concluded that DA-receptor blockade impairs sweet reward by blunting its "unconditioned incentive value". By this terminology we refer to the property of sweet reward to activate an appetitive behavioural response that results in contact between the sweet solution and taste receptors in the tongue; this behaviour consists of approach, snout contact, tongue protrusion and licking. Therefore, incentive value is the result of the coupling of hedonic value with mechanisms that transduce that value into a response that results in ingestion. Within this context, incentive value of a reward is not exactly superimposable to the notion of "incentive salience attribution" of BERRIDGE and ROBINSON, as this terminology refers to the expression of incentive properties by conditioned stimuli and is, therefore, dependent on a previous associative learning step; incentive value, in the present case, is likely to be innate, as the response it elicits takes place in rat pups naive to the independent ingestion paradigm (HALL and BRYAN 1980, 1981). A similar example of reduction of the incentive value of unconditioned but not of conditioned stimuli by DA-receptor blockade has been reported by LOPEZ and ETTENBERG (2001).

In studies of the role of DA in reward, taste reactivity has been utilized as a direct expression of the hedonic value of rewards. In this paradigm, solutions are infused directly in the mouth of the subject through intraoral cannulas. Sweet solutions evoke the emission of a characteristic pattern of hedonic reactions (frontal tongue protrusion; lateral tongue protrusion, paw licking) while bitter solutions evoke aversive reactions (gapes, forelimb fails, head shakes) (GRILL and NORGREN 1978). These reactions are affected by drive state, by drugs and by brain lesions in a manner compatible with the notion that they reflect the motivational valence and value of the taste. Using this paradigm, TREIT and BERRIDGE (1990) failed to show changes in hedonic and aversive reactions after 1 mg/kg of haloperidol. Negative results were also obtained after 6-OHDA lesions by BERRIDGE et al. (1989) and by BERRIDGE and ROBINSON (1998). LEEB et al. (1991), however, reported that pimozide reduces hedonic reactions to sucrose; moreover, PARKER and LOPEZ JR. (1990) reported that pimozide enhances the aversiveness of quinine solutions. In order to investigate the reason for these discrepancies, BERRIDGE and PARKER and their collaborators joined together in a collaborative study (PECINA et al. 1997) reaching the following conclusions: pimozide reduces hedonic taste reactions to sucrose, but this effect takes place slowly and, in any case, after the first minute of the trial, which explains the negative results of TREIT and BERRIDGE (1990) who utilized 1-min infusions. Because of this and since aversive reactions to quinine were also found to be reduced, the authors attributed the effect of pimozide to a sensorimotor impairment rather than to blunting of taste hedonia. This explanation, however, is not fully convincing. In fact, on the second and third trial, hedonic reactions are impaired from the beginning of the trial (see, for example, frontal tongue protrusions and lateral tongue protrusions in Experiment 1B). As drug trials were performed every 48h and alternated every 24h with no-drug trials, a motor impairment or a fatigue effect from previous drug trials cannot account for the between-trial reduction of hedonic taste reactivity by pimozide. As to the aversive reactions, no reduction was observed after correction of the results by PARKER and LOPEZ JR. (1990) for the reduction in scoring time due to reduction of motor activity induced by pimozide. These negative findings, in turn, contrast with the reduction of aversive reactions reported by PECINA et al. (1997). Another aspect of this study is the fact that a single dose level (0.5 mg/kg) of pimozide was investigated; this dose level in turn, is twice what was shown by XENAKIS and SCLAFANI (1981) and by GEARY and SMITH (1985) to be effective in reducing sucrose intake. Further studies with lower doses of pimozide and with other DA-receptor antagonists might clarify this point. As to 6-OHDA lesions, the fact that these lesions do not affect the reactivity to hedonic as well as aversive taste stimuli (BERRIDGE and ROBINSON 1998) does not necessarily exclude the possibility that acute blockade of DA transmission would. Thus, blockade of D_1 receptors by SCH23390 or SCH398166 impair taste aversion learning (FENU et al. 2001) in spite of the fact that 6-OHDA lesions are ineffective (BERRIDGE et al. 1989; BERRIDGE and ROBINSON 1998).

From the above discussion it appears that even in taste reactivity tests it is difficult to avoid the confounding influence of a performance effect of neuroleptics.

As already pointed out by BENINGER (1989) and by ETTENBERG (1989), one way to distinguish the influence of a performance effect in the action of neuroleptics on behaviour is to test for neuroleptic effects in their absence. In this case the neuroleptic is administered during separate acquisition sessions of stimulus–reward or stimulus–response associations. Although this artifice eliminates the possibility of a performance effect of neuroleptics during response expression, it does not prevent an effect of performance impairment on the efficiency of conditioning during acquisition nor the possibility of state-dependent learning.

HSIAO and SMITH (1995) have taken advantage of this principle to test the hypothesis that D_2 receptor blockade impairs the reinforcing properties of sucrose. To this end, raclopride or saline were administered during intake of sucrose solutions flavoured with a different flavour, depending whether subjects were given raclopride or saline. Although raclopride reduced sucrose intake, control subjects were allowed to drink a volume of sucrose correspondent to that taken under raclopride. In this way, any reduction of the efficiency of conditioning under raclopride due to reduced sucrose-flavour pairing was controlled for. On a two-bottle test against water, the intake of the flavoured sucrose solution paired with raclopride was reduced compared to the intake of the flavoured solution paired with saline, and this difference was not due to a conditioned taste aversion. The possibility of an impairment in the ability to discriminate the gustatory stimuli was excluded since it had been demonstrated by WILLNER et al. (1990a) that pimozide does not impair the efficiency of sucrose solutions of different concentrations to act as discriminative stimuli in a T-maze. Therefore it was concluded that raclopride impairs the reinforcing properties of sucrose. Although this effect can be understood as the result of a reduction of the hedonic properties of sucrose it would be equally consistent with an impairment of Pavlovian incentive learning.

II. Operant Responding for Sweet Reward

In operant paradigms of responding for sucrose solutions, pimozide reduced responding for weak but not for strong sucrose solutions, its concentration-response function being strikingly superimposable to that of sucrose solutions adulterated with a fixed concentration of quinine (BAILEY et al. 1986). The similarity between the concentration/response function for sucrose under pimozide and for sucrose adulterated by quinine (BAILEY et al. 1986) raises the possibility that under neuroleptics low sucrose concentrations acquire aversive properties. Operant responding for 45 mg sucrose pellets of different concentrations (1%, 10%, 95%) had an inverted U-shape concentration-response function; raclopride affected responding according to a shift to the right of the concentration-response function, decreasing responding for low (1%) and increasing responding for high (35%) sucrose concentration in the pellets (PHILLIPS et al. 1991a). Thus, observations obtained from operant responding for sucrose are fairly consistent with those obtained from consumption of sucrose solutions of different concentrations (see above).

Summing up, studies of the effect of DA-receptor blockade on consumption of and on responding for sweet reward are difficult to interpret on the basis of a unitary explanatory framework. Thus, a simple performance impairment does not explain the specific reduction of consumption and of responding for low sucrose concentrations. A performance impairment also does not account for the apparent shift in the overall concentration/response curve for sucrose consumption and responding (PHILLIPS et al. 1991a,b). For this, a more likely explanation would be a reduction of the rewarding properties of sucrose, consistent with the anhedonia hypothesis, or an impairment of the unconditioned incentive properties of sweet reward.

E. The Motor Deficit Issue

As we have pointed out, a fundamental argument for excluding a motor impairment as the basis for neuroleptic-induced reduction of instrumental responding has been the circumstance that in continuous reinforcement (CRF) schedules of responding this effect takes place within and not at the beginning of the session (WISE et al. 1978b). Thus, in contrast to Parkinson's-like motor impairment, that typically affects movement initiation, the impairment induced by neuroleptics on instrumental responding seems to involve response maintenance rather than initiation. This argument, however, is weakened by the fact that not only this does not apply to variable interval (VI) schedules of reinforcement, where responding is reduced from the start of the session but, even in the case of CRF schedules, it depends on the amount of training. Thus, contrary to previous studies (WISE et al. 1978a), GRAY and WISE (1980) in Experiment 2 observed that 0.5 and 1.0 mg/kg of pimozide dose-dependently reduced bar pressing from the first series of test trials. More recently DICKINSON et al. (2000) have shown that 0.5 mg/kg of pimozide reduces bar pressing from the beginning of the session in rats tested 1 h and 2 h after the drug. Thus, under certain conditions, probably in relation to the degree of training, pimozide impairs responding in a manner compatible with a motor impairment.

A motor impairment was also advocated by a number of studies showing that neuroleptics affect operant responding in a manner incompatible with blunting of reinforcement. Thus, the observation by PHILLIPS and FIBIGER (1979), replicated by various studies (GRAY and WISE 1980; MASON et al. 1980; TOMBAUGH et al. 1980; FELDON et al. 1988), that neuroleptics reduce responding also when given under extinction, is not only incompatible with the idea that neuroleptics produce effects homologous to extinction (original anhedonia hypothesis) but also with the idea that they blunt the impact of incentives (revised anhedonia hypothesis). In fact, as argued by PHILLIPS and FIBIGER (1979), if indeed neuroleptics impair both primary reinforcement (abolished under extinction) as well as secondary reinforcement (preserved under extinction), they should similarly impair reinforced and non-reinforced responding; instead, neuroleptics impair non-reinforced responding to a larger extent than reinforced responding (PHILLIPS and FIBIGER 1979). Moreover, under short intertrial intervals, neuroleptics administered in acquisition did not induce an increased resistance to extinction as would be expected if, by reducing reward, they induce a partial reward extinction effect (FELDON and WEINER 1991). A further problem for the anhedonia hypothesis comes from transfer studies from extinction to neuroleptic treatment and vice versa. Consistent with the anhedonia hypothesis, successful transfer from extinction to neuroleptic (i.e. the reduction of responding induced by extinction is maintained and continued when subjects are switched from non-reinforcement to neuroleptics+reinforcement) was reported by WISE et al. (1978a,b) and by GERBER et al. (1981) but was not replicated by other laboratories (MASON et al. 1980; TOMBAUGH et al. 1980; BENINGER 1982) and was actually lost (WISE et al. 1978a; GERBER et al. 1981) if reinforced responding under neuroleptic is compared with responding on the last extinction session (see discussion by WISE 1982, p 44). On the other hand, the effect of neuroleptics on reinforced responding does not transfer to extinction. Thus, when rats are shifted from neuroleptic plus reinforcement to non-reinforcement (extinction), responding resumes for the whole session (BENINGER 1982). Apparently, rats behaved on the extinction test as if the neuroleptic, in spite of the reduction of responding, had not prevented reinforcement from taking place (BENINGER 1982). This "asymmetry" of the transfer between extinction and neuroleptics (GERBER et al. 1981) cast further doubts on the simple hypothesis that neuroleptics reduce responding by impairing reinforcement and suggest that additional effects of these drugs play a major role in their effect on instrumental responding.

Among the early studies that are commonly quoted as providing evidence against the anhedonia hypothesis is that of ETTENBERG et al. (1981). In this study it was reported that flupenthixol differentially affects responding for ICSS, depending on the type operant response; thus, doses that completely block bar pressing only partially reduce nose poking. This study, however, contrasts with a study by GERHARDT and LIEBMAN (1981) showing that haloperidol (and clonidine) dose-dependently reduce bar pressing and nose poking for

ICSS in a superimposable manner. The same authors point out a number of differences between their study and that of ETTENBERG et al. (1981); thus, in the study by ETTENBERG et al. (1981) nose-poking sessions always preceded bar-pressing sessions while in the study by GERHARDT and LIEBMAN (1981) the order was counterbalanced; moreover, in the ETTENBERG et al. (1981) study each session lasted 40 min, during which responding was tested on a full range of current intensities, while in the study by GERHARDT and LIEBMAN (1981) each session lasted 15 min. Thus, the conditions of the ETTENBERG et al. (1981) study could have motorically and motivationally favoured nose poking over bar pressing. WAUQUIER and NIEMEGEERS (1979), on the other hand, had shown earlier that four different neuroleptics (pimozide, haloperidol, azaperone and pipamperone) all reduced licking at lower doses than bar pressing for ICSS. It is notable that the threshold current for responding with licking is higher than for responding with bar pressing, while that for nose poking is lower than for bar pressing (WAUQUIER and NIEMEGEERS 1979; GERHARDT and LIEBMAN 1981). It would appear, therefore, that to the extent that reinforcement is expressed by the strength of motivation for responding, the reduction of responding induced by neuroleptics can depend on the kind of response by which reinforcement is obtained. The above observations, therefore, are equally compatible with an effect on performance or on reinforcement or on both.

More recently it has been reported that in animals bar pressing for sucrose reward, a within-session reduction of responding by neuroleptics is obtained under extinction not only in the presence of conditioned cues but also in their absence (PHILLIPS et al. 1991a). Notably, removal of the cues resulted in decrease of responding within the first 5 min, while in the presence of the cues reduction of responding did not take place until 11–15 min into the session. As the final level of responding obtained under raclopride was the same in the presence or in the absence of cues, it appears that the slower reduction of responding in the presence of cues is consistent with a contrasting influence of incentive cues on neuroleptic-induced impairment of responding (see Fig. 6 in PHILLIPS et al. 1991a). These observations, therefore, are consistent with a role of motor impairment in the effect of neuroleptics on instrumental responding.

Further evidence for such a role comes from comparative studies of the impact of extinction and neuroleptics on temporal and force aspects of performance during instrumental responding (FAUSTMAN and FOWLER 1981; ASIN and FIBIGER 1984; FOWLER et al. 1986). Thus, while extinction induced a progressive reduction of the force with no change in the duration of bar pressing, neuroleptics increased response duration and inter-response interval without decreasing or even increasing peak force (FOWLER et al. 1986; FOWLER and KIRKPATRICK 1989). The effects of neuroleptics on response duration appear to have a similar within-session profile as the reduction of rate of responding and the increase in inter-response interval (FOWLER and KIRKPATRICK 1989; LIAO and FOWLER 1990).

Further information on the nature of the response decrements induced by neuroleptics can be derived from studies of their effects on the kinematics of licking behaviour and from a comparison with shift-in-reward value. Haloperidol reduces at low doses (<0.3 mg/kg) the peak force of licks and lick rate without modifying lick rhythm (FOWLER and DAS 1994). Olanzapine and clozapine, two atypical neuroleptics devoid of extrapyramidal side-effects, reduced peak force and lick rate but also duration and rhythm of licks (DAS and FOWLER 1996). The effects of olanzapine and clozapine on peak force and number of licks, however, did not have the within-session distribution of the effects of haloperidol. Furthermore, anticholinergic treatment attenuated the effects of haloperidol on licking rate and force (FOWLER and DAS 1994) and additionally abolished their within-session distribution (DAS and FOWLER 1995).

These observations with atypical neuroleptics complement those of SANGER (1986) and of SANGER and PERRAULT (1995) that clozapine and olanzapine and other atypical neuroleptics do not induce within-session decrements of operant responding. Collectively, the observations summarized in this section could be interpreted to indicate that the within-session effect of neuroleptics on responding is the result of an impairment of extrapyramidal function (FOWLER 1990). The precise mechanism by which this impairment results in the within-session effect varies. Some authors interpreted the within-session effect as an expression of fatigue (FIBIGER et al. 1976; PHILLIPS and FIBIGER 1979). Related to this explanation is that of ETTENBERG et al. (1981) that referred neuroleptic-induced performance deficit to the "kinetic requirements" of the response; thus, neuroleptics would be more prone to impair responses that have a higher "motor demand". This indeed seems to be the case for bar pressing versus nose poking (FOWLER and KIRKPATRICK 1989). Related to this hypothesis are those that take into account the economical reward/effort ratio. Neuroleptics would progressively decrease responding by increasing the perceived effort of the task in the face of a constant reward, thereby reducing behavioural output (SINNAMON 1982). However, contrary to these hypotheses, if force required for reinforcement is increased, the impairment of responding by neuroleptics is reduced rather than worsened (FOWLER and KIRKPATRICK 1989). These observations in turn exclude the possibility that the time-dependent effect of neuroleptics on responding is the result of fatigue. TOMBAUGH et al. (1982), instead, assumed that the within-session decrease of responding was the result of an aversive conditioning due to the frustrative effect of the uncoupling between the subject intention and its motor expression due to the performance deficit. This hypothesis is consistent with the observation that the kinematic effects of neuroleptics on performance, namely lengthening of response duration and increase in peak force precede the decrease in rate [and in lengthening of inter-response time (IRT)] that is typically manifested late in the session (FOWLER and KIRKPATRICK 1989) and therefore might be the result of aversion learning. If this hypothesis were correct, however, neuroleptics should not affect independently performance and motivation/reinforcement efficiency as shown by matching law experiments (see below).

From this analysis it appears that the effects of neuroleptics on instrumental responding cannot be fully accounted for by a motor deficit, although such action certainly contributes, depending on the paradigm, to the impairment of responding.

F. Response-Reinforcement Functions

The relationship between response rate and reinforcer value is described by a rectangular hyperbola: response rate saturates at high reinforcer value reaching an asymptote that corresponds to maximal response capacity. The reinforcer value at which 50% of maximal response rate is obtained is a measure of the impact of the reinforcer on responding. Changes in this impact would result in parallel shifts of the response-reinforcer function to the right or to the left depending on whether reinforcing impact is reduced or increased. On the other hand, changes in response capacity would result in changes in asymptotic responding (STELLAR and STELLAR 1985).

This apparently simple construct has been utilized to investigate the effect of drugs acting on the DA system on performance as separate from reinforcer impact. Different reinforcement and response modalities have been utilized in these studies. In ICSS studies, the relationship between frequency (in Hz, pulses/second), intensity (in A) or duration (or number of pulses) of stimulation and responding (expressed as rate of bar-pressing or running speed in a maze) has been studied so as to obtain frequency-response, intensity-response and reward summation functions (LIEBMAN 1983; MILIARESSIS et al. 1986a; STELLAR et al. 1988; EDMONDS and GALLISTEL 1974; HUNT and ATRENS 1992a). In instrumental responding for food, water or drug-reward, the relationship between rate of responding and rate of reinforcement among different VI or random interval (RI) schedules is utilized (HERRNSTEIN 1970; DEVILLIERS and HERRNSTEIN 1976; HEYMAN et al. 1987).

I. Reward Summation Studies

In studies of maze running for ICSS, pimozide shifted to the right the reward summation function and also reduced asymptotic running speed (FRANKLIN 1978; STELLAR et al. 1983); similar effects were obtained by infusion of *cis*-flupenthixol (0.5g per side) in the nucleus accumbens (NAc) (STELLAR et al. 1983; STELLAR and CORBETT 1989). In bar pressing for ICSS, pimozide and molindone shifted to the right, while amphetamine shifted to the left, the reward summation function (GALLISTEL and KARRAS 1984; HAMILTON et al. 1985). In fixed-interval schedules of bar pressing for ICSS, however, pimozide (HUNT and ATRENS 1992b) and spiroperidol (HUNT et al. 1994) reduced responding abruptly along a dose-response curve by an effect on performance rather than reinforcer impact. SCH23390, instead, changed the response-frequency function of bar pressing for ICSS in a dose-related manner as a result of a combined effect on performance and on reinforcer impact (HUNT et al. 1994). In studies of maze running and bar pressing for ICSS relating the frequency of stimulation to response rate, both SCH23390 and raclopride shifted the reward summation function to the right but also reduced its slope, suggesting a reduction of reinforcing impact and an impairment of performance (NAKAJIMA and MCKENZIE 1986; NAKAJIMA and O'REGAN 1991; NAKAJIMA and PATTERSON 1997).

In maze running for ICSS, shift in the response-frequency function by *cis*-flupenthixol was more pronounced in self-stimulation of the medial forebrain bundle (MFB) than of the prefrontal cortex (PFCX) (CORBETT 1990). Moreover, intra-accumbens *cis*-flupenthixol was more effective in shifting to the right the response-frequency function for MFB than for caudate-putamen or medial prefrontal cortex self-stimulation (STELLAR and CORBETT 1989).

In bar pressing for ICSS, pimozide, at low doses, induced a parallel shift to the right of the rate-frequency function in four subjects, while in two subjects it decreased the slope. Higher doses completely blocked responding (MILIARESSIS et al. 1986b).

These studies show that DA-receptor blockers affect both the slope of the reward-summation function, the asymptotic response capacity and the reinforcement magnitude necessary for half-maximal response. In some studies (e.g. HUNT and ATRENS 1992a; HUNT et al. 1994) utilizing fixed-interval (FI) schedules of bar pressing for ICSS, D_2 receptor antagonists do not show any effect on reinforcement, while in studies utilizing continuous reinforcement (CRF) (i.e. fixed ratio of 1) schedules (e.g. NAKAJIMA and O'REGAN 1991) they do; the reason for this discrepancy might reside in the fact that under CRF schedules response and reinforcement are interdependent; thus, higher rates of responding increase the reinforcement rate per se in CRF schedules, while in FI schedules, the constant inter-reinforcement interval allows the estimation of reinforcement and response as independent factors of the response-reinforcement function. Under these conditions only the D_1 antagonist SCH23390 fulfils the criteria for reduction of the reinforcing impact of ICSS (HUNT et al. 1994). These observations, while consistent with the possibility that ICSS at the MFB is mediated by release of DA acting on D_1 receptors, also indicate that, at least for ICSS, the effect of D_2 receptor antagonists on response rate is mostly explained by an impairment of performance. Thus, even if one cannot exclude that DA acting on D_2 receptors contributes to reinforcement by ICSS, its eventual contribution is obscured by the performance impairment.

II. Intensity-Threshold Studies

According to WISE (1982), studies of the effect of neuroleptics on the intensity threshold of electrical current at which ICSS is obtained provide the best evidence for a role of DA in the rewarding impact of ICSS. Indeed, neuroleptics increase threshold intensity for ICSS (ESPOSITO et al. 1979; SCHAEFER and HOLTZMAN 1979; ZAREVICS and SETLER 1979; LYNCH and WISE 1985; SCHAEFER and MICHAEL 1985; BIRD and KORNETSKI 1990), even in paradigms that depend on active titration of current intensity by the subject through a second lever (SCHAEFER and HOLTZMAN 1979; ZAREVICS and SETLER 1979). However, SCHAEFER and MICHAEL (1980) reported that chlorpromazine does not affect reinforcement thresholds at doses that reduce response rate while clozapine increases reinforcement thresholds without reducing response rate. Haloperidol and loxapine, instead, affect both performance and reinforcement (SCHAEFER and MICHAEL 1980).

More recently it has been shown that also in the case of the reward threshold for ICSS, CRF schedules result in an artefactually large increase in the reward threshold after neuroleptics, due to the fact that, by decreasing response rate, neuroleptics indirectly reduce reinforcement. Control over reward magnitude by the use of a fixed-interval schedule results in a lesser increase of intensity threshold and in a decrease of maximal rate of responding by neuroleptics, consistent with a combined effect on reward and on performance (BOYE and ROMPRE 1996). In spite of their effects on the reinforcement threshold, neuroleptics did not impair the discriminative stimulus properties (SCHAEFER and MICHAEL 1985) nor the detection threshold of ICSS (BIRD and KORNETSKY 1990).

III. Response-Reinforcement Matching Studies

Response-reinforcement matching studies take advantage of the fact that in VI and RI schedules, the operant responding response rate (R) adjusts itself in a regular pattern in relation to the reinforcement rate (r). In these conditions, R and r obtained under different, concurrent or alternate schedules, or under multiple sequential schedules fit a rectangular hyperbola, the matching law of HERRNSTEIN (1970), where R is a negatively accelerated function of r according to the equation:

$$R = R_{max} \times 2/2 + K_h$$

where the two fitting parameters are the maximal response rate (R_{max}) corresponding to asymptotic R and the reinforcement rate at which corresponds half-maximal R (K_h). These two parameters are differentially affected by experimental manipulation; thus, while R_{max} is affected by changes in the force needed to perform the response, K_h is affected by changes in drive state, reinforcer value and amount of reinforcer (DEVILLIERS and HERRNSTEIN 1976; HEYMAN and MONAGHAN 1987).

These observations indicate that K_h is an estimate of reinforcement efficiency and degree of motivation while R_{max} is an estimate of performance efficiency and motor capacity (MCSWEENEY 1978; MCDOWELL and WOOD 1984;

HEYMAN and MONAGHAN 1987). Therefore, an increase in K_h (an homolog of K_d and K_m) corresponds to a reduction of motivation and reinforcer efficiency, while a decrease in R_{max} corresponds to a decrease in motor capacity and performance.

HEYMAN (1983) first reported on a five-component multiple VI schedule that while pimozide increases K_h and decreases R_{max}, intermediate doses of amphetamine elicit the opposite effect. Different conclusions were reached by MORLEY et al. (1984a). These authors utilized only two different VI reinforcement schedules, a low-frequency (10s) and a high-frequency (100s) one. According to Herrnstein's equation, an increase in K_h would have been reflected in a larger reduction of R in the low-frequency than in the high-frequency reinforcement schedule. Instead, pimozide similarly reduced R on both components suggesting that pimozide reduces performance without affecting reinforcement. It is arguable, however, that the failure of MORLEY et al. (1984b) to detect a reinforcement deficit under pimozide was the result of the insufficient range of reinforcement frequencies utilized as compared to HEYMAN (1983). Subsequent studies by HEYMAN et al. (1986), WILLNER et al. (1987, 1989, 1990b) and PHILLIPS et al. (1991d) have confirmed that indeed pimozide affects both motivation/reinforcer efficacy and performance capacity. Chlorpromazine behaved similarly to pimozide (HEYMAN et al. 1986), while SCH23390 reduced motivation at doses that did not affect performance. Sulpiride slightly but significantly decreased motivation without affecting performance (WILLNER et al. 1990b). PHILLIPS et al. (1991d) also showed that, depending on the time into the session at which responding was tested, pimozide reduced R_{max} from the beginning of the session while increased K_h only late in the session. These observations indicate that, as previously suggested by WISE (1982), DA-receptor blockers reduce responding by an immediate effect on performance and by a delayed effect on incentive-motivation.

This interpretation, which is consistent with the revised anhedonia hypothesis, is more in line with the above results than an interpretation in terms of blunting of the primary rewarding impact of the reinforcer (original anhedonia hypothesis). In fact, a blockade of the hedonic properties by neuroleptics would have reduced K_h from the beginning of the session. It is possible, however, that K_h is not a measure of local reinforcement efficacy but of the strength of motivation that results from reinforcement efficiency. Thus, these observations are equally compatible with the possibility that reduction of the hedonic properties of the primary reinforcer results in reduction of the motivational value of conditioned stimuli that maintain responding.

Finally, according to HEYMAN et al. (1987), cis-flupenthixol, a D_1/D_2 antagonist, has selective effects on performance as, even in low doses, selectively decreases R_{max} without affecting K_h.

The above studies provide important evidence that DA-receptor blockers exert multiple effects on responding as a result of two separate effects: a reduction of the motivational impact of reinforcement and a reduction of performance. In most cases both these effect are present. Selective D_1 and D_2

antagonists can, at lower doses, selectively affect reinforcement efficacy while D_1/D_2 antagonists have a more pronounced affect on performance, probably as a result of a more pervasive D_1/D_2 interaction in the motor terminal areas of the DA system.

G. Dissociating Reinforcement from Incentive-Motivation and Performance

One way to circumvent the problem of the performance effects of DA-receptor blockers is to separate exposure to the antagonist and testing of its effects (BENINGER 1989; ETTENBERG 1989). In a first series of studies, haloperidol (0.075 and 0.15 mg/kg) was given intermittently on 10 (33%) out of 30 single daily sessions of running for food or water reinforcement in a straight runway (ETTENBERG and CAMP 1986a,b). During the following 12 days, responding was tested in single daily sessions under extinction conditions. No impairment in movement initiation nor in performance was observed under haloperidol as indicated by the unchanged latency to leave the start box and the marginal increase in the time to reach the goal box. On the extinction phase, rats intermittently exposed to haloperidol showed a significant resistance to extinction compared to controls not given the drug; this effect, in turn, was similar to that observed in a group in which reinforcement was omitted on the same proportion of trials (33%). Thus, haloperidol did not impair maze-running performance for food or water reward but slowed down the rate of extinction of the motivated response on subsequent drug-free test sessions, much in the same way as intermittent non-reinforcement. Failure of haloperidol to affect maze running on trial suggests that at the doses given, haloperidol differentially affects the expression of the response-eliciting (incentive) properties of conditional stimuli predictive of reward and the reinforcing properties of reward: while the first ones are intact the second ones are impaired.

These results have been confirmed by FELDON et al. (1988), who also tested the effect of haloperidol given daily during reinforced and non-reinforced sessions of a partially reinforcement paradigm (50% of the responses unrewarded). Under these conditions haloperidol, contrary to the predictions of the anhedonia hypothesis, did not facilitate extinction. Further studies by FELDON and WEINER (1991), performed on multiple-trial sessions, show that, contrary to the observations in single-trial sessions, haloperidol fails to impair reinforcement as indicated by failure to produce a resistance to extinction when given during continuous reinforcement and a facilitation of extinction when given during partial reinforcement. These observations, coupled to the fact that haloperidol increases the rate of extinction when given during extinction, have been taken to indicate that haloperidol reduces the impact of reinforcement only on single-trial reinforcement schedules while increasing the impact of non-reinforcement both on single- and on multiple-trial reinforcement schedules (FELDON and WEINER 1991). The difference between the impact of haloperidol on reinforcement in single versus multiple schedules has been explained by the different learning processes operative in the two conditions (FELDON and WEINER 1991). Thus, while responding on multiple trial schedules utilizes response-outcome relationships, this is not the case of single-trial schedules, which depend on the acquisition of incentive properties by stimuli that precede responding. In single schedules, neuroleptics might reduce the impact of reinforcement by impairing incentive learning.

In a further series of studies by ETTENBERG and associates, stimuli were explicitly paired (CS+) or unpaired (CS–) with reinforcement, thus becoming predictive of reinforcement and of non-reinforcement, respectively, in a straight runway. Haloperidol (0.15–0.30 mg/kg) failed to increase run times in response to the CSs, while it strongly increased run times in a drug-free test performed on the next day. Similar results were obtained with conventional reinforcers such as food (HORVITZ and ETTENBERG 1991) and sex (LOPEZ and ETTENBERG 2001) and drug reinforcers (i.v. heroin) (MCFARLAND and ETTENBERG 1995). Results consistent with an impairment of reinforcement independently from motor impairment have been obtained by the same group on the response-reinstating properties of reinforcement by conventional and drug reinforcers. In this paradigm, subjects are first trained to run the maze in response to reward (food, HORVITZ and ETTENBERG 1988; water, ETTENBERG and HORVITZ 1990) or to drug reward (i.v. amphetamine, ETTENBERG 1990; i.v. heroin, ETTENBERG et al. 1996). Once the response is extinguished by a series of non-reinforced sessions, responding is reinstated by a single re-exposure to the reward in the goal box after haloperidol or saline administration. On the next day, testing for maze running in the absence of haloperidol shows a reduction of response in the haloperidol as compared to the saline exposed group.

These observations could be explained either by an impairment of response-reinforcement (haloperidol impairs the ability of the reinforcer to strengthen extinguished S-R associations) or of stimulus–reinforcement (haloperidol impairs the ability of the reinforcer to strengthen the incentive properties of the goal box). These studies, however, have been performed in a straight runway and the response measured (run time to the goal box) is a natural and elementary incentive response, such as approach behaviour This response may not be equivalent to an unnatural and complex response, such as bar pressing. Because of this, some authors do not regard maze running paradigms as expression of instrumental behaviour but instead of Pavlovian and incentive-motivational responding, being based on learning of stimulus-contingencies rather than response-contingencies (DICKINSON and BALLEINE 1994). This might be the reason why the motor impairment demonstrated in operant paradigms involving an explicit instrumental response, such as bar pressing, does not apply to paradigms involving maze running. Therefore, the apparent similarity between two effects of neuroleptics, the within-session impairment of bar pressing shown by WISE and colleagues and the delayed reduction of maze running in primary reinforcement paradigms shown by

ETTENBERG and colleagues, may not be a reflection of their homology but rather of their analogy, that is, of a commonality in a phenomenological aspect rather than in a basic aspect. Therefore, although for the principle of parsimony one would favour a unitary mechanism of the effect of neuroleptics in operant responding and in maze running paradigms, the differences inherent to them make this principle not readily applicable to this specific case.

An additional reason for considering the impairment induced by neuroleptics on reinforcement by bar pressing as not homologous to that obtained on reinforcement by maze running is the fact that while D_1 and D_2 receptor antagonists are similarly effective in producing within-session reduction of bar pressing, only D_2 antagonists impair reinstatement of responding in maze-running paradigms (CHAUSMER and ETTENBERG 1997). This difference is particularly puzzling given the circumstance that D_1 receptor antagonists have been indicated to be more specific than D_2 antagonists in reducing reinforcement as compared to their ability to impair performance as estimated from their ability to induce microcatalepsy (FOWLER and LIOU 1994; FOWLER and LIOU 1998) and to modify the reward summation function for ICSS (HUNT and ATRENS 1992b). Failure of D_1 receptor blockade to impair reinforcement in the paradigm of ETTENBERG et al., however, is also inconsistent with the idea that this paradigm involves Pavlovian stimulus reinforcement rather than response reinforcement, and that D_2 antagonists given on trial act on the acquisition of stimulus–reward association. For these reasons, further studies are needed to clarify this issue (see below). Allowing the above caveats, we favour the interpretation of the effects of neuroleptics in the paradigm of ETTENBERG and colleagues as due to an impairment of Pavlovian incentive learning rather than of response reinforcement. These studies also indicate that, once acquired by Pavlovian learning, the expression of the incentive properties of stimuli are resistant to neuroleptics. Viewed from this perspective, the delayed, within-session impairment of responding induced by neuroleptics can be explained by a progressive loss of incentive properties of Pavlovian stimuli on instrumental responding as a result of impairment of Pavlovian incentive learning.

H. Incentive Accounts of the Role of Dopamine in Behaviour

An important place in the current understanding of the role of DA in behaviour is occupied by incentive accounts. These hypotheses assume that DA mediates or modulates the expression of the incentive properties of stimuli. Apart from their common "incentive" label, however, these hypotheses differ substantially in certain mechanistic aspects that are critical for their testing and from which depends their "working" character. Some authors (SALAMONE et al. 1997) have favoured the idea of a response-energizing role of DA. Others (BLACKBURN et al. 1987), instead, have attributed to DA a preparatory role.

The response-energizing hypothesis is based on a series of studies by SALAMONE and colleagues showing that in a choice condition impairment of DA transmission by 6-OHDA lesions and DA-receptor blockers biases response selection to the less demanding response (see SALAMONE et al. 1997 for review).

On the other hand, the preparatory hypothesis comes from the observation that pimozide reduces preparatory responses (the number of entries into a niche where food is expected) at doses that do not reduce food consumption when the food itself is available (BLACKBURN et al. 1987).

This observation is homologous to a classic effect of neuroleptics, that of disrupting active avoidance responses to the CS at doses that do not impair aversive responses to the US (DEWS and MORSE 1961). More recently it has been reported that neuroleptics, at doses that induce within-session reduction of instrumental responding, impair the property of a Pavlovian CS to facilitate responding for the US (transfer from Pavlovian to instrumental) (DICKINSON et al. 2000). These effects, however, can be also attributed to a subtle motor impairment that differentially affects responding in relation to the strength of the stimulus. Thus, when the sensory salience of the CS relative to that of the US is increased, the CS selectivity of the response-impairing effect of the neuroleptic is abolished (BIGNAMI 1978). Evidence from other studies, on the other hand, does not support a role of DA in the response-eliciting properties of incentives. Thus, presentation of a novel CS reinstates responding for ICSS blocked by pimozide (FRANKLIN and McCoy 1979), an observation that contrasts with the idea that neuroleptics specifically impair the incentive effects of stimuli. Moreover, studies by ETTENBERG and associates reviewed in the previous section show that neuroleptics, at doses that prevent reinforcement learning, do not impair incentive responses to CSs (HORVITZ and ETTENBERG 1991; McFARLAND and ETTENBERG 1995; LOPEZ and ETTENBERG 2001). Further studies from the same group have addressed the issue of the activational and directional properties of incentives (McFARLAND and ETTENBERG 1999). Rats were previously trained to run a straight maze in response to olfactory stimuli predictive of the occurrence (S+) or absence (S−) of food or of i.v. heroin reward in the goal box. Haloperidol (0.075–0.30 mg/kg) reduced basal locomotor activity but not the locomotion induced by exposure to the CS+ in a different environment. Moreover, haloperidol (0.15–0.30 mg/kg) did not affect preference for the compartment correspondent to the CS+ over the one correspondent to the CS− (McFARLAND and ETTENBERG 1999). Therefore, haloperidol, at doses that impair reinforcement, did not affect the activational (CS− induced locomotion) nor the directional properties (discrimination between CS+ and CS−) of incentives. These observations challenge the observations of BLACKBURN et al. (1987) that pimozide specifically reduces incentive/preparatory responses (visits to the niche where food is expected) in response of a food-predictive CS. McFARLAND and ETTENBERG (1999) attribute this discrepancy to failure of BLACKBURN et al. (1987) to account for the effect of pimozide on basal responding in the absence of the CS+.

Further studies by ETTENBERG and colleagues show a notable difference between the effect of neuroleptics on primary, unconditioned, and on secondary, conditioned, incentives. Thus, haloperidol, at doses that do not affect run times for food (HORVITZ and ETTENBERG 1991) or in response to drug-conditioned CSs (McFARLAND and ETTENBERG 1995), did increase runtime of sexually naive male rats in response to oestrus female cues but not to non-oestrus female cues or to the empty goal box (LOPEZ and ETTENBERG 2001). Moreover, the same or even lower doses of haloperidol prevented the ejaculation-induced decrease in runtime in response to oestrus and non-oestrus female cues (LOPEZ and ETTENBERG 2000).

Therefore, DA-receptor blockade impairs sexual reward as well as approach responses to primary incentives (olfactory sexual stimuli) but fails to block approach responses to secondary, conditioned incentives.

I. Stimulus-Bound Incentive Role of Dopamine?

As we have already pointed out, the within-session impairment of responding induced by neuroleptics in VI schedules of reinforcement led WISE (1982) to envision an incentive role of DA in his revised anhedonia hypothesis. In this hypothesis, however, neuroleptics were still understood to blunt hedonia, except that this effect was not regarded as restricted to primary reinforcers but was extended to conditioned incentive stimuli (WISE 1982). This concept, on the other hand, was in line with incentive theories of the time assuming that incentives acquire all the motivational properties of the rewards to which they have been conditioned, including their hedonic properties (BINDRA 1974). More recently, however, a distinction between hedonic and response-eliciting properties of conditioned stimuli has been made, and DA has been assigned a role in response-eliciting but not in hedonic properties (ROBINSON and BERRIDGE 1993; BERRIDGE 1996; BERRIDGE and ROBINSON 1998). The mechanism by which DA exerts this function has been termed "incentive salience attribution", and has been thought to be part of a two step process: first, a Pavlovian stimulus–reward association resulting in the acquisition of conditional directional properties by the reward-associated stimulus; second, an incentive salience attribution, by which the conditioned stimulus is imbued with response-eliciting (incentive) properties as a result of its conditioned ability of releasing DA. According to BERRIDGE and ROBINSON (1998), an incentive stimulus derives its ability to elicit a response from the property of triggering a burst of spikes in DA neurons and therefore a phasic release of DA in the striatum.

This hypothesis, however, suffers from a number of difficulties. The first is terminological in nature. Thus, the expression "incentive salience attribution" is inadequate and possibly misleading because terms normally referred to attentional (salience) and explicit aspects (attribution) are utilized to indicate a process of behavioural response expression (incentive) regarded by the same authors as implicit/procedural in nature. On the other hand, the assumption, made by ROBINSON and BERRIDGE (1993), that the attribution process is the result of a phasic stimulus-bound activity of DA neurons, is in contrast with available evidence on the time-relationship between stimulus-bound burst activity in DA neurons and movement-related activity in basal ganglia output neurons. These studies show that, by the time presentation of a stimulus results in activation of DA neurons (100ms) (SCHULTZ 1998) and DA starts to elicit its post synaptic effects (>150ms) (GONON et al. 1997), responsive units along the efferent pathway of the basal ganglia have already initiated their discharge sequence that leads to inhibition of output neurons in the substantia nigra (SN) and globus pallidus (GP) by fast γ-aminobutyric acid (GABA) receptors (HIKOSAKA and WURTZ 1983). Therefore, by the time stimulus-bound activity of DA neurons takes place, the stimulus has already been translated down the basal ganglia output. These observations make it unlikely that, as proposed by BERRIDGE and ROBINSON (1998), phasic DA transmission is on-line with action. A direct relationship between release of DA and action is also incompatible with the circumstance that stimuli effective in activating DA neurons are not necessarily action triggers but might serve instead as instruction signals predictive of action-trigger stimuli that may not themselves stimulate DA neurons; eventually, activation of DA neurons follows responding, being elicited by response outcome (reward) (SCHULTZ 1998). Therefore, if indeed release of DA plays a role in the expression of incentive responding, this role cannot be envisioned, as assumed by BERRIDGE and ROBINSON (1998), in series between the stimulus and the response but instead as secondary to stimuli that triggered the initial response.

II. Dopamine and Incentive Arousal

Although not essential in general for the expression of the incentive properties of stimuli, DA might play an incentive role under special circumstances related to the experimental conditions of specific behavioural paradigms. One such condition might be schedule-induced adjunctive behaviour. In this paradigm, cumulative arousal (KILLEEN et al. 1978b) related to expectancy of a food pellet, insufficient per se to reduce food drive, induced by an intermittent (1–4min) schedule of food presentation, results in a steady increase of DA throughout the whole striatum (CHURCH et al. 1987; MCCULLOUGH and SALAMONE 1992). This tonic increase of DA transmission might be instrumental for adjunctive behaviour to take place. A similar mechanism might be operative in VI and in CRF schedules. In both these conditions, build-up of DA in the extracellular fluid induces a state of incentive arousal. Under CRF schedules, blockade of DA transmission would not impair responding at the beginning of the session but after a certain delay, consistent with the within-session effect of neuroleptics on instrumental behaviour. Accordingly, DA would be the substrate of an arousal state (incentive arousal) that non-specifically increases the response-eliciting properties of incentives. Motivational stimuli have DA-independent incentive properties that are amplified under arousal states as a result of heightened DA transmission. This one is a major difference between our hypothesis and that of BERRIDGE and ROBINSON (1998) who regard DA as the critical substrate for any incentive property of stimuli; this view, however, is untenable in the light of the observations of ETTENBERG and associates (see above). Another major difference between our hypothesis and that of BERRIDGE and ROBINSON is the notion that the incentive role of DA is not linked to its phasic, stimulus-bound release (BERRIDGE and ROBINSON 1998) but to a state (incentive arousal) elicited by a prolonged, most likely tonic, increase of DA in the extracellular compartment of terminal DA areas.

The notion of incentive arousal here described is much like the incentive state of some early incentive theorists, particularly COFER (1972) and KILLEEN (1975). In turn, the role here attributed to DA has many similarities with that envisioned by WISE (1982, p 52) in his revised anhedonia hypothesis (see quotations in the Introduction to this chapter). Two aspects, however, weaken the incentive connotation of the hypothesis by WISE (1982): the first, as pointed out by SALAMONE (1992), is the coexistence of two principles theoretically mutually exclusive, the reinforcement principle, derived from the original anhedonia hypothesis, linking DA to response reinforcement, and the incentive principle, linking DA to stimulus reinforcement; the second aspect is that even in the revised anhedonia hypothesis, the main function of DA remains that of mediating hedonia, consistent with the notion that incentives acquire also the hedonic properties of the rewards they are conditioned to (BINDRA 1974, 1978).

III. Incentive Role of Drug-Stimulated Dopamine Transmission

While a role of DA in the incentive properties of stimuli can be demonstrated only under specific conditions, the specific incentive properties of psychostimulants can be easily shown. Indeed, the notion of an incentive role of endogenous DA is largely derived from the role attributed to DA as the substrate of the effect of psychostimulants on reinforcement and instrumental responding (ROBINSON and BERRIDGE 1993; DI CHIARA 1995). It is far from our intention to review here the immense literature on the behavioural properties of psychostimulants. Suffice it to say that psychostimulants elicit typical unconditional incentive effects in the form of approach towards stimuli and exploratory behaviour related to novelty of the context (ROBINSON and BERRIDGE 1993; DI CHIARA 1995). Recently, a facilitation of transfer from Pavlovian to instrumental responding after intra-shell infusions of amphetamine has been reported, but it is not known if this effect takes place also after systemic administration (WYVELL and BERRIDGE 2000).

Psychostimulants also facilitate conditioned reinforcement (the ability of a Pavlovian CS to elicit responding instrumental to its presentation), an effect that, given its origin from Pavlovian learning, can be considered as based on the incentive properties of the stimulus.

Here, however, a clear-cut distinction should be made: neuroleptics differentially affect drug-induced and basal activity of responding. Thus, in secondary reinforcement, neuroleptics prevent the stimulant effect of psychostimulants while leaving intact basal responding. Similarly, neuroleptics block amphetamine locomotion at doses that do not reduce the expression of place-preference or the behavioural activating effects of CSs, taken as examples of the directional and activational properties of incentives, respectively.

This lack of symmetry between the effects of neuroleptics on the incentive effects of psychostimulants and on the incentive properties of non-drug stimuli is critical for a correct interpretation of the role of DA in incentive-motivation. Failure to acknowledge this has led to the erroneous extension to conventional reward of a role of DA that seems to apply mostly or exclusively to psychostimulants.

I. Associative Learning Accounts

Associative learning can be distinguished in Pavlovian and instrumental learning (see Sect. B., "Terminology"). A role of DA in instrumental incentive learning has been excluded in view of the observation that systemic neuroleptics do not affect this form of learning (DICKINSON et al. 2000). Recently, SMITH-ROE and KELLEY (2000) reported that intra-accumbens core co-infusion of the D_1 antagonist SCH23390 and the N-methyl-D-aspartate (NMDA) antagonist AP-5 slows the acquisition of bar pressing and of nose poking for food. The same treatment failed to reduce locomotor activity and feeding. On this basis the authors concluded that NMDA and D_1 receptors are involved in appetitive instrumental learning.

I. Pavlovian Incentive Learning

The ability of a stimulus, conditioned to a reward or punisher (US), to elicit a "consummatory" (KONORSKI 1967) conditioned response is not impaired by the administration of DA-receptor blockers during CS-US pairing. Large doses of chlorpromazine given during shock-tone pairing trials did not prevent the ability of the tone to elicit conditioned emotional aversive responses on a subsequent test (BENINGER et al. 1980). Similarly, pimozide failed to impair conditioned prod burying when administered during prod-shock pairings (BENINGER and PHILLIPS 1980). Moreover, neuroleptics do not impair the acquisition of an operant discrimination (TOMBAUGH et al. 1980) and 6-OHDA lesions do not impair learning of brightness discrimination in an electrified U maze (PRICE and FIBIGER 1975). The acquisition of discrimination in an underwater Y maze is impaired by administration of spiroperidol and by 6-OHDA lesions (RANJE and UNGERSTEDT 1977a,b) but this effect has been explained by performance impairment during the learning phase resulting in delay of stimulus–reward association (BENINGER 1983). Haloperidol and pimozide reduce classical conditioning of the rabbit nictitating membrane, but this effect has been explained by a reduction of CS salience rather than by an impairment of CS-US association (HUNT 1956; HARVEY and GORMEZANO 1981). Findings generally consistent with a lack of impairment of Pavlovian association have been reported by Berridge and Robinson (BERRIDGE and ROBINSON 1998) in a conditioned taste aversion learning paradigm utilizing taste reactivity as a means to estimate the affective properties of the taste stimulus. In this paradigm, a novel taste (sucrose, saccharin, chocolate, etc.) is associated with visceral malaise produced by intraperitoneal lithium. 6-OHDA lesions that reduced by more than 98% DA in the neostriatum and by 85%–99% DA in the NAc did not impair the acquisition of aversive taste reactions to intraoral sucrose previously paired with intraperitoneal lithium-induced malaise.

Recent studies utilizing acute blockade of DA transmission by DA-receptor antagonists rather than chronic lesions, which might result in compensatory changes, have provided evidence for a role of DA D_1 receptors in conditioned taste aversion learning. In contrast to classical Pavlovian learning that tolerates only short delays (2s) between CS/US presentation, in conditioned taste aversion the US (lithium) can be administered up to 3h after the to-be-conditioned taste, consistent with the function of this associative mechanism, which relates to avoidance of foods with harmful post-ingestive effects. Due to this delayed association with the US, a representation of the CS has to be stored in short-term memory for the time necessary to be efficiently associated with the US. Systemic administration of the D_1 receptor antagonist SCH23390 5min after exposure to the CS (sucrose or saccharin) results in reduction of conditioned taste aversion (CTA) on a subsequent test performed in the absence of the D_1 antagonist (FENU et al. 2001). The effect of the D_1 antagonist was time-dependent, since it did not take place if the D_1 antagonists were given 45min instead of 5min after the CS or at various time intervals before it. These characteristics are consistent with the idea that the antagonist is acting at a time critical for the formation and consolidation of the short-term memory trace of the CS. These effects of SCH23390 could be reproduced by local infusion of the more selective D_1 antagonist SCH23390 in the NAc shell and to a lesser extent in the lateral hypothalamus, a DA-rich area that receives direct projections from the shell. No effects were obtained from the NAc core nor from the bed nucleus of stria terminalis. These observations are consistent with a role of DA in the formation and consolidation of a short-term memory trace of the novel gustatory stimuli (FENU et al. 2001). This mechanism might be coupled to release of DA in the NAc shell by novel appetitive stimuli. Thus, appetitive taste stimuli release of DA in the shell and this response undergoes single-trial habituation (BASSAREO and DI CHIARA 1997). DA has been implicated in consolidation into long-term memory of Pavlovian stimulus–reward associations. These studies have been reviewed by ROBBINS and EVERITT (see Chap. 19, this volume) and will not be further discussed here, except for pointing out that these observations are quite different from those obtained in CTA studies (FENU et al. 2001) since they refer to consolidation into long-term memory of the CS-US association rather than into formation and consolidation of a short-term memory trace of the CS.

These negative studies contrast with a number of other studies showing that DA-receptor blockers impair the acquisition of secondary reinforcing properties and of the ability of exerting incentive influences on instrumental behaviour in drug-free tests if administered during CS-US pairings. The earliest reports of these effects are from BENINGER and PHILLIPS. In 1980 these authors (BENINGER and PHILLIPS 1980) first pre-exposed rats to a two-lever operant box, depression of one of which produced a 3-s tone, rats were then conditioned in the absence of the levers to tone–food pairings, and finally they were tested for responding on the tone lever. Pimozide (0.5 or 1.0 mg/kg) was administered in conjunction with the Pavlovian conditioning session (tone-food pairings); in this way the ability of DA-receptor blockade to impair the acquisition of secondary reinforcing properties by the tone was tested in drug-free instrumental sessions. Conditioned reinforcement, as indicated by an increase of responding for the tone in the test session compared to responding on pre-exposure sessions, was obtained in the group conditioned under saline or under 0.5 mg/kg pimozide but not under 1.0 mg/kg pimozide. The possibility that failure to increase responding on test was due to difficulty to retrieve the tone–food association due to learning under the pimozide state (state-dependent learning) was excluded by the fact administration of pimozide on test reduced responding to a greater extent on the inactive lever than on the active one, indicating secondary reinforcement; No such difference, on the other hand, was observed when pimozide was administered both on test and in the conditioning phase, thus excluding that the effect of pimozide 1.0 mg/kg was due to state-dependent learning. A more difficult possibility to rule out is that pimozide impaired conditioning by impairing feeding and therefore degrading the tone-food contingency. Indeed, delay of reward is known to impair Pavlovian stimulus–reinforcement (JENKINS 1950; BERSH 1951). In fact, while undrugged controls did eat the pellets within 3 s of delivery on 99% of the occasions, this figure decreased to 80% in the case of pimozide-administered animals. This effect of pimozide, however, is consistent with a deficit of movement initiation, quite common in neuroleptic-treated rats (FIBIGER et al. 1975). The authors, while acknowledging those difficulties, excluded, however, an influence of this effect of pimozide on the efficiency of conditioning on the basis of the observation that presentation of food pellets non-contingently upon tones (random) did not impair conditioning. Thus, the high rate of tones and pellets presentation (one of each every 45 s) did provide a sufficient degree of causal pairing to ensure conditioning. Alternatively, presentation of the tone under the state induced by the scheduled exposure to the food (one pellet every 5 s) could have provided efficient conditioning. This second possibility is particularly notable since this effect of pimozide might be an example of interaction between neuroleptics and a schedule-induced state. Under this state, build-up of DA release might be instrumental for acquisition and expression of motivated behaviour.

In a further study, HOFFMAN and BENINGER (1985) addressed the specificity of the effect of pimozide on the acquisition of secondary reinforcement. Thus, it was hypothesized that the effect of pimozide was due to an action on the strength of conditioning. This issue, in turn, tapped into the role of performance impairment on the efficiency of conditioning under pimozide. Thus, a range of doses of pimozide (0.5, 1.0, 2.0 and 4.0 mg/kg) was tested for its effects on 2-day and 4-day conditioning. Groups of rats administered 1 h after each conditioning session with pimozide in their home cage were run to control for cumulative drug effects unrelated to an action on conditioning. The results showed a reciprocal interaction between duration of conditioning and dose of pimozide: the longer the conditioning the higher the dose of pimozide needed to impair its efficiency. After 2.0 mg/kg pimozide and 2 days of conditioning, home-cage controls did show secondary reinforcement in spite of the fact that their feeding latencies were in the same range as those of the group given pimozide during conditioning which however failed to show secondary reinforcement on test. In the 4-day conditioning group, no significant differences in latency of feeding between saline and pimozide groups were observed. This study, therefore, seems to exclude a performance deficit during conditioning as the mechanism of the effect of pimozide on secondary reinforcement and also provide an explanation for the failure of previous studies (TOMBAUGH et al. 1983) to show an impairment of acquisition of discriminated responding and on food-conditioned place-preference by pimozide administration.

The same approach utilized in the above studies was applied by BENINGER and PHILLIPS (1981) to study the role of DA in the acquisition of the transfer of classical conditioning to an operant discrimination. This phenomenon, also termed transfer from Pavlovian to instrumental responding (PIT), consists in the ability of response non-contingent presentation of a CS to specifically facilitate responding instrumental to the presentation of the US to which the CS has been previously conditioned by Pavlovian association (ESTES 1943; BOWER and GRUSEC 1964; TRAPOLD et al. 1968; MELLGREN and OST 1969; LOVIBOND 1983). In the transfer study by BENINGER and PHILLIPS (1981), differently from the previous one (BENINGER and PHILLIPS 1980), operant boxes were equipped with only one lever. In those conditions, non-contingent presentation of the food-conditioned tone increased the rate of acquisition of operant discrimination during the test, and this effect was significantly impaired in the group given pimozide during Pavlovian pairing. The effect was significant in the first three sessions considered (sessions 2 to 4), marginal ($p < 0.056$) in the second three sessions (5 to 7) and non-significant in the third three sessions (8 to 10). No differences in the latency to eat the pellets during tone–food pairing were observed. State-dependency was excluded on the basis of the observation of the previous study (BENINGER and PHILLIPS 1981).

These observations have been confirmed by a recent study by DICKINSON et al. (2000) of the effect of pimozide (0.25 mg/kg) and *cis*-flupenthixol (0.5 mg/kg), given during Pavlovian pairing of a CS+ with food or sucrose, on the ability of the same CS+ to increase responding for the relative US over the rate obtained under presentation of a CS–. Both pimozide and *cis*-flupenthixol reduced PIT when given during Pavlovian training. Although the drugs did not affect the rate of magazine entries during conditioning, thus excluding a role of non-specific impairment of conditioning due to a performance effect, they did reduce responding when given during the instrumental sessions, thus precluding the possibility of excluding a state-dependent effect; this effect, however, seems unlikely given the observation of BENINGER and PHILLIPS (1980) on the acquisition of secondary reinforcement. This study, on the other hand, adds important evidence on the CS+ specificity of the effect of pimozide, ruling out attentional mechanisms mediated by salient stimuli non-associated to the US (CS–).

The conclusion of these series of studies is that impairment of DA transmission by neuroleptics during Pavlovian conditioning of an arbitrary stimulus impairs the incentive effects of the stimulus on instrumental responding and its ability to acquire conditional reinforcing properties.

These observations might provide an explanation for much of the effects of neuroleptics on instrumental behaviour. In relation to this, it is important to point out the critical difference between the role of DA in the acquisition of incentive properties of stimuli and its eventual role in expression. Thus, the role attributed to DA by all incentive/activational theories thus far posited, from the first one by WISE (1982) (behavioural arousing) to that of BLACKBURN et al. (1992) (preparatory), SALAMONE et al. (1997) (energizing), BERRIDGE and ROBINSON (1998) (incentive salience attribution) and ROBBINS et al. (1989b) (gain-amplifying) has been always referred to as an action on the expression phase of responding. It was for BENINGER and PHILLIPS (1980 and 1981) and BENINGER (1983) to posit a role of DA in the acquisition of incentive properties of conditioned stimuli during Pavlovian learning. This concept, however, has been confused and eventually weakened in later accounts by the failure to distinguish between learning (i.e. acquisition) and expression of incentive properties of stimuli and by the related practice of referring to an action on incentive learning virtually any effect of DA agonists and antagonists on instrumental behaviour (BENINGER and MILLER 1998; SUTTON and BENINGER 1999). Thus, SUTTON and BENINGER (1999) apply the term incentive learning to any approach response to reward-associated stimuli (SUTTON and BENINGER 1999, p 95). Even DICKINSON et al. (2000) do not seem to be immune from this tendency when dealing with the effect of neuroleptics on Pavlovian incentive learning. Here we will assume that, without further connotation, Pavlovian incentive learning refers specifically to *acquisition* of incentive influences of stimuli. Accordingly, in the study by DICKINSON et al. (2000), the reduction of instrumental performance by administration of pimozide and *cis*-flupenthixol on test should not be taken as indicative of an impairment of Pavlovian incentive learning but rather of an impairment of its expression, i.e. of the expression of the incentive properties of the CS. This, in turn, is the weakest aspect of the effect of neuroleptics on PIT, since one cannot exclude a role of a performance effect at doses of pimozide that induce a within-session impairment of responding independent from the interval between the drug and the test of instrumental performance (DICKINSON et al. 2000). In fact, this observation simply indicates that testing takes place during steady-state levels of fractional occupation of DA receptors by the drug but tells us nothing about the nature of the impairment of responding.

II. Place-Conditioning Studies

Evidence for a role of DA in Pavlovian incentive learning comes from place-conditioning studies. This paradigm (see CARR et al. 1989; HOFFMAN 1989; CALCAGNETTI and SCHECHTER 1994; TZSCHENTKE 1998 for reviews) involves pairing of a specific context with a reward or a punisher (US) and testing the appetitive or aversive properties of the place (CS) under extinction. As pairing is not contingent upon a response, this learning is Pavlovian in nature; however, the CR is, unlike the response to the US (UR), an approach response towards the context paired with the reward (place-preference) or away from the context paired with the punisher (place-aversion). Therefore in place-preference, the conditioned response is an incentive response to a distal CS much like the preparatory CR of KONORSKI (1967). We maintain that place conditioning can be understood as a Pavlovian incentive response and that, therefore, its acquisition involves Pavlovian incentive learning. For this reason, place conditioning is well suited as a paradigm for the study of the role of DA in the acquisition and expression of Pavlovian incentive responding. The information obtained from place-conditioning studies is therefore similar to that obtained from studies on the transfer from Pavlovian to instrumental responding, except that the conditioned approach or avoidance response to a Pavlovian stimulus (context) rather than the facilitation of instrumental responding by the non-contingent presentation of a Pavlovian stimulus is considered.

An advantage of such a paradigm is that a performance effect of DA-receptor blockers on the expression of the conditioned response can be excluded by administering the drug only during acquisition. This arrangement does not exclude the possibility that the effects of the drug are due to failure to retrieve, in the absence of the drug state, the learned association formed under the drug state (state dependency). This, however, can be controlled by the administration of the drug both in the acquisition and in the expression phase.

Place conditioning has been widely utilized to investigate the role of DA in the action of drugs and by non-drug stimuli. DA-receptor antagonists effectively impair place conditioning elicited by appetitive stimuli when given during conditioning. Thus, SPYRAKI et al. (1982) reported that haloperidol (0.1 and 0.2 mg/kg) given during conditioning to hungry rats blocked the establishment of preference for the food-paired compartment. At variance with these observations, TOMBAUGH et al. (1982) reported that pimozide (1.0 mg/kg) failed to impair acquisition of incentive properties by a light or by a distinct compartment paired with food. A procedural difference between these studies is that in the study by Tombaugh and colleagues' (1982) rats were food deprived on test, while they were fed ad libitum in the study of Spyraki et al. (1982).

It is possible that in the study of Tombaugh et al. (1982), a deprivation state had enhanced the incentive properties of the food-paired environment to a degree sufficient to overcome any impairment of Pavlovian incentive learning during acquisition. Impairment of the acquisition of place preference by DA-receptor blockade could be due to reward devaluation or to impairment of Pavlovian association. The study of Agmo (1995) shows that cis-flupenthixol blocks the acquisition of preference for a compartment paired to drinking of 18% sucrose solution without reducing sucrose consumption. These results were interpreted to indicate that DA is essential for Pavlovian incentive learning but not for the impact of reward. Further studies show that raclopride, while not impairing lordosis behaviour in female hamsters during sexual activity, prevents the establishment of preference for the place where sexual activity took place. If lordosis behaviour is taken as a measure of the hedonic impact of sexual activity, it appears that raclopride impairs Pavlovian incentive learning without reducing the rewarding impact of sexual stimulation.

Similar conclusions were reached in studies of place-preference conditioned water drinking (Agmo et al. 1993). In this case both SCH23390 (a D_1 receptor blocker) and raclopride (a D_2 receptor blocker) were able to impair place preference acquisition at doses that did not impair water drinking. Finally, SCH23390 impaired at very low doses (0.01 and 0.03 mg/kg) the acquisition of place-preference conditioned by novel objects while did not impair the interaction with novel objects (Besheer et al. 1999). Under certain conditions, D_2-specific neuroleptics, while ineffective per se, are able to facilitate place preference induced by food. These neuroleptics are sulpiride, pimozide and amisulpride while chlorpromazine, haloperidol and metoclopramide were ineffective (Guyon et al. 1993). These results can be explained by assuming that DA can inhibit its own activity via D_2-like DA receptors. Consistent with this, SCH23390 prevented this effect. In this study, amisulpride, given on test attenuated the effect of the same drug given during conditioning. Guyon et al. (1993) interpreted this observation as indicating that the impairment of associative learning was in part related to state-dependency. However, a more likely explanation is that amisulpride, given on test, impairs the expression of preference by impairment of performance. A further example of the property of neuroleptics to impair the acquisition of incentive properties by stimuli paired with rewards is the observation that haloperidol (0.3 mg/kg) given during non-contingent electrical stimulation of the lateral hypothalamus prevented the establishment of conditioned preference for the compartment paired to the hypothalamic stimulation (Ettenberg and Duvauchelle 1988). It is notable that in this study, hypothalamic stimulation was not contingent upon a subject response but was instead administered by the experiments.

The relative paucity of the studies that have utilized the place-conditioning paradigm for investigating the role of DA in the incentive properties of natural stimuli contrasts with the abundance of studies that have utilized this paradigm for investigating the incentive properties of drugs. These studies, with few exceptions, show that neuroleptics and DA D_1 antagonists impair the acquisition of drug-conditioned place preference (see HOFFMAN 1989; ROTHMAN et al. 1989; CALCAGNETTI and SCHECHTER 1994; TZSCHENTKE 1998). This property has been taken by BENINGER and associates (BENINGER 1991; BENINGER and MILLER 1998; SUTTON and BENINGER 1999) as evidence for a role of DA in incentive learning. However, as most if not all drugs inducing place-preference also increase extracellular DA in the NAc shell (DI CHIARA 1999), one cannot exclude that in this case DA antagonists act by directly blunting reward rather than by impairing context-reward association. This possibility applies in particular to psychostimulants that depend from the ability to increase DA in the NAc for most of their unconditioned effects, including the rewarding ones. For this reason, any impairment of the acquisition of drug-induced place preference by DA-receptor antagonists cannot be taken as evidence for a role of DA in Pavlovian incentive learning. An exception to this, however, is provided by aversive drugs such as naloxone, lithium and picrotoxin for which an increase of DA in the NAc has not been observed (BASSAREO et al. 1996). These drugs elicit place aversion that is blocked by the administration of the D_1 receptor blockers SCH23390 and SCH39166 given during pairing with a specific compartment (ACQUAS et al. 1989; ACQUAS and DI CHIARA 1994). This effect cannot be explained by a role of DA in the aversive impact of the drug, since it is highly unlikely that these drugs elicit aversion by releasing DA. A more likely explanation is, therefore, that this effect is the result of an impairment of Pavlovian incentive learning.

Similar considerations can be made for the finding that haloperidol impaired the place aversion induced by a benzodiazepine inverse agonist (FG7142) known to induce anxiety but not convulsions in naive rats (DI SCALA and SANDNER 1989). Moreover, SHIPPENBERG and HERZ (1987) reported that SCH23390 blocks the establishment of place aversion to a κ-opioid agonist. In relation to these studies, it is notable that SCH23390, given in low doses, (12.5–25 mg/kg s.c.) induced place-aversion for the compartment to which it had been paired (ACQUAS and DI CHIARA 1994). This observation might seem incompatible with the idea that blockade of D_1 receptors impairs Pavlovian incentive learning. However, a higher dose of SCH39166 (50 μg/kg s.c.) paired with both compartments prevented the establishment of place aversion induced by a dose of 12.5 μg/kg of the same drug (ACQUAS and DI CHIARA 1994). Thus, lower doses of SCH39166 are needed to induce an aversive state than to impair Pavlovian incentive learning. This conclusion is consistent with the observation that low doses of SCH39166 (12.5–25 μg/kg s.c.) are sufficient to impair conditioning to amphetamine while higher doses (50–100 μg/kg) are needed to impair place preference to morphine and place-aversion to lithium (ACQUAS and DI CHIARA 1994). Again, lower doses were needed to block DA-dependent reward (amphetamine) than Pavlovian incentive learning (morphine and lithium).

J. An Interpretative Framework of the Role of Dopamine in Reward

If one considers that most hypotheses on the a role of DA in reward have been built on the basis of the effects of DA-receptor blockers on rewarded behaviour, it might seem contradictory that, in spite of the effort devoted to this issue in the past 25 years, the precise mechanism of the effect of these drugs on reward remains elusive. The main reason for this is probably that DA-receptor blockers can act at various stages and on different aspects of the process by which organisms respond to stimuli (including rewarding ones) in a motivationally meaningful fashion.

A general consideration that can be drawn from studies on the effect of DA-receptor blockers on behaviour, and in particular on instrumental behaviour, is that the effect obtained is critically dependent upon the paradigm utilized. Thus, the fact that in a given paradigm neuroleptics affect responding by one mechanism does not exclude that they can affect it by a different mechanism in another paradigm. Operant paradigms are quite complex, and because of this the interpretation of the effect of neuroleptics on instrumental responding is particularly demanding. However, bar pressing, chain pulling and other unnatural behaviours, as opposed to such species-specific behaviours as maze-running or nose-poking, are thought to more likely fulfil the criteria of goal-directed action and act-outcome relationships that distinguish instrumental from Pavlovian responding (DICKINSON and BALLEINE 1994).

Among the various steps and phases into which the effect of neuroleptics on instrumental responding can be dissected, it is useful to distinguish acquisition from expression phases and Pavlovian/incentive-motivational from instrumental mechanisms. Instrumental responding, however, heavily depends on Pavlovian stimuli that exert incentive-motivational influences on its expression (DICKINSON 1994; RESCORLA 1994).

Concerning acquisition, associative learning is operative in Pavlovian and incentive responding, in the form of stimulus–reward associations, and in instrumental responding in the form of stimulus–response and act-outcome associations (see Sect. B., "Terminology"). However, the difference between the associative mechanisms involved in Pavlovian as compared to instrumental responding should not be overlooked. Thus, Pavlovian associations are long lasting and rigid, while instrumental associations are flexible and reversible in order to adapt response to the changing needs of the outside world. Thus, response reinforcement, according to behaviourist accounts, is dynamically controlled by response outcome through its influence on the strength of the S–R associations (WATSON 1913; HULL 1943). Thus, when describing the effect of neuroleptics on reinforcement, it is necessary to specify which kind of reinforcement is meant by this term, i.e. response- or stimulus–reinforcement and, in the case of response-reinforcement, if one refers to incentive or to habit learning.

The first hypothesis on the role of DA in reward to be considered here is the original anhedonia hypothesis (WISE 1982). A role of DA in the hedonic properties of sweet reward, although consistent with various observations, does not explain the whole evidence available (see Sect. D.I., "Sweet Reward"). Thus, a role of DA in the incentive, as distinguished from the hedonic properties of sweet reward, has been proposed as a way out from the inadequacies of a purely hedonic interpretation (BERRIDGE 1996). Pharmacological evidence, however, suggests that DA could mediate euphoria (DREVETS et al. 2001) and a role of DA in euphoria is the tenet of current hypotheses on the role of DA in normal and abnormal mood states (euthymia, dysthymia, depression, mania) (PAPP et al. 1991). These observations might be interpreted to mean that stimulus-hedonia (i.e. taste hedonia) is DA independent while state-hedonia (i.e. euphoria) is DA dependent.

A more operational version of the anhedonia hypothesis is that of a role of DA in response reinforcement. For this, however, a major difficulty is constituted by the inextricable relationship of the response-reinforcement construct with performance, and the fact that DA plays an important role in extrapyramidal motor functions. On the basis of the within-session character of the effect of neuroleptics on responding, WISE (1982) did exclude a primary role of motor impairment in this effect. However, some aspects of the effect of neuroleptics on instrumental responding are readily explained by a motor impairment. This is the case of atypical neuroleptics. These drugs have a reduced liability for inducing Parkinson's disease-like symptoms but retain the antipsychotic potential of classic neuroleptics and are apparently unable to induce the within-session decrease of responding typical of classic neuroleptics (DAS and FOWLER 1995, 1996; SANGER and PERRAULT 1995). On the other hand, antimuscarinic drugs abolish the within-session effect of classic neuroleptics (FOWLER and DAS 1994). These observations would favour an interpretation of the effect of neuroleptics on instrumental responding in terms of a performance effect and in addition cast much doubts on the relevance of these effects for the antipsychotic action of neuroleptics. However, strong evidence for a role of DA in reinforcement and/or motivation comes from reward-summation and response-reinforcement matching studies. Thus, SCH23390, at doses that impair reinforcement, produces little impairment of performance (HUNT et al. 1994). This observation in turn is consistent with the reduced ability of SCH23390 to induce bradykinesia (increase in movement duration) and other signs of motor slowing when compared with raclopride, a D_2 antagonist (FOWLER and LIOU 1998). In matching law experiments, it has been observed that DA-receptor blockers reduce maximal response rate (K_s) (an index of performance impairment) from the beginning of the session but increase reinforcement needed for maintaining half-maximal response rate (K_h) (an index of reduced reinforcement impact/motivational strength) only late in the session (PHILLIPS et al. 1991d; WILLNER et al. 1990b). In the case of SCH23390 and sulpiride, reduction in motivational strength takes place late in the session at doses that do not affect performance (PHILLIPS et al. 1991a). This observation is particularly relevant since SCH23390 and sulpiride have been reported to be unable to elicit within-session reductions of responding in conventional operant schedules (SANGER 1987; SANGER and PERRAULT 1995). In view of this, the observation that atypical neuroleptics fail to induce within-session reduction of responding on conventional schedules (SANGER and PERRAULT 1995) does not exclude that they induce a within-session reduction of motivation on multiple schedules. Matching law studies with atypical neuroleptics on multiple schedules is needed to test this possibility.

One way to overcome the confounding influence of motor impairment in studies of the effect of DA receptor blockers on behaviour has consisted of testing for the action of these drugs in their absence (BENINGER 1989; ETTENBERG 1989). This idea is in principle ingenious but in practice does not necessarily allow testing of the role of DA in instrumental responding. In fact, a basic requirement of this experimental approach is the temporal separation between application of the drug and testing of its effects on operant responding. This wide temporal dissociation is extraneous to instrumental associations that, as pointed out above, are on-line with action. As a result of this, such a principle can only be applied to Pavlovian associations, i.e. to stimulus contingencies rather than to response contingencies. It is also debatable to what extent reinforcement by maze running, largely utilized in these studies, is homologous to reinforcement by bar pressing. Thus, it has been suggested that in maze running experiments, stimulus-reward rather than stimulus-response reinforcement is at work (DICKINSON and BALLEINE 1994). Thus, in the experiments of ETTENBERG et al. the effect of neuroleptics on test should be related to an impairment of Pavlovian rather than instrumental incentive learning (DI CHIARA 1999). If this interpretation is correct, the observations of ETTENBERG et al. cannot be utilized as evidence for a role of DA in instrumental reinforcement but rather in Pavlovian incentive learning (DI CHIARA 1999).

The difficulty in distinguishing non-specific performance effects from specific effects on response reinforcement after neuroleptics has led some authors to interpret the role of DA in terms of a complex sensory-motor function (SALAMONE 1992). This account, however, is not clearly distinguishable from one involving a motor-deficit mechanism. Thus, the effects of DA-receptor blockers on response selection observed by SALAMONE et al. (1997) can be explained in terms of a motor bias that redirects choice among two rewards as a result of increased response cost of the larger reward. These observations, however, could also be taken to suggest an "energizing" role of DA on responding (SALAMONE et al. 1997). However, impairment of DA transmission rather than reducing, actually increases the force necessary for maintaining response performance (FOWLER and KIRKPATRICK 1989). Therefore, if anything, such energizing role would be not dissimilar from an incentive role of DA. Evidence provided by ETTENBERG and colleagues (HORVITZ and ETTENBERG

1991; McFarland and Ettenberg 1995; McFarland and Ettenberg 1999; Lopez and Ettenberg 2001), however, tends to exclude that DA is essential for the expression of the response-eliciting properties of conditioned incentive stimuli. It is possible, however, that two kinds of incentive influences by Pavlovian stimuli on instrumental responding can be distinguished: a stimulus-bound, DA-dependent influence and a state-like, DA-dependent one. This latter one builds up with time and becomes relevant only within the session rather that at its beginning, being related to the accumulation of DA in the extracellular fluid and onto its receptor. Blockade of such an influence mechanism can account for the within-session reduction of responding induced by neuroleptics.

However, as maintenance of response reinforcement in already-trained subjects involves learning of act-outcome relationships (instrumental incentive learning) or of S–R (habit) learning, an impairment of response-reinforcement learning is also compatible with the above evidence.

Indeed, a role of DA in response-reinforcement learning is at the root of current computational models of instrumental responding (Montague et al. 1996; Schultz et al. 1997). According to these models, DA released in response to an unexpected reward would strengthen synaptic connections in neural chains that promote actions whose consequence is reinforcement (i.e. presentation of the reward); as this goal is accomplished and reward presentation becomes reliably predicted, the reward itself progressively loses the ability to stimulate DA neurons and this property is acquired by unpredictable stimuli that reliably predict the occurrence of the reward (Schultz 1998). Under these conditions, reward omission results in phasic inhibition of DA release that, if repeated over time, would result in extinction of the instrumental response.

This simple interpretative scheme has been criticized by Redgrave et al. (1999) and others on the ground that if indeed DA carries a reinforcement signal, it remains unexplained why the property of activating DA transfers to a stimulus that precedes response outcome. However, instrumental responding is controlled by consequences rather than by premises; therefore, the fact that release of DA precedes reinforcement does not mean that release of DA cannot control responding; it can, except that this control takes place post-hoc, i.e. after response emission and in relation to its outcome. Therefore, if responding fails to produce the expected outcome, release of DA is phasically depressed in conjunction with that failure and this depression would be instrumental for extinction of learned habits (Schultz 1998). At the same time, release of DA by the reward-predictive stimulus will progressively lose its ability to activate DA neurons and to release DA. These characteristics are in turn consistent with the possibility that activation of DA transmission by secondary reward-predictive stimuli serves to recruit new instrumental responses by which reward can be reliably obtained. One problem with the above model, however, is that it can be reduced to a role of DA in response-reinforcement and therefore brings us back to the debated issue of the role of DA in response reinforcement. Thus, when providing evidence for a role of DA in response reinforcement, we are again confronted with the confounding influence of the effect of neuroleptics on motor performance. As already discussed, studies dissociating performance effects of neuroleptics from reinforcement effects have utilized approach responses in mazes and therefore cannot be taken as rigorous evidence for a role of DA in instrumental reinforcement (DICKINSON and BALLEINE 1994). These paradigms, however, can be also viewed as models of Pavlovian incentive learning. Thus, an alternative interpretation of the role of DA in instrumental responding is one that still implicates incentive stimuli, except that it would not take place at the level of the expression of incentive influences but of their acquisition, i.e. on Pavlovian incentive learning. BENINGER (1983) defined incentive learning as "the acquisition by environmental stimuli of the ability to elicit responding" (p 178) and regarded it as distinct from Pavlovian learning, defined as learning of "the association of environmental stimuli with the stimulus aspects of reinforcement"; however, since in both forms of learning the stimulus aspect of reinforcement is expressed by a behavioural response, the difference between Pavlovian learning and incentive learning is unclear. A further occasion for confusion has arisen from the fact that the same author, in later reviews (BENINGER and MILLER 1998) labelled any impairment of instrumental responding by neuroleptics as impairment of incentive learning, thus leading the reader to assume that the kind of incentive learning under consideration was indeed instrumental incentive learning. Indeed, the definition formulated by BENINGER (1983) fails to capture the real essence of incentive learning and its difference from other forms of learning, including instrumental learning and classical conditioning. Clearly, the kind of incentive learning dealt with by BENINGER is learning of stimulus contingencies (BENINGER and PHILLIPS 1980, 1981), i.e. learning according to Pavlovian rules. The differences between classical Pavlovian conditioning and Pavlovian incentive learning are eventually appreciated in their consequences for expression; thus, classical conditioning is expressed in consummatory responses while Pavlovian incentive learning is expressed in preparatory/appetitive responses (see Sect. B., "Terminology"). This distinction in turn corresponds to the Konorskian analysis of Pavlovian conditioned responses (KONORSKI 1967).

As we have seen in the specific analysis of the literature, the evidence that neuroleptics impair Pavlovian incentive learning survives the ability of neuroleptics to induce non-specific performance impairment during conditioning or state-dependent learning. In turn, impairment of Pavlovian incentive learning can provide an explanation for the observations of ETTENBERG and colleagues (see Sect. G., "Dissociating Reinforcement from Incentive-Motivation and Performance").

From a more general point of view, impairment of Pavlovian influences on instrumental responding by an action on Pavlovian incentive learning might also provide an interpretative key to the delayed (within-session) effects of neuroleptics on instrumental responding (see above). Immediate effects, however, are not accounted for by this hypothesis but instead by a more general action on performance consistently with the early suggestion of PHILLIPS and FIBIGER (1979) or by a general arousing effect of incentives related to release of DA in the forebrain (PFCX, NAc core) and related to an incentive arousal function of DA (see Sect. H,II., "Dopamine and Incentive Arousal").

Acknowledgements. I wish to thank Dr. Elio Acquas for establishing the Reference Manager Database and Ms. Adelaide Ciuti for typing the manuscript from my unreadable handwriting. I also thank my collaborators whose fresh enthusiasm and genuine interest has been a constant source of invention and energy for research.

The studies from our laboratory reported in this review have been funded by Ministero dell'Università e della Ricerca Scientifica e Tecnologica (COFIN and Centro di Eccellenza per gli Studi sulle Dipendenze), by Consiglio Nazionale delle Ricerche (Centro per la Neurofarmacologia), by Università degli Studi di Cagliari and by Regione Autonoma della Sardegna, Assessorato alla Sanità.

References

Acquas E, Carboni E, Leone P, Di Chiara G (1989) SCH 23390 blocks drug-conditioned place-preference and place-aversion: anhedonia (lack of reward) or apathy (lack of motivation) after dopamine-receptor blockade? Psychopharmacology (Berl) 99:151–155

Acquas E, Di Chiara G (1994) D1 receptor blockade stereospecifically impairs the acquisition of drug-conditioned place preference and place aversion. Behav Pharmacol 5:555–569

Agmo A, Federman I, Navarro V, Padua M, Velazquez G (1993) Reward and reinforcement produced by drinking water: role of opioids and dopamine receptor subtypes. Pharmacol Biochem Behav 46:183–194

Agmo A, Galvan A, Talamantes B (1995) Reward and reinforcement produced by drinking sucrose: two processes that may depend on different neurotransmitters. Pharmacol Biochem Behav 52:403–414

Asin KE, Fibiger HC (1984) Force requirements in lever-pressing and responding after haloperidol. Pharmacol Biochem Behav 20:323–326

Bailey CS, Hsiao S, King JE (1986) Hedonic reactivity to sucrose in rats:modification by pimozide. Physiol Behav 38:447–452

Bassareo V, Di Chiara G (1997) Differential influence of associative and nonassociative learning mechanisms on the responsiveness of prefrontal and accumbal dopamine transmission to food stimuli in rats fed ad libitum. J Neurosci 17:851–861

Bassareo V, Tanda G, Petromilli P, Giua C, Di Chiara G (1996) Non-psychostimulant drugs of abuse and anxiogenic drugs activate with differential selectivity dopamine transmission in the nucleus accumbens and in the medial prefrontal cortex of the rat [published erratum appears in Psychopharmacology (Berl) 1996 Oct;127(3): 289–90]. Psychopharmacology (Berl) 124:293–299

Beninger RJ (1982) A comparison of the effects of pimozide and nonreinforcement on discriminated operant responding in rats. Pharmacol Biochem Behav 16:667–669

Beninger RJ (1983) The role of dopamine in locomotor activity and learning. Brain Res 287:173–196

Beninger RJ (1989) Dissociating the effects of altered dopaminergic function on performance and learning. Brain Res Bull 23:365–371

Beninger RJ (1991) Receptor subtype-specific dopamine agonists and antagonists and conditioned behavior. In: Willner P, Skeel-Kruger J (eds) The mesolimbic dopamine system: from motivation to action. John Wiley and Sons, New York, pp 273–299

Beninger RJ, Mason ST, Phillips AG, Fibiger HC (1980) The use of extinction to investigate the nature of neuroleptic-induced avoidance deficits. Psychopharmacology (Berl) 69:11–18

Beninger RJ, Miller R (1998) Dopamine D1-like receptors and reward-related incentive learning. Neurosci Biobehav Rev 22:335–345

Beninger RJ, Phillips AG (1980) The effect of pimozide on the establishment of conditioned reinforcement. Psychopharmacology (Berl) 68:147–153

Beninger RJ, Phillips AG (1981) The effects of pimozide during pairing on the transfer of classical conditioning to an operant discrimination. Pharmacol Biochem Behav 14:101–105

Berger B, Tassin JP, Blanc G, Moyne MA, Thierry AM (1974) Histochemical confirmation for dopaminergic innervation of the rat cerebral cortex after destruction of the noradrenergic ascending pathways. Brain Res 81:332–337

Berridge KC (1996) Food reward: brain substrates of wanting and liking. Neurosci Biobehav Rev 20:1–25

Berridge KC (2000) Measuring hedonic impact in animals and infants: microstructure of affective taste reactivity patterns. Neurosci Biobehav Rev 24:173–198

Berridge KC, Robinson TE (1998) What is the role of dopamine in reward: hedonic impact, reward learning, or incentive salience? Brain Res Brain Res Rev 28:309–369

Berridge KC, Venier IL, Robinson TE (1989) Taste reactivity analysis of 6-hydroxydopamine-induced aphagia: implications for arousal and anhedonia hypotheses of dopamine function. Behav Neurosci 103:36–45

Berridge MJ (1989) The Albert Lasker Medical Awards. Inositol trisphosphate, calcium, lithium, and cell signaling. JAMA 262:1834–1841

Bersh PJ (1951) The influence of two variables upon the esthablishment of a secondary reinforcer for operant responding. J Exp Psychol 41:62–73

Besheer J, Jensen HC, Bevins RA (1999) Dopamine antagonism in a novel-object recognition and a novel-object place conditioning preparation with rats. Behav Brain Res 103:35–44

Bignami G (1978) Effects of neuroleptics, ethanol, hypnotic-sedatives, tranquilizers, narcotics and minor stimulants in aversive paradigms. In: Anisman A, Bignami G (eds) Psychopharmacology of aversively motivated behaviour. Plenum Press, New York, pp 385–402

Bindra D (1974) A motivational view of learning, performance, and behavior modification. Psychol Rev 81:199–213

Bindra D (1978) How adaptive behavior is produced: a perceptual-motivation alternative to response reinforcement. Behavioral and Brain Sciences 1:41–91

Bird M, Kornetsky C (1990) Dissociation of the attentional and motivational effects of pimozide on the threshold for rewarding brain stimulation. Neuropsychopharmacology 3:33–40

Blackburn JR, Pfaus JG, Phillips AG (1992) Dopamine functions in appetitive and defensive behaviours. Prog Neurobiol 39:247–279

Blackburn JR, Phillips AG, Fibiger HC (1987) Dopamine and preparatory behavior: I. Effects of pimozide. Behav Neurosci 101:352–360

Bolles RC (1972) Reinforcement, expectancy and learning. Psychol Rev 79:394–409

Boring EG (1945) The use of operational definitions in science. Psychol Rev 52:243–245

Bower G, Grusec T (1964) Effect of prior Pavlovian discrimination training upon learning an operant discrimination. J Exp Anal Behav 7:401–404

Boye SM, Rompre PP (1996) Effect of pimozide on self-stimulation threshold under a continuous and fixed-interval schedule of reinforcement. Behav Brain Res 78:243–245

Brauer LH, de Wit H (1996) Subjective responses to d-amphetamine alone and after pimozide pretreatment in normal, healthy volunteers. Biol Psychiatry 39:26–32

Calcagnetti DJ, Schechter MD (1994) Nicotine place preference using the biased method of conditioning. Prog Neuropsychopharmacol Biol Psychiatry 18:925–933

Carr GD, Fibiger HC, Phillips AG (1989) Conditioned place preference as a measure of drug reward. In: Liebman JM, Cooper SJ (eds) The neuropsychopharmacological basis of reward. Oxford University Press, New York, pp 264–319

Chausmer AL, Ettenberg A (1997) A role for D2, but not D1, dopamine receptors in the response-reinstating effects of food reinforcement. Pharmacol Biochem Behav 57:681–685

Church WH, Justice JB, Jr., Neill DB (1987) Detecting behaviorally relevant changes in extracellular dopamine with microdialysis. Brain Res 412:397–399

Cofer CN (1972) Motivation and Emotion. Glenview, IL Scott, Foresman

Colwill RM, Rescorla RA (1990) Effect of reinforcer devaluation on discriminative control of instrumental behavior. J Exp Psychol Anim Behav Process 16:40–47

Corbett D (1990) Differences in sensitivity to neuroleptic blockade: medial forebrain bundle versus frontal cortex self-stimulation. Behav Brain Res 36:91–96

Crow TJ (1972) Catecholamine-containing neurones and electrical self-stimulation: 1. A review of some data. Psychological Medicine 2:414–421

Das S, Fowler SC (1995) Acute and subchronic effects of clozapine on licking in rats: tolerance to disruptive effects on number of licks, but no tolerance to rhythm slowing. Psychopharmacology (Berl) 120:249–255

Das S, Fowler SC (1996) Similarity of clozapine's and olanzapine's acute effects on rats' lapping behavior. Psychopharmacology (Berl) 123:374–378

deVilliers PA, Herrnstein RJ (1976) Towards a law of response strength. Psychol Bull 83:1131–1153

Dews PB, Morse WH (1961) Behavioral pharmacology. Annual Reviews of Pharmacology 1:145–174

Di Chiara G (1995) The role of dopamine in drug abuse viewed from the perspective of its role in motivation [published erratum appears in Drug Alcohol Depend 1995 Aug;39(2):155]. Drug Alcohol Depend 38:95–137

Di Chiara G (1999) Drug addiction as dopamine-dependent associative learning disorder. Eur J Pharmacol 375:13–30

Di Scala G, Sandner G (1989) Conditioned place aversion produced by FG 7142 is attenuated by haloperidol. Psychopharmacology (Berl) 99:176–180

Dickinson, A (1994) Instrumental conditioning. In: Animal learning and cognition MacKintosh N (eds), San Diego, Academic Press, pp 45–79

Dickinson A, Balleine B (1994) Motivational control of goal-directed action. Animal Learning & Behavior 22:1–18

Dickinson A, Smith J, Mirenowicz J (2000) Dissociation of Pavlovian and instrumental incentive learning under dopamine antagonists. Behav Neurosci 114:468–483

Drevets WC, Gautier C, Price JC, Kupfer DJ, Kinahan PE, Grace AA, Price JL, Mathis CA (2001) Amphetamine-induced dopamine release in human ventral striatum correlates with euphoria. Biol Psychiatry 49:81–96

Eckman G (1967) The measurement of subjective reactions. Forsvarsedicine 33:27–41

Edmonds DE, Gallistel CR (1974) Parametric analysis of brain stimulation reward in the rat: III. Effect of performance variables on the reward summation function. J Comp Physiol Psychol 87:876–883

Esposito RU, Faulkner W, Kornetsky C (1979) Specific modulation of brain stimulation reward by haloperidol. Pharmacol Biochem Behav 10:937–940

Estes WK (1943) Discriminative conditioning. Journal of Experimental Psychology 32:150–155

Ettenberg A (1989) Dopamine, neuroleptics and reinforced behavior. Neurosci Biobehav Rev 13:105–111

Ettenberg A (1990) Haloperidol prevents the reinstatement of amphetamine-rewarded runway responding in rats. Pharmacol Biochem Behav 36:635–638

Ettenberg A, Camp CH (1986b) A partial reinforcement extinction effect in water-reinforced rats intermittently treated with haloperidol. Pharmacol Biochem Behav 25:1231–1235

Ettenberg A, Camp CH (1986a) Haloperidol induces a partial reinforcement extinction effect in rats: implications for a dopamine involvement in food reward. Pharmacol Biochem Behav 25:813–821

Ettenberg A, Duvauchelle CL (1988) Haloperidol blocks the conditioned place preferences induced by rewarding brain stimulation. Behav Neurosci 102:687–691

Ettenberg A, Horvitz JC (1990) Pimozide prevents the response-reinstating effects of water reinforcement in rats. Pharmacol Biochem Behav 37:465–469

Ettenberg A, Koob GF, Bloom FE (1981) Response artifact in the measurement of neuroleptic-induced anhedonia. Science 213:357–359

Ettenberg A, MacConell LA, Geist TD (1996) Effects of haloperidol in a response-reinstatement model of heroin relapse. Psychopharmacology (Berl) 124:205–210

Falk JL (1977) The origin and function of adjunctive behaviour. Animal Learning & Behavior 5:325–335

Faustman WO, Fowler SC (1981) Use of operant response duration to distinguish the effects of haloperidol from nonreward. Pharmacol Biochem Behav 15:327–329

Feigl (1945) Operationism and scientific method. Psychol Rev 250–259

Feldon J, Katz Y, Weiner I (1988) The effects of haloperidol on the partial reinforcement extinction effect (PREE): implications for neuroleptic drug action on reinforcement and nonreinforcement. Psychopharmacology (Berl) 95:528–533

Feldon J, Weiner I (1991) Effects of haloperidol on the multitrial partial reinforcement extinction effect (PREE): evidence for neuroleptic drug action on nonreinforcement but not on reinforcement. Psychopharmacology (Berl) 105:407–414

Fenu S, Bassareo V, Di Chiara G (2001) A role for dopamine d1 receptors of the nucleus accumbens shell in conditioned taste aversion learning. J Neurosci 21:6897–6904

Fibiger HC (1978) Drugs and reinforcement mechanisms: a critical review of the catecholamine theory. Annu Rev Pharmacol Toxicol 18:37–56

Fibiger HC, Carter DA, Phillips AG (1976) Decreased intracranial self-stimulation after neuroleptics or 6-hydroxydopamine: evidence for mediation by motor deficits rather than by reduced reward. Psychopharmacology (Berl) 47:21–27

Fibiger HC, Phillips AG (1986) Reward, motivation, cognition: Psychobiology of mesotelencephalic systems. 647–675

Fibiger HC, Zis AP, Phillips AG (1975) Haloperidol-induced disruption of conditioned avoidance responding: attenuation by prior training or by anticholinergic drugs. Eur J Pharmacol 30:309–314

Fowler SC (1990) Neuroleptics produce within-session response decrements: Facts and theories. Drug Development Research 20:101–116

Fowler SC, Das S (1994) Haloperidol-induced decrements in force and duration of rats' tongue movements during licking are attenuated by concomitant anticholinergic treatment. Pharmacol Biochem Behav 49:813–817

Fowler SC, Kirkpatrick MA (1989) Behavior-decrementing effects of low doses of haloperidol result from disruptions in response force and duration. Behav Pharmacol 1:123–132

Fowler SC, LaCerra MM, Ettenberg A (1986) Effects of haloperidol on the biophysical characteristics of operant responding: implications for motor and reinforcement processes. Pharmacol Biochem Behav 25:791–796

Fowler SC, Liou JR (1994) Microcatalepsy and disruption of forelimb usage during operant behavior: differences between dopamine D1 (SCH-23390) and D2 (raclopride) antagonists. Psychopharmacology (Berl) 115:24–30

Fowler SC, Liou JR (1998) Haloperidol, raclopride, and eticlopride induce microcatalepsy during operant performance in rats, but clozapine and SCH 23390 do not. Psychopharmacology (Berl) 140:81–90

Franklin KB (1978) Catecholamines and self-stimulation: reward and performances effects dissociated. Pharmacol Biochem Behav 9:813–820

Franklin KB, McCoy SN (1979) Pimozide-induced extinction in rats: stimulus control of responding rules out motor deficit. Pharmacol Biochem Behav 11:71–75

Gallagher M, Graham PW, Holland PC (1990) The amygdala central nucleus and appetitive Pavlovian conditioning: lesions impair one class of conditioned behavior. J Neurosci 10:1906–1911

Gallistel CR, Karras D (1984) Pimozide and amphetamine have opposing effects on the reward summation function. Pharmacol Biochem Behav 20:73–77

Geary N, Smith GP (1985) Pimozide decreases the positive reinforcing effect of sham fed sucrose in the rat. Pharmacol Biochem Behav 22:787–790

Gerber GJ, Sing J, Wise RA (1981) Pimozide attenuates lever pressing for water reinforcement in rats. Pharmacol Biochem Behav 14:201–205

Gerhardt S, Liebman JM (1981) Differential effects of drug treatments on nose-poke and bar-press self- stimulation. Pharmacol Biochem Behav 15:767–771

Glickman SE, Schiff BB (1967) A biological theory of reinforcement. Psychol Rev 74:81–109

Gonon F (1997) Prolonged and extrasynaptic excitatory action of dopamine mediated by D1 receptors in the rat striatum in vivo. J Neurosci 17:5972–5978

Gramling SE, Fowler SC (1986) Some effects of pimozide and of shifts in sucrose concentration on lick rate, duration, and interlick interval. Pharmacol Biochem Behav 25:219–222

Gramling SE, Fowler SC, Collins KR (1984) Some effects of pimozide on nondeprived rats licking sucrose solutions in an anhedonia paradigm. Pharmacol Biochem Behav 21:617–624

Gray T, Wise RA (1980) Effects of pimozide on lever pressing behavior maintained on an intermittent reinforcement schedule. Pharmacol Biochem Behav 12:931–935

Grill HJ, Norgren R (1978) The taste reactivity test. I. Mimetic responses to gustatory stimuli in neurologically normal rats. Brain Res 143:263–279

Guyon A, Assouly-Besse F, Biala G, Puech AJ, Thiebot MH (1993) Potentiation by low doses of selected neuroleptics of food-induced conditioned place preference in rats. Psychopharmacology (Berl) 110:460–466

Hall WG, Bryan TE (1980) The ontogeny of feeding in rats: II. Independent ingestive behavior. J Comp Physiol Psychol 94:746–756

Hall WG, Bryan TE (1981) The ontogeny of feeding in rats: IV. Taste development as measured by intake and behavioral responses to oral infusions of sucrose and quinine. J Comp Physiol Psychol 95:240–251

Hamilton AL, Stellar JR, Hart EB (1985) Reward, performance, and the response strength method in self-stimulating rats: validation and neuroleptics. Physiol Behav 35:897–904

Harvey JA, Gormezano I (1981) Effects of haloperidol and pimozide on classical conditioning of the rabbit nictitating membrane response. J Pharmacol Exp Ther 218:712–719

Herrnstein RJ (1970) On the law of effect. J Exp Anal Behav 13:243–266

Heyman GM (1983) A parametric evaluation of the hedonic and motoric effects of drugs: pimozide and amphetamine. J Exp Anal Behav 40:113–122

Heyman GM, Kinzie DL, Seiden LS (1986) Chlorpromazine and pimozide alter reinforcement efficacy and motor performance. Psychopharmacology (Berl) 88:346–353

Heyman GM, Monaghan MM (1987) Effects of changes in response requirement and deprivation on the parameters of the matching law equation: new data and review. J Exp Psychol Anim Behav Process 13:384–394

Heyman GM, Monaghan MM, Clody DE (1987) Low doses of cis-flupentixol attenuate motor performance. Psychopharmacology (Berl) 93:477–482

Hikosaka O, Wurtz RH (1983) Visual and oculomotor functions of monkey substantia nigra pars reticulata. I. Relation of visual and auditory responses to saccades. J Neurophysiol 49:1230–1253

Hoffman DC (1989) The use of place conditioning in studying the neuropharmacology of drug reinforcement. Brain Res Bull 23:373–387

Hoffman DC, Beninger RJ (1985) The effects of pimozide on the establishment of conditioned reinforcement as a function of the amount of conditioning. Psychopharmacology (Berl) 87:454–460

Horvitz JC, Ettenberg A (1988) Haloperidol blocks the response-reinstating effects of food reward: a methodology for separating neuroleptic effects on reinforcement and motor processes. Pharmacol Biochem Behav 31:861–865

Horvitz JC, Ettenberg A (1991) Conditioned incentive properties of a food-paired conditioned stimulus remain intact during dopamine receptor blockade. Behav Neurosci 105:536–541

Hsiao S, Smith GP (1995) Raclopride reduces sucrose preference in rats. Pharmacol Biochem Behav 50:121–125

Hull, CL (1943) Principles of behavior, an introduction of behavior theory. New York D. Appleton-Century

Hunt GE, Atrens DM (1992a) Parametric manipulations and fixed-interval self-stimulation. Physiol Behav 51:1009–1020

Hunt GE, Atrens DM (1992b) Reward summation and the effects of pimozide, clonidine, and amphetamine on fixed-interval responding for brain stimulation. Pharmacol Biochem Behav 42:563–577

Hunt GE, Atrens DM, Jackson DM (1994) Reward summation and the effects of dopamine D1 and D2 agonists and antagonists on fixed-interval responding for brain stimulation. Pharmacol Biochem Behav 48:853–862

Hunt HF (1956) Some effects of drugs on classical (type s) conditioning. Ann NY Acad Sci 65:258–267

Jenkins WO (1950) A temporal gradient of derived reinforcement. Am J Psychology 63:237–243

Killcross S, Robbins TW, Everitt BJ (1997) Different types of fear-conditioned behaviour mediated by separate nuclei within amygdala. Nature 388:377–380

Killeen P (1975) On the temporal control of behavior. Psychol Rev 82:89–115

Killeen P, Hanson S, Osbourne S (1978a) Arousal: Its genesis and manifestation as response rate. Psychol Rev 85:571–581

Killeen PR, Hanson SJ, Osborne SR (1978b) Arousal: its genesis and manifestation as response rate. Psychol Rev 85:571–581

Konorski, J (1967) Integrative activity of the brain. Chicago University of Chigaco Press

Le Moal M, Simon H (1991) Mesocorticolimbic dopaminergic network: functional and regulatory roles. Physiol Rev 71:155–234

Leeb K, Parker L, Eikelboom R (1991) Effects of pimozide on the hedonic properties of sucrose: analysis by the taste reactivity test. Pharmacol Biochem Behav 39: 895–901

Liao RM, Fowler SC (1990) Haloperidol produces within-session increments in operant response duration in rats. Pharmacol Biochem Behav 36:191–201

Liebman JM (1983) Discriminating between reward and performance: a critical review of intracranial self-stimulation methodology. Neurosci Biobehav Rev 7:45–72

Lindvall O, Bjorklund A (1974) The organization of the ascending catecholamine neuron systems in the rat brain as revealed by the glyoxylic acid fluorescence method. Acta Physiol Scand Suppl 412:1–48

Lindvall O, Bjorklund A, Nobin A, Stenevi U (1974) The adrenergic innervation of the rat thalamus as revealed by the glyoxylic acid fluorescence method. J Comp Neurol 154:317–347

Lopez HH, Ettenberg A (2000) Haloperidol challenge during copulation prevents subsequent increase in male sexual motivation. Pharmacol Biochem Behav 67: 387–393

Lopez HH, Ettenberg A (2001) Dopamine antagonism attenuates the unconditioned incentive value of estrous female cues. Pharmacol Biochem Behav 68:411–416

Lovibond PF (1983) Facilitation of instrumental behavior by a Pavlovian appetitive conditioned stimulus. J Exp Psychol Anim Behav Process 9:225–247

Lynch MR, Wise RA (1985) Relative effectiveness of pimozide, haloperidol and trifluoperazine on self-stimulation rate-intensity functions. Pharmacol Biochem Behav 23:777–780

Mason ST, Beninger RJ, Fibiger HC, Phillips AG (1980) Pimozide-induced suppression of responding: evidence against a block of food reward. Pharmacol Biochem Behav 12:917–923

McCullough LD, Salamone JD (1992) Involvement of nucleus accumbens dopamine in the motor activity induced by periodic food presentation:a microdialysis and behavioral study. Brain Res 592:29–36

McDowell JJ, Wood HM (1984) Confirmation of linear system theory prediction:changes in Herrnstein's K as a function of changes in reinforcer magnitude. J Exp Anal Behav 41:183–192

McFarland K, Ettenberg A (1995) Haloperidol differentially affects reinforcement and motivational processes in rats running an alley for intravenous heroin. Psychopharmacology (Berl) 122:346–350

McFarland K, Ettenberg A (1999) Haloperidol does not attenuate conditioned place preferences or locomotor activation produced by. Pharmacol Biochem Behav 62:631–641

McNair DM, Lorr M, Droppleman LF (1971) Manual for the profile of mood states. San Diego San Diego, Educational and Industrial Testing Service

McSweeney FK (1978) Prediction of concurrent key-peck and treadle press responding from simple schedule performance. Animal Learning & Behavior 6:444–450

Mellgren RL, Ost JW (1969) Transfer of Pavlovian differential conditioning to an operant discrimination. J Comp Physiol Psychol 67:390–394

Miliaressis E, Malette J, Coulombe D (1986b) The effects of pimozide on the reinforcing efficacy of central grey stimulation in the rat. Behav Brain Res 21:95–100

Miliaressis E, Rompre PP, Laviolette P, Philippe L, Coulombe D (1986a) The curve-shift paradigm in self-stimulation. Physiol Behav 37:85–91

Miller R, Wickens JR, Beninger RJ (1990) Dopamine D-1 and D-2 receptors in relation to reward and performance: a case for the D-1 receptor as a primary site of therapeutic action of neuroleptic drugs. Prog Neurobiol 34:143–183

Montague PR, Dayan P, Sejnowski TJ (1996) A framework for mesencephalic dopamine systems based on predictive Hebbian learning. J Neurosci 16:1936–1947

Morley MJ, Bradshaw CM, Szabadi E (1984a) The effect of pimozide on variable-interval performance: a test of the "anhedonia" hypothesis of the mode of action of neuroleptics. Psychopharmacology (Berl) 84:531–536

Morley MJ, Bradshaw CM, Szabadi E (1984b) The effects of pimozide on variable-interval performance: a test of the anhedonia hypothesis of the mode of action of neuroleptics. Psychopharmacology 84:531–536

Moss FA, Thorndike EL (1934) Comparative Psychology. New York Prentice-Hall

Muscat R, Willner P (1989) Effects of dopamine receptor antagonists on sucrose consumption and preference. Psychopharmacology (Berl) 99:98–102

Nakajima S, McKenzie GM (1986) Reduction of the rewarding effect of brain stimulation by a blockade of dopamine D1 receptor with SCH 23390. Pharmacol Biochem Behav 24:919–923

Nakajima S, O'Regan NB (1991) The effects of dopaminergic agonists and antagonists on the frequency- response function for hypothalamic self-stimulation in the rat. Pharmacol Biochem Behav 39:465–468

Nakajima S, Patterson RL (1997) The involvement of dopamine D2 receptors, but not D3 or D4 receptors, in the rewarding effect of brain stimulation in the rat. Brain Res 760:74–79

Newton TF, Ling W, Kalechstein AD, Uslaner J, Tervo K (2001) Risperidone pretreatment reduces the euphoric effects of experimentally administered cocaine. Psychiatry Res 102:227–233

Papp M, Willner P, Muscat R (1991) An animal model of anhedonia: attenuation of sucrose consumption and place preference conditioning by chronic unpredictable mild stress. Psychopharmacology (Berl) 104:255–259

Parker LA, Lopez N, Jr. (1990) Pimozide enhances the aversiveness of quinine solution. Pharmacol Biochem Behav 36:653–659

Parkinson J, Fudge J, Hurd Y, Pennartz C, Peoples L (2000) Finding motivation at Seabrook Island: the ventral striatum, learning and plasticity. Trends Neurosci 23:383–384

Pavlov IP (1927) Conditioned reflexes: an Investigation of the Physiological Activity of the Cerebral Cortex. London Oxford University Press

Pecina S, Berridge KC, Parker LA (1997) Pimozide does not shift palatability: separation of anhedonia from sensorimotor suppression by taste reactivity. Pharmacol Biochem Behav 58:801–811

Phillips AG, Fibiger HC (1979) Decreased resistance to extinction after haloperidol: implications for the role of dopamine in reinforcement. Pharmacol Biochem Behav 10:751–760

Phillips G, Willner P, Muscat R (1991a) Suppression or facilitation of operant behaviour by raclopride dependent on concentration of sucrose reward. Psychopharmacology (Berl) 105:239–246

Phillips G, Willner P, Muscat R (1991b) Reward-dependent suppression or facilitation of consummatory behaviour by raclopride. Psychopharmacology (Berl) 105: 355–360

Phillips G, Willner P, Muscat R (1991c) Anatomical substrates for neuroleptic-induced reward attenuation and neuroleptic-induced response decrement. Behavioural Pharmacology 2:129–141

Phillips G, Willner P, Sampson D, Nunn J, Muscat R (1991d) Time-, schedule-, and reinforcer-dependent effects of pimozide and amphetamine. Psychopharmacology (Berl) 104:125–131

Price MT, Fibiger HC (1975) Discriminated escape learning and response to electric shock after 6-hydroxydopamine lesions of the nigro-neostriatal dopaminergic projection. Pharmacol Biochem Behav 3:285–290

Ranje C, Ungerstedt U (1977a) Discriminative and motor performance in rats after interference with dopamine neurotransmission with spiroperidol. Eur J Pharmacol 43:39–46

Ranje C, Ungerstedt U (1977b) Lack of acquisition in dopamine denervated animals tested in an underwater Y-maze. Brain Res 134:95–111

Redgrave P, Prescott TJ, Gurney K (1999) Is the short-latency dopamine response too short to signal reward error? Trends Neurosci 22:146–151

Rescorla RA (1994) Control of instrumental performance by Pavlovian and instrumental stimuli. J Exp Psychol Anim Behav Process 20:44–50

Rescorla RA, Solomon RL (1967) Two-process learning theory: Relationships between Pavlovian conditioning and instrumental learning. Psychol Rev 74:151–182

Robbins TW, Cador M, Taylor JR, Everitt BJ (1989a) Limbic-striatal interactions in reward-related processes. Neurosci Biobehav Rev 13:155–162

Robbins TW, Everitt BJ, Ryan CN, Marston HM, Jones GH, Page KJ (1989b) Comparative effects of quisqualic and ibotenic acid-induced lesions of the substantia innominata and globus pallidus on the acquisition of a conditional visual discrimination: differential effects on cholinergic mechanisms. Neuroscience 28:337–352

Robinson TE, Berridge KC (1993) The neural basis of drug craving: an incentive-sensitization theory of addiction. Brain Res Brain Res Rev 18:247–291

Rolls ET, Rolls BJ, Kelly PH, Shaw SG, Wood RJ, Dale R (1974) The relative attenuation of self-stimulation, eating and drinking produced by dopamine receptor blockade. Psychopharmacologia 38:219–230

Rothman RB, Mele A, Reid AA, Akunne H, Greig N, Thurkauf A, Rice KC, Pert A (1989) Tight binding dopamine reuptake inhibitors as cocaine antagonists. A strategy for drug development [published erratum appears in FEBS Lett 1990 Jan 15;260(1):152]. FEBS Lett 257:341–344

Salamone JD (1987) The actions of neuroleptic drugs on appetitive instrumental behaviors. In: Iversen LL, Iversen SD, Snyder SH (eds) Handbook of psychopharmacology. Plenum Press, New York, pp 575–608

Salamone JD (1992) Complex motor and sensorimotor functions of striatal and accumbens dopamine: involvement in instrumental behavior processes. Psychopharmacology (Berl) 107:160–174

Salamone JD, Cousins MS, Snyder BJ (1997) Behavioral functions of nucleus accumbens dopamine: empirical and conceptual problems with the anhedonia hypothesis. Neurosci Biobehav Rev 21:341–359

Sanger DJ (1986) Drug taking as adjunctive behaviour. In: Behavioural analyis of drug dependence Academic Press, Inc., pp 123–160

Sanger DJ (1987) The actions of SCH 23390, a D1 receptor antagonist, on operant and avoidance behavior in rats. Pharmacol Biochem Behav 26:509–513

Sanger DJ, Perrault G (1995) Effects of typical and atypical antipsychotic drugs on response decrement patterns in rats. J Pharmacol Exp Ther 272:708–713

Schaefer GJ, Holtzman SG (1979) Free-operant and auto-titration brain self-stimulation procedures in the rat: a comparison of drug effects. Pharmacol Biochem Behav 10:127–135

Schaefer GJ, Michael RP (1980) Acute effects of neuroleptics on brain self-stimulation thresholds in rats. Psychopharmacology (Berl) 67:9–15

Schaefer GJ, Michael RP (1985) The discriminative stimulus properties and detection thresholds of intracranial self-stimulation: effects of d-amphetamine, morphine, and haloperidol. Psychopharmacology (Berl) 85:289–294

Schneider LH, Davis JD, Watson CA, Smith GP (1990) Similar effect of raclopride and reduced sucrose concentration on the microstructure of sucrose sham feeding. Eur J Pharmacol 186:61–70

Schneider LH, Gibbs J, Smith GP (1986) D-2 selective receptor antagonists suppress sucrose sham feeding in the rat. Brain Res Bull 17:605–611

Schultz W (1998) Predictive reward signal of dopamine neurons. J Neurophysiol 80:1–27

Schultz W, Dayan P, Montague PR (1997) A neural substrate of prediction and reward. Science 275:1593–1599

Shippenberg TS, Herz A (1987) Place preference conditioning reveals the involvement of D1-dopamine receptors in the motivational properties of mu- and kappa-opioid agonists. Brain Res 436:169–172

Sinnamon HM (1982) The reward-effort model: an economic framework for examining the mechanism of neuroleptic action. Behavioral and Brain Sciences 5:73–75

Skinner BF (1938) The Behavior of Organisms: an experimental analysis. New York D. Appleton-Century Company Inc.

Smith-Roe SL, Kelley AE (2000) Coincident activation of NMDA and dopamine D1 receptors within the nucleus accumbens core is required for appetitive instrumental learning. J Neurosci 20:7737–7742

Smith GP, Gibbs J (1979) Postprandial satiety. In: Sprague J, Epstein A (eds) Progress in Psychobiology and Physiological Psychology. New York, Academic Press

Spyraki C, Fibiger HC, Phillips AG (1982) Cocaine-induced place preference conditioning: lack of effects of neuroleptics and 6-hydroxydopamine lesions. Brain Res 253:195–203

Stellar JR, Corbett D (1989) Regional neuroleptic microinjections indicate a role for nucleus accumbens in lateral hypothalamic self-stimulation reward. Brain Res 477:126–143

Stellar JR, Kelley AE, Corbett D (1983) Effects of peripheral and central dopamine blockade on lateral hypothalamic self-stimulation: evidence for both reward and motor deficits. Pharmacol Biochem Behav 18:433–442

Stellar JR, Stellar E (1985) The neurobiology of motivation and reward. New York Springer-Verlag

Stellar JR, Waraczynski M, Wong K (1988) The reward summation function in hypothalamic self-stimulation. In: Commons M, Church RM, Stellar JR, Wagner AR (eds) Quantitative analysis of behavior, Vol. 7. Biological determinants of reinforcement. Lawrence Erlbaum Associates, Hillsdale, pp 31–57

Sutton MA, Beninger RJ (1999) Psychopharmacology of conditioned reward: evidence for a rewarding signal at D1-like dopamine receptors. Psychopharmacology (Berl) 144:95–110

Thierry AM, Blanc G, Sobel A, Stinus L, Golwinski J (1973) Dopaminergic terminals in the rat cortex. Science 182:499–501

Thorndike, EL (1898) Animal Intelligence: An Experimental Study of the Associative Processes in Animals. New York Mc Millan

Toates F (1998) The interaction of cognitive and stimulus-response processes in the control of behaviour. Neurosci Biobehav Rev 22:59–83

Tolman EC (1959) Principles of purposive behavior. In: Koch S (ed) Psychology: a study of a science. New York, McGraw-Hill

Tombaugh TN, Anisman H, Tombaugh J (1980) Extinction and dopamine receptor blockade after intermittent reinforcement training: failure to observe functional equivalence. Psychopharmacology (Berl) 70:19–28

Tombaugh TN, Szostak C, Mills P (1983) Failure of pimozide to disrupt the acquisition of light-dark and spatial discrimination problems. Psychopharmacology (Berl) 79:161–168

Tombaugh TN, Szostak C, Voorneveld P, Tombaugh JW (1982) Failure to obtain functional equivalence between dopamine receptor blockade and extinction: evidence supporting a sensory-motor conditioning hypothesis. Pharmacol Biochem Behav 16:67–72

Towell A, Muscat R, Willner P (1987) Effects of pimozide on sucrose consumption and preference. Psychopharmacology (Berl) 92:262–264

Trapold MA, Lawton GW, Dick RA, Gross DM (1968) Transfer of training from differential classical to differential instrumental conditioning. J Exp Psychol 76:568–573

Treit D, Berridge KC (1990) A comparison of benzodiazepine, serotonin, and dopamine agents in the taste-reactivity paradigm. Pharmacol Biochem Behav 37:451–456

Tyrka A, Gayle C, Smith GP (1992) Raclopride decreases sucrose intake of rat pups in independent ingestion tests. Pharmacol Biochem Behav 43:863–869

Tyrka A, Smith GP (1991) Potency of SCH 23390 for decreasing sucrose intake in rat pups depends on mode of ingestion. Pharmacol Biochem Behav 39:955–961

Tyrka A, Smith GP (1993) SCH23390, but not raclopride, decreases intake of intraorally infused 10% sucrose in adult rats. Pharmacol Biochem Behav 45:243–246

Tzschentke TM (1998) Measuring reward with the conditioned place preference paradigm: a comprehensive review of drug effects, recent progress and new issues. Prog Neurobiol 56:613–672

Watson JB (1913) Psychology as the behaviourist views it. Psychol Rev 20:158–177

Wauquier A, Niemegeers CJ (1979) A comparison between lick or lever-pressing contingent reward and the effects of neuroleptics thereon. Arch Int Pharmacodyn Ther 239:230–240

Weatherford SC, Greenberg D, Gibbs J, Smith GP (1990) The potency of D-1 and D-2 receptor antagonists is inversely related to the reward value of sham-fed corn oil and sucrose in rats. Pharmacol Biochem Behav 37:317–323

Weingarten HP, Watson SD (1982) Sham feeding as a procedure for assessing the influence of diet palatability on food intake. Physiol Behav 28:401–407

Willner P, Towell A, Muscat R (1987) Effects of amphetamine and pimozide on reinforcement and motor parameters in variable-interval performance. Journal of Psychopharmacology 1:140–153

Willner P, Phillips G, Sampson D, Muscat R (1989) Time-dependent and schedule-dependent effects of dopamine receptor blockade. Behavioural Pharmacology 1:169–176

Willner P, Papp M, Phillips G, Maleeh M, Muscat R (1990a) Pimozide does not impair sweetness discrimination. Psychopharmacology (Berl) 102:278–282

Willner P, Sampson D, Phillips G, Muscat R (1990b) A matching law analysis of the effects of dopamine receptor antagonists. Psychopharmacology (Berl) 101:560–567

Wise RA (1978) Catecholamine theories of reward: a critical review. Brain Res 152:215–247

Wise RA (1982) Neuroleptics and operant behavior: The anhedonia hypothesis. The Behavioral and Brain Sciences 5:39–87

Wise RA, Bozarth MA (1987) A psychomotor stimulant theory of addiction. Psychol Rev 94:469–492

Wise RA, Spindler J, deWit H, Gerberg GJ (1978b) Neuroleptic-induced "anhedonia" in rats: pimozide blocks reward quality of food. Science 201:262–264

Wyvell CL, Berridge KC (2000) Intra-accumbens amphetamine increases the conditioned incentive salience of sucrose reward: enhancement of reward "wanting" without enhanced "liking" or response reinforcement. J Neurosci 20:8122–8130

Xenakis S, Sclafani A (1981) The effects of pimozide on the consumption of a palatable saccharin- glucose solution in the rat. Pharmacol Biochem Behav 15:435–442

Xenakis S, Sclafani A (1982) The dopaminergic mediation of a sweet reward in normal and VMH hyperphagic rats. Pharmacol Biochem Behav 16:293–302

Zarevics P, Setler PE (1979) Simultaneous rate-independent and rate-dependent assessment of intracranial self-stimulation: evidence for the direct involvement of dopamine in brain reinforcement mechanisms. Brain Res 169:499–512

CHAPTER 23
Molecular and Cellular Events Regulating Dopamine Neuron Survival

G.U. CORSINI, R. MAGGIO, and F. VAGLINI

A. Introduction

The progressive degeneration of dopamine (DA)-melanized neurons of the substantia nigra pars compacta (SNpc) represents the pathological hallmark of Parkinson's disease (PD), although nigral lesion is not the only cell derangement present in the disease. In addition to DA neurons, other neuronal pathways are involved in the plurifocal damage typical of the idiopathic form. Nigral cell loss, however, and the resulting dopaminergic denervation of the corpus striatum, the recipient area of nigral projections, triggers a cascade of functional changes in the whole basal ganglia circuitry, which leads ultimately to the expression of PD motor symptoms. The disease is characterized by a clinical triad of cardinal symptoms including tremor, bradykinesia and muscle rigidity.

The main pathological feature of PD is a typical hyaline inclusion in the cytoplasm of neurons, the so-called "Lewy body". These inclusions occur in several brain areas but most prominently in the substantia nigra. Lewy bodies are round-shaped entities with a dense eosinophilic core and a pale surrounding halo. Although these pathological inclusions are also observed in other neurodegenerative disorders, the presence of Lewy bodies in the substantia nigra, in combination with nigrostriatal cell loss, is generally considered to be the specific pathological condition that defines idiopathic PD. Other neurodegenerative diseases that also include this condition are consistently considered PD-plus or PD-like motor disorders primarily affecting other sub-cortical nuclei.

Due to the discovery of new pathological entities, clinical identification of the idiopathic form is becoming a more and more difficult task. Consistently, drug response and clinical outcome only, before brain autopsy, are the two main features of the disease which may help diagnosis. "Parkinsonism" is a term which was originally coined to define PD-like motor disturbances with a known aetiology, assumed to be different from the one responsible for idiopathic PD. Several forms of parkinsonism have been ascertained up to now. In the 1980s, the discovery that 1-methyl-4-phenyl-1,2,3,6-tetrahydropyridine (MPTP) causes a parkinsonism that is indistinguishable from the idiopathic disease, has not only provided new insight into the mechanisms of DA cell death, but has also added new support for the "toxic theory" of idiopathic PD. As a result, a definition of the idiopathic entity has become more and more complicated, even if there have been enormous advances in scientific knowledge about the whole disease.

Although idiopathic PD is usually sporadic, it has been reported that a genetic factor may contribute to the increase in the incidence in relatives of some affected people. The recent discovery of gene linkage in a few families with strong patterns of inheritance has provided new insight into the role of specific proteins (α-synuclein, parkin) in the pathophysiology of the disease.

The aim of the present chapter is to describe and discuss the current status of our understanding of the molecular and cellular events regulating DA neuron survival. For a comprehensive review on PD, see this volume or a recent report by DUNNET and BJÖRKLUND (1999).

B. Mechanisms of DA Cell Death

The primary cause of nigrostriatal pathway degeneration in idiopathic PD remains obscure. Despite the difficulty in studying basic mechanisms of cell death of this uncertain clinical entity, most information about the pathophysiology of the disease derives from studies in human and experimental parkinsonism. Selective neurotoxins, viral agents or genetic changes, as known aetiological factors in parkinsonism, provide useful models in order to understand the intra- and extra-neuronal events regulating DA cell viability. The principal source of direct evidence, however, comes from brain analysis in post-mortem studies. With this technique in pivotal experiments, OLEH HORNYKIEWICZ showed a marked loss of DA and its major metabolite in the basal ganglia of PD patients (EHRINGER and HORNYKIEWICZ 1960; BERNHEIMER and HORNYKIEWICZ 1964; BERNHEIMER and HORNYKIEWICZ 1965) and provided a proper understanding of the pathological and clinical correlates of the disease (BERNHEIMER et al. 1973). However, this generally accepted concept of loss of striatal DA as responsible for motor disabilities has very recently been confuted (WILLIS and ARMSTRONG 1998).

Physiologically, a progressive decline in neurons of the SNpc and DA content of the striatum with aging has been demonstrated (McGEER et al. 1989). The rate of nigral cell degeneration during the aging process is about 5% per decade (McGEER et al. 1989; FEARNLEY and LEES 1991). In PD there is a tenfold higher rate of cell death (45% reduction per decade) up to a critical threshold of about 50% of nigral cell counts, when the first motor symptoms appear, compared with age-matched controls (McGEER et al. 1989). Striatal DA is accordingly reduced to about 80% at the onset of symptoms (BERNHEIMER et al. 1973) and this evaluation has recently been confirmed with positron emission tomography (PET) studies in vivo (BROOKS 1998). In parkinsonism, instead, as observed in MPTP cases, the acute insult induced by the aetiological factor may cause rapid but partial loss of nigral cells, which is then followed by the natural decline due to the physiological aging process (CALNE et al. 1985). All this evidence suggests that in idiopathic PD an active disease process of accelerated nigral cell death is taking place (CALNE 1994).

One of the most interesting pathological issues about the primary lesion of nigrostriatal degeneration is whether it originates in the striatum or in the SN. Solving this problem would aid the identification of aetiological factors and pathogenetic mechanisms as well. Some authors have dealt with this difficult subject and they have suggested that the anatomical genesis of degeneration is the striatum, on the basis of certain animal models of neurotoxin-induced parkinsonism (HERKENHAM et al. 1991; ICHITANI et al. 1991). Actually, it is easy to accept this suggestion as for the case of MPTP- or 6-hydroxydopamine (6-OHDA)-induced damage in primates or in rats respectively. These neurotoxins are avidly taken up by the high-affinity DA transporter (DAT) at the synaptic membranes in terminal boutons of DA neurons (JONSSON 1980; JAVITCH and SNYDER 1984) where they primarily induce a derangement of the synaptosomal metabolism (BAUMGARTEN and ZIMMERMANN 1992a; FORNAI et al. 1997a).

However, a primary lesion in the striatum has also been postulated in PD (HORNYKIEWICZ 1991) in spite of the traditional assumption that the degenerative process starts to affect perikarya in the SN (BERNHEIMER et al. 1973; OLANOW and TATTON 1999). The crucial solution of this issue might draw the attention of research from one brain area (midbrain) to another (forebrain) in order to study also the extraneuronal environment modulating the primary lesions.

Several molecular and cellular events involved in the progressive decline of DA nigrostriatal neurons have recently been identified. The underlying mechanisms of these events may be identified as extra- and intra-neuronal, which may all interact leading to neuronal dysfunction and cell death. Among the extra-neuronal events, we will focus on the role of the noradrenergic system, excitotoxicity, trophic factors and selective neurotoxins. Among the intra-neuronal mechanisms, we will discuss oxidative stress, mitochondrial dysfunction, P450 involvement, cellular vulnerability and apoptosis.

C. Extraneuronal Events

I. Noradrenergic System

The potential role of norepinephrine (NE) in the pathogenetic mechanisms of PD was already suggested by the classic study of Hornykiewicz in 1960 (EHRINGER and HORNYKIEWICZ 1960).

In the CNS, the main nucleus of NE neurons is located in the pons (locus coeruleus) and an extensive cell loss has been observed in PD, precisely in this brain area (FORNO 1996). Actually, in addition to DA cell loss in the SN, the concomitant lesion of the locus coeruleus (LC) represents the cardinal feature of morphological analysis in parkinsonian brains (ALBIN et al. 1989). The accompanying pathological elements found in the LC of patients range from the classic Lewy bodies to neurofibrillary tangles (ALBIN et al. 1989). However, the fortuitous finding of these elements in the LC of normal aging people has been extensively reported (FORNO and ALVORD 1971; ALVORD and FORNO 1992).

Recent studies in PD brains indicate that the degree of LC cell loss compared with age-matched controls is about 70% (BERTRAND et al. 1997), with a homogeneous degeneration involving both the caudal and the rostral portion of the nucleus (CHAN-PALAY and ASAN 1989a; GERMAN et al. 1992). In contrast, in PD complicated with dementia, the rostral portion of the nucleus that projects to the forebrain is particularly affected (GERMAN et al. 1992).

As stated above, the biochemical finding of DA loss occurring in PD is not only a statistically significant neurochemical change, but it can also be considered as a constant and necessary neurochemical feature of idiopathic PD. Similarly to DA, NE contents in several brain areas of PD patients are constantly and significantly reduced indicating, differently from what has been observed with other neurotransmitters, that NE loss must be considered as a fundamental biochemical marker of the disease (HORNYKIEWICZ and KISH 1986; HORNYKIEWICZ and PIFL 1994). The brain areas maximally affected by the disease are the motor cortex, the SNpc and the hippocampus, but also the cerebellum and lumbar spinal cord are often involved (KISH et al. 1984; SCATTON et al. 1986).

The loss of NE neurons in PD may result in some clinical correlates which have been more recently analysed and widely discussed (GERLACH and RIEDERER 1993).

Besides the classic motor symptoms, PD patients constantly present a long sequela of non-motor alterations (BIRKMAYER et al. 1987). Among these, depression is a frequent mental disorder which affects PD patients, often even preceding by several years the appearance of motor disabilities (DOONEIEF et al. 1992). It is well known that NE is one of the major neurotransmitters involved in the pathophysiology and treatment of several forms of depression (ZUBENKO et al. 1990) and in PD, NE impairment might yield this mental disorder (CHAN-PALAY and ASAN 1989a,b). In PD, autonomic failure of some extent has frequently been observed, and this has been related to the NE reduction occurring in the spinal cord (SCATTON et al. 1986). Actually, GOTO and HIRANO (1991) provided further evidence that a massive cell loss in the parabrachial nucleus, within the LC complex, is responsible for the autonomic impairment in PD patients. Among the non-motor symptoms occurring in PD patients, dementia deserves special attention; this is considered as an adjunctive feature defining a distinct clinical entity as a part of a continuous spectrum of neurodegenerative disorders (PERL et al. 1998). Dementia is always related to the finding of pathological elements and NE loss in the neocortex (PERL et al. 1998; ALBIN et al. 1989). Indeed, it has been reported that in PD with dementia, the LC is more markedly affected in its rostral pole and this relationship further strengthens the hypothesis of NE involvement in cognitive disorders (GERMAN et al. 1992).

Among motor disabilities in PD, "freezing" has been attributed to cell loss in the LC (MIZUNO et al. 1994). This abrupt fit of immobility which affects the patient during his motor activity is considered a phenomenon clinically distinct from akinesia and not unrelated to drug treatment (POEWE and GRANATA 1997).

1. NE in Experimental Parkinsonism

The use of selective neurotoxins in experimental parkinsonism suggests the necessity of lesioning the nigrostriatal DA pathway more and more specifically in order to reproduce a simple animal model closer to the human disease. This has been currently achieved through the local or systemic administration of 6-OHDA, MPTP or methamphetamine, particularly in rodents and primates (HERKEN and HUCHO 1992).

However, the concept of "selective lesioning" does not take into account the involvement of the LC, as mentioned above. Indeed, if we consider the real selectivity of these neurotoxins we notice that they all produce lesions to both the DA and NE systems when systemically injected (HERKEN and HUCHO 1992). Under these conditions, 6-OHDA is able to lesion even peripheral NE neurons (BAUMGARTEN and ZIMMERMANN 1992a). Similarly, MPTP, which is considered to be the most selective neurotoxin for DA neurons, often produces an extensive lesion to the NE system both in primates (FORNO 1996) and in mice (SENIUK et al. 1990; PIFL et al. 1991). However, the regional distribution of NE loss which is obtained with MPTP in monkeys does not match that occurring in idiopathic PD (PIFL et al. 1991). For this reason, in order to reproduce the natural progression of PD, a selective NE depletion before lesioning DA neurons was achieved both in primates and in mice (MAVRIDIS et al. 1991; MARIEN et al. 1993; BING et al. 1994). The current non-invasive method to reduce NE content in the brain consists of systemic administration of N-(2-chloroethyl)-N-ethyl-2-bromobenzylamine (DSP-4), an experimental tool which has been used since the 1970s (Ross and RENYI 1976). This neurotoxin selectively affects NE axon terminals arising from LC neurons, which have an affinity uptake for DSP-4 2.7 times higher than extra-LC terminals (ZACZEK et al. 1990). DSP-4 has also been used in some experimental studies to demonstrate that the impairment of LC neurons produces a significant worsening of the subsequent nigrostriatal degeneration, suggesting that NE cells of LC might have a protective role on DA neurons (MAVRIDIS et al. 1991; MARIEN et al. 1993; BING et al. 1994; FORNAI et al. 1995a). In detail, in a recent experiment, MPTP was administered to LC-lesioned monkeys and the animals were observed for a period of 9 weeks, and then brain analysis was performed. Between the LC-lesioned and unlesioned monkeys, there was a marked difference in the recovery process from motor impairment. As a matter of fact, at the end of the study, the control monkeys had almost completely recovered from the experimental parkinsonism, whereas the LC-lesioned primates were still markedly affected. Brain analysis in these animals proved that the fall in striatal DA and the damage to nigral neurons were more marked in comparison with the monkeys that recovered. The authors concluded that the NE system of the LC represents a critical factor in promoting the recovery of lesioned DA neurons (MAVRIDIS et al. 1991). Alternatively, these data could be interpreted as indicating that the regular functioning of LC may mitigate the neurotoxin-induced lesion *ab initio*, rather than favouring recovery: in the sham-treated animals even a partial lesion of DA terminals would have elicited a marked parkinsonian syndrome indistinguishable from that observed in the LC-lesioned monkeys. In any case, these data suggest that the NE system may have some neurotrophic properties towards DA neurons. This hypothesis is supported by the recent observations that NE agonists increase fibroblast growth factor (FGF) mRNA in glial cells (FOLLESA and MOCCHETTI 1993), and that FGF has a protective role towards toxin-induced lesions to DA neurons (OTTO and UNSICKER 1990). Similar findings have been obtained in rodents where either MPTP or methamphetamine has been used as the lesioning tool.

In mice, MARIEN et al. (1993) reported that pretreatment with DSP-4 enhances the decrease in striatal DA caused by MPTP. These findings were confirmed by a subsequent study in which 6-OHDA was unilaterally injected into the LC and sub-threshold doses of MPTP followed after a 10-day period. In these animals, a dramatic reduction was found in the number of tyrosine hydroxylase (TH)-positive cells in the SN ipsilaterally to the LC lesion (BING et al. 1994). Further studies confirmed these results by using methamphetamine as a neurotoxic agent. It is in fact widely reported that methamphetamine induces a long-lasting depletion of striatal DA due to the lesion to striatal DA terminals (BAUMGARTEN and ZIMMERMANN 1992a). In our recent studies, pretreatment with DSP-4 markedly potentiated the striatal DA fall induced by the administration of a dose of methamphetamine which causes an intermediate lesion to nigrostriatal terminals, both in mice and in rats (FORNAI et al. 1995a, 1996a). All the above-mentioned studies confirm the concept of a protective role of the NE system on SNpc neurons. However, other considerations must be taken into account before reaching conclusions about such a role of the NE system in these pivotal studies. Recently, SONSALLA et al. (SPECIALE et al. 1998) reported that 1-methyl-4-phenylpyridinium (MPP$^+$) is sequestered within neurons that contain vesicular monoamine transporters (VMAT), suggesting that the NE system is also responsible for the uptake of endogenously formed MPP$^+$, a phenomenon which had already been previously observed (JAVITCH et al. 1985). This might explain the increased acute toxicity of MPTP on DA neurons when the NE system is impaired. However, it has been well demonstrated that selective NE uptake inhibitors protect against NE depletion by MPTP without affecting its lesioning action on the DA neurons in the striatum (SUNDSTROM and JONSSON 1985; MAYER et al. 1986). Therefore, it may be ruled out that a simple buffering activity is responsible for the protective role of the NE system against MPTP toxicity. This conclusion is further confirmed by a more recent report by us in which pretreatment with DSP-4 does not change the striatal kinetics of MPTP/MPP$^+$ in mice (FORNAI et al. 1997b).

An alternative hypothesis for the protective role of the NE system is suggested by the observation that glutamate may have an important role in nigrostriatal neurotoxicity in experimental parkinsonism (see Sect. C.II.). As far as this problem is concerned, alpha-2-adrenoceptor stimulation inhibits the release of glutamate in several brain regions (KAMISAKI et al. 1992). This may represent the molecular event by which endogenous NE physiologically protects DA neurons. Accordingly, it has been recently reported by us that the alpha-2 agonist, clonidine, completely prevents MPTP-induced toxicity in mice, whereas the antagonist yohimbine potentiates such toxicity and reduces the protective effects of clonidine (FORNAI et al. 1995b). However, it has to be pointed out that other mechanisms related to alpha-2 receptors may be involved as well. As in previous experiments with MPTP, methamphetamine toxicity is potentiated by the LC lesion as an acute effect rather than a chronic modification of the recovery process. In a very recent study, FORNAI et al. (1999) were able to perform a time course (as long as 90 days) of the recovery process after methamphetamine-induced damage to DA neurons in LC-lesioned mice. No difference was observed in the rate of recovery of striatal DA levels between intact and LC-lesioned animals during the time-period studied. We concluded that the NE system does not affect the recovery of the remaining DA terminals, but rather the early molecular events leading to neurotoxin-induced DA degeneration. One of the early molecular events that is induced by all DA neurotoxins is DA release from nerve terminals. 6-OHDA, MPTP and methamphetamine all have a marked potency in releasing DA, which is fundamental for the subsequent cell lesion (ROLLEMA et al. 1988; BAUMGARTEN and ZIMMERMANN 1992a; O'DELL et al. 1993). We have recently demonstrated, by using in vivo striatal brain dialysis, that methamphetamine-induced DA release increases in animals pretreated with DSP-4 (FORNAI et al. 1999).

The role of the LC in PD is far from being fully understood, but innovative hypotheses and new insights have been provided with the help of experimental animal models. The data presented above indicate that the NE system may influence motor and non-motor symptoms in PD and modulate the vulnerability of nigrostriatal DA neurons. The LC integrity and its physiological functioning might be a crucially decisive factor for the onset and progression of some neurodegenerative diseases.

II. Excitatory Amino Acids

1. Excitotoxicity in PD

Excitatory amino acids (EAA), such as glutamate and aspartate, are endogenous neurotransmitters which produce neuronal depolarization. These acidic amino acids and their analogues may exert, under certain circumstances, an excessive hyperstimulation of their physiological receptors, which may lead to post-synaptic neuronal dysfunction and finally cell death. This destructive event has been termed as "excitotoxicity" (OLNEY 1980).

Currently, the study of this neurotoxic phenomenon represents one of the most active areas of interest in neuroscience, especially for the clinical implications in neurodegenerative disorders (CHOI 1988a,b; LIPTON and ROSENBERG 1994). Among these amino acids, glutamic acid is the most widespread in the CNS where, under extreme conditions, it could become frankly toxicant. Excessive or prolonged release of glutamate into the synaptic cleft, impaired clearance of glutamate from the extracellular space and impairment of surrounding γ-aminobutyric acid (GABA)ergic inhibition are all conditions which may transform a physiological neurotransmitter into a selective neurotoxin (OLNEY 1990). Excitotoxicity is a glutamate receptor-mediated phenomenon that explains the selective vulnerability of different neuronal pathways (BAUMGARTEN and ZIMMERMANN 1992b). After repeated acute insults, there is a neuronal loss with a progressive recruitment of synaptically linked cells. This trans-synaptic recruitment is similar to the one observed in some neurodegenerative diseases, where the cell loss follows functional patterns of the downstream neurons. In these neurodegenerative disorders, often called "multisystemic", the pathways affected overlap the circuitry enrolled during the acute insult (BAUMGARTEN and ZIMMERMANN 1992b). In PD, in spite of the plurifocal damage, it is difficult to identify a specific and consistent circuitry recruitment, as observed, instead, in multisystemic atrophy (MSA) or in progressive supranuclear palsy (PSP). However, the glutamatergic pathways from both the neocortex and the subthalamic nucleus to the SN and to the striatum, through the activation of N-methyl-D-aspartate (NMDA) and α-amino-3-hydroxy-5-methyl-4-isoxazolepropionic acid (AMPA) receptors could play an important role in sustaining excitotoxicity towards the nigrostriatal DA neurons (COTMAN et al. 1987; VILA et al. 1999).

Recently, pharmacological and toxicological evidence strongly suggests an involvement of excitotoxicity in PD (HENNEBERRY et al. 1989; CARLSSON and CARLSSON 1990; ALBIN and GREENAMYRE 1992; COYLE and PUTTFARCKEN 1993). In addition, one of the consequences of the degeneration of nigrostriatal DA neurons is an overactivity of glutamatergic subthalamopallidal, subthalamonigral and corticostriatal pathways in the basal ganglia, as demonstrated in rats and monkeys rendered parkinsonian, or in humans with PD (MILLER and DELONG 1987; BERGMAN et al. 1990; HOLLERMAN and GRACE 1992; WÜLLNER et al. 1992; CALABRESI et al. 1993; BERGMAN et al. 1994; ANGLADE et al. 1996; VILA et al. 1997). This increased glutamatergic tone induces an overactivity of the basal ganglia output structures, which appears to be critical for the development and the sustaining of the clinical symptoms of the disease: it has been demonstrated that subthalamotomy or the reduction of subthalamic activity by high-frequency electrical stimulation alleviates the major symptoms of PD (BERGMAN et al. 1990; AZIZ et al. 1991; SELLAL et al. 1992; BENAZZOUZ et al. 1993; LIMOUSIN et al. 1995; GURIDI et al. 1996; BLANDINI et al. 1997). Similarly, it has been reported that pharmacological EAA receptor blockade, in order to slow down the overactivity of the glutamatergic transmission, relieves the motor impairments induced by DA striatal depletion (KLOCKGETHER and TURSKI 1990; BROTCHIE et al. 1991; KLOCKGETHER et al. 1991; GREENAMYRE et al. 1994). In line with this, clinical trials using non-competitive NMDA antagonists have suggested that these drugs improve the parkinsonian symptoms (BERNHEIMER and HORNYKIEWICZ 1965; BONUCCELLI et al. 1992; MONTASTRUC et al. 1992). In addition to the symptomatic approach to the study of the glutamatergic system in PD, excitotoxicity has been investigated with the use of NMDA or AMPA receptor antagonists in experimental parkinsonism. Under these toxin-induced conditions, excitotoxicity might represent a final common pathway that operates apart from the specific aetiological factors of neuronal damage. It is therefore appropriate to analyse the experimental models of PD in order to carefully consider how excitotoxicity contributes to the nigrostriatal degeneration induced by neurotoxic agents.

2. Excitotoxicity in Experimental Parkinsonism

Experimental models of parkinsonism, as obtained with MPTP, methamphetamine or 6-OHDA, should help us to reveal the underlying mechanisms responsible for DA neuronal damage. However, the further these models are from the human disease, the more questions are raised. Indeed, one of the main weaknesses of these models consists in an artificial acute onset of the lesion, which does not reproduce the natural course of the human disease. PD, like other neurodegenerative disorders, is by nature slow and progressive. Furthermore, a reliable model should reproduce the same behavioural features as the natural disease. In rodents, for instance, in spite of their ability to lesion the DA system, these neurotoxins do not reproduce any behavioural aspect resembling human parkinsonism. On the contrary, the effects of MPTP administration in non-human primates are much more similar to what was observed by William Langston in young drug abusers after inadvertent self-administration of MPTP (LANGSTON et al. 1983; LANGSTON et al. 1984b) The resemblance between MPTP-induced parkinsonism in humans and idiopathic PD was so striking that this not only added new weight to the toxic hypothesis of causation, but also provided the most accurate primate model of the disease (LANGSTON 1998).

3. The MPTP Model

The discovery that a contaminant found in an illicit drug, MPTP, was able to induce a severe state of parkinsonism in humans gave considerably new impulse to research on PD (DAVIS et al. 1979; LANGSTON et al. 1983). Soon afterwards, intense investigation was carried out on the MPTP mechanism of action, in an effort to gain insight into the aetiology, as well as to halt the progression of the disease. The discovery of MPTP renewed interest in the toxic hypothesis of PD and stressed the role of free radicals and mitochondrial impairment as determining factors (KOPIN 1992). Furthermore, it was immediately clear that a marked species difference in sensitivity to MPTP toxicity was present. Monkeys treated with the toxin showed a behavioural syndrome resembling PD (BURNS et al. 1983). On the contrary, rats were almost completely insensitive (CHIUEH et al. 1984a), whereas mice could be lesioned only by using high doses (HALLMAN et al. 1984; HEIKKILA et al. 1984a).

MPTP is metabolized to MPP$^+$ via an intermediate 1-methyl-4-phenyl-2,3-dihydropyridinium (MDPD+), and this oxidation is catalysed by monoamine-oxidase type B (MAO-B) (CHIBA et al. 1984; MARKEY et al. 1984). This crucial sequence in the mechanism of toxicity was further confirmed by the fact that MAO-B-inhibitors administered prior to MPTP provided complete protection from MPTP toxicity both in primates and in mice (HEIKKILA et al. 1984b; LANGSTON et al. 1984a). Consequently, in order to slow down the progression of the idiopathic degenerative process, an irreversible MAO-B inhibitor, L-deprenyl, has been used in PD patients (TETRUD and LANGSTON 1989; THE PARKINSON'S STUDY GROUP 1989). The subsequent step in MPTP research focused on the mechanism of selectivity of the lesion. Soon, it became clear that DA neurons accumulate MPP$^+$ through the high-affinity DAT (JAVITCH and SNYDER 1984; JAVITCH et al. 1985). This was confirmed by studies indicating that selective DA uptake inhibitors were able to prevent MPTP toxicity (PILEBLAD and CARLSSON 1985; RICAURTE et al. 1985; SUNDSTROM and JONSSON 1985) as well as lesions due to intracerebroventricularly administered MPP$^+$ (SUNDSTROM et al. 1986).

It is widely accepted that the biochemical steps reported above are fundamental for MPTP toxicity. It is, however, important to mention that, during MPTP/MPP$^+$ biotransformation, an MAO-B-dependent adduct which binds covalently to macromolecules is formed (CORSINI et al. 1986, 1988). The covalent binding of a product of MPTP oxidation to both MAO-A and MAO-B is consistent with the progressive irreversible "suicide" or "mechanism-based" inhibition of both forms of MAO during incubation with MPTP (SALACH et al. 1984). In any case, once MPP$^+$ is formed and it is selectively taken up by DA neurons, a cascade of partially known intracellular events takes place, leading to nerve terminal derangement (SUNDSTROM et al. 1994). In order to clarify some MPP$^+$-induced cellular events, DEL ZOMPO et al. (1986) described the existence of reversible MPP$^+$ binding sites within discrete brain areas. This binding was displaced by several compounds, the most potent of which is the hypotensive and MAO-A inhibitor, debrisoquine (DEL ZOMPO et al. 1990). This MPP$^+$ site was initially interpreted as the substrate recognition site on the MAO type A enzyme, but subsequently the same authors demonstrated that MPP$^+$ binds with a high affinity to ^{3}H-tyramine sites on synaptic vesicles (DEL ZOMPO et al. 1991; VACCARI et al. 1991).

Other authors focused their attention on the formation of free radicals or hydrogen peroxides (H_2O_2) during the oxidative process of MPTP to MPP$^+$. This was initially suggested by PERRY et al. (1985) in studies showing that α-tocopherol, β-carotene, L-ascorbic acid or N-acetylcysteine, well-known as free radical scavengers, partially protect against MPTP toxicity in mice. Although a toxic role for free radicals was also suggested in other studies (PERRY et al. 1982; JOHANNESSEN et al. 1986), some investigators failed to find a significant protection of antioxidants against MPTP toxicity in monkeys (PERRY et al. 1987). Furthermore, lipid peroxidation was not apparent in MPTP-treated mice (CORONGIU et al. 1987). On the contrary, evidence regarding the role played by MPP$^+$ in inhibiting the mitochondrial function is more consistent, at least in rodents. In this animal species, MPP$^+$ is taken up by mitochondria, where it is accumulated, directly inhibiting complex I (NICKLAS et al. 1985; RAMSAY and SINGER 1986; SONSALLA and NICKLAS 1992). As a consequence of the mitochondrial impairment, the reduction in oxidized nicotinamide adenine dinucleotide (NAD) was subsequently demonstrated (SONSALLA et al. 1992a). Accordingly, MPP$^+$ toxicity has been directly associated with a failure of energy supplies in some in vitro models (DI MONTE et al. 1986; DENTON and HOWARD 1987). Similar findings were obtained in a preparation of synaptosomes, where MPP$^+$ decreased the adenosine triphosphate (ATP) content in a dose-dependent manner (SCOTCHNER et al. 1990). Subsequently, in vivo studies also confirmed that DA depletion induced by MPTP administration is preceded by a fall in ATP contents in mouse striatum (CHAN et al. 1993). The failure of the energy supply directly affects the nigrostriatal DA system, since the ventral mesencephalon and the striatum are the most affected areas, and this phenomenon is prevented by DA uptake inhibitors (CHAN et al. 1991). It is worth noting that ATP depletion is obtained with a concentration of MPP$^+$ of about $10^{-4}M$, which is reached in mice with large doses. This overall evidence regarding the mitochondrial function promoted several investigations aimed at discovering the presence of a genetic/acquired mitochondrial complex I deficiency in parkinsonian patients (SCHAPIRA et al. 1990; LESTIENNE et al. 1990). Although contradictory findings were obtained from these and other studies, the issue has broadened remarkably and it constitutes a separate section of this chapter.

Although free radical formation and mitochondrial impairment may support apparently contrasting hypotheses concerning the mechanism of action of MPP$^+$, they are not necessarily inconsistent. Actually, as suggested by SUNDSTROM et al. (1994), the impairment of the mitochondrial function may indeed trigger an increase in intracellular levels of free radicals. Accordingly, the inhibition of complex I by rotenone or MPP$^+$ has been found to increase free radical production (HASEGAWA et al. 1990).

a) Species Differences in MPTP Toxicity

Marked species differences in MPTP toxicity have been widely described: the toxin has been administered to several animal strains and species such as monkeys, dogs, rodents, amphibians and even fish (CHIUEH et al. 1984a,b; BARBEAU et al. 1985a; SCHNEIDER et al. 1986; RAPISARDI et al. 1990; YOUDIM et al. 1992). In C57 Black mice, a dose of 30mg/kg of MPTP produces an intermediate degree of striatal DA depletion which recovers almost completely in about 3 months with a minimal cell loss in the SN (ZUDDAS et al. 1989a). The same dose of MPTP is less effective in Swiss-Webster mice, which also show a wider range of variability in DA fall. In rats, a massive dose of MPTP is required to produce a minimal reduction in striatal DA (GIOVANNI et al. 1994a). On the contrary, primates are the most susceptible species, especially old world monkeys. In *Macaca fascicularis*, in particular, a dose as low as 1–3 mg/kg is sufficient to produce a series of motor symptoms which are similar to the ones observed in severely affected parkinsonian patients (BURNS et al. 1983). In these animals, a complete striatal DA depletion and a massive degeneration of DA cell bodies are observed in the SNpc (LANGSTON et al. 1984c). This led to the conclusion that the monkey model of MPTP-induced parkinsonism represents the best experimental condition to study both the pathophysiology and the symptomatic treatment of the disease. Instead, the mouse model not only does not accurately reproduce the pathochemical and behavioural features of PD but it also shows less selective neuronal lesions. This strongly suggests that the mechanism of MPTP toxicity in mice might be different from the one operating in monkeys, and that the findings observed in the mouse model should be interpreted with more accurate considerations.

Currently, although it is generally accepted that MPTP-treated non-human primates represent the best animal model for PD, due to both ethical limitations and high costs, rodents are far more used than monkeys. This important issue needs further consideration in order to investigate phenomena such as sensitivity and resistance to neurotoxins and the underlying mechanisms. In particular, it has been suggested that different strains of mice displaying a different degree of sensitivity to MPTP toxicity present a significant direct correlation with neostriatal MPP^+ levels measured ex vivo (GIOVANNI et al. 1991). In addition, species differences have also been correlated with different striatal levels of MPP^+.

b) MPP^+ Kinetics

Comparative studies among mice, rats and monkeys have suggested that the higher sensitivity of monkeys compared with rodents is reflected in a much longer striatal retention of MPP^+ (half-life about 10 days) compared with mice (half-life about 4 h) (JOHANNESSEN 1991). This evidence led to the concept that striatal MPP^+ levels represent a predictive factor for MPTP toxicity (GIOVANNI et al. 1991; JOHANNESSEN 1991). For this reason, the different sensitivity among animal species has been considered to depend on metabolic steps located upstream to the entry of MPP^+ within the DA terminals, whereas the mechanism of action of intradopaminergic MPP^+, once inside the DA cells, has been thought to be fairly similar across the different species. This assumption is partly correct. Striatal MPP^+ half-life is almost completely dependent on enzymatic metabolism inside the neuron, since the uptake of terminals is a powerful mechanism of MPP^+ clearance from extraneuronal medium. Therefore, the different half-lives, and hence the sensitivity among animal species, are probably due to different enzyme patterns within the DA neurons. However, several findings must be considered before attributing species differences in MPTP toxicity to differences in the availability of striatal MPP^+.

In particular, ZUDDAS et al. (1994) reported that species differences (i.e. mice vs rats) cannot be explained by a parallel difference in synthesis, accumulation or retention of MPP^+ within the basal ganglia. However, it could be argued that the ex vivo striatal levels of MPP^+ did not account for the MPP^+ content which is really retained within the DA synaptic terminals. Indeed, MPP^+ can be accumulated in various striatal neurons and glial cells as well (HERKENHAM et al. 1991; DI MONTE et al. 1992; RUSS et al. 1992). In vivo brain dialysis of extracellular striatal MPP^+ across different animal species might provide clues to the amount of the toxin that is really available for DA terminals. In addition, in vitro measurement of the different uptake of MPP^+ within the striatal dopaminergic synaptosomes could provide conclusive information about the "bioavailability" of striatal MPP^+ for the dopaminergic terminals. Using this pluri-methodological approach, GIOVANNI et al. (1994a,b) recently reported that post-mortem striatal MPP^+ levels, in vivo striatal extracellular MPP^+ concentrations and in vitro MPP^+ uptake into dopaminergic synaptosomes do not correlate with the higher sensitivity to MPTP of mice compared with rats. Unexpectedly, but confirming ZUDDAS's previous findings (ZUDDAS et al. 1994), ex vivo striatal MPP^+ levels were higher in rats than in mice. All these findings, obtained using various experimental approaches, rule out differences in the amount of intradopaminergic MPP^+ among different species (mice and rats) and suggest that species sensitivity may depend on the molecular effects of intraneuronal MPP^+. In line with this, GIOVANNI et al. (1994a,b) reported that similar amounts of intradopaminergic MPP^+ increased DA release 40-fold in mice, whereas this increase was much less pronounced (threefold) in rats.

If we consider striatal MPP^+ kinetics in the same animal species (mouse), important conclusions emerge suggesting contradictory issues. In this regard, a more recent study indicates that striatal MPP^+ levels do not necessarily correlate with MPTP-induced striatal DA depletion (toxicity) after different combined treatments (VAGLINI et al. 1996). This study offered confirmation of previous findings (COHEN et al. 1984) that an MAO-B inhibitor (–) deprenyl and a DA uptake blocker, GBR-12909, prevent MPTP-induced striatal DA decrease. This protective effect was accompanied by a time course of striatal MPP^+ levels which is consistent with the lack of appearance of MPP^+ due to MAO-B inhibition (deprenyl) or an accelerated clearance of MPP^+ within the striatum due to the inhibition of MPP^+ uptake from dopaminergic terminals (GBR-12909). Indeed, by preventing the storage of the toxic metabolite within the dopaminergic axons, GBR induced an accelerated clearance of MPP^+ from the striatum. Consistently with the classic correlative hypothesis (IRWIN et al. 1987a), and confirming previous data (CORSINI et al. 1985; IRWIN et al. 1987b; VAGLINI et al. 1994), the MPTP enhancer diethyldithiocarbamate (DDC) produced a potentiation of MPTP toxicity which was accompanied by increased striatal MPP$^+$ levels. These data confirmed previous findings (IRWIN et al. 1987b) showing that DDC enhances striatal MPP$^+$ levels both by increasing the maximum amount of striatal MPP$^+$ and by slowing its clearance from the striatum.

Similarly, as already reported (CORSINI et al. 1987; ZUDDAS et al. 1989a), acetaldehyde also enhanced MPTP toxicity and prolonged MPP$^+$ half-life, although unlike DDC, it did not increase MPP$^+$ levels at the peak time. By contrast, the co-administration of MK-801 with MPTP, although ineffective in preventing the long-term MPTP-induced striatal DA decrease, caused an increased striatal amount of MPP$^+$, confirming previous data (VAGLINI et al. 1994), which could be related to a decreased firing rate in the dopaminergic neurons. Similarly, nicotine in combination with MPTP produced a significant increase in the amount of striatal MPP$^+$, which did not produce any effect on striatal DA levels (FORNAI et al. 1996b). Interestingly, although MK-801 and nicotine increased the total amount of striatal MPP$^+$, they did not prolong the MPP$^+$ half-life. Indeed, after treatment with these substances, the time course of MPP$^+$ was modified by an increase in the duration of the peak levels which were still on a plateau 4h after MPTP administration. Conversely, after this time, MPP$^+$ levels rapidly fell to reach the same amount as controls 6h after treatment. Strikingly, the alpha-2 agonist clonidine caused a complete protection of MPTP toxicity, in conjunction with an increased retention of MPP$^+$ in the striatum (FORNAI et al. 1995b). Remarkably, although clonidine produced a striatal time course of MPP$^+$ that was opposite to that occurring after GBR, it produced a similar effect (complete prevention) as GBR on DA striatal levels. It is well known that clonidine reduces striatal DA utilization (ANDÉN and GRABOWSKA 1976); in this way clonidine might prolong the storage time of MPP$^+$ within the striatal dopaminergic terminals. Similarly, MK-801 and nicotine, acting as non-competitive NMDA receptor antagonists, might reduce the excitation of nigrostriatal DA terminals (OVERTON and CLARK 1991) thereby prolonging striatal MPP$^+$ storage.

Apart from the different mechanisms of action that could account for these results, our data in mice are in sharp contrast with the current belief that a direct relationship exists between striatal MPP$^+$ concentrations and the degree of MPTP-induced depletion of striatal DA (IRWIN et al. 1987a; RIACHI et al. 1988; RIACHI et al. 1989). In order to explain these conflicting results, we may, therefore, assume that striatal levels of MPP$^+$ do not reflect the exact amount of the toxin inside the DA terminals. Different metabolic pathways of MPP$^+$ outside versus inside the DA neurons of the striatum may account for the discrepancies observed under treatment with different drugs. It is likely that only DDC or acetaldehyde, which increase MPTP toxicity in mice, might affect MPP$^+$ kinetics inside the DA neurons, thus effectively prolonging its toxic property.

As mentioned above, MPP$^+$, once taken up by nigrostriatal DA terminals, causes a permanent inhibition of the respiratory chain within mitochondria in

Molecular and Cellular Events 335

mice. This effect, similarly to the above-mentioned findings about ATP depletion, is obtained at a very high concentration ($10^{-4} M$), and, even considering both the neuronal and the mitochondrial uptake of MPP⁺, it is unlikely that the toxic metabolite will reach comparably critical brain levels in the monkey, in which MPTP produces parkinsonism at a dose of a few milligrams per kilogram. Moreover, in vitro studies have also shown that MPP⁺ is toxic to all cells at concentrations >$10^{-3} M$ (SANCHEZ-RAMOS et al. 1988). It is unlikely, therefore, that MPTP toxicity in monkeys is due to mitochondrial impairment alone, and further mechanisms should be sought. On the contrary, high doses are necessary in mice to induce minimal damage to DA terminals, and huge, repeated administration is necessary to lesion cell bodies (RICAURTE et al. 1986; ZUDDAS et al. 1989a). Under these extreme circumstances, a less-selective lesion is obtained, and it is likely that MPP⁺ levels will reach the critical concentrations needed to inhibit mitochondrial function. If this is the case, it would be appropriate to carefully criticize all the data obtained in non-primate animal species.

In agreement with the above-reported findings of GIOVANNI et al. (1994a), and in an attempt to understand species sensitivity in MPTP toxicity, a massive DA release induced by intradopaminergic MPP⁺ may be assumed to be a crucial intraneuronal event leading to the toxic insult in vulnerable animals. This mechanism is further confirmed at the molecular level by VACCARI et al. (1991), who found that MPP⁺ binds with a high affinity to tyramine-labelled sites on synaptic vesicles, thus conceivably displacing DA from its storage sites. The massive DA release, under concomitant MAO inhibition, may constitute the basic mechanism of toxicity common to all the toxins that affect DA neurons.

c) Excitotoxicity in the MPTP Model

MPP⁺ alters intracellular calcium homeostasis (FREI and RICHTER 1986) and this effect is likely to be due to its modulating action on a subclass of calcium channels (CHIEUH and HUANG 1991). In 1989, SONSALLA et al. explored the possible role of EAA in contributing to methamphetamine or MPTP toxicity (SONSALLA et al. 1989). This paper suggested quite clearly that the non-competitive NMDA receptor antagonist, MK-801, does not protect against MPTP toxicity in mice, and this conclusion was subsequently confirmed in a more recent study (SONSALLA et al. 1992b). However, opposite conclusions were reached later on after intranigral MPP⁺ administration in rats (TURSKY et al. 1991). Actually, findings concerning the role played by EAAs on MPTP toxicity are controversial (TIPTON and SINGER 1993; OSSOWSKA 1994). Some authors have suggested an involvement of EAA during MPP⁺ toxicity when injected directly into the SN (TURSKY et al. 1991) or the striatum (STOREY et al. 1992) of the rat. Others could not confirm this involvement (SONSALLA et al. 1992b), even after systemic administration of MPTP in mice (SONSALLA et al. 1989; KUPSCH et al. 1992; SONSALLA et al. 1992b). Such contradictory issues appear, once again, to be related to the different animal species that have been from time to time considered and an extensive review on this specific subject has recently been published (FORNAI et al. 1997a).

Although data regarding the role of EAA in MPTP-induced parkinsonism in rodents were conflicting, further studies in primates demonstrated a clear protective effect of EAA antagonists (ZUDDAS et al. 1992; LANGE et al. 1993). These authors found that apart from preventing degeneration of DA neurons, MK-801 also prevented the appearance of parkinsonian symptoms after MPTP, although this NMDA receptor antagonist was unable to inhibit the parkinsonism that has already developed (CROSSMAN et al. 1989; CLOSE et al. 1990; RUPNIAK et al. 1992).

These data strongly support the concept of a crucial role of glutamate in MPTP toxicity and, together with the classic hypothesis about the mechanisms of action of MPTP with marked species differences, highlight the inadequacy of MPTP models in rodents to represent a reliable experimental condition to study PD.

4. Methamphetamine Toxicity

Methamphetamine is another neurotoxin which damages nigrostriatal DA pathways in a fairly selective manner. This lesion is characterized by a long-lasting depletion of striatal DA, a decrease in tyrosine-hydroxylase activity, a reduction in the density of high-affinity DA uptake sites and the histochemical observation of nerve terminal degeneration in the striatum of rodents and monkeys as well (SEIDEN et al. 1975; ELLISON et al. 1978; WAGNER et al. 1980).

The mechanisms underlying methamphetamine neurotoxicity have been studied primarily with respect to the dopaminergic system and suggest that DA contributes to the neurotoxic effects of methamphetamine. In fact, α-methyl-p-tyrosine (αMPT), a catecholamine synthesis inhibitor, prevents the toxic effects of methamphetamine (WAGNER et al. 1983; SCHMIDT et al. 1985; AXT et al. 1990), whereas L-3,4-dihydroxyphenylalanine administration reverses the protective effects of αMPT (SCHMIDT 1992; WEIHMULLER et al. 1993).

Two major hypotheses have been proposed to explain methamphetamine-induced neurotoxicity. First, it has been suggested that the ability of the drug to mobilize DA from intraneuronal pools to the extracellular space by outward transport through DAT may allow extraneuronal DA oxidation to highly reactive molecules, resulting in subsequent neurotoxicity (SEIDEN and VOSMER 1984; DE VITO and WAGNER 1989; AXT et al. 1990; MAREK et al. 1990a,b; O'DELL et al. 1991, 1993). Alternatively, redistribution of DA from synaptic vesicles to cytoplasmic compartments and consequent elevation of oxidizable DA concentrations has been postulated to be primarily responsible for DA terminal injury by amphetamines (CUBELLS et al. 1994; LIU and EDWARDS 1997; WRONA et al. 1997; UHL 1998). Thus, although DA clearly plays a role in methamphetamine neurotoxicity, the DA pool responsible for this toxicity remains unclear. Recently, elegant studies of molecular biology by FUMAGALLI et al. (1998) demonstrated that mice lacking DAT are protected against the toxic effects of methamphetamine. More recently, the same authors found that mice heterozygous for VMAT2$^{+/-}$ display an attenuated striatal extracellular DA overflow after methamphetamine treatment compared with wild-type mice. In addition, indices of hydroxyl radical (OH˙) formation were elevated by methamphetamine markedly less in VMAT2$^{+/-}$ mice than in wild-type animals. Nevertheless, more prominent DA and metabolite depletion and decrease in DAT expression were observed in heterozygous mice. These results suggest a dissociation between the ability of the drug to modulate extraneuronal DA dynamics and the degree of methamphetamine-induced neurotoxicity in VMAT2$^{+/-}$ mice. Furthermore, these data suggest that alterations in intraneuronal DA compartmentalization, rather than elevation in extraneuronal levels, may represent the primary cause for the increased vulnerability of the cell to the neurotoxic action of methamphetamine (FUMAGALLI et al. 1999).

In the pivotal study mentioned above, SONSALLA et al. (1989) reported that MK-801 prevents the fall in striatal DA and tyrosine hydroxylase activity induced by methamphetamine in mouse striatum. This finding, which has since then been confirmed and extended by further studies (FULLER et al. 1992; MURAKI et al. 1992), not only indicates that EAAs are directly involved in the mechanism of methamphetamine neurotoxicity, but also confirms that excitotoxicity represents a crucial event in nigrostriatal degeneration processes. This issue is soundly grounded on a subsequent study providing the most articulate pharmacological evidence (SONSALLA et al. 1991). In this study, both noncompetitive NMDA receptor antagonists (MK-801, phencyclidine, ketamine), as well as ifenprodil and SL 82.0715, and competitive NMDA receptor antagonists (CGS 19755, NPC 126126) markedly prevented the DA lesion induced by methamphetamine. These results have been confirmed by further investigations (MARSHALL et al. 1992; O'DELL et al. 1992). In contrast with the NMDA receptor, whose activation by glutamic acid seems to be involved in the neurotoxic effect of methamphetamine, the AMPA receptor seems not to be involved. As a matter of fact, neither 2,3-dihydroxy-6-nitro-7-sulphamoylbenzo(F)quinoxaline (NBQX) peripherally administered, nor another agonist of AMPA receptors, quisqualic acid injected directly into the striatum, affected methamphetamine toxicity (SONSALLA et al. 1992a). These findings clearly indicate that glutamic acid is directly involved in the mechanism of methamphetamine neurotoxicity through a selective NMDA-mediated activity.

5. 6-OHDA Toxicity

In 1968, MALMFORS and SACHS confirmed previous findings of THOENEN and TRANZER demonstrating that 6-OHDA is neurotoxic towards adrenergic nerve fibres, and in this connection coined the term of "chemical axotomy" (THOENEN and TRANZER 1968). Soon after, several studies from different laboratories indicated that this toxin is able to destroy central noradrenergic and dopaminergic projections, provided it is either injected directly into the brain parenchyma or into the ventricular CSF (UNGERSTEDT 1968; BLOOM et al. 1969; URETSKY and IVERSEN 1969, 1970). These pioneer studies highlighted the enormous potential of 6-OHDA as a selective tool in experimental neurobiology, pharmacology and toxicology, and methods of application have been widely reviewed (THOENEN and TRANZER 1973; BREESE 1975; JONSSON 1980; SCHALLERT and WILCOX 1985; KOSTRZEWA 1988; BAUMGARTEN and ZIMMERMANN 1992a). 6-OHDA is taken up in a fairly selective manner by catecholaminergic neurons, where it induces neurotransmitter release and metabolic cell derangement due to its rapid autoxidation to quinoidal electrophilic systems and formation of secondary reaction products. The generation of a variety of free radicals and reactive products, apart from participating in the oxidative breakdown of 6-OHDA, must be considered as the main cytotoxic mechanism mediating various aspects of the in vivo neurotoxicity of 6-OHDA (BAUMGARTEN and ZIMMERMANN 1992a).

More recently, GLINKA and YOUDIM (1995) reported that 6-OHDA inhibits the enzymes of the mitochondrial respiratory chain, NADH dehydrogenase (complex I) and cytochrome c oxidase (complex IV). These authors conclude that 6-OHDA itself, and not its oxidation products, is responsible for the neurotoxicity via inhibition of respiratory chain enzymes. A further study from this laboratory confirmed these results and concluded that free radicals are not involved in the interaction between 6-OHDA and the respiratory chain (GLINKA et al. 1997).

Closely related to this effect is the finding that 6-OHDA, at very low concentrations, induces apoptosis in the rat pheochromocytoma cell line, PC 12 (OCHU et al. 1998). This study demonstrates the involvement of a caspase-3-like protease in 6-OHDA-induced apoptosis, and that caspase inhibition is sufficient to rescue PC 12 cells from the apoptotic but not the necrotic component of 6-OHDA neurotoxicity. In line with these conclusions, DODEL et al. (1999) showed that exposure to relatively low concentrations of 6-OHDA induces apoptosis of cerebellar granule neurons and this effect is associated with activation of a caspase-3-like protease.

Accordingly, primary cultures of neocortical neurons from transgenic mice over-expressing human Bcl-2 were consistently protected against 6-OHDA toxicity in a dose-dependent manner (OFFEN et al. 1998). These authors support the concept that 6-OHDA-induced cell death is apoptotic in nature and indirectly confirm a previous report that neurons of mice deficient in Bcl-2 are more susceptible to neurotoxins.

All these recent findings suggest that the neurotoxicity induced by 6-OHDA, which was believed to be due, in part, to the production of reactive oxygen species (ROS) and/or an inhibition of mitochondrial function, might be the result of more specific cellular events carried out at very low concentrations of the toxin.

In vitro studies indicate that L-dopa, the natural precursor of DA and the commonly used antiparkinsonian drug, evokes weak excitatory responses (OLNEY et al. 1990). Moreover, its orthohydroxylated derivative 6-hydroxydopa (6-OHDOPA) has been reported to be a potent depolarizing agent (AIZENMAN et al. 1990; ROSENBERG et al. 1991). The non-NMDA receptor antagonist 6-cyano-7-nitroquinoxaline-2,3-dione (CNQX), but not 2-amino-5-phosphonopentanoic acid (AP-5), a competitive NMDA receptor antagonist, antagonized the excitatory responses evoked by both compounds which clearly proved to be neurotoxic (OLNEY et al. 1990). L-Dopa, but more potently 6-OHDOPA, produced degeneration of embryonic retinal neurons in chickens. Furthermore, 6-OHDOPA similarly to 6-OHDA, caused neuronal lesions when injected into the SN, striatum and frontal cortex of rats and when applied to rat cortical cell cultures (OLNEY et al. 1990; ROSENBERG et al. 1991; AIZENMAN et al. 1992). The non-NMDA receptor antagonist, CNQX, but not MK-801, counteracted the neurotoxic effects of L-dopa.

Indirect findings on the role of EAAs in 6-OHDA neurotoxicity emerged from in vivo lesioned animals (PIALLAT et al. 1995). The selective lesion of the subthalamic nucleus provided complete protection against the DA nigral cell loss induced by the intrastriatal injection of 6-OHDA. All these studies suggested that these compounds may exert an excitotoxic action directly or indirectly via the glutamate receptors. A biochemical study, showing that 6-OHDOPA displaced 80% of [^{3}H]AMPA from striatal binding sites, further supported this conclusion (CHA et al. 1991). In the same study, [^{3}H]kainate binding sites were also displaced by 6-OHDOPA and L-dopa to an even lower extent. These results support the concept that at least 6-OHDOPA is a potent agonist at AMPA receptors.

6. Conclusions on Excitotoxicity

In order to conclude for a specific role of EAA in DA neuron viability, however, different experimental approaches are necessary. The pharmacological approach to study the involvement of glutamate receptors in experimental parkinsonism represents only one of these multidisciplinary methodologies. Anatomical evidence, in fact, is necessary to demonstrate that glutamatergic neurons impinge on the nigrostriatal pathway and endogenous EAAs actually induce cell death and that the integrity of glutamatergic bundles really contributes to DA cell damage. Electrophysiological investigation should provide further information about whether DA cell death is produced by experimentally increasing the activity of the upstream neuronal pathway. Finally, a biochemical approach is essential in order to demonstrate that cell damage is obtained by microinfusing EAAs into the same brain regions and that, during experimental parkinsonism, an acute increase in EAA levels is observed in these areas.

SNpc receives a rich glutamatergic innervation from the cerebral cortex (FONNUM 1984) and the subthalamic nucleus (MEREU et al. 1991). Consistently, DA neurons in this area possess NMDA receptors and are stimulated, via these receptors, by glutamic acid (MEREU et al. 1991; OVERTON and CLARK 1991). The caudate-putamen nucleus also receives a glutamatergic pathway from the cerebral cortex and shows a high density of NMDA, AMPA and kainate receptors (FONNUM 1984; ALBIN et al. 1992; TALLAKSEN-GREENE et al. 1992). In the striatum, furthermore, the stimulation of NMDA receptors increases the excitability of nigrostriatal DA terminals (OVERTON and CLARK 1991). Hence, anatomical, pharmacological and biochemical evidence indicates that glutamic acid profoundly influences nigrostriatal neurons at the level of both their cell bodies and axon terminals. Accordingly, MPP$^+$ induced a massive release of glutamate and aspartate from rat striatum and this effect was antagonized by MK-801 (CARBONI et al. 1990). A similar increase in striatal glutamate release was found after repeated administration of methamphetamine to rats (NASH and YAMAMOTO 1992). Consistently, a recent study reported similar results with lesions induced by 6-OHDA (PIALLAT et al. 1995). In this study, the authors prevented 6-OHDA-induced parkinsonism in rats by lesioning the subthalamic nucleus, thus interrupting the glutamatergic activity to the SN.

Conversely, focal administrations within rat striatum of glutamate receptor agonists did not reproduce the pathochemical features of parkinsonism (OLNEY and DE GUBAREFF 1978). However, the unilateral administration of NMDA into the striatum potentiated the selective DA lesion induced by methamphetamine (SONSALLA et al. 1992a). In the light of all these findings, we may conclude that excitotoxicity has a precise role in contributing to DA cell death and that the glutamatergic tone represents an extra-neuronal event precipitating DA cell derangement.

III. Neurotrophic Factors

The progressive nature of PD and the fact that DA neuron degeneration in the SN is slow and protracted present opportunities for therapeutic intervention aimed at blocking or slowing down the degenerative process. Neurotrophic factors are good candidates for this task, and they can rescue injured neurons before they enter the irreversible death pathway in the adult brain (MUFSON et al. 1999). A large number of growth factors stimulate dopaminergic neuron survival and differentiation in cell culture systems; these include: basic FGF-2, insulin-like growth factor (IGF)-1 and -2, interleukin (IL)-6, epidermal growth factor (EGF), transforming growth factor (TGF)-α, brain-derived neurotrophic factor (BDNF), neurotrophins (NT)-3 and -4/5, neurturin, ciliary neurotrophic factor (CTNF), TGF-β1, plasminogen and lastly glial cell line-derived neurotrophic factor (GDNF) (for a review, see HEFTI 1994). While cell cultures have become an easy experimental setting to assess the efficacy of trophic factors on dopaminergic neuron survival and differentiation, it has now become clear that these data have to be validated in animal models of experimental parkinsonism. In line with this concept, few of the growth factors mentioned above have been shown to be effective in animal models of experimental parkinsonism (HEFTI 1997). Some of them have been studied in detail and will be reviewed in this chapter.

In the early years of the past decade much emphasis was placed on FGF-2; this neurotrophin appeared to be particularly important for dopaminergic cell survival and differentiation (Otto and Unsicker 1993; Casper and Blum 1995). Adult dopaminergic neurons contain FGF-2 (Bean et al. 1991; Cintra et al. 1991), which is anterogradely transported by rat nigrostriatal dopaminergic neurons (McGeer et al. 1992). Pathological analysis of parkinsonian brains revealed that FGF-2 is severely depleted in the SN (Tooyama et al. 1993). Otto and Unsicker (1990) have shown that intracerebral administration of FGF-2 to adult mice treated with MPTP promotes the recovery of the dopaminergic function. Despite these results, interest in FGF-2 has dropped in recent years because of the discovery of new and more potent factors. Nevertheless, the importance of FGF-2 in dopaminergic cell survival is still to be established, especially in relation to recent findings that demonstrated how this factor can be induced in the striatum of rodents by compounds like nicotine and MK-801, which prevent experimental parkinsonism (Belluardo et al. 1998; Maggio et al. 1998). Although these results are suggestive, a cause-effect relationship between the induction of FGF-2 and neuroprotection has not been demonstrated yet, and more experiments need to be conducted.

NT-4/5 and in particular BDNF are two other factors that have been extensively studied for their property to protect dopaminergic cells from toxic insults. Although effective in culture, both NT-4/5 and BDNF failed to prevent neuronal loss after nigrostriatal transection in adult rats and the diminution of DA levels in mice treated with MPTP (Knusel et al. 1992). Experiments have also been performed with the chronic infusion or repeated injection of BDNF into the SN of intact rats, but they gave conflicting results: dopaminergic hyperfunction in the first case (Altar et al. 1992) and dopaminergic hypofunction in the second (Lapchak et al. 1993). These data, together with the fact that TrkB, the receptor for NT-4/5 and BDNF, is expressed at a very low level in the SN (Ringstedt et al. 1993), suggest that these neurotrophins would not be able to attenuate nigrostriatal degeneration in PD.

GDNF is the first member of a new family of trophic factors distantly related to the transforming growth factor-β superfamily. It is synthesized by many cell types and affects the survival and development of a varied set of neuronal and non-neuronal cells (Lin et al. 1993; Henderson et al. 1994; Mount et al. 1995; Arenas et al. 1995; Hellmich et al. 1996; Suvanto et al. 1996). GDNF exerts its activity through the glycosyl-phosphatidylinositol-linked protein (designated GDNFR-α) which binds GDNF with a high affinity and which is expressed on GDNF-responsive cells like nigrostriatal dopaminergic neurons (Treanor et al. 1996; Trupp et al. 1996). A related trophic factor, neurturin, which shares a 42% homology with mature GDNF, has been identified and cloned (Kotzbauer et al. 1996).

In the standard 6-OHDA and MPTP lesion models in rodents, recombinant GDNF has three different effects on DA neurons: (1) direct rescue of injured or axotomized neurons when given before or shortly after the insult; (2) promotion of axonal sprouting or regeneration in chronically lesioned animals; (3) stimulation of DA turnover and function in lesioned, and possibly also intact, neurons (BJORKLUND et al. 1997; GASH et al. 1998). Recovery from nigrostriatal injury is induced by GDNF also in non-human primates. Rhesus monkeys, infused with MPTP in the right carotid artery to create a hemi-parkinsonian model, had a behavioural improvement after GDNF administration; this improvement was found in relation to three of the cardinal features of PD: bradykinesia, rigidity and postural stability (MIYOSHI et al. 1997). This functional recovery was maintained by monthly injection of GDNF in the lateral ventricle and it correlated well with the neurochemical and immunohistological marker of dopaminergic neuron survival. These promising results have prompted researchers to begin clinical trials using intraventricular injection of GDNF (KORDOWER et al. 1999).

While GDNF emerges as a potent factor in the survival of dopaminergic neurons, data from knockout animals suggest that a GDNF homologue, rather than GDNF itself, represents the natural factor that promotes survival and differentiation of dopaminergic neurons. This conclusion is supported by the fact that mice lacking GDNF develop normal nigrostriatal dopaminergic neurons (MOORE et al. 1996; PICHEL et al. 1996; SANCHEZ et al. 1996). As mentioned above, new homologues of GDNF (neurturin) are being identified and it is likely that in the near future a new GDNF-like factor that regulates dopaminergic cell development will be identified. Whereas knockout mice for GDNF do not show any impairment of nigrostriatal cells, they lack the kidney as well as the enteric nervous system (MOORE et al. 1996; PICHEL et al. 1996; SANCHEZ et al. 1996). This implies that pharmacological administration of this factor could have a potent action upon these organs, and raises the problem that many neurotrophins have multiple effects on the neuronal and non-neuronal systems and they have to be considered before systemic administration.

Another practical difficulty in the use of neurotrophic factors to halt the neurodegenerative process in PD, as well as in other neurodegenerative diseases, is that these proteins do not easily cross the blood–brain barrier. While biotechnology has provided delivery systems that allow neurotrophic factors to cross the blood–brain barrier, an alternative approach to the use of these peptides could be the use of compounds that can stimulate the synthesis of neurotrophins in specific areas of the brain. Many compounds with these characteristics have been identified so far, including convulsant agents (RIVA et al. 1992) antipsychotic drugs (RIVA et al. 1999), MK-801 (MAGGIO et al. 1998) and nicotine (BELLUARDO et al. 1998; MAGGIO et al. 1998). Few of these drugs are of practical use; nicotine and nicotinic agonists could be good candidates. It has been shown that nicotine increases FGF-2 and BDNF in the striatum and protects dopaminergic neurons in rodent models of parkinsonism. Nicotine acts by stimulating the heteropentameric nicotinic acetylcholine receptor. The heterogeneity in the sub-unit composition of this receptor in different areas of the brain has led to the discovery of drugs that selectively recognize specific nicotinic receptors. It will be of interest to study whether compounds with a high selectivity for certain sub-units could reproduce the neuroprotective effect of nicotine, without retaining its detrimental effects.

Another alternative to neurotrophic agents is offered by the so-called immunophilin ligands like cyclosporin A and FK506. These compounds have neurite-growth-promoting and neuroprotective effects in vitro on many neuronal cell types, including mesencephalic DA neurons (COSTANTINI et al. 1998). Experiments with non-immunosuppressive analogues of cyclosporin A and FK506, which are thought to act by a different mechanism (regulation of intracellular Ca^{2+} release), have shown that the neurotrophic effect can be dissociated from the immunosuppressive one (SNYDER et al. 1998). Initial experiments in 6-OHDA and MPTP-lesioned rats and mice indicate that systemically administered immunophilins may be promising therapeutic agents in neurodegenerative diseases (STEINER et al. 1997; COSTANTINI et al. 1998).

D. Intraneuronal Events

I. Oxidative Stress

Oxidative stress is perhaps the causative mechanism that has received most consideration in PD because of the oxidative metabolism of DA potentially yielding ROS (HALLIWELL and GUTTERIDGE 1985; OLANOW and TATTON 1999). Formation of ROS by DA occurs either by autoxidation with the formation of quinone and semiquinone or by enzymatic reaction through MAO with the formation of H_2O_2, which by way of the Fenton reaction can react with iron and form the highly reactive OH⁻ (OLANOW and TATTON 1999). A variety of critical biomolecules can then be damaged by ROS, leading to neurodegeneration. The hallmark of oxidative damage in the cell is the increased level of lipid peroxidation products (i.e. malondialdehyde, lipid hydroperoxide). These products have been found to be enhanced in the SN of parkinsonian patients but not in their cerebellum, indicating the anatomical localization of the oxidative damage (DEXTER et al. 1989a, 1994). Additional evidence has been provided in favour of the importance of the oxidative damage (ALAM et al. 1997a,b).

Substantially, two pathogenetic modalities are often considered when oxidative stress is accepted to be the cause of cell death: (1) an increased production of ROS; (2) a diminished brain capacity to buffer the production of ROS. If this distinction can be acceptable from the descriptive point of view, it is likely that in vivo, at least at a certain point of the degenerative process, both of the components concur to the same extent.

The most compelling evidence that an increase in ROS production can induce dopaminergic cell death in PD comes from animal studies using the selective neurotoxin 6-OHDA. This compound, the use of which has become the canonical way to induce a hemilateral lesion of the nigrostriatal tract in rodents, is actively accumulated in catecholamine neurons, and there it reaches a concentration which is high enough to destroy the neurons by oxidizing to toxic species. As we mentioned above, DA has the same oxidative potential as 6-OHDA; and, in conditions of increased turnover, diminished metabolism or reduced compartmentalization, it can reach a dangerous concentration in the cell.

The toxicity of a widely abused drug, methamphetamine, supports this view. Methamphetamine enters the neuron through the DAT, or in part by passive diffusion due to its lipophilicity (SEIDEN and SABOL 1996). Inside the neuron it increases the cytoplasmic level of DA, which, together with its reduced metabolism by MAO, causes the damage (LAVOIE and HASTINGS 1999). It is also important to mention that knockout mice heterozygous for VMAT2 showed an increase in sensitivity of the DA system to methamphetamine, most probably due to an altered intracellular compartmentalization of DA (MILLER et al. 1999). While much experimental evidence points to DA as potentially harmful to the same cells synthesizing it, no clinical data exist that support an altered homeostasis of DA as the *primum movens* in the pathogenesis of PD. However, it should be considered that the amplified variation of DA turnover that we observe in the brief period of an acute experimental intoxication should be diluted out in the time scale of a chronic disease such as PD. Therefore, a minimal alteration of DA turnover could pass undetected, even using the most sophisticated diagnostic tool.

The potential toxicity of DA leads to the controversial issue of the possible harmful effect of L-dopa therapy. L-Dopa has been suspected of accelerating the course of PD and, while it has been shown to be clearly toxic in vitro for mesencephalic cell cultures (TANAKA et al. 1991), no definitive evidence of toxicity has been provided in experimental animal models (HEFTI et al. 1981; PERRY et al. 1984) or in clinical trials (QUINN et al. 1986).

Numerous studies, using brain imaging and analytical techniques, have shown that the level of iron is elevated in the SNpc of parkinsonian patients (RIEDERER et al. 1989; DEXTER et al. 1989b; SOFIC et al. 1991; OLANOW 1992) and this increase is localized in neuromelanin-containing granules (HIRSCH et al. 1991; GOOD et al. 1992; YOUDIM et al. 1993). As we have mentioned above, the ferrous state of iron can catalyse the transformation of H_2O_2 in the highly reactive OH^-. The ferritin complex (apoferritin+iron), deputed to store iron in the tissues and to reduce the actual concentration of free iron in the cytoplasm, has been found to increase, decrease or remain unchanged (RIEDERER et al. 1989; DEXTER et al. 1990; MANN et al. 1994). As the toxicity of iron depends on the amount of its unbound form, it is evident that the level of the ferritin complex determines the extent to which an excess of iron is rendered less reactive. It is not clear how iron accumulates in the SN, or whether this is primary or secondary to the neurodegenerative process. FAUCHEUX et al. (1995) have reported an increase in lactoferrin receptor in the SN of parkinsonian patients, which may account for the increased accumulation of iron in these neurons. On the other hand, iron accumulation has been observed in other neurodegenerative diseases and in experimental conditions like 6-OHDA or MPTP lesions (VALBERG et al. 1989; CONNOR et al. 1992; OLANOW and YOUDIM 1996).

While the role of iron in the aetiopathogenesis of PD remains to be defined, it is likely that it contributes to the neurodegenerative process once this is established.

If we consider the diminution of the anti-oxidative defences, we find that glutathione (GSH), which is in part deputed to clear the excess of H_2O_2, is reduced by about 30%–40% in the SN of parkinsonian patients (RIEDERER et al. 1989; SOFIC et al. 1992; SIAN et al. 1994a). Although this reduction seems to be peculiar to PD brain, since no similar findings have been reported in other neurodegenerative diseases, its significance is not clear for at least two reasons: (1) the level of reduction does not seem to be high enough to cause degeneration in dopaminergic neurons in culture (MITHÖFER et al. 1992); (2) chronic depletion of GSH in rat does not lead to a decrease in the number of dopaminergic cells in the SN (SCHAPIRA et al. 1990). Furthermore, a corresponding increase in GSSG, the oxidized form of GSH, has not been reported in parkinsonian brains (SOFIC et al. 1992), suggesting that the reduction of GSH might not be due to oxidative stress. The cause of GSH depletion is not clear; the synthesizing enzyme does not seem to be affected. Consideration should be given to mitochondrial dysfunction; a decrease in complex I activity (MIZUNO et al. 1989; SCHAPIRA et al. 1990) could account for GSH reduction (MITHÖFER et al. 1992). Another possibility could be that the GSH reduction derives from L-dopa therapy rather than from the disease process itself; GSH could react with semiquinone radicals formed by L-dopa (or DA) and decrease progressively as the therapy proceeds (SPENCER et al. 1995).

In conclusion, considering all the data reviewed above, we can say that while the primary role of oxidative stress in PD remains unproved, it is likely that at a certain point in the natural progression of the disease, oxidative stress will become an important, if not the most important factor in the neurodegeneration of DA neurons.

II. Nitric Oxide

Intracellular calcium levels are known to predict excitotoxic neuronal death, and the increased levels following activation of NMDA receptors are buffered by mitochondria (SCHAPIRA et al. 1989). Accumulation of calcium within mitochondria, followed by mitochondrial depolarization, are critical features of excitotoxic cell death and are associated with an increased free radical production and activation of nitric oxide synthase (NOS) (DAWSON et al. 1991; BECKMAN et al. 1990). The increased generation of superoxide as well as NO radicals can lead to the production of peroxynitrite via chemical interaction of superoxide with NO. Peroxynitrite appears to be a critical mediator of cell death in both in vitro and in vivo models of excitotoxicity.

The role of NO in excitotoxicity has been demonstrated both in vitro and in vivo. NOS inhibitors block glutamate neurotoxicity in cultured striatal and cortical neurons (DAWSON et al. 1993); the selective neuronal NOS inhibitor 7-nitroindazole (7-NI) dose-dependently protects against MPTP-induced DA depletion and tyrosine hydroxylase-positive neuronal loss in the SN (SCHULZ et al. 1995; HANTRAYE et al. 1996).

Furthermore, mice deficient in neuronal NOS were resistant to MPTP neurotoxicity (PRZEDBORSKI et al. 1996). In baboons, an acute dosing regimen of MPTP results in a 94%–98% depletion of DA in the putamen and caudate nucleus. 7-NI alone had no effect on DA levels, but when co-administered with MPTP, it completely protected against MPTP-induced DA depletion in the striatum and tyrosine hydroxylase-positive neuronal loss in the SN (HANTRAYE et al. 1996). Furthermore, the administration of 7-NI protected against motor and cognitive deficits (HANTRAYE et al. 1996). It is important to note that a recent study has demonstrated that 7-NI also inhibits MAO-B; this effect might therefore be the one responsible for its ability to provide neuroprotection against MPTP toxicity (CASTAGNOLI et al. 1997). While there is no doubt that NO has a critical role in neurodegeneration, it is not clear whether its altered production represents a pathogenetic step in PD.

III. Apoptosis and Mitochondria

Apoptosis is a form of death in which genetic programs intrinsic to the cell are activated to induce its destruction. It is different from necrosis, which is a passive process that the cell simply suffers without contributing to it. Several reports have now indicated the presence of apoptotic cells in post-mortem PD brains (AGID 1995; MOCHIZUKI et al. 1996; ANGLADE et al. 1997; TATTON et al. 1998).

The key features of apoptosis involve DNA fragmentation. In the late state of the apoptotic process, there is an activation of endonucleases that cleave the DNA into fragments of different lengths, which distribute in gel electrophoresis as a classic ladder. As neurons in PD die over a prolonged period of time, it is likely that they enter the degenerative process at different times; consequently, the possibility of seeing an apoptotic cell in the SN is very limited. Nevertheless, a recent technique using in situ 3′-end labelling (ISEL), in which the cut ends of the DNA are labelled with a chromagen or a fluorochrome, has allowed the detection of apoptotic cells in parkinsonian brains. Approximately 1%–2% of SN neurons were ISEL positive in the study of AGID (1995) and MOCHIZUKI et al. (1996). If we consider the short half-life of a cell when it enters the apoptotic process, the percentage of ISEL-positive neurons seems to be too high, and does not account for the number of DA neurons that undergo degeneration as a consequence of PD. A rational explanation for these findings has been given by OLANOW and TATTON (1999) who suggested that these apoptotic cells may reflect an altered vulnerability of neurons in the SN of patients in a pre-agonal period rather than the effective number of neurons committed to death by the disease.

What makes dopaminergic neurons more vulnerable and susceptible to apoptosis is still a matter of discussion. Mitochondria could play a critical role in this respect. It has been shown that homogenates of these organelles can induce nuclear changes characteristic of apoptosis in a cell-free system (NEWMEYER et al. 1994). This finding suggests that factors contained in mitochondria can induce apoptosis, and their release in the cytoplasm could start the process. In the apoptotic scenario, the membrane potential across the inner mitochondrial membrane collapses, indicating the opening of a large conductance channel (PETIT et al. 1996). This channel, which spans the inner and outer mitochondrial membrane, is known as the permeability transition pore (LEMASTERS et al. 1997). The opening of this pore results in a volume dysregulation of mitochondria due to the hyperosmolarity of the matrix, which causes the expansion of the matrix itself. This volume expansion can eventually cause membrane rupture and extrusion of apoptotic factors such as cytochrome C from the internal mitochondrial matrix. It is interesting to note that this pore can be regulated by compounds, like ROS, which can be generated in the course of oxidative stress; therefore, mitochondria could be the actual target of ROS. Agents that keep this channel closed, such as cyclosporine-A, block apoptosis (SUSIN et al. 1996). The mitochondrion is the cell organelle devoted to energy formation by means of five respiratory chain complexes. Complex I represents the first chain of cytochromes leading to the synthesis of ATP from adenosine diphosphate (ADP), the high-level energy supply of the cell. As reported above, MPP$^+$, the toxic metabolite of MPTP, inhibits complex I, thus reducing mitochondrial respiration, resulting in cell derangement due to energy failure (DI MONTE et al. 1986; DENTON and HOWARD 1987). This important suggestion from experimental parkinsonism has allowed the investigation of complex I activity in PD. At present, after conflicting results, complex I activity is reported to be reduced in the SN and selectively in PD (SCHAPIRA et al. 1989). This reduction is moderate in entity and does not account for the extensive lesion present in the SN of these patients (MIZUNO et al. 1998). While the role of mitochondria in the pathogenesis of DA cell death remains to be established, it is important to note that mitochondrial dysfunction may be responsible for several metabolic alterations contributing to the derangement of intracellular calcium homeostasis and excessive oxidative stress. This organelle is a new target for drugs directed at the reduction of the progression of PD.

IV. Cytochrome P450 System

Cytochrome P450 enzymes are heme-containing proteins that are responsible for the oxidative metabolism of many endogenous substances as well as foreign chemicals. It is generally believed that the drug-metabolizing enzymes have evolved due to the interaction between plants and animals and that the ancestral gene of cytochrome P450 already existed 3.5 billion years ago, i.e. before the divergence of prokaryotes and eukaryotes. However, the number of new isozymes of this superfamily within eukaryotes has increased considerably during the past 800 million years, and this is believed to be the consequence of the so-called "plant-animal warfare" (GONZALEZ and NEBERT 1990).

As animals began to diverge from plants and to ingest them, plants developed toxins that protected them. In their turn, animals with new isoforms of P450 were favoured by being able to metabolize and detoxify those compounds. It is worth noting that many drugs are either extracted from plants or are derivatives of plant products.

The vast majority of drugs and other xenobiotics are degraded into more hydrophilic metabolites via a small number of metabolic pathways, mainly by P450 enzymes localized in the liver, although every tissue has some metabolic activity. The classification of these enzymes is based on gene similarity (cytochromes within families have more than 40% homology in their protein sequence) (DALY et al. 1996). CYP 1, CYP 2 and CYP 3 are the major families of the CYP system involved in drug metabolism. CYP 3A4 seems to be the most important (50% of drugs metabolized); it is followed by CYP 2D6 (20%), CYP 2C9 and CYP 2C19 (15%), whereas the remaining metabolism is carried out by CYP 2E1, CYP 2A6 and CYP 1A2 (BERTZ and GRANNEMAN 1997). Variability in drug metabolism recognizes four main causes: (1) genetic polymorphisms; (2) induction or inhibition due to concomitant drug therapies or environmental factors; (3) physiological status; (4) disease states. The genes encoding CYP 2A6, CYP 2C9, CYP 2C19 and CYP 2D6 are functionally polymorphic, therefore at least 40% of P450-dependent drug metabolism is performed by polymorphic enzymes (INGELMAN-SUNDBERG et al. 1999).

1. Cytochrome P450 in the CNS

Cytochrome P450-dependent activities have been identified in the CNS of several species including human brain (NORMAN and NEAL 1976; COHN et al. 1977; PAUL et al. 1977; SASAME et al. 1977; ROSELLI et al. 1985; LICCIONE and MAINES 1989; RAVINDRANATH et al. 1989). Apart from other components of the electron transport chain (NADPH cytochrome P450 reductase), seven different P450 isoenzymes have been detected in this tissue (WARNER et al. 1994).

Each isozyme has its own distribution in the brain, with a specific localization in one or more cell-types; isozymes are present in neurons, as well as in glial and endothelial cells (WARNER et al. 1994). In 1987, GHERSI-EGEA et al. reported that the P450 system was located predominantly in the mitochondrial fraction, and to a lesser extent in the microsomal fraction, of rat brain. Subsequently, this was confirmed also in the monkey brain (ISCAN et al. 1990). Concerning cellular localization, a preferential association with synaptic structures has been identified for some isozymes by several authors (SCHLINGER and CALLARD 1989; HANSSON et al. 1990). Synthesized in the cell body, P450 enzymes are transported to the nerve endings by the axonal flow (HANSSON et al. 1990).

It is well accepted that P450 levels in the brain are much lower than those in the liver, and that they are about 1% of the hepatic content (GHERSI-EGEA et al. 1989). Cytochrome P450 from the CNS has been purified, and its activity in reconstituted systems has been evaluated (ANANDATHEERTHAVARADA et al. 1992; BERGH and STROBEL 1992).

In mouse brain, the total amount of P450 was estimated to be 2% of the liver content (NABESHIMA et al. 1981). In this study, the authors administered phenobarbital or morphine, in an attempt to induce the P450 system in this animal species. Unfortunately, neither treatment succeeded in providing any measurable induction. In the brain of the rat, a species which is by far the most extensively studied, several isozymes have been identified using immunohistochemical and Western blot techniques, as well as molecular genetic assays and measurement of catalytic activities. The following isoenzymes have been detected: 1A1, 2B1, 2D1, 2E1, aromatase, 3β-diol hydroxylase and cytochrome P450 reductase (WARNER et al. 1994). In 1987, FONNE-PFISTER et al. reported the detection of bufuralol-hydroxylase activity in human brain microsomes, which was inhibited by antibodies toward rat liver CYP 2D1. Later on, a cDNA for CYP 2D6 from a human cDNA library was identified and sequenced (TYNDALE et al. 1991). In an autoptic study, RAVINDRANATH et al. measured the P450 content in the human brain (BHAMRE et al. 1992). Furthermore, isozyme identification, purification and catalytic activities have been performed by the same authors (RAVINDRANATH et al. 1990). It is difficult today to suggest a functional meaning for P450 in the CNS. Several reports indicate that the P450 system in this organ may contribute to the metabolism of foreign compounds and endogenous substrates as well (WARNER et al. 1994).

2. P450 System and DA Neurons

The presence of the P450 system in DA neurons has only recently been studied more in detail. The use of new modern techniques of double staining have made it possible to directly and selectively detect within DA neurons the presence of P450 enzymes, or their messenger RNAs, co-localized with tyrosine hydroxylase, the specific marker of catecholaminergic neurons. Extensive studies have been performed in rat brain as well as in human post-mortem samples, and the SN and the basal ganglia are the two most widely investigated areas in this connection. One crucial issue in identifying different isozymes is the specificity of the antibodies and antisera employed. Cross-reactivity commonly occurs due to the high degree of similarity among the various enzymes, and this partly depends on the different techniques of antibody production. Most studies have employed antisera raised against P450 systems from rat liver microsomes, where even the minor presence of different isozymes as contaminants will raise cross-reactive antibodies. To overcome all these problems of specificity, in recent studies highly specific anti-peptide antisera raised against unique peptide sequences of P450 enzymes were developed and double staining was performed with anti-tyrosine hydroxylase in rat SN and basal ganglia (WATTS et al. 1998; RIEDL et al. 1999). Co-localization was assessed by confocal laser scanning microscopy, which allowed the detection of P450 enzyme expression in DA neurons. Among the different enzymes studied, only CYP 2E1 and CYP 2C13/2C6 were found in tyrosine-hydroxylase-positive neurons in the SN (WATTS et al. 1998). Previous findings indicated the occurrence of CYP 2E1 in rat SN (HANSSON et al. 1990; SOHDA et al. 1993) and of other isozymes as well: CYP 1A1 (KOHLER et al. 1988; ANANDATHEERTHAVARADA et al. 1993), CYP 2D1 (NORRIS et al. 1996), CYP 2D4 (HEDLUND et al. 1996) and NADPH-P450 oxidoreductase (HAGLUND et al. 1984). Discrepancies exist regarding the presence of the CYP 2D family and its distribution within the brain. Although several data indicate its presence in DA neurons of the SN (RIEDL et al. 1999), conclusive results for these enzyme members still have to be achieved (NORRIS et al. 1996; WATTS et al. 1998). However, CYP 2D5 is extensively expressed in rat basal ganglia and lesioning of the nigrostriatal pathway with 6-OHDA reduced the number of neurons expressing CYP 2D5 by 50%, suggesting that this enzyme is present in DA neurons (RIEDL et al. 1999).

In man, CYP 2D6 has been reported to be associated with the DA transporter in the brain (NIZNIK et al. 1990). Consistently, many ^{3}H-GBR 12935 binding sites have been found in rat (ANDERSON 1987) and human brain (HIRAI et al. 1988; MARCUSSON and ERIKSSON 1988). One of these binding sites has been identified as the DAT, and the second has been termed the "piperazine acceptor" site (ANDERSON 1987). ALLARD et al. (1994) reported the presence of this site in the human brain and suggested its identification as CYP 2D6, which was found in several regions, including the SN and basal ganglia. The presence of CYP 2E1 in human SN was also detected by identifying mRNA for this enzyme (FARIN and OMIECINSKI 1993). In order to understand the role of these enzymes within DA neurons under physiological and pathological conditions, it seems useful to summarize their principal functions and gene regulations.

a) CYP 2D6

A genetic polymorphism has been described for CYP 2D6; this enzyme is absent in about 5%–10% of the European population (BROSEN and GRAM 1989). As a result, these people do not metabolize several drugs and are, therefore, called "poor metabolizers". Currently, there are more than 80 drugs whose metabolism depends on this enzyme, including anti-hypertensive, β-blockers and anti-depressive agents (KROEMER and EICHELBAUM 1995). CYP 2D6 is not inducible and its activity can be strongly inhibited by selective serotonin uptake inhibitors such as paroxetine or fluoxetine (HARVEY and PRESKORN 1996). The determination of a poor metabolizer can be obtained by phenotyping or by genotyping. The former consists of the oral intake of a test drug (dextromethorphan), the collection of urine for 8h and the evaluation of the parent drug/metabolite ratio (BAUMANN and JONZIER-PEREY 1988). The latter consists of a direct analysis of the mutations leading to reduced CYP 2D6 expression in the DNA extracted from leucocytes (HEIM and MEYER 1990). It has been shown that a small percentage of people have a very high

CYP 2D6 activity due to the multiduplication of the CYP 2D6 gene, resulting in an increased CYP 2D6 activity (JOHANSSON et al. 1993). The percentage of these so-called "ultrarapid metabolizers" is variable, ranging from 1.5% in Germany (GRIESE et al. 1998) to 7% in Spain (AGUNDEZ et al. 1995a) and to 29% in Ethiopia (AKLILLU et al. 1996). Genotyping the ultrarapid metabolizers consists in determining the presence of gene duplication in the DNA extracted from leucocytes (JOHANSSON et al. 1993).

b) CYP 2E1

The expression of CYP 2E1 may vary as a result of polymorphism in CYP 2E1 promoters; consequently, the levels of this enzyme are by no means constant among individuals, but they do not exhibit the marked interindividual variation characteristic of other P450 enzymes (PARKINSON 1996). CYP 2E1 was first identified as MEOS, the microsomal ethanol oxidizing system (LIEBER 1990). In addition to ethanol, CYP 2E1 catalyses the biotransformation of a large number of halogenated alkanes (GUENGERICH et al. 1991). This enzyme is inducible by ethanol and isoniazid, and is inhibited by several compounds including DDC and aldehydes (PARKINSON 1996).

3. The P450 System in PD

In 1985, BARBEAU et al. elegantly presented evidence for an association of a CYP 2D6 defect with PD (BARBEAU et al. 1985b). Indeed, they postulated that subjects with a reduced CYP 2D6 enzyme (poor metabolizers) are vulnerable for PD because of the impaired capacity to detoxify those neurotoxins that are harmful for DA neurons. It is worth noting that recent studies have actually indicated CYP 2D6 as the major detoxifying enzyme for the PD-inducing neurotoxin, MPTP (COLEMAN et al. 1996; GILHAM et al. 1997). After this bold pioneer report, however, many conflicting results were obtained in phenotypic and genotypic CYP 2D6 studies, which have recently been outlined in a comprehensive review by RIEDL et al. (1998).

As a matter of fact, several enzymes involved in the metabolism of endogenous compounds and xenobiotics have been studied in relation to PD. However, cytochrome P450 in particular drew attention, due to its ability to defend the body against xenobiotic aggression. In particular, six P450 enzymes have been examined with respect to PD: CYP 1A1 (KURTH 1993; BENNET et al. 1994; KURTH and TAKAKUBO et al. 1996), CYP 2C9 (FERRARI et al. 1990; PEETERS et al. 1994), CYP 2C19 (GUDJONSSON et al. 1990; TSUNEOKA et al. 1996), CYP 1A2, CYP 2E1 (FACTOR et al. 1989; STEVENTON et al. 1989) and CYP 2D6 (RIEDL et al. 1998). Since the first enthusiastic claim, more than 50 reports have debated the role of CYP 2D6 in the pathogenesis of PD. Subsequent phenotypic studies have failed to support a link between this isozyme and PD. Similarly, the most extensive genetic studies initially confirmed this link, but a critical analysis of the recent studies from different groups again failed to draw any definitive conclusion (RIEDL et al. 1998). Indeed, with respect to CYP 2D6, no laboratories have succeeded in replicating the initial report of SMITH et al. (1992), according to which the frequency of poor metabolizers significantly increased in a PD population. Subsequent reports have been conflicting, although some groups have claimed differences in the allelic frequency of CYP 2D6*4 and other CYP 2D6 allelic variants in PD. Two recent meta-analyses failed to find an increased frequency of poor metabolizers among PD patients (CHRISTENSEN et al. 1998; ROSTAMI-HODJEGAN et al. 1998). On the contrary, an earlier meta-analysis suggested a weak association, but this included fewer studies (MCCANN et al. 1997). As a result of their inability to observe any association, other authors performed sub-group analyses, thus suggesting a possible link with "young onset PD" (AGUNDEZ et al. 1995b) or PD with prominent tremor (AKHMEDOVA et al. 1995). Unfortunately, these findings have not been replicated either (SANDY et al. 1996). Although most studies have been negative, there are some critical issues that have recently been addressed by LE COUTEUR and MCCANN (1998) in connection with this problem. First, it is impossible, on the basis of current studies, to completely refute CYP 2D6, as, in order to have a definitive study of a statistical power, one would need almost 3,000 subjects to exclude a 50% increase in the frequency of poor metabolizers among PD patients. The second issue is that studies should consider only patients who have had neurotoxin exposure. If CYP 2D6 polymorphism influences vulnerability to PD by affecting the metabolism of an environmental neurotoxin, then studies should include only those subjects who have undergone this kind of neurotoxin exposure. The authors concluded that this stratification for toxin exposure is necessary in order to rule out the role of CYP 2D6 in the pathogenesis of PD.

This last concept of an environmental toxin and CYP 2D6, as its metabolizing enzyme, opens an old issue regarding the toxic hypothesis of PD, which originated from the incidental discovery of MPTP as a widespread impurity (LANGSTON et al. 1983). Indeed, MPTP is metabolized by some P450 enzymes and by CYP 2D6 in particular (COLEMAN et al. 1996; GILHAM et al. 1997) and has recently been discovered to be a synthetic impurity of heterocyclic drugs (KRAMER et al. 1998). In this study, the authors assessed the risk of administering MPTP orally and reported that compounds containing less than 5 ppm MPTP do not involve any neurotoxicological health risk. They concluded surprisingly that it may be assumed that MPTP is also present as a yet undiscovered minor impurity in various existing drugs (KRAMER et al. 1998). If this were true, MPTP or one of its analogues would represent the toxin probably responsible not for idiopathic PD, but for a specific subgroup of parkinsonism. In this case, CYP 2D6-related metabolism would be of extreme importance, and phenotypic and genotypic studies should be carried out on different and selected types of subjects.

One class of drugs which is extensively used and which has recently been reported to induce reversible and irreversible extrapyramidal side effects is SSRI (selective serotonin reuptake inhibitor) (LANE 1998). A recent review lists all the numerous reports on this matter and states that neurological side effects vary from parkinsonism to akathisia and tardive dyskinesia. Minor extrapyramidal symptoms, such as dystonia, myoclonus or tremor, and major disorders, such as parkinsonism and choreic movements, are elicited variably by paroxetine, fluoxetine, sertraline and fluvoxamine. However, most reports focus on the role of fluoxetine and paroxetine in inducing these motor disorders (LANE 1998). In particular, if we consider parkinsonism and related symptoms as a major toxic target of these compounds, we may conclude that SSRI occasionally induces a parkinsonian syndrome (BOUCHARD et al. 1989; JIMENEZ et al. 1994; AL ADWANI 1995; SINGH et al. 1995) or a resting tremor (JIMENEZ et al. 1994; COULTER and PILLANS 1995; SINGH et al. 1995) or a deterioration of PD patients under L-dopa treatment (BOUCHARD et al. 1989; BROD 1989; CHOUINARD and SULTAN 1992; DARIC et al. 1993; STEUR 1993; JIMENEZ et al. 1994; MECO et al. 1994; AL ADWANI 1995; ORENGO et al. 1996; SIMONS 1996; GORMLEY et al. 1997). Among the various pharmacological actions of these compounds, it is unlikely that these untoward effects are due to their selective capacity to inhibit serotonin uptake, since tricyclic antidepressants, which share the same mechanism of action, do not show such a marked predisposition to elicit extrapyramidal disorders, although some evidence exists in this connection (for a review see BOYER and FEIGHNER 1991). What is worth noting, besides, is that paroxetine, fluoxetine and nor-fluoxetine are the most potent inhibitors of CYP 2D6 activity (ALDERMAN et al. 1994; PRESKORN and MAGNUS 1994), while sertraline, fluvoxamine, and citalopram are less effective (PRESKORN and MAGNUS 1994; ERESHEFSKY 1996). Tricyclic antidepressants are also considered to be inhibitors of CYP 2D6, but with a far lower potency (PARKINSON 1996).

It has been suggested that the induction of extrapyramidal side effects by SSRI might be related to the deficient cytochrome P450 isoenzyme status (LANE 1998). However, the fact that potent inhibitors of CYP 2D6 may induce Parkinson-related disorders further strengthens the role of this enzyme in DA neuron viability. Whether this inhibition takes place in the liver or in specific brain areas is difficult to ascertain at the moment. Currently, it is also difficult to rule out, again, that the presence of a causative toxin metabolized by CYP 2D6 may trigger the motor disorders induced by SSRI, unless these drugs are under the conditions reported by KRAMER et al. (1998), i.e. a synthetic toxic impurity is present.

It is interesting to note that another example of irreversible drug-induced parkinsonism has been reported since the 1980s; that is the one observed after flunarizine treatment (CHOUZA et al. 1986; MONTASTRUC et al. 1994; NEGROTTI and CALZETTI 1997). Flunarizine has an abnormally long half-life in humans and inhibits CYP 2D6 (KARIYA et al. 1992). This property might represent, again, the fundamental mechanism leading to the death of DA neurons.

4. P450 in Experimental Parkinsonism

As reported above, in 1985 CORSINI et al. unexpectedly found that DDC markedly enhanced MPTP toxicity in mice (CORSINI et al. 1985). This effect was initially interpreted as due to the inhibition of superoxide dismutase leading to an increase in oxidative stress induced by the toxin. Subsequently, among numerous compounds tested, other enhancers of MPTP toxicity (ethanol and acetaldehyde) were found by the same authors (CORSINI et al. 1987). After this further discovery, this group suggested that these compounds could increase the potency of the toxin via an inhibition of aldehyde dehydrogenase within the striatum. The "enhancers", at the same time, prolonged the striatal half-life of MPP$^+$, the toxic metabolite of MPTP (IRWIN et al. 1987b; ZUDDAS 1989b), and this was interpreted as the causative factor of this enhancement.

However, a more recent paper by VAGLINI et al. (1996) demonstrated that striatal MPP$^+$ levels do not necessarily correlate with MPTP toxicity in the same animal species (mouse), and they further on suggested, as previously reported, that DDC-increased toxicity was probably due to an independent action on glutamate receptors (VAGLINI et al. 1996). However, as reported above, it is likely that the prolonged storage of MPP$^+$ inside the DA neurons is crucial for its toxic effects. According to this interpretation, the enzymes, which may metabolize MPP$^+$ inside the DA neurons, have a cardinal role in MPTP toxicity. As reported in the previous section, CYP 2E1 and the CYP 2D family are the most widely represented isozymes within the DA neurons (WATTS et al. 1998; RIEDL et al. 1999), and it is likely that these two P450 enzymes are responsible for MPP$^+$ clearance. As a matter of fact, DDC, ethanol and acetaldehyde have recently been discovered to be specific inhibitors of CYP 2E1 when they are acutely administered (STOTT et al. 1997). This specific inhibition inside the DA neuron may account for the increase in MPP$^+$ striatal half-life, and thus toxicity.

In general, MPP$^+$ metabolism, unlike MPTP, has been poorly investigated. JOHANNESSEN et al. (1985) postulated that MPP$^+$ may be transformed into free radical species, and other authors provided evidence for CYP 2D isoform involvement (FONNE-PFISTER and MEYER 1988; JOLIVALT et al. 1995). It is interesting to note that CYP 2E1 is associated with the metabolism of several small planar molecules, such as nitrosoamines, benzene, alcohol and 3-hydroxypiridine (PARKINSON 1996), and is present in a functional form because its levels can be induced by prior treatment with isoniazid (PARK et al. 1993). CYP 2E1 therefore may represent, in this particular case, a detoxification pathway of MPP$^+$, whose inhibition by DDC leads to an increased toxicity. A similar conclusion can be drawn for CYP 2D isozymes. CYP 2D6, the isoform present in humans and monkeys, metabolizes MPTP and MPP$^+$ probably to harmless compounds (FONNE-PFISTER and MEYER 1988; JOLIVALT et al. 1995; COLEMAN et al. 1996; GILHAM et al. 1997). Therefore, "CYP 2D6-poor metabolizers", or the drugs that inhibit this isoenzyme, may represent susceptible factors favouring the neurotoxicity induced by MPTP (BARBEAU et al. 1985b; LANE 1998).

It is worth noting that MPP$^+$-binding sites, as described by DEL ZOMPO et al. (1986), may partly correspond in mouse brain to the substrate recognition sites of CYP 2D isozymes. This MPP$^+$ binding, indeed, is displaced potently by debrisoquine and its analogues, which are good substrates for the P450 system (DEL ZOMPO et al. 1990).

MPP⁺ binding has also been studied in post-mortem brain of PD patients and, among the several brain areas analysed, only the SN showed a reduction in this binding in comparison with age-matched controls (CORSINI et al. 1988). This reduction may be interpreted as a result of CYP 2D6 loss in the SNpc following DA neuron degeneration, a finding that is similar to that observed by RIEDL et al. (1999) in rat brain after 6-OHDA lesion of DA neurons. Furthermore, CYP 2D isoforms not only metabolize the neurotoxins MPTP and/or MPP⁺, but also markedly participate in the metabolism of methamphetamine and its analogues (LIN et al. 1995). Actually, similar conclusions must be drawn for these toxic compounds that are widely abused by humans. The role of CYP 2D6-mediated metabolism of amphetamines must be considered not only for the hepatic enzyme, but also for the one present in DA neurons. At present, it is difficult to suggest the effective physiological role of this enzyme in the DA neuron. It is likely that it behaves like a guard towards endogenous or exogenous harmful intruders (false transmitters) that may affect DA metabolism. The concept of a "false transmitter" implies that endogenous chemicals may be handled within the neurons like the natural transmitter, thereby influencing the intraneuronal disposition and release of the natural transmitter (THOENEN 1969). Among the various false transmitters that affect DA neurons, tryptamine is one of the most widely studied (BAUMGARTEN and ZIMMERMAN 1992a). Tryptamine is an endogenous substrate of CYP 2D6 (MARTINEZ et al. 1997) and its implication in PD and in schizophrenia as well has been evaluated since the 1960s (BRUNE and HIMWHICH 1962; KEUHL et al. 1968; HERKERT and KEUP 1969; SMITH and KELLOW 1969).

E. Toxicity of Dopamine

Several in vitro and in vivo studies have demonstrated that DA is a toxic compound that may contribute to neurodegenerative disorders such as PD. Its toxicity consists primarily of its ability to produce ROS, such as H_2O_2, superoxide radical and OH⁻ (HALLIWELL 1992). This toxic event may occur via MAO through DA metabolism (MAKER et al. 1981) or through direct conversion of DA into reactive metabolites, which may covalently bind cell macromolecules (STOKES et al. 1999). This second pathway consists of a spontaneous oxidation of the catechol moiety to a reactive quinone, and this conversion is accelerated by the presence of transition metal ions (manganese or iron) or different enzymes (WICK et al. 1977; GRAHAM 1978; DONALDSON et al. 1982; HALLIWELL and GUTTERIDGE 1984; HASTINGS 1995; NAPOLITANO et al. 1995; STOKES et al. 1996). This property is rare among other catecholamines and suggests that DA has a higher toxic potential, as demonstrated in a neuroblastoma cell model (GRAHAM 1978). The cytotoxicity of catecholamines and of DA, in particular, has also been confirmed in dissociated rat neural cell systems (ROSENBERG 1988; MICHEL and HEFTI 1990).

In vivo studies have also indicated that DA has neurotoxic properties. In rats, intrastriatal injections of DA produced dose-dependent lesions in this brain area (FILLOUX and TOWNSEND 1993; HASTINGS et al. 1996). These lesions, after focal administration, were associated with the formation of protein–catechol conjugates, and were reduced by the co-administration of antioxidants (HASTINGS et al. 1996). Actually, DA with its oxidation product, DA quinone, readily participates in nucleophilic addition reactions, thus forming conjugates with sulphydryl groups, which are the strongest nucleophiles in the cell at physiological pH (TSE et al. 1976). This reaction takes place predominantly at position 5 on the ring, thus forming 5-S-cysteinyl-DA, and to a lesser extent at position 2 (KATO et al. 1986). Similarly, other catechol-containing molecules such as L-dopa and DOPAC may form quinones that react with the sulphydryl groups of cysteine (FORNSTEDT et al. 1986; HASTINGS and ZIGMOND 1994).

In the cell, instead, sulphydryl groups are mainly represented as cysteine, which exists largely as a free amino acid or as part of the tripeptide GSH and of proteins. GSH is the most widespread free radical scavenger in nature, and is the main detoxifying agent of living cells. This tripeptide participates in several conjugation reactions with quinones or epoxides, thus removing the toxic species and protecting cell proteins from harmful insults (SIES and KETTERER 1988). When GSH and other detoxifying agents are depleted in the cell, the toxic agent may react with the sulphydryl groups of proteins and macromolecules (BAINS and SHAW 1997). This reaction implies the covalent binding of active sites and the inactivation of the functional state of macromolecules (GILLETTE 1982). This disruptive event ultimately results in a critical derangement of cell homeostasis leading to cell death. Consistently, it has been demonstrated that the free (not protein-bound) conjugates of DA increase in guinea-pig striatum with age, under ascorbate deficiency and after reserpine exposure (FORNSTEDT and CARLSSON 1989; FORNSTEDT et al. 1990; FORNSTEDT and CARLSSON 1991). Instead, the catechols bound to proteins have been measured by HASTINGS and ZIGMOND (1994) by using an HPLC assay following isolation and acid hydrolysis of the proteins. With this procedure, the authors were able to detect a dramatic increase in the free and protein-bound conjugates during the selective toxicity of DA terminals, in response to the intrastriatal administration of DA (HASTINGS et al. 1996). In this study, a positive correlation was observed between the extent of toxicity and the amount of protein conjugates. At the same time, when ascorbate or GSH were administered during this treatment, both the lesion size and the amount of the conjugates were markedly reduced. It is interesting to note that protein-DA conjugates were observed during methamphetamine toxicity, indicating that also endogenous DA may be converted into toxic species (LAVOIE and HASTINGS 1999). This recent study is of particular interest, since it points out that neurotoxin-induced DA lesions might be due to the endogenous leakage of the neurotransmitter, which in turn reaches critical levels in the cytoplasmic medium, where it triggers protein binding.

Several studies have been performed with the aim of detecting vulnerable proteins, whose inactivation by DA quinones may be responsible for the toxic event. DA quinone has been shown to inactivate DAT and glutamate transporter (BERMAN et al. 1996; BERMAN and HASTINGS 1997a), tyrosine hydroxylase (Xu et al. 1998) and tryptophan hydroxylase (KUHN and ARTHUR 1998). It is worth noting that exposure to DA quinones alters the mitochondrial function, suggesting further speculations about the molecular basis of DA toxicity (BERMAN and HASTINGS 1997b). The hypothesis of DA toxicity as a result of quinone formation and its reactivity, despite extensive experimental support, needs further evidence. As a matter of fact, it is difficult to accept that a selective lesion leading to programmed cell death is due to an aspecific mechanism, such as covalent protein binding by reactive species. In the liver, covalent protein binding by reactive quinones or epoxides leads to the necrotic type of degenerative process (POTTER et al. 1974). It is conceivable, nevertheless, that the brain activates complex mechanisms in order to limit the affected area, and, therefore, before reaching a massive spread of the toxic agents, selective pro-apoptotic molecules are triggered.

In this connection, very recent studies try to suggest alternative mechanisms to explain DA toxicity, even though quinone formation cannot be completely ruled out (PARDINI et al. 1999; MAGGIO et al. 2000). In these studies, apomorphine has been used as an experimental tool in order to investigate DA toxicity, besides its property to stimulate DA receptors. Apomorphine retains in its molecule the chemical structure of DA and, like this neurotransmitter, is easily oxidized to a quinone, but due to its liposolubility and unlike DA, it readily crosses the membrane of all cells and tissues. This alkaloid has been tested on Chinese hamster ovary cells (CHO-K1), a cell line lacking DA receptors. Under specific conditions of culture, apomorphine and different DA-related compounds induced an antiproliferative effect that is not due to production of ROS (MAGGIO et al. 2000). On the same cell line, apomorphine at higher doses affected cell viability and induced apoptotic-type death, as measured by different methodological approaches (PARDINI et al. 1999). The authors suggest that apomorphine effects are similar to the ones exerted by some anti-cancer agents like quercetin (KANG and LIANG 1997) and genistein (MATSUKAWA et al. 1993). The cytostatic and cytotoxic effect of quercetin could be mediated in part by inhibition of protein kinase C (PKC) (FERRIOLA et al. 1989), and remarkably apomorphine has recently been found to inhibit PKC and PKA (WANG et al. 1997). It is worth noting that the IC_{50} values found to inhibit PKC ($8\mu M$) and PKA ($1\mu M$) are similar to the EC_{50} values responsible for the antiproliferative effects. As these results were obtained with an experimental tool different from DA on specific cell cultures, alternative explanations for the mechanisms of DA toxicity could be considered.

As mentioned above, the elegant study by FUMAGALLI et al. (1999) indicated that in heterozygous VMAT2 knockout mice, the increased methamphetamine neurotoxicity was accompanied by a less-pronounced increase in extracellular DA and indices of free radical formation, compared with wild-type mice. This evidence, despite the methodological limitations of this study, further points out different key mechanisms of DA toxicity. In PD, the formation of conjugates of DA was reported by SPENCER et al. (1998). In particular, free cysteinyl- and GSH-catechol derivatives were found to increase in post-mortem brain samples of PD patients, compared with controls. Protein-bound DA, however, has not been determined yet in these brain autopsies. GSH levels, on the contrary, have been measured and found to be significantly lower in the SN of parkinsonian brains (PERRY et al. 1982; SIAN et al. 1994b). This strongly supports the concept that oxidative stress and DA quinones, in particular, take place in DA neurons during the disease, and that DA adducts to proteins and other macromolecules are necessarily formed.

In the cell cultures, DA binds covalently to DNA (STOKES et al. 1996). In human cell lines, including human promyelocytic leukemia (HL-60) and glioblastoma, DA has been shown to form DNA adducts which increase with H_2O_2 exposure and are prevented by ascorbic acid (LEVAY and BODELL 1993). Furthermore, DA induces strand breaks in cultured human fibroblasts and non-enzymatic strand scissions in circular DNA (MOLDEUS et al. 1983). These findings indicate that DA has a potential genotoxic effect and this property might be exerted in living organisms.

I. DA and Apoptosis

In PD, the number of apoptotic nuclei in the SN is greater than those seen in normal aging (OLANOW and TATTON 1989). With a more sophisticated histological technique, TATTON et al. (1998) confirmed previous observations indicating that apoptosis is likely to be the type of cell death occurring in nigrostriatal DA neurons (AGID 1995; MOCHIZUKI et al. 1996). If DA itself is supposed to be the main putative cause of cell degeneration in PD, the cell death induced by this neurotransmitter should be of the same type as that occurring in the disease. Actually, DA elicits typical markers of apoptosis in several cell lines of neural origin, including cultured chick sympathetic neurons (ZIV et al. 1994), PC 12 (WALKINSHAW and WATERS 1995), a clonal catecholaminergic cell line (MASSERANO et al. 1996) and primary cultures (HOYT et al. 1997). DA-induced apoptosis is significantly inhibited by cocaine or antisense oligonucleotides that impair DA transport into the cell (SIMANTOV et al. 1996). In this connection, apomorphine, which shares similarities in its chemical structure with DA, freely crosses the cell membranes, due to its liposolubility, and induces apoptosis in several cell lines also of non-neural origin (PARDINI et al. 1999). In particular, the CHO cell line, which does not express either DA receptors or DAT, shows biochemical and histological features of apoptosis when exposed to a concentration of apomorphine of $10\mu M$. This further confirms that DA and DA-like compounds also produce cell death of an apoptotic type. It is worth noting that antioxidants such as GSH but not vitamin E or C protect neuroblastoma cells or PC 12 cells from DA-induced apoptosis (GABBAY et al. 1996; OFFEN et al. 1996). In cultured rat forebrain neurons, this protecting effect has been observed with other thiol-containing compounds that inhibited also the covalent binding of DA to proteins (HOYT et al. 1997). It is likely that scavengers act directly on DA quinones, thus preventing their reactivity against key protective proteins; however, a more selective mechanism of these thiol-containing compounds cannot be ruled out. Several studies, indeed, have reported that GSH and other thiol compounds promote the activity of growth factor receptors that antagonize the cascade of apoptotic events (KRAKER et al. 1992; CLARK and KONSTANTOPOULOS 1993; ENGL et al. 1994; SHOWALTER et al. 1997).

Furthermore, the apoptotic property of DA is confirmed by the fact that PC 12 cells overexpressing Bcl-2 show a significant resistance to DA toxicity (OFFEN et al. 1997) and that DA activates the c-Jun N-terminal kinase (JNK) pathway and increases the c-Jun protein (LUO et al. 1998).

In conclusion, DA toxicity in in vitro systems is well documented. In vivo, however, several abnormal conditions must occur in order to reach the critical levels of DA that may exert its cytotoxic potential. Huge amounts of the neurotransmitter are stored in vesicles where pH and ascorbic acid maintain the catechol residue in the reduced state. This compartmentalization of DA in the neuron represents the primary protecting event. Outside the vesicles, the enzymes which promptly metabolize DA (MAO), the transporters which take care of its clearance (DAT and VMAT), and thiol-containing compounds (GSH) are all powerful systems counteracting its toxicity. It is conceivable that acute or chronic alterations in some of these systems might lead to conditions supporting DA toxicity.

F. Conclusions About the Pathogenesis of PD

Currently, it is unthinkable to search for "the primary cause" of idiopathic Parkinson's disease. It is now clear that the common final lesion of the nigrostriatal DA pathway, leading to the clinical symptoms typical of PD, is the result of different aetiological insults. The majority of these insults, today, are known and many others will probably be discovered in the near future. Consequently, this review of the molecular and cellular events regulating DA cell viability should provide new information about the aetiology of parkinsonism.

In this chapter, we started by describing the protective role of the noradrenergic system on DA neurons. It is likely that a chronic disorder of this system may lead to the disease. We have also reported the effects of the glutamatergic system on DA cell viability and its possible implication, as a cofactor, in determining the disease. Indeed, we do not know exactly to what extent it may be held directly responsible for DA cell death, but it is likely that any chronic insult increasing the overall activity of the glutamatergic system might lead to progressive cell loss. Furthermore, in this chapter we have dealt with the main neurotoxins for the DA nigrostriatal pathway and their fundamental mechanisms responsible for acute neuronal damage. Besides opening up a new field of research into the aetiology of PD, these neurotoxins have highlighted the weakest aspects of the metabolism of DA neurons and the basic conditions necessary for the pathogenesis of the disease. Furthermore, the study of these neurotoxins, which are simple, widespread molecules, has renewed the hypothesis of a toxic parkinsonism, which nowadays we still believe to be relevant.

The discovery that MPTP is an impurity present during the synthesis of several compounds and even drugs (KRAMER et al. 1998) rouses old worries about the diffusion of the toxin and its environmental impact. Amphetamines, too, are widespread molecules and their abuse by youngsters may be responsible for early onset PD. We wonder, therefore, whether toxin-induced parkinsonism represents a real nosological entity, apart from the outbreak observed by LANGSTON et al. in 1983. On the grounds of this, it is likely that vulnerable subjects, who are "slow metabolizers" (BARBEAU et al. 1985b) in the liver and/or in DA neurons, may slowly proceed towards the disease, under chronic exposure to the toxin. This possibility further suggests that studies on the association of metabolic enzymes with PD must be performed on homogeneous populations with putative neurotoxin exposure (LE COUTEUR and MCCANN 1998). Furthermore, in this connection, the concomitant use of compounds or drugs, which may inhibit or induce the metabolizing enzymes, should be taken into consideration. This enzymatic variability, due to exposure to these agents, may markedly affect the results obtained in genotypic association studies. In particular, it is worth noting that the inhibition of CYP 2D6, as observed after selective serotonin uptake inhibitor (SSRI) treatment, may be responsible for the extrapyramidal disorders elicited by these drugs (LANE 1998). We are persuaded that toxin-induced parkinsonism is a type of disease that still has to be properly classified.

Another section of this chapter concerns the trophic factors for DA neurons. Research has indicated the factors which specifically protect DA neurons and whose deficiency may lead to an increased vulnerability. These factors protect against oxidative stress, NO toxicity and apoptotic processes, indicating that the fragile homeostasis of DA neurons is finely regulated by them. However, one should wonder why DA neurons are more vulnerable in comparison with other types of neurons. A possible answer has been put forward in the last section of this chapter: DA neurons contain dopamine, which is a highly toxic compound! Several detoxifying factors operate actively within the neuron in order to prevent the toxic potential of this neurotransmitter. DA is safely stored and packed in compartmentalized structures, i.e. vesicles. In this connection, the vesicular pH, binding sites and dehydroascorbic acid content all help to maintain these vesicles operating fully and well. The vesicular transporter (VMAT) must be intact. When these systems are affected, or other molecules interfere with them, DA itself leaks out, flooding the synapse and thus leading to toxicity. The mitochondrial MAO are crucial for inactivating DA and glutathione, as well as essential to buffer the oxidative potential of the catechol which progressively turns into a quinone (STOKES et al. 1999). DA quinone binds irreversibly to key macromolecules, which are crucial for neuronal survival (LAVOIE and HASTINGS 1999). This is the basic mechanism responsible for the vulnerability of DA neurons and for the process triggering the cascade of events leading to apoptosis. In order to search for possible aetiological factors affecting this homeostasis, we should also consider that many endogenous substances may take part in this process, including the so-called "false transmitters" (THOENEN 1969). Phenylethylamines, such as tyramine, and indoleamines, such as tryptamine, are well-known false transmitters that affect DA neurons (BAUMGARTEN and ZIMMERMAN 1992a). These indirect amines deceive DAT, by sneaking into the neuron and interfering with cytochrome P450 (MARTINEZ et al. 1997), which is supposed to inactivate foreign compounds as well as classic neurotoxins.

The identification by GOLBE (1999) of an Italo-American family from Contursi (Salerno, Italy), in which five generations suffered from PD, allowed the discovery of the first genetic alteration in this neurodegenerative disease. The substitution of the guanine 209 with an adenine in the gene of α-synuclein has proved to be the alteration responsible for parkinsonism in the Contursi family (POLYMEROPOULOS et al. 1997). As regards the hypothesis about the mechanisms regulating DA cell viability, we should ask ourselves how this theory relates to the discovery that a defective α-synuclein is responsible for some cases of familial parkinsonism. α-Synuclein is a small brain-specific protein, which is expressed to a high degree in pre-synaptic structures of neuronal pathways of the brain, since it is a component of the vesicular apparatus (MAROTEAUX and SCHELLER 1991; CLAYTON and GEORGE 1998; LAVEDAN 1998). It is likely that defective α-synucleins affect the complex machinery involved in DA storage.

If we all agree that idiopathic PD is caused by multiple factors, it is not clear how these multiple factors converge toward a common pathological picture, as is shown by the presence of Lewy bodies. Again, the parkinsonism due to the genetic alteration of α-synuclein gives us a clue to understand this apparent discrepancy. A misfolded α-synuclein, which can precipitate in the form of aggregates, has been indicated as one of the possible pathogenetic steps of the disease (CONWAY et al.1998; EL-AGNAF et al. 1998; ENGELENDER et al. 1999; GIASSON et al. 1999; NARHI et al. 1999). α-Synuclein has been detected in Lewy bodies (SPILLANTINI et al. 1997, 1998), suggesting that precipitates of this protein could contribute to the formation of these neuronal inclusions. If this were the case, processes that modify protein folding and alter their solubility could result in the formation of nuclei of protein aggregates, around which neuronal inclusions could take shape. As discussed previously, the pathogenesis of PD recognizes several conditions that can alter the structure of proteins, including the oxidative stress induced by DA. The imbalance of the oxido-reductive status of SH groups could easily disrupt the proper three-dimensional structure of proteins, inducing their precipitation in the form of aggregates. It is likely that proteins with a high turnover (for instance vesicular proteins) are the most affected by this process. We could then conclude that, if this analysis of the pathogenesis proves to be correct, hopefully we will soon have a therapeutic approach aimed at halting or even preventing the natural progression of the disease.

Acknowledgements. This work was supported by grants from the MURST 1998 Prot. 9805409813-003. We wish to thank Miss. Manuela Corsini for preparing the manuscript and Prof. Packham for reviewing the language.

References

Agid Y (1995) Aging, disease and nerve cell death. Bull Acad Natl Med 179:1193–1203
Agundez JA, Ledesma MC, Ladero JM, Benitez J (1995a) Prevalence of CYP2D6 gene duplication and its repercussion on the oxidative phenotype in a white population. Clin Pharmacol Ther 57:265–269
Agundez JA, Jimenez-Jimenez FJ, Luengo A, Bernal ML, Molina JA, Ayuso L, Vazquez A, Parra J, Duarte J, Coria F et al. (1995b) Association between the oxidative polymorphism and early onset of Parkinson's disease. Clin Pharmacol Ther 57:291–298
Aizenman E, White WF, Loring RH, Rosenberg PA (1990) A 3-4-dihydroxyphenylalanine oxidation product is a non-*N*-methyl-D-aspartate glutamayergic agonist in rat cortical neurons. Neurosci Lett 116:168–171
Aizenman E, Boeckman FA, Rosenberg PA (1992) Glutathione prevents 2,4,5-trihydroxyphenylalanine excitotoxicity by maintaining it in a reduced, non-active form. Neurosci Lett 44:233–236
Akhmedova SN, Pushnova EA, Yakimovsky AF, Avtonomov VV, Schwartz EI (1995) Frequency of a specific cytochrome P4502D6B (CYP2D6B) mutant allele in clinically differentiated groups of patients with Parkinson disease. Biochem Mol Med 54:88–90
Aklillu E, Persson I, Bertilsson L, Johansson I, Rodrigues F, Ingelman-Sundberg M (1996) Frequent distribution of ultrarapid metabolizers of debrisoquine in an Ethiopian population carrying duplicated and multiduplicated functional CYP2D6 alleles. J Pharmacol Exp Ther 278:441–446
Al Adwani A (1995) Brain damage and tardive dyskinesia. Br J Psychiatry 167:410–411
Alam ZI, Daniel SE, Lees AJ, Marsden DC, Jenner P, Halliwell B (1997a) A generalised increase in protein carbonyls in the brain in Parkinson's but not incidental Lewy body disease. J Neurochem 69:1326–1329
Alam ZI, Jenner A, Daniel SE, Lees AJ, Cairns N, Marsden CD, Jenner P, Halliwell B (1997b) Oxidative DNA damage in the parkinsonian brain: an apparent selective increase in 8-hydroxyguanine levels in substantia nigra. J Neurochem 69:1196–1203
Albin RL, Greenamyre JT (1992) Alternative excitotoxic hypothesis. Neurology 42:733–738
Albin RL, Young AB, Penney JB (1989) The functional anatomy of basal ganglia disorders. Trends Neurosci 12:366–375
Albin RL, Makowiec RL, Hollingsworth ZR, Dure LS 4th, Penney JB, Young AB (1992) Excitatory amino acid binding sites in the basal ganglia of the rat: a quantitative autoradiographic study. Neuroscience 46:35–48
Alderman J, Greenblatt DJ, Allison J, Chung M, Harrison W (1994) Desipramine pharmacokinetics with serotonin reuptake inhibitors (SSRI), paroxetine or sertraline. Neuropsychopharmacology 10 [Suppl]:263S
Allard P, Marcusson JO, Ross SB (1994) [3H]GBR-12935 binding to cytochrome P450 in the human brain. J Neurochem 62:342–348

Altar CA, Boylan CB, Jackson C, Hershenson S, Miller J, Wiegand SJ, Lindsay RM, Hyman C (1992) Brain-derived neurotrophic factor augments rotational behavior and nigrostriatal dopamine turnover in vivo. Proc Natl Acad Sci USA 89:11347–11351

Alvord EC, Forno LS (1992) Pathology. In: Koller WC (ed) Handbook of Parkinson's Disease. Dekker, New York, pp 258–259

Anandatheerthavarada HK, Boyd MR, Ravindranath V (1992) Characterization of a phenobarbital-inducible cytochrome P-450, NADPH-cytochrome P-450 reductase and reconstituted cytochrome P-450 mono-oxygenase system from rat brain. Evidence for constitutive presence in rat and human brain. Biochem J 288:483–488

Anandatheerthavarada HK, Williams JF, Wecker L (1993) Differential effect of chronic nicotine administration on brain cytochrome P4501A1/2 and P4502E1. Biochem Biophys Res Commun 194:312–318

Andén NE, Grabowska M (1976) Pharmacological evidence for a stimulation of dopamine neurons by noradrenaline neurons in the brain. Eur J Pharmacol 39:275–282

Anderson PH (1987) Biochemical and pharmacological characterization of [^{3}H]GBR-12935 binding in vitro to rat striatal membranes: labelling of the dopamine uptake complex. J Neurochem 48:1887–1896

Anglade P, Mouatt-Prigent A, Agid Y, Hirsch EC (1996) Synaptic plasticity in the striatum of patients with Parkinson's disease. Neurodegeneration 5:121–128

Anglade P, Vyas S, Javoy-Agid F, Herrero MT, Michel PP, Marquez J, Mouatt-Prigent A, Ruberg M, Hirsch EC, Agid Y (1997) Apoptosis and autophagy in nigral neurons of patients with Parkinson's disease. Histol Histopathol 12:25–31

Arenas E, Trupp M, Akerud P, Ibanez CF (1995) GDNF prevents degeneration and promotes the phenotype of brain noradrenergic neurons in vivo. Neuron 15:1465–1473

Axt KJ, Commins DL, Vosmer G, Seiden LS (1990) α-Methyl-p-tyrosine pretreatment partially prevents methamphetamine-induced endogenous neurotoxin formation. Brain res 515:269–276

Aziz TZ, Peggs D, Sambrook MA, Crossman AR (1991) Lesions of the subthalamic nucleus for the alleviation of 1-methyl-4-phenyl-1,2,3,6-tetrahydropyridine (MPTP)-induced parkinsonism in the primate. Mov Disord 4:288–292

Bains JS, Shaw CA (1997) Neurodegenerative disorders in humans: the role of glutathione in oxidative stress-mediated neuronal death. Brain Res Brain Res Rev 25:335–358

Barbeau A, Dallaire L, Buu NT, Veilleux F, Boyer H, DeLanney LE, Irwin I, Langston EB, Langston JW (1985a) New amphibian models for the study of 1-methyl-4-phenyl-1,2,3,6-tetrahydropyridine (MPTP). Life Sci 36:1125–1134

Barbeau A, Cloutier T, Roy M, Plasse L, Paris S, Poirier J (1985b) Ecogenetics of Parkinson's disease: 4-hydroxylation of debrisoquine. Lancet 2:1213–1216

Baumann P, Jonzier-Perey M (1988) GC and GC-MS procedures for simultaneous phenotyping with dextromethorphan and mephenytoin. Clin Chim Acta 171:211–222

Baumgarten HG, Zimmermann B (1992a) Neurotoxic Phenylalkylamines and Indolealkylamines. In: Herken H, Hucho F (eds) Selective Neurotoxicity. Springer, Berlin Heidelberg New York (Handbook of Experimental Pharmacology vol 102, Ch VIII, pp 225–291)

Baumgarten HG, Zimmermann B (1992b) Cellular and subcellular targets of neurotoxins: the concept of selective vulnerability. In: Herken H, Hucho F (eds) Selective Neurotoxicity. Springer, Berlin Heidelberg New York (Handbook of Experimental Pharmacology vol 102, Ch I, pp 1–27)

Bean AJ, Elde R, Cao Y, Oellig C, Tamminga C, Goldstein M, Petterson RF, Hokfelt T (1991) Expression of acidic and basic fibroblast growth factors in the substantia nigra of rat, monkey, and human. Proc Natl Acad Sci USA 88:10237–10241

Beckman JS, Beckman TW, Chen J, Marshall PA, Freeman BA (1990) Apparent hydroxyl radical production by peroxynitrite: implications for endothelial injury from nitric oxide and superoxide. Proc Natl Acad Sci USA 87:1620–1624

Belluardo N, Blum M, Mudo G, Andbjer B, Fuxe K (1998) Acute intermittent nicotine treatment produces regional increases of basic fibroblast growth factor messenger RNA and protein in the tel- and diencephalon of the rat. Neuroscience 83:723–740

Benazzouz A, Gross Ch, Féger J, Boraud T, Bioulac B (1993) Reversal of rigidity and improvement in motor performance by subthalamic high-frequency stimulation in MPTP-treated monkeys. Eur J Neurosci 5:382–389

Bennet P, Ramsden DB, Williams AC, Ho SL (1994) Cytochrome P450 1A1 (CYP1A1) gene in familial and sporadic idiopathic Parkinson's disease (IPD). Mov Disord [Suppl] 9:33

Bergh AF, Strobel HW (1992) Reconstitution of the brain mixed function oxidase system: purification of NADPH-cytochrome P450 reductase and partial purification of cytochrome P450 from whole rat brain. J Neurochem 59:575–581

Bergman H, Wichmann T, DeLong MR (1990) Reversal of experimental parkinsonism by lesions of the subthalamic nucleus. Science 249:1436–1438

Bergman H, Wichmann T, Karmon B, DeLong MR (1994) The primate subthalamic nucleus. II. Neuronal activity in the MPTP model of parkinsonism. J Neurophysiol 72:507–520

Berman SB, Hastings TG (1997a) Inhibition of glutamate transport in synaptosomes by dopamine oxidation and reactive oxygen species. J Neurochem 69:1185–1195

Berman SB, Hastings TG (1997b) Effects of dopamine oxidation products on mitochondrial function: implications for Parkinson's disease. Soc Neurosci Abstr 23:1370

Berman SB, Zigmond MJ, Hastings TG (1996) Modification of dopamine transporter function: effect of reactive oxygen species and dopamine. J Neurochem 67:593–600

Bernheimer H, Hornykiewicz O (1964) Das verhalten des dopamin metaboliten homovanillinsäure im gehirn von normalen und Parkinson kranken menschen. Arc Exp. Path Pharmakol 247:305–306

Bernheimer H, Hornykiewicz O (1965) Herabgesetzti konzentration homo-vanillansäure im gehirn von parkinson ranken menschen als ausdruk der störung des zentralen dopaminstoffwechsels. Klin Wochenschr 43:711–715

Bernheimer H, Birkmayer W, Hornykiewicz O, Jellinger K, Seitelberger F (1973) Brain dopamine and syndromes of Parkinson and Huntington: clinical, morphological and neurochemical correlations. J Neurol Sci 20:415–455

Bertrand E, Lechowicz W, Szpak GM, Dynecki J (1997) Qualitative and quantitative analysis of locus coeruleus neurons in Parkinson's disease. Folia Neuropathol 35:80–86

Bertz RJ, Granneman GR (1997) Use of in vitro and in vivo data to estimate the likelihood of metabolic pharmacokinetic interactions. Clin Pharmacokinet 32:210–258

Bhamre S, Anandatheerthavarada HK, Shankar SK, Ravindranath V (1992) Microsomal cytochrome P450 in human brain regions. Biochem Pharmacol 44:1223–1225

Bing G, Zhang Y, Watanabe Y, McEwen BS, Stone EA (1994) Locus coeruleus lesions potentiate neurotoxic effects of MPTP in dopaminergic neurons of the substantia nigra. Brain Res 668:261–265

Birkmayer W, Danielczyk W, Riederer P (1987) Symptoms and side effects in the course of Parkinson's disease. New Trends in Clinical Neuropharmacology 1:37–48

Bjorklund A, Rosenblad C, Winkler C, Kirik D (1997) Studies on neuroprotective and regenerative effects of GDNF in a partial lesion model of Parkinson's disease. Neurobiol Dis 4:186–200

Blandini F, Garcia-Osuna M, Greenamyre JT (1997) Subthalamic ablation reverses changes in basal ganglia oxidative metabolism and motor response to apomorphine induced by nigrostriatal lesion in rats. Eur J Neurosci 9:1407–1413

Bloom FE, Algeri S, Gropetti A, Revuelta A, Costa E (1969) Lesions of central norepinephrine terminals with 6-OH-dopamine: biochemistry and fine structure. Science 166:1284–1286

Bonuccelli U, Del Dotto P, Piccini P, Beghé F, Corsini GU, Muratorio A (1992) Dextromethorphan and Parkinsonism. Lancet 340:53

Bouchard RH, Pourcher E, Vincent P (1989) Fluoxetine and extrapyramidal side effects. Am J Psychiat 146:1352–1353

Boyer WF, Feighner JP (1991) Side effects of the selective serotonin re-uptake inhibitors. In: Feighner JP, Boyer WF (eds) Selective serotonin re-uptake inhibitors. John Wiley, Chichester, pp 133–152

Breese GR (1975) Chemical and immunochemical lesions by specific neurotoxin substances and antisera. In: Iversen LL, Iversen SD, Synder SH (eds) Handbook of psychopharmacology, vol 1. Plenum, New York, pp 137–189

Brod TM (1989) Fluoxetine and extrapyramidal side effects. Am J Psychiat 146: 1352–1353

Brooks DJ (1998) The early diagnosis of Parkinson's disease. Ann Neurol [Suppl] 44:510–518

Brosen K, Gram LF (1989) Clinical significance of the sparteine/debrisoquine oxidation polymorphism. Eur J Clin Pharmacol 36:537–547

Brotchie JM, Mitchell IJ, Sambrook MA, Crossman AR (1991) Alleviation of parkinsonism by antagonism of excitatory amino acid transmission in the medial segment of the globus pallidus in rat and primate. Mov Disord 6:133–138

Brune GG, Himwhich HE (1962) Indole metabolites in schizophrenic patients. Arch Gentile Psychiat 6:324–328

Burns RA, Chiueh CC, Markey SP, Ebert MH, Jacobowicz DM, Kopin IJ (1983) A primate model of parkinsonism: selective destruction of dopaminergic neurons in the pars compacta of the substantia nigra by 1-methyl-4-phenyl-1,2,3,6-tetrahydropyridine. Proc Natl Acad Sci USA 80:4546–4550

Calabresi P, Mercuri NB, Sancesario G, Bernardi G (1993) Electrophysiology of dopamine-denervated striatal neurons. Implications for Parkinson's disease. Brain 116:433–452

Calne DB (1994) Is idiopathic Parkinsonism the consequence of an event or process? Neurology 44:5–10

Calne DB, Langston JW, Martin WR, Stoessl AJ, Ruth TJ, Adam MJ, Pate BD, Schulzer M (1985) Positron emission tomography after MPTP: observations relating to the cause of Parkinson's disease. Nature 317:246–248

Carboni S, Melis F, Pani L, Hadjiconstantinou M, Rossetti ZL (1990) The non-competitive NMDA-receptor antagonist MK-801 prevents the massive release of glutamate and aspartate from rat striatum induced by 1-methyl-4-phenylpyridinium (MPP+). Neurosci Lett 117:129–133

Carlsson M, Carlsson A (1990) Interactions between glutamatergic and monoaminergic systems within the basal ganglia – implications for schizophrenia and Parkinson's disease. Trends Neurosci 13:272–277

Casper D, Blum M (1995) Epidermal growth factor and basic fibroblast growth factor protect dopaminergic neurons from glutamate toxicity in culture. J Neurochem 65:1016–1026

Castagnoli K, Palmer S, Anderson A, Bueters T, Castagnoli N Jr (1997) The neuronal nitric oxide synthase inhibitor 7-nitroindazole also inhibits the monoamine oxidase-B-catalyzed oxidation of 1-methyl-4-phenyl-1,2,3,6-tetrahydropyridine. Chem Res Toxicol 10:364–368

Cha JH, Dure LSIV, Sakurai SY, Penney JB, Young AB (1991) 2,4,5 trihydroxyphenylalanine (6-hydroxy-DOPA) displaces [^{3}H]AMPA binding in rat striatum. Neurosci Lett 132:55–58

Chan P, DeLanney LE, Irwin I, Langston JW, Di Monte DA (1991) Rapid ATP loss caused by 1-methyl-4-phenyl-1,2,3,6-tetrahydropyridine in mouse brain. J Neurochem 57:348–351

Chan P, Langston JW, Irwin I, DeLanney LE, Di Monte DA (1993) 2-deoxyglucose enhances 1-methyl-4-phenyl-1,2,3,6-tetrahydropyridine-induced ATP loss in the mouse brain. J Neurochem 61:610–616

Chan-Palay V, Asan E (1989a) Quantification of catecholamine neurons in the locus coeruleus in the human brain of normal young and older adults an in depression. J Comp Neurol 287:357–372

Chan-Palay V, Asan E (1989b) Alterations in catecholamine neurons of the locus coeruleus in senile dementia of the Alzheimer's type and in Parkinson's disease with and without dementia and depression. J Comp Neurol 287:373–392

Chiba K, Trevor A, Castagnoli N Jr (1984) Metabolism of the neurotoxic tertiary amine, MPTP, by brain monoamine oxidase. Biochem Biophys Res Commun 120:1228–1232

Chieuh CC, Huang SJ (1991) MPP+ enhances potassium-evoked striatal dopamine release through a omega-conotoxin-insensitive, tetrodotoxin- and nimodipine-sensitive calcium-dependent mechanism. Ann NY Acad Sci 635:393–396

Chiueh CC, Markey SP, Burns RS, Johannessen JN, Jacobowitz DM, Kopin I (1984a) Neurochemical and behavioral effects of 1-methyl-4-phenyl-1,2,3,6-tetrahydropyridine in rat, guinea pig and monkey. Psychopharm Bull 20:548–553

Chiueh CC, Markey SP, Burns RS, Johannessen JN, Pert A, Kopin IJ (1984b) Neurochemical and behavioral effects of systemic and intranigral administration of N-methyl-4-phenyl-1,2,3,6-tetrahydropyridine in the rat. Eur J Pharmacol 100:189–194

Choi DW (1988a) Glutamate neurotoxicity and diseases of the nervous system. Neuron 1:623–634

Choi DW (1988b) Calcium-mediated neurotoxicity: a relationship to specific channel types and role in ischemic damage. Trends Neurosci 11:456–469

Chouinard G, Sultan S (1992) A case of Parkinson's Disease exacerbated by fluoxetine. Hum Psychopharmac 7:63–66

Chouza C, Scaramelli A, Caamano JL, De Medina O, Aljanati R, Romero S (1986) Parkinsonism, tardive dyskinesia, akathisia, and depression induced by flunarizine. Lancet 1:1303–1304

Christensen PM, Gotzsche PC, Brosen K (1998) The sparteine/debrisoquine (CYP2D6) oxidation polymorphism and the risk of Parkinson's disease: a meta-analysis. Pharmacogenetics 8:473–479

Cintra A, Cao YH, Oellig C, Tinner B, Bortolotti F, Goldstein M, Pettersson RF, Fuxe K (1991) Basic FGF is present in dopaminergic neurons of the ventral midbrain of the rat. Neuroreport 2:597–600

Clark S, Konstantopoulos N (1993) Sulphydryl agents modulate insulin- and epidermal growth factor (EGF)-receptor kinase via reaction with intracellular receptor domains: differential effects on basal versus activated receptors. Biochem J 292:217–223

Clayton DF, George JM (1998) The synucleins: a family of proteins involved in synaptic function, plasticity, neurodegeneration and disease. Trends Neurosci 21:249–254

Close SP, Elliot PJ, Hayes AG, Marriot AS (1990) Effects of classical and novel agent in a MPTP-induced reversible model of Parkinson's disease. Psychopharmacology 102:295–300

Cohen G, Pasik P, Cohen B, Leist A, Mytilineou C, Yahr MD (1984) Pargyline and deprenyl prevent the neurotoxicity of 1-methyl-4-phenyl-1,2,3,6-tetrahydropyridine (MPTP) in monkeys. Eur J Pharmacol 106:209–210

Cohn JA, Alvares AP, Kappas A (1977) On the occurrence of cytochrome P-450 and aryl hydrocarbon hydroxylase activity in rat brain. J Exp Med 145:1607–1611

Coleman T, Ellis SW, Martin IJ, Lennard MS, Tucker GT (1996) 1-Methyl-4-phenyl-1,2,3,6-tetrahydropyridine (MPTP) is N-demethylated by cytochromes P450 2D6, 1A2 and 3A4 – implications for susceptibility to Parkinson's disease. J Pharmacol Exp Ther 277:685–690

Connor JR, Menzies SL, St Martin SM, Mufson EJ (1992) A histochemical study of iron, transferrin, and ferritin in Alzheimer's diseased brains. J Neurosci Res 31:75–83

Conway KA, Harper JD, Lansbury PT (1998) Accelerated in vitro fibril formation by a mutant alpha-synuclein linked to early-onset Parkinson disease. Nat Med 4: 1318–1320

Corongiu FP, Dessì MA, Banni S, Bernardi F, Piccardi MP, Del Zompo M, Corsini GU (1987) MPTP fails to induce lipid peroxidation in vivo. Biochem Pharmacol 36: 2251–2253

Corsini GU, Pintus S, Chiueh CC, Weiss JF, Kopin IJ (1985) 1-methyl-4-phenyl-1,2,3,6-tetrahydropyridine (MPTP) neurotoxicity in mice is enhanced by pretreatment with diethyldithiocarbamate. Eur J Pharmacol 119:127–128

Corsini GU, Pintus S, Bocchetta A, Piccardi MP, Del Zompo M (1986) A reactive metabolite of 1-methyl-4-phenyl-1,2,3,6-tetrahydropyridine is formed in rat brain in vitro by type B monoamine oxidase. J Pharmacol Exp Ther 238:648–652

Corsini GU, Zuddas A, Bonuccelli U, Schinelli S, Kopin IJ (1987) 1-methyl-4-phenyl-1,2,3,6-tetrahydropyridine (MPTP) neurotoxicity in mice is enhanced by ethanol or acetaldehyde. Life Sci 40:827–832

Corsini GU, Bocchetta A, Zuddas A, Piccardi MP, Del Zompo M (1988) Covalent protein binding of a metabolite of 1-methyl-4-phenyl-1,2,3,6-tetrahydropyridine to mouse and monkey brain in vivo and in vitro. Biochem Pharmacol 37:4163–4169

Costantini LC, Chaturvedi P, Armistead DM, McCaffrey PG, Deacon TW, Isacson O (1998) A novel immunophilin ligand: distinct branching effects on dopaminergic neurons in culture and neurotrophic actions after oral administration in an animal model of Parkinson's disease. Neurobiol Dis 5:97–106

Cotman CW, Monaghan DT, Ottersen OP, Storm-Mathisen J (1987) Anatomical organization of excitatory amino acid receptors and their pathways. Trends Neurosci 10:273–280

Coulter DM, Pillans PI (1995) Fluoxetine and extrapyramidal side effects. Am J Psychiatry 152:122–125

Coyle JY, Puttfarcken P (1993) Oxidative stress, glutamate, and neurodegenerative disorders. Science 262:689–695

Crossman AR, Peggs D, Boyce S, Luquin MR, Sambrook MA (1989) Effect of the NMDA antagonist MK801 on MPTP-induced parkinsonism in the monkey. Neuropharmacology 28:1271–1273

Cubells JF, Rayport S, Rajindron G, Sulzer D (1994) Methamphetamine neurotoxicity involves vacuolation of endocytic organelles and dopamine-dependent intracellular stress. J Neurosci 14:2260–2271

Daly AK, Brockmöller J, Broly F, Eichelbaum M, Evans WE, Gonzalez FJ, Huang JD, Idle JR, Ingelman-Sundberg M, Ishizaki T, Jacqz-Aigrain E, Meyer UA, Nebert DW, Steen VM, Wolf CR, Zanger UM (1996) Nomenclature for human CYP2D6 alleles. Pharmacogenetics 6:193–201

Daric C, Dollfus S, Mihout B, Omnient Y, Petit M (1993) Fluoxetine et symptomes extrapyramidaux. L'Encephale 19:61–62

Davis CG, Williams AC, Markey SP, Ebert MH, Calne ED, Reichert CM, Kopin IJ (1979) Chronic parkinsonism secondary to intravenous injection of meperidine analogues. Psychiat Res 1:249–254

Dawson VL, Dawson TM, London ED, Bredt DS, Snyder SH (1991) Nitric oxide mediates glutamate neurotoxicity in primary cortical cultures. Proc Natl Acad Sci USA 88:6368–6371

Dawson VL, Dawson TM, Bartley DA, Uhl GR, Snyder SH (1993) Mechanisms of nitric oxide-mediated neurotoxicity in primary brain cultures. J Neurosci 13:2651–2661

Del Zompo M, Bernardi F, Maggio R, Piccardi MP, Johannessen JN, Corsini GU (1986) High affinity binding sites for 1-methyl-4-phenyl-pyridinium ion (MPP+) are present in mouse brain. European J Pharmacol 129:87–92

Del Zompo M, Ruiu S, Maggio R, Piccardi MP, Corsini GU (1990) [3H]1-methyl-4-phenyl-2,3-dihydropyridinium ion binding sites in mouse brain: pharmacological and biological characterization. J Neurochem 54:1905–1910

Del Zompo M, Piccardi MP, Ruiu S, Corsini GU, Vaccari A (1991) High-affinity binding of [3H]1-methyl-4-phenyl-2,3-dihydropyridinium ion to mouse striatal membranes: putative vesicular location. Eur J Pharmacol 202:293–294

Denton T, howard BD (1987) A dopaminergic cell line variant resistant to the neurotoxin 1-methyl-4-phenyl-1,2,3,6-tetrahydropyridine. J Neurochem 49:622–629

De Vito MJ, Wagner GC (1989) Methamphetamine-induced neuronal damage: a possible role for free radicals. Neuropharmacology 28:1145–1150

Dexter DT, Carter CJ, Wells FR, Javoy-Agid F, Agid Y, Lees A, Jenner P, Marsden CD (1989a) Basal lipid peroxidation in substantia nigra is increased in Parkinson's disease. J Neurochem 52:381–389

Dexter DT, Wells FR, Lees AJ, Agid F, Agid Y, Jenner P, Marsden CD (1989b) Increased nigral iron content and alterations in other metal ions occurring in brain in Parkinson's disease. J Neurochem 52:1830–1836

Dexter DT, Carayon A, Vidailhet M, Ruberg M, Agid F, Agid Y, Lees AJ, Wells FR, Jenner P, Marsden CD (1990) Decreased ferritin levels in brain in Parkinson's disease. J Neurochem 55:16–20

Dexter DT, Holley AE, Flitter WD, Slater TF, Wells FR, Daniel SE, Lees AJ, Jenner P, Marsden CD (1994) Increased levels of lipid hydroperoxides in the parkinsonian substantia nigra: an HPLC and ESR study. Mov Disord 9:92–97

Di Monte D, Jewell SA, Ekstromm G, Sandy MS, Smith MT (1986) 1-methyl-4-phenyl-1,2,3,6-tetrahydropyridine (MPTP) and 1-methyl-4-phenylpyridine (MPP+) cause rapid ATP depletion in isolated hepatocytes. Biochem Biophys Res Commun 137:310–315

Di Monte DA, Wu EY, Irwin I, DeLanney LE, Langston JW (1992) Production and disposition of 1-methyl-4-phenylpyridinium in primary cultures of mouse astrocytes. Glia 5:48–55

Dodel RC, Du Y, Bales KR, Ling Z, Carvey PM, Paul SM (1999) Caspase-3-like proteases and 6-hydroxydopamine induced neuronal cell death. Brain Res Mol Brain Res 64:141–148

Donaldson J, McGregor D, LaBella F (1982) Manganese neurotoxicity: a model for free radical mediated neurodegeneration? Can J Physiol Pharmacol 60:1398–1405

Dooneief G, Mirabello E, Bell K, Marder K, Stern Y, Mayeux R (1992) An estimate of the incidence of depression in idiopathic Parkinson's disease. Arch Neurol 49:305–307

Dunnet SB, Bjorklund A (1999) Prospects for new restorative and neuroprotective treatments in Parkinson's disease. Nature [Suppl] 399:A32–A39

Ehringer H, Hornykiewicz O (1960) Verteilung von Noradrenalin und Dopamin (3-Hydroxytyramin) im Gehirn des Menschen und ihr verhalten bei Erkrankungen des Extrapyramidalen Systems. Klin Wochenschr 38:1236–1239

El-Agnaf OM, Jakes R, Curran MD, Wallace A (1998) Effects of the mutations Ala30 to Pro and Ala53 to Thr on the physical and morphological properties of alpha-synuclein protein implicated in Parkinson's disease. FEBS Lett 440:67–70

Ellison G, Eison MS, Huberman HS, Daniel F (1978) Long-term changes in dopaminergic innervation of caudate nucleus after continuous amphetamine administration. Science 201:276–278

Engelender S, Kaminsky Z, Guo X, Sharp AH, Amaravi RK, Kleiderlein JJ, Margolis RL, Troncoso JC, Lanahan AA, Worley PF, Dawson VL, Dawson TM, Ross CA (1999) Synphilin-1 associates with alpha-synuclein and promotes the formation of cytosolic inclusions. Nat Genet 22:110–114

Engl J, Moule M, Yip CC (1994) Dithiothreitol stimulates insulin receptor autophosphorylation at the juxtamembrane domain. Biochem Biophys Res Commun 201:1439–1444

Ereshefsky L (1996) Drug-drug interactions involving antidepressants: focus on venlafaxine. J Clin Psychopharmacol 16 [Suppl]:37S–53S

Factor SA, Weiner WJ, Hefti F (1989) Acetaminophen metabolism by cytochrome P450 monooxygenases in Parkinson's disease. Ann Neurol 26:286–288

Farin FM, Omiecinski CJ (1993) Regiospecific expression of cytochrome P-450s and microsomal epoxide hydrolase in human brain tissue. J Toxicol Environ Health 40:317–335

Faucheux BA, Nillesse N, Damier P, Spik G, Mouatt-Prigent A, Pierce A, Leveugle B, Kubis N, Hauw JJ, Agid Y et al. (1995) Expression of lactoferrin receptors is increased in the mesencephalon of patients with Parkinson disease. Proc Natl Acad Sci USA 92:9603–9607

Fearnley J, Lees AJ (1991) Parkinson's disease: neuropathology. Brain 114:2283–2301

Ferrari MD, Peeters EA, Haan J, Roos RA, Vermey P, De Wolff FA, Buruma OJ (1990) Cytochrome P450 and Parkinson's disease. Poor parahydroxylation of phenytoin. J Neurol Sci 96:153–157

Ferriola PC, Cody V, Middleton E Jr (1989) Protein kinase C inhibition by plant flavonoids. Kinetic mechanisms and structure-activity relationships. Biochem Pharmacol 38:1617–1624

Filloux F, Townsend JJ (1993) Pre- and postsynaptic neurotoxic effects of dopamine demonstrated by intrastriatal injection. Exp Neurol 119:79–88

Follesa P, Mocchetti I (1993) Regulation of basic fibroblast growth factor and nerve growth factor mRNA by beta-adrenergic receptor activation and adrenal steroids in rat central nervous system. Mol Pharmacol 43:132–138

Fonne-Pfister R, Meyer UA (1988) Xenobiotic and endobiotic inhibitors of cytochrome P-450dbl function, the target of the debrisoquine/sparteine type polymorphism. Biochem Pharmacol 37:3829–3835

Fonne-Pfister R, Bargetzi MJ, Meyer UA (1987) MPTP, the neurotoxin inducing Parkinson's disease, is a potent competitive inhibitor of human and rat cytochrome P450 isozymes (P450bufI, P450db1) catalyzing debrisoquine 4-hydroxylation. Biochem Biophys Res Commun 148:1144–1150

Fonnum F (1984) Glutamate: a neurotransmitter in the mammalian brain. J Neurochem 42:1–11

Fornai F, Bassi L, Torracca MT, Scalori V, Corsini GU (1995a) Norepinephrine loss exacerbates methamphetamine-induced striatal dopamine depletion in mice. Eur J Pharmacol 283:99–102

Fornai F, Alessandri MG, Fascetti F, Vaglini F, Corsini GU (1995b) Clonidine suppresses 1-methyl-4-phenyl-1,2,3,6-tetrahydropyridine-induced reductions of striatal dopamine and tyrosine hydroxylase activity in mice. J Neurochem 65:704–709

Fornai F, Torracca MT, Bassi L, D'Errigo DA, Scalori V, Corsini GU (1996a) Norepinephrine loss selectively enhances chronic nigro-striatal dopamine depletion in mice and rats. Brain Res 735:349–353

Fornai F, Vaglini F, Maggio R, Bonuccelli U, Corsini GU (1996b) Excitatory Amino Acids and MPTP toxicity. In: Battistin L, Scarlato G, Caraceni T, Ruggeri S (eds) Advances in Neurology. Vol. 69. Lippincot-Raven Publishers, Philadelphia, pp 167–176

Fornai F, Vaglini F, Maggio R, Bonuccelli U, Corsini GU (1997a) Species differences in the role of excitatory amino acids in experimental parkinsonism. Neurosci Biobehav Rev 21:401–415

Fornai F, Alessandri MG, Torracca MT, Bassi L, Corsini GU (1997b) Effects of noradrenergic lesions on MPTP/MPP+ kinetics and MPTP-induced nigrostriatal dopamine depletions. J Pharmacol Exp Ther 283:100–107

Fornai F, Giorgi FS, Alessandri MG, Giusiani M, Corsini GU (1999) Effects of pretreatment with DSP-4 on methamphetamine-induced striatal dopamine losses and pharmacokinetics. J Neurochem 72:777–784

Forno LS (1996) Neuropathology of Parkinson's Disease. J Neuropathol Exp Neurol 55:259–272

Forno LS, Alvord EC Jr (1971) The pathology of parkinsonism, Part I. Some new observations and correlations. J Comp Neurol 8:120–130

Fornstedt B, Carlsson A (1989) A marked rise in 5-S-cysteinyl-dopamine levels in guinea-pig striatum following reserpine treatment. J Neural Transm 76:155–161

Fornstedt B, Carlsson A (1991) Vitamin C deficiency facilitates 5-S-cysteinyldopamine formation in guinea pig striatum. J Neurochem 56:407–414

Fornstedt B, Rosengren E, Carlsson A (1986) Occurrence and distribution of 5-S-cysteinyl derivatives of dopamine, dopa and dopac in the brains of eight mammalian species. Neuropharmacology 25:451–454

Fornstedt B, Pileblad E, Carlsson A (1990) In vivo autoxidation of dopamine in guinea pig striatum increases with age. J Neurochem 55:655–659

Frei B, Richter C (1986) N-methyl-4-phenyl-pyridine (MPP+) together with 6-hydroxydopamine or dopamine stimulates calcium release from mitochondria. FEBS Lett 198:99–102

Fuller RW, Heimrick-Luecke SK, Ornstein PL (1992) Protection against amphetamine-induced neurotoxicity toward striatal dopamine neurons in rodents by LY274614, an excitatory amino acid antagonist. Neuropharmacology 31:1027–1032

Fumagalli F, Gainetdinov RR, Valenzano KJ, Caron MG (1998) Role of dopamine transporter in methamphetamine-induced neurotoxicity: evidence from mice lacking the transporter. J Neurosci 18:4861–4869

Fumagalli F, Gainetdinov RR, Wang YM, Valenzano KJ, Miller GW, Caron MG (1999) Increased methamphetamine neurotoxicity in heterozygous vesicular monoamine transporter 2 knock-out mice. J Neurosci 19:2424–2431

Gabbay M, Tauber M, Porat S, Simantov R (1996) Selective role of glutathione in protecting human neuronal cells from dopamine-induced apoptosis. Neuropharmacology 35:571–578

Gash DM, Zhang Z, Gerhardt G (1998) Neuroprotective and neurorestorative properties of GDNF. Ann Neurol [Suppl] 44:S121–S125

Gerlach M, Riederer P (1993) The pathophysiological basis of Parkinson's disease. In: Szlenyi I (ed) Series of New Drugs. Vol 1. Inhibitors of Monoamine Oxidase. Birkhauser, Basel, pp 25–50

German DC, Manaye KF, White CL (1992) Disease-specific pattern of locus coeruleus cell loss. Ann Neurol 32:667–676

Ghersi-Egea JF, Walther B, Minn A, Siest G (1987) Quantitative measurement of cerebral cytochrome P-450 by second derivative spectrophotometry. J Neurosci Methods 20:261–269

Ghersi-Egea JF, Minn A, Daval JL, Jayyosi Z, Arnould V, Souhaili-El Amri H, Siest G (1989) NADPH:cytochrome P-450(c) reductase: biochemical characterization in rat brain and cultured neurons and evolution of activity during development. Neurochem Res 14:883–887

Giasson BI, Uryu K, Trojanowski JQ, Lee VM (1999) Mutant and wild type human alpha-synucleins assemble into elongated filaments with distinct morphologies in vitro. J Biol Chem 274:7619–7622

Gilham DE, Cairns W, Paine MJ, Modi S, Poulsom R, Roberts GC, Wolf CR (1997) Metabolism of MPTP by cytochrome P4502D6 and the demonstration of 2D6 mRNA in human foetal and adult brain by in situ hybridization. Xenobiotica 27: 111–125

Gillette JR (1982) The problem of chemically reactive metabolites. Drug Metab Rev 13:941–961

Giovanni A, Sieber BA, Heikkila RE, Sonsalla PK (1991) Correlation between the neostriatal content of the 1-methyl-4-phenyl-pyridinium species and dopaminergic neurotoxicity following 1-methyl-4-phenyl-1,2,3,6-tetrahydropyridine administration to several strains of mice. J Pharmacol Exp Ther 257:691–697

Giovanni A, Sieber BA, Heikkila RE, Sonsalla PK (1994a) Studies on species sensitivity to the dopaminergic neurotoxin 1-methyl-4-phenyl 1-methyl-4-phenyl-1,2,3,6-tetrahydropyridine. Part 1: Systemic administration. J Pharmacol Exp Ther 270:1000–1007

Giovanni A, Sonsalla PK, Heikkila RE (1994b) Studies on species sensitivity to the dopaminergic neurotoxin 1-methyl-4-phenyl 1-methyl-4-phenyl-1,2,3,6-tetrahydropyridine. Part 2: Central administration of 1-methyl-4-phenylpyridinium. J Pharmacol Exp Ther 270:1008–1014

Glinka YY, Youdim MB (1995) Inhibition of mitochondrial complexes I and IV by 6-hydroxydopamine. Eur J Pharmacol 292:329–332

Glinka YY, Gassen M, Youdim MB (1997) Mechanism of 6-hydroxydopamine neurotoxicity. J Neural Transm (Suppl) 50:55–66

Golbe LI (1999) Alpha-synuclein and Parkinson's disease. Mov Disord 14:6–9

Gonzalez FJ, Nebert DW (1990) Evolution of the P450 gene superfamily: animal-plant warfare, molecular drive and human genetic differences in drug oxidation. TIGS 6:182–186

Good PF, Olanow CW, Perl DP (1992) Neuromelanin-containing neurons of the substantia nigra accumulate iron and aluminum in Parkinson's disease: a LAMMA study. Brain Res 593:343–346

Gormley N, Watters L, Lawlor BA (1997) Extrapyramidal side-effects in elderly patients exposed to selective serotonin reuptake inhibitors. Hum Psychopharmac: Clini Exp 12:139–143

Goto S, Hirano A (1991) Catecholaminergic neurons in the parabrachial nucleus of normal individuals and patients with idiopathic Parkinson's disease. Ann Neurol 30:192–196

Graham DG (1978) Oxidative pathways for catecholamines in the genesis of neuromelanin and cytotoxic quinones. Mol Pharmacol 14:633–643

Greenamyre JT, Eller RV, Zhang Z, Ovadia A, Kurlan R, Gash DM (1994) Antiparkinsonian effects of remacemide hydrochloride, a glutamate antagonist, in rodent and primate models of Parkinson's disease. Ann Neurol 35:655–661

Griese EU, Zanger UM, Brudermanns U, Gaedigk A, Mikus G, Morike K, Stuven T, Eichelbaum M (1998) Assessment of the predictive power of genotypes for the in-vivo catalytic function of CYP2D6 in a German population. Pharmacogenetics 8:15–26

Gudjonsson O, Sanz E, Alvan G, Aquilonius SM, Reviriego J (1990) Poor hydroxylator phenotypes of debrisoquine and S-mephenytoin are not over-represented in a group of patients with Parkinson's disease. Br J Clin Pharmacol 30:301–302

Guengerich FP, Kim DH, Iwasaki M (1991) Role of human cytochrome P-450 IIE1 in the oxidation of many low molecular weight cancer suspects. Chem Res Toxicol 4:168–179

Guridi J, Herrero MT, Luquin MR, Guillen J, Ruberg M, Laguna J, Vila M, Javoy-Agid F, Agid Y, Hirsch EC, Obeso JA (1996) Subthalamotomy in parkinsonian monkeys. Behavioural and biochemical analysis. Brain 119:1717–1727

Haglund L, Kohler C, Haaparanta T, Goldstein M, Gustafsson JA (1984) Presence of NADPH-cytochrome P450 reductase in central catecholaminergic neurones. Nature 307:259–262

Halliwell B (1992) Reactive oxygen species and the central nervous system. J Neurochem 59:1609–1623

Halliwell B, Gutteridge JM (1984) Oxygen toxicity, oxygen radicals, transition metals and disease. Biochem J 219:1–14

Halliwell B, Gutteridge JM (1985) Oxygen radicals and the nervous system. Trends Neurosci 8:22–29

Hallman H, Olson L, Jonsson G (1984) Neurotoxicity of the meperidine analog 1-methyl-4-phenyl-1,2,3,6-tetrahydropyridine on brain catecholamine neurons in the mouse. Eur J Pharmacol 97:133–136

Hansson T, Tindberg N, Ingelman-Sundberg M, Kohler C (1990) Regional distribution of ethanol-inducible cytochrome P450 IIE1 in the rat central nervous system. Neuroscience 34:451–463

Hantraye P, Brouillet E, Ferrante R, Palfi S, Dolan R, Matthews RT, Beal MF (1996) Inhibition of neuronal nitric oxide synthase prevents MPTP-induced parkinsonism in baboons. Nat Med 2:1017–1021

Harvey AT, Preskorn SH (1996) Cytochrome P450 enzymes: interpretation of their interactions with selective serotonin reuptake inhibitors. Part I. J Clin Psychopharmacol 16:273–285

Hasegawa E, Takeshige K, Oishi T, Murai Y, Minikami S (1990) 1-methyl-4-phenyl-pyridinium (MPP+) induces NADH-dependent superoxide formation, and enhances NADH-dependent lipid peroxidation in bovine heart submitochondrial particles. Biochem Biophys Res Commun 170:1049–1055

Hastings TG (1995) Enzymatic oxidation of dopamine: the role of prostaglandin H synthase. J Neurochem 64:919–924

Hastings TG, Zigmond MJ (1994) Identification of catechol-protein conjugates in neostriatal slices incubated with [3H]dopamine: impact of ascorbic acid and glutathione. J Neurochem 63:1126–1132

Hastings TG, Lewis DA, Zigmond MJ (1996) Role of oxidation in the neurotoxic effects of intrastriatal dopamine injections. Proc Natl Acad Sci USA 93:1956–1961

Hedlund E, Wyss A, Kainu T, Backlund M, Kohler C, Pelto-Huikko M, Gustafsson JA, Warner M (1996) Cytochrome P4502D4 in the brain: specific neuronal regulation by clozapine and toluene. Mol Pharmacol 50:342–350

Hefti F (1994) Neurotrophic factor therapy for nervous system degenerative diseases. J Neurobiol 25:1418–1435

Hefti F (1997) Pharmacology of neurotrophic factors. Annu Rev Pharmacol Toxicol 37:239–267

Hefti F, Melamed E, Bhawan J, Wurtman R (1981) Long-term administration of L-Dopa does not damage dopaminergic neurons in the mouse. Neurol 31:1194–1195

Heikkila RE, Hess A, Duvosin RC (1984a) Dopaminergic neurotoxicity of 1-methyl-4-phenyl-1,2,3,6-tetrahydropyridine in mice. Science 224:1451–1453

Heikkila RE, Manzino L, Cabbat FS, Duvoisin RC (1984b) Protection against the dopaminergic toxicity of 1-methyl-4-phenyl-1,2,3,6-tetrahydropyridine by monoamine oxidase inhibitors. Nature 311:467–469

Heim M, Meyer UA (1990) Genotyping of poor metabolisers of debrisoquine by allele-specific PCR amplification. Lancet 336:529–532

Hellmich HL, Kos L, Cho ES, Mahon KA, Zimmer A (1996) Embryonic expression of glial cell-line derived neurotrophic factor (GDNF) suggests multiple developmental roles in neural differentiation and epithelial-mesenchymal interactions. Mech Dev 54:95–105

Henderson CE, Phillips HS, Pollock RA, Davies AM, Lemeulle C, Armanini M, Simmons L, Moffet B, Vandlen RA et al. (1994) GDNF: a potent survival factor for motoneurons present in peripheral nerve and muscle. Science 266:1062–1064

Henneberry RC, Novelli A, Cox JA, Lysko PG (1989) Neurotoxicity of the N-methyl-D-aspartate receptor in energy-compromised neurons. An hypothesis for cell death in ageing and disease. Ann NY Acad Sci 568:225–233

Herken H, Hucho F (eds) (1992) Selective Neurotoxicity. Vol 102. Springer, Berlin Heidelberg New York (Handbook of Experimental Pharmacology)

Herkenham M, Little MD, Bankiewicz K, Yang SC, Markey SP, Johannessen JN (1991) Selective retention of MPP^+ within the monoaminergic systems of the primate brain following MPTP administration: an in vivo autoradiographic study. Neuroscience 40:133–158

Herkert EE, Keup W (1969) Excretion patterns of tryptamine, indoleacetic acid, and 5-hydroxyindoleacetic acid, and their correlation with mental changes in schizophrenic patients under medication with alpha-methyldopa. Psychopharmacologia 15:48–59

Hirai M, Kitamura N, Hashimoto T, Nakai T, Mita T, Shirakawa O, Yamadori T, Amano T, Noguchi-Kuno SA, Tanaka C (1988) [^{3}H]GBR-12935 binding sites in human striatal membranes: binding characteristics and changes in parkinsonians and schizophrenics. Jpn J Pharmacol 47:237–243

Hirsch EC, Brandel JP, Galle P, Javoy-Agid F, Agid Y (1991) Iron and aluminum increase in the substantia nigra of patients with Parkinson's disease: an X-ray microanalysis. J Neurochem 56:446–451

Hollerman JR, Grace AA (1992) Subthalamic nucleus cell firing in the 6-OHDA-treated rat. Basal activity and response to haloperidol. Brain Res 590:291–299

Hornykiewicz O (1991) Neurochemical pathology of Parkinson's disease-anatomical, functional and pathological aspects. Proc 10th Int Symp Park Dis Oct 7

Hornykiewicz O, Kish SJ (1986) Biochemical pathophysiology of Parkinson's disease. Adv Neurol 45:19–34

Hornykiewicz O, Pifl C (1994) The validity of the MPTP primate model for neurochemical pathology of idiopathic Parkinson's disease. In: Briley M, Marien M (eds) Noradrenergic mechanisms in Parkinson's disease. CRC Press, Boca Raton, pp 59–71

Hoyt KR, Reynolds IJ, Hastings TG (1997) Mechanisms of dopamine-induced cell death in cultured rat forebrain neurons: interactions with and differences from glutamate-induced cell death. Exp Neurol 143:269–281

Ichitani Y, Okamura H, Nagatsu I, Ibata Y (1991) Evidence for degeneration of nigral dopamine neurones after 6-hydroxydopamine into the rat striatum. Proc 10th Int Symp Park Dis Oct 158

Ingelman-Sundberg M, Oscarson M, McLellan RA (1999) Polymorphic human cytochrome P450 enzymes: an opportunity for individualized drug treatment. TIPS 20:342–349

Irwin I, Langston JW, DeLanney LE (1987a) 4-Phenylpyridine (4PP) and MPTP: The relationship between striatal MPP+ concentrations and neurotoxicity. Life Sci 40:731

Irwin I, Wu EY, DeLanney LE, Trevor L, Langston JW (1987b) The effect of diethyldithiocarbamate on the biodisposition of MPTP: an explanation for enhanced neurotoxicity. Eur J Pharmacol 141:209–217

Iscan M, Reuhl K, Weiss B, Maines MD (1990) Regional and subcellular distribution of cytochrome P-450-dependent drug metabolism in monkey brain: the olfactory bulb and the mitochondrial fraction have high levels of activity. Biochem Biophys Res Commun 169:858–863

Javitch JA, Snyder SH (1984) Uptake of MPP$^+$ by dopamine neurons explains selectivity of parkinsonism-inducing neurotoxin, MPTP. Eur J Pharmacol 106:455–456

Javitch JA, D'Amato RJ, Strittmatter SM, Snyder SH (1985) Parkinsonism-inducing neurotoxin, N-methyl-4-phenyl-1,2,3,6-tetrahydropyridine: uptake of the metabolite N-methyl-4-phenylpyridine by dopamine neurons explains selective toxicity. Proc Natl Acad Sci USA 82:2173–2177

Jimenez FJ, Tejeiro J, Martinez-Junquera G, Cabrera-Valdivia F, Alarcon J, Garcia-Albea E (1994) Parkinsonism exarcebated by paroxetine. Neurology 44:2046

Johannessen JN (1991) A model for chronic neurotoxicity: long-term retention of the neurotoxin 1-methyl-4-phenyl-pyridinium (MPP$^+$) within catecholaminergic neurons. Neurotoxicology 12:285–302

Johannessen JN, Kelner L, Hanselman D, Shih MC, Markey S (1985) In vitro oxidation of MPTP by primate neural tissue: a potential model of MPTP neurotoxicity. Neurochem Int 7:169–176

Johannessen JN, Adams JP, Shuller HM, Bacon JP, Markey AP (1986) 1-methyl-4-phenyl-pyridinium ion (MPP$^+$) induces oxidative stress in the rodent. Life Sci 38:743–749

Johansson I, Lundqvist E, Bertilsson L, Dahl ML, Sjoqvist F, Ingelman-Sundberg M (1993) Inherited amplification of an active gene in the cytochrome P450 CYP2D locus as a cause of ultrarapid metabolism of debrisoquine. Proc Natl Acad Sci USA 90:11825–11829

Jolivalt C, Minn A, Vincent-Viry M, Galteau MM, Siest G (1995) Dextromethorphan O-demethylase activity in rat brain microsomes. Neurosci Lett 187:65–68

Jonsson G (1980) Chemical neurotoxins as denervation tools in neurobiology. Annu Rev Neurosci 3:169–187

Kamisaki Y, Hamahashi T, Hamada T, Maeda K, Itoh T (1992) Presynaptic inhibition by clonidine of neurotransmitter amino acid release in several brain regions. Eur J Phamacol 217:57–63

Kang TB, Liang NC (1997) Studies on the inhibitory effects of quercetin on the growth of HL-60 leukemia cells. Biochem Pharmacol 54:1013–1018

Kariya S, Isozaki S, Narimatsu S, Suzuki T (1992) Oxidative metabolism of flunarizine in rat liver microsomes. Res Commun Chem Pathol Pharmacol 78:85–95

Kato T, Ito S, Fujita K (1986) Tyrosinase-catalyzed binding of 3,4-dihydroxyphenylalanine with proteins through the sulfhydryl group. Biochim Biophys Acta 881: 415–421

Keuhl FA, Vanden-Heuval WJA, Ormond RE (1968) Urinary metabolites in Parkinson's disease. Nature 217:471–490

Kish SJ, Shannak KS, Rajput AH, Gilbert JJ, Hornykiewicz O (1984) Cerebellar norepinephrine in patients with Parkinson's disease and control subjects. Arch Neurol 41:612–614

Klockgether T, Turski L (1990) NMDA antagonists potentiate antiparkinsonian actions of L-DOPA in monoamine-depleted rats. Ann Neurol 28:539–546

Klockgether T, Turski L, Honoré T, Zhang Z, Gash DM, Kurlan R, Greenamyre JT (1991) The AMPA receptor antagonist NBQX has antiparkinsonian effects in monoamine depleted rats and MPTP treated monkeys. Ann Neurol 30:717–723

Knusel B, Beck KD, Winslow JW, Rosenthal A, Burton LE, Widmer HR, Nikolics K, Hefti F (1992) Brain-derived neurotrophic factor administration protects basal forebrain cholinergic but not nigral dopaminergic neurons from degenerative changes after axotomy in the adult rat brain. J Neurosci 12:4391–4402

Kohler C, Eriksson LG, Hansson T, Warner M, Ake-Gustafsson J (1988) Immunohistochemical localization of cytochrome P-450 in the rat brain. Neurosci Lett 84: 109–114

Kopin IJ (1992) Mechanisms of 1-methyl-4-phenyl-1,2,3,6-tetrahydropyridine induced destruction of dopaminergic neurons. In: Herken H, Hucho F (eds) Selective Neurotoxicity. Springer, Berlin Heidelberg New York (Handbook of Experimental Pharmacology, vol 102, Ch X, pp 333–356)

Kordower JH, Palfi S, Chen EY, Ma SY, Sendera T, Cochran EJ, Cochran EJ, Mufson EJ, Penn R, Goetz CG, Comella CD (1999) Clinicopathological findings following intraventricular glial-derived neurotrophic factor treatment in a patient with Parkinson's disease. Ann Neurol 46:419–424

Kostrzewa RM (1988) Reorganization of noradrenergic neuronal systems following neonatal chemical and surgical injury. Prog Brain Res 73:405–423

Kotzbauer PT, Lampe PA, Heuckeroth RO, Golden JP, Creedon DJ, Johnson EM Jr, Milbrandt J (1996) Neurturin, a relative of glial-cell-line-derived neurotrophic factor. Nature 384:467–470

Kraker AJ, Wemple MJ, Moore CW (1992) Effect of sulphydryl reagents on the inhibition of epidermal growth factor (EGF) receptor tyrosine kinase by erbstatin. Proceedings of the American Association for Cancer Research 33:512

Kramer PJ, Caldwell J, Hofmann A, Tempel P, Weisse G (1998) Neurotoxicity risk assessment of MPTP (N-methyl-4-phenyl-1,2,3,6-tetrahydropyridine) as a synthetic impurity of drugs. Hum Exp Toxicol 17:283–293

Kroemer HK, Eichelbaum M (1995) "It's the genes, stupid". Molecular bases and clinical consequences of genetic cytochrome P450 2D6 polymorphism. Life Sci 56:2285–2298

Kuhn DM, Arthur R Jr (1998) Dopamine inactivates tryptophan hydroxylase and forms a redox-cycling quinoprotein: possible endogenous toxin to serotonin neurons. J Neurosci 18:7111–7117

Kupsch A, Loshmann PA, Sauer H, Arnold G, Renner P, Pufal D, Burg M, Wachtel H ten Bruggencate G, Oertel WH (1992) Do NMDA receptor antagonists protect against MPTP toxicity? Biochemical and immunocytochemical analysis in black mice. Brain Res 592:74–83

Kurth MC, Kurth JH (1993) Variant cytochrome P450 CYP2D6 allelic frequencies in Parkinson's disease. Am J Med Genet 48:166–168

Lane RM (1998) SSRI-induced extrapyramidal side-effects and akathisia: implications for treatment. J Psychopharmacol 12:192–214

Lange KW, Loschmann PA, Sofic E, Burg M, Horowski R, Kalveram KT, Wachtel H, Riederer P (1993) The competitive NMDA antagonist CPP protects substantia nigra neurons from MPTP-induced degeneration in primates. Naunyn Schmiedeberg's Arch Pharmacol 348:586–592

Langston JW (1998) Epidemiology versus genetics in Parkinson's disease: progress in resolving an age-old debate. Ann Neurol [Suppl] 44:S45–S52

Langston JW, Ballard P, Tetrud JW, Irwin I (1983) Chronic parkinsonism in humans due to a product of meperidine-analog synthesis. Science 219:979–980

Langston JW, Irwin I, Langston EB (1984a) Pargyline prevents MPTP-induced parkinsonism in primates. Science 225:1480–1482

Langston JW, Forno LS, Rebert CS, Irwin I (1984b) Selective nigral toxicity after systemic administration of 1-methyl-4-phenyl-1,2,3,6-tetrahydropyridine (MPTP) in the squirrel monkey. Brain Res 292:390–394

Langston JW, Irwin I, Langston EB, Forno LS (1984c) 1-methyl-4-phenyl-pyridinium ion (MPP$^+$): identification of a metabolite of MPTP, a toxin selective to the substantia nigra. Neurosci Lett 48:87–92

Lapchak PA, Beck KD, Araujo DM, Irwin I, Langston JW, Hefti F (1993) Chronic intranigral administration of brain-derived neurotrophic factor produces striatal dopaminergic hypofunction in unlesioned adult rats and fails to attenuate the decline of striatal dopaminergic function following medial forebrain bundle transection. Neuroscience 53:639–650

Lavedan C (1998) The synuclein family. Genome Res 8:871–880

LaVoie MJ, Hastings TG (1999) Dopamine quinone formation and protein modification associated with striatal neurotoxicity of methamphetamine: evidence against a role for extracellular dopamine. J Neurosci 19:1484–1491

Le Couteur DG, McCann SJ (1998) P450 enzymes and Parkinson's disease. Mov Disord 13:851–852

Lemasters JJ, Nieminen AL, Qian T, Trost LC, Herman B (1997) The mitochondrial permeability transition in toxic, hypoxic and reperfusion injury. Mol Cell Biochem 174:159–165

Lestienne P, Nelson I, Riederer P, Jellinger K, Reichmann H (1990) Normal mitochondrial genome in brain from patients with Parkinson's disease and complex I defect. J Neurochem 55:1810–1812

Levay G, Bodell WJ (1993) Detection of dopamine – DNA adducts: potential role in Parkinson's disease. Carcinogenesis 14:1241–1245

Liccione JJ, Maines MD (1989) Manganese-mediated increase in the rat brain mitochondrial cytochrome P-450 and drug metabolism activity: susceptibility of the striatum. J Pharmacol Exp Ther 248:222–228

Lieber CS (1990) Mechanism of ethanol induced hepatic injury. Pharmacol Ther 46:1–41

Limousin P, Pollak P, Benazzouz A, Hoffmann D, Lebas JF, Broussolle E, Perret JE, Benabid AL (1995) Effect on parkinsonian signs and symptoms of bilateral subthalamic nucleus stimulation. Lancet 345:91–95

Lin LF, Doherty DH, Lile JD, Bektesh S, Collins F (1993) GDNF: a glial cell line-derived neurotrophic factor for midbrain dopaminergic neurons. Science 260:1130–1132

Lin LY, Kumagai Y, Hiratsuka A, Narimatsu S, Suzuki T, Funae Y, Distefano EW, Cho AK (1995) Cytochrome P4502D isozymes catalyze the 4-hydroxylation of methamphetamine enantiomers. Drug Metab Dispos 23:610–614

Lipton SA, Rosenberg PA (1994) Excitatory amino acids as a final common pathway for neurologic disorders. New Engl J Med 330:613–622

Liu Y, Edwards RH (1997) The role of vesicular transport proteins in synaptic transmission and neural degeneration. Annu Rev Neurosci 20:125–156

Luo Y, Umegaki H, Wang X, Abe R, Roth GS (1998) Dopamine induces apoptosis through an oxidation-involved SAPK/JNK activation pathway. J Biol Chem 273: 3756–3764

Maggio R, Riva M, Vaglini F, Fornai F, Molteni R, Armogida M, Racagni G, Corsini GU (1998) Nicotine prevents experimental parkinsonism in rodents and induces striatal increase of neurotrophic factors. J Neurochem 71:2439–2446

Maggio R, Armogida M, Scarselli M, Salvadori F, Longoni BM, Pardini C, Chiarenza A, Chiacchio S, Vaglini F, Bernardini R, Colzi A, Corsini GU (2000) Dopamine agonists and analogues have an antiproliferative effect on CHO-K1 cells. Neurotoxicity Res 1:285–297

Maker HS, Weiss C, Silides DJ, Cohen G (1981) Coupling of dopamine oxidation (monoamine oxidase activity) to glutathione oxidation via the generation of hydrogen peroxide in rat brain homogenates. J Neurochem 36:589–593

Malmfors T, Sachs C (1968) Degeneration of adrenergic nerves produced by 6-hydroxydopamine. Eur J Pharmacol 3:89–92

Mann VM, Cooper JM, Daniel SE, Srai K, Jenner P, Marsden CD, Schapira AH (1994) Complex I, iron, and ferritin in Parkinson's disease substantia nigra. Ann Neurol 36:876–881

Marcusson J, Eriksson K (1988) [3H]GBR-12935 binding to dopamine uptake sites in the human brain. Brain Res 457:122–129

Marek CJ, Vosmer G, Seiden LS (1990a) Dopamine uptake inhibitors block long-term neurotoxic effects of methamphetamine upon dopaminergic neurons. Brain Res 513:274–279

Marek CJ, Vosmer G, Seiden LS (1990b) The effects of monoamine uptake inhibitors and methamphetamine on neostriatal 6-hydroxydopamine (6-OHDA) formation, short-term monoamine depletions and locomotor activity in the rat. Brain Res 516:1–7

Marien M, Briley M, Colpaert F (1993) Noradrenaline depletion exacerbates MPTP-induced striatal dopamine loss in mice. Eur J Pharmacol 236:487–489

Markey SP, Johannessen JN, Chiueh CC, Burns RS, Herkenham MA (1984) Intraneuronal generation of a pyridinium metabolite may cause drug-induced parkinsonism. Nature 311:464–467

Maroteaux L, Scheller RH (1991) The rat brain synucleins; family of proteins transiently associated with neuronal membrane. Brain Res Mol Brain Res 11:335–343

Marshall JF, O'Dell SJ, Weihmuller FB (1992) Dopamine-glutamate interactions in methamphetamine-induced neurotoxicity. J Neural Transm (Gen Sec) 91:241–254

Martinez C, Agundez JA, Gervasini G, Martin R, Benitez J (1997) Tryptamine: a possible endogenous substrate for CYP2D6. Pharmacogenetics 7:85–93

Masserano JM, Gong L, Kulaga H, Baker I, Wyatt RJ (1996) Dopamine induces apoptotic cell death of a catecholaminergic cell line derived from the central nervous system. Mol Pharmacol 50:1309–1315

Matsukawa Y, Marui N, Sakai T, Satomi Y, Yoshida M, Matsumoto K, Nishino H, Aoike A (1993) Genistein arrests cell cycle progression at G2-M. Cancer Res 53:1328–1331

Mavridis M, Degryse AD, Lategan AJ, Marien MR, Colpaert FC (1991) Effects of locus coeruleus lesions on parkinsonian signs, striatal dopamine and substantia nigra cell loss after of 1-methyl-4-phenyl-1,2,3,6-tetrahydropyridine in monkeys: a possible role for the locus coeruleus in the progression of Parkinson's disease. Neuroscience 41:507–523

Mayer RA, Kindt MV, Heikkila RE (1986) Prevention of the nigrostriatal toxicity of 1-methyl-4-phenyl-1,2,3,6-tetrahydropyridine by inhibitors of 3,4-dihydroxyphenyl-ethylamine transport. J Neurochem 47:1073–1079

McCann SJ, Pond SM, James KM, Le Couteur DG (1997) The association between polymorphisms in the cytochrome P-450 2D6 gene and Parkinson's disease: a case-control study and meta-analysis. J Neurol Sci 153:50–53

McGeer EG, Singh EA, McGeer PL (1992) Apparent anterograde transport of basic fibroblast growth factor in the rat nigrostriatal dopamine system. Neurosci Lett 148:31–33

McGeer PL, Itagaki S, Akiyama H, McGeer EG (1989) Rate of cell death in parkinsonism indicates active neuropathological process. Ann Neurol 24:574–576

Meco G, Bonifati V, Fabrizio E, Vanacore N (1994) Worsening of Parkinsonism with fluvoxamine-two cases. Hum Psychopharmac 9:439–441

Mereu G, Costa E, Armstrong DM, Vicini S (1991) Glutamate receptors subtypes mediate excitatory synaptic currents of dopamine neurons in midbrain slices. J Neurosci 11:1359–1366

Michel PP, Hefti F (1990) Toxicity of 6-hydroxydopamine and dopamine for dopaminergic neurons in culture. J Neurosci Res 26:428–435

Miller GW, Gainetdinov RR, Levey AI, Caron MG (1999) Dopamine transporters and neuronal injury. Trends Pharmacol Sci 20:424–429

Miller W, Delong MR (1987) Altered tonic activity of neurons in the globus pallidus and subthalamic nucleus in the primate MPTP model of parkinsonism. In: Carpenter MB, Jayaraman A (eds) The basal ganglia. Plenum, New York, pp 414–427

Mithöfer K, Sandy MS, Smith MT, Di Monte D (1992) Mitochondrial poisons cause depletion of reduced glutathione in isolated hepatocytes. Arch Biochem Biophys 295:132–136

Miyoshi Y, Zhang Z, Ovadia A, Lapchak PA, Collins F, Hilt D, Lebel C, Kryscio R, Gash DM (1997) Glial cell line-derived neurotrophic factor-levodopa interactions and reduction of side effects in parkinsonian monkeys. Ann Neurol 42:208–214

Mizuno Y, Ohta S, Tanaka M, Takamiya S, Suzuki K, Sato T, Oya H, Ozawa T, Kagawa Y (1989) Deficiencies in complex I subunits of the respiratory chain in Parkinson's disease. Biochem Biophys Res Commun 163:1450–1455

Mizuno Y, Kondo T, Mori H (1994) Various aspects of motor fluctuations and their management in Parkinson's disease. Neurology 44:S29–S34

Mizuno Y, Yoshino H, Ikebe S, Hattori N, Kobayashi T, Shimoda-Matsubayashi S, Matsumine H, Kondo T (1998) Mitochondrial dysfunction in Parkinson's disease. Ann Neurol 44[Suppl]:S99–S109

Mochizuki H, Goto K, Mori H, Mizuno Y (1996) Histochemical detection of apoptosis in Parkinson's disease. J Neurol Sci 137:120–123

Moldeus P, Nordenskjold M, Bolcsfoldi G, Eiche A, Haglund U, Lambert B (1983) Genetic toxicity of dopamine. Mutat Res 124:9–24

Montastruc JL, Rascol O, Senard JM, Rascol A (1992) A pilot study of N-methyl-D-aspartate (NMDA) antagonist in Parkinson's disease. J Neurol Neurosurg Psychiatr 55:630–631

Montastruc JL, Llau ME, Rascol O, Senard JM (1994) Drug-induced parkinsonism: a review. Fundam Clin Pharmacol 8:293–306

Moore MW, Klein RD, Farinas I, Sauer H, Armanini M, Phillips H, Reichardt LF, Ryan AM, Carver-Moore K, Rosenthal A (1996) Renal and neuronal abnormalities in mice lacking GDNF. Nature 382:76–79

Mount HT, Dean DO, Alberch J, Dreyfus CF, Black IB (1995) Glial cell line-derived neurotrophic factor promotes the survival and morphologic differentiation of Purkinje cells. Proc Natl Acad Sci USA 92:9092–9096

Mufson EJ, Kroin JS, Sendera TJ, Sobreviela T (1999) Distribution and retrograde transport of trophic factors in the central nervous system: functional implications for the treatment of neurodegenerative diseases. Prog Neurobiol 57:451–484

Muraki A, Koyama T, Nakayama M, Ohmori T, Yamashita I (1992) MK-801, a noncompetitive antagonist of NMDA receptor prevents metamphetamine-induced decrease of striatal dopamine uptake sites in the rat striatum. Neurosci Lett 136:39–42

Nabeshima T, Fontenot J, Ho IK (1981) Effects of chronic administration of pentobarbital or morphine on the brain microsomal cytochrome P-450 system. Biochem Pharmacol 30:1142–1145

Napolitano A, Crescenzi O, Pezzella A, Prota G (1995) Generation of the neurotoxin 6-hydroxydopamine by peroxidase/H_2O_2 oxidation of dopamine. J Med Chem 38:917–922

Narhi L, Wood SJ, Steavenson S, Jiang Y, Wu GM, Anafi D, Kaufman SA, Martin F, Sitney K, Denis P, Louis JC, Wypych J, Biere AL, Citron M (Both familial Parkinson's disease mutations accelerate alpha-synuclein aggregation. J Biol Chem 274:9843–9846

Nash JF, Yamamoto BK (1992) Methamphetamine neurotoxicity and striatal glutamate release: comparison to 3,4-methylenedioxymethamphetamine. Brain Res 581:237–243

Negrotti A, Calzetti S (1997) A long-term follow-up study of cinnarizine- and flunarizine-induced parkinsonism. Mov Disord 12:107–110

Newmeyer DD, Farschon DM, Reed JC (1994) Cell-free apoptosis in Xenopus egg extracts: inhibition by Bcl-2 and requirement for an organelle fraction enriched in mitochondria. Cell 79:353–364

Nicklas WJ, Vyas I, Heikkila RE (1985) Inhibition of NADH-linked oxidation in brain mitochondria by 1-methyl-4-phenylpyridine, a metabolite of the neurotoxin 1-methyl-4-phenyl-1,2,3,6-tetrahydropyridine. Life Sci 36:2503–2508

Niznik HB, Tyndale RF, Sallee FR, Gonzalez FJ, Hardwick JP, Inaba T, Kalow W (1990) The dopamine transporter and cytochrome P45OIID1 (debrisoquine 4-hydroxylase) in brain: resolution and identification of two distinct [3H]GBR-12935 binding proteins. Arch Biochem Biophys 276:424–432

Norman BJ, Neal RA (1976) Examination of the metabolism in vitro of parathion (diethyl p-nitrophenyl phosphorothionate) by rat lung and brain. Biochem Pharmacol 25:37–45

Norris PJ, Hardwick JP, Emson PC (1996) Regional distribution of cytochrome P450 2D1 in the rat central nervous system. J Comp Neurol 366:244–258

Ochu EE, Rothwell NJ, Waters CM (1998) Caspases mediate 6-hydroxydopamine-induced apoptosis but not necrosis in PC12 cells. J Neurochem 70:2637–2640

O'Dell SJ, Weihmuller FB, Marshall JF (1991) Multiple methamphetamine injections induce marked increases in extracellular striatal dopamine which correlate with subsequent neurotoxicity. Brain Res 564:256–260

O'Dell SJ, Weihmuller FB, Marshall JF (1992) MK-801 prevents methamphetamine-induced striatal dopamine damage and reduces extracellular dopamine overflow. In: Neurotoxins and neurodegenerative disease. Ann NY Acad Sci 648:317–319

O'Dell SJ, Weihmuller FB, Marshall JF (1993) Methamphetamine-induced dopamine overflow and injury to striatal dopamine terminals: attenuation by dopamine D1 and D2 antagonists. J Neurochem 60:1792–1799

Offen D, Ziv I, Sternin H, Melamed E, Hochman A (1996) Prevention of dopamine-induced cell death by thiol antioxidants: possible implications for treatment of Parkinson's disease. Exp Neurol 141:32–39

Offen D, Ziv I, Panet H, Wasserman L, Stein R, Melamed E, Barzilai A (1997) Dopamine-induced apoptosis is inhibited in PC12 cells expressing Bcl-2. Cell Mol Neurobiol 17:289–304

Offen D, Beart PM, Cheung NS, Pascoe CJ, Hochmen A, Gorodin S, Melamed E, Bernard O (1998) Transgenic mice expressing human Bcl-2 in their neurons are resistant to 6-hydroxydopamine and 1-methyl-4-phenyl-1,2,3,6-tetrahydropyridine neurotoxicity. Proc Natl Acad Sci USA 95:5789–5794

Olanow CW (1992) Magnetic resonance imaging in parkinsonism. Neurol Clin 10:405–420

Olanow CW, Tatton WG (1999) Etiology and pathogenesis of Parkinson's disease. Annu Rev Neurosci 22:123–144

Olanow CW, Youdim MHB (1996) Iron and neurodegeneration: prospects for neuroprotection. In: Olanow CW, Jenner P, Youdim M (eds) Neurodegeneration and Neuroprotection In Parkinson's disease. Academic Press, London, pp 55–67

Olney JW (1980) Excitotoxic mechanisms of neurotoxicity. In: Spencer P, Shaumberg HH (eds) Experimental and clinical neurotoxicology. Williams and Wilkins, Baltimore, pp 272–294

Olney JW (1990) Excitotoxic amino acids and neuropsychiatric disorders. A Rev Pharmacol Toxicol 30:47–71

Olney JW, de Gubareff T (1978) Glutamate neurotoxicity and Huntington's chorea. Nature 271:557–559

Olney JW, Zorumsky CF, Stewart GR, Price MT, Wang G, Labruyere J (1990) Excitotoxicity of L-DOPA and 6-OHDA; implications for Parkinson's and Huntington's disease. Exp Neurol 108:269–272

Orengo CA, Kunik ME, Molinari V, Workman RH (1996) The use and tolerability of fluoxetine in geropsychiatric inpatients. J Clin Psychiatry 57:12–16

Ossowska K (1994) The role of excitatory amino acids in experimental models of Parkinson's disease. J Neural Transm (P-D Sect) 8:39–71

Otto D, Unsicker K (1990) Basic FGF reverses chemical and morphological deficits in the nigrostriatal system of MPTP-treated mice. J Neurosci 10:1912–1921

Otto D, Unsicker K (1993) FGF-2 mediated protection of cultured mesencephalic dopaminergic neurons against MPTP and MPP$^+$: specificity and impact of culture conditions, non-dopaminergic neurons, and astroglial cells. J Neurosci Res 34: 382–393

Overton P, Clark D (1991) N-methyl-D-aspartate increases the excitability of nigrostriatal dopamine terminals. Eur J Pharmacol 201:117–120

Pardini C, Vaglini F, Gesi M, Martini F, Corsini GU (1999) Apomorphine induces apoptotic cell death in Chinese Hamster Overy Cell line. Parkinsonism & Related Disorders 5[Suppl]:S30

Park KS, Sohn DH, Veech RL, Song BJ (1993) Translational activation of ethanol-inducible cytochrome P450 (CYP2E1) by isoniazid. Eur J Pharmacol 248:7–14

Parkinson A (1996) Biotransformation of xenobiotics. In: Klaassen (ed) Casarett and Doull's Toxicology: The Basic Science of Poison, 5th edn. McGraw-Hill, New York St. Louis San Francisco, pp 113–186

Paul SM, Axelrod J, Diliberto EJ Jr (1977) Catechol estrogen-forming enzyme of brain: demonstration of a cytochrome p450 monooxygenase. Endocrinology 101:1604–1610

Peeters EA, Bloem BR, Kuiper MA, Vermeij P, de Wolff FA, Wolters EC, Roos RA, Ferrari MD (1994) Phenytoin parahydroxylation is not impaired in patients with young-onset Parkinson's disease. Clin Neurol Neurosurg 96:296–299

Perl DP, Olanow CW, Calne DB (1998) Alzheimer's disease and Parkinson's disease; distinct entities or extremes of a spectrum of neurodegeneration? Ann Neurol [Suppl] 44:S19–S31

Perry TL, Godin DV, Hansen S (1982) Parkinson's disease: a disorder due to nigral glutathione deficiency? Neurosci Lett 33:305–310

Perry TL, Yong VW, Ito M, Foulks JG, Wall RA, Godin DV, Clavier RM (1984) Nigrostriatal dopaminergic neurons remain undamaged in rats given high doses of L-DOPA and carbidopa chronically. J Neurochem 43:990–993

Perry TL, Yong VW, Clavier RM, Jones K, Wright JM, Foulks JG, Wall RA (1985) Partial protection from the dopaminergic neurotoxin N-methyl-4-phenyl-1,2,3,6-tetrahydropyridine by four different antioxidants in the mouse. Neurosci Lett 60: 109–114

Perry TL, Yong VW, Hansen S, Jones K, Bergeron C, Foulks JG, Wright JM (1987) Alpha-tocopherol and beta-carotene do not protect marmosets against the dopaminergic neurotoxicity of N-methyl-4-phenyl-1,2,3,6-tetrahydropyridine. J Neurol Sci 81:321–331

Petit PX, Susin SA, Zamzami N, Mignotte B, Kroemer G (1996) Mitochondria and programmed cell death: back to the future. FEBS Lett 396:7–13

Piallat B, Bernazzouz A, Bressand K, Vercueil L, Benabid AL (1995) Subthalamic nucleus lesion in rats prevents dopaminergic nigral neuron degeneration after striatal 6-OHDA injection. Soc Neurosci Abs 230:5

Pichel JG, Shen L, Sheng HZ, Granholm AC, Drago J, Grinberg A, Lee EJ, Huang SP, Saarma M, Hoffer BJ, Sariola H, Westphal H (1996) Defects in enteric innervation and kidney development in mice lacking GDNF. Nature 382:73–76

Pifl CH, Schingnitz G, Hornykiewicz O (1991) Effect of 1-methyl-4-phenyl-1,2,3,6-tetrahydropyridine on the regional distribution of the monoamines in rhesus monkeys. Neuroscience 44:591–605

Pileblad E, Carlsson A (1985) Catecholamine-uptake inhibitors prevent the neurotoxicity of 1-methyl-4-phenyl-1,2,3,6-tetrahydropyridine (MPTP) in mouse brain. Neuropharmacology 24:689–692

Poewe W, Granata R (1997) Neurological Principles and Practice. In: Watts RL, Koller WC (eds) Movement Disorders. McGraw-Hill, New York, pp 201–219

Polymeropoulos MH, Lavedan C, Leroy E, Ide SE, Dehejia A, Dutra A, Pike B, Root H, Rubenstein J, Boyer R, Stenroos ES, Chandrasekharappa S, Athanassiadou A, Papapetropoulos T, Johnson WG, Lazzarini AM, Duvoisin RC, Di Iorio G, Golbe LI, Nussbaum RL (1997) Mutation in the alpha-synuclein gene identified in families with Parkinson's disease. Science 276:2045–2047

Potter WZ, Thorgeirsson SS, Jollow DJ, Mitchell JR (1974) Acetaminophen-induced hepatic necrosis. V. Correlation of hepatic necrosis, covalent binding and glutathione depletion in hamsters. Pharmacology 12:129–143

Preskorn SH, Magnus RD (1994) Inhibition of hepatic P-450 isoenzymes by serotonin selective reuptake inhibitors: in vitro and in vivo findings and their implications for patient care. Psychopharmacol Bull 30:251–259

Przedborski S, Jackson-Lewis V, Yokoyama R, Shibata T, Dawson VL, Dawson TM (1996) Role of neuronal nitric oxide in 1-methyl-4-phenyl-1,2,3,6-tetrahydropyridine (MPTP)-induced dopaminergic neurotoxicity. Proc Natl Acad Sci USA 1996 93:4565–4571

Quinn N, Parkes D, Janota I, Marsden CD (1986) Preservation of the substantia nigra and locus coeruleus in a patient receiving levodopa (2kg) plus decarboxylase inhibitor over a four-year period. Mov Disord 1:65–68

Ramsay RR, Singer TP (1986) Energy-dependent uptake of 1-methyl-4-phenylpyridinium, the neurotoxic metabolite of 1-methyl-4-phenyl-1,2,3,6-tetrahydropyridine, by mitochondria. J Biol Chem 261:7885–7887

Rapisardi SC, Warrington VOP, Wilson JS (1990) Effects of MPTP on fine structure of neurons in substantia nigra of dogs. Brain Res 512:147–154

Ravindranath V, Anandatheerthavarada HK, Shankar SK (1989) Xenobiotic metabolism in human brain – presence of cytochrome P-450 and associated monooxygenases. Brain Res 496:331–335

Ravindranath V, Anandatheerthavarada HK, Shankar SK (1990) NADPH cytochrome P-450 reductase in rat, mouse and human brain. Biochem Pharmacol 39:1013–1018

Riachi NJ, Harik SI, Kalaria RN, Sayre LM (1988) On the mechanisms underlying 1-methyl-4-phenyl-1,2,3,6-tetrahydropyridine neurotoxicity II: Susceptibility among mammalian species correlate with the toxin's metabolic patterns in rat microvessels and liver. J Pharmacol Exp Ther 244:443–448

Riachi NJ, Lamanna JC, Harik SI (1989) Entry of 1-methyl-4-phenyl-1,2,3,6-tetrahydropyridine into the rat brain. J Pharmacol Exp Ther 249:744–748

Ricaurte GA, Langston JW, DeLannery LE, Irwin I, Brooks JD (1985) Dopamine uptake blockers protect against the dopamine depleting effect of 1-methyl-4-phenyl-1,2,3,6-tetrahydropyridine (MPTP) in the mouse striatum. Neurosci Lett 59:259–264

Ricaurte GA, Langston JW, DeLannery LE, Irwin I, Peroutka SJ, Forno LS (1986) Fate of nigrostriatal neurons in young mature mice given 1-methyl-4-phenyl-1,2,3,6-tetrahydropyridine: a neurochemical and morphological reassessment. Brain Res 376:117–124

Riederer P, Sofic E, Rausch WD, Schmidt B, Reynolds GP, Jellinger K, Youdim MB (1989) Transition metals, ferritin, glutathione, and ascorbic acid in parkinsonian brains. J Neurochem 52:515–520

Riedl AG, Watts PM, Jenner P, Marsden CD (1998) P450 enzymes and Parkinson's disease: the story so far. Mov Disord 13:212–220

Riedl AG, Watts PM, Edwards RJ, Schulz-Utermoehl T, Boobis AR, Jenner P, Marsden CD (1999) Expression and localisation of CYP2D enzymes in rat basal ganglia. Brain Res 822:175–191

Ringstedt T, Lagercrantz H, Persson H (1993) Expression of members of the trk family in the developing postnatal rat brain. Brain Res Dev Brain Res 72:119–131

Riva MA, Gale K, Mocchetti I (1992) Basic fibroblast growth factor mRNA increases in specific brain regions following convulsive seizures. Brain Res Mol Brain Res 15:311–318

Riva MA, Molteni R, Tascedda F, Massironi A, Racagni G (1999) Selective modulation of fibroblast growth factor-2 expression in the rat brain by the atypical antipsychotic clozapine. Neuropharmacology 38:1075–1082

Rollema H, Kuhr WG, Kranenborg G, De Vries J, Van den Berg C (1988) MPP+-induced efflux of DA and lactate from rat striatum have similar time courses as shown by in vivo brain dialysis. J Pharmacol Exp Ther 245:858–866

Roselli CE, Horton LE, Resko JA (1985) Distribution and regulation of aromatase activity in the rat hypothalamus and limbic system. Endocrinology 117:2471–2477

Rosenberg PA (1988) Catecholamine toxicity in cerebral cortex in dissociated cell culture. J Neurosci 8:2887–2894

Rosenberg PA, Loring R, Xie Y, Zaleskas V, Aizenman E (1991) 2,3,5, Trihydroxyphenylalanine in solution forms a non-N-methyl-D-aspartate glutamatergic agonist and neurotoxin. Proc Natl Acad Sci USA 88:4865–4869

Ross SB, Renyi AL (1976) On the long-lasting inhibitory effect of N-(-2-chloroethyl)-N-ethyl-2-bromobenzylamine (DSP-4) on the active uptake of noradrenaline. J Pharm Pharmacol 28:458–459

Rostami-Hodjegan A, Lennard MS, Woods HF, Tucker GT (1998) Meta-analysis of studies of the CYP2D6 polymorphism in relation to lung cancer and Parkinson's disease. Pharmacogenetics 8:227–238

Rupniak NMJ, Boyce S, Steventon MJ, Iversen SD, Marsden CD (1992) Dystonia induced by combined treatment with L-dopa and MK-801 in parkinsonian monkeys. Ann Neurol 32:103–105

Russ H, Gliese M, Sonna J, Shomig E (1992) The extraneuronal transport mechanism for noradrenaline (uptake 2) avidly transports 1-methyl-4-phenylpyridinium (MPP$^+$). Naunyn Schmiedebergs Arch Pharmacol 346:158–165

Salach JI, Singer TP, Castagnoli N Jr, Trevor A (1984) Oxidation of the neurotoxic amine 1-methyl-4-phenyl-1,2,3,6-tetrahydropyridine (MPTP) by monoamine oxidases A and B and suicide inactivation of the enzymes by MPTP. Biochem Biophys Res Commun 125:831–835

Sanchez MP, Silos-Santiago I, Frisen J, He B, Lira SA, Barbacid M (1996) Renal agenesis and the absence of enteric neurons in mice lacking GDNF. Nature 382:70–73

Sanchez-Ramos JR, Michel P, Weiner WJ, Hefti F (1988) Selective destruction of cultured dopaminergic neurons from rat mesencephalon by 1-methyl-4-phenylpyridinium: cytochemical and morphological evidence. J neurochem 50:1934–1944

Sandy MS, Armstrong M, Tanner CM, Daly AK, Di Monte DA, Langston JW, Idle JR (1996) CYP2D6 allelic frequencies in young-onset Parkinson's disease. Neurology 47:225–230

Sasame HA, Ames MM, Nelson SD (1977) Cytochrome P-450 and NADPH cytochrome c reductase in rat brain: formation of catechols and reactive catechol metabolites. Biochem Biophys Res Commun 78:919–926

Scatton B, Dennis T, L'Hereux R, Monfort J, Duyckaerts C, Javoy-Agid F (1986) Degeneration of noradrenergic and serotoninergic but not dopaminergic neurons in the lumbar spinal cord of parkinsonian patients. Brain Res 380:181–185

Schallert T, Wilcox RE (1985) Neurotransmitter-selective brain lesions. In: Boulton AA, Baker GB (eds) Neuromethods, vol 1. Humana, Clifton, pp 343–387

Schapira AHV, Cooper JM, Dexter D (1989) Mitochondrial complex 1 deficiency in Parkinson's disease. Lancet 1:1269

Schapira AHV, Cooper JM, Dexter D, Clark JB, Jenner P, Marsden CD (1990) Mitochondrial complex 1 deficiency in Parkinson's disease. J Neurochem 54:823–827

Schlinger BA, Callard GV (1989) Localization of aromatase in synaptosomal and microsomal subfractions of quail (Coturnix coturnix japonica) brain. Neuroendocrinology 49:434–441

Schmidt CJ (1992) L-DOPA potentiates the neurotoxicity of some amphetamine analogues. Ann NY Acad Sci 648:343–344

Schmidt CJ, Ritter JK, Sonsalla PK, Hanson GR, Gibb JW (1985) Role of dopamine in the neurotoxic effects of methamphetamine. J Pharmacol Exp Ther 233:539–544

Schneider JS, Juviler A, Markham CH (1986) Production of a Parkinson-like syndrome in the cat with 1-methyl-4-phenyl-1,2,3,6-tetrahydropyridine (MPTP): behavior, histology and biochemistry. Exp Neurol 91:293–307

Schulz JB, Matthews RT, Muqit MM, Browne SE, Beal MF (1995) Inhibition of neuronal nitric oxide synthase by 7-nitroindazole protects against MPTP-induced neurotoxicity in mice. J Neurochem 64:936–939

Scotchner KP, Irwin I, DeLanney LE, Langston JW, Di Monte D (1990) Effects of 1-methyl-4-phenyl-1,2,3,6-tetrahydropyridine and 1-methyl-4-phenylpyridinium ion on ATP levels of mouse brain synaptosomes. J Neurochem 54:1295–1301

Seiden LS, Sabol KE (1996) Methamphetamine and methylenedioxymethamphetamine neurotoxicity: possible mechanisms of cell destruction. NIDA Res Monogr 163:251–276

Seiden LS, Vosmer G (1984) Formation of 6-hydroxydopamine in caudate nucleus of the rat brain after a single large dose of methylamphetamine. Pharmacol Biochem Behav 21:29–31

Seiden LS, Fischman MW, Shuster CR (1975) Long-term methamphetamine-induced changes in brain catecholamines in tolerant rhesus monkeys. Drug Alcohol Depend 1:215–219

Sellal F, Hirsch E, Lisovoski F, Mutschler V, Collard M, Marescaux C (1992) Contralateral disappearance of parkinsonian signa after subthalamic hematoma. Neurology 42:255–256

Seniuk NA, Tatton WG, Greenwood CE (1990) Dose-dependent destruction of the coeruleus-cortical and nigral-striatal projections by MPTP. Brain Res 27:7–20

Showalter HDH, Sercel AD, Boguslawa ML, Wolfangel CD, Ambroso LA, Elliot WL, Fry DW, Kraker AJ, Howard CT, Lu GH, Moore CW, Nelson JM, Roberts BJ, Vincent PW, Denny WA, Thompson AM (1997) Tyrosine kinase inhibitors. 6. Structure-activity relationships among N- and 3-substitute 2,2-dislenobis (1H-indoles) for inhibition of protein tyrosine kinases and comparative in vitro and in vivo studies against selected sulfur congeners. J Med Chem 40:413–426

Sian J, Dexter DT, Lees AJ, Daniel S, Agid Y, Javoy-Agid F, Jenner P, Marsden CD (1994a) Alterations in glutathione levels in Parkinson's disease and other neurodegenerative disorders affecting basal ganglia. Ann Neurol 36:348–355

Sian J, Dexter DT, Lees AJ, Daniel S, Jenner P, Marsden CD (1994b) Glutathione-related enzymes in brain in Parkinson's disease. Ann Neurol 36:356–361

Sies H, Ketterer B (1988) Glutathione conjugation mechanisms and biological significance. Academic, London

Simantov R, Blinder E, Ratovitski T, Tauber M, Gabbay M, Porat S (1996) Dopamine-induced apoptosis in human neuronal cells: inhibition by nucleic acids antisense to the dopamine transporter. Neuroscience 74:39–50

Simons JA (1996) Fluoxetine in Parkinson's disease. Mov Disord 11:581–582

Singh RK, Gupta AK, Singh B (1995) Acute organic brain syndrome after fluoxetine treatment. Am J Psychiatry 152:295–296

Smith CA, Gough AC, Leigh PN, Summers BA, Harding AE, Maraganore DM, Sturman SG, Schapira AH, Williams AC et al. (1992) Debrisoquine hydroxylase gene polymorphism and susceptibility to Parkinson's disease. Lancet 339:1375–1377

Smith I, Kellow AH (1969) Aromatic amines and Parkinson's disease. Nature 221:1261

Snyder SH, Sabatini DM, Lai MM, Steiner JP, Hamilton GS, Suzdak PD (1998) Neural actions of immunophilin ligands. Trends Pharmacol Sci 19:21–26

Sofic E, Paulus W, Jellinger K, Riederer P, Youdim MB (1991) Selective increase of iron in substantia nigra zona compacta of parkinsonian brains. J Neurochem 56:978–982

Sofic E, Lange KW, Jellinger K, Riederer P (1992) Reduced and oxidized glutathione in the substantia nigra of patients with Parkinson's disease. Neurosci Lett 142:128–130

Sohda T, Shimizu M, Kamimura S, Okumura M (1993) Immunohistochemical demonstration of ethanol-inducible P450 2E1 in rat brain. Alcohol Alcohol 28:69–75

Sonsalla PK, Nicklas WJ (1992) MPTP and animal models of Parkinson's disease. In: Koller WC (ed) Handbook of Parkinson's disease, 2nd edn. Marcel Dekker, New York, pp 319–340

Sonsalla PK, Nicklas WJ, Heikkila RE (1989) Role for excitatory amino acids in methamphetamine-induced nigrostriatal dopaminergic toxicity. Science 243:398–400

Sonsalla PK, Riordan DE, Heikkila RE (1991) Competitive and non-competitive antagonists at N-methyl-$\underline{d}$-aspartate receptors protect against methamphetamine-induced dopaminergic damage in mice. J Pharmacol Exp Ther 256:506–512

Sonsalla PK, Giovanni A, Sieber BA, Delle Donne K, Manzino L (1992a) Characteristics of dopaminergic neurotoxicity produced by MPTP and methamphetamine. In: Neurotoxins and neurodegenerative disease. Ann NY Acad Sci 648:229–238

Sonsalla PK, Zeevalk GD, Manzino L, Giovanni A, Nicklas WJ (1992b) MK-801 fails to protect against the dopaminergic neuropathology produced by systemic 1-methyl-4-phenyl-1,2,3,6-tetrahydropyridine in mice or intranigral 1-methyl-4-phenylpyridium in rats. J Neurochem 58:1979–1982

Speciale SG, Liang CL, Sonsalla PK, Edwards RH, German DC (1998) The neurotoxin 1-methyl-4-phenylpyridium is sequestered within neurons that contain vesicular monoamine transporter. Neuroscience 84:1177–1185

Spencer JP, Jenner P, Halliwell B (1995) Superoxide-dependent depletion of reduced glutathione by L-DOPA and dopamine. Relevance to Parkinson's disease. Neuroreport 6:1480–1484

Spencer JP, Jenner P, Daniel SE, Lees AJ, Marsden DC, Halliwell B (1998) Conjugates of catecholamines with cysteine and GSH in Parkinson's disease: possible mechanisms of formation involving reactive oxygen species. J Neurochem 71:2112–2122

Spillantini MG, Schmidt ML, Lee VM, Trojanowski JQ, Jakes R, Goedert M (1997) Alpha-synuclein in Lewy bodies. Nature 388:839–840

Spillantini MG, Crowther RA, Jakes R, Hasegawa M, Goedert M (1998) alpha-Synuclein in filamentous inclusions of Lewy bodies from Parkinson's disease and dementia with Lewy bodies. Proc Natl Acad Sci USA 95:6469–6473

Steiner JP, Hamilton GS, Ross DT, Valentine HL, Guo H, Connolly MA, Liang S, Ramsey C, Li JH, Huang W, Howorth P, Soni R, Fuller M, Sauer H, Nowotnik AC, Suzdak PD (1997) Neurotrophic immunophilin ligands stimulate structural and functional recovery in neurodegenerative animal models. Proc Natl Acad Sci USA 94:2019–2024

Steur EN (1993) Increase of Parkinson disability after fluoxetine medication. Neurology 43:211–213

Steventon GB, Heafield MT, Waring RH, Williams AC (1989) Xenobiotic metabolism in Parkinson's disease. Neurology 39:883–887

Stokes AH, Brown BG, Lee CK, Doolittle DJ, Vrana KE (1996) Tyrosinase enhances the covalent modification of DNA by dopamine. Brain Res Mol Brain Res 42:167–170

Stokes AH, Hastings TG, Vrana KE (1999) Cytotoxic and genotoxic potential of dopamine. J Neurosci Res 55:659–665

Storey E, Hyman BT, Jenkins B, Brouillet E, Miller JM, Rosen BR, Beal MF (1992) 1-methyl-4-phenylpyridium produces excitotoxic lesion in rat striatum as a result of impairment of oxidative metabolism. J Neurochem 58:1975–1978

Stott I, Murthy A, Robinson A, Thomas NW, Fry JR (1997) Low-dose diethyldithiocarbamate attenuates the hepatotoxicity of 1,3-dichloro-2-propanol and selectively inhibits CYP2E1 activity in the rat. Hum Exp Toxicol 16:262–266

Sundstrom E, Jonsson G (1985) Pharmacological interference with the neurotoxic action of 1-methyl-4-phenyl-1,2,3,6-tetrahydropyridine (MPTP) on central catecholamine neurons in the mouse. Eur J Pharmacol 110:293–299

Sundstrom E, Goldstein M, Jonsson G (1986) Uptake inhibition protects nigro-striatal dopamine neurons from the neurotoxicity of 1-methyl-4-phenylpyridine (MPP$^+$) in mice. Eur J Pharmacol 131:289–292

Sundstrom E, Henriksson BG, Mohammed AH, Souverbie F (1994) MPTP-treated mice: a useful model for Parkinson's disease? In: Woodruff ML, Nonneman AJ (eds) Toxin-induced models of neurological disorders. Plenum Press, New York, pp 121–137

Susin SA, Zamzami N, Castedo M, Kroemer G (1996) The cell biology of apoptosis: evidence for the implication of mitochondria. Apoptosis 1:231–242

Suvanto P, Hiltunen JO, Arumae U, Moshnyakov M, Sariola H, Sainio K, Saarma M (1996) Localization of glial cell line-derived neurotrophic factor (GDNF) mRNA in embryonic rat by in situ hybridization. Eur J Neurosci 8:816–822

Takakubo F, Yamamoto M, Ogawa N, Yamashita Y, Mizuno Y, Kondo I (1996) Genetic association between cytochrome P450IA1 gene and susceptibility to Parkinson's disease. J Neural Transm Gen Sect 103:843–849

Tallaksen-Greene SJ, Wiley RG, Albin RL (1992) Localization of striatal excitatory amino acid binding site subtypes to striatonigral projection neurons. Brain Res 594:165–170

Tanaka M, Sotomatsu A, Kanai H, Hirai S (1991) Dopa and dopamine cause cultured neuronal death in the presence of iron. J Neurol Sci 101:198–203

Tatton NA, Maclean-Fraser A, Tatton WG, Perl DP, Olanow CW (1998) A fluorescent double-labeling method to detect and confirm apoptotic nuclei in Parkinson's disease. Ann Neurol [Suppl] 44:S142–S148

Tetrud JW, Langston JW (1989) Effects of L-deprenyl (selegiline) on the natural history of Parkinson's disease. Science 245:519–522

The Parkinson's Study Group (1989) Effect of deprenyl on the progression and disability in early Parkinson's disease. N Engl J Med 20:1364–1371

Thoenen H (1969) Bildung und funktionelle Bedeutung adrenerger Ersatztransmitter. Exp Med Pathol Klin 27:1–85

Thoenen H, Tranzer JP (1968) Chemical sympathectomy by selective destruction of adrenergic nerve endings with 6-hydroxydopamine. Naunyn-Schmiedebergs Arch Pharmacol 261:271–288

Thoenen H, Tranzer JP (1973) The pharmacology of 6-hydroxydopamine. Annu Rev Pharmacol 13:169–180

Tipton KF, Singer TP (1993) Advances in our understanding of the mechanisms of the neurotoxicity of MPTP and related compounds. J Neurochem 61:1191–1206

Tooyama I, Kawamata T, Walker D, Yamada T, Hanai K, Kimura H, Iwane M, Igarashi K, McGeer EG, McGeer PL (1993) High molecular weight basic fibroblast growth factor-like protein is localized to a subpopulation of mesencephalic dopaminergic neurons in the rat brain. Neurology 43:372–376

Treanor JJ, Goodman L, de Sauvage F, Stone DM, Poulsen KT, Beck CD, Gray C, Armanini MP, Pollock RA, Hefti F, Phillips HS, Goddard A, Moore MW, Buj-Bello

A, Davies AM, Asai N, Takahashi M, Vandlen R, Henderson CE, Rosenthal A (1996) Characterization of a multicomponent receptor for GDNF. Nature 382: 80–83
Trupp M, Arenas E, Fainzilber M, Nilsson AS, Sieber BA, Grigoriou M, Kilkenny C, Salazar-Grueso E, Pachnis V, Arumae U (1996) Functional receptor for GDNF encoded by the c-ret proto-oncogene. Nature 381:785–789
Tse DC, McCreery RL, Adams RN (1976) Potential oxidative pathways of brain catecholamines. J Med Chem 19:37–40
Tsuneoka Y, Fukushima K, Matsuo Y, Ichikawa Y, Watanabe Y (1996) Genotype analysis of the CYP2C19 gene in the Japanese population. Life Sci 59:1711–1715
Tursky L, Bressler K, Rettig KJ, Loshmann PA, Wachtel H (1991) Protection of substantia nigra from MPP+ neurotoxicity by N-methyl-D-aspartate antagonists. Nature 349:414–418
Tyndale RF, Sunahara R, Inaba T, Kalow W, Gonzalez FJ, Niznik HB (1991) Neuronal cytochrome P450IID1 (debrisoquine/sparteine-type): potent inhibition of activity by (−)-cocaine and nucleotide sequence identity to human hepatic P450 gene CYP2D6. Mol Pharmacol 40:63–68
Uhl GR (1998) Hypothesis: the role of dopaminergic transporters in selective vulnerability of cells in Parkinson's disease. Ann Neurol 43:555–560
Ungerstedt U (1968) 6-hydroxydopamine induced degeneration of central monoamine neurons. Eur J Pharmacol 5:107–110
Uretsky NJ, Iversen LL (1969) Effects of 6-hydroxydopamine on noradrenaline-containing neurons in the rat brain. Nature 221:557–559
Uretsky NJ, Iversen LL (1970) Effects of 6-hydroxydopamine on catecholamine containing neurones in the rat brain. J Neurochem 17:269–278
Vaccari A, Del Zompo M, Melis F, Gessa GL, Rossetti ZL (1991) Interaction of 1-methyl-4-phenylpyridinium ion and tyramine with a site putatively involved in the striatal vesicular release of dopamine. Br J Pharmacol 104:573–574
Vaglini F, Fascetti F, Fornai F, Maggio R, Corsini GU (1994) (+)MK-801 prevents the DDC-induced enhancement of MPTP toxicity in mice. Brain Res 668:194–203
Vaglini F, Fascetti F, Tedeschi D, Cavalletti M, Fornai F, Corsini GU (1996) Striatal MPP+ levels do not necessarily correlate with striatal dopamine levels after MPTP treatment in mice. Neurodegeneration 5:129–136
Valberg LS, Flanagan PR, Kertesz A, Ebers GC (1989) Abnormalities in iron metabolism in multiple sclerosis. Can J Neurol Sci 16:184–186
Vila M, Levy R, Herrero MT, Ruberg M, Faucheux B, Obeso JA, Agid Y, Hirsch EC (1997) Consequences of nigrostriatal denervation on the functioning of the basal ganglia in human and non-human primates: an in situ hybridization study of cytochrome oxidase submit I mRNA. J Neurosci 17:765–773
Vila M, Marin C, Ruberg M, Jimenez A, Raisman-Vozari R, Agid Y, Tolosa E, Hirsch EC (1999) Systemic administration of NDMA and AMPA receptor antagonists reverses the neurochemical changes induced by nigrostriatal denervation in basal ganglia. J Neurochem 73:344–352
Wagner GC, Ricaurte GA, Seiden LS, Schuster CR, Miller RJ, Westley J (1980) Long-lasting depletion of striatal dopamine and loss of dopamine uptake sites following repeated administration of methamphetamine. Brain Res 181:151–160
Wagner GC, Lucot JB, Schuster CR, Seiden LS (1983) α-methyltyrosine attenuates and reserpine increases methamphetamine-induced neuronal changes. Brain Res 270:285–288
Walkinshaw G, Waters CM (1995) Induction of apoptosis in catecholaminergic PC12 cells by L-DOPA. Implications for the treatment of Parkinson's disease. J Clin Invest 95:2458–2464
Wang BH, Lu ZX, Polya GM (1997) Inhibition of eukaryote protein kinases by isoquinoline and oxazine alkaloids. Planta Med 63:494–498
Warner M, Wyss A, Yoshida S, Gustafsson JA (1994) Cytochrome P450 enzymes in brain. Methods in Neurosciences, vol. XXII. Academic Press, New York, pp 51–66

Watts PM, Riedl AG, Douek DC, Edwards RJ, Boobis AR, Jenner P, Marsden CD (1998) Co-localization of P450 enzymes in the rat substantia nigra with tyrosine hydroxylase. Neuroscience 86:511–519

Weihmuller FB, O'Dell SJ, Marshall JF (1993) L-DOPA pretreatment potentiates striatal dopamine overflow and produces dopamine terminal injury after a single dose of methamphetamine. Brain Res 623:303–307

Wick MM, Byers L, Frei F (1977) L-dopa: selective toxicity for melanoma cells in vitro. Science 197:468–469

Willis GL, Armstrong SM (1998) Orphan neurones and amine excess: the functional neuropathology of Parkinsonism and neuropsychiatric disease. Brain research Reviews 27:177–242

Wrona MZ, Yang Z, Zhang F, Dryhurst G (1997) Potential new insights into the molecular mechanisms of methamphetamine-induced neurodegeneration. Natl Inst Drug Abuse res Monogr 173:146–174

Wüllner U, Kupsch A, Arnold G, Renner P, Scheid R, Oertel W, Klockgether T (1992) The competitive NMDA antagonist CGP 40116 enhances L-DOPA response in MPTP-treated marmosets. Neuropharmacology 31:713–715

Xu Y, Stokes AH, Roskoski R Jr, Vrana KE (1998) Dopamine, in the presence of tyrosinase, covalently modifies and inactivates tyrosine hydroxylase. J Neurosci Res 54:691–697

Youdim MB, Dhariwal K, Levine M, Markey CJ, Markey S, Caohuy H, Adeyemo OM, Pollard HB (1992) MPTP-induced "parkinsonism" in the goldfish. Neurochem Int (Suppl) 20:275S–278S

Youdim MBH, Ben-Shachar D, Riederer P (1993) Iron-melanin interaction and Parkinson's disease. NIPS 8:45–49

Zaczek R, Fritschy JM, Culp S, De Souza EB, Grzanna R (1990) Differential effects of DSP-4 on noradrenaline axons in cerebral cortex and hypothalamus may reflect heterogeneity of noradrenaline uptake sites. Brain Res 522:308–314

Ziv I, Melamed E, Nardi N, Luria D, Achiron A, Offen D, Barzilai A (1994) Dopamine induces apoptosis-like cell death in cultured chick sympathetic neurons – a possible novel pathogenetic mechanism in Parkinson's disease. Neurosci Lett 170: 136–140

Zubenko GS, Moossy J, Kopp U (1990) Neurochemical correlates of major depression and primary dementia. Arch Neurol 47:209–214

Zuddas A, Corsini GU, Schinelli S, Johannessen JN, di Porzio U, Kopin IJ (1989a) MPTP treatment combined with ethanol or acetaldehyde selectively destroys dopaminergic neurons in mouse substantia nigra. Brain Res 501:1–10

Zuddas A, Corsini GU, Schinelli S, Barker JL, Kopin IJ, di Porzio U (1989b) Acetaldehyde directly enhances MPP$^+$ neurotoxicity and delays its elimination from the striatum. Brain Res 501:11–22

Zuddas A, Oberto G, Vaglini F, Fascetti F, Fornai F, Corsini GU (1992) MK-801 prevents 1-methyl-4-phenyl-1,2,3,6-tetrahydropyridine-induced Parkinsonism in primates. J Neurochem 59:733–739

Zuddas A, Fascetti F, Corsini GU, Piccardi P (1994) In Brown Norway rats, MPP+ is accumulated in nigrostriatal dopaminergic terminals but it is not neurotoxic: a model of neural resistance t MPTP toxicity. Exp Neurol 127:54–61

CHAPTER 24
Dopamine and Depression

P. WILLNER

A. Introduction

Depression has been described as "the common cold of psychiatry". Unlike the common cold, the symptoms of depression may vary greatly from patient to patient, so much so that two patients diagnosed as suffering from a major depression may show no overlap in their symptoms (FIBIGER 1991). In these circumstances, it is prudent, when attempting to understand this protean disorder, to focus on its cardinal symptoms, which are (1) depressed mood and (2) loss of interest or pleasure in usually pleasurable activities (AMERICAN PSYCHIATRIC ASSOCIATION 1994). These symptoms are associated with characteristic abnormalities in the way in which depressed people process information. On the one hand, their cognitions and perceptions are biased towards the pessimistic: they selectively abstract and remember information consistent with a negative view of themselves, their place in the world, and the future (BECK 1987). On the other hand, they think and act more slowly, and experience particular difficulty in initiating actions (WILLNER 1985; BERMANZOHN and SIRIS 1992). A simple hypothesis to explain both of these central features of depression is that they reflect an impairment of incentive motivation. Incentives are stimuli associated with rewards, which serve to confirm that behaviour is on track to attain its goal and to increase the vigour of goal-directed behaviour (BINDRA 1974). A decrease in the impact of positive incentives would explain both a negative cognitive bias (because responses to positive events are weakened) and a decrease in response initiation (because behaviour is no longer energized). This, in turn, explains the core symptoms of low mood, loss of pleasure and loss of interest.

This simple functional hypothesis has an equally simple structural counterpart. The mesolimbic dopamine (DA) system has been recognized as a crucial substrate for rewarded behaviour since the pioneering studies of WISE and colleagues over 20 years ago (WISE 1982). The "DA hypothesis of reward" proposed that rewarding events, irrespective of their modality, shared the common property of activating the mesocorticolimbic DA system, and in particular, the DA projection to the nucleus accumbens; conversely, inactivation of DA function would lead to anhedonia, the inability to experience pleasure

(WISE 1982). This hypothesis, with its obvious implications for a potential role for the mesocorticolimbic DA system as a substrate for affective disorders, has stimulated an extensive body of behavioural pharmacological research, which is reviewed elsewhere in this volume (see Chap. 22, this volume), aimed at clarifying its precise role(s) in reward, reward-related learning and affect. While the hypothesis that the DA projection to the nucleus accumbens functions as a "reward pathway" remains controversial (e.g. SALAMONE et al. 1997), it is now indisputable that this pathway plays a crucial role in the selection and orchestration of goal-directed behaviours, particularly those elicited by incentive stimuli, and in reward-related learning (WILLNER and SCHEEL-KRUGER 1991; LE MOAL 1995; BENINGER and NAKONECHNY 1996). These properties arise from the fact that DA functions to gate the flow of information through the nucleus accumbens, which serves as the major interface through which information in limbic structures gains access to motor output systems (MOGENSON and YIM 1991). The major non-DA afferent projections to the nucleus accumbens, which represent the major output pathways of the limbic system, are from amygdala, hippocampus and prefrontal cortex (GROENEWEGEN et al. 1991). It is noteworthy that all of these structures are also implicated in affective psychopathology (WILLNER 1985).

Traditional accounts of the biochemical basis of depression have focussed largely on noradrenaline (NA) and serotonin (5-HT), and although most of the evidence that coalesced into the "catecholamine hypothesis of depression" does not distinguish clearly between NA and DA, the potential role of DA was at first overlooked. Following two influential reviews that drew attention to this oversight (RANDRUP et al. 1975; WILLNER 1983), there has been an upsurge of interest in the possible involvement of DA in affective disorders. In fact, as will be seen below, there is little in the recent clinical evidence to justify this change of fashion; the pressure to reconsider the role of DA in depression arises largely from preclinical developments. One is the theoretical argument, summarized above, developed from a growing understanding of the role of DA in motivated behaviour, and the recognition that some of the major symptoms of depression are consistent with a decreased level of functional activity in the mesolimbic DA system. The other is the now substantial body of work (reviewed below) demonstrating that antidepressant drugs enhance the functioning of mesolimbic DA synapses.

This "dopamine hypothesis of depression" differs somewhat from earlier "biochemical theories" in that it not only proposes a relationship between a biochemical entity (DA) and a mental disorder (depression), but also defines explicitly the nature of the relationship in terms of the functional properties of the relevant DA neurons. The hypothesis also defines certain boundary conditions: it involves a limited set of DA projections (the mesocorticolimbic system), and a limited set of depressive symptoms (anhedonia and lack of interest). This chapter reviews critically the recent clinical and preclinical evidence pertaining to the involvement of the mesocorticolimbic DA system in depression. It also presents a further development of the hypothesis, based on differential effects of positive and negative incentive stimuli on mesocorticolimbic DA release.

B. DA Function in Affective Disorders
I. DA Turnover

Numerous studies have attempted to assess forebrain DA function in depressed patients by measuring levels of the DA metabolite homovanillic acid (HVA) in cerebrospinal fluid (CSF). In some studies patients were pretreated with probenecid to block the transport of HVA out of the CSF; this procedure, which measures the accumulation of HVA, is considered to give a better estimate of DA turnover. Most studies have tended to report a decrease in CSF HVA in depressed patients, and this relationship holds strongly in studies using the probenecid technique. Decreases in CSF HVA are particularly pronounced in patients with marked psychomotor retardation. In fact, a 1983 review of this area concluded that "The consistent finding of decreased post-probenecid CSF HVA accumulation in depressed patients, particularly those with psychomotor retardation, is probably the most firmly established observation in the neurochemistry of depression" (WILLNER 1983). More recent studies have not altered this conclusion (JIMERSON 1987; REDDY et al. 1992; BROWN and GERSHON 1993).

DA turnover, measured post-mortem, is also reduced in the caudate nucleus and nucleus accumbens of depressed suicides (BOWDEN et al. 1997b). As DA uptake is unchanged in depressed suicides (ALLARD and NORLEN 1997; BOWDEN et al. 1997c), the decreased turnover apparently reflects decreased DA release. There are also many reports of decreased CSF HVA in depressed suicide attempters (BROWN and GERSHON 1993). Consistent with these findings, a decrease in 24-h urinary excretion of HVA and DOPAC has been reported in depressed suicide attempters (ROY et al. 1992). As abnormalities of DA metabolism are not observed in non-depressed suicide attempters (BROWN and GERSHON 1993), these data provide further evidence that decreased DA turnover is a correlate of depression.

Nevertheless, the interpretation of these data is far from straightforward. Although one study has reported that CSF HVA was lower in melancholic than in non-melancholic patients (ROY et al. 1985), this relationship is probably explained by the association between low CSF HVA and psychomotor retardation (WILLNER 1983; BROWN and GERSHON 1993), which is a prominent feature of melancholia. In fact, low CSF HVA has been associated with psychomotor slowing (bradyphrenia) not only in depressed patients, but also in Parkinson's disease and Alzheimer's disease (WOLFE et al. 1990). In agitated patients, however, CSF HVA levels are normal or slightly elevated (WILLNER 1983). CSF HVA levels (as well as plasma DA: SCHATZBERG and ROTHSCHILD 1988) are also elevated in delusional patients (WILLNER 1983). Again, this finding may reflect psychomotor change: in a study of psychotic patients, CSF

HVA levels were elevated in those with delusions and agitation, but normal in those with delusions but no agitation (VAN PRAAG et al. 1975). CSF HVA levels are usually found to be elevated in mania (JIMERSON 1987). These data suggest that CSF HVA levels may reflect motor activity rather than mood, and further raise the problem of whether a reduction in HVA level is the primary cause or a secondary reflection of psychomotor retardation. This latter problem was first posed in an early study in which a group of depressed patients were asked to simulate mania: the exercise did increase DA turnover, but also elevated mood (POST et al. 1973). Recently, however, decreased DA turnover in depressed patients has been demonstrated using the arteriovenous HVA concentration gradient. As participants were supine throughout the sampling period, motor activity differences can be excluded as an explanation of these differences. This study also reported a significant correlation between DA turnover and severity of depression (LAMBERT et al. 2000).

It is hardly surprising that CSF HVA levels are associated with level of motor activity, since CSF HVA derives largely from the caudate nucleus, on account of its large size and its periventricular location. In schizophrenic patients, decreased CSF HVA concentrations are associated with ventricular enlargement (DAVIS et al. 1991), which is equally common in major depressive disorder (JESTE et al. 1988). Indeed, positron emission tomography (PET) imaging studies have reported hypometabolism of the head of the caudate nucleus in unipolar and bipolar depressed patients, which may reflect a decreased DA activity in this structure (BAXTER et al. 1989). Similar findings have been reported in ventromedial caudate and nucleus accumbens in patients diagnosed with familial pure depressive disease (DREVETS et al. 1992). However, the contribution to CSF HVA of DA release in mesolimbic structures such as the nucleus accumbens and frontal cortex is relatively minor. There is therefore no reason to expect that changes in mesolimbic DA function would be apparent in studies measuring HVA levels in lumbar CSF; it is far more likely that any such changes would be obscured by alterations in nigrostriatal DA function associated with changes in motor output. Thus, although most reviewers have tended to interpret the HVA data as evidence for a DA dysfunction in depression (RANDRUP et al. 1975; WILLNER 1983; JIMERSON 1987), these data are actually silent with respect to the important question of the state of activity in the mesocorticolimbic DA system.

In one series of studies, increased levels of DA itself were observed in CSF of melancholic patients, with a tendency towards higher concentrations in patients who were delusional (GJERRIS et al. 1987). CSF DA levels have been found to correlate with extraversion in depressed patients (KING et al. 1986). However, the proportion of DA in lumbar CSF originating in the forebrain is unknown.

The possibility that abnormalities of DA turnover may be a feature of depression per se, rather than a consequence of altered motor activity, prompts the question of whether there are relevant genetic differences in depressed individuals. Following an initial report of a genetic marker for depression on the short arm of chromosome 11 (EGELAND et al. 1987), many studies have examined the tyrosine hydroxylase gene and other markers located within this region While some studies have reported positive findings (e.g. SERRETTI et al. 1998), most have not. Also, no abnormalities of the DA transporter gene have been detected in either unipolar or bipolar depressed patients (GOMEZ-CASERO et al. 1996; MANKI et al. 1996).

II. DA Receptors

Five different DA receptors are currently recognized, which fall into two families, D1-like (D_1 and D_5) and D2-like (D_2, D_3 and D_4). D_1 and D_2 receptors are present in all brain regions that receive a DA projection; both subtypes are expressed at a high level in the dorsal and ventral striatum, but D_1 receptors predominate in prefrontal cortex. DA autoreceptors are of the D_2 subtype, with a possible D_3 contribution; there are no D_1 autoreceptors. D_3 and D_4 receptors are localized almost exclusively within "limbic" areas, particularly the nucleus accumbens shell, and so are of particular interest in relation to affective disorders (see Chaps. 5, 7 and 8, Vol. I).

There have as yet been relatively few studies of DA receptors in depressed patients. Post-mortem studies have reported no change in D_1 or D_2 receptor binding in depressed suicides (BOWDEN et al. 1997a) or D_4 receptors in patients with major depression (SUMIYOSHI et al. 1995), relative to matched controls. One PET imaging study has reported a decrease in D_1 receptor binding in the frontal cortex, but not the striatum, of a small number of bipolar patients (only one of whom was depressed at the time of the study); however, the ligand used in this study (SCH-23390) also binds to 5HT receptors (SUHARA et al. 1992). PET studies of D_2 receptors suggest that D_2 receptor numbers may be elevated in manic but not in non-delusional depressed patients (WONG et al. 1985). Single photon emission computed tomography (SPECT) imaging studies have reported either no change (EBERT et al. 1996), or unilateral (SHAH et al. 1997) or bilateral (D'HAENEN and BOSSUYT 1994) increases in D_2 receptor binding in the basal ganglia. The latter findings are compatible with a decrease in DA turnover, but are subject to similar problems of interpretation: the relative contributions of mood and motor activity and the inability of current techniques to image the nucleus accumbens independently of the dorsal striatum.

Although there is strong evidence for a genetic contribution to bipolar affective disorder, molecular genetic studies have so far provided no evidence that bipolar disorder is associated with abnormalities of the genes coding for the D_1, D_2, D_3 or D_5 receptors. Both positive and negative findings have been reported with respect to D_4 receptor polymorphisms (reviewed by WILLNER 1999). An association between a D_4 receptor polymorphism and the personality trait of sensation seeking has been reported in several studies (e.g. EBSTEIN et al. 1997), but others have disputed this observation (e.g. POGUE-GEILE et al. 1998).

III. Neuroendocrine Studies

The tuberoinfundibular DA system has neuroendocrine functions, inhibiting the release of prolactin and stimulating the release of growth hormone (GH). Thus, basal levels of these hormones have been examined as potential markers of DA function in affective disorders, and their responses to DA agonists have been used to evaluate DA receptor responsiveness. These studies suffer two serious limitations: the inability to generalize any conclusions to the forebrain DA systems, and the involvement of many other neurotransmitters in neuroendocrine regulation; in particular, a stimulatory role of 5HT in prolactin secretion and a stimulatory role of alpha-adrenergic receptors in GH secretion.

Abnormal prolactin levels have frequently been reported in depressed patients, but there is no consistency in the direction of change: low, normal and high values have been reported in different studies (WILLNER 1983; JIMERSON 1987; BROWN and GERSHON 1993). Prolactin responses were also normal in depressed patients following DA agonist (JIMERSON 1987; BROWN and GERSHON 1993) or antagonist (ANDERSON and COWEN 1991) challenges. However, two studies have reported a decrease in prolactin levels in seasonal affective disorder (SAD), which was seen in both unipolar and bipolar patients, and was present during both winter depression and summer euthymia (DEPUE et al. 1989, 1990). This apparent trait abnormality in SAD patients is consistent either with increased DA function or with decreased 5HT function. The former interpretation is supported by the observation that SAD patients also showed a seasonally independent increase in spontaneous eye blinking: this behaviour is thought to be under dopaminergic control, being increased by D_2 agonists and suppressed by D_2 antagonists (DEPUE et al. 1989, 1990). However, blink rate appears to be normal in patients with major depression (EBERT et al. 1996).

Studies of GH are similarly inconclusive. Basal GH levels have been reported to be decreased (BOYER et al. 1986), normal (ANSSEAU et al. 1988) or increased (MENDLEWICZ et al. 1985) in major depression, and normal in mania (JIMERSON 1987). One study reported a blunting of the GH response to apomorphine in major depression, relative to patients with minor depression or normal controls (ANSSEAU et al. 1988), but no differences were observed in many earlier studies, using either a slightly higher dose of apomorphine (0.75 vs 0.5 mg), or L-dopa (WILLNER 1983; JIMERSON 1987). The group reporting blunted GH responses to apomorphine have reported a difference between major and minor depressives in two further studies (ANSSEAU et al. 1987; PITCHOT et al. 1992), and have also reported blunted responses in manic patients (ANSSEAU et al. 1987) and in suicide attempters (PITCHOT et al. 1992b). The same group also reported that blunted apomorphine responses in depressed patients were associated with low introversion and anxiety scores on the Minnesota Multiphase Personality Inventory (MMPI), but not with severity of depression (PITCHOT et al. 1990); others have reported a negative correlation between GH response and severity of delusions (MELTZER et al.

1984)]. Together, these observations suggest that there may be some subsensitivity to apomorphine in a subgroup of depressed patients. If these findings are confirmed, the question remains of whether they reflect DA receptor subsensitivity, or a more general decrease in GH responsivity [it is well established that the GH response to alpha-adrenergic challenges is subsensitive in major depression (Siever and Uhde 1984)]. The relevance of GH changes for forebrain DA function also remains to be determined. Nevertheless, a recent study reported that the growth hormone response to apomorphine was lower in patients who later responded to SSRI treatment, which was interpreted to mean that the patients who responded were those with lower pre-treatment DA receptor sensitivity (Healy and McKeon 2000).

IV. Summary

These clinical data may be summarized as follows:

1. There is clear evidence of decreased DA turnover in depressed patients, but whether this is a primary abnormality or secondary to decreased motor activity, and whether there are abnormalities of DA turnover in the nucleus accumbens, is unclear.
2. There is no consistent evidence of any genetic abnormality of DA receptors in depressed patients; there is a hint of an increase in D_2 receptor binding, but studies are inconsistent and do not provide information on the state of D_2 receptors within the nucleus accumbens.
3. Neuroendocrine studies suggest that DA function may be elevated in SAD patients, but these studies do not permit any clear conclusions with respect to major depressive disorder.

All in all, this picture is somewhat disappointing in its lack of clarity. This reflects primarily on the extreme difficulty of measuring mesoaccumbens DA function in human subjects.

C. Mood Effects of DA Agonists and Antagonists

I. Psychostimulants

The psychostimulants amphetamine and methylphenidate cause activation and euphoria in normal volunteers. Although these drugs enhance activity at both DA and NA synapses, the psychostimulant effects are mediated at DA synapses, since they are antagonized by DA receptor blockers, but not by adrenergic receptor blockers (Nurnberger et al. 1982, 1984; Jacobs and Silverstone 1988). The euphoric effects of psychostimulants at low doses closely parallel the symptomatology of hypomania, while high doses, particularly when taken repeatedly or chronically, can cause grandiosity, delusions, dysphoria, and all the other symptoms of a full-blown manic episode (Jacobs and Silverstone 1988; Post et al. 1991).

Single doses of amphetamine or methylphenidate also cause a transient mood elevation in a high proportion (>50%) of depressed patients (LITTLE 1988); the response in depressed patients appears similar, both in size and in the proportion of subjects responding, to that seen in non-depressed volunteers (NURNBERGER et al. 1982; CANTELLO et al. 1989). Following an initial report by FAWCETT and SIMONPOULOUS (1971), a number of studies have used the acute mood response to psychostimulants to predict the clinical response to chronic antidepressant therapy. A review of this literature confirmed that the response to antidepressants was well predicted by the result of an amphetamine challenge (85% improvement in responders vs 43% in non-responders), but questioned the predictive value of a methylphenidate challenge (66% improvement in responders vs 68% in non-responders) (LITTLE 1988). However, the amphetamine and methylphenidate studies differ in that the former involved mainly patients treated with imipramine and desipramine, while the latter also included a high proportion of patients treated with "serotonergic" antidepressants. A reanalysis of the same literature showed that the acute response to methylphenidate does predict antidepressant efficacy, provided that the analysis is restricted to patients treated with "noradrenergic" antidepressants (GWIRTSMAN and GUZE 1989).

Psychostimulants are not themselves considered to be efficacious as antidepressants. In early trials, the catecholamine precursor L-dopa produced a modest global improvement, primarily in retarded patients, but the effect was largely one of psychomotor activation with little effect on mood; in bipolar patients, dopa frequently caused a switch into hypomania (GOODWIN and SACK 1974). These data have been interpreted as evidence against a prominent role for DA in depression. However, the effects of dopa were greatest in patients with the lowest pretreatment CSF HVA levels (VAN PRAAG and KORF 1975). This suggests that the effect of dopa might primarily be to increase DA release in the caudate nucleus, perhaps causing motor side-effects that could mask any potentially therapeutic effects of an increase in mesolimbic DA release. A more recent open study reported striking effects of methylphenidate when added to an SSRI in treatment-resistant patients (STOLL et al. 1996).

Amphetamine is known to be effective in the treatment of old age depression (AYD and ZOHAR 1987), but efficacy in major depressive disorder has not been demonstrated in younger patients. Nonetheless, despite the absence of clinical trial data, amphetamine continues to find widespread, if little publicized, use in the treatment of depression (AYD and ZOHAR 1987). In contrast to L-dopa, low doses of amphetamine increase synaptic DA levels preferentially within the nucleus accumbens and prefrontal cortex, relative to the dorsal striatum (DI CHIARA et al. 1993; TANDA et al. 1997; DREVETS et al. 1999).

II. DA-Active Antidepressants

More convincing antidepressant effects have been reported with the directly acting DA agonists piribedil and bromocriptine. These were largely open trials, but there are also controlled studies, including a double-blind trial showing piribedil to be superior to placebo, particularly in patients with low pretreatment CSF HVA, and two large trials which found no difference in antidepressant efficacy between bromocriptine and imipramine (WILLNER 1983). The antidepressant response to bromocriptine may be greater in bipolar patients (SILVERSTONE 1984), and one study suggests a preferential effect of bromocriptine on emotional blunting (AMMAR and MARTIN 1991). Hypomanic responses during bromocriptine therapy have been reported (JOUVENT et al. 1983; SILVERSTONE 1984). Striking and rapid therapeutic effects of piribedil have been observed in previously non-responsive patients whose sleep EEG showed signs characteristic for Parkinson's disease; in patients not showing these signs, piribedil was ineffective (MOURET et al. 1989).

Trials of DA agonists in depression are not currently fashionable, but a recent double-blind study found effects superior to placebo and comparable to fluoxetine for pramipexole, a very selective D_3-preferring D_2/D_3 receptor agonist (BENNETT and PIERCEY 1999). It is also notable that DA uptake inhibition is a prominent feature of a number of newer antidepressants, including nomifensine, bupropion, and amineptine (BROWN and GERSHON 1993). The mechanism of action of bupropion, which is widely used both as monotherapy for depression and in combination with selective serotonin reuptake inhibitors (SSRIs), appears to involve both DA and NA components (ASCHER et al. 1995). Amineptine, which is a relatively selective DA uptake inhibitor, was more efficacious than clomipramine, and had a faster onset of antidepressant action, in a double-blind trial in retarded patients; another dopaminomimetic agent, minaprine, was also more effective than clomipramine in retarded patients (RAMPELLO et al. 1991). However, trials of a very selective DA uptake inhibitor, GBR-12909 (ANDERSEN 1989) were aborted because the drug appeared to be ineffective.

Contrary to expectations, given the antidepressant effects of DA agonists, there is also clear evidence that under certain circumstances, neuroleptics, which are DA receptor antagonists, are also active as antidepressants (ROBERTSON and TRIMBLE 1981; NELSON 1987). One potential resolution of this apparent paradox (which will be discussed further below) is that neuroleptics may be antidepressant only at low doses, which act preferentially as DA autoreceptor antagonists and so increase DA turnover. This hypothesis has been advanced in particular in relation to certain atypical antidepressants, such as sulpiride, which are said to have "activating" properties. Antidepressant effects of sulpiride are seen in a dose range of 100–300mg/day (e.g. RUTHER et al. 1999), which is considerably lower than the typical antipsychotic dose of 800–1,000mg/day. A DA-activating effect of sulpiride at low doses is supported by the finding that low doses of sulpiride antagonized the sedative actions of apomorphine in human subjects (SERRA et al. 1990). The antidepressant properties of a related compound, amisulpride, are also seen only at low doses (LECRUBIER et al. 1997), which act presynaptically to increase DA release. Interestingly, while the DA receptor-blocking effects of sulpiride and amisulpride are seen in all terminal areas, the increase in DA turnover seen at low doses is specific to the nucleus accumbens (SCATTON et al. 1994). The mechanisms underlying this anatomical specificity are unknown.

Antidepressant effects have also been reported for roxindole, a putatively selective DA autoreceptor agonist. In an open trial, roxindole caused rapid improvements in 8 of 12 patients suffering from a major depressive episode, as well as reducing depression and anergia in schizophrenic patients (BENKERT et al. 1992). Roxindole possesses 5HT uptake-inhibiting and 5HT agonist actions, both of which could contribute to an antidepressant effect, but neuroendocrine data (suppression of prolactin secretion: BENKERT et al. 1992) suggest that DA agonism is the predominant action of this drug. If, as claimed, roxindole is a selective autoreceptor agonist, the effect should be to decrease DA function. However, it is questionable whether roxindole is antidepressant by virtue of decreasing DA function: the drug also appears to be effective in negative schizophrenia (BENKERT et al. 1992), which is compatible with a DA-activating effect.

III. Neuroleptic-Induced Depression

Depression is frequently encountered as a side-effect of neuroleptic therapy in schizophrenia (RANDRUP et al. 1975; SIRIS 1991). This is a complex issue, with debates about whether "neuroleptic-induced depression" is a side-effect of treatment, a part of schizophrenia, a secondary effect of having schizophrenia, or the unmasking of a pre-existing depression when psychotic symptoms are brought under control. However, schizophrenic patients on neuroleptics are more likely to show full depressive syndromes than those not on neuroleptics, with a strong association between neuroleptic use and anhedonia, and this relationship holds up after controlling for level of psychosis (HARROW et al. 1994). This suggests that "neuroleptic-induced depression" is genuine, and there are strong grounds for believing that the effect is caused by antagonism of DA receptors. Conversely, neuroleptic drugs also decrease manic symptomatology. Although classical neuroleptics act at a variety of receptor sites, antimanic effects are also observed with drugs that act relatively specifically as DA receptor antagonists (JIMERSON 1987). In normal volunteers neuroleptics induce feelings of dysphoria, paralysis of volition and fatigue (BELMAKER and WALD 1977).

It is still widely believed that the catecholamine-depleting drug reserpine causes depression, on the basis of a series of reports in the 1950s, despite the findings of GOODWIN et al. (1972), on reanalysis of these data, that the great majority of "reserpine depression" patients had been incorrectly diagnosed. Patients treated with reserpine tended to display a "pseudodepression" characterized by psychomotor slowing, fatigue and anhedonia but lacking cognitive features of depression such as hopelessness or guilt. Only a small proportion of patients (5%–9%) showed symptoms analogous to major depression, and these patients usually had a prior history of mood disorders

(GOODWIN et al. 1972). It remains unclear whether the doses of reserpine administered in the "reserpine depression" studies were sufficient to decrease DA function. However, it may be significant that in the GOODWIN et al. (1972) reanalysis, major depression was considered to be the correct diagnosis in almost 50% of patients who developed marked psychomotor retardation.

IV. Parkinson's Disease

More convincing evidence of an association between DA depletion and depression is seen in the high incidence of depression in Parkinson's disease (RANDRUP et al. 1975; MURRAY 1996 – but see TAYLOR and SAINT-CYR 1990). At the level of symptomatology, there is substantial overlap between Parkinsonian akinesia and depressive psychomotor retardation (TAYLOR and SAINT-CYR 1990; BERMANZOHN and SIRIS 1992). It is difficult to determine whether Parkinsonian depression should be considered a secondary response to loss of motor function, rather than a direct consequence of DA depletion. There is no agreement in the literature as to whether the severity of depression is correlated with the extent of physical impairment. However, Parkinsonian depression is more severe than would be expected from the physical symptoms alone, and the onset of depression can precede the physical disabilities (GUZE and BARRIO 1991; MURRAY 1996).

It is now recognized that Parkinson's disease can not be considered as a pure DA deficiency syndrome: NA, 5HT, acetylcholine (ACh), somatostatin and neurotensin are also abnormal (PERRY 1987). Nevertheless, there are good reasons to relate the symptoms of Parkinsonian depression to DA depletion. In one well-designed study, depressed patients showed profound attenuation of the euphoric response to methylphenidate, relative to non-depressed Parkinsonian patients, depressed non-Parkinsonian patients, and normal controls (CANTELLO et al. 1989). The antidepressant effect of dopa in Parkinson's disease (GOODWIN and SACK 1974; RANDRUP et al. 1975; AMMAR and MARTIN 1991) also points towards a dopaminergic substrate of Parkinsonian depression. In some cases there is clear evidence that mood improvement precedes the improvement in physical symptoms (MURPHY 1972), suggesting that the antidepressant effect cannot be simply explained away as secondary to an improvement in physical symptoms. Antidepressant effects of bupropion (GOETZ et al. 1984) and bromocriptine (JOUVENT et al. 1983) have also been reported in Parkinsonian patients.

V. Neuroleptics as Antidepressants

The clinical pharmacology literature reviewed in this section is broadly consistent with the hypothesis that increases in DA function elevate mood and decreases in DA function induce symptoms of depression. However, not all of the data are compatible with this formulation. In particular, the fact that neuroleptics are used to treat depression (ROBERTSON and TRIMBLE 1981; NELSON

1987) strikes at the heart of the dopamine hypothesis. This phenomenon therefore requires careful consideration.

One hypothesis, discussed above, is that neuroleptics are administered in depression at low doses that interact selectively with presynaptic autoreceptors. However, while an autoreceptor hypothesis might explain some of the data, particularly those pertaining to sulpiride, it is not necessarily the case that low doses are used when neuroleptics are prescribed as antidepressants. Doses below the antipsychotic range have usually been prescribed in studies of mild, non-endogenous depression, but in delusional depression, neuroleptics are more commonly prescribed at normal antipsychotic doses (NELSON 1987). However, it is not certain that DA antagonism is the mechanism of antidepressant action. Indeed, in one study, antidepressant effects of *cis*-flupenthixol were negatively correlated with increase in serum prolactin levels, suggesting that DA blockade might actually antagonize the antidepressant effect (ROBERTSON and TRIMBLE 1981). In a similar vein, antidepressant effects on withdrawal of neuroleptics are well documented, though the evidence tends to arise from case reports rather than formal studies (RANDRUP et al. 1975). In a controlled trial, DEL ZOMPO et al. (1990) treated depressed patients with a cocktail of haloperidol and clomipramine, and reported marked improvement, relative to a group treated with clomipramine alone, when the haloperidol component was withdrawn after three weeks treatment. It was assumed that the improvement resulted from the unmasking of DA receptors rendered supersensitive by chronic neuroleptic treatment. Clearly, more trials of this kind are needed, and the proposed mechanism of action requires confirmation.

It is also questionable whether neuroleptics are truly antidepressant, and examination of the pattern of symptomatic improvement may provide the clearest resolution to the paradox of the antidepressant action of neuroleptics. In brief, there is no evidence that neuroleptics can improve either psychomotor retardation or anhedonia, the core symptom of depression most closely associated with the DA hypothesis. The antidepressant potential of neuroleptics is most firmly established in delusional depression, which responds well to combined therapy with a neuroleptic/tricyclic mixture, but responds poorly if at all to tricyclics alone. However, neuroleptics alone are also ineffective in delusional depression: they produce a substantial global improvement, but this arises almost entirely from a decrease in agitation and delusional thinking; motor retardation, lack of energy and anhedonia do not respond to neuroleptic treatment, and indeed, may become worse (NELSON 1987). In endogenous depressions, while neuroleptics have been claimed to be as effective as tricyclics, or nearly so, this appearance may be spurious, insofar as the studies in question may have seriously underestimated the true effectiveness of tricyclics (owing to a failure to attain adequate plasma drug levels, and other factors) (NELSON 1987). On the basis of the findings in delusional depression, it seems likely that the global improvement seen in endogenous depressives treated with neuroleptics results from the preponderance in these studies of agitated and delusional patients (ROBERTSON and TRIMBLE 1981; NELSON 1987). This analysis of the place of neuroleptics in the treatment of depression implies that retardation and delusions are mediated by different sets of DA terminals, which may be activated independently (FIBIGER 1991). In support of this assumption, it is well established that different components of the mesocorticolimbic DA projection are differentially regulated (LE MOAL 1995).

VI. Summary

The effects of DA agonists and antagonists on mood are intriguing:

1. Depression is associated with conditions in which DA function is decreased, including Parkinson's disease and neuroleptic drug treatment of schizophrenia.
2. Neuroleptics have also been claimed to be antidepressant, at DA-blocking doses, but probably are not.
3. Clear evidence of antidepressant efficacy is seen with specific D_2/D_3 agonists and with certain D_2/D_3 antagonists that act presynaptically to increase DA release within the nucleus accumbens specifically.
4. Antidepressant efficacy is also seen with agents that include DA uptake inhibition as a prominent component of their spectrum of activity. However, neither psychostimulants nor a pure DA uptake inhibitor have been shown to be effective in major depressive disorder.

A hypothesis that integrates these data is that (1) DA-active agents are antidepressant if they increase DA function specifically within the nucleus accumbens, but (2) if they also increase DA function at critical sites elsewhere in the brain, they must also possess other neurochemical properties that counteract these effects. This hypothesis awaits investigation.

D. Dopaminergic Consequences of Antidepressant Treatment

I. DA autoreceptor Desensitization

Most antidepressant drugs have little effect on DA function following acute administration. In particular, tricyclic antidepressants do not act as potent DA uptake inhibitors (WILLNER 1983), in contrast to their well-known effects at adrenergic and serotonergic synapses [though some data suggest that antidepressants may cause significant inhibition of DA uptake within the nucleus accumbens and frontal cortex (CARBONI et al. 1990; DE MONTIS et al. 1990)]. Nevertheless, there is now considerable evidence that antidepressants do enhance dopaminergic function following chronic administration.

In one of the earliest studies to demonstrate an antidepressant-induced increase in DA function, SERRA et al. (1979) reported that imipramine, amitriptyline and mianserin all decreased the sedative effect of a low dose of apomorphine. Since it was assumed that this latter effect was mediated by stimulation of DA autoreceptors, the results were interpreted as a decrease in autoreceptor sensitivity. However, the evidence that antidepressants desensitize DA autoreceptors is equivocal. There are a number of supportive studies, using a variety of techniques, but equally, there have been failures to replicate all of these data (WILLNER 1983). Some studies have reported that clear evidence of DA autoreceptor subsensitivity was not present until 3–7 days following withdrawal from chronic antidepressant treatment (SCAVONE et al. 1986; TOWELL et al. 1986). Another reason to question the relevance of DA autoreceptor desensitization for the clinical action of antidepressants is that these data were obtained in "normal" rats. However, rats exposed to chronic mild stress, which has been proposed as an animal model of depression (see below), also show evidence of DA autoreceptor desensitization similar to that sometimes seen following chronic antidepressant treatment in "normal" animals (WILLNER and PAPP 1997). Finally, changes in apomorphine-induced sedation do not necessarily imply changes in DA autoreceptor function. High doses of apomorphine cause locomotor stimulation, so a decrease in apomorphine-induced sedation might equally well indicate an increase in postsynaptic responsiveness, rather than autoreceptor subsensitivity.

II. Sensitization of D_2/D_3 Receptors

In fact, a substantial body of literature now demonstrates that following chronic treatment, antidepressants do increase the responsiveness of postsynaptic D_2/D_3 receptors in the mesolimbic system. These effects are seen irrespective of the primary neurochemical action of the drug (WILLNER 1989; MAJ 1990). The majority of studies have examined the locomotor stimulant response to moderate doses of apomorphine or amphetamine, which is consistently elevated following chronic administration of antidepressants. Similar effects have been observed using the specific D_2/D_3 agonist quinpirole (MAJ 1990). There are well-known pharmacokinetic interactions between antidepressants and amphetamine. However, antidepressants also increased the psychomotor stimulant effect when amphetamine, or DA itself, was administered directly to the nucleus accumbens (MAJ 1990), confirming a true pharmacodynamic interaction. Furthermore, these effects were present within a short time (2h) of the final antidepressant treatment, confirming that, unlike DA autoreceptor desensitization, the increase in responsiveness of postsynaptic D_2/D_3 receptors is not simply a withdrawal effect. The potentiation of D_2/D_3 receptor function by chronic antidepressant treatment is confined to mesolimbic terminal regions: antidepressants do not increase the intensity of stereotyped behaviours caused by high doses of amphetamine, which are mediated by DA release within the dorsal striatum (WILLNER 1989). Neither did chronic antidepressant treatment potentiate a DA-mediated neuroendocrine response (PRZEGALINSKI et al. 1990).

Receptor binding studies have usually failed to detect any alterations in the binding parameters of D_2/D_3 receptors that would explain the increased functional responses. The majority of these studies are of limited relevance, as they assayed DA receptors in samples of dorsal striatum. Nevertheless, negative findings have also been reported in nucleus accumbens. However, D_2/D_3 receptors in limbic forebrain (but not dorsal striatum) have an increased affinity for the agonist ligand quinpirole following chronic antidepressant administration to rats, and an increase in receptor number has recently been reported, in ventral but not dorsal striatum, using an agonist ligand (MAJ et al. 1996). Increased D_3 receptor binding in ventral striatal regions, following chronic antidepressant treatment, has also been recently reported (MAJ et al. 1998).

While it has proved difficult, using conventional antagonist ligands, to demonstrate structural changes in D_2 receptors corresponding to their increased functional sensitivity, this may reflect the use of inappropriate experimental methods. Rats subjected to chronic mild stress, which reproduce many of the symptoms of depression (see below), show a decrease in D_2/D_3 receptor numbers in limbic forebrain. This decrease was completely reversed by chronic treatment with imipramine (WILLNER and PAPP 1997). In the same study, non-stressed animals treated with imipramine failed to show an increase in D_2/D_3 receptor binding, consistent with earlier data.

In addition to increasing the responsiveness of D_2/D_3 receptors, antidepressants also decrease the number of D_1 receptors, following chronic treatment (MAJ 1990). This effect is associated with a decrease in the ability of DA to stimulate adenylyl cyclase (MAJ 1990), and a decreased behavioural response (grooming) to D_1 receptor stimulation (MAJ et al. 1989), consistent with the binding data. A role for D_1 receptor changes in the sensitization of D_2/D_3 receptors has been proposed (SERRA et al. 1990), but this seems unlikely, as the downregulation of D_1 receptors is species specific: D_1 receptors were downregulated by chronic imipramine in rats but not in mice (NOWAK et al. 1991). Furthermore, D_1 receptors were not downregulated by chronic imipramine in chronically stressed rats, which did show D_2/D_3 receptor upregulation (PAPP et al. 1994). In both of these studies, functionally relevant behavioural effects of chronic antidepressant treatment were seen in the absence of D_1 receptor changes.

III. Clinical Evidence

Three recent clinical studies have reported increased D_2/D_3 receptor binding following chronic antidepressant treatment. One study reported increased binding post-mortem in antidepressant-treated depressed patients who had committed suicide. Increases were observed in caudate, putamen and nucleus accumbens of antidepressant-treated suicides, relative to non-depressed controls, but not in depressed suicide victims who had not been treated with antidepressants (BOWDEN et al. 1997a). A second study, using SPECT imaging, reported that the extent of clinical recovery following SSRI treatment of depression was significantly correlated with the size of the increase in D_2 receptor binding in striatum and anterior cingulate gyrus (LARISCH et al. 1997). The third study, also using SPECT imaging, also found increased D_2 receptor binding in the basal ganglia of patients who responded to SSRIs (D'HAENEN et al. 1999). Consistent with the animal data, however, chronic SSRI treatment did not increase D_2 receptor binding in non-depressed volunteers (TIIHONEN et al. 1996).

IV. Summary

In contrast to the traditional focus on NA and 5HT systems, chronic treatment with antidepressant drugs also reliably affects transmission at DA synapses, via several mechanisms:

1. Chronic antidepressant treatment desensitizes both presynaptic DA autoreceptors and postsynaptic D_1 receptors. However, in both cases there is reason to doubt the clinical relevance of these effects.
2. Chronic antidepressant treatment increases the sensitivity of D_2/D_3 receptors; these effects appear to be specific to the nucleus accumbens.
3. Increases in D_2/D_3 receptor binding following chronic antidepressant treatment have been relatively difficult to demonstrate in normal animals, and appear to require the use of agonist ligands. However, increased D_2/D_3 binding is readily observed following chronic antidepressant treatment, using conventional methods, in an animal model of depression (chronic mild stress) and in depressed patients.

These data point to increased transmission at D_2/D_3 receptors in the nucleus accumbens as a potentially important mechanism of antidepressant action.

E. Dopaminergic Mechanisms in Animal Models of Depression

I. D_2/D_3 Receptor Sensitization as a Mechanism of Antidepressant Action

Although these data confirm that antidepressants change the functional status of DA receptors in the nucleus accumbens, they give little insight into the role that these changes play in the clinical action of antidepressants. In particular, the data reviewed demonstrate only that antidepressants change the properties of DA receptors. These changes could be responsible for, correlated with, irrelevant to, or even counteractive of the clinical effects.

Animal models of depression provide one means of addressing this question, albeit indirectly. The mechanisms by which antidepressants act to bring about their functional effects have been analysed most extensively in the Porsolt forced swim test. In this model, rats or mice are required to swim in a confined space, and antidepressants prolong the period in which the animal displays active escape behaviour. Immobility in the swim test may be reversed not only by antidepressants, but also by D_2/D_3 receptor agonists, applied systemically or to the nucleus accumbens (BORSINI and MELI 1990). Conversely, a number of studies have reported that antidepressant effects in the swim test were reversed by DA antagonists (BORSINI and MELI 1990); these include studies in which antidepressants were administered chronically (PULVIRENTI and SAMANIN 1986; DELINA-STULA et al. 1988). The effects of chronically administered tricyclic antidepressants were reversed by the administration of sulpiride in the nucleus accumbens, but not in the dorsal striatum (CERVO and SAMANIN 1988). Despite these positive findings, BORSINI and MELI (1990) urge caution in accepting that the data demonstrate a DA mechanism of antidepressant action in the swim test, and suggest that the effects of intra-accumbens sulpiride could be related to the presence in the mesolimbic system of non-dopaminergic sulpiride binding sites that also bind antidepressants. The swim test has been criticized on a number of counts, most prominently that it responds to acute administration of antidepressants, unlike the clinical situation, which requires chronic treatment. This criticism is not entirely justified, since the test only responds acutely to extremely high drug doses, but becomes slowly more sensitive with repeated treatment (WILLNER 1989). However, the validity of the test as a model of depression is extremely weak.

Dopaminergic mechanisms have also been analysed in animal models of depression more valid and realistic than the swim test. For example, a decrease in D_2 receptor binding has been reported in socially subordinate female cynomolgous monkeys, which display many features reminiscent of affective pathology (SHIVELY et al. 1997).

The most extensive investigations of this type have employed the chronic mild stress (CMS) procedure, in which rats or mice are exposed chronically (weeks or months) to a variety of mild unpredictable stressors. This causes the appearance of almost all of the behavioural symptoms and many physiological changes characteristic of depression, including a generalized decrease in responsiveness to rewards (anhedonia). Normal behaviour in this model is restored by chronic, but not by acute, administration of a wide range of tricyclic or atypical antidepressants (WILLNER et al. 1992; WILLNER 1997a; WILLNER and PAPP 1997). These behavioural changes are accompanied by a decrease in D_2/D_3 receptor binding and D_2 mRNA expression in the nucleus accumbens, and a pronounced functional subsensitivity to the rewarding and locomotor stimulant effects of the D_2/D_3 agonist quinpirole, administered systemically or within the nucleus accumbens. All of these effects are also reversed by chronic antidepressant treatment (DZIEDZICKA-WASYLEWSKA et al. 1997; WILLNER and PAPP 1997).

The question of whether these changes in D_2/D_3 receptor function are responsible for the therapeutic action of antidepressant drugs in the CMS model has been investigated in studies that asked whether the effect of antidepressant treatment would be reversed by interfering with transmission at DA synapses. In these studies, animals successfully treated with antidepressants were treated acutely with D_2/D_3 receptor antagonists, at low doses that were without effect in non-stressed animals or in untreated stressed animals. This treatment reversed the effects of a wide variety of antidepressants on rewarded behaviour (including tricyclics, specific 5HT or NA uptake inhibitors, or mianserin) (WILLNER and PAPP 1997). Chronic stress also causes an antidepressant-reversible decrease in aggressive behaviour, and this effect of chronic antidepressant treatment was also reversed by acute administration of DA antagonists (ZEBROWSKA-LUPINA et al. 1992). These data argue strongly that an increase in D_2/D_3 receptor responsiveness may be responsible for the therapeutic action of antidepressants in this model (WILLNER and PAPP 1997).

II. Clinical Evidence

The antagonist challenge strategy can also be applied in clinical studies to investigate the involvement of specific neurotransmitter systems in antidepressant action. This method has been used recently to demonstrate the involvement of 5HT systems in the action of serotonergic antidepressants (using the tryptophan depletion technique to antagonize 5HT transmission) and the involvement of NA systems in the action of noradrenergic antidepressants (using the catecholamine synthesis inhibitor alpha-methyl-*p*-tyrosine to antagonize NA transmission) (SALOMON et al. 1993; MILLER et al. 1996).

The role of D_2/D_3 receptor sensitization has been studied using the same method that has been used in rodents. Patients who had recovered from depression after chronic treatment with SSRIs were treated acutely with the D_2/D_3 receptor antagonist sulpiride, at a low dose that was shown to be almost without effect in non-depressed volunteers. All of the patients showed a marked increase in symptoms of depression (WILLNER 1997b; P. Willner et al., in preparation). This study would appear to confirm that the increase in D_2/D_3 receptor binding that has been observed in antidepressant-treated patients (BOWDEN et al. 1997a; LARISCH et al. 1997; D'HAENEN et al. 1999) may actually be responsible for their recovery from depression.

III. Reciprocal Changes in DA Responses to Reward and Stress

In addition to decreasing D_2 receptor function in the nucleus accumbens, CMS also decreases DA release in the same region. For example, a variant of the CMS procedure has been shown to disrupt appetitive behaviour maintained by a highly palatable reward (GHIGLIERI et al. 1997) and to decrease basal levels of extracellular DA in the nucleus accumbens shell (GAMBARANA et al. 1999), both of these changes being reversed by chronic antidepressant treatment.

Particularly cogent data have recently been presented by DI CHIARA and colleagues. In this study, basal DA concentrations were not decreased significantly in animals exposed to CMS, but there were marked changes in responses to a palatable reward and to a stressor (tail pinch), and these changes were regionally specific and opposite in direction. In control animals, tail pinch increased DA release primarily in the prefrontal cortex, while consumption of a palatable food increased DA release both in prefrontal cortex and in the nucleus accumbens shell. CMS markedly inhibited the responses to rewards, but potentiated the response to stress. Both of these effects of CMS were reversed by chronic treatment with the antidepressant desipramine. Thus, in both prefrontal cortex and nucleus accumbens shell, desmethylimipramine (DMI) reversed both the inhibition of DA release in response to reward, and the enhancement of DA release in response to stress (DI CHIARA et al. 1999).

These findings appear to provide a potential explanation and mechanism for the negative information processing bias characteristic of depression. The simultaneous inhibition of DA responses to reward and enhancement of responses to stress both serve to bias information processing in the direction of increased salience of emotionally negative stimuli; and both halves of this equation were normalized by antidepressant treatment. The mechanisms whereby CMS causes opposite changes in DA release in response to appetitive and aversive stimulation are currently unknown. CMS has been shown to increase the size of the releasable pool of DA in the nucleus accumbens (STAMFORD et al. 1991; WILLNER et al. 1991). This could account for the elevation of DA release in response to stress: if there is no change in the ability of stressors to activate DA neurons, the larger pool of releasable DA would lead to an increase in DA release. However, this factor cannot explain the attenuation of DA release in response to rewards. This discrepancy suggests that CMS may act to prevent reward-related information from activating DA neurons in the ventral tegmental area (VTA) by mechanisms that remain to be determined.

The parallel with depression is even more striking when pre- and postsynaptic mechanisms are considered together. In the case of reward, CMS decreases both presynaptic (release) and postsynaptic (D_2 receptor sensitivity) measures of DA function, and these two effects combine to decrease the behavioural response to rewards. In the case of stress, CMS increases DA release, but at the same time decreases D_2 receptor sensitivity, so these two effects will tend to cancel one another out, leaving the response to an acute stressor relatively unchanged. This is consistent with old data (reviewed by WILLNER 1984) showing that while depressed patients are greatly distressed by psychic pain, their response to acute physical pain is relatively normal.

IV. Summary

These studies provide evidence that changes in DA function may be central to an understanding of depression and antidepressant drug action:

1. The therapeutic effect of chronic antidepressant treatment is temporarily reversed by acute blockade of D_2/D_3 receptors, both in a realistic animal model of depression (CMS) and in patients.
2. Rewarding events release DA in both the nucleus accumbens and the prefrontal cortex, while aversive events release DA preferentially in the prefrontal cortex. CMS inhibits the response to rewards, while increasing the response to aversive events. Both of these effects are normalized by chronic antidepressant treatment.

These data confirm the relevance of DA synapses as a crucial substrate for antidepressant action and provide a potential neural substrate for the information-processing biases characteristic of depression.

F. Conclusions

I. Limitations of the Dopamine Hypothesis

The data reviewed in the preceding section present a strong case that elevation of DA transmission in the nucleus accumbens may represent a "final common pathway" responsible for at least part of the spectrum of behavioural actions of antidepressant drugs. The mechanisms by which antidepressants bring about these changes are not well understood, but the best guess at present is that the effects are indirectly mediated via primary actions at NA or 5HT terminals. Nevertheless, the evidence supporting a dopaminergic mechanism of antidepressant action is largely preclinical: clinical studies evaluating the role of DA mechanisms in the action of classical antidepressants are sparse.

Similarly, there is relatively little clinical evidence that unambiguously supports the hypothesis of DA hypofunction in depression. Indeed, some evidence runs directly counter to the DA hypothesis, particularly the clinical use of neuroleptics in depression. As discussed above, there are a number of potential resolutions of this troublesome paradox, including the possibility of autoreceptor-selective actions of neuroleptics at low doses, the possibility that neuroleptics control delusions but actually worsen depressive symptoms, and the possibility that DA hypofunction in some terminal fields coexists with DA hyperfunction in other regions (NELSON 1987; FIBIGER 1991). [The latter hypothesis has also been advanced to explain the coexistence of negative and positive symptoms in schizophrenia (DAVIS et al. 1991; FIBIGER 1991)]. There has been little research directed specifically at understanding the place of neuroleptics in the treatment of depression: more is urgently needed.

Setting aside the question of neuroleptics as antidepressants, the effects of pharmacological interventions in human subjects lead broadly to the conclusion that inhibiting DA transmission is therapeutic in mania and induces depressive symptomatology in normal volunteers, while stimulation of DA transmission has antidepressant effects and induces manic symptoms.

However, the extent of overlap between these pharmacological effects and clinical changes is far from complete. While the effects of psychostimulants provide a good match to the symptoms of mania, the primary effects of neuroleptics or reserpine in normal subjects are fatigue, apathy and dysphoria (GOODWIN et al. 1972; BELMAKER and WALD 1977). Conversely, while L-dopa readily induces hypomania in depressed patients, there is little evidence of mood improvement (GOODWIN and SACK 1974). Thus, the pharmacological evidence for DA involvement appears rather stronger in mania than in depression. However, this conclusion overlooks the anatomical non-specificity of these drugs: they are of limited value as research tools for evaluating whether depression is associated with a dysfunction of mesolimbic DA specifically. In contrast to L-dopa, directly acting DA agonists do appear to be effective antidepressants, though the number of controlled trials remains unacceptably low. The clinical efficacy of these agents may reflect a preferential action within the nucleus accumbens, but it is not yet possible to evaluate this hypothesis in human subjects.

Similarly, the inability to measure DA activity within the nucleus accumbens seriously limits the value of virtually all of the correlative studies of DA function in depression and mania. The fact that there are no reliable neuroendocrine changes in affective disorder patients simply tells us that there are no generalized abnormalities of D_2/D_3 receptors, not that such abnormalities are absent within the nucleus accumbens specifically. Similarly, we have no useful information on the release of DA from mesocorticolimbic terminals in human subjects. The clearest evidence implicating DA in depression, the decrease in CSF HVA concentrations in retarded depression, is intriguing but appears to relate primarily to changes in motor function. Discovering the direction of causality in this relationship remains an important objective. However, the priority for understanding the role of DA in depression must be to redress the imbalance between the preclinical and the clinical evidence. This requires the development of research tools for human use with sufficient anatomical precision to evaluate DA function within distinct terminal fields. Neuroimaging techniques are already very close to achieving this objective.

II. Syndromes or Symptoms?

As noted in the introduction to this chapter, our emerging understanding of the behavioural functions of forebrain DA systems suggests that the involvement of DA in affective disorders might profitably be analysed at the level of symptoms rather than syndromes. The clinical literature contains a number of findings that support this position. Thus, there is some evidence that emotional blunting responds more rapidly and more completely than other symptoms in depressed patients treated with bromocriptine (AMMAR and MARTIN 1991), and that DA-uptake inhibiting antidepressants may be superior to tricyclics in retarded patients (RAMPELLO et al. 1991). Conversely, Parkinsonian or pre-Parkinsonian (MOURET et al. 1989) depressions, which respond to treatment with DA agonists (RANDRUP et al. 1975; MOURET et al. 1989; AMMAR and MARTIN 1991), are characterized by decreased motivation and drive, but not by feelings of guilt, self-blame and worthlessness (BROWN et al. 1988); these characteristic depressive cognitions are also conspicuously absent from descriptions of neuroleptic- or reserpine-induced depressive states (GOODWIN et al. 1972; BELMAKER and WALD 1977).

From a psychobiological standpoint, it seems obvious that the major psychiatric syndromes are likely to involve multiple neurotransmitter systems, which contribute to different syndromes to differing degrees. A research strategy that follows from this observation is to investigate, as a first step, the involvement of specific pathways in specific behavioural processes, which need not, on a priori grounds, bear any obvious relationship to nosological boundaries. It is clear that features of what might be termed a DA-deficiency syndrome, involving low CSF HVA, anhedonia, psychomotor slowing and a good response to DA agonist treatment, are characteristic not only of depression, but also of Parkinson's disease (VAN PRAAG et al. 1975; TAYLOR and SAINT-CYR 1990; WOLFE et al. 1990; BERMANZOHN and SIRIS 1992) and negative schizophrenia (VAN PRAAG et al. 1975; KIRKPATRICK and BUCHANAN 1990; FIBIGER 1991; BENKERT et al. 1992). At the other extreme, there is considerable overlap in symptoms between positive schizophrenia and mania, and these common symptoms are reliably reproduced in psychostimulant-induced psychoses (FIBIGER 1991; POST et al. 1991). The extent to which these similar functional outcomes reflect common underlying mechanisms remains to be determined, and represents a major challenge for future research. However, the difficulties of pursuing a research agenda that cuts across DSM-IV diagnostic categories should not be underestimated.

III. The Wider Picture

While the present chapter has focussed on the role of the mesocorticolimbic DA system in depression and antidepressant drug action, it is clear that this role can only be fully understood in the context of a broader picture of the functioning of the neuroanatomical systems with which the mesocorticolimbic DA system interacts. Information processing in the forebrain is based on a set of parallel cortical-striatal-pallidal-thalamic loops (ALEXANDER and CRUTCHER 1990). The mesocorticolimbic DA system innervates primarily the limbic cortex-ventral striatum-ventral pallidum-mediodorsal thalamus loop, which has been implicated in pathophysiology of depression (SWERDLOW and KOOB 1987; MCHUGH 1989). A decrease in DA activity within the nucleus accumbens should disinhibit outputs from the ventral striatum (SMITH and BOLAM 1990), leading to increased activity within prefrontal cortex and amygdala (SWERDLOW and KOOB 1987). This pattern of activity is indeed observed in functional imaging studies of depressed patients (DREVETS et al. 1992), providing important indirect evidence to support the DA hypothesis advanced in this chapter.

References

Alexander GE, Crutcher MD (1990) Functional architecture of basal ganglia circuits: neural substrates of parallel processing. Trends Neurosci 13:266–271

Allard P, Norlen M (1997) Unchanged density of caudate nucleus dopamine uptake sites in depressed suicide victims. J Neural Transms 104:1353–1360

American Psychiatric Association (1994) DSM IV-Diagnostic and Statistical Manual of Psychiatric Disorders, 4th Edition. American Psychiatric Association, Washington D.C.

Andersen PH (1989) The dopamine uptake inhibitor GBR 12909: selectivity and molecular mechanism of action. Eur J Pharmacol 166:493–504

Ascher JA, Cole JO, Colin JN, Feighner JP, Ferris RM, Fibiger HC, Golden RN, Martin P, Potter WZ, Richelson E et al. (1995) Bupropion: A review of its mechanism of antidepressant activity. J Clin Psychiatr 56:395–401

Ammar S, Martin P (1991) Modelisation des effets des agonistes dopaminergiques en psychopharmacologie: vers une homothetie clinique et experimentale. Psychol Francaise 36:221–232

Anderson IM, Cowen PJ (1991) Prolactin response to the dopamine antagonist, metoclopramide, in depression. Biol Psychiatr 30:313–316

Ansseau M, von Frenckell R, Cerfontaine JL, Papart P, Franck G, Timset-Berthier M, Geenen V, Legros J-J (1987) Neuroendocrine evaluation of catecholaminergic neurotransmission in mania. Psychiatr Res 22:193–206

Ansseau M, von Frenckell R, Cerfontaine JL, Papart P, Franck G, Timset-Berthier M, Geenen V, Legros J-J (1988) Blunted response of growth hormone to clonidine and apomorphine in endogenous depression. Brit J Psychiatr 153:65–71

Ayd FJ Jr, Zohar J (1987) Psychostimulant (amphetamine or methylphenidate) therapy for chronic and treatment-resistant depression. In: Zohar J, Belmaker RH (eds) Treating Resistant Depression. New York: PMA Corp, 343–355

Baxter LR, Schwartz JM, Phelps ME, Mazziotta JC, Guze BH, Selin CE, Gerner RH, Sumida RM (1989) Reduction of prefrontal cortex glucose metabolism common to three types of depression. Arch Gen Psychiatr 46:243–250

Beck AR (1987) Cognitive models of depression. J Cog Psychother 1:5–37

Belmaker RH, Wald D (1977) Haloperidol in normals. Brit J Psychiatr 131:222–223

Beninger RL, Nakonechny PL (1996) Dopamine D1-like receptors and molecular mechanisms of incentive learning. In: Beninger RJ, Paolomo T, Archer T (eds) Dopamine Disease States, Editorial CYM, Madrid, pp 407–431

Benkert O, Brunder G, Wetzel H (1992) Dopamine autoreceptor agonists in the treatment of schizophrenia and major depression. Pharmacopsychiatry 25:254–260

Bennett JP, Piercey MF (1999) Pramipexole – A new kind of dopamine agonist for the treatment of Parkinson's Disease. J Neurosci Res 163:25–31

Bermanzohn PC, Siris G (1992) Akinesia: A syndrome common to Parkinsonism, retarded depression and negative symptoms of schizophrenia. Compr Psychiatr 33:221–232

Bindra D (1974) A motivational view of learning, performance, and behavior modification. Psychol Rev 81:199–213

Borsini F, Meli A (1990) The forced swimming test: Its contribution to the understanding of the mechanisms of action of antidepressants. In: Gessa GL, Serra G, (eds) Dopamine and Mental Depression. Oxford: Pergamon Press, 63–76

Bowden C, Theodorou AE, Cheetham SC, Lowther S, Katona CL, Crompton MR, Horton RW (1997a) Dopamine D1 and D2 receptor binding sites in brain samples from depressed suicides and controls. Brain Res 752:227–233

Bowden C, Cheetham SC, Lowther S, Katona CL, Crompton MR, Horton RW (1997b) Reduced dopamine turnover in the basal ganglia of depressed suicides. Brain Res 769:135–140

Bowden C, Cheetham SC, Lowther S, Katona CL, Crompton MR, Horton RW (1997c) Dopamine uptake sites, labelled with [^{3}H]GBR12935, in brain samples from depressed suicides and controls. Eur Neuropsychopharmacol 7:247–252

Boyer P, Davila M, Schaub C, Nassiet J (1986) Growth hormone response to clonidine stimulation in depressive states. Part I. Psychiatr Psychobiol 1:189–195

Brown AS, Gershon S (1993) Dopamine and depression J Neural Transm [Gen Sect] 91:75–109

Brown RG, MacCarthy B, Gotham A-M, Der GJ, Marsden CD (1988) Depression and disability in Parkinson's disease: A follow-up of 132 cases. Psychol Med 18:49–55

Cantello R, Aguggia M, Gilli M, Delsedime M, Chiardo Cutin I, Riccio A, Mutani R (1989) Major depression in Parkinson's disease and the mood response to intravenous methylphenidate: possible role of the "hedonic" dopamine synapse. J Neurol Neurosurg Psychiatr 52:724–731

Carboni E, Tanda GL, Frau R, Di Chiara G (1990) Blockade of the noradrenaline carrier increases extracellular dopamine concentrations in the prefrontal cortex. Evidence that dopamine is taken up in vivo by noradrenergic terminals. J Neurochem 55:1067–1070

Cervo L, Samanin R (1988) Repeated treatment with imipramine and amitriptyline reduces the immobility of rats in the swimming test by enhancing dopamine mechanisms in the nucleus accumbens. J Pharm Pharmacol 1940:155–156

Davis KL, Kahn RS, Ko G, Davidson M (1991) Dopamine in schizophrenia: A review and reconceptualization. Am J Psychiatr 148:1474–1486

De Montis MG, Devoto P, Gessa GL, Porcella A, Serra G, Tagliamonte A (1990) Possible role of DA receptors in the mechanism of action of antidepressants. In: Gessa GL, Serra G (eds) Dopamine and Mental Depression. Oxford: Pergamon Press, 147–157

Del Zompo M, Boccheta A, Bernardi F, Corsini GU (1990) Clinical evidence for a role of dopaminergic system in depressive syndromes. In: Gessa GL, Serra G (eds) Dopamine and Mental Depression. Oxford: Pergamon Press, 177–184

Delina-Stula A, Radeke E, van Riezen H (1988) Enhanced functional responsiveness of the dopaminergic system: The mechanism of anti-immobility effects of antidepressants in the behavioural despair test in the rat. Neuropharmacology 27:943–947

Depue RA, Arbisi P, Spoont MR, Krauss S, Leon A, Ainsworth B (1989) Seasonal and mood independence of low basal prolactin secretion in premenopausal women with seasonal affective disorder. Am J Psychiatr 146:989–995

Depue RA, Arbisi P, Krauss S, Iacono WG, Leon A, Muir R, Allen J (1990) Seasonal independence of low prolactin concentration and high spontaneous eye blink rates in unipolar and bipolar II seasonal affective disorder. Arch Gen Psychiatr 47:356–364

D'haenen H, Bossuyt A (1994) Dopamine D2 receptors in the brain measured with SPECT. Biol Psychiatr 15:128–132

D'haenen H, Steensens D, Bossuyt A (1999) Successful antidepressant treatment with SSRIs is associated with an increased D2 binding. Int J Neuropsychopharmacol 2:S14

Di Chiara G, Tanda G, Frau R, Carboni E (1993) On the preferential release of dopamine in the nucleus accumbens by amphetamine: further evidence obtained by vertically implanted concentric probes. Psychopharmacology 112:398–402

Di Chiara G, Loddo P, Tanda G (1999) Reciprocal changes in prefrontal and limbic dopamine responsiveness to aversive and rewarding stimuli after chronic mild stress: Implications for the psychobiology of depression. Biol Psychiatr 46:1624–1633

Drevets WC, Videen TO, Price JL, Preskorn SH, Carmichael ST, Raichle ME (1992) A functional anatomical study of unipolar depression. J Neurosci 12:3628–3636

Drevets WC, Price JC, Kupfer DJ, Kinahan PE, Lopresti B, Holt D, Mathis C (1999) PET measures of amphetamine-induced dopamine release in ventral versus dorsal striatum. Neuropsychopharmacology 21:694–709

Dziedzicka-Wasylewska M, Willner P, Papp M (1997) Changes in dopamine receptor mRNA expression following chronic mild stress and chronic antidepressant treatment. Behav Pharmacol 8:607–618

Ebert D, Albert R, Hammon G, Strasser B, May A, Merz A (1996) Eye-blink rates and depression. Is the antidepressant effect of sleep deprivation mediated by the dopamine system? Neuropsychopharmacology 15:332–339

Ebert D, Feistel H, Loew T, Priner A (1996) Dopamine and depression: Striatal dopamine D2 receptor SPECT before and after antidepressant therapy. Pschopharmacology 126:91–94

Ebstein RP, Nemanov L, Klotz I, Gritsenko I, Belmaker RH (1997) Additional evidence for an association between the dopamine D4 receptor (DRD4) exon III repeat polymorphism and the human personality trait of Novelty Seeking. Mol Psychiatr 2:472–477

Egeland JA, Gerhard DS, Pauls DL, Sussex JN, Kidd KK, Allen CR, Hostetter AM, Housman DE (1987) Bipolar affective disorders linked to DNA markers on chromosome 11. Nature 325:783–787

Fawcett J, Simonopoulos V (1971) Dextroamphetamine response as a possible predictor of improvement with tricyclic therapy in depression. Arch Gen Psychiatr 25:247–255

Fibiger HC (1991) The dopamine hypotheses of schizophrenia and depression: Contradictions and speculations. In: Willner P, Scheel-Kruger J (eds) The Mesolimbic Dopamine System: From Motivation to Action. Chichester: John Wiley and Sons, 1991:615–637

Gambarana C, Masi F, Tagliamonte A, Scheggi S, Ghiglieri O, De Montis MG (1999) A chronic stress that impairs reactivity in rats also decreases dopaminergic transmission in the nucleus accumbens: A microdialysis study. J Neurochem 72:2039–46

Ghiglieri O, Gambarana C, Scheggi S, Tagliamonte, A, Willner P, De Montis G (1997) Palatable food induces an appetitive behaviour in satiated rats which can be inhibited by chronic stress. Behav Pharmacol 8:619–628

Gjerris A, Werdelin L, Rafaelson OJ, Alling C, Christensen NJ (1987) CSF dopamine increased in depression: CSF dopamine, noradrenaline and their metabolites in depressed patients and in controls. J Affect Disord 13:279–286

Goetz CG, Tanner CM, Klawans HL (1984) Bupropion in Parkinson's disease. Neurology 34:1092–1094

Gomez-Casero E, Perez de Castro I, Saiz-Ruiz J, Llinares C, Fernandez-Piqueras J (1996) No association between particular DRD3 and DAT gene polymorphisms and manic-depressive illness in a Spanish sample. Psychiatr Genet 6:209–212

Goodwin FK, Ebert MH, Bunney WE (1972) Mental effects of reserpine in man: A review. In: Shader RI (ed) Psychiatric Complications of Medical Drugs. New York: Raven Press, 73–101

Goodwin FK, Sack RL (1974) Central dopamine function in affective illness: evidence from precursors, enzyme inhibitors, and studies of central dopamine turnover. In: Usdin E (ed) Neuropsychopharmacology of Monoamines and their Regulatory Enzymes. New York: Raven Press, 261–279

Groenewegen HJ, Berendse HW, Meredith GE, Haber SN, Voorn P, Wolters JG, Lohman AHM (1991) Functional anatomy of the ventral, limbic system-innervated striatum. In: Willner P, Scheel-Kruger J (eds) The Mesolimbic Dopamine System: From Motivation to Action. Wiley, Chichester, pp 19–59

Guze BH, Barrio JC (1991) The etiology of depression in Parkinson's disease patients. Psychosomatics 32:390–394

Gwirtsman HE, Guze BH (1989) Amphetamine, but not methylphenidate, predicts antidepressant response. J Clin Psychopharmacol 9:453

Harrow M, Yonan CA, Sands JR, Marengo J (1994) Depression in schizophrenia: Are neuroleptics, akinesia, or anhedonia involved. Schizophr Bull 120:327–338

Healy E, McKeon P (2000) Dopaminergic sensitivity and prediction of antidepressant response. J Psychopharmacol 14:152–156

Jacobs D, Silverstone T (1988) Dextroamphetamine-induced arousal in human subjects as a model for mania. Psychol Med 16:323–329

Jeste DV, Lohr JB, Goodwin FK (1988) Neuroanatomical studies of major affective disorders: A review and suggestions for future research. Brit J Psychiatr 153:444–459

Jimerson DC (1987) Role of dopamine mechanisms in the affective disorders. In: Meltzer HY (ed) Psychopharmacology: The Third Generation of Progress. New York: Raven Press, 515–511

Jouvent R, Abensour P, Bonnet AM, Widlocher D, Ajid Y, Lhermitte F (1983) Antiparkinson and antidepressant effects of high doses of bromocriptine. J Affect Disord 5:141–145

King RJ, Mefford IN, Wang C, Murchison A, Caligari EJ, Berger PA (1986) CSF dopamine levels correlate with extraversion in depressed patients. Psychiatr Res 19:305–310

Kirkpatrick B, Buchanan RW (1990) Anhedonia and the deficit syndrome of schizophrenia. Psychiatr Res 31:25–30

Lambert G, Johansson M, Agren H, Friberg P (2000) Reduced brain norepinephrine and dopamine turnover in depressive illness: Evidence in support of the catecholamine hypothesis of affective disorders. Arch Gen Psychiatr 57:787–793

Larisch R, Klimke A, Vosberg H, Loffler S, Gaebel W, Muller-Gartner H-W (1997) In vivo evidence for the involvement of dopamine-D2 receptors in striatum and anterior cingulate gyrus in major depression. Neuroimage 5:251–260

Lecrubier Y, Boyer P, Turjanski S, Rein W (1997) Amisulpride versus imipramine and placebo in dysthymia and major depression. J Affect Disord 43:95–103

Little KY (1988) Amphetamine, but not methylphenidate, predicts antidepressant response. J Clin Psychopharmacol 8:177–183

Le Moal M (1995) Mesocorticolimbic dopaminergic neurons: Functional and regulatory roles. In: Bloom FE, Kupfer DJ (eds) Psychopharmacology: The Fourth Generation of Progress. Raven, New York, pp 283–294

Maj J (1990) Behavioral effects of antidepressant drugs given repeatedly on the dopaminergic system. In: Gessa GL, Serra G (eds) Dopamine and Mental Depression. Oxford: Pergamon Press, 139–146

Maj J, Papp M, Skuza G, Bigajska K, Zazula M (1989) The influence of repeated treatment with imipramine, (+)- and (−)-oxaprotiline on behavioural effects of dopamine D-1 and D-2 agonists. J Neural Transm 76:29–38

Maj J, Dziedzicka-Wasylewska M, Rogoz R, Rogoz Z, Skuza G (1996) Antidepressant drugs given repeatedly change the biding of the dopamine D2 receptor agonist [3H]N-0437, to dopamine D2 receptors in the rat brain. Eur J Pharmacol 23:49–54

Maj J, Dziedzicka-Wasylewska M, Rogoz R, Rogoz Z (1998) Effect of antidepressant drugs administered repeatedly on the dopamine D3 receptors in the rat brain. Eur J Pharmacol 351:31–37

Manki H, Kanabe S, Muramatsu T, Higuchi S, Suzuki E, Matsushita S, Onon Y, Chiba H, Shintani F, Nakamura M, Yagi G, Asai M (1996) Dopamine D2, D3 and D4 receptor and transporter gene polymorphisms and mood disorders. J Affect Disord 40:7–13

McHugh PR (1989) The neuropsychiatry of basal ganglia disorders. Neuropsychiatr. Neuropsychol Behav Neurol 2:239–247

Meltzer HY, Kolakowska T, Fang VS, Fogg L, Robertson A, Lewine R, Strahilevitz M, Busch D (1984) Growth hormone and prolactin response to apomorphine in schizophrenia and major affective disorders. Relation to duration of illness and affective symptoms. Arch Gen Psychiatr 41:512–519

Mendlewicz J, Linkowski P, Kerkhofs M, Desmedt D, Goldstein J, Copinschi G, Van Cauter E (1985) Diurnal hypersecretion of growth hormone in depression. J Clin Endocrinol Metab 60:505–512

Miller HL, Delgado PL, Salomon RM, Berman R, Krystal JH, Heninger GR, Charney DS (1996) Clinical and biochemical effects of catecholamine depletion on antidepressant-induced remission of symptoms. Arch Gen Psychiatr 53:117–128

Mogenson GJ, Yim CC (1991) Neuromodulatory functions of the mesolimbic dopamine system: Electrophysiological and behavioural studies. In: Willner P, Scheel-Kruger J (eds) The Mesolimbic Dopamine System: From Motivation to Action. Wiley, Chichester, pp 105–130

Mouret J, LeMoine P, Minuit M-P (1989) Marqueurs polygraphiques, cliniques et therapeutiques des depressions dopamino-dependantes (DDD). Confront Psychiatr, Special Issue, 430–437

Murphy DL (1972) L-dopa, behavioral activation and psychopathology. Res Publ Ass Res Nerv Ment Dis 50:472–493

Murray JB (1996) Depression in Parkinson's disease. J Psychol 130:659–667

Nelson JC (1987) The use of antipsychotic drugs in the treatment of depression. In: Zohar J, Belmaker RH (eds) Treating Resistant Depression. New York: PMA Corp, 131–146

Nowak G, Skolnick P, Paul IA (1991) Downregulation of dopamine1 (D1) receptors is species-specific. Pharmaol Biochem Behav 39:769–771

Nurnberger JJ Jr, Gershon ES, Simmons S, Ebert M, Kessler LR, Dibble ED, Jimerson SS, Brown GM, Gold P, Jimerson DC, Guroff JJ, Storch FI (1982) Behavioral, biochemical and neurochemical responses to amphetamine in normal twins and "well-state" bipolar patients. Psychoneuroendocrinology 7:163–176

Nurnberger JJ Jr, Simmons-Alling S, Kessler L, Jimerson S, Schreiber J, Hollander E, Tamminga CA, Suzan Nadi N, Goldstein DS, Gershon ES (1984) Separate mechanisms for behavioral, cardiovascular and hormonal responses to dextroamphetamine in man. Psychopharmacology 84:200–204

Papp M, Klimek V, Willner P (1994) Parallel changes in dopamine D2 receptor binding in limbic forebrain associated with chronic mild stress-induced anhedonia and its reversal by imipramine. Psychopharmacology 115:441–446

Perry EK (1987) Cortical neurotransmitter chemistry in Alzheimer's disease. In: Meltzer HY (ed) Psychopharmacology: The Third Generation of Progress. New York: Raven Press, 887–895

Pitchot W, Ansseau M, Gonzalez Moreno A, Hansenne M, von Frenckell R (1992a) Dopaminergic function in panic disorder: Comparison with major and minor depression. Biol Psychiatr 32:1004–1011

Pitchot W, Hansenne M, Gonzalez Moreno A, Ansseau M (1992b) Suicidal behavior and growth hormone response to apomorphine test. Biol Psychiatr 31:1213–1219

Pitchot W, Hansenne M, Gonzalez Moreno A, von Frenckell R, Ansseau M (1990) Psychopathological correlates of dopaminergic disturbances in major depression. Neuropsychobiology 24:169–172

Pogue-Geile M, Ferrell R, Deka R, Debski T, Manuck S (1998) Human novelty-seeking personality traits and dopamine D4 receptor polymorphisms: A twin and genetic association study. Am J Med Genet 81:44–48

Post RM, Kotin J, Goodwin FK, Gordon E (1973) Psychomotor activity and cerebrospinal fluid metabolites in affective illness. Am J Psychiatr 130:67–72

Post RM, Weiss SRB, Pert A (1991) Animal models of mania. In: Willner P, Scheel-Kruger J (eds) The Mesolimbic Dopamine System: From Motivation to Action. Chichester: John Wiley and Sons, 443–472

Przegalinski E, Budziszewska B, Blaszcztnska E (1990) Repeated treatment with antidepressant drugs and/or electroconvulsive shock (ECS) does not affect the quinpirole-induced elevation of serum corticosterone concentration in rats. J Psychopharmacol 4:198–203

Pulvirenti L, Samanin R (1986) Antagonism by dopamine, but not noradrenaline receptor blockers of the anti-immobility activity of desipramine after different treatment schedules in the rat. Pharmacol Res Comm 18:73–80

Rampello L, Nicoletti G, Raffaele R (1991) Dopaminergic hypothesis for retarded depression: A symptom profile for predicting therapeutical responses. Acta Psychiatr. Scand 84:552–554

Randrup A, Munkvad I, Fog R, Gerlach J, Molander L, Kjellberg B, Scheel-Kruger J (1975) Mania, depression and brain dopamine. In: Essman WB, Valzelli L (eds) Current Developments in Psychopharmacology (Vol. 2). New York: Spectrum Press, 206–248

Reddy PL, Khanna S, Subhash MN, Channabasavanna SM, Sridhara Rama Rao BS (1992) CSF amine metabolites in depression. Biol Psychiatr 31:112–118

Robertson MM, Trimble MR (1981) Neuroleptics as antidepressants. Neuropharmacology 20:1335–1336

Roy A, Karoum F, Pollack S (1992) Marked reduction in indexes of dopamine transmission among patients with depression who attempt suicide. Arch Gen Psychiatr 49:447–450

Roy A, Pickar D, Linnoila M, Doran AR, Ninan P, Paul SM (1985) Cerebrospinal fluid monoamine and monoamine metabolite concentrations in melancholia. Psychiatr Res 15:281–290

Ruther E, Degner D, Munzel U, Brunner E, Lenhard G, Biehl J, Vogtle-Junkert U (1999) Antidepressant action of sulpiride. Results of a placebo-controlled double-blind trial. Pharmakopsychiatry 32:127–135

Salomon, RM, Miller HL, Delgado PL, Charney D (1993) The use of tryptophan depletion to evaluate serotonin function in depression and other neuropsychiatric disorders. Int Clin Psychopharmacol 8 [Suppl 2]:41–46

Salamone JD, Cousins MS, Snyder BJ (1997) Behavioral functions of nucleus accumbens dopamine: empirical and conceptual problems with the anhedonia hypothesis. Neurosci Biobehav Rev 21:341–359

Scatton B, Perrault G, Sanger DJ, Shoemaker H, Carter C, Fage D, Gonon F, Chergui K, Cudennec A, Benavides J (1994) Pharmacological profile of amisulpride, an atypical neuroleptic which preferentially blocks presynaptic D2/D3 receptors. Neuropsychopharm 10 [Suppl, Part 1]:242 S

Scavone C, Aizenstein ML, De Lucia R, Da Silva Planeta C (1986) Chronic imipramine administration reduces apomorphine inhibitory effects. Eur J Pharmacol 132:263–267

Schatzberg AF, Rothschild AJ (1988) The roles of glucocorticoid and dopaminergic systems in delusional (psychotic) depression. Ann NY Acad Sci 537:462–471

Serra G, Argiolas A, Fadda F, Melis MR, Gessa GL (1979) Chronic treatment with antidepressants prevents the inhibitory effect of small doses of antidepressants on dopamine synthesis and motor activity. Life Sci 25:415–424

Serra G, Collu M, D'Aquila PS, de Montis GM, Gessa GL (1990) Possible role of dopamine D1 receptor in the behavioural supersensitivity to dopamine agonists induced by chronic treatment with antidepressants. Brain Res 527:234–243

Serra G, Forgione A, D'Aquila PS, Collu M, Fratta W, Gessa GL (1990) Possible mechanism of antidepressant effect of L-sulpiride. Clin Neuropharmacol 13 [Suppl 1]:S76–S83

Serretti A, Macciardi F, Verga M, Cusin C, Pedrini S, Smeraldi E (1998) Tyrosine hydroxylase gene associated with depressive symptomatology in mood disorder. Am J Med Genet 81:127–130

Shah PJ, Ogilvie AD, Goodwin GM, Ebmeier KP (1997) Clinical and psychometric correlates of dopamine D2 binding in depression. Psychol Med 27:1247–1256

Shively CA, Grant KA, Ehrenkaufer RL, Mach RH, Nader MA (1997) Stress, depression, and brain dopamine in female cynomolgous monkeys. Ann NY Acad Sci 807:574–577

Siever LJ, Uhde TW (1984) New studies and perspectives on the noradrenergic receptor system in depression: Effects of the alpha2-adrenergic agonist clonidine. Biol Psychiatr 19:131–156

Silverstone T (1984) Response to bromocriptine distinguishes bipolar from unipolar depression. Lancet 1:903–904

Siris S (1991) Diagnosis of secondary depression in schizophrenia. Implications for DSM-IV. Schizophr Bull 17:75–98

Smith AD, Bolam JP (1990) The neural network of the basal ganglia as revealed by the study of synaptic connections of identified neurones. Trends Neurosci 5:776–794

Stamford JA, Muscat R, O'Connor JJ, Patel J, Trout SJ, Wieczorek WJ, Zruk ZL, Willner P (1991) Subsensitivity to reward following chronic mild stress is associated with increased release of mesolimbic dopamine. Psychopharmacology 105:275–282

Stoll AL, Pillay SS, Diamond L, Workum SB, Cole JO (1996) Methylphenidate augmentation of serotonin selective reuptake inhibitors: a case series. J Clin Psychopharmacol 57:72–76

Suhara T, Nakayama K, Inoue O, Fukuda H, Shimizu M, Mori A, Tateno Y (1992) D1 dopamine receptor binding in mood disorders measured by positron emission tomography. Psychopharmacology 106:14–18

Sumiyoshi T, Stockmeier CA, Overholser JC, Thompson PA, Meltzer HY (1995) Dopamine D4 receptors and effects of guanine nucleotides on [3H]raclopride binding in postmortem caudate nucleus of subjects with schizophrenia or major depression. Brain Res 681:109–116

Swerdlow NR, Koob GF (1987) Dopamine, schizophrenia, mania and depression: toward a unified hypothesis of cortico-striato-pallido-thalamic function. Behav Brain Sci 10:197–245

Tanda G, Pontieri FF, Frau R, Di Chiara G (1997) Contribution of blockade of the noradrenaline carrier to the increase of extracellular dopamine in the rat prefrontal cortex by amphetamine and cocaine. Eur J Neurosci 9:2077–2085

Taylor AE, Saint-Cyr JA (1990) Depression in Parkinson's disease: Reconciling physiological and psychological perspectives. Neuropsychiatr Pract Opin 2:92–98

Tiihonen J, Kuoppamaki M, Nagre K, Bergmen JM Eronen E, Syvalahti E, Hietala J (1996) Serotonergic modulation of striatal D2 dopamine receptor binding in humans measured with positron emission tomography. Psychopharmacology 126:277–280

Towell A, Willner P, Muscat R (1986) Behavioural evidence for autoreceptor subsensitivity in the mesolimbic dopamine system during withdrawal from antidepressant drugs. Psychopharmacology 90:64–71

Van Praag HM, Korf J (1975) Central monoamine deficiency in depression: Causative or secondary phenomenon. Pharmacopsychiatry 8:321–326

Van Praag HM, Korf J, Lakke JPWF, Schut T (1975) Dopamine metabolism in depression, psychoses, and Parkinson's disease: The problem of specificity of biological variables in behavior disorders. Psychol Med 5:138–146

Willner P (1983) Dopamine and depression: A review of recent evidence. Brain Res Rev 6:211–246

Willner P (1984) Cognitive functioning in depression: A review of theory and research. Psychol Med 14:807–823

Willner P (1985) Depression: A Psychobiological Synthesis. Wiley, New York

Willner P (1989) Sensitization to antidepressant drugs. In: Emmett-Oglesby MV, Goudie AJ (eds) Psychoactive drugs: Tolerance and Sensitization. Clifton, N.J.: Humana Press, 407–459

Willner P (1997a) Validity, reliability and utility of the chronic mild stress (CMS) model of depression: A ten-year review and evaluation. Psychopharmacology 134:319–329

Willner P (1997b) The chronic mild stress procedure as an animal model of depression: Valid, reasonably reliable, and useful. Psychopharmacology 134:371–377

Willner P (1999) Dopaminergic mechanisms in depression and mania. In: Watson (ed) Psychopharmacology: The Fourth Generation of Progress, On-Line Edition, http://www.acnp.org/citations/GN401000093. Lippincott Williams & Wilkins, New York

Willner P, Papp M (1997) Animal models to detect antidepressants: Are new strategies necessary to detect new antidepressants? In: Skolnick P (ed) Antidepressants: New Pharmacological Strategies, Humana, Totowa NJ, pp 213–234

Willner P, Scheel-Kruger J (eds) (1991) The Mesolimbic Dopamine System: From Motivation to Action. Chichester: John Wiley and Sons

Willner P, Klimek V, Golembiowska K, Muscat R (1991) Changes in mesolimbic dopamine may explain stress-induced anhedonia. Psychobiology 19:79–84

Willner P, Muscat R, Papp M (1992) Chronic mild stress-induced anhedonia: A realistic animal model of depression. Neurosci Biobehav Rev 16:525–534

Wise RA (1982) Neuroleptics and operant behaviour: The anhedonia hypothesis. Behav Brain Sci 5:39–88

Wolfe N, Katz DI, Albert ML, Almozlino A, Durso R, Smith MC, Volicer L (1990) Neuropsychological profile linked to low dopamine in Alzheimer's disease, major depression, and Parkinson's disease. J Neurol Neurosur Psychiatr 53:915–917

Wong DF, Wagner HN Jr, Pearlson G, Dannals RF, Links JM, Ravert HT, Wilson AA, Suneja S, Bjorvvinssen E, Kuhar MJ, Tune L (1985) Dopamine receptor binding of C-11-3-N-methyl-spiperone in the caudate of schizophrenic and bipolar disorder: A preliminary report. Psychopharmacol Bull 21:595–598

Zebrowska-Lupina I, Ossowska G, Klenk-Majewska B (1992) The influence of antidepressants on aggressive behavior in stressed rats: The role of dopamine. Pol J Pharmacol Pharm 44:325–335

CHAPTER 25

Dopamine in Schizophrenia

Dysfunctional Information Processing in Basal Ganglia – Thalamocortical Split Circuits

I. WEINER and D. JOEL

A. The Dopamine Hypothesis of Schizophrenia

Schizophrenia is a major mental disorder with about 0.85%–1% lifetime prevalence world wide (JABLENSKY et al. 1992). The course of schizophrenia is characterized by the onset of clinical symptoms after puberty and a high symptom heterogeneity. Schizophrenia symptoms are considered to fall into two major classes: positive and negative. According to the Diagnostic and Statistical Manual of Mental Disorders (DSM)-IV, the former include "distortions or exaggerations of inferential thinking (delusions), perception (hallucinations), language, and communication (disorganized speech), and behavioral monitoring (grossly disorganized or catatonic behavior)", and the latter include "restrictions in the range and intensity of emotional expression (affective flattening), in the fluency and production of thought and speech (alogia), and in the initiation of goal-directed behavior (avolition)" (pp. 274–275). While additional classifications of symptoms have been proposed (e.g., CROW 1980; ANDREASEN 1982; LIDDLE 1987; CARPENTER et al. 1988, 1999; LIDDLE et al. 1989; KAY 1990; BUCHANAN and CARPENTER 1994; ANDREASEN et al. 1995; TANDON 1995), the dopamine (DA) hypothesis of schizophrenia has been primarily related to the positive–negative classification (see below).

For about three decades, the DA hypothesis of schizophrenia has been the reigning biological hypothesis of the neural mechanisms underlying this disorder (CARLSSON and LINDQUIST 1963; MATTHYSSE 1973; SNYDER 1973, 1974, 1976; MELTZER and STAHL 1976; BURT et al. 1977; OWEN et al. 1978; MCKENNA 1987; SEEMAN 1987; LIEBERMAN et al. 1990, 1997; WILLNER 1997). The DA hypothesis has undergone numerous revisions, but has proven remarkably resistant to obliteration. In its original formulation, the hypothesis stated that schizophrenia is due to a central hyperdopaminergic state. This was based on two complementary lines of indirect pharmacological evidence: the DA releaser amphetamine as well as other DA-enhancing agents such as the DA precursor L-dopa or methylphenidate, produced and exacerbated schizophrenic symptoms (JENKINS and GROH 1970; ANGRIST et al. 1971, 1974, 1980; SNYDER 1973; JANOWSKY and DAVIS 1976; VAN KAMMEN et al. 1982; LIBERMAN et al. 1984, 1987; DAVIDSON et al. 1987), whereas drugs that were effective in the treatment of amphetamine-induced psychosis and schizophrenia [neuroleptics or antipsychotic drugs (APDs)] decreased DA activity, and their clinical potency was correlated with their potency in blocking D_2 receptors (CARLSSON and LINDQUIST 1963; CREESE et al. 1976; HYTTEL et al. 1985; FARDE et al. 1988, 1992; NORDSTROM et al. 1993).

Initially, the focus was on hyperdopaminergia in the mesostriatal DA system [DA projections from the substantia nigra pars compacta, retrorubral area and ventral tegmental area (VTA) to the neostriatum), because the neostriatum has the highest concentration of D_2 receptors in the brain, and because prolonged treatment with neuroleptics produced motor disturbances (extrapyramidal side effects), which are associated with the mesostriatal DA system (SEDVALL 1996; ARNT et al. 1997; JOYCE et al. 1997). Indeed, the most widely accepted definition of a neuroleptic drug had been that it has antipsychotic activity and induces extrapyramidal side effects (ARNT and SKRSFELDT 1998). In parallel, repeated high-dose amphetamine administration was deemed necessary for producing schizophrenia-like symptoms in humans (SEGEL 1975; GROVES and REBEC 1976; KOKKINIDIS and ANISMAN 1980), and studies in rodents have revealed that the effects of comparable regimens of amphetamine administration, which led to stereotypy, are mediated by the mesostriatal DA system (CREESE and IVERSEN 1975; COSTALL et al. 1977; STATON and SOLOMON 1984). By corollary, the most essential preclinical test of antipsychotic activity was antagonism of amphetamine-induced stereotypy in rodents (ARNT and SKARSFELDT 1998).

The first wave of challenges to the original formulation of the DA hypothesis sprung from its very cornerstones, namely, the effects of amphetamine and neuroleptic drugs, as well as from the growing recognition of the importance of negative symptoms in schizophrenia. Thus, on the one hand it became apparent that amphetamine does not produce the entire spectrum of schizophrenic symptoms but only those considered to belong to the "positive" category; moreover, it improved negative symptoms (OGURA et al. 1976; ANGRIST et al. 1980, 1982, 1985; DAVIDSON et al. 1987; DANIEL et al. 1990; SANFILIPO et al. 1996). On the other hand, while neuroleptics were effective in treating positive symptoms, their efficacy in treating negative symptoms turned out to be limited (JOHNSTONE et al. 1978; MELTZER et al. 1986; 1994; KANE 1995; BREIER et al. 1987; BREIER and BERG 1999), and these drugs themselves could lead to a syndrome similar to the negative symptomatology of schizophrenia (BELMAKER and WALD 1977; CHATTERJEE et al. 1995; HEINZ et al. 1998). These problems were reinforced by studies of the main DA metabolite, homovanillic acid (HVA), in the cerebro-spinal fluid (CSF) and plasma of schizophrenic patients which yielded inconclusive results regarding changes in DA turnover (POST et al. 1975; VAN KAMMEN et al. 1986; REYNOLDS, 1989; HSIAO et al. 1993; KAHN and DAVIS, 1995; BEUGER et al. 1996), and indeed have pointed to DA hypofunction in some schizophrenic patients (BJERKENSTEDT et al. 1985; LINDSTROM 1985; WIESELGREN and LINDSTROM 1998). Taken together, these findings have led to the proposition that positive symptoms are associated with an increased DA function, whereas negative symptoms are associated with a decreased DA function (MELTZER 1985; WYAT 1986; DAVIS et al. 1991).

The postulated site of DA dysfunction has been re-conceptualized as well. The advent of the "atypical" neuroleptic clozapine, which had superior efficacy against positive as well as negative symptoms, has undermined the connection between antipsychotic efficacy and extrapyramidal side effects, as this drug had high antipsychotic efficacy at doses that did not produce extrapyramidal side effects (HOGBERG et al. 1987; ARNT and SKARSFELDT 1998; KINON and LIEBERMAN 1996). The latter was consistent with findings that relatively to the typical APD haloperidol, clozapine produced a much weaker striatal D_2 blockade (FARDE et al. 1989, 1992; KERWIN 1994; MELTZER et al. 1994; ARNT and SKARSFELDT 1998). Subsequently, a wide separation between the doses used to control psychosis and those that induce extrapyramidal side effects has become a major characteristic of the novel or "atypical" APDs (KINON and LIEBERMAN 1996; ARNT and SKARSFELDT 1998). Furthermore, electrophysiological, biochemical, and behavioral studies of typical and atypical APDs in rodents have revealed that clozapine and other atypical APDs exhibit selectivity for the mesolimbic DA system [originating in the VTA and terminating in the nucleus accumbens (NAC)] and reverse amphetamine-induced activity, mediated primarily by the mesolimbic DA system (PIJNENBURG et al. 1975; STATON and SOLOMON 1984) but not stereotypy (mediated primarily by the mesostriatal DA system; STATON and SOLOMON 1984). In addition, Fos immunohistochemistry studies have shown that the NAC might be the common site of action of all APDs (DEUTCH et al. 1992; ROBERTSON and FIBIGER 1992). These findings have led to the hypothesis that antipsychotic activity is mediated by inhibition of DA function in limbic regions, whereas extrapyramidal side effects are mediated by inhibition of the mesostriatal DA function, and, by inference, that schizophrenia may be related to excessive activity in the mesolimbic DA system. As evidence has accumulated from rodent studies that the mesolimbic DA system plays a central role in complex motivational and cognitive processes (e.g., TAGHZOUTI et al. 1985; MOGENSON et al. 1988; ANNETT et al. 1989; CADOR et al. 1989, 1991; COLE and ROBBINS 1989; Van DEN BOS and COOLS 1989; LEMOAL and SIMON 1991; VAN DEN BOS et al. 1991; PENNARTZ et al. 1994; SALAMONE 1994, 1997; SEAMANS and PHILLIPS 1994; IKEMOTO and PANKSEPP 1999, see also Chap. 19, this volume), and that the NAC receives, in addition to its DA input from the VTA, input from all the brain regions implicated in the pathophysiology of schizophrenia (SESACK et al. 1989; GROENEWEGEN et al. 1990, 1991; BERENDSE et al. 1992, see below), the locus of the subcortical DA dysfunction in this disorder has been shifted to the mesolimbic DA system (SWERDLOW and KOOB 1987; CSERNANSKY et al. 1991; GRACE 1991, 1993; GRAY et al. 1991; DEUTCH 1992; JOYCE 1993, although recently, there has been a comeback of the mesostriatal system, see GRAYBIEL 1997).

While pharmacology and results of animal studies have increasingly implicated the mesolimbic DA system, the results of neuropathological and neuroimaging studies in schizophrenia patients have increasingly pointed to a dysfunction of cortical areas in schizophrenia. Thus, findings revealed a functional abnormality of the frontal cortex in schizophrenia ("hypofrontality", e.g., WOLKIN et al. 1985, 1988, 1992; GUR et al. 1987; VOLKOW et al. 1987; WEINBERGER 1988; WEINBERGER et al. 1988; BERMAN and WEINBERGER 1990; BUCHSBAUM et al. 1990; ANDREASEN et al. 1992, 1996) and structural abnormalities were found in frontal and temporal brain regions in schizophrenia, including the prefrontal cortex, the entorhinal cortex, the hippocampus, and the amygdala (e.g., KOVELMAN and SCHEIBEL 1984; BOGERTS et al. 1985; JAKOB and BECKMAN 1986; BOGERTS 1991, 1993; KNABLE and WEINBERGER 1995; HARRISON 1995, 1999; SELEMON et al. 1995, 1998; WEINBERGER and LIPSKA 1995; ARNOLD and TROJANOWSKI 1996; RAJKOWSKA et al. 1998; WEICKERT and WEINBERGER 1998; BENES 1999). These findings have resurrected the proposition of KRAEPELIN (1919) that schizophrenia symptoms resulted from pathology of the frontal and temporal lobes, but were unrelated to the evidence of mesolimbic DA dysfunction. In addition, while the initial wave of DA hypothesis revision implied that positive and negative symptoms characterize distinct subgroups of patients, it has become clear that positive and negative symptoms coexist in schizophrenic patients (ANDREASEN 1982; TANDON 1995; WILLNER 1997). Therefore, models of schizophrenia that could link the pharmacological and neuropathological/neuroimaging lines of evidence as well as accommodate the coexistence of positive and negative symptoms have become imperative.

The formulation of such hypotheses has been made possible by the results of rodent studies which showed that: (1) the NAC is a site of convergence and interaction between the ascending mesolimbic DA system and the glutamatergic inputs from all the cortical regions whose dysfunction has been implicated in schizophrenia (KRAYNIAK et al. 1981; KELLEY and DOMESIK 1982; LOPES DA SILVA et al. 1984; FULLER et al. 1987; GROENEWEGEN et al. 1987, 1991, 1996, 1999; SESACK et al. 1989; MCDONALD 1991; BROG et al. 1993; JOYCE 1993; PENNARTZ et al. 1994; O'DONNEL and GRACE 1995; FINCH 1996; WRIGHT and GROENEWEGEN 1996); and (2) perturbations of these cortical regions can modify mesolimbic DA function (KELLY and ROBERTS 1983; ISAACSON 1984; MOGENSON and NIELSEN 1984; LOUILOT et al. 1985; YIM and MOGENSON 1988; JASKIW et al. 1990; CADOR et al. 1991; LE MOAL and SIMON 1991; LIPSKA et al. 1992; BURNS et al. 1993; LIPSKA et al. 1993; WILKINSON et al. 1993; LE MOAL 1995; KARREMAN and MOGHADDAM 1996).

Two major versions of what can be termed the revised DA hypothesis have received prominence. One version, based on rodent experiments showing that the mesocortical DA system can regulate the activity of the mesolimbic DA system (e.g., CARTER and PYCOCK 1980; PYCOCK et al. 1980; LOUILOT et al. 1989; DEUTCH et al. 1990; JASKIW et al. 1991; LEMOAL and SIMON 1991; DEUTCH 1992; ROSIN et al. 1992; LE MOAL 1995; KING et al. 1997), states that schizophrenia is associated with hypodopaminergia in the prefrontal cortex and a consequent hyperdopaminergia in the mesolimbic DA system, leading to negative and positive symptoms, respectively (MELTZER 1985; WYAT 1986; WEINBERGER 1987; DAVIS et al. 1991; DEUTCH 1992; DEUTCH et al. 1992).

The second, more general, hypothesis posits that schizophrenia reflects a dysfunction of fronto-temporolimbic-mesolimbic DA circuitry in which mesolimbic DA hyperactivity is either primary or secondary to a disrupted/reduced cortical input to the mesolimbic DA system (FRITH 1987; SWERDLOW and KOOB 1987; WEINBERGER 1987; CARLSSON 1988, 1995; WEINBERGER et al. 1988; CARLSSON and CARLSSON 1990; ROBBINS 1990, 1991; CSERNANSKY et al. 1991; GRACE 1991; GRAY et al. 1991; DEUTCH 1992; JOYCE 1993; WEINBERGER and LIPSKA 1995; O'DONNEL and GRACE 1998; MOORE et al. 1999). GRACE (1991, 1993) has advanced an influential hypothesis according to which reduced cortical input to the NAC leads to both decreased (tonic) and increased (phasic) striatal DA function leading to negative and positive symptoms, respectively (GRACE 1991, 2000; O'DONNEL and GRACE 1998; MOORE et al. 1999, see below). CSERNANSKY et al. (1991) have proposed an additional model which combines decreased and increased striatal DA function, albeit via different mechanisms (see below).

Direct evidence for a DA dysfunction in schizophrenia has been lacking for years. As noted above, studies of HVA concentration in the CSF and in plasma have yielded inconsistent results (see DAVIS et al. 1991; KAHN and DAVIS 1995 for reviews of this literature). Early postmortem studies reported elevated levels of brain DA and DA metabolites (BOWERS, 1974; BIRD et al. 1977, 1979a,b,c) as well as significantly elevated numbers of D_2 receptors (LEE and SEEMAN 1977; LEE et al. 1978; CROSS et al. 1981; MACKAY 1982; SEEMAN et al. 1984, 1987) in schizophrenic brains. The possibility that such an increase is related to antipsychotic treatment (which has been shown in rats to elevate striatal D_2 receptors) has been contested by findings of elevated D_2 receptors in never-medicated patients (LEE et al. 1978; OWEN et al. 1978; CROSS et al. 1981; FARDE et al. 1987). However, later studies of D_2 receptor densities in neuroleptic-naïve patients using neuroimaging techniques have yielded conflicting results (WONG et al. 1986, 1997a,b; MARTINOT et al. 1989, 1990, 1991; FARDE et al. 1990; TUNE et al. 1993, 1996; HIETALA et al. 1994, PILOWSKY et al. 1994; NORDSTROM et al. 1995; BREIER et al. 1997; LARUELLE et al. 1997). Two recently published meta-analyses (LARUELLE et al. 1998; ZAKZANIS and HANSEN 1998) and several reviews (DAVIS et al. 1991; SOARES and INNIS 1999) have concluded that striatal D_2 receptor levels are moderately elevated in a substantial portion but not in all patients with schizophrenia, although FARDE et al. (1995, 1997) concluded that the weight of the evidence does not point to such an elevation. Interestingly, the elevation of D_2 receptors in drug-naïve schizophrenics appears largely in the limbic region of the striatum (MITA et al. 1986; JOYCE et al. 1988, 1997). In any event, it should be evident that elevation/upregulation of D_2 receptors is not indicative of higher DA activity but rather would be in agreement with DA hypoactivity (GRACE 1991; CSERNANSKY et al. 1991).

In recent years, the development of sophisticated imaging techniques has finally allowed the demonstration of DA dysfunction in the living brain, which may with further developments become substantiated as a direct support for the DA hypothesis. A series of studies using D_2 radioreceptor imaging have found larger displacement of the ligand from striatal D_2 receptors following amphetamine challenge in untreated and neuroleptic-naïve schizophrenics compared to healthy controls, pointing to a greater stimulation of these receptors (LARUELLE et al. 1996, 1999; BREIER et al. 1997; ABI-DARGHAM et al. 1998). Importantly, excessive DA function was found in patients who were in an active phase of the illness but not in patients in remission, suggesting that DA abnormality is not stable but is related to the clinical stage of the disease and may subserve or at least contribute to the transition to the active phase. Furthermore, amphetamine challenge and the concomitant increase in DA transmission were correlated with an exacerbation of positive symptoms and an improvement in negative symptoms, indicating that increased DA transmission is related to positive symptoms, whereas reduced DA function may be associated with negative symptoms.

Positron emission tomography (PET) studies assessing striatal DA synthesis as measured by the uptake of labeled dopa have yielded comparable findings. Thus, increased rate of DA synthesis, consistent with increased presynaptic activity, was found in both first-admission and more chronic psychotic schizophrenic patients (HIETALA et al. 1994, 1995, 1999; REITH et al. 1994; DAO-COSTELLANA et al. 1997; HAGBERG et al. 1998; LINDSTROM et al. 1999), whereas decreased DA synthesis appears to characterize schizophrenic patients with psychomotor slowing and depressive symptoms, i.e., primarily negative symptomatology (HIETALA et al. 1995, 1999; DAO-COSTELLANA et al. 1997).

Recently, BERTOLINO et al. (2000) found a selective negative correlation between a measure of neuronal integrity in dorsolateral prefrontal cortex (assessed as N-acetylaspartate relative concentrations measured with MRS imaging) and amphetamine-induced release of striatal DA (assessed as changes in striatal raclopride binding measured with PET) in schizophrenia patients but not in healthy controls. These results show that increased release of striatal DA after amphetamine in schizophrenia might be related to reduced glutamatergic activity of prefrontal cortex neurons, and are consistent with the hypothesis that DA dysregulation in schizophrenia may be prefrontally determined.

There have been some findings suggesting DA dysfunction in the cortex of schizophrenic patients. Using immunocytochemical methods, AKIL et al. (1999) have found morphological alterations in DA axons in some areas of the prefrontal cortex, suggesting disturbance in DA neurotransmission. LINDSTROM et al. (1999) have found increased uptake of L-dopa in the medial prefrontal cortex, pointing to an elevated DA synthesis. There are also some reports on receptor abnormalities in the cortex of schizophrenic subjects, including decreased density of D_1 receptors in the prefrontal cortex (OKUBO et al. 1997), increased density of D_4 receptors in the entorhinal cortex (LAHTI et al. 1998), and a higher (STEFANIS et al. 1998) or lower (MEADOR-WOODRUFF et al. 1997) level of D_4 mRNA in the frontal cortex.

Taken together, the results obtained with the newly developed methods of neurochemical brain imaging (SOARES and INNIS 1999) support the DA hypothesis and moreover, suggest that hyperresponsivity of striatal DA is an important component of the positive symptomatology of schizophrenia, whereas reduced striatal DA function is involved in the pathophysiology of negative symptoms. However, the mechanisms underlying the DA abnormality remain unknown. For example, abnormal DA responsiveness to amphetamine could reflect either a presynaptic mechanism, i.e., increased DA release, or a postsynaptic mechanism, i.e., increased affinity of D_2 receptors. Another important issue relates to the basal levels of endogenously released DA in schizophrenia; thus, some authors suggested that different basal DA levels in patients may account for conflicting PET measurements of D_2 receptors (WONG et al. 1986; FARDE et al. 1990, 1997), and LARUELLE et al. (1999) suggested that the response to amphetamine challenge may be associated with a dysregulation of baseline DA activity. Direct measure of baseline DA levels will be necessary to unravel the nature of abnormal DA transmission in schizophrenia. Finally, due to limitations in the anatomic resolution, most imaging studies have measured the neostriatum, and the contribution of the ventral (limbic) striatum has remained largely unknown. This is a major limitation given the present emphasis on the abnormality of the mesolimbic DA system in schizophrenia. The same problem is evident with PET and single photon emission computed tomography (SPECT) studies of D_2 receptor occupancy by APDs. Only few studies exist and only for striatal D_2 receptors (ARNT and SKARSFELD 1998).

Finally, there is some evidence that D_2 receptor gene polymorphism affects susceptibility to schizophrenia (OHARA et al. 1998; SERRETTI et al. 1998; JONSSON et al. 1999, but see KANESHIMA et al. 1997; TALLERICO et al. 1999), and that allelic variation in the D_3 receptor gene may play a role in the pathophysiology of schizophrenia (DUBERTRET et al. 1998; SCHARFETTER et al. 1999, but see MALHOTRA et al. 1998).

B. Schizophrenia as a Dopamine-Dependent Dysfunctional Information Processing in Basal Ganglia–Thalamocortical Circuits

Several paths have been taken in attempting to link DA dysfunction to schizophrenia symptomatology. As noted above, most often DA dysfunction is related at the gross level to positive vs negative symptoms, either in a region-specific manner, e.g., hypodopaminergia of the prefrontal cortex is responsible for negative symptoms, whereas mesolimbic DA overactivity is responsible for positive symptoms (DAVIS et al. 1991; MELTZER 1985), or in relation to the mode of striatal DA release, i.e., increased phasic and decreased tonic release are responsible for positive and negative symptoms, respectively (GRACE 1991;

O'DONNEL and GRACE 1998; MOORE et al. 1999). Recently, it has been suggested that schizophrenia may involve a hypodopaminergic state in the dorsal striatum coupled with a hyperdopaminergic state in the ventral striatum (O'DONNEL and GRACE 1998; LARUELLE et al. 2000), or an imbalance between modes of DA activity within the prefrontal cortex, i.e., decreased phasic and increased tonic release (BRAVER et al. 1999; COHEN et al. 1999).

Other approaches include the selection of a "basic" cognitive deficit which presumably underlies many schizophrenic symptoms (e.g., lack of contextual modulation; BRAVER et al. 1999; lack of influence of past regularities on current perception; GRAY et al. 1991; disruption of working memory; GOLDMAN-RAKIC 1999), and/or endowing mesolimbic or cortical DA with a "basic" function (gain, gating, switching) whose impairment produces the schizophrenic deficit (e.g., SWERDLOW and KOOB 1987).

One of the major advances in the understanding of information processing in the forebrain has been the discovery that anatomically and functionally associated regions of the striatum and the frontal cortex are linked within several limbic, associative and motor basal ganglia–thalamocortical circuits (DELONG and GEORGOPOULOS 1981; PENNEY and YOUNG 1983, 1986; ALEXANDER et al. 1986; MARSDEN 1986; GROENEWEGEN and BERENDSE 1994; JOEL and WEINER 1994; PARENT and HAZRATI 1995; WISE et al. 1996). Each circuit receives glutamatergic input from several separate but functionally related cortical areas, traverses specific regions of the striatum, the internal segment of the globus pallidus, substantia nigra pars reticulata (SNR), ventral pallidum and thalamus, and projects back upon a frontocortical area. Within each circuit, striatal output reaches the basal ganglia output nuclei (SNR, internal segment of globus pallidus, and ventral pallidum) via a "direct" pathway and via an "indirect pathway," which traverses the external segment of the globus pallidus and the subthalamic nucleus (DELONG et al. 1985; PENNEY and YOUNG 1986; ALBIN et al. 1989; ALEXANDER and CRUTCHER 1990; ALEXANDER et al. 1990; WISE et al. 1996; JOEL and WEINER 1997). Striatal neurons of the direct pathway contain γ-aminobutyric acid (GABA) and substance P and preferentially express D_1 receptors, while neurons of the indirect pathway contain GABA and enkephalin and preferentially express D_2 receptors (ALBIN et al. 1989; GERFEN et al. 1990; REINER and ANDERSON 1990; GERFEN and WILSON 1996; WISE et al. 1996; Chap. 11, Vol. I). Given the preponderance of DA innervation of the striatum, it has been proposed that the understanding of the role of DA in schizophrenia might profit from an understanding of the nature of information processing within the basal ganglia–thalamocortical circuits and its modulation by DA (FRITH 1987; SWERDLOW and KOOB 1987; CARLSSON 1988; ROBBINS 1990; 1991; GRAY et al. 1991; GRAYBIEL 1997; CARLSSON et al. 1999, 2000; MOORE et al. 1999). Guided by this rationale, several "circuit models" of schizophrenia have been described. These models typically include a description of the circuit contribution to normal behavior; the modulating effect of DA on the circuit functioning; and the effects of dysfunctional DA on circuit functioning and the resulting symptomatology.

I. Circuit Models of Schizophrenia

PENNEY and YOUNG (1983, 1986; ALBIN et al. 1989) were the first to describe how abnormalities of striatal DA disrupt the functioning of basal ganglia–thalamocortical circuitry. These authors focused on movement disorders (e.g., Parkinson's disease), and thus on the motor circuit. In their model, the direct pathway determines which sensory stimuli are used to initiate motor action, whereas the indirect pathway suppresses unwanted responses to sensory stimuli or determines which stimuli are disregarded. DA decreases the activity of striatal neurons of the indirect pathway and potentiates the activity of striatal neurons of the direct pathway. Therefore, reduced DA input to the striatum in Parkinson's disease results in underactivity of the direct pathway and overactivity of the indirect pathway, leading to reduced initiation and increased suppression of movement, manifested clinically as bradykinesia and hypokinesia.

Based on the model of PENNEY and YOUNG, SWERDLOW and KOOB (1987) presented a circuit model of schizophrenia in which the pathophysiology of this disorder was attributed to a malfunctional limbic cortico-striato-pallido-thalamo-cortical circuit. In this circuit, the limbic striatum (NAC) selects and maintains specific sets of impulses originating in limbic structures and frontal cortex, which form the basis of emotional and cognitive processes, by inhibiting pallidal cells and thus disinhibiting the transfer of information from the thalamus to the cortex. An important component of this selection process is the sharpening of cortical information that is achieved by the dense collateral inhibitory network within the NAC. DA modulates the capacity of NAC neurons to filter out irrelevant patterns and initiate new patterns or switch existing patterns by inhibiting these neurons and thus disrupting lateral inhibition. Overactivity of DA input to the NAC results in the loss of lateral inhibition causing inhibition of pallidothalamic efferents; this in turn causes rapid changes and a loss of focused corticothalamic activity in cortical regions controlling cognitive and emotional processes. This results in rapid switching and an inability to filter inappropriate cognitive and emotional cortical information at the NAC level, manifested clinically as "flight of ideas" (rapid switching) and "loose associations" (unfiltered information) characteristic of psychosis.

GRAY et al. (1991) have extended SWERDLOW and KOOB's (1987) model to include also a dorsal cortico-striato-pallido-thalamo-cortical circuit which is responsible for executing the steps of goal-directed motor programs. The function of the NAC system in this model is to monitor the smooth running of the motor program in terms of progress toward the intended goal and to switch between steps in the program, guided by the projections to the striatum from the prefrontal cortex, the amygdala, and the septohippocampal system. The latter is responsible for checking whether the actual outcome of a particular motor step matches the expected outcome, and this information is transmitted from the subiculum to the NAC. Positive symptoms of schizophrenia result from a disruption in the subiculo-NAC projection that leads to a failure to integrate past regularities with ongoing motor programs. The role of DA in this model is identical to that described by SWERDLOW and KOOB, namely enabling switching of striatal activity into a new pattern by inhibiting striatal neurons and consequently disrupting lateral inhibition within the NAC. GRAY et al. added a mechanism for a topographical specificity of DA effects, achieved by local DA increase in the region of active cortical (particularly subicular) glutamatergic terminals. Excess DA overcomes this topographical specificity, thus inhibiting striatal neurons indiscriminately, leading to a disruption of the running of all steps in all motor programs indiscriminately. Essentially the same process will be caused by loss of the subicular input to the NAC. Behavioral control will revert to new stimuli or familiar stimuli will be treated as novel. In psychological terms, this will be manifested in the weakening of the influence of past regularities on current behavior, and in a failure to monitor willed intentions correctly, which are considered by GRAY et al. to be basic to the schizophrenic condition and to account for most of the positive symptoms of this disorder.

Both SWERDLOW and KOOB's and GRAY and colleagues' models incorporate only mesolimbic DA hyperfunction and relatedly account only for positive symptoms of schizophrenia. In SWERDLOW and KOOB's model, negative symptoms were suggested to result from NAC cell loss, which would limit the amount of cortical information passing through the NAC, leading to paucity of affect and behavior. SWERDLOW and KOOB did describe the effects of DA underactivity in their limbic circuit and suggested that this would result in an inability to initiate or switch sets of cortical activity, leading to psychomotor retardation, paucity of affect, cognitive perseveration, and anhedonia, but this was proposed as a model of depression.

CARLSSON (1988) proposed that striatal neurons can inhibit thalamocortical neurons and thus filter off part of the sensory input to the thalamus to protect the cortex from sensory overload and hyperarousal. Since DA inhibits (and glutamate excites) these striatal neurons, excess DA (or glutamatergic deficiency) should reduce this striatal protective influence and thus lead to psychosis. Recently, CARLSSON et al. (1999, 2000) have updated the model by referring to the distinction between striatal neurons of the direct and indirect pathways. Specifically, they attributed the protective striatal function to neurons of the indirect pathway, and further suggested that neurons of the direct pathway exert an opposite, excitatory influence on thalamocortical neurons. Underactivity of the latter, induced for example by glutamatergic deficiency, is suggested to contribute to the negative symptoms of schizophrenia, whereas underactivity of the indirect pathway, induced by hyperactivity of DA or glutamatergic deficiency, is suggested to contribute to psychosis.

CSERNANSKY et al. (1991) have also provided a model in which alterations in the limbic basal ganglia–thalamocortical circuit lead to the positive and negative symptoms of schizophrenia. According to their model, hippocampal activation of the circuit, by activating NAC neurons, results in inhibition of behavioral output. DA can modulate this inhibition by inhibiting NAC neurons, thus inhibiting the circuit. In schizophrenia, abnormalities in limbic structures result in chronic increase in glutamatergic input to the NAC. Secondary to this increase, the level of NAC DA becomes reduced. Increased glutamatergic and decreased DA input to the NAC lead to overactivity of NAC neurons and thus overactivity of the circuit, and provide the pathophysiological basis of the prodromal/residual state of schizophrenia. The chronic decrease in DA in turn leads to an increase in the density of D_2 receptors in the NAC. As a result, an acute increase in NAC DA (e.g., by environmental stressors) will have an abnormally large disinhibitory effect on NAC neurons, leading to pathological release of behavior, i.e., to psychosis.

It should be noted that SWERDLOW and KOOB, CARLSSON et al. and CSERNANSKY et al. emphasize the loss of the inhibitory effects of the limbic circuit on behavior as the core abnormality of psychosis. In terms of current views of the organization of the basal ganglia–thalamocortical circuits, and as acknowledged by CARLSSON et al. (1999, 2000), these three models can be reformulated as implicating underactivity of the indirect pathway in the pathophysiology of psychosis. The contribution of dysfunction of the indirect pathway to the florid state of schizophrenia is strengthened by the fact that in the early stages of Huntington's disease, which are characterized primarily by degeneration of striatal neurons of the indirect pathway, patients often show schizophrenia-like symptoms and are sometimes incorrectly diagnosed as suffering from schizophrenia (JOEL and WEINER 1997; for an elaborated account see JOEL 2001).

O'DONNEL and GRACE (1998) have recently described three different dysfunctional circuitries postulated to underlie the three clusters of schizophrenic symptoms as delineated by LIDDLE et al. (1992): reality distortion (positive), psychomotor poverty (negative), and disorganization. Positive symptoms are attributed to disrupted hippocampal and prefrontal inputs to the NAC shell, combined with an increase in phasic DA release and a decrease in prefrontal cortex-dependent tonic DA levels. These lead to a decrease in the overall cell activity in the NAC shell, resulting in abnormally depressed activity in the mediodorsal–prefrontal loop, leading to both hypofrontality and the emergence of positive symptoms, possibly via the orbitofrontal cortex. Psychomotor retardation is attributed to decreased input from the dorsolateral prefrontal cortex to the associative striatum, combined with the resultant decrease in tonic DA levels. These result in reduced activity of striatal neurons, leading to increased inhibition of the mediodorsal/ventroanterior–prefrontal loop, which is reflected in a perseverative state and an overall psychomotor retardation. The disorganization syndrome is attributed to disrupted input from the cingulate cortex and the ventrolateral prefrontal cortex, which lead to decreased activity of neurons of the NAC core. This will lead to increased inhibition of the reticular thalamic nucleus, resulting in a breakdown of thalamic filtering, as manifested in schizophrenics' inability to focus attention or maintain a coherent line of thought.

II. The Split Circuit Model of Schizophrenia

We have recently presented a new model of basal ganglia–thalamocortical organization, namely, the split circuit scheme, which emphasizes the open interconnected nature of the circuits (JOEL and WEINER 1994, 1997, 2000), as opposed to the common view that these circuits are structurally and functionally segregated (e.g., ALEXANDER et al. 1986, 1990; ALEXANDER and CRUTCHER 1990). The model describes three open–interconnected split circuits, a motor, an associative, and a limbic, each containing a closed circuit through which information is channeled from a frontocortical subregion through specific subregions of the basal ganglia and thalamus back to the frontocortical area of origin, as well as several types of pathways connecting it to the other split circuits. The open–interconnected model is the first to explicitly incorporate the striatal connections with the dopaminergic system into a scheme of basal ganglia–thalamocortical circuits, in the form of closed and open loops (see Fig. 1 for a detailed description of the circuits and their interconnecting pathways).

In functional terms, we proposed that the motor, the associative, and the limbic split circuits provide the brain machinery for the selection and execution of goal-directed routine behavior, with the connections within each circuit subserving the selection of circuit-specific elements (motor acts, motor programs, and goals, respectively), and the connections between the circuits serving to coordinate their actions in order to produce complex goal-directed behavior (JOEL and WEINER 1994, for a detailed exposition of the model see

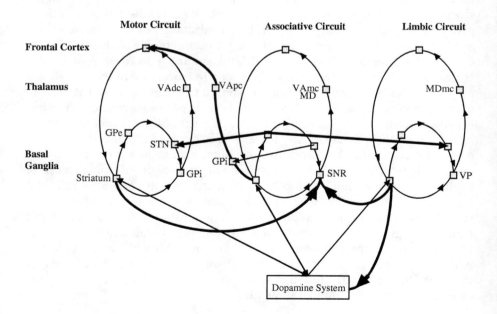

JOEL and WEINER 1999). In line with the widely held premise that the frontal cortex has a central role in flexible behavior, planning and decision making (e.g., MILNER 1963; LURIA 1973; PRIBRAM 1973; SHALLICE 1982; NORMAN and SHALLICE 1986; GOLDMAN-RAKIC 1987; FUSTER 1990; KOLB and WHISHAW 1990; ROBBINS 1990, 1991; LEVINE et al. 1992; STUSS 1992), and the striatum subserves routine or automatic aspects of behavior (e.g., COOLS 1980; MARSDEN 1982; IVERSEN 1984; NORMAN and SHALLICE 1986; ROLLS and WILLIAMS 1987; ROBBINS 1990, 1991; ROBBINS and BROWN 1990; MILLER and WICKENS 1991; BERRIDGE and WHISHAW 1992; LEVINE et al. 1992; GRAYBIEL et al. 1994; HIKOSAKA 1994;

◀────────────────────────────────────

Fig. 1. A summary diagram of the structural organization of the motor, associative, and limbic split circuits. Each split circuit contains a closed circuit and an open route or an open pathway. The associative split circuit: The closed associative circuit comprises the associative striatum, SNR, VAmc, and MD thalamic nuclei and the associative prefrontal cortex (including the frontal eye field and dorsolateral prefrontal cortex). The open associative pathway arises from the associative striatum, traverses the associative GPi and VApc, and terminates in the premotor cortex, which projects to the motor striatum. The motor split circuit: The closed motor circuit comprises the motor striatum, motor GPi, VAdc, and the primary motor cortex and supplementary motor area. The open motor route consists of motor striatal projections to SNR. The limbic split circuit: The closed limbic circuit comprises the limbic striatum, ventral (limbic) pallidum, MDmc, and the limbic prefrontal cortex (including the orbitofrontal cortex and anterior cingulate area). The open limbic route consists of limbic striatal projections to SNR. There may also be an open limbic pathway arising from the limbic striatum and projecting via the rostromedial GPi to motor/premotor cortices. Included within each of the closed circuits as well as within the open associative pathway are a direct and a closed indirect pathway. In addition, the associative split circuit contains an open indirect pathway that connects it with the motor split circuit, and possibly an open indirect pathway that connects it with the limbic split circuit. Each split circuit has a closed loop with the DA system, and in addition, there are two open loops connecting the limbic split circuit with the motor and the associative split circuits. Abbreviations: *GPe*, external segment of the globus pallidus; *GPi*, internal segment of the globus pallidus; *MD*, mediodorsal thalamic nucleus; *MDmc*, mediodorsal thalamic nucleus, magnocellular subdivision; *STN*, subthalamic nucleus; *PFC*, prefrontal cortex; *VAdc*, ventral anterior thalamic nucleus, densocellular subdivision; *VAmc*, ventral anterior thalamic nucleus, magnocellular subdivision; *VApc*, ventral anterior thalamic nucleus, parvocellular subdivision. Definitions: Closed circuit: A striato-frontocortical pathway that reenters the frontocortical area which is the source of cortical input to this striatal subregion; Open pathway: A striato-frontocortical pathway that terminates in a frontocortical area which innervates a different striatal subregion; Open route: The striatonigral portion of an open pathway; Closed indirect pathway: An indirect pathway (striatum-GPe-STN-GPi/SNR) that connects functionally corresponding subregions of the basal ganglia, that is, which terminates in the same subregion of the basal ganglia output nuclei as the direct pathway; Open indirect pathway: An indirect pathway (striatum-GPe-STN-GPi/SNR) which connects functionally non-corresponding subregions of the basal ganglia, that is, which terminates in a different subregion of the basal ganglia output nuclei than the direct pathway; Closed loop: A loop (striatum-DA system-striatum) which terminates in the striatal subregion from which it originates; Open loop: A loop (striatum-DA system-striatum) which terminates in a different striatal subregion than that from which it originates. Pathways connecting between circuits are demarcated in *thick lines*

MARSDEN and OBESO 1994), within each circuit, the frontal component subserves a non-routine or non-automatic selection [similar to the supervisory attentional system in NORMAN and SHALLICE'S (1986) scheme], whereas the striatal component acts as an automatic selection device [similar to the contention scheduling mechanism in NORMAN and SHALLICE'S (1986) scheme; see ROBBINS 1990, 1991 and WISE et al. 1996].

Below we combine the split circuit scheme with the physiological and behavioral functions of DA as they have emerged from animal research, to present a detailed description of the circuits' contribution to normal goal-directed behavior and its modulation by DA. This model is then used to account for some symptoms of schizophrenia, based on the postulated dual DA dysfunction in this disorder as detailed by GRACE and colleagues (GRACE 1991, 1993, 2000; O'DONNEL and GRACE 1998; MOORE et al. 1999). It is postulated that (1) routine goal-related information is actively maintained in the limbic striatum, and serves to direct and propel the selection and execution of motor plans by the associative and motor split circuits; (2) Striatal DA plays a fundamental role in the establishment, selection, and maintenance of routine goals as well as in limbic control of the associative and motor circuits because of its critical role in motivation, learning, and selection; and (3) The primary deficit of schizophrenia lies in an impairment of goal-directed control of routine behavior due to abnormal cortical and dopaminergic inputs to the limbic striatum. The latter is a revision and an elaboration of the idea that the cardinal deficit of schizophrenia lies in a failure to develop and maintain coherent patterns of goal-directed behavior (BLEULER 1911; KRAEPELIN 1919; ANSCOMBE 1987; FRITH 1987, 1992; STRAUSS 1987; COHEN and SERVAN-SCHREIBER 1992; HEMSLEY 1994; LIDDLE 1995; ZEC 1995; COHEN et al. 1996, 1999; FRISTON 1998; JAHANSHANI and FRITH 1998; BRAVER et al. 1999).

1. Striatum as a Contention Scheduling Device

Several characteristics of the striatum are important for understanding its functioning as a selection device. Striatal neurons are found in one of two stable subthreshold membrane potential states; a "down" state, in which the neuron is very hyperpolarized and does not generate action potentials, and an "up" state, in which the neuron is depolarized and a relatively weak excitatory synaptic input can trigger action potentials. The transition from the down to the up state depends on a temporally and spatially synchronized input from a relatively large subset of the neuron's cortical glutamatergic afferents (PENNARTZ et al. 1994; HOUK 1995; WILSON 1995; FINCH 1996; GERFEN and WILSON 1996; see Chap. 11, Vol. I). Given that the organization of corticostriatal projections is such that (1) different combinations of cortical inputs converge on different zones within a given striatal subregion and (2) each striatal neuron receives only few synapses from each of the thousands of cortical neurons innervating it (FLAHERTY and GRAYBIEL 1993, 1994; GRAYBIEL et al. 1994; PENNARTZ et al. 1994; GRAYBIEL and KIMURA 1995; WICKENS and KOTTER

1995; WILSON 1995; FINCH 1996; GERFEN and WILSON 1996; GRAYBIEL 1998), a striatal neuron's reaction to specific cortical inputs is likely to depend upon the current cortical context, i.e,. the pattern of activity in the different cortical regions which innervate this striatal neuron (LIDSKY et al. 1985; ROLLS and WILLIAMS 1987; ROLLS and JOHNSTONE 1992; AOSAKI et al. 1994; PENNARTZ et al. 1994; PLENZ and AERTSEN 1994; HOUK 1995; HOUK and WISE 1995; HOUK et al. 1995; KIMURA 1995; SCHULTZ et al. 1995a; GRAYBIEL 1998). Moreover, these characteristics suggest that cortical context determines a set of possible striatal outputs corresponding to the set of striatal neurons that have been driven into the up state. The specific striatal output is determined by the firing induced in a subset of these neurons by specific cortical inputs, which together with a winner-take-all mechanism subserved by inhibitory axon collaterals or feed-forward inhibition by striatal interneurons (GROVES 1983; PENNEY and YOUNG 1983; SWERDLOW and KOOB 1987; MILLER and WICKENS 1991; PENNARTZ et al. 1994; WICKENS and KOTTER 1995; WILSON 1995; KITA 1996), restricts the number of activated neurons and serves to select one specific output.

An additional important characteristic of the striatal selection mechanism is its ability to be molded by experience, manifested in long-term changes in corticostriatal synaptic efficacy (KIMURA 1987; MILLER and WICKENS 1991; FLAHERTY and GRAYBIEL 1994; PENNARTZ et al. 1994; SCHULTZ et al. 1995a; WICKENS and KOTTER 1995). Furthermore, it has been suggested that striatal neurons of the direct pathway "learn" to select the most appropriate element in response to specific stimuli, whereas neurons of the indirect pathway "learn" to suppress inappropriate elements (HOUK and WISE 1995, JOEL and WEINER 1999), as well as contribute to the termination of behavioral elements (BROTCHIE et al. 1991a,b; OBESO et al. 1994; WICHMAN et al. 1994a).

Under conditions requiring the selection of a new response strategy (e.g., during the initial stages of learning), the prefrontal cortex, interacting with different association and limbic regions, selects and directs behavior. Concurrently, the corticostriatal projections arising from these different cortical regions may drive a set of striatal neurons into the up state. Specific cortical activity patterns may then activate a subset of these neurons, of both the direct and indirect pathways, encoding the facilitation and suppression of specific behavioral elements, respectively. If the actual behavior leads to favorable outcomes to the organism, the activated corticostriatal synapses (both those responsible for the transition to the up state and those which induced firing) onto the activated striatal neurons of both pathways are strengthened, so that in the future this subset of neurons is more likely to be driven into the up state by the same or a similar cortical context and to be activated by the specific input.

Repeated occurrences of this sequence of events, i.e., reinforcement of a specific behavior in a specific context, will lead to the following: First, whenever the cortical context occurs it will drive a set of striatal neurons into the up state, thus determining a set of behaviors appropriate to that context as well as a set of behaviors inappropriate to that context. Second, the actual behavior will be selected according to the specific cortical inputs to the neurons in the up state which will induce firing in a subgroup of direct pathway neurons, whereas inappropriate behaviors which may be triggered by these same cortical inputs will be suppressed due to the activation of indirect pathway neurons. Specific changes of the cortical input will activate other indirect and direct pathway neurons leading to the termination of the current behavior and to the initiation of a new behavior. Third, the behavior selected at any moment will be the most appropriate for the current situation according to past experience.

2. The Interaction Between the Striatum and the Frontal Cortex

Information regarding the selection of the most appropriate behavioral element in the current context is continuously channeled from the striatum to the frontal cortex via the basal ganglia output nuclei (SNR, internal segment of globus pallidus, ventral pallidum) and thalamus, and acts to bias the activity patterns of cortical neurons towards the selection of this behavior. Striatal information, however, does not necessarily translate into behavioral output since the frontal cortex receives in addition information about the current context from other cortical regions. Whether the actual behavioral output is the one selected by the striatum depends on the strength of the striatal biasing effect on the frontal cortex (which is a function of the degree of activation of striatal neurons), and on the degree of correspondence between the striatal and cortical biasing effects on the activity pattern in the frontal cortex.

In well-learned/highly familiar situations, striatal neurons which encode the most appropriate behavior, as well as those encoding incompatible behaviors, are expected to be strongly activated, and their strong biasing effect is expected to have a high degree of correspondence with the cortical biasing effect. The result is an effortless production of routine behavior. In novel or ill-learned situations, striatal neurons are expected to be weakly activated, and in addition, their (weaker) biasing effect on the activity of frontal neurons is unlikely to coincide with the biasing effects of other inputs to the frontal cortex. Under such circumstances the selection of behavioral output cannot be achieved automatically, but requires a supervisory process, subserved by the interaction of the frontal cortex with other brain regions (e.g., posterior association regions and high-order limbic cortices), which will yield alternative ways of action. In intermediate situations in which the cortical context is only partly familiar, striatal neurons will be activated, albeit less strongly than they would be in the fully familiar context, but their biasing effect is less likely to coincide with the cortical biasing effect. Two outcomes can ensue: (1) The cortical biasing effect may be sufficiently strong to counteract the striatal biasing effect; the routine behavior will not be performed, and the supervisory process will intervene. (2) The striatal biasing effect is sufficiently strong to lead to the execution of the routine behavior.

3. Contention Scheduling of Goals by the Limbic Striatum

The striato-frontal interaction described above takes place within each of the circuits, but on different types of information, namely, motor, cognitive, or limbic. Specifically, the motor striatum subserves the contention scheduling of simple motor acts, the associative striatum subserves the contention scheduling of motor programs (which include cognitive and motor components) and the limbic striatum subserves the contention scheduling of goals (JOEL and WEINER 1999).

The proposition that the limbic striatum subserves the contention scheduling of goals, i.e., directs an organism's behavior toward specific end-points, e.g., obtaining appetitive stimuli (such as food, warmth, affection, the recognition of others), avoiding aversive stimuli (such as shock, anger, rejection), exploring novel stimuli, etc., is consistent with the current view, pioneered by NAUTA et al. (1978) and MOGENSON et al. (1980), that the limbic striatum plays a fundamental role in the translation of limbic information to action, that is, in the "directional" aspects of motivation (e.g., ROBBINS and EVERITT 1982, 1992; BENINGER 1983; CADOR et al. 1989, 1991; EVERITT et al. 1991; SCHEEL-KRUGER and WILLNER 1991; LAVOIE and MIZUMORI 1992; SCHULTZ et al. 1992, 1995a; KALIVAS et al. 1993; PENNARTZ et al. 1994; SALAMONE 1994; BENINGER and MILLER 1998; DEPUE and COLLINS 1999; IKEMOTO and PANKSEPP 1999).

The proposition that goals are selected by a contention scheduling mechanism has the following implications: (1) Goals are selected according to their activation level which is determined by external and internal information (provided by the inputs to the limbic striatum); (2) As a result of a reinforcement-driven learning mechanism, the most appropriate goal is selected, that is, the goal that according to past experience is expected to maximize reward in the present situation; (3) In routine situations, the selection of goals is automatic and effortless; (4) In novel, ill-learned or dangerous situations, in which automatic selection of goals is not possible, a supervisory mechanism, residing in the limbic prefrontal cortex, selects a goal. These characteristics are in line with current views of goal-directed behavior according to which goals are activated by environmental and internal factors and selected in a way that maximizes expected value. In addition, it is accepted that when activity is well organized and routine, action moves from goal to goal fairly smoothly without requiring a deliberate "choice" or "decision" to change goals. The effortful process of selecting a goal is required under unusual internal or external stimulation (for review, see PERVIN 1983, 1996).

4. The Role of Tonic and Phasic DA in the Contention Scheduling of Goals

What is needed for adaptive goal-directed behavior? The "goal system" must select (sometimes among several competing goals) the goal most appropriate to a given situation; maintain it throughout the course of the behavior; be resistant to interference; terminate it as soon as it is fulfilled or turns out to be inadequate for the situation, and switch to a different goal. Current data and theories suggest that striatal DA is critically involved in these aspects of contention scheduling of goals.

While the physiological and behavioral consequences of striatal DA have been extensively documented (see Chap. 11, Vol. I; Chap. 19, this volume), they have been seldomly related specifically to the mode of DA release. However, it has been increasingly recognized that the two modes of DA transmission may play distinct roles in the modulation of corticostriatal synaptic transmission and plasticity, as well as in behavioral and cognitive processes (GRACE 1991, 1993, 2000; SCHULTZ 1998; MOORE et al. 1999; Chap. 19, this volume).

a) Tonic and Phasic DA Release

DA cells exhibit two spontaneously occurring electrophysiological states: single spiking, in which the majority of cells are found, and burst firing (BUNNEY et al. 1991; WHITE 1991; KALIVAS 1993; JOHNSON et al. 1994; GRACE 1995; TEPPER et al. 1995). DA levels at the terminal fields depend on the firing mode of DA cells as well as on other factors acting directly on DA terminals. Specifically, there are two modes of DA release, phasic and tonic. The former refers to the release of DA during an action potential, which is rapidly inactivated via reuptake into presynaptic terminals and diffusion. The level of phasic release depends primarily on the mode of DA cell activity and is markedly enhanced when cells fire in bursts (GRACE and BUNNEY 1984; GONON 1988; MURASE et al. 1992; NISSBRANDT et al.1994; GARRIS et al. 1997). Tonic DA transmission represents the steady state DA level in the extracellular space. It is relatively constant and tightly regulated (GRACE 1991, 1993). Tonic DA release may be driven by the presynaptic actions of glutamatergic cortical inputs onto DA terminals in the striatum, and is also affected by "spillover" from synaptic release (phasic) as well as by DA released from non-synaptic sites along the axons. As such, tonic DA would be secondarily affected by DA cell activity and directly affected by how much DA escapes from sites of release (e.g., via changes in release or reuptake; for a detailed description of phasic and tonic DA transmission, see GRACE 1991, 1993, 2000; MOORE et al. 1998).

Electrophysiological studies in behaving animals (SCHULTZ 1986, 1998; KIAYTKIN 1988; SCHULTZ and ROMO 1990; MILLER et al. 1991; SCHULTZ et al. 1992) have shown that DA cells switch to burst firing following the occurrence of salient, novel, or reinforcing (unconditioned and conditioned) stimuli. A critical feature of DA response is its dependence on event unpredictability: DA neurons respond to stimuli which have an innate or acquired (via learning) significance as long as they are unpredictable, but stop responding when they become predictable; during learning, DA responses transfer from primary rewards to reward-predicting stimuli. The responsiveness of striatal DA to significant stimuli has been supported also by in vivo microdialysis studies (SALAMONE et al. 1997; see Chap. 19, this volume).

Based on the above and the evidence for dopamine-dependent long-term changes in corticostriatal synaptic efficacy (CALABRESI et al. 1992, 1996; PENNARTZ et al. 1993; WICKENS et al. 1996; CHARPIER and DENIAU 1997, see Chap. 11, Vol. I), it has been suggested that DA governs associative learning in the striatum by providing a "teaching" or an "error" signal which modulates corticostriatal synaptic transmission (WICKENS 1990; MILLER and WICKENS 1991; GRAYBIEL et al. 1994; PENNARTZ et al. 1994; GROVES et al. 1995; HOUK 1995; HOUK et al. 1995; KIMURA 1995; PENNARTZ 1995; SCHULTZ et al. 1995a,b; WICKENS and KOTTER 1995; GRAYBIEL 1998). The dependence of DA response on event unpredictability is also consistent with the theoretical positions and behavioral data that learning takes place only as long as the to-be-associated events are unpredictable (RESCORLA and WAGNER 1972; PEARCE and HALL 1980).

While phasic DA plays a significant role in the processing of significant and unpredicted events, this cannot account for the wide range of behavioral deficits following injury to the DA system by means of lesions or pharmacological treatments (SCHULTZ 1998). Overall, DA depletion or DA blockade result in a greatly impoverished behavioral repertoire and a disorganization of motivated, adaptive goal-directed interaction with the environment, ranging from locomotor activity, through species-specific behaviors like food hoarding and maternal nursing, to a wide variety of positively and negatively reinforced instrumental responding (ROBBINS and EVERITT 1982; BENINGER 1983; LEMOAL and SIMON 1991; BLACKBURN et al. 1992; SALAMONE 1994; LE MOAL 1995; BENINGER and MILLER 1998; DI CHIARA 1998; SCHULTZ 1998; IKEMOTO and PANKSEPP 1999). Importantly, many of the lost functions are still present but not expressed without DA, since they can be reinstated by DA agonists or strong environmental stimulation (LYNCH and CAREY 1987; KEEFE et al. 1989; LEMOAL and SIMON 1991; SCHULTZ 1998). These data have been interpreted as demonstrating that DA has a general enabling, activating, energizing, or invigorating function, attributed primarily to mesolimbic DA (TAYLOR and ROBBINS 1984, 1986; COLE and ROBBINS 1987; LE MOAL and SIMON 1991; ROBBINS and EVERITT 1992, 1996; SALAMONE 1994; SALAMONE et al. 1997; BERRIDGE and ROBINSON 1998; DI CHIARA 1998; SCHULTZ 1998; IKEMOTO and PANKSEPP 1999; Chap. 19, this volume). Although gross manipulations of the DA system affect both phasic and tonic DA transmission, the observations that (1) DA alterations affect a wide range of behaviors, many of which do not trigger burst activity in DA cells; (2) DA agonists can reverse the effects of DA loss, although they do not restore DA phasic transmission (LEMOAL and SIMON 1991; SCHULTZ 1998), and (3) artificial increases in DA level (e.g., by amphetamine) invigorate a wide range of behaviors (LYON and ROBBINS 1975; EVENDEN and ROBBINS 1983; TAYLOR and ROBBINS 1984, 1986; LJUNGBERG and ENQUIST 1987), suggest that the enabling/energizing function of DA depends on tonic DA levels rather than on phasic DA release (SCHULTZ 1998).

In view of the above, it has been suggested that the two modes of DA neurotransmission subserve different functions. Thus, SCHULTZ (1998) concluded that phasic DA subserves the signaling of significant alerting stimuli, and tonic DA subserves the enabling of a wide range of behaviors without temporal coding. Similarly, MOORE et al. (1999) suggested that tonic DA provides sufficient level of DA receptor stimulation necessary for the initiation and execution of well-learned behavior, while phasic transmission is necessary for novelty-induced behaviors and learning. These authors (GRACE 1991, 1993, 2000; SCHULTZ 1998; MOORE et al. 1999) have also described the modulation of corticostriatal synaptic transmission and plasticity by phasic and tonic DA which may mediate the functional role of DA in the striatum. Below we describe the role of striatal DA in contention scheduling with a focus on the contention scheduling of goals. DA is assumed to play the same role in the contention scheduling of motor programs and motor acts.

b) The Establishment of Goals in the Limbic Striatum

In the present model, the phasic increase in striatal DA following the unpredicted occurrence of conditioned and unconditioned reinforcers acts as a reinforcement signal which serves to strengthen synapses between active corticostriatal terminals and active striatal neurons. The strengthening of corticostriatal synapses on active direct pathway neurons will result in a greater likelihood that the encoded goal, which has led to favorable outcomes to the organism, is selected again in the same or a similar context. Similarly, the strengthening of corticostriatal synapses on active indirect pathway neurons will result in a greater likelihood that inappropriate goals are suppressed in the same or a similar context.

It should be noted that conditioned and unconditioned stimuli may act not only as reinforcers of the goal-directed behavior which precedes them, but also as stimuli guiding behavior. During learning, these stimuli acquire the capacity to activate direct pathway neurons encoding the next sub-goal as well as indirect pathway neurons encoding the previous sub-goal. Consequently, they will be able to trigger the termination of the previous sub-goal and the initiation of the next sub-goal, thus enabling a smooth transition between the different components of a routine goal-directed behavior. Thus, although these stimuli lose their ability to increase DA cell firing as they become predicted and therefore lose their ability to support further learning, they do not lose their ability to direct behavior.

c) Goal Selection

Striatal DA also serves to modulate the selection process (SCHULTZ 1998; MOORE et al. 1999). It has been suggested that by inhibiting striatal neurons or attenuating their responses to excitatory and inhibitory inputs (via activation of D_2 receptors; UCHIMURA et al. 1986; O'DONNEL and GRACE 1994; CEPEDA et al. 1995; YAN et al. 1997, see Chap. 11, Vol. I), DA may restrict striatal output to the most strongly activated neurons (YANG and MOGENSON 1984; MOGENSON et al. 1993; PENNARTZ et al. 1994; SCHULTZ et al. 1995; O'DONNEL and GRACE 1998; DEPUE and COLLINS 1999; MOORE et al. 1999). Furthermore, based on findings that activation of D_1 receptors enhances the response to excitatory glutamatergic input of striatal neurons in the up state but reduces the response of neurons in the down state (KAWAGUCHI et al. 1989; CEPEDA et al. 1993, 1998; HERNANDEZ-LOPEZ et al. 1997; SCHULTZ 1998; Chap. 11, Vol. I), it has been suggested that DA increases the contrast gradient between weak and strong cortical inputs (O'DONNEL and GRACE 1998; SCHULTZ 1998).

Since D_1 and D_2 receptors are expressed preferentially on striatal neurons of the direct and indirect pathways, respectively (ALBIN et al. 1989; GERFEN et al. 1990; REINER and ANDERSON 1990; DELONG and WICHMANN 1993; GERFEN and WILSON 1996), DA may simultaneously act (1) to suppress the activity of indirect pathway neurons, except for the most active ones, thus enabling the initiation (by direct pathway neurons) of a wide variety of goals while concomitantly preventing the selection of goals inappropriate to the current context, and (2) to enhance the contrast between neurons transferred to the up state by the cortical context and those which were not, thus ensuring that only context-appropriate goals will compete for behavioral expression. Since the selection process is modulated by the level of striatal DA, it is affected by both tonic and phasic DA release (see also section B.II.4.e below).

d) Goal Maintenance and Energizing

Once a goal is selected, tonic DA serves to maintain it at a sufficient level of activation and protect it from interference as well as to determine the effort which will be invested in attaining it. These functions are suggested to be subserved by the differential effects of activation of D_1 receptors on striatal neurons, depending on their membrane potential. Thus, tonic DA maintains the selected goal by facilitating firing of neurons already in the up state, and simultaneously provides protection from interference by suppressing firing or transition to the up state of neurons in the down state. Since most (about 80%) of D_1 receptors are in the low-affinity state (RICHFIELD et al. 1989), these actions may be achieved either by activating the high-affinity D_1 receptors, or by local increases in DA level at the region of the active striatal cells which will suffice for activating low-affinity receptors. Such an increase may be achieved either via the stimulating effects of glutamate released from the active corticostriatal terminals on DA release (GRACE 1991; MOORE et al. 1999) or through the indirect facilitatory effects exerted by striatal neurons on DA cells (see below; see KALIVAS et al. 1993 for a related view). The higher the levels of tonic DA, the higher the response of striatal cells to a given cortical input, and thus the higher is striatal facilitation of the selected output. In this way, the level of tonic DA determines the energy level of the selected goal.

An additional effect of the local increase in DA may be to activate D_2 autoreceptors. This activation may lead to reduced phasic DA release in response to DA cell firing. The consequences of such a decrease are detailed below.

e) Switching Between Goals

Based on lesion and drug studies showing that DA loss produces inflexible and perseverative behavior whereas increase in DA promotes behavioral switching, DA has been attributed a central role in switching (LYONS and ROBBINS 1975; ROBBINS and EVERITT 1982; OADES 1985; TAGHZOUTI et al. 1985; SWERDLOW and KOOB 1987; VAN DEN BOS and COOLS 1989; WEINER 1990; GRAY et al. 1991; LE MOAL and SIMON 1991; VAN DEN BOS et al. 1991; PENNARTZ et al. 1994). While the relationship between switching and tonic/phasic DA has not received attention, switching may be subserved by both modes of release depending on the conditions which elicit switching (REDGRAVE et al. 1999).

Goals are changed either in the course of routine chains of behaviors, namely, in response to predicted events, or when the situation unexpectedly changes, namely, in response to unpredicted events. Therefore, we suggest that the former depends on tonic DA whereas the latter depends on phasic changes in DA level. During routine behavior, the attainment of each sub-goal of a routine motor program is predicted, and thus not accompanied by phasic changes in striatal DA. Rather, switching between goals during the performance of routine goal-directed behaviors is subserved by the mechanisms detailed above for goal selection, namely, the attainment of a sub-goal triggers both its termination and the initiation of the subsequent sub-goal by activating neurons of the indirect and direct pathways, respectively. In this way, tonic DA enables smooth transition between successive sub-goals.

The phasic increase in striatal DA accompanying the unexpected occurrence of significant events depresses the activity of indirect pathway neurons, retarding the suppression of all goals, except for the inappropriate ones (see section B.II.4.c), thus reducing constraints on the concomitant goal selection by direct pathway neurons. The DA effects on direct pathway neurons are more selective. Specifically, via its differential action on neurons that are in the up vs down state, DA facilitates the selection of a context-appropriate goal. Moreover, it biases the selection away from the goal that had been active just before the unexpected occurrence of a significant event. This is achieved by the attenuation of the phasic DA increase in the region of striatal neurons that have just been active, as a result of activation of DA autoreceptors in this region (see section B.II.4.d). Such attenuation may serve to favor the selection of new sets of neurons, i.e., of new goals, as well as to prevent switching back to the set which has just been active, thereby preventing dithering between goals (REDGRAVE et al. 1999).

It should be pointed out that phasic increase in striatal DA is hypothesized to mediate both switching and learning. Thus, increased DA input to the striatum following unexpected significant stimuli, both facilitates a switch in the set of active striatal neurons from the set which has just been active to a new set, thus favoring a change in behavior, and strengthens active corticostriatal synapses of neurons which have just been active, thus raising the likelihood that these neurons will be activated again by this cortical context, and thus that the behavior will occur again in the same or a similar situation.

Finally, phasic changes in striatal DA may also contribute to the termination of a goal-directed behavior which has proved to be ineffective. DA neurons were found to decrease their firing rate in response to the omission of an expected stimulus (SCHULTZ et al. 1993, 1995b). This may lead to a transient decrease in tonic DA levels, which will affect particularly high affinity D_2 receptors (SCHULTZ 1998). The decreased activation of these receptors may lead to increased activity of neurons of the indirect pathway, including those which encode the termination of the current goal, thus leading to a behavioral arrest which enables a reevaluation of the situation and a reselection of a goal.

5. The Translation of Goals to Behavior

In the present scheme, the limbic split circuit selects goals, without specifying the specific motor program by means of which these goals are to be achieved. The latter is suggested to be the function of the associative split circuit acting together with the motor split circuit. However, via its connections with these circuits, the limbic split circuit directs the selection and execution of motor programs towards achieving the selected goal.

Specifically, information regarding the most appropriate goal in the current context:

1. Is channeled from the limbic striatum via the open limbic route to SNR, where it acts to bias nigral output according to the current goal of the organism. In this way selection of goals in the limbic striatum can affect the transfer of information in the striato-nigro-thalamo-cortical pathway to the associative prefrontal cortex, which is involved in the selection and execution of motor programs. It can also affect the nigral output to the superior colliculus and in this way contribute to the reallocation of attention when the goal is changed or when a novel or surprising stimulus appears.
2. Modulates the dopaminergic input to the limbic striatum as well as to the motor and associative striatum, via the closed and open loops originating from the limbic striatum. We have recently suggested that via each of the loops, closed or open, the striatum exerts a direct inhibitory effect on DA cells as well as an indirect disinhibitory effect, i.e., facilitation of burst firing in DA cells (JOEL and WEINER 2000). Thus, the activation of a set of limbic striatal neurons encoding a specific goal is expected to directly inhibit dopaminergic neurons. This inhibition can counteract the excitatory input to the dopaminergic cells when the goal is attained and thus prevent the firing of dopaminergic neurons to predicted rewards (SCHULTZ et al. 1993, 1995b; WICKENS and KOTTER 1995; BROWN et al. 1999). In this way, the limbic striatum can prevent switching following the attainment of sub-goals in the course of performing a routine goal-directed behavior, as well as restrict learning in all striatal subregions in well-learned situations. The indirect

facilitatory effect of the limbic striatum on DA cells stems from the projections of striatal neurons onto GABAergic neurons in SNR and VTA and their subsequent projections onto DA cells as well as onto the GABAergic neurons of the limbic pallidum, which also project onto DA cells. These disinhibitory effects may provide a mechanism whereby the limbic striatum can adjust DA levels in the different striatal subregions according to the motivational state of the organism, and thus modulate the degree of effort invested in the execution of the encoded goal-directed behavior (JOEL and WEINER 2000; for a related view see KALIVAS et al. 1993).
3. Is continuously channeled to the limbic prefrontal cortex where it acts to bias the activity patterns of cortical neurons towards the selection of this goal and contributes to sustained activity of limbic prefrontal cortex neurons, which maintain active goals and intentions in the absence of the external and internal stimuli that arouse them. From the limbic prefrontal cortex the information can be channeled to the associative split circuit via corticocortical connections between the limbic prefrontal cortex and the associative prefrontal cortex. The projections from the limbic prefrontal cortex to the associative prefrontal cortex may directly bias the selection of motor programs by the associative prefrontal cortex according to the current goal, encoded in the limbic prefrontal cortex. As both prefrontal regions subserve a supervisory mechanism, this link may be particularly important for the effortful and deliberate process of goal selection. This is in contrast with the transfer of information via the different open pathways that subserve an automatic, effortless process by which goals can affect different aspects of behavior.

6. Schizophrenia

A failure to exert control over thoughts and actions has been often considered to be central to schizophrenia (KRAEPELIN 1919; ANSCOMBE 1987; FRITH 1987, 1992; STRAUSS 1987; COHEN and SERVAN-SCHREIBER 1992; HEMSLEY 1994; LIDDLE 1995; ZEC 1995; COHEN et al. 1996, 1999; FRISTON 1998; JAHANSHANI and FRITH 1998; BRAVER et al. 1999). Given that a core characteristic of coherent and flexible behavior is that it is goal-directed or purposeful, the failure of control in schizophrenia has been attributed to a disruption of a system which allows the generation of efficient goal-directed behaviors. Influenced by KRAEPELIN's (1919) view that schizophrenia is a disorder of volition, SHALLICE and NORMAN's (1986) supervisory attentional system and FRITH's (1987, 1992) powerful exposition of schizophrenia as a disorder caused by a breakdown in the monitoring of willed intentions, there has been an increasing trend to argue that schizophrenic symptomatology may reflect a failure of high-level cognitive control system or of a central executive mechanism which guides and coordinates behavior in a flexible fashion, particularly in novel and complex situations. Such control has been typically envisaged as a top–down process, with the level of control being "higher" to lower level selection (REDGRAVE et al. 1999), and most typically residing in the prefrontal cortex (KRAEPELIN 1919; FRITH 1987, 1992; STRAUSS 1987; WEINBERGER et al. 1988; ROBBINS 1990, 1991; COHEN and SERVAN-SCHREIBER 1992; LIDDLE 1995; WEINBERGER and LIPSKA 1995; ZEC 1995; COHEN et al. 1996, 1999; CRIDER 1997; JAHANSHANI and FRITH 1998; BRAVER et al. 1999).

We would like to forward a different view: Most of human behavior, whether internally or externally driven, is routine, and most of the time people do not face novel and complex situations. Indeed, what makes the normal adult behavior smooth, flexible, and adaptive is that most of the time people transact with fairly familiar internal and external environments, which elicit routine goal states that give rise to routine behavioral programs. Moreover, when the "executive" comes into play, its role in most cases is to stop the ongoing inappropriate routine behavior and aid in choosing an alternative behavior from the existing repertoire; it is only in very unfamiliar and unexpected situations that a dramatic re-appraisal and re-learning are needed. Finally, when normal persons face unfamiliar and unexpected situations for which they do not have a routine behavioral program or one that is easily adaptable to the situation, they do not fair out very well either.

While we accept the position that schizophrenia involves a disturbance in executive functions, the pervasiveness of the schizophrenic deficits in almost all aspects of functioning points in our opinion to a profound disturbance also in the routine aspects of behavior. Indeed, in these patients "problems may be noted in any form of goal-directed behavior leading to difficulties in performing activities of daily living such as organizing meals or maintaining hygiene" (DSM-IV, p 276). We propose that precisely such a disturbance of routine goal-directed behavior results from a disruption of the contention scheduling of goals in the limbic striatum due to cortical dysfunction and a dysregulation of phasic and tonic DA, and the resultant dysfunction of the limbic, associative, and motor split circuits. On this view, schizophrenic symptomatology, rather than reflecting a failure of top-down control, reflects an impaired interplay between top-down and bottom-up control processes within each circuit, as well as impaired "medial-to-lateral" control processes between the circuits.

We want to note that our account of DA dysfunction is limited to the striatal portion of the circuits and does not include the well-documented DA role in the frontal component of the circuit, e.g., in working memory and many other executive functions (Chap. 19, this volume). Likewise, while our model retains the notion of an "impaired executive" or deficient supervisory processes in schizophrenia, it is silent with regard to the direct contribution of prefrontal and temporal dysfunction to schizophrenic symptomatology, which has been described in detail by others (e.g., WEINBERGER 1987, 1988; COHEN and SERVAN-SCHREIBER 1992; COHEN et al. 1996, 1999; GOLDMAN-RAKIC 1999; Chap. 19, this volume). This is because the dysfunction of these regions is considered to disrupt supervisory processes which are important in novel or non-routine situations, whereas our model focuses on routine behavior and therefore on the basal ganglia.

Finally, we adhere to the notion that schizophrenia is a neurodevelopmental disorder (e.g., MURRAY and LEWIS 1987; BOGERTS 1991, 1993; MEDNICK et al. 1991; MURRAY et al. 1991; HARRISON 1995, 1999; KNABLE and WEINBERGER 1995; WEINBERGER and LIPSKA 1995; TURNER et al. 1997; WEICKERT and WEINBERGER 1998; KESHAVAN 1999; KESHAVAN and HOGARTY 1999), in which an early damage (occurring in utero or in early neonatal period) to prefrontal and/or temporolimbic cortices interacts with the development of the brain to lead via as yet unknown (but widely speculated; e.g. KNABLE and WEINBERGER 1995; WEINBERGER and LIPSKA 1995; FRISTON 1998; KESHAVAN 1999; KESHAVAN and HOGARTY 1999) mechanisms to the late appearance of symptoms.

a) Fronto-temporo-limbic Cortical Dysfunction and Dysregulation of Tonic and Phasic DA Transmission in Schizophrenia

As noted in the introduction, based on extensive evidence of morphometric abnormalities in frontal and temporal cortices, and on neuroimaging studies of brain function in patients with schizophrenia pointing to an abnormal pattern of fronto-temporal activation/interaction, it has been increasingly accepted that schizophrenia involves an abnormality in prefrontal-temporal neuronal and/or functional connectivity (LIDDLE 1987, 1995; FRISTON et al. 1992; WEINBERGER et al. 1992; FRISTON and FRITH 1995; FRITH et al. 1995; GAREY et al. 1995; GLANTZ and LEWIS 1995; KNABLE and WEINBERGER 1995; SPENCE et al. 1997; SELEMON et al. 1995, 1998; FLETCHER et al. 1996; DOLAN and FLETCHER 1997; JAHANSHANI and FRITH 1998; RAJKOWSKA et al. 1998; GOLDMAN RAKIC 1999), and that this abnormality leads to a dysregulation of mesolimbic DA.

GRACE (1991, 2000; O'DONNEL and GRACE 1998; MOORE et al. 1999) has advanced a refined hypothesis of mesolimbic DA dysfunction based on the dual control of DA release in the NAC. In this model, tonic DA levels regulate phasic DA release via activation of DA synthesis- and release-modulating autoreceptors, so that the amount of phasically released DA is an inverse function of the basal level of tonic DA present in the extrasynaptic space. A pathological decrease in the activity of cortical inputs to the NAC leads to a reduction in tonic DA release, leading to a decrease in the basal extracellular levels of DA in the NAC. The resultant decrease of DA terminal autoreceptor stimulation leads to abnormal enhancement of spike-dependent DA release. Consequently, cell firing, and in particular, bursting of DA cells would lead to a release of abnormally large amounts of DA, and produce pathologically high degrees of postsynaptic receptor stimulation (for a detailed description see GRACE 1991, 1993, 2000; O'DONNEL and GRACE 1998; MOORE et al. 1999).

b) The Consequences of Fronto-temporo-limbic Cortical Dysfunction: Disrupted Establishment of Goals

The abnormal functioning of fronto-temporo-limbic cortical regions and the resulting disorganized fronto-temporo-limbic input to the limbic striatum is expected to lead to an abnormal establishment of goals in the limbic striatum.

As a consequence of disrupted functioning of and information flow between fronto-temporo-limbic cortical regions, the effortful goal selection process in non-routine situations that takes place in the limbic prefrontal cortex in concert with temporo-limbic regions will be abnormal. Goal selection will be less determined by information in temporo-limbic regions, e.g., about one's own emotions and those of others, the significance of stimuli and events, and memories/knowledge related to the current situation, so that many of the selected goals will be unrelated or inappropriate to the context. Since reinforcement of most behaviors is context dependent, i.e., the same behavior is reinforced in some situations but not in others, many of the individual's behaviors will be inconsistently reinforced or punished.

Since the establishment of goals in the limbic striatum progresses concurrently with goal selection in the limbic prefrontal cortex, and depends on repeated reinforcement of the selected goal, it will also be abnormal. Specifically, the striatum will learn to select only those goals which are reinforced or at least not punished under most situations familiar to the individual, leading to the establishment of a limited repertoire of goals, mostly avoidant in nature. Moreover, goals established in the striatum will be less context-dependent than normal, i.e. their activation will be more dependent on specific information derived mostly from the limbic prefrontal cortex, and less dependent on the cortical context derived from temporo-limbic regions.

It is, therefore, hypothesized that many of the persisting deficit symptoms in schizophrenia result from the individual's inability to acquire through life experiences a rich repertoire of goals which can be automatically selected in a context-appropriate fashion and lead to behaviors which are appropriate and thus reinforced. This will lead in general to poverty of behavior as well as to inappropriate behavior and withdrawal. In addition, since interpersonal interaction and communication are probably the most context-sensitive human behaviors, often requiring complex processes of inferring the right context (see SPERBER and WILSON 1987), they are likely to be most adversely affected by a dysfunction in the mechanism responsible for the selection of context-appropriate goals. This may account for the pervasive impairment of schizophrenic individuals in the social domain, characterized by poor social relations and social skills, lack of interpersonal competence, and lack of the ability to engage in socially appropriate behaviors (DWORKIN 1992).

The dysfunctional process described above presumably takes place throughout the life of an individual destined to become schizophrenic, consistent with the observation that some dysfunction may appear already in the prodromal stage. Such dysfunction is mainly characterized by negative symptoms, such as social withdrawal and isolation, although they are much milder than they are after the schizophrenic illness begins (DAVIS et al. 1991; FAUSTMAN and HOFF 1995). The variability of presenting symptoms in the prodromal stage is likely to reflect differences in the severity of cortical abnormalities and in the life experiences of each individual. This is in line with the observation that individuals with more evidence of structural brain abnormalities have a poorer premorbid adjustment, more prominent negative signs, symptoms, and cognitive impairments, and a poorer outcome (DSM-IV).

It should also be pointed out that the above account does not incorporate a DA dysfunction, since it is not clear whether such a dysfunction is expressed prior to the first psychotic episode. However, in some cases it may be present already at the prodromal stage, as evidenced by the presence of mild positive-like symptoms (e.g., odd beliefs but not of delusional proportion; DSM-IV).

c) The Consequences of Dysregulation of the DA Input to the Limbic Striatum

α) Reduced Tonic DA: Goal Selection, Activation and Maintenance

In familiar, routine situations, tonic DA provides a sufficient level of DA receptor activation, permitting the selection and maintenance of goals. A reduction of tonic DA release and of tonic DA levels in the limbic striatum of schizophrenic patients will thus lead to deficits in goal selection, maintenance, and energizing. Specifically, a reduced DA level will lead to insufficient activation of neurons of the direct pathway, and thus to difficulties in the initiation of goals. This will be compounded by an insufficient inhibition of neurons of the indirect pathway and thus an overinhibition of all goals, which will further impair the initiation (by direct pathway neurons) of the most appropriate goal. In addition to difficulties in the selection of an appropriate goal, the loss of DA-energizing effect will result in a weak activation of the selected goal. Low motivation, apathy, loss of interest or pleasure (anhedonia), restriction of the range and intensity of emotional expression and reactivity (flat or blunted affect) will follow.

Sufficient DA levels are needed not only for "energizing" the selected goal, but also for preventing the activation of competing goals. Therefore, weak activation of the selected goal may lead to difficulties in maintaining the selected goal in the face of relatively minor changes in the situation, i.e., to increased sensitivity to interference. It should be noted that since such minor changes are by definition not accompanied by a rise in DA level, the newly selected (interfering) goal is also of low energy. Therefore, the patient is expected to switch repeatedly between different low-energy goals. Reduced goal activation may also result in a gradual decay of goal representation, which may eventually result in the cessation of goal representation in the limbic striatum.

Weak activity of striatal neurons of the direct pathway will result in a weak biasing effect on the activity of the limbic prefrontal cortex. This will disrupt the automatic selection and maintenance of goals in the limbic prefrontal cortex in routine situations, thus requiring a supervisory mechanism for the selection of goals, as normally happens in ill-learned situations. Moreover, since the supervisory process depends on interactions of the limbic prefrontal cortex with other association and limbic cortical regions, and these interactions are dysfunctional in schizophrenia, the goal selection process will not only cease to be automatic and effortless but will also be impaired (as described in the previous section).

The dysfunction of the limbic striatum as a consequence of reduced tonic DA will not only affect the functioning of the closed limbic circuit, as described above, but also its modulation of the functioning of the motor and associative circuits, enacted via the open limbic route and the open loops. Thus, reduced goal activation in the limbic striatum may lead, via reduced activity of the open loops, to a reduction in the facilitating, i.e., disinhibiting, effects of the active limbic striatal neurons encoding a goal on tonic DA levels in the associative and motor split circuits. As a consequence, the degree of effort invested in performing the relevant goal-directed behavior, which depends on tonic DA levels in the motor and associative split circuits, will be lowered. Reduced DA input to these circuits will also lead to difficulties in initiating goal-directed activities (avolition), manifested in decreased behavioral output and reduction in the production of thought and speech (alogia).

Reduced goal activation will also lead to a reduction in the inhibitory effect of the active limbic striatal neurons encoding a goal on the phasic response of DA neurons to the occurrence of this goal, and thus to an abnormally high phasic response of DA neurons. The consequences of the resultant abnormal phasic DA release in the three striatal regions are detailed in the next section. In addition, such an abnormal phasic response of DA cells may disrupt the functioning of the closed associative circuit by interfering with the throughput of associative striatal information via SNR. We (JOEL and WEINER 2000) have recently suggested that striatal input to SNR leads to local increases in dendritically released DA in the regions of SNR neurons that were inhibited by the striatal input. This local DA increase acts to increase signal to noise ratio in striatonigral transmission because it (1) increases (via D_1 presynaptic receptors) GABA release from the active striatal terminals in the regions of inhibited SNR cells but not in other SNR regions innervated by the active striatal neurons, and (2) excites (via D_2 postsynaptic receptors) SNR neurons in the vicinity of the inhibited SNR cells, thus increasing the contrast between the inhibited SNR cells which transmit striatal information and other SNR cells. Since one of the factors increasing dendritic release is the switch of DA cells from the single spiking mode to the bursting mode, loss of the regulation of DA neurons burst firing by the limbic striatum will lead to an unregulated dendritic release and thus loss of the spatially restricted increase in dendritic release. Consequently, the sharpening of striatal neurotransmission will be lost, disrupting associative striatal throughput via SNR to the associative prefrontal cortex and the superior colliculus, thus impairing the selection and execution of motor plans as well as the allocation of attention.

Loss of an active goal representation in the limbic striatum may also lead, via reduced activity of the open limbic route, to a disintegration of the modulating effect of the limbic striatum on the selection and execution of motor programs in the associative split circuit. As a consequence, behavior will be triggered by any event which can activate motor programs in the associative striatum, including stimuli or thoughts that have established strong stimulus-response associations in the associative striatum as well as motor or cognitive components of well-learned motor programs. Behavior will be either stimulus-bound or disorganized, as each of the executed elements may lead to the next element in the same motor program or to elements of other motor programs. These would be reflected in a wide range of stereotypic behaviors, ranging from simple motor acts such as pacing and rocking, to more complex ritualistic behaviors documented in schizophrenic patients; disorganized speech or loosening of associations, i.e., slipping off the track from one topic to another; as well as disorganization of any form of goal-directed behavior.

It should be pointed out that at the behavioral level, it would be difficult to distinguish between abnormal behavior which results from repeated switching between low-energy goals and that resulting from reduced modulation of behavior by goals, because both would be reflected in a failure to persist in goal-directed behaviors. We presume, however, that the two deficits would be accompanied by different subjective experience. Thus, premature switching of goals should still allow subjective perception that behavior is related to one's goals, whereas a dissociation between goals and behavior may lead to feelings of loss of control, and even to a feeling of alienation towards one's behavior.

β) Abnormal Phasic DA Release: Learning and Switching

Normally, phasic DA, occurring following the encounter of unexpected significant events (external or internal), facilitates the acquisition of new goals as well as switching between already established goals. The dysregulation of phasic DA will lead to exaggerated phasic DA release in response to stimuli which normally lead to phasic DA release, such as novel or unexpected reinforcing stimuli (exaggerated phasic release), as well as to phasic DA release in response to stimuli which normally would not lead to such release, such as weak novel stimuli, repeatedly presented stimuli, and predictable reinforcing stimuli (inappropriate phasic release).

Since phasic increase of striatal DA facilitates switching between goals, increased phasic DA release will lead to switching following the occurrence of events which are not relevant to the current goal and which normally would not have led to phasic DA release, as well as following the expected occurrence of goal-related events (i.e., achieving a sub-goal), which although expected will lead to phasic DA release. Both will lead to repeated premature abortion of current goals and re-selection of different goals, leading to high distractibility. Importantly, the patient may be distracted not only by task-irrelevant stimuli, as is widely documented, but also following the completion of each step of the goal-directed behavioral sequence. This should lead to profound difficulties in persisting in any goal-directed behavior. Moreover, since DA inhibits neurons of the indirect pathway, increased phasic DA release may lead to abnormal suppression of indirect pathway neurons, including those encoding the suppression of inappropriate goals. Thus, the patient will not only be highly distractible, but will also be more likely to switch to inappropriate goals, leading to inappropriate or bizarre behaviors.

In addition to disrupting goal-directed behavior, increased phasic DA release may charge events with a particular intensity and give rise to spurious sense of significance (ANSCOMBE 1987) at the experiential level; moreover, phasic DA release in response to task-irrelevant and task-relevant events will give rise to different subjective interpretations/experience. The former will lead to the attribution of heightened significance to insignificant stimuli, resulting in the widely documented attraction of schizophrenics to irrelevant stimuli (KRAEPELIN 1919; ZEC 1995), whereas the latter will lead to attribution of heightened significance to one's own actions. Thus, whereas normally the attainment of an expected goal as a result of performing the routine goal-directed behavior is not accompanied by changes in striatal DA levels and thus remains "unnoticed," abnormal phasic DA increase following the attainment of a goal after performing the relevant goal-directed behavior, may lead to inappropriate feelings of achievement, excitement or surprise, or an excessive sense of personal agency in schizophrenic patients. This may contribute to grandiosity delusions. A similar misattribution of significance/achievement to other people's routine actions may contribute to suspiciousness, hostility, and paranoid delusions.

The difficulties in performing goal-directed behaviors resulting from inappropriate switching between goals in the limbic striatum may be compounded by the consequences of exaggerated and inappropriate phasic DA release in the associative and motor striatum, namely, over-switching between motor programs and between components of motor programs. The over-responsiveness of the DA system may also lead to excessive triggering of motor programs by current stimuli and thoughts so that motor programs will be under less control by goals selected in the limbic striatum (normally exerted via the open limbic route). This may lead to a gross disorganization in the performance of activities of daily living such as organizing meals or maintaining hygiene as well as disorganized speech or loosening of associations. At the experiential level, dissociation between goals and behavior will lead to feelings of loss of control and alienation towards one's behavior. In the extreme case, delusions of alien control, i.e., attribution of one's actions to an external agent, may appear.

During psychotic episodes, increased phasic DA release is likely to lead to periods of increased tonic DA (MOORE et al. 1999). Under these conditions selected goals will be over-activated, leading to a disproportional effort in attaining them. In addition, since DA has a focusing effect, increased tonic DA will lead to a reduction in the number of alternative goals which are activated enough to be selected, leading to reduced variability of behavioral output (LYON and ROBBINS 1975), so that the patient will alternate between relatively few behaviors, each executed with great effort. Since the degree of activation depends on the degree a specific goal has been learned in the current context, only well-learned goals will be activated enough to be selected. Moreover, since DA acting on D_1 receptors facilitates the activity of already active neurons, and since the normal mechanism by which already active goals have less chances of being reselected, is likely to be overwhelmed by the high DA level, the current goal will be not only highly activated but also hard to replace. These may be further exaggerated by DA inhibitory effects on neurons of the indirect pathway that are responsible for suppressing or terminating the current goal as well as for suppressing inappropriate goals. Therefore, prolonged periods of increased phasic DA may lead to highly motivated, inappropriate, stereotypic, and perseverative behavior.

Increased tonic DA levels may result in an additional problem. Since striatal neurons of the direct pathway are highly active, their biasing effect on the limbic prefrontal cortex is expected to be abnormally high. Under such conditions, the biasing effect exerted on the limbic prefrontal cortex by other cortical regions might not be sufficient to counteract the strong striatal biasing effect, resulting in great difficulties in resisting the performance of routine goal-directed behavior. This may be reflected in a high rate of "capture errors," i.e., performing the routine behavior instead of a behavior one intended to, and may be experienced as being forcefully driven to perform specific behaviors in spite of intentions to behave differently. At the extreme the patient may feel as if he has no free will, or as if his free will has been overtaken by some strong and alien force.

In addition to facilitating switching, phasic increase in striatal DA levels governs striatal learning. Therefore, increased phasic DA release will lead to a rapid learning of new goals, motor programs, and motor acts, as well as to over-learning of routine goals, motor programs, and motor acts in the limbic, associative, and motor circuits, respectively. Moreover, the inappropriate phasic DA release to incidental/insignificant stimuli and to predicted reinforcers will lead to inappropriate learning, i.e., to the establishment of goals with odd or bizarre content, and to the acquisition of superstitious behaviors that are performed as a part of a goal-directed sequence, although they are not necessary for attaining the goal. During a psychotic episode, this will be reflected in the development of highly energized bizarre behaviors that gradually replace previous behaviors. At the experiential level, abnormally rapid and redundant associations, seeing relationships where they do not exist, and excessive perception of a correspondence between one's goals and chance occurrences of external events, may lead to magical thinking, ideas of references, exaggerated inferential thinking (delusions), and the breaking of boundaries between the inner and the outer worlds.

Even more critically, the faulty learning occurring during each psychotic episode will increasingly broaden the patient's repertoire of inadequate and bizarre goals and behaviors. This may account for the findings that considerable proportion of patients experience some progression of their illness, with recurrent psychotic episodes resulting in lower levels of recovery and higher levels of residual symptoms, and that the longer the period of psychosis experienced prior to receiving APD treatment, the poorer the treatment response and the outcome (HUBER et al. 1980; MAY et al. 1981; WYATT 1991; LOEBEL et al. 1992; McGLASHAN and FENTON 1993; LIEBERMAN et al. 1996, 1997; McGLASHAN 1999). Likewise, cumulative defective learning experience is consistent with findings that assertive rehabilitation efforts appear to improve long-term outcome (DAVIDSON and McGLASHAN 1997).

d) Summary: Phasic and Tonic DA Dysregulation and Schizophrenia Symptoms

As may be evident from the discussion thus far, both abnormally low tonic DA and high phasic DA are hypothesized to lead to similar deficits, including excessive and immature switching, perseveration, disorganization, and a dissociation between goals and behavior. The two states are suggested to differ in what may be termed the "energy level" accompanying the observed deficit: low energy with low tonic DA and high energy with increased phasic DA. It is precisely such a difference in energy level that seems to distinguish productive from deficit symptoms, and indeed may be discerned in the symptom description of DSM-IV. Thus, positive symptoms are said to include "grossly disorganized behavior: problems may be noted in any form of goal-directed behavior leading to difficulties in performing activities of daily living such as organizing meals or maintaining hygiene," whereas under negative symptoms, the description appears as "Avolition: is characterized by inability to initiate and persist in goal-directed activities. The person may sit for long periods of time and show little interest in participating in work or social activities." Likewise, positive symptoms include "Catatonic motor behaviors ... which range from extreme degree of catatonic stupor to purposeless and unstimulated excessive motor activity," while negative symptoms include "Abnormal psychomotor activity, e.g., pacing, rocking or apathetic immobility, odd mannerisms, posturing, ritualistic or stereotyped behavior"; and positive symptoms include "loosening of associations, disorganized speech," while negative symptoms include "problems with focusing attention, distractibility." In general, boundary problems in classification and diagnosis of schizophrenia symptoms are widely acknowledged (STRAUSS et al. 1974; FRITH 1987; ANDREASEN 1982; BILDER et al. 1985; CORNBLATT et al. 1985; FRITH 1987; CARPENTER et al. 1988; CARPENTER and BUCHANAN 1989; KAY 1990; LYON 1991; ROBBINS 1991; TANDON and GREDEN 1991; ANDREASEN et al. 1995; TANDON 1995; CRIDER 1997). As pointed out by LYON (1991), one of the reasons for such problems may stem from an excessive focus on the "content" of the aberrant behaviors rather than on its "structure"; Indeed, our account resonates with that of LYON (1991) who suggested that schizophrenia symptoms may be grouped under four major types of behavioral change: switching, focusing, fragmentation, and stereotypy.

Differences in "energy level" will be reflected in the accompanying subjective (and therefore communicated) experience. In the low energy state, the patient will primarily feel unenergetic, unable to carry out his intentions and plans, passive, apathetic, withdrawn, and displaced. In the high energy state, the patient may feel highly energetic, overwhelmed with a sense of personal significance, meaning and control, or controlled by great powers, culminating in delusions. As summarized by ANSCOMBE (1987), "some patients describe an animated world full of significance while others describe experience that is empty and null" (p. 242).

Energy level may also be reflected in the severity of the symptoms. In particular, in low energy the processes involved are relatively slow and weak, enabling the supervisory systems to correct at least some of the deviance; in high energy, the supervisory systems, which are by themselves malfunctional, collapse, which will be reflected in a more extreme behavioral disorganization.

The most devastating consequence of either abnormally low or abnormally high DA in the limbic striatum is the splitting between goals and behavior. In both cases, the patients become disconnected from the motivational and intentional origins of their behavior, cannot give coherence to their behavior, loose sense of control, and increasingly become observers of their behavior rather than its initiators. Moreover, it is the "routineness" of one's goals and actions, i.e., the rapid and efficient choice of well-known courses of action in different situations, and the correspondence between purpose and outcomes which render one's behavior coherent to oneself and to others and link the person's inner world with the objective outer world. One can say that I know myself because I am familiar with the actions I take in different situations. In addition, since most adult individuals belonging to the same class, culture, etc., share many routine goals and actions, this ensures social coherence and approval. Repeated activation of goals and actions that lack routineness and coherence and are situation-inadequate may lead to a loss of sense of self, depersonalization, disturbances of ego and identity, perception of the outside world as alien and uncontrollable/incomprehensible, as well as to social alienation. These should lead to attempts to explain such an incoherent world, and a delusional framework might be just such an attempt (e.g., JASPERS et al. 1959; BOWERS 1974; MAHER 1974; MILLER 1984; ANSCOMBE 1987; SHANER 1999).

Finally, a note is in order with regard to the most prominent symptom of psychosis, hallucinations (BREIER and BERG 1999; EPSTEIN et al. 1999), which are apparently associated with DA hyperfunction since they are most efficiently treated by D_2 antagonists (BREIER and BERG 1999). While the present model can accommodate the development of delusions, it does not relate at all to hallucinations. However, as pointed out by EPSTEIN et al. (1999), in schizophrenia hallucinations are related to concurrent delusions, and both were shown by these authors to be associated with altered blood flow in the ventral striatum, medial temporal, and frontal regions, i.e., in the limbic circuit. Indeed, EPSTEIN et al. suggested that hallucinations and delusions result from disrupted balance between frontal and temporal inputs to the ventral striatum, which is normally used for maintaining a coherent stream of goal-directed behavior, and that this imbalance leads to aberrant representations of the external world. Thus, it is possible that the disruption of routine goal-directed behavior stemming from distorted processing in the limbic split circuit as described here could lead also to hallucinations.

In sum, we have suggested that dysregulation of mesolimbic DA in schizophrenia culminates in a dissociation between the activity of the limbic, associative, and motor basal ganglia–thalamocortical split circuits. This dissociation may provide the neurophysiological basis for the "splitting of mental faculties" which is conveyed in BLEULER's (1911) name schizophrenia, and has retained a central position in leading recent formulations of the psychopathology of this disorder (e.g., FRITH 1992; ZEC 1995; ANDREASEN et al. 1996, 1999; GRAYBIEL 1997; FRISTON 1998). In addition, the present proposition, that schizophrenia symptomatology results from the effects of DA dysregulation on both the direct and indirect pathways, implies that the full understanding of the action of APDs, as well as the development of new drugs, should take into account their effects on both pathways. An ideal antipsychotic treatment should normalize the functioning of both pathways.

References

Abi-Dargham A, Gil R, Krystal J, Baldwin RM, Seibyl JP, Bowers M, van Dyck CH, Charney DS, Innis RB, Laruelle M (1998) Increased striatal dopamine transmission in schizophrenia: confirmation in a second cohort. Am J Psychiatry 155: 761–767

Akil M, Pierri JN, Whitehead RE, Edgar CL, Mohila C, Sampson AR, Lewis DA (1999) Lamina-specific alterations in the dopamine innervation of the prefrontal cortex in schizophrenic subjects. Am J Psychiatry 156:1580–1589

Albin RL, Young AB, Penney JB (1989) The functional anatomy of basal ganglia disorders. Trends Neurosci 12:366–375

Alexander GE, Crutcher MD (1990) Functional architecture of basal ganglia circuits: neural substrates of parallel processing. Trends Neurosci 13:266–271

Alexander GE, Crutcher MD, Delong MR (1990) Basal ganglia–thalamocortical circuits: parallel substrates for motor, oculomotor, "prefrontal" and "limbic" functions. Prog Brain Res 85:119–146

Alexander GE, Delong MR, Strick PL (1986) Parallel organization of functionally segregated circuits linking basal ganglia and cortex. Ann Rev Neurosci 9:357–381

American Psychiatric Association (1994) Diagnostic and statistical manual of mental disorders. 4th edition. Washington, DC. American Psychiatric Press.

Andreasen NC (1982) Negative symptoms in schizophrenia. Definition and reliability. Arch Gen Psychiatry 39:784–788

Andreasen NC, Arndt S, Alliger R, Miller D, Flaum M (1995) Symptoms of schizophrenia. Methods, meanings, and mechanisms. Arch Gen Psychiatry 52:341–351

Andreasen NC, Nopoulos P, O'Leary DS, Miller DD, Wassink T, Flaum M (1999) Defining the phenotype of schizophrenia: cognitive dysmetria and its neural mechanisms. Biol Psychiatry 46:908–920

Andreasen NC, O'Leary DS, Cizadlo T, Arndt S, Rezai K, Ponto LL, Watkins GL, Hichwa RD (1996) Schizophrenia and cognitive dysmetria: a positron-emission tomography study of dysfunctional prefrontal-thalamic-cerebellar circuitry. Proc Natl Acad Sci USA 93:9985–9990

Andreasen NC, Rezai K, Alliger R, Swayze VWd, Flaum M, Kirchner P, Cohen G, O'Leary DS (1992) Hypofrontality in neuroleptic-naive patients and in patients with chronic schizophrenia. Assessment with xenon 133 single-photon emission computed tomography and the Tower of London. Arch Gen Psychiatry 49:943–958

Angrist B, Peselow E, Rubinstein M, Corwin J, Rotrosen J (1982) Partial improvement in negative schizophrenic symptoms after amphetamine. Psychopharmacology 78: 128–130

Angrist B, Peselow E, Rubinstein M, Wolkin A, Rotrosen J (1985) Amphetamine response and relapse risk after depot neuroleptic discontinuation. Psychopharmacology 85:277–283

Angrist B, Rotrosen J, Gershon S (1980) Responses to apomorphine, amphetamine, and neuroleptics in schizophrenic subjects. Psychopharmacology (Berl) 67:31–38

Angrist B, Sathananthan G, Wilk S, Gershon S (1974) Amphetamine psychosis: behavioral and biochemical aspects. J Psychiatr Res 11:13–23

Angrist BM, Shopsin B, Gershon S (1971) Comparative psychotomimetic effects of stereoisomers of amphetamine. Nature 234:152–153

Annett LE, McGregor A, Robbins TW (1989) The effects of ibotenic acid lesions of the nucleus accumbens on spatial learning and extinction in the rat. Behavioural-Brain-Research 31:231–242

Anscombe F (1987) The disorder of consciousness in schizophrenia. Schiz Bull 13: 241–260

Aosaki T, Graybiel AM, Kimura M (1994) Effect of the nigrostriatal dopamine system on acquired neural responses in the striatum of behaving monkeys. Science 265: 412–415

Arnold SE, Trojanowski JQ (1996) Recent advances in defining the neuropathology of schizophrenia. Acta Neuropathol (Berl) 92:217–231

Arnt J, Skarsfeldt T (1998) Do novel antipsychotics have similar pharmacological characteristics? A review of the evidence. Neuropsychopharmacology 18:63–101

Arnt J, Skarsfeldt T, Hyttel J (1997) Differentiation of classical and novel antipsychotics using animal models. Int Clin Psychopharmacol 12 [Suppl 1]:S9–S17

Belmaker RH, Wald D (1977) Haloperidol in normals [letter]. Br J Psychiatry 131: 222–223

Benes FM (1999) Alterations of neural circuitry within layer II of anterior cingulate cortex in schizophrenia. J Psychiatr Res 33:511–512

Beninger RJ (1983) The role of dopamine in locomotor activity and learning. Brain Res 287:173–196

Beninger RJ, Miller R (1998) Dopamine D1-like receptors and reward-related incentive learning. Neurosci Biobehav Rev 22:335–345

Berendse HW, Galis-de Graaf Y, Groenewegen HJ (1992) Topographical organization and relationship with ventral striatal compartments of prefrontal corticostriatal projections in the rat. J Comp Neurol 316:314–347

Berman KF, Weinberger DR (1990) The prefrontal cortex in schizophrenia and other neuropsychiatric diseases: in vivo physiological correlates of cognitive deficits. Prog Brain Res 85:521–536

Berridge KC, Robinson TE (1998) What is the role of dopamine in reward: hedonic impact, reward learning, or incentive salience? Brain Res Rev 28:309–369

Berridge KC, Whishaw IQ (1992) Cortex, striatum and cerebellum – control of serial order in a grooming sequence. Exp Brain Res 90:275–290

Bertolino A, Breier A, Callicott JH, Adler C, Mattay VS, Shapiro M, Frank JA, Pickar D, Weinberger DR (2000) The relationship between dorsolateral prefrontal neuronal N-acetylaspartate and evoked release of striatal dopamine in schizophrenia. Neuropsychopharmacology 22:125–132

Beuger M, van Kammen DP, Kelley ME, Yao J (1996) Dopamine turnover in schizophrenia before and after haloperidol withdrawal. CSF, plasma, and urine studies. Neuropsychopharmacology 15:75–86

Bilder RM, Mukherjee S, Rieder RO, Pandurangi AK (1985) Symptomatic and neuropsychological components of defect states. Schizophr Bull 11:409–419

Bird ED, Crow TJ, Iversen LL, Longden A, Mackay AV, Riley GJ, Spokes EG (1979b) Dopamine and homovanillic acid concentrations in the post-mortem brain in schizophrenia [proceedings]. J Physiol (Lond) 293:36P–37P

Bird ED, Spokes EG, Barnes J, MacKay AV, Iversen LL, Shepherd M (1977) Increased brain dopamine and reduced glutamic acid decarboxylase and choline acetyl transferase activity in schizophrenia and related psychoses. Lancet 2:1157–1158

Bird ED, Spokes EG, Iversen LL (1979a) Brain norepinephrine and dopamine in schizophrenia [letter]. Science 204:93–94

Bird ED, Spokes EG, Iversen LL (1979c) Increased dopamine concentration in limbic areas of brain from patients dying with schizophrenia. Brain 102:347–360

Bjerkenstedt L, Edman G, Hagenfeldt L, Sedvall G, Wiesel FA (1985) Plasma amino acids in relation to cerebrospinal fluid monoamine metabolites in schizophrenic patients and healthy controls. Br J Psychiatry 147:276–282

Blackburn JR, Pfaus JG, Phillips AG (1992) Dopamine functions in appetitive and defensive behaviours. Prog Neurobiol 39:247–279

Bleuler E (1911) Dementia Praecox or the Group of Schizophrenias. International Universities Press, New York

Bogerts B (1991) The neuropathology of schizophrenia: Pathophysiological and neurodevelopmental implications. In: Mednick SA, Cannon TD, Barr CE, Lyon M (eds) Fetal neural development and adult schizophrenia. Cambridge University Press, Cambridge, pp 153–173

Bogerts B (1993) Recent advances in the neuropathology of schizophrenia. Schizophr Bull 19:431–445

Bogerts B, Meertz E, Schonfeldt-Bausch R (1985) Basal ganglia and limbic system pathology in schizophrenia. A morphometric study of brain volume and shrinkage. Arch Gen Psychiatry 42:784–791

Bowers MB, Jr (1974) Central dopamine turnover in schizophrenic syndromes. Arch Gen Psychiatry 31:50–54

Braver TS, Barch DM, Cohen JD (1999) Cognition and control in schizophrenia: a computational model of dopamine and prefrontal function. Biol Psychiatry 46:312–328

Breier A, Berg PH (1999) The psychosis of schizophrenia: prevalence, response to atypical antipsychotics, and prediction of outcome. Biol Psychiatry 46:361–364

Breier A, Su TP, Saunders R, Carson RE, Kolachana BS, de Bartolomeis A, Weinberger DR, Weisenfeld N, Malhotra AK, Eckelman WC, Pickar D (1997) Schizophrenia is associated with elevated amphetamine-induced synaptic dopamine concentrations: evidence from a novel positron emission tomography method. Proc Natl Acad Sci USA 94:2569–2574

Brog JS, Salyapongse A, Deutch AY, Zahm DS (1993) The patterns of afferent innervation of the core and shell in the "accumbens" part of the rat ventral striatum: immunohistochemical detection of retrogradely transported fluoro-gold. J Comp Neurol 338:255–278

Brown J, Bullock D, Grossberg S (1999) How the basal ganglia use parallel excitatory and inhibitory learning pathways to selectively respond to unexpected rewarding cues. J Neurosci 19:10502–10511

Buchanan RW, Carpenter WT (1994) Domains of psychopathology – an approach to the reduction of heterogeneity in schizophrenia. J Nerv Ment Dis 182:193–204

Buchsbaum MS, Nuechterlein KH, Haier RJ, Wu J, Sicotte N, Hazlett E, Asarnow R, Potkin S, Guich S (1990) Glucose metabolic rate in normals and schizophrenics during the Continuous Performance Test assessed by positron emission tomography. Br J Psychiatry 156:216–227

Bunney WE, Bunney BG (2000) Evidence for a compromised dorsolateral prefrontal cortical parallel circuit in schizophrenia. Brain Res Rev 31:138–146

Burns LH, Annett L, Kelley AE, Everitt BJ, Robbins TW (1996) Effects of lesions to amygdala, ventral subiculum, medial prefrontal cortex, and nucleus accumbens on the reaction to novelty: implication for limbic-striatal interactions. Behav Neurosci 110:60–73

Burt DR, Creese I, Snyder SH (1977) Antischizophrenic drugs: chronic treatment elevates dopamine receptor binding in brain. Science 196:326–328

Cador M, Robbins TW, Everitt BJ (1989) Involvement of the amygdala in stimulus-reward associations: interaction with the ventral striatum. Neuroscience 30:77–86

Cador M, Taylor JR, Robbins TW (1991) Potentiation of the effects of reward-related stimuli by dopaminergic- dependent mechanisms in the nucleus accumbens. Psychopharmacology 104:377–385

Calabresi P, Maj R, Pisani A, Mercuri NB, Bernardi G (1992) Long-term synaptic depression in the striatum: physiological and pharmacological characterization. J Neurosci 12:4224–4233

Calabresi P, Pisani A, Mercuri NB, Bernardi G (1996) The corticostriatal projection: from synaptic plasticity to dysfunctions of the basal ganglia. Trends Neurosci 19:19–24

Carlsson A (1988) The current status of the dopamine hypothesis of schizophrenia. Neuropsychopharmacology 1:179–186

Carlsson A, Lindqvist M (1963) Effect of chlorpromazine or haloperidol on formation of 3-methoxytyramine and normetanephrine in mouse brain. Acta Pharmacologica et Toxicologica 20:140–144

Carlsson A, Waters N, Carlsson ML (1999) Neurotransmitter interactions in schizophrenia–therapeutic implications. Biol Psychiatry 46:1388–1395

Carlsson M, Carlsson A (1990) Interactions between glutamatergic and monoaminergic systems within the basal ganglia—implications for schizophrenia and Parkinson's disease. Trends Neurosci 13:272–276

Carpenter WT, Jr, Buchanan RW (1989) Domains of psychopathology relevant to the study of etiology and treatment of schizophrenia. In: Shulz SC, Tamminga CT (eds) Schizophrenia: Scientific progress. Oxford University press, New York, pp 13–22

Carpenter WT, Jr, Heinrichs DW, Wagman AM (1988) Deficit and nondeficit forms of schizophrenia: the concept. Am J Psychiatry 145:578–583

Carpenter WT, Kirkpatrick B, Buchanan RW (1999) Schizophrenia: syndromes and diseases. J Psychiatr Res 33:473–475

Carter CJ, Pyccock CJ (1980) Behavioral and biochemical effects of dopamine and noradrenaline depletion within the medial prefrontal cortex of the rat. Brain Res 192:163–176

Cepeda C, Buchwald NA, Levine MS (1993) Neuromodulatory actions of dopamine in the neostriatum are dependent upon the excitatory amino acid receptor subtypes activated. Proc Natl Acad Sci USA 90:9576–9580

Cepeda C, Chandler SH, Shumate LW, Levine MS (1995) Persistent Na+ conductance in medium-sized neostriatal neurons: characterization using infrared videomicroscopy and whole cell patch- clamp recordings. J Neurophysiol 74:1343–1348

Cepeda C, Colwell CS, Itri JN, Chandler SH, Levine MS (1998) Dopaminergic modulation of NMDA-induced whole cell currents in neostriatal neurons in slices: contribution of calcium conductances. J Neurophysiol 79:82–94

Charpier S, Deniau JM (1997) In vivo activity-dependent plasticity at cortico-striatal connections: evidence for physiological long-term potentiation. Proc Natl Acad Sci USA 94:7036–7040

Chatterjee A, Chakos M, Koreen A, Geisler S, Sheitman B, Woerner M, Kane JM, Alvir J, Lieberman JA (1995) Prevalence and clinical correlates of extrapyramidal signs and spontaneous dyskinesia in never-medicated schizophrenic patients. Am J Psychiatry 152:1724–1729

Cohen JD, Barch DM, Carter C, Servan-Schreiber D (1999) Context-processing deficits in schizophrenia: converging evidence from three theoretically motivated cognitive tasks. J Abnorm Psychol 108:120–133

Cohen JD, Braver TS, O'Reilly RC (1996) A computational approach to prefrontal cortex, cognitive control and schizophrenia: recent developments and current challenges. Philos Trans R Soc Lond B Biol Sci 351:1515–1527

Cohen JD, Servan-Schreiber D (1992) Context, cortex, and dopamine: a connectionist approach to behavior and biology in schizophrenia. Psychol Rev 99:45–77

Cohen JD, Servan-Schreiber D (1993) A theory of dopamine function and its role in cognitive deficits in schizophrenia. Schizophr Bull 19:85–104

Cole BJ, Robbins TW (1987) Amphetamine impairs the discriminative performance of rats with dorsal noradrenergic bundle lesions on a 5-choice serial reaction time task: new evidence for central dopaminergic-noradrenergic interactions. Psychopharmacology 91:458–466

Cole BJ, Robbins TW (1989) Effects of 6-hydroxydopamine lesions of the nucleus accumbens septi on performance of a 5-choice serial reaction time task in rats: implications for theories of selective attention and arousal. Behav Brain Res 33:165–179

Cools AR (1980) Role of the neostriatal dopaminergic activity in sequencing and selecting behavioral strategies: facilitation of processes involved in selecting the best strategy in a stressful situation. Behav Brain Res 1:361–378

Cornblatt BA, Lezenweger MF, Dworkin RH, Erlenmeyer-Kimling L (1985) Positive and negative schizophrenic symptoms, attention, and information processing. Schiz Bull 11:397–408

Costall B, Naylor RJ, Cannon JG, Lee T (1977) Differentiation of the dopamine mechanisms mediating stereotyped behaviour and hyperactivity in the nucleus accumbens and caudate-putamen. J Pharm Pharmacol 29:337–342

Creese I, Burt DR, Snyder SH (1976) Dopamine receptor binding predicts clinical and pharmacological potencies of antischizophrenic drugs. Science 192:481–483

Creese I, Iversen SD (1975) The pharmacological and anatomical substrates of the amphetamine response in the rat. Brain Res 83:419–436

Crider A (1997) Perseveration in schizophrenia. Schizophr Bull 23:63–74

Cross AJ, Crow TJ, Owen F (1981) 3H-Flupenthixol binding in post-mortem brains of schizophrenics: evidence for a selective increase in dopamine D2 receptors. Psychopharmacology 74:122–124

Crow TJ (1980) Molecular pathology of schizophrenia: more than one disease process? Br Med J 280:66–68

Csernansky JG, Murphy GM, Faustman WO (1991) Limbic/mesolimbic connections and the pathogenesis of schizophrenia. Biol Psychiatry 30:383–400

Daniel DG, Breslin N, Clardy J, Goldberg T, Gold J, Kleinman J, Weinberger DR (1990) The effect of L-DOPA on negative symptoms: cognitive performance and regional cerebral blood flow in schizophrenia. Biol Psychiatry 27:118A

Dao-Castellana MH, Paillere-Martinot ML, Hantraye P, Attar-Levy D, Remy P, Crouzel C, Artiges E, Feline A, Syrota A, Martinot JL (1997) Presynaptic dopaminergic function in the striatum of schizophrenic patients. Schizophr Res 23:167–174

Davidson L, McGlashan TH (1997) The varied outcomes of schizophrenia. Can J Psychiatry 42:34–43

Davidson M, Keefe RS, Mohs RC, Siever LJ, Losonczy MF, Horvath TB, Davis KL (1987) L-dopa challenge and relapse in schizophrenia. Am J Psychiatry 144:934–938

Davis KL, Kahn RS, Ko G, Davidson M (1991) Dopamine in schizophrenia: a review and reconceptualization. Am J Psychiatry 148:1474–1486

DeLong MR, Crutcher MD, Georgopoulos AP (1985) Primate globus pallidus and subthalamic nucleus: functional organization. J Neurophysiol 53:530–543

DeLong MR, Georgopoulos AP (1981) Motor functions of the basal ganglia. In: Brookhart JM, Mountcastle VB, Brooks VB (eds) Hanbook of Physiology. American Physiological Society, Bethseda, pp 1017–1061

DeLong MR, Wichmann T (1993) Basal ganglia–thalamocortical circuits in parkinsonian signs. Clinical Neuroscience 1:18–26

Depue RA, Collins PF (1999) Neurobiology of the structure of personality: Dopamine, facilitation of incentive motivation, and extraversion. Behav Brain Sci 22:491–569

Deutch AY (1992) The regulation of subcortical dopamine systems by the prefrontal cortex: interactions of central dopamine systems and the pathogenesis of schizophrenia. J Neural Transm Suppl 36:61–89

Deutch AY, Clark WA, Roth RH (1990) Prefrontal cortical dopamine depletion enhances the responsiveness of mesolimbic dopamine neurons to stress. Brain Res 521:311–315

Deutch AY, Lee MC, Iadarola MJ (1992) Regionally specific effects of atypical antipsychotic drugs on striatal fos expression: The nucleus accumbens shell as a locus of antipsychotic action. Mol Cell Neurosci 3:332–341

Di Chiara G (1998) A motivational learning hypothesis of the role of mesolimbic dopamine in compulsive drug use. J Psychopharmacol 12:54–67

Dolan RJ, Fletcher PC (1997) Dissociating prefrontal and hippocampal function in episodic memory encoding. Nature 388:582–585

Dubertret C, Gorwood P, Ades J, Feingold J, Schwartz JC, Sokoloff P (1998) Meta-analysis of DRD3 gene and schizophrenia: ethnic heterogeneity and significant association in Caucasians. Am J Med Genet 81:318–322

Dworkin RH (1992) Affective deficits and social deficits in schizophrenia: what's what? Schizophr Bull 18:59–64

Epstein J, Stern E, Silbersweig D (1999) Mesolimbic activity associated with psychosis in schizophrenia. Symptom-specific PET studies. Ann N Y Acad Sci 877:562–574

Evenden JL, Robbins TW (1983) Increased response switching, perseveration and perseverative switching following d-amphetamine in the rat. Psychopharmacology 80:67–73

Everitt BJ, Morris KA, O'Brien A, Robbins TW (1991) The basolateral amygdala-ventral striatal system and conditioned place preference: further evidence of limbic-striatal interactions underlying reward-related processes. Neuroscience 42:1–18

Farde L (1997) Brain imaging of schizophrenia – the dopamine hypothesis. Schizophr Res 28:157–162

Farde L, Halldin C, Stone-Elander S, Sedvall G (1987) PET analysis of human dopamine receptor subtypes using 11C-SCH 23390 and 11C-raclopride. Psychopharmacology 92:278–284

Farde L, Nordstrom AL, Wiesel FA, Pauli S, Halldin C, Sedvall G (1992) Positron emission tomographic analysis of central D1 and D2 dopamine receptor occupancy in patients treated with classical neuroleptics and clozapine. Relation to extrapyramidal side effects. Arch Gen Psychiatry 49:538–544

Farde L, Nyberg S, Oxenstierna G, Nakashima Y, Halldin C, Ericsson B (1995) Positron emission tomography studies on D2 and 5-HT2 receptor binding in risperidone-treated schizophrenic patients. J Clin Psychopharmacol 15:19S-23S

Farde L, Weisel FA, Nordstroem AL, Sedvall G (1989) D1- and D2-dopamine receptor occupancy during treatment with conventional and atypical neuroleptics. Psychopharmacology 99:S28-S31

Farde L, Wiesel FA, Halldin C, Sedvall G (1988) Central D2-dopamine receptor occupancy in schizophrenic patients treated with antipsychotic drugs. Arch Gen Psychiatry 45:71–76

Farde L, Wiesel FA, Stone-Elander S, Halldin C, Nordstrom AL, Hall H, Sedvall G (1990) D2 dopamine receptors in neuroleptic-naive schizophrenic patients. A positron emission tomography study with [11C]raclopride. Arch Gen Psychiatry 47:213–219

Faustman WO, Hoff AL (1995) Effects of antipsychotic drugs on neuropsychological measures. In: Csernansky JG (ed) Handbook of Experimental Pharmacology: Antipsychotics. Springer, Berlin, pp 445–478

Finch DM (1996) Neurophysiology of converging synaptic inputs from the rat prefrontal cortex, amygdala, midline thalamus, and hippocampal formation onto single neurons of the caudate/putamen and nucleus accumbens. Hippocampus 6:495–512

Flaherty AW, Graybiel AM (1993) Two input systems for body representations in the primate striatal matrix: experimental evidence in the squirrel monkey. J Neurosci 13:1120–1137

Flaherty AW, Graybiel AM (1994) Input–output organization of the sensorimotor striatum in the squirrel monkey. J Neurosci 14:599–610

Fletcher PC, Frith CD, Grasby PM, Friston KJ, Dolan RJ (1996) Local and distributed effects of apomorphine on fronto-temporal function in acute unmedicated schizophrenia. J Neurosci 16:7055–7062

Friston KJ (1998) The disconnection hypothesis. Schizophr Res 30:115–125

Friston KJ (1999) Schizophrenia and the disconnection hypothesis. Acta Psychiatr Scand Suppl 395:68–79

Friston KJ, Frith CD (1995) Schizophrenia: a disconnection syndrome? Clin Neurosci 3:89–97

Friston KJ, Liddle PF, Frith CD, Hirsch SR, Frackowiak RS (1992) The left medial temporal region and schizophrenia. A PET study. Brain 115:367–382

Frith CD (1987) The positive and negative symptoms of schizophrenia reflect impairments in the perception and initiation of action. Psychol Med 17:631–648

Frith CD (1992) The cognitive neuropsychology of schizophrenia. Lawrence Erlbaum Associates Ltd, Hillsdale

Frith CD, Friston KJ, Herold S, Silbersweig D, Fletcher P, Cahill C, Dolan RJ, Frackowiak RS, Liddle PF (1995) Regional brain activity in chronic schizophrenic patients during the performance of a verbal fluency task. Br J Psychiatry 167: 343–349

Fuller TA, Russchen FT, Price JL (1987) Sources of presumptive glutamergic/aspartergic afferents to the rat ventral striatopallidal region. J Comp Neurol 258:317–338

Fuster JM (1990) Behavioral electrophysiology of the prefrontal cortex of the primate. Prog Brain Res 85:313–324

Garey LJ, Ong WY, Patel TS, Kanani M, Davis A, Hornstein C, Bauer M (1995) Reduction in dendritic spine number on cortical pyramidal neurons in schizophrenia. Soc Neurosci Abstr 21:237

Garris PA, Christensen JR, Rebec GV, Wightman RM (1997) Real-time measurement of electrically evoked extracellular dopamine in the striatum of freely moving rats. J Neurochem 68:152–161

Gerfen CR, Engber TM, Mahan LC, Susel Z, Chase TN, Monsma FJ, Jr, Sibley DR (1990) D1 and D2 dopamine receptor-regulated gene expression of striatonigral and striatopallidal neurons. Science 250:1429–1432

Gerfen CR, Wilson CJ (1996) The basal ganglia. In: Swanson LW, Bjorklumd A, Hokfelt T (eds) Handbook of chemical neuroanatomy. Elsevier Science BV, pp 371–468

Glantz LA, Lewis DA (1995) Assessment of spine density on layer III pyramidal cells in the prefrontal cortex of schizophrenic subjects. Soc Neurosci Abstr 21:239

Goldman-Rakic PS (1987) Circuitry of the prefrontal cortex and the regulation of behavior by representational memory. In: Planum F, Mountcastle VB (eds) Handbook of Physiology. American Physiological Society, Bethesda, pp 373–417

Goldman-Rakic PS (1999) The physiological approach: functional architecture of working memory and disordered cognition in schizophrenia. Biol Psychiatry 46:650–661

Gonon FG (1988) Nonlinear relationship between impulse flow and dopamine released by rat midbrain dopaminergic neurons as studied by in vivo electrochemistry. Neuroscience 24:19–28

Grace AA (1991) Phasic versus tonic dopamine release and the modulation of dopamine system responsivity: a hypothesis for the etiology of schizophrenia. Neuroscience 41:1–24

Grace AA (1993) Cortical regulation of subcortical dopamine systems and its possible relevance to schizophrenia. J Neural Transm Gen Sect 91:111–134

Grace AA (1995) The tonic/phasic model of dopamine system regulation: its relevance for understanding how stimulant abuse can alter basal ganglia function. Drug Alcohol Depend 37:111–129

Grace AA (2000) Gating of information flow within the limbic system and the pathophysiology of schizophrenia. Brain Res Rev 31:330–341

Grace AA, Bunney BS (1984) The control of firing pattern in nigral dopamine neurons: single spike firing. J Neurosci 4:2866–2876

Gray JA, Feldon J, Rawlins JNP, Hemsley DR, Smith AD (1991) The neuropsychology of schizophrenia. Behav Brain Sci 14:1–84

Graybiel AM (1997) The basal ganglia and cognitive pattern generators. Schizophr Bull 23:459–469

Graybiel AM (1998) The basal ganglia and chunking of action repertoires. Neurobiol Learn Mem 70:119–136

Graybiel AM, Aosaki T, Flaherty AW, Kimura M (1994) The basal ganglia and adaptive motor control. Science 265:1826–1831

Graybiel AM, Kimura M (1995) Adaptive neural networks in the basal ganglia. In: Houk JC, Davis JL, Beiser DG (eds) Models of information processing in the basal ganglia. MIT Press, Cambridge, pp 103–116

Groenewegen HJ, Berendse HW (1994) Anatomical relationships between the prefrontal cortex and the basal ganglia in the rat. In: Thierry A-M, Glowinski J, Goldman-Rakic P, Christen Y (eds) Motor and Cognitive Functions of the Prefrontal Cortex. Fondation IPSEN, Springer-Verlag, pp 51–77

Groenewegen HJ, Berendse HW, Meredith GE, Haber SN, Voorn P, Wolters JG, Lohman AHM (1991) Functional anatomy of the ventral, limbic system-innervated striatum. In: Wilner P, Scheel-Kruger J (eds) The mesolimbic dopamine system: from motivation to action. John Wiley, Chinchester, pp 19–59

Groenewegen HJ, Berendse HW, Wolters JG, Lohman AHM (1990) The anatomical relationship of the prefrontal cortex with the striatopallidal system, the thalamus and the amygdala: evidence for a parallel organization. Prog Brain Res 85:95–118

Groenewegen HJ, Vermeulen-Van der Zee E, Te Kortschot A, Witter MP (1987) Organization of the projections from the subiculum to the ventral striatum in the rat. A study using anterograde transport of Phaseolus vulgaris leucoagglutinin. Neuroscience 23:103–120

Groenewegen HJ, Wright CI, Beijer AV (1996) The nucleus accumbens: gateway for limbic structures to reach the motor system? Prog Brain Res 107:485–511

Groenewegen HJ, Wright CI, Uylings HB (1997) The anatomical relationships of the prefrontal cortex with limbic structures and the basal ganglia. J Psychopharmacol 11:99–106

Groves PM (1983) A theory of the functional organization of the neostriatum and the neostriatal control of voluntary movement. Brain Res Rev 5:109–132

Groves PM, Garcia-Munoz M, Linder JC, Manley MS, Martone ME, Young SJ (1995) Elements of the intrinsic organization and information processing in the neostriatum. In: Houk JC, Davis JL, Beiser DG (eds) Models of information processing in the basal ganglia. MIT Press, Cambridge, pp 51–96

Groves PM, Rebec GV (1976) Biochemistry and behavior: Some central actions of amphetamines and antipsychotic drugs. Ann Rev Psychol 27:91–127

Gur RE, Resnick SM, Alavi A, Gur RC, Caroff S, Dann R, Silver FL, Saykin AJ, Chawluk JB, Kushner M et al. (1987) Regional brain function in schizophrenia. I. A positron emission tomography study. Arch Gen Psychiatry 44:119–125

Hagberg G, Gefvert O, Lidstrom L, Bergstrom M, Langstrom B (1998) L-[beta-11C]DOPA utilization in schizophrenia. Neuroimage 7:A45

Harrison PJ (1995) On the neuropathology of schizophrenia and its dementia: neurodevelopmental, neurodegenerative, or both? Neurodegeneration 4:1–12

Harrison PJ (1999) The neuropathology of schizophrenia. A critical review of the data and their interpretation. Brain 122:593–624

Heinz A, Knable MB, Coppola R, Gorey JG, Jones DW, Lee KS, Weinberger DR (1998) Psychomotor slowing, negative symptoms and dopamine receptor availability – an IBZM SPECT study in neuroleptic-treated and drug-free schizophrenic. Schizophr Res 31:19–26

Hemsley DR (1994) A cognitive model for schizophrenia and its possible neural basis. Acta Psychiatr Scand Suppl 384:80–86

Hernandez-Lopez S, Bargas J, Surmeier DJ, Reyes A, Galarraga E (1997) D1 receptor activation enhances evoked discharge in neostriatal medium spiny neurons by modulating an L-type Ca2+ conductance. J Neurosci 17:3334–3342

Hietala J, Syvalahti E, Vilkman H, Vuorio K, Rakkolainen V, Bergman J, Haaparanta M, Solin O, Kuoppamaki M, Eronen E, Ruotsalainen U, Salokangas RK (1999) Depressive symptoms and presynaptic dopamine function in neuroleptic-naive schizophrenia. Schizophr Res 35:41–50

Hietala J, Syvalahti E, Vuorio K, Nagren K, Lehikoinen P, Ruotsalainen U, Rakkolainen V, Lehtinen V, Wegelius U (1994) Striatal D2 dopamine receptor characteristics in neuroleptic-naive schizophrenic patients studied with positron emission tomography. Arch Gen Psychiatry 51:116–123

Hietala J, Syvalahti E, Vuorio K, Rakkolainen V, Bergman J, Haaparanta M, Solin O, Kuoppamaki M, Kirvela O, Ruotsalainen U et al. (1995) Presynaptic dopamine function in striatum of neuroleptic-naive schizophrenic patients. Lancet 346: 1130–1131

Hikosaka O (1994) Role of basal ganglia in control of innate movements, learned behavior and cognition. In: Percheron G, McKenzie JS, Feger J (eds) The basal ganglia IV: New ideas and data on structure and function. Penum Press, New York, pp 589–596

Hogberg T, Ramsby S, Ogren SO, Norinder U (1987) New selective dopamine D-2 antagonists as antipsychotic agents. Pharmacological, chemical, structural and theoretical considerations. Acta Pharm Suec 24:289–328

Houk JC (1995) Information processing in modular circuits linking basal ganglia and cerebral cortex. In: Houk JC, Davis JL, Beiser DG (eds) Models of information processing in the basal ganglia. MIT Press, Cambridge, pp 3–9

Houk JC, Adams JL, Barto AG (1995) A model of how the basal ganglia generate and use reward signals that predict reinforcement. In: Houk JC, Davis JL, Beiser DG (eds) Models of information processing in the basal ganglia. MIT Press, Cambridge, pp 249–270

Houk JC, Wise SP (1995) Distributed modular architectures linking basal ganglia, cerebellum, and cerebral cortex: their role in planning and controlling action. Cerebral Cortex 2:95–110

Hsiao JK, Colison J, Bartko JJ, Doran AR, Konicki PE, Potter WZ, Pickar D (1993) Monoamine neurotransmitter interactions in drug-free and neuroleptic-treated schizophrenics. Arch Gen Psychiatry 50:606–614

Huber G, Gross G, Schuttler R, Linz M (1980) Longitudinal studies of schizophrenic patients. Schizophr Bull 6:592–605

Hyttel J, Larsen JJ, Christensen AV, Arnt J (1985) Receptor-binding profiles of neuroleptics. Psychopharmacology Suppl 2:9–18

Ikemoto S, Panksepp J (1999) The role of nucleus accumbens dopamine in motivated behavior: a unifying interpretation with special reference to reward-seeking. Brain Res Rev 31:6–41

Iversen SD (1984) Behavioral effects of manipulation of basal-ganglia neurotransmitters. In: Evered D, O'Connor M (eds) Functions of the Basal Ganglia. Pitman, London, pp 183–200

Jablensky A, Sartorius N, Ernberg G, Anker M, Korten A, Cooper JE, Day R, Bertelsen A (1992) Schizophrenia: manifestations, incidence and course in different cultures. A World Health Organization ten-country study. Psychol Med Monogr Suppl 20:1–97

Jahanshahi M, Frith CD (1998) Willed action and its impairments. Cognitive Neuropsychology 15:483–533

Jakob H, Beckmann H (1986) Prenatal developmental disturbances in the limbic allocortex in schizophrenics. J Neural Transm 65:303–326

Janowsky DS, Davis JM (1976) Methylphenidate, dextroamphetamine, and levamfetamine. Effects on schizophrenic symptoms. Arch Gen Psychiatry 33:304–308

Jaskiw GE, Karoum F, Freed WJ, Phillips I, Kleinman JE, Weinberger DR (1990) Effect of ibotenic acid lesions of the medial prefrontal cortex on amphetamine-induced locomotion and regional brain catecholamine concentrations in the rat. Brain Res 534:263–272

Jaskiw GE, Weinberger DR, Crawley JN (1991) Microinjection of apomorphine into the prefrontal cortex of the rat reduces dopamine metabolite concentrations in microdialysate from the caudate nucleus. Biol Psychiatry 29:703–706

Jaspers K (1959 translated 1963) General psychopathology. Manchester University Press, Manchester

Jenkins RB, Groh RH (1970) Mental symptoms in Parkinsonian patients treated with L-dopa. Lancet 2:177–179

Joel D, Weiner I (1994) The organization of the basal ganglia–thalamocortical circuits: open interconnected rather than closed segregated. Neuroscience 63:363–379

Joel D, Weiner I (1997) The connections of the primate subthalamic nucleus: indirect pathways and the open-interconnected scheme of basal ganglia–thalamocortical circuitry. Brain Res Rev 23:62–78

Joel D, Weiner I (1999) Striatal contention scheduling and the split circuit scheme of basal ganglia–thalamocortical circuitry: From anatomy to behaviour. In: Miller R, Wickens JR (eds) Conceptual Advances in Brain Research: Brain dynamics and the striatal complex. Harwood Academic Publishers, pp 209–236

Joel D, Weiner I (2000) The connections of the dopaminergic system with the striatum in rats and primates: An analysis with respect to the functional and compartmental organization of the striatum. Neuroscience 96:451–474

Joel D (2001) The open interconnected model of basal ganglia-thalamocortical circuitry and its relevance to the clinical syndrome of Huntington's disease. Movement Disorders 16:407–423

Johnson LR, Aylward RLM, Hussain Z, Totterdell S (1994) Input from the amygdala to the rat nucleus accumbens: its relationship with tyrosine hydroxylase immunoreactivity and identified neurons. Neuroscience 61:851–865

Johnstone EC, Crow TJ, Frith CD, Carney MW, Price JS (1978) Mechanism of the antipsychotic effect in the treatment of acute schizophrenia. Lancet 1:848–851

Jonsson EG, Nothen MM, Neidt H, Forslund K, Rylander G, Mattila-Evenden M, Asberg M, Propping P, Sedvall GC (1999) Association between a promoter polymorphism in the dopamine D2 receptor gene and schizophrenia. Schizophr Res 40:31–36

Joyce JN (1993) The dopamine hypothesis of schizophrenia: limbic interactions with serotonin and norepinephrine. Psychopharmacology (Berl) 112:S16–S34

Joyce JN, Lexow N, Bird E, Winokur A (1988) Organization of dopamine D1 and D2 receptors in human striatum: receptor autoradiographic studies in Huntington's disease and schizophrenia. Synapse 2:546–557

Joyce JN, Meador-Woodruff JH (1997) Linking the family of D2 receptors to neuronal circuits in human brain: insights into schizophrenia. Neuropsychopharmacology 16:375–384

Kahn RS, Davis KL (1995) New developments in dopamine and schizophrenia. In: Bloom FE, Kupfer DJ (eds) Psychopharmacology: the forth generation of progress. Raven Press, New York, pp 1193–1203

Kalivas PW (1993) Neurotransmitter regulation of dopamine neurons in the ventral tegmental area. Brain Res Rev 18:75–113

Kalivas PW, Churchill L, Klitenick MA (1993) The circuitry mediating the translation of motivational stimuli into adaptive motor responses. In: Kalivas PW, Barnes CD (eds) Limbic motor circuits and neuropsychiatry. CRC Press, Boca Raton, pp 237–287

Kane JM (1995) Current problems with the pharmacotherapy of schizophrenia. Clin Neuropharmacol 18:S154–S161

Kaneshima M, Higa T, Nakamoto H, Nagamine M (1997) An association study between the Cys311 variant of dopamine D2 receptor gene and schizophrenia in the Okinawan population. Psychiatry Clin Neurosci 51:379–381

Karreman M, Moghaddam B (1996) The prefrontal cortex regulates the basal release of dopamine in the limbic striatum: an effect mediated by ventral tegmental area. J Neurochem 66:589–98

Kawaguchi Y, Wilson CJ, Emson PC (1989) Intracellular recording of identified neostriatal patch and matrix spiny cells in a slice preparation preserving cortical inputs. J Neurophysiol 62:1052–1068

Kay SR (1990) Pyramidical model of schizophrenia. Schizophr Bull 16:537–545

Keefe KA, Salamone JD, Zigmond MJ, Stricker EM (1989) Paradoxical kinesia in parkinsonism is not caused by dopamine release. Studies in an animal model. Arch Neurol 46:1070–1075

Kelley AE, Domesick VB (1982) The distribution of the projection from the hippocampal formation to the nucleus accumbens in the rat: an anterograde- and retrograde-horseradish peroxidase study. Neuroscience 7:2321–2335

Kelly PH, Roberts DCS (1983) Effects of amphetamine and apomorphine on locomotor activity after 6-OHDA and electrolytic lesions of the nucleus accumbens septi. Pharmacol Biochem Behav 19:137–143

Kerwin RW (1994) The new atypical antipsychotics. A lack of extrapyramidal side-effects and new routes in schizophrenia research. Br J Psychiatry 164:141–148

Keshavan MS (1999) Development, disease and degeneration in schizophrenia: a unitary pathophysiological model. J Psychiatr Res 33:513–521

Keshavan MS, Hogarty GE (1999) Brain maturational processes and delayed onset in schizophrenia. Dev Psychopathol 11:525–543

Kiaytkin EA (1988) Functional properties of presumed dopamine-containing and other ventral tegmental area neurons in conscious rats. Int J Neurosci 42:21–43

Kimura M (1987) The putamen neuron: activity and the association of a sensory stimulus with movement in the monkey. In: Carpenter MB, Jayaraman A (eds) The basal ganglia II: Structure and function-current concepts. Planum Press, New York, pp 337–347

Kimura M (1995) Role of basal ganglia in behavioral learning. Neurosci Res 22:353–358

King D, Zigmund MJ, Finlay JM (1997) Effects of dopamine depletion in the medial prefrontal cortex on the stress-induced increase in extracellular dopamine in the nucleus accumbens core and shell. Neuroscience 77:141–153

Kinon BJ, Lieberman JA (1996) Mechanisms of action of atypical antipsychotic drugs: A critical analysis. Psychopharmacology 124:2–34

Kita H (1996) Two pathways between the cortex and the basal ganglia output nuclei and the globus pallidus. In: Ohye C, Kimura M, McKenzie JS (eds) The basal ganglia V. Plenum Press, New York, pp 72–94

Knable MB, Weinberger DR (1995) Are mental diseases brain diseases? The contribution of neuropathology to understanding of schizophrenic psychoses. Eur Arch Psychiatry Clin Neurosci 245:224–230

Kokkinidis L, Anisman H (1980) Amphetamine models of paranoid schizophrenia – an overview and elaboration of animal experimentation. Psychol Bull 88:551–579

Kolb B, Whishaw IQ (1990) Fundamentals of human neuropsychology. In: Atkinson RC, Lindzey G, Thompson RF (eds). W.H. Freeman and Company, New York

Kovelman JA, Scheibel AB (1984) A neurohistological correlate of schizophrenia. Biol Psychiatry 19:1601–1621

Kraepelin E (1919, 1971) Dementia praecox and paraphrenia. Robert E. Krieger Publishing Co. Inc., New York

Krayniak PF, Meibach RC, Siegel A (1981) A projection from the entorhinal cortex to the nucleus accumbens in the rat. Brain Res 209:427–431

Lahti RA, Roberts RC, Cochrane EV, Primus RJ, Gallager DW, Conley RR, Tamminga CA (1998) Direct determination of dopamine D4 receptors in normal and schizophrenic postmortem brain tissue: a [3H]NGD-94-1 study. Mol Psychiatry 3:528–533

Laruelle M (1998) Imaging dopamine transmission in schizophrenia. A review and meta-analysis. Q J Nucl Med 42:211–221

Laruelle M, Abi-Dargham A, Gil R, Kegeles L, Innis R (1999) Increased dopamine transmission in schizophrenia: relationship to illness phases. Biol Psychiatry 46: 56–72

Laruelle M, Abi-Dargham A, van Dyck C, Gil R, D'Souza DC, Krystal J, Seibyl J, Baldwin R, Innis R (2000) Dopamine and serotonin transporters in patients with schizophrenia: an imaging study with [(123)I]beta-CIT. Biol Psychiatry 47: 371–379

Laruelle M, Abi-Dargham A, van Dyck CH, Gil R, D'Souza CD, Erdos J, McCance E, Rosenblatt W, Fingado C, Zoghbi SS, Baldwin RM, Seibyl JP, Krystal JH, Charney DS, Innis RB (1996) Single photon emission computerized tomography imaging of amphetamine-induced dopamine release in drug-free schizophrenic subjects. Proc Natl Acad Sci USA 93:9235–9240

Laruelle M, D'Souza CD, Baldwin RM, Abi-Dargham A, Kanes SJ, Fingado CL, Seibyl JP, Zoghbi SS, Bowers MB, Jatlow P, Charney DS, Innis RB (1997) Imaging D2 receptor occupancy by endogenous dopamine in humans. Neuropsychopharmacology 17:162–174

Lavoie AM, Mizumori SJY (1992) Spatial, movement- and reward-sensitive discharge by medial nucleus accumbens. Soc Neurosci Abstr 19:707

Le Moal M (1995) Mesocorticolimbic dopaminergic neurons: Functional and regulatory roles. In: Bloom FE, Kupfer DJ (eds) Psychopharmacology: the forth generation of progress. Raven Press, New York, pp 283–294

Le Moal M, Simon H (1991) Mesocorticolimbic dopaminergic network: functional and regulatory roles. Physiol Rev 71:155–234

Lee T, Seeman P, Tourtellotte WW, Farley IJ, Hornykeiwicz O (1978) Binding of 3H-neuroleptics and 3H-apomorphine in schizophrenic brains. Nature 274:897–900

Levine DS, Leven SJ, Prueitt PS (1992) Integration, disintergration and the frontal lobes. In: Levine DS, Leven SJ (eds) Motivation, emotion, and goal direction in neural networks. Lawrence Erlbaum Associates, Publishers, Hilsdale, pp 301–335

Liddle PF (1987) Schizophrenic syndromes, cognitive performance and neurological dysfunction. Psychol Med 17:49–57

Liddle PF (1995) Inner connections within domain of dementia praecox: role of supervisory mental processes in schizophrenia. Eur Arch Psychiatry Clin Neurosci 245:210–215

Liddle PF, Barnes TR, Morris D, Haque S (1989) Three syndromes in chronic schizophrenia. Br J Psychiatry [Suppl]:119–122

Liddle PF, Friston KJ, Frith CD, Hirsch SR, Jones T, Frackowiak RS (1992) Patterns of cerebral blood flow in schizophrenia. Br J Psychiatry 160:179–186

Lidsky TI, Manetto C, Schneider JS (1985) A consideration of sensory factors involved in motor functions of the basal ganglia. Brain Res Rev 9:133–146

Lieberman JA, Kane JM, Alvir J (1987) Provocative tests with psychostimulant drugs in schizophrenia. Psychopharmacology 91:415–433

Lieberman JA, Kane JM, Gadaleta D, Brenner R, Lesser MS, Kinon B (1984) Methylphenidate challenge as a predictor of relapse in schizophrenia. Am J Psychiatry 141:633–638

Lieberman JA, Kinon BJ, Loebel AD (1990) Dopaminergic mechanisms in idiopathic and drug-induced psychoses. Schizophr Bull 16:97–110

Lieberman JA, Koreen AR, Chakos M, Sheitman B, Woerner M, Alvir JM, Bilder R (1996) Factors influencing treatment response and outcome of first-episode schizophrenia: implications for understanding the pathophysiology of schizophrenia. J Clin Psychiatry 57:5–9

Lieberman JA, Sheitman BB, Kinon BJ (1997) Neurochemical sensitization in the pathophysiology of schizophrenia: deficits and dysfunction in neuronal regulation and plasticity. Neuropsychopharmacology 17:205–29

Lindstrom LH (1985) Low HVA and normal 5HIAA CSF levels in drug-free schizophrenic patients compared to healthy volunteers: correlations to symptomatology and family history. Psychiatry Res 14:265–273

Lindstrom LH, Gefvert O, Hagberg G, Lundberg T, Bergstrom M, Hartvig P, Langstrom B (1999) Increased dopamine synthesis rate in medial prefrontal cortex and striatum in schizophrenia indicated by L-(beta-11C) DOPA and PET. Biol Psychiatry 46:681–688

Lipska BK, Jaskiw GE, Chrapusta S, Karoum F, Weinberger DR (1992) Ibotenic acid lesion of the ventral hippocampus differentially affects dopamine and its metabolites in the nucleus accumbens and prefrontal cortex in the rat. Brain Res 585:1–6

Lipska BK, Jaskiw GE, Weinberger DR (1993) Postpubertal emergence of hyperresponsiveness to stress and to amphetamine after neonatal excitotoxic hippocampal damage – a potential animal model of schizophrenia. Neuropsychopharmacology 9:67–75

Ljungberg T, Enquist M (1987) Disruptive effects of low doses of d-amphetamine on the ability of rats to organize behaviour into functional sequences. Psychopharmacology 93:146–151

Loebel AD, Lieberman JA, Alvir JM, Mayerhoff DI, Geisler SH, Szymanski SR (1992) Duration of psychosis and outcome in first-episode schizophrenia. Am J Psychiatry 149:1183–1188

Lopes da Silva FH, Arnolds DE, Neijt HC (1984) A functional link between the limbic cortex and ventral striatum: physiology of the subiculum accumbens pathway. Exp Brain Res 55:205–214

Louilot A, Le Moal M, Simon H (1989) Opposite influences of dopaminergic pathways to the prefrontal cortex or the septum on the dopaminergic transmission in the nucleus accumbens. An in vivo voltammetric study. Neuroscience 29:45–56

Louilot A, Simon H, Taghzouti K, Le Moal M (1985) Modulation of dopaminergic activity in the nucleus accumbens following facilitation or blockade of the dopaminergic transmission in the amygdala: a study by in vivo differential pulse voltammetry. Brain Res 346:141–145

Luria AR (1973) The frontal lobes and the regulation of behavior. In: Pribram KH, Luria AR (eds) Psychophysiology of the frontal lobes. Academic Press, New York, pp 3–26

Lynch MR, Carey RJ (1987) Environmental stimulation promotes recovery from haloperidol-induced extinction of open field behavior in rats. Psychopharmacology 92:206–209

Lyon M, Robbins TW (1975) The action of central nervous system stimulant drugs: a general theory concerning amphetamine effects Current developments in psychopharmacology. Spectrum, New York, pp 80–163

Mackay AV, Iversen LL, Rossor M, Spokes E, Bird E, Arregui A, Creese I, Synder SH (1982) Increased brain dopamine and dopamine receptors in schizophrenia. Arch Gen Psychiatry 39:991–997

Maher BA (1974) Delusional thinking and perceptual disorder. J Individ Psychol 30:98–113

Malhotra AK, Goldman D, Buchanan RW, Rooney W, Clifton A, Kosmidis MH, Breier A, Pickar D (1998) The dopamine D3 receptor (DRD3) Ser9Gly polymorphism and schizophrenia: a haplotype relative risk study and association with clozapine response. Mol Psychiatry 3:72–75

Marsden CD (1982) The mysterious motor function of the basal ganglia. Neurology 32:514–539

Marsden CD (1986) Movement disorders and the basal ganglia. Trends Neurosci 9:512–515

Marsden CD, Obeso JA (1994) The functions of the basal ganglia and the paradox of stereotaxic surgery in Parkinson's disease. Brain 117:877–897

Martinot JL, Huret JD, Peron-Magnan P, Mazoyer BM, Baron JC, Caillard V, Syrota A, Loo H (1989) Striatal D2 dopaminergic receptor status ascertained in vivo by positron emission tomography and 76Br-bromospiperone in untreated schizophrenics. Psychiatry Res 29:357–358

Martinot JL, Paillere-Martinot ML, Loc'h C, Hardy P, Poirier MF, Mazoyer B, Beaufils B, Maziere B, Allilaire JF, Syrota A (1991) The estimated density of D2 striatal receptors in schizophrenia. A study with positron emission tomography and 76Br-bromolisuride. Br J Psychiatry 158:346–350

Martinot JL, Peron-Magnan P, Huret JD, Mazoyer B, Baron JC, Boulenger JP, Loc'h C, Maziere B, Caillard V, Loo H et al. (1990) Striatal D2 dopaminergic receptors assessed with positron emission tomography and [76Br]bromospiperone in untreated schizophrenic patients. Am J Psychiatry 147:44–50

Matthysee S (1973) Antipsychotic drug actions: a clue to the neuropathology of schizophrenia? Fed Proc 32:200–205

May PR, Tuma AH, Dixon WJ, Yale C, Thiele DA, Kraude WH (1981) Schizophrenia. A follow-up study of the results of five forms of treatment. Arch Gen Psychiatry 38:776–784

McDonald AJ (1991) Topographical organization of amygdaloid projections to the caudatoputamen, nucleus accumbens, and related striatal-like areas of the rat brain. Neuroscience 44:15–33

McGlashan TH (1999) Duration of untreated psychosis in first-episode schizophrenia: marker or determinant of course? Biol Psychiatry 46:899–907

McGlashan TH, Fenton WS (1993) Subtype progression and pathophysiologic deterioration in early schizophrenia. Schizophr Bull 19:71–84

McKenna PJ (1987) Pathology, phenomenology and the dopamine hypothesis of schizophrenia. Br J Psychiatry 151:288–301

Meador-Woodruff JH, Haroutunian V, Powchik P, Davidson M, Davis KL, Watson SJ (1997) Dopamine receptor transcript expression in striatum and prefrontal and occipital cortex. Focal abnormalities in orbitofrontal cortex in schizophrenia. Arch Gen Psychiatry 54:1089–1095

Mednick SA, Canon TD, Barr CE, Lyon M (1991) Fetal neural development and adult schizophrenia. Cambridge University Press, Cambridge

Meltzer HY (1985) Dopamine and negative symptoms in schizophrenia: Critique of the Type I–Type II hypothesis. In: Alpert M (ed) Controversies in Schizophrenia: Changes and Constancies. Guilford Press, New York, pp 110–136

Meltzer HY (1994) An overview of the mechanism of action of clozapine. J Clin Psychiatry 55 [Suppl B]:47–52

Meltzer HY, Sommers AA, Luchins DJ (1986) The effect of neuroleptics and other psychotropic drugs on negative symptoms in schizophrenia. J Clin Psychopharmacol 6:329–338

Meltzer HY, Stahl SM (1976) The dopamine hypothesis of schizophrenia: a review. Schizophr Bull 2:19–76

Miller R (1984) Major psychosis and dopamine: controversial features and some suggestions. Psychol Med 14:779–789

Miller R, Wickens JE (1991) Corticostriatal cell assemblies in selective attention and in representation of predictable and controllable events. Concepts in Neuroscience 2:65–95

Milner B (1963) Effects of different brain lesions on card sorting, the role of the frontal lobes. Arch Neurol 9:90–100

Mita T, Hanada S, Nishino N, Kuno T, Nakai H, Yamadori T, Mizoi Y, Tanaka C (1986) Decreased serotonin S2 and increased dopamine D2 receptors in chronic schizophrenics. Biol Psychiatry 21:1407–1414

Mogenson GJ, Brudzynski SM, Wu M, Yang CR, Yim CCY (1993) From motivation to action: a review of dopaminergic regulation of limbic, nucleus accumbens, ventral pallidum, peduncolopontine nucleus circuitries involved in limbic-motor integration. In: Kalivas PW, Barnes CD (eds) Limbic motor circuits and neuropsychiatry. CRC Press, Boca Raton, pp 193–236

Mogenson GJ, Jones DL, Yim CY (1980) From motivation to action: functional interface between the limbic system and the motor system. Prog Neurobiol 14:69–97

Mogenson GJ, Nielsen M (1984) A study of the contribution of hippocampal-accumbens-subpallidal projections to locomotor activity. Behav Neural Biol 42: 38–51

Mogenson GJ, Yang CR, Yim CY (1988) Influence of dopamine on limbic inputs to the nucleus accumbens. Ann N Y Acad Sci 537:86–100

Moore H, West AR, Grace AA (1999) The regulation of forebrain dopamine transmission: relevance to the pathophysiology and psychopathology of schizophrenia. Biol Psychiatry 46:40–55

Murray RM, Jones P, O'Callaghan E (1991) Fetal brain development and later schizophrenia. In: Bock GR, Whelan J (eds) The childhood environment and adult disease. John Wiley & Sons, Chichester, pp 155–170

Murray RM, Lewis SW (1987) Is schizophrenia a nerodevelopmental disorder? Br Med J 295:681–682

Nauta WJH, Smith GP, Faull RLM, Domesick VB (1978) Efferent connections and nigral afferents of the nucleus accumbens septi in the rat. Neuroscience 3:385–401

Nissbrandt H, Elverfors A, Engberg G (1994) Pharmacologically induced cessation of burst activity in nigral dopamine neurons: significance for the terminal dopamine efflux. Synapse 17:217–224

Nordstrom AL, Farde L, Eriksson L, Halldin C (1995) No elevated D2 dopamine receptors in neuroleptic-naive schizophrenic patients revealed by positron emission tomography and [11C]N-methylspiperone. Psychiatry Res 61:67–83

Nordstrom AL, Farde L, Wiesel FA, Forslund K, Pauli S, Halldin C, Uppfeldt G (1993) Central D2-dopamine receptor occupancy in relation to antipsychotic drug effects: a double-blind PET study of schizophrenic patients. Biol Psychiatry 33:227–235

Norman DA, Shallice T (1986) Attention to action: willed and automatic control of behavior. In: Davidson RJ, Schwartz GE, Shapiro D (eds) Consciousness and self-regulation: advances in research. Plenum Press, New York, pp 1–18

O'Donnell P, Grace AA (1994) Tonic D2-mediated attenuation of cortical excitation in nucleus accumbens neurons recorded in vitro. Brain Res 634:105–112

O'Donnell P, Grace AA (1995) Synaptic interactions among excitatory afferents to nucleus accumbens neurons: hippocampal gating of prefrontal cortical input. J Neurosci 15:3622–3639

O'Donnell P, Grace AA (1998) Dysfunctions in multiple interrelated systems as the neurobiological bases of schizophrenic symptom clusters. Schizophr Bull 24: 267–83

Oades RD (1985) The role of noradrenaline in tuning and dopamine in switching between signals in the CNS. Neurosci Biobehav Rev 9:261–282

Obeso JA, Guridi J, Herrero M-T (1994) Role of the subthalamic nucleus in normal and pathological conditions. In: Percheron G, McKenzie JS, Feger J (eds) The basal ganglia IV: New ideas and data on structure and function. Plenum Press, New York, pp 365–370

Ogura C, Kishimoto A, Nakao T (1976) Clinical effect of L-dopa on schizophrenia. Curr Ther Res Clin Exp 20:308–318

Ohara K, Nagai M, Tani K, Nakamura Y, Ino A (1998) Functional polymorphism of −141C Ins/Del in the dopamine D2 receptor gene promoter and schizophrenia. Psychiatry Res 81:117–123

Okubo Y, Suhara T, Suzuki K, Kobayashi K, Inoue O, Terasaki O, Someya Y, Sassa T, Sudo Y, Matsushima E, Iyo M, Tateno Y, Toru M (1997) Decreased prefrontal dopamine D1 receptors in schizophrenia revealed by PET. Nature 385:634–636

Owen F, Cross AJ, Crow TJ, Longden A, Poulter M, Riley GJ (1978) Increased dopamine-receptor sensitivity in schizophrenia. Lancet 2:223–226

Parent A, Hazrati LN (1995) Functional anatomy of the basal ganglia. I. The cortico-basal ganglia-thalamo-cortical loop. Brain Res Rev 20:91–127

Pearce JM, Hall G (1980) A model for Pavlovian learning: Variations in the effectiveness of conditioned but not unconditioned stimuli. Psych Rev 87:532–552

Pennartz CM, Ameerun RF, Groenewegen HJ, Lopes da Silva FH (1993) Synaptic plasticity in an in vitro slice preparation of the rat nucleus accumbens. Eur J Neurosci 5:107–117

Pennartz CM, Groenewegen HJ, Lopes da Silva FH (1994) The nucleus accumbens as a complex of functionally distinct neuronal ensembles: an integration of behavioural, electrophysiological and anatomical data. Prog Neurobiol 42:719–761

Pennartz CMA (1995) The ascending neuromodulatory systems in learning by reinforcement: comparing computational conjectures with experimental findings. Brain Res Rev 21:219–245

Penney JB, Young AB (1983) Speculations on the functional anatomy of basal ganglia disorders. Ann Rev Neurosci 6:73–97

Penney JB, Young AB (1986) Striatal inhomogeneities and basal ganglia function. Movement Disorders 1:3–15

Pervin LA (1983) The stasis and flow of behavior: towards a theory of goals. In: Page MM (ed) Personality: current theory and research. University of Nebraska Press, Lincoln, pp 1–53

Pervin LA (1996) The science of personality. John Wiley & Sons, Inc., New York

Pijnenburg AJ, Honig WM, Van-Rossum JM (1975) Inhibition of d-amphetamine-induced locomotor activity by injection of haloperidol into the nucleus accumbens of the rat. Psychopharmacologia 41:87–95

Pilowsky LS, Costa DC, Ell PJ, Verhoeff NP, Murray RM, Kerwin RW (1994) D2 dopamine receptor binding in the basal ganglia of antipsychotic-free schizophrenic patients. An 123I-IBZM single photon emission computerised tomography study. Br J Psychiatry 164:16–26

Plenz D, Aertsen A (1994) The basal ganglia: "minimal coherence detection" in cortical activity distribution. In: Percheron G, McKenzie JS, Feger J (eds) The basal ganglia IV: New ideas and data on structure and function. Penum Press, New York, pp 579–588

Post RM, Fink E, Carpenter WTJ, Goodwin FK (1975) Cerebrospinal fluid amine metabolites in acute schizophrenia. Arch Gen Psychiatry 32:1063–1069

Pribram KH (1973) The primate frontal cortex – executive of the brain. In: Pribram KH, Luria AR (eds) Psychophysiology of the frontal lobes. Academic Press, New York, pp 293–314

Pycock CJ, Kerwin RW, Carter CJ (1980) Effect of lesion of cortical dopamine terminals on subcortical dopamine receptors in rats. Nature 286:74–76

Rajkowska G, Selemon LD, Goldman-Rakic PS (1998) Neuronal and glial somal size in the prefrontal cortex: a postmortem morphometric study of schizophrenia and Huntington disease. Arch Gen Psychiatry 55:215–224

Redgrave P, Prescott TJ, Gurney K (1999) The basal ganglia: a vertebrate solution to the selection problem? Neuroscience 89:1009–1023

Reiner A, Anderson KD (1990) The patterns of neurotransmitter and neuropeptide co-occurrence among striatal projection neurons: conclusions based on recent findings. Brain Res Rev 15:251–265

Reith J, Benkelfat C, Sherwin A, Yasuhara Y, Kuwabara H, Andermann F, Bachneff S, Cumming P, Diksic M, Dyve SE et al. (1994) Elevated dopa decarboxylase activity in living brain of patients with psychosis. Proc Natl Acad Sci USA 91: 11651–11654

Rescorla RA, Wagner AR (1972) A theory of Pavlovian conditioning: Variations in the effectiveness of reinforcement and nonreinforcement. In: Black AH, Prokasy WF (eds) Classical Conditioning II: Current Research and Theory. Appleton Century-Crofts, New York

Reynolds GP (1989) Beyond the dopamine hypothesis. The neurochemical pathology of schizophrenia. Br J Psychiatry 155:305–316

Richfield EK, Penney JB, Young AB (1989) Anatomical and affinity state comparisons between dopamine D1 and D2 receptors in the rat nervous system. Neuroscience 30:767–777

Robbins TW (1990) The case of frontostriatal dysfunction in schizophrenia. Schizophr Bull 16:391–402
Robbins TW (1991) Cognitive deficits in schizophrenia and Parkinson's disease: Neural basis and the role of dopamine. In: Willner P, Scheel-Kruger J (eds) The mesolimbic dopamine system: From motivation to action. John Wiley & Sons, pp 497–529
Robbins TW, Brown VJ (1990) The role of the striatum in the mental chronometry of action: a theoretical review. Rev Neurosci 2:181–213
Robbins TW, Everitt BJ (1982) Functional studies of the central catecholamines. Int Rev Neurobiol 23:303–365
Robbins TW, Everitt BJ (1992) Functions of dopamine in the dorsal and ventral striatum. Semin Neurosci 4:119–128
Robbins TW, Everitt BJ (1996) Neurobehavioural mechanisms of reward and motivation. Curr Opin Neurobiol 6:228–236
Robbins TW, Everitt BJ (in press) Dopamine- its role in behaviour and cognition in experimental animals and humans. In: Di Chiara G (ed) Dopamine. Springer Verlag
Robertson GS, Fibiger HC (1992) Neuroleptics increase c-fos expression in the forebrain: contrasting effects of haloperidol and clozapine. Neuroscience 46:315–328
Rolls ET, Johnstone S (1992) Neurophysiological analysis of striatal function. In: Wallesch C, Vallar G (eds) Neuropsychological Disorders with Subcortical Lesions. University Press, Oxford, pp 61–97
Rolls ET, Williams GV (1987) Sensory and movement-related neuronal activity in different regions of the primate striatum. In: Schneider JS, Lidsky TT (eds) Basal Ganglia and Behavior: Sensory aspects and motor functioning. Hans Huber, Bern, pp 37–59
Rosin DL, Clark WA, Goldstein M, Roth RH, Deutch AY (1992) Effects of 6-hydroxydopamine lesions of the prefrontal cortex on tyrosine hydroxylase activity in mesolimbic and nigrostriatal dopamine systems. Neuroscience 48:831–839
Salamone JD (1994) The involvement of nucleus accumbens dopamine in appetitive and aversive motivation. Behav Brain Res 61:117–133
Salamone JD, Cousins MS, Snyder BJ (1997) Behavioral functions of nucleus accumbens dopamine: empirical and conceptual problems with the anhedonia hypothesis. Neurosci Biobehav Rev 21:341–359
Sanfilipo M, Wolkin A, Angrist B, van Kammen DP, Duncan E, Wieland S, Cooper TB, Peselow ED, Rotrosen J (1996) Amphetamine and negative symptoms of schizophrenia. Psychopharmacology (Berl) 123:211–214
Scharfetter J, Chaudhry HR, Hornik K, Fuchs K, Sieghart W, Kasper S, Aschauer HN (1999) Dopamine D3 receptor gene polymorphism and response to clozapine in schizophrenic Pakastani patients. Eur Neuropsychopharmacol 10:17–20
Scheel-Kruger J, Willner P (1991) The mesolimbic system: principles of operation. In: Willner P, Scheel-Kruger J (eds) The Mesolimbic Dopamine System: From Motivation to Action. Wiley J. & Sons, Chichester, pp 559–597
Schultz W (1986) Responses of midbrain dopamine neurons to behavioral trigger stimuli in the monkey. J Neurophysiol 56:1439–1461
Schultz W (1998) Predictive reward signal of dopamine neurons. J Neurophysiol 80:1–27
Schultz W, Apiccela P, Romo R, Scarnati E (1995a) Context-dependent activity in primate striatum reflecting past and future behavioral events. In: Houk JC, Davis JL, Beiser DG (eds) Models of information processing in the basal ganglia. MIT Press, Cambridge, pp 11–27
Schultz W, Apicella P, Ljungberg T (1993) Responses of monkey dopamine neurons to reward and conditioned stimuli during successive steps of learning a delayed response task. J Neurosci 13:900–913
Schultz W, Apicella P, Scarnati E, Ljungberg T (1992) Neuronal activity in monkey ventral striatum related to the expectation of reward. J Neurosci 12:4595–4610

Schultz W, Romo R (1990) Dopamine neurons of the monkey midbrain: contingencies of responses to stimuli eliciting immediate behavioral reactions. J Neurophysiol 63:607–624

Schultz W, Romo R, Ljungberg T, Mirenowicz J, Hellerman JR, Dickinson A (1995b) Reward-related signals carried by dopaminergic neurons. In: Houk JC, Davis JL, Beiser DG (eds) Models of information processing in the basal ganglia. MIT Press, Cambridge, pp 233–248

Seamans JK, Phillips AG (1994) Selective memory impairments produced by transient lidocaine-induced lesions of the nucleus accumbens in rats. Behav Neurosci 108: 456–468

Sedvall GC (1996) Neurobiological correlates of acute neuroleptic treatment. Int Clin Psychopharmacol 11 [Suppl 2]:41–46

Seeman P (1987) Dopamine receptors and the dopamine hypothesis of schizophrenia. Synapse 1:133–152

Seeman P, Bzowej NH, Guan HC, Bergeron C, Reynolds GP, Bird ED, Riederer P, Jellinger K, Tourtellotte WW (1987) Human brain D1 and D2 dopamine receptors in schizophrenia, Alzheimer's, Parkinson's, and Huntington's diseases. Neuropsychopharmacology 1:5–15

Seeman P, Ulpian C, Bergeron C, Riederer P, Jellinger K, Gabriel E, Reynolds GP, Tourtellotte WW (1984) Bimodal distribution of dopamine receptor densities in brains of schizophrenics. Science 225:728–731

Segal DS (1975) Behavioral and neurochemical correlates of repeated d-amphetamine administration. In: Mandell AJ (ed) Neurobiological mechanisms of adaptation and behavior. Raven Press, New York

Selemon LD, Rajkowska G, Goldman-Rakic PS (1995) Abnormally high neuronal density in the schizophrenic cortex. A morphometric analysis of prefrontal area 9 and occipital area 17. Arch Gen Psychiatry 52:805–818

Selemon LD, Rajkowska G, Goldman-Rakic PS (1998) Elevated neuronal density in prefrontal area 46 in brains from schizophrenic patients: application of a three-dimensional, stereologic counting method. J Comp Neurol 392:402–412

Serretti A, Macciardi F, Smeraldi E (1998) Dopamine receptor D2 Ser/Cys311 variant associated with disorganized symptomatology of schizophrenia. Schizophr Res 34: 207–210

Sesack SR, Deutch AY, Roth RH, Bunney BS (1989) Topographical organization of the efferent projections of the medial prefrontal cortex in the rat: an anterograde tract-tracing study with Phaseolus vulgaris leucoagglutinin. J Comp Neurol 290:213–242

Shallice T (1982) Specific impairments of planning. Phil Trans R Soc Lond b 298: 199–209

Shaner A (1999) Delusions, superstitious conditioning and chaotic dopamine neurodynamics. Med Hypotheses 52:119–123

Snyder SH (1973) Amphetamine psychosis: a "model" schizophrenia mediated by catecholamines. Am J Psychiatry 130:61–67

Snyder SH (1976) The dopamine hypothesis of schizophrenia: focus on the dopamine receptor. Am J Psychiatry 133:197–202

Snyder SH, Banerjee SP, Yamamura HI, Greenberg D (1974) Drugs, neurotransmitters, and schizophrenia. Science 184:1243–1253

Soares JC, Innis RB (1999) Neurochemical brain imaging investigations of schizophrenia. Biol Psychiatry 46:600–615

Spence SA, Brooks DJ, Hirsch SR, Liddle PF, Meehan J, Grasby PM (1997) A PET study of voluntary movement in schizophrenic patients experiencing passivity phenomena (delusions of alien control). Brain 120:1997–2011

Sperber D, Wilson D (1987) Precis of relevance: Communication and cognition. Behav Brain Sci 10:697–754

Staton DM, Solomon P (1984) Microinjections of d-amphetamine into the nucleus accumbens and caudate-putamen differentially affect stereotypy and locomotion in the rat. Physiol Psychol 12:159–162

Stefanis NC, Bresnick JN, Kerwin RW, Schofield WN, McAllister G (1998) Elevation of D4 dopamine receptor mRNA in postmortem schizophrenic brain. Brain Res Mol Brain Res 53:112–129

Strauss JS (1987) Processes of healing and chronicity in schizophrenia. In: Hafner H, Gattaz WF (eds) Search for the causes of schizophrenia. Springer, Berlin, pp 75–87

Strauss JS, Carpenter WT, Jr, Bartko JJ (1974) The diagnosis and understanding of schizophrenia. Summary and conclusions. Schizophr Bull 70–80

Stuss DT (1992) Biological and Psychological Development of Executive Functions. Brain Cognition 20:8–23

Surmeier DJ, Calabresi P (in press) Cellular actions of dopamine. In: Di Chiara G (ed) Dopamine. Springer Verlag

Swerdlow NR, Koob GF (1987) Dopamine, schizophrenia, mania and depression: toward a unified hypothesis of cortico-striato-pallido-thalamic function. Behav Brain Res 10:215–217

Taghzouti K, Simon H, Louilot A, Herman J, Le Moal M (1985) Behavioral study after local injection of 6-hydroxydopamine into the nucleus accumbens in the rat. Brain Res 344:9–20

Tallerico T, Ulpian C, Liu IS (1999) Dopamine D2 receptor promoter polymorphism: no association with schizophrenia. Psychiatry Res 85:215–219

Tandon R (1995) Neurobiological substrate of dimensions of schizophrenic illness. J Psychiatr Res 29:255–260

Tandon R, Greden JF (1991) Negative symptoms of schizophrenia: the need for conceptual clarity. Biol Psychiatry 30:321–325

Taylor JR, Robbins TW (1984) Enhanced behavioural control by conditioned reinforcers following microinjections of d-amphetamine into the nucleus accumbens. Psychopharmacology 84:405–412

Taylor JR, Robbins TW (1986) 6-Hydroxydopamine lesions of the nucleus accumbens, but not of the caudate nucleus, attenuate enhanced responding with reward-related stimuli produced by intra-accumbens d-amphetamine. Psychopharmacology 90:390–397

Tepper JM, Martin LP, Anderson DR (1995) GABAA receptor-mediated inhibition of rat substantia nigra dopaminergic neurons by pars reticulata projection neurons. J Neurosci 15:3092–3103

Tune L, Barta P, Wong D, Powers RE, Pearlson G, Tien AY, Wagner HN (1996) Striatal dopamine D2 receptor quantification and superior temporal gyrus: volume determination in 14 chronic schizophrenic subjects. Psychiatry Res 67:155–158

Tune LE, Wong DF, Pearlson G, Strauss M, Young T, Shaya EK, Dannals RF, Wilson AA, Ravert HT, Sapp J et al. (1993) Dopamine D2 receptor density estimates in schizophrenia: a positron emission tomography study with 11C-N-methylspiperone. Psychiatry Res 49:219–237

Turner EE, Fedtsova N, Jeste DV (1997) Cellular and molecular neuropathology of schizophrenia: new directions from developmental neurobiology. Schizophr Res 27:169–180

Uchimura N, Higashi H, Nishi S (1986) Hyperpolarizing and depolarizing actions of dopamine via D-1 and D-2 receptors on nucleus accumbens neurons. Brain Res 375:368–372

van den Bos R, Charria Ortiz GA, Bergmans AC, Cools AR (1991) Evidence that dopamine in the nucleus accumbens is involved in the ability of rats to switch to cue-directed behaviours. Behav Brain Res 42:107–114

van den Bos R, Cools AR (1989) The involvement of the nucleus accumbens in the ability of rats to switch to cue-directed behaviours. Life Sciences 44:1697–1704

van Kammen DP, Docherty JP, Bunney WE, Jr (1982) Prediction of early relapse after pimozide discontinuation by response to d-amphetamine during pimozide treatment. Biol Psychiatry 17:233–242 van Kammen DP, van Kammen WB, Mann LS, Seppala T, Linnoila M (1986) Dopamine metabolism in the cerebrospinal fluid of drug-free schizophrenic patients with and without cortical atrophy. Arch Gen Psychiatry 43:978–983

Volkow ND, Wolf AP, Van Gelder P, Brodie JD, Overall JE, Cancro R, Gomez-Mont F (1987) Phenomenological correlates of metabolic activity in 18 patients with chronic schizophrenia. Am J Psychiatry 144:15115–15118

Weickert CS, Weinberger DR (1998) A candidate molecule approach to defining developmental pathology in schizophrenia. Schizophr Bull 24:303–316

Weinberger DR (1987) Implications of normal brain development for the pathogenesis of schizophrenia. Arch Gen Psychiatry 44:660–669

Weinberger DR (1988) Schizophrenia and the frontal lobe. Trends Neurosci 11:367–370

Weinberger DR, Berman KF, Illowsky BP (1988) Physiological dysfunction of dorsolateral prefrontal cortex in schizophrenia. III. A new cohort and evidence for a monoaminergic mechanism. Arch Gen Psychiatry 45:609–615

Weinberger DR, Berman KF, Suddath R, Torrey EF (1992) Evidence of dysfunction of a prefrontal-limbic network in schizophrenia: a magnetic resonance imaging and regional cerebral blood flow study of discordant monozygotic twins. Am J Psychiatry 149:890–897

Weinberger DR, Lipska BK (1995) Cortical maldevelopment, anti-psychotic drugs, and schizophrenia: a search for common ground. Schizophr Res 16:87–110

Weiner I (1990) Neural substrates of latent inhibition: The switching model. Psychol Bull 108:442–461

White FJ (1991) Neurotransmission in the mesoaccumbens dopamine system. In: Willner P, Scheel-Kruger J (eds) The Mesolimbic Dopamine System: From Motivation to Action. Wiley J. & Sons, Chichester, pp 61–103

Wichman T, Bergman H, DeLong MR (1994) The primate subthalamic nucleus. I. Functional properties in intact animals. J Neurophysiol 72:494–506

Wickens J (1990) Striatal dopamine in motor activation and reward-mediated learning: Steps towards a unifying model. J Neural Transm 80:9–31

Wickens J, Kotter R (1995) Cellular models of reinforcement. In: Houk JC, Davis JL, Beiser DG (eds) Models of information processing in the basal ganglia. MIT Press, Cambridge, pp 187–214

Wickens JR, Begg AJ, Arbuthnott GW (1996) Dopamine reverses the depression of rat corticostriatal synapses which normally follows high-frequency stimulation of cortex in vitro. Neuroscience 70:1–5

Wieselgren IM, Lindstrom LH (1998) CSF levels of HVA and 5-HIAA in drug-free schizophrenic patients and healthy controls: a prospective study focused on their predictive value for outcome in schizophrenia. Psychiatry Res 81:101–110

Wilkinson LS, Mittleman G, Torres E, Humby T, Hall FS, Robbins TW (1993) Enhancement of amphetamine-induced locomotor activity and dopamine release in nucleus accumbens following excitotoxic lesions of the hippocampus. Behav Brain Res 55:143–150

Willner P (1997) The dopamine hypothesis of schizophrenia: current status, future prospects. Int Clin Psychopharmacol 12:297–308

Wilson CJ (1995) The contribution of cortical neurons to the firing pattern of striatal spiny neurons. In: Houk JC, Davis JL, Beiser DG (eds) Models of information processing in the basal ganglia. MIT Press, Cambridge, pp 29–50

Wise SP, Murray EA, Gerfen CR (1996) The frontal cortex-basal ganglia system in primates. Crit Rev Neurobiol 10:317–356

Wolkin A, Angrist B, Wolf A, Brodie JD, Wolkin B, Jaeger J, Cancro R, Rotrosen J (1988) Low frontal glucose utilization in chronic schizophrenia: a replication study. Am J Psychiatry 145:251–253

Wolkin A, Jaeger J, Brodie JD, Wolf AP, Fowler J, Rotrosen J, Gomez-Mont F, Cancro R (1985) Persistence of cerebral metabolic abnormalities in chronic schizophrenia as determined by positron emission tomography. Am J Psychiatry 142:564–571

Wolkin A, Sanfilipo M, Wolf AP, Angrist B, Brodie JD, Rotrosen J (1992) Negative symptoms and hypofrontality in chronic schizophrenia. Arch Gen Psychiatry 49: 959–965

Wong DF, Wagner HN, Jr, Tune LE, Dannals RF, Pearlson GD, Links JM, Tamminga CA, Broussolle EP, Ravert HT, Wilson AA et al. (1986) Positron emission tomography reveals elevated D2 dopamine receptors in drug-naive schizophrenics. Science 234:1558–1563

Wong DF, Young D, Wilson PD, Meltzer CC, Gjedde A (1997a) Quantification of neuroreceptors in the living human brain: III. D2- like dopamine receptors: theory, validation, and changes during normal aging. J Cereb Blood Flow Metab 17:316–330

Wong WF, Pearlson GD, Tune LE, Young LT, Meltzer CC, Dannals RF, Ravert HT, Reith J, Kuhar MJ, Gjedde A (1997b) Quantification of neuroreceptors in the living human brain: IV. Effect of aging and elevations of D2-like receptors in schizophrenia and bipolar illness. J Cereb Blood Flow Metab 17:331–342

Wright CI, Groenewegen HJ (1995) Patterns of convergence and segregation in the medial nucleus accumbens of the rat: relationships of prefrontal cortical, midline thalamic, and basal amygdaloid afferents. J Comp Neurol 361:383–403

Wyatt RJ (1986) The dopamine hypothesis: variations on a theme (II). Psychopharmacol Bull 22:923–927

Wyatt RJ (1991) Neuroleptics and the natural course of schizophrenia. Schizophr Bull 17:325–351

Yan Z, Song WJ, Surmeier J (1997) D2 dopamine receptors reduce N-type Ca^{2+} currents in rat neostriatal cholinergic interneurons through a membrane-delimited, protein-kinase-C-insensitive pathway. J Neurophysiol 77:1003–1015

Yang CR, Mogenson GJ (1984) Electrophysiological responses of neurones in the nucleus accumbens to hippocampal stimulation and the attenuation of the excitatory responses by the mesolimbic dopaminergic system. Brain Res 324:69–84

Yim CY, Mogenson GJ (1988) Neuromodulatory action of dopamine in the nucleus accumbens: an in vivo intracellular study. Neuroscience 26:403–415

Zakzanis KK, Hansen KT (1998) Dopamine D2 densities and the schizophrenic brain. Schizophr Res 32:201–206

Zec RF (1995) Neuropsychology of schizophrenia according to Kraepelin: disorders of volition and executive functioning. Eur Arch Psychiatry Clin Neurosci 245:216–223

CHAPTER 26
Atypical Antipsychotics

J.E. LEYSEN

A. Introduction

Neuroleptic action was first discovered and defined in 1952 with the clinical use of chlorpromazine, known before as an antihistamine (DELAY et al. 1952). The identification of dopamine as a neurotransmitter in the brain and neurochemical and pharmacological studies revealed that neuroleptic activity involved dopamine antagonism (CARLSSON and LINDQVIST 1963; VAN ROSSUM 1966). The butyrophenone, haloperidol, discovered in 1958, became the prototype of a neuroleptic with selective dopamine antagonistic action (DIVRY et al. 1958; JANSSEN et al. 1959). Following these discoveries, a first generation of neuroleptics was developed between the 1960s and mid-1980s, which were designed to be dopamine antagonists. Over 70 neuroleptics, belonging to more than 10 different chemical classes, were brought to the European market (LEYSEN and NIEMEGEERS 1985). All these compounds appeared to block dopamine D_2 receptors in the brain and a correlation was shown between their affinity for D_2 receptors and dosages used for treating positive symptoms of schizophrenia (CREESE et al. 1976; SEEMAN et al. 1976). However, a direct relationship also exists between blockade of D_2 receptors and the induction of extrapyramidal symptoms and the elevation of plasma prolactin levels (VAN WIELINCK and LEYSEN 1983; KUENSTLER et al. 1999). Although in the early years neuroleptics were successfully used at moderate doses, treatment dosages were markedly increased over the years and side-effects became a major problem. Moreover, the first generation of neuroleptics could suppress the positive symptoms of schizophrenia, but did not treat the negative symptoms. Later, when used at high dose, secondary negative symptoms were precipitated (CARPENTER 1995).

Pharmacological and receptor studies revealed that in the different classes of neuroleptics there were compounds which interacted with several different neurotransmitter receptors. The blockade of different receptors gave rise to different types of side-effects or could bring certain therapeutic benefit (LEYSEN 1984; LEYSEN and NIEMEGEERS 1985; LEYSEN et al. 1993).

The identification in 1978 (LEYSEN et al. 1978) of the S_2 or 5-HT_2 receptors in the brain and the finding that certain neuroleptics had high affinity for these receptors led to the development of a second generation of antipsychotics with predominant 5-HT$_2$ and more moderate D$_2$ antagonism. The term antipsychotic became preferred since the clinical picture that was aimed at with the new compounds differed from the classical definition of a neuroleptic, which in fact included also the side-effect profile. Existing antipsychotics, of which predominant 5-HT$_2$ antagonism was recognised, were the butyrophenone, pipamperone and the 6,7,6-membered-ring tricyclic, clozapine (LEYSEN et al. 1978). These compounds were noted in clinical use for atypical actions. Pipamperone was reported to have anti-agitation properties, to normalise disturbed sleep rhythms in psychiatric patients and to improve social interaction. Clozapine was noted for its therapeutic efficacy in treatment resistant patients with schizophrenic-like symptoms, its low incidence of induction of EPS and the absence of elevation of plasma prolactin levels (LINDSTROM 1989). Clinical studies with a potent and relatively selective 5-HT$_2$ antagonist, ritanserin, revealed anti-dysthymic action and potential to alleviate negative symptoms of schizophrenia (LEYSEN et al. 1985; REYNTJENS et al. 1986; DUINKERKE et al. 1993). 5-HT$_2$ receptors were further subclassified into 5-HT$_{2A}$, 5-HT$_{2B}$ and 5-HT$_{2C}$ receptors (HOYER and MARTIN 1996, 1997). It appeared that 5-HT$_{2A}$ receptor blockade was of particular importance for the beneficial action in patients suffering from schizophrenia (SORENSEN et al. 1993; SCHMIDT et al. 1995; CARLSSON et al. 1997). It was proposed that a balanced 5-HT$_{2A}$/D$_2$ receptor blockade, with an affinity ratio of at least tenfold, and careful dosing to maintain an appropriate moderate D$_2$ receptor blockade (40%–75%) in the basal ganglia, could provide a more optimal treatment of positive and negative symptoms of schizophrenia with reduced side-effect liability (MELTZER and NASH 1991; NYBERG et al. 1996). New compounds were developed to meet this goal: risperidone (JANSSEN et al. 1988; LEYSEN et al. 1988), its active metabolite 9-OH risperidone (VAN BEIJSTERVELDT et al. 1994), ziprasidone (SEEGER et al. 1995), zotepine (NEEDHAM et al. 1996), olanzapine (MOORE et al. 1992; BYMASTER et al. 1996), and quetiapine (MIGLER et al. 1993; SALLER and SALAMA 1993) came onto the worldwide market between 1992 and 2000. The compounds have different chemical structures (Fig. 1), and although the balanced 5-HT$_{2A}$/D$_2$ antagonism is a key feature, they interact with various different biogenic amine receptors, the profiles being different for each of the compounds.

In this chapter, we describe the receptor profiles of the second generation of antipsychotics, "the balanced 5-HT$_{2A}$/D$_2$ antagonists", compared to the prototype compounds: haloperidol, pipamperone and clozapine. The interaction of the compounds with a comprehensive list of biogenic amine receptors is reported and implications for therapeutic and side-effects are discussed (for extensive reviews see ARNT and SKARSFELDT 1998; LEYSEN 2000).

B. Receptor Binding Profile of Antipsychotics

Various antipsychotics were found to have high to moderate affinity for biogenic amine receptors. The currently known subtypes of dopamine, 5-HT,

Atypical Antipsychotics

Butyrophenones

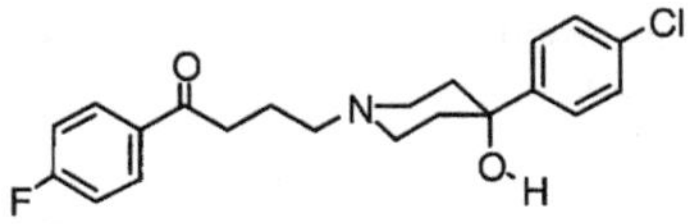

haloperidol

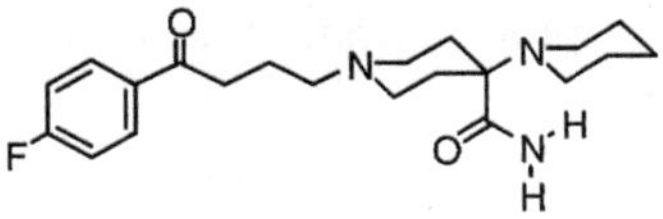

pipamperone

Benzisoxazoles

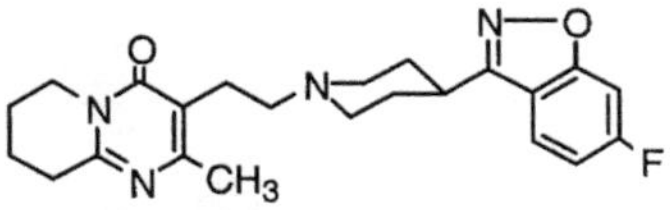

risperidone

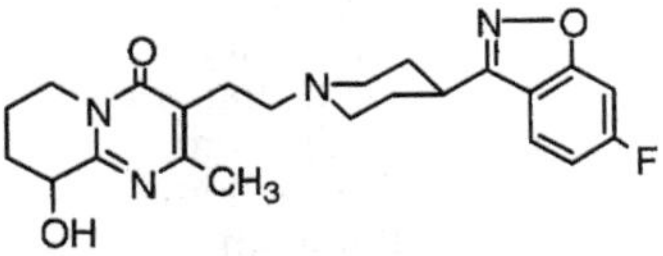

9-OH risperidone

Pyrimidine benzthiazole

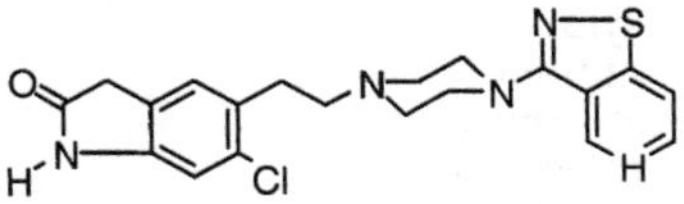

ziprasidone

6,7,6- or 6,7,5-membered-ring tricyclics

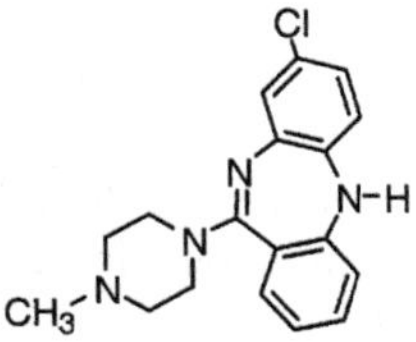

clozapine

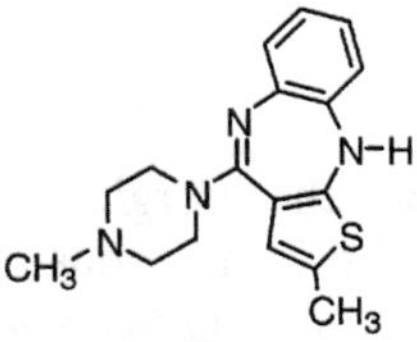

olanzapine

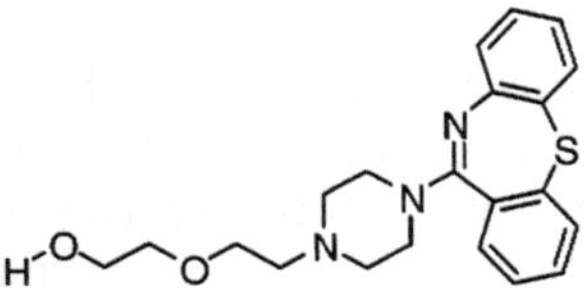

quetiapine

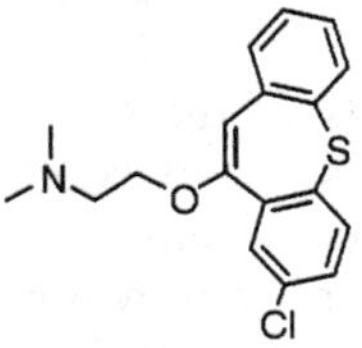

zotepine

Fig. 1. Chemical structure of antipsychotics

Table 1. Dopamine receptor subtypes and clinical applications

Receptor	Brain areas with high density	Second messenger response	Therapeutic effects	Side-effects
D_1	Putamen – caudate N. accumbens Frontal cortex	Increase cAMP	Agonist: improvement cognitive and motoric functions	Antagonist: impairment of cognitive functions
D_5	Cortex, basal ganglia Hippocampus Diencephalon Brain stem Cerebellum	Increase cAMP	To be explored	To be explored
D_2	Putamen – caudate N. accumbens Substantia nigra	Decrease cAMP	Agonist: improvement motoric function; antagonist: treatment positive symptoms, anti-emetic	Agonist: hallucinations, emesis; antagonist: EPS, prolactin elevation
D_3	Islands of Calleja N. accumbens Cerebellum	Decrease cAMP	Antagonist: hypothesised reduction of dystonia	
D_4	Hippocampus Entorhinal cortex	Decrease cAMP	Antagonist: hypothesised treatment of negative symptoms	

For distribution studies see CILIAX et al. 2000; BERGSON et al. 1995; LAHTI et al. 1998; SEEMAN 1995.

α-adrenergic, cholinergic muscarinic and histamine receptors are listed in Tables 1–3; the brain areas of high densities and possible therapeutic effects and side-effects of agonists and antagonists are presented.

The receptor binding profiles of haloperidol, pipamperone, risperidone, 9-OH risperidone, ziprasidone, zotepine, olanzapine, clozapine and quetiapine are shown in Table 4.

The affinity of the compounds for the receptors is indicated by the pK_i-value = $-\log$ (K_i-value as molar concentration). The K_i indicates the concentration of the compound producing 50% occupancy of the receptor. K_i-values are measured in vitro by inhibition of radioligand binding to receptors in cell membrane preparations. Table 4 shows the radioligands employed for labelling the receptors respectively, and mostly cloned human receptors expressed in cells were used.

Table 2. 5-HT receptor subtypes and clinical applications

Receptor	Brain areas with high density	Second messenger response	Therapeutic effects	Side-effects
5-HT$_{1A}$	Hippocampus, septum, amygdala, raphe n.	Decrease cAMP, open K$^+$ channels	Agonist: possible anxiolytic effect	Strong agonism, possible excitation
5-HT$_{1B}$	Substantia nigra, globus pallidus, Tuberculum olfactorium, superior colliculus	Decrease cAMP	Antagonist: possible antidepressant; agonist: anti migraine	Agonist: coronary constriction
5-HT$_{1D}$	Trigeminal ganglia, trigeminal sensory neurons	Decrease cAMP	Agonist: anti migraine	
5-ht$_{1E}$	Striatum, amygdala, cortex	Decrease cAMP	To be explored	To be explored
5-ht$_{1F}$	Striatum, hippocampus, cortex	Decrease cAMP	Agonist: possible anti migraine	
5-HT$_{2A}$	Frontal and cingulate cortex, striatum, n. accumbens, pedunculopontine n., laterodorsal tegmental n.	Increase inositol phosphate, intracellular Ca^{++}, arachidonic acid	Antagonist: treatment of dysthymia, negative symptoms	Agonist: hallucinations, tremors, convulsions
5-HT$_{2B}$	Amygdala	Increase inositol phosphate, intracellular Ca^{++}	To be explored	To be explored
5-HT$_{2C}$	Choroid plexus, widespread throughout brain	Increase inositol phosphate, intracellular Ca^{++}, arachidonic acid	Antagonist: anxiolytic, increased food intake	Agonist: anorectic; antagonist: weight gain
5-HT$_3$	Area postrema, n. tractus solitarius, substantia gelatinosa, trigeminal n.	Ion channel opening, permeable to Na$^+$, K$^+$, Ca^{++}	Antagonist: anti-emetic	Agonist: emesis, sensory pain
5-HT$_4$	Basal ganglia, hippocampus	Increase cAMP	Agonist: possible improved cognitive function, gastrokinetic	
5-ht$_5$	Glial cells	Decrease cAMP	To be explored	To be explored
5-HT$_6$	T. olfactorium, n. accumbens, striatum, frontal, entorhinal cortex, hippocampus, cerebellum	Increase cAMP	Antagonist: possible improvement of cognitive function; to be further explored	To be explored
5-HT$_7$	Medial thalamic nuclei, dentate gyrus, cortex, amygdala	Increase cAMP	Agonist: phase shift, circadian rhythm	To be explored

For review see HOYER and MARTIN 1996, 1997; BARNES and SHARP 1999.

Table 3. α-Adrenoceptor, cholinergic muscarinic and histamine receptor subtypes and clinical applications

Receptor	Second messenger response	Therapeutic effects	Side-effects
α_1-Adrenoceptor[a,b] ($\alpha_{1A}, \alpha_{1B}, \alpha_{1D}$)	increase inositol phosphate, intracellular Ca^{++}	Agonist: treatment of narcolepsy, day-time sleepiness	Antagonist: orthostatic hypotension, reflex tachycardia, sedation
α_2-Adrenoceptor[a] ($\alpha_{2A}, \alpha_{2B}, \alpha_{2C}$)	Decrease cAMP	Antagonist: possible antidepressant, increased drive and motivation; agonist: analgesic	Antagonist: increased cardiac output; agonist: hypotension
Cholinergic muscarinic[c]			
m1, m3, m5	Increase inositol phosphate, intracellular Ca^{++}	Antagonist: anti-ulcer, gastrointestinal spasmolytic	Antagonist: dry mouth, blurred vision, urinary retention, constipation, confusion, hallucinations
m2, m4	Decrease cAMP, opening K^+ channel		
Histamine			
H_1	Increase inositol phosphate	Antagonist: anti-allergic	Antagonist: sedation, weight gain
H_2	Increase cAMP	Antagonist: anti-gastric acid, suggested treatment of negative symptoms	
H_3	Decrease cAMP	Antagonist: possible improvement of attention and vigilance	

[a] For review see DOCHERTY 1998.
[b] SIRVIÖ and MACDONALD 1999.
[c] For review see CAULFIELD 1993.

Table 4. Receptor binding profiles of antipsychotics, pK$_i$ values ± SD (n), –log M

Receptor	Radioligand	Tissue	Temp.	Haloperidol	Pipamperone	Risperidone	9-OH-risperidone	Ziprasidone
Dopamine								
rD$_1$	[^{3}H]SCH23390	Rat striatum	37	6.56 ± 0.08 (3)	5.61 ± 0.03 (3)	6.21 ± 0.08 (3)	6.19 ± 0.11 (5)	6.47 ± 0.13 (3)
hD$_{2L}$	[^{3}H]Spiperone	Human D$_{2L}$-CHO	37	8.69 ± 0.09 (4)	6.71 ± 0.23 (4)	8.39 ± 0.23 (5)	8.37 ± 0.08 (11)	8.17 ± 0.10 (3)
hD$_3$	[^{125}I]Iodosulpride	Human D$_3$-CHO	37	8.25 ± 0.19 (3)	6.58 ± 0.17 (3)	7.85 ± 0.12 (3)	8.15 ± 0.10 (11)	7.61 ± 0.12 (4)
hD$_4$	[^{3}H]Spiperone	Human D$_{4.2}$-L929	37	7.93 ± 0.18 (5)	7.95 ± 0.37 (5)	7.82 ± 0.11 (5)	7.55 ± 0.09 (5)	6.98 ± 0.02 (3)
Serotonin								
h5-HT$_{1A}$	[^{3}H]8OHDPAT	Human 5-HT$_{1A}$-Hela	37	5.79 ± 0.13 (3)	5.46 ± 0.30 (3)	6.37 ± 0.13 (3)	6.23 ± 0.04 (3)	8.26 ± 0.18 (5)
h5-HT$_{1B}$	[^{3}H]Alniditan	Human 5-HT$_{1B}$-HEK293	37	<5 (3)	5.54 ± 0.15 (3)	6.84 ± 0.04 (2)	6.82 ± 0.24 (2)	8.06 ± 0.03 (3)
h5-HT$_{1D}$	[^{3}H]Alniditan	Human 5-HT$_{1D}$-C6 glioma	37	6.35 ± 0.23 (3)	6.14 ± 0.45 (3)	7.80 ± 0.38 (5)	7.98 ± 010 (3)	8.50 ± 0.11 (4)
h5-HT$_{1E}$	[^{3}H]5-HT	Human 5-HT$_{1E}$-CHO	37	<5 (3)	5.44 ± 0.01 (3)	5.68 ± 0.08 (3)	5.80 ± 0.04 (3)	6.12 ± 0.21 (4)
h5-ht$_{1F}$	[^{3}H]5-HT	Human 5-HT$_{1F}$-COS7	37	<5 (3)	<5 (3)	<5 (3)	<5 (4)	6.86 ± 0.21 (2)
h5-HT$_{2A}$	[^{125}I]R093274	Human 5-HT$_{2A}$-L929	37	6.52 ± 0.10 (3)	8.19 ± 0.08 (3)	9.39 ± 0.26 (3)	9.10 ± 0.22 (6)	8.68 ± 0.09 (5)
h5-HT$_{2B}$	[^{3}H]5-HT	Human 5-HT$_{2B}$-CHO	25	5.57 ± 0.14 (3)	7.37 ± 0.13 (3)	7.74 ± 0.14 (3)	7.49 ± 0.16 (3)	8.40 ± 0.08 (2)
h5-HT$_{2C}$	[^{3}H]Mesulergine	Human 5-HT$_{2C}$-sf9	37	5.70 ± 0.14 (3)	7.30 ± 0.07 (3)	7.80 ± 0.17 (3)	7.71 ± 0.03 (4)	8.57 ± 0.25 (3)
m5-HT$_3$	[^{3}H]GR65630	NXG108CC15 cells	37	<5 (3)	<5 (2)	<5 (2)	<5 (5)	5.88 ± 0.31 (4)
h5-HT$_{4L}$	[^{3}H]R116712	Human 5-HT$_4$-COS7	37	<5 (3)	<5 (3)	<5 (3)	5.46 ± 0.07 (5)	<5 (3)
h5-ht$_5$	[^{3}H]5-Carboxamido tryptamine	Human 5-HT$_5$-HEK293	37	<5 (3)	<5 (3)	6.49 ± 0.06 (4)	6.00 ± 0.36 (3)	6.23 ± 0.25 (4)
h5-HT$_6$	[^{3}H]LSD	Human 5-HT$_6$-HEK293	37	<5 (3)	6.22 ± 0.03 (3)	5.53 ± 0.18 (2)	5.68 ± 0.15 (2)	7.43 ± 0.11 (2)
h5-HT$_7$	[^{3}H]LSD	Human 5-HT$_7$-CHO	37	6.22 ± 0.16 (3)	6.54 ± 0.18 (3)	8.35 ± 0.12 (3)	8.21 ± 0.13 (5)	8.15 ± 0.14 (4)
Adrenaline								
hα$_{1A}$	[^{3}H]Prazosin	Human α$_{1A}$-CHO	25	7.12 ± 0.20 (3)	6.19 ± 0.20 (3)	8.25 ± 0.19 (3)	8.00 ± 0.20 (3)	7.22 ± 0.20 (3)
hα$_{2A}$	[^{3}H]Rauwolscine	Human α$_{2A}$-CHO	25	5.98 ± 0.08 (4)	6.15 ± 0.13 (4)	7.66 ± 0.10 (4)	7.55 ± 0.07 (5)	6.73 ± 0.13 (4)
hα$_{2B}$	[^{3}H]Rauwolscine	Human α$_{2B}$-CHO	25	6.29 ± 0.12 (4)	7.26 ± 0.05 (4)	8.06 ± 0.06 (3)	8.04 ± 0.12 (5)	7.08 ± 0.22 (3)
hα$_{2C}$	[^{3}H]Rauwolscine	Human α$_{2C}$-CHO	25	6.36 ± 0.10 (4)	6.25 ± 0.12 (3)	8.08 ± 0.22 (4)	8.02 ± 0.11 (4)	7.10 ± 0.12 (3)
Histamine								
hH$_1$	[^{3}H]Pyrilamine	Human H$_1$-CHO	25	5.92 ± 0.10 (3)	5.74 ± 0.06 (3)	7.48 ± 0.56 (4)	7.47 ± 0.10 (5)	6.57 ± 0.05 (3)
Acetylcholine Muscarinic	[^{3}H]Dexetimide	Rat striatum	37	5.46 ± 0.38 (3)	5.23 ± 0.52 (3)	<5 (5)	5.46 ± 0.12 (4)	5.61 ± 0.01 (2)
				Zotepine	Olanzapine	Clozapine	Quetiapine	
Dopamine								
rD$_1$	[^{3}H]SCH23390	Rat striatum	37	6.99 ± 0.16 (3)	6.93 ± 0.32 (3)	6.27 ± 0.04 (3)	5.35 ± 0.18 (3)	
hD$_{2L}$	[^{3}H]Spiperone	Human D$_{2L}$-CHO	37	7.90 ± 0.47 (4)	7.20 ± 0.52 (4)	6.75 ± 0.06 (3)	6.14 ± 0.05 (3)	
hD$_3$	[^{125}I]Iodosulpride	Human D$_3$-CHO	37	8.29 ± 0.14 (3)	7.30 ± 0.24 (3)	6.62 ± 0.14 (3)	6.03 ± 0.15 (3)	
hD$_4$	[^{3}H]Spiperone	Human D$_{4.2}$-L929	37	7.56 ± 0.28 (4)	7.53 ± 0.10 (4)	7.26 ± 0.14 (4)	5.65 ± 0.13 (7)	

Table 4. Continued

Receptor	Radioligand	Tissue	Temp.	Haloperidol	Pipamperone	Risperidone	9-OH-risperidone	Ziprasidone
Serotonin								
h5-HT$_{1A}$	[^{3}H]8OHDPAT	Human 5-HT$_{1A}$-Hela	37	6.50 ± 0.07 (3)	5.49 ± 0.06 (3)	6.72 ± 0.19 (3)	6.38 ± 0.21 (3)	
h5-HT$_{1B}$	[^{3}H]Alniditan	Human 5-HT$_{1B}$-HEK293	37	6.36 ± 0.05 (3)	5.56 ± 0.07 (3)	5.53 ± 0.07 (3)	5.34 ± 0.01 (3)	
h5-HT$_{1D}$	[^{3}H]Alniditan	Human 5-HT$_{1D}$-C6 glioma	37	6.93 ± 0.33 (4)	5.73 ± 0.24 (4)	5.96 ± 0.19 (4)	<5 (6)	
h5-ht$_{1E}$	[^{3}H]5-HT	Human 5-HT$_{1E}$-CHO	37	6.24 ± 0.12 (5)	5.59 ± 0.03 (5)	6.06 ± 0.02 (4)	5.80 ± 0.04 (3)	
h5-ht$_{1F}$	[^{125}I]R093274	Human 5-HT$_{1F}$-COS7	37	5.93 ± 0.06 (2)	6.21 ± 0.24 (2)	6.76 ± 0.64 (3)	5.82 ± 0.25 (4)	
h5-HT$_{2A}$	[^{3}H]5-HT	Human 5-HT$_{2A}$-L929	37	8.58 ± 0.15 (3)	8.63 ± 0.09 (3)	8.20 ± 0.13 (7)	6.64 ± 0.47 (4)	
h5-HT$_{2B}$	[^{3}H]5-HT	Human 5-HT$_{2B}$-CHO	25	9.01 ± 0.20 (3)	8.08 ± 0.03 (3)	8.18 ± 0.08 (3)	6.77 ± 0.17 (3)	
h5-HT$_{2C}$	[^{3}H]Mesulergine	Human 5-HT2C-sf9	37	8.61 ± 0.34 (4)	7.26 ± 0.41 (3)	7.98 ± 0.44 (3)	5.49 ± 0.11 (3)	
m5-HT$_3$	[^{3}H]GR65630	NXG108CC15 cells	37	6.62 ± 0.05 (3)	6.87 ± 0.19 (3)	7.00 ± 0.23 (3)	5.58 ± 0.38 (4)	
h5-HT$_{4L}$	[^{3}H]R116712	Human 5-HT$_4$-COS7	37	5.94 ± 0.06 (3)	<5 (3)	<5 (3)	<5 (4)	
h5-ht$_5$	[^{3}H]5-Carboxamidotryptamine	Human 5-HT$_5$-HEK293	37	6.74 ± 0.35 (3)	5.86 ± 0.01 (3)	5.96 ± 0.12 (2)	<5 (5)	
h5-HT$_6$	[^{3}H]LSD	Human 5-HT$_6$-HEK293	37	8.09 ± 0.01 (2)	8.25 ± 0.30 (2)	8.04 ± 0.22 (4)	5.86 ± 0.46 (3)	
h5-HT$_7$	[^{3}H]LSD	Human 5-HT$_7$-CHO	37	8.40 ± 0.26 (3)	6.53 ± 0.22 (3)	7.41 ± 0.27 (3)	6.46 ± 0.28 (3)	
Adrenaline								
hα$_1$	[^{3}H]Prazosin	Human α$_{1A}$-CHO	25	8.43 ± 0.02 (3)	6.51 ± 0.14 (3)	7.54 ± 0.15 (3)	7.67 ± 0.06 (3)	
hα$_{2A}$	[^{3}H]Rauwolscine	Human α$_{2A}$-CHO	25	6.81 ± 0.14 (5)	6.37 ± 0.12 (5)	7.27 ± 0.13 (4)	5.75 ± 0.08 (4)	
hα$_{2B}$	[^{3}H]Rauwolscine	Human α$_{2B}$-CHO	25	8.28 ± 0.05 (3)	6.78 ± 0.13 (3)	7.65 ± 0.10 (4)	7.05 ± 0.06 (4)	
hα$_{2C}$	[^{3}H]Rauwolscine	Human α$_{2C}$-CHO	25	6.89 ± 0.04 (3)	6.67 ± 0.40 (3)	8.11 ± 0.10 (3)	6.54 ± 0.13 (4)	
Histamine								
hH$_1$	[^{3}H]Pyrilamine	Human H$_1$-CHO	25	9.25 ± 0.70 (3)	8.84 ± 0.08 (3)	8.97 ± 0.09 (4)	8.08 ± 0.27 (4)	
Acetylcholine								
Muscarinic	[^{3}H]Dexetimide	Rat striatum	37	6.03 ± 0.36 (3)	7.26 ± 0.40 (3)	7.48 ± 0.06 (4)	5.97 ± 0.07 (3)	
		Human m1		7.74[a]	8.60[b]	8.51[a]–8.85[b]	6.87[b]	
		Human m2		6.85[a]	7.89[b]	7.32[a]–8.15[b]	6.15[b]	
		Human m3		7.14[a]	8.00[b]	7.70[b]–8.22[b]	6.65[b]	
		Human m4[c]		7.11[a]	8.22[b]	7.96[b]–8.30[b]	5.52[b]	
		Human m5		6.59[a]		7.95[b]		

Up to a concentration of 10 μM the compounds did not bind to β$_1$, β$_2$, β$_3$ adrenoceptors, neuropeptide receptors (cholecystokinin CCKA, CCKB; neurokinin NK1, NK2, NK3; bradykinin BK$_2$), NMDA receptors (MK801 site, glycine site), AMPA receptors.

Some compounds bound at μM concentrations to opiate receptors (μ, δ, κ), dihydropyridine labelled Ca^{++} channel sites, batrachotoxin-labelled Na$^+$ channels sites, DA, 5-HT, NE transporter. Ziprasidone showed pK$_i$ = 6.72 at the 5-HT transporter and pK$_i$ = 6.42 at the NE transporter. Zotepine showed pK$_i$ = 8.3 on the NE transporter.

Haloperidol had nM affinity for haloperidol labelled σ sites, the other compounds had μM affinity.

Example of reading a pK$_i$-value: pK$_i$ = 8.69 = −log K$_i$; K$_i$ = 10$^{-8.69}$ = 10$^{-9\cdot0.31}$ = 2.10^{-9} M = 2 nM.

Data from SCHOTTE et al. (1996) and LEYSEN et al. (2000), except [a]BOLDEN et al. (1991) and [b]BYMASTER et al. (1996). [c] Agonistic activity at m4 receptors was reported for clozapine EC$_{50}$ 60 nM and olanzapine EC$_{50}$ 1900 nM (BOLDEN et al. 1993).

Table 5 shows an overview of the receptors to which each of the compounds bind with an affinity higher than or equal to the D_2 receptors. The relative binding affinity for D_2 receptors and the applied clinical dose range is indicated as well as the ratio in affinity for 5-HT_{2A} versus D_2 receptors. This overview table facilitates the discussion and comparison of the receptor binding profiles of the compounds.

C. Interaction with Dopamine Receptors

The key feature for treating positive symptoms of schizophrenia is dopamine D_2 receptor blockade. All the antipsychotics indeed bind to D_2 receptors, but with a potency difference of over 350-fold between haloperidol, the most potent and most selective D_2 antagonist, and quetiapine, the least potent of the compounds. Studies on the effect of the compounds on D_2 receptor signalling showed that they all are full antagonists and have the potential of reducing the signalling below basal levels, indicating that they have inverse agonist properties. In general, there is a relatively good correlation between the D_2 receptor binding affinity, the potency to block D_2 receptor signalling in vitro and the potency of the compounds to block D_2 receptor-mediated behaviour in vivo. The currently investigated antipsychotics do not differentiate between D_2 and D_3 receptor binding and inhibition of signalling. A detailed study and discussion of the effects of the recent and reference antipsychotics at human D_2 and D_3 receptor signalling is reported in VANHAUWE et al. (2000). The contribution of D_3 receptor blockade in the therapeutic or side-effects of the antipsychotics is as yet unknown.

The discovery that clozapine has a higher affinity for D_4 receptors than for D_2 receptors prompted the hypothesis that D_4 receptors blockade may contribute antipsychotic properties. Also pipamperone has higher affinity for D_4 than for D_2 receptors. Several selective D_4 antagonists were developed and clinically investigated. However, the role of D_4 receptors has been disputed (ROTH et al. 1995). Most selective D_4 antagonists have been abandoned because of lack of efficacy (BRISTOW et al. 1997).

As mentioned above, the degree of central D_2 receptor occupancy is of prime importance for differentiating between therapeutic and side-effects of the antipsychotics. Careful dose-titration studies with the new antipsychotics, in particular risperidone (DAVIS and JANICAK 1996; JONES 1997), and studies of the in vivo occupancy of striatal D_2 receptors in humans using positron emission tomography with isotope labelled D_2 receptor ligands, indicated that one should aim for 40%–75% occupancy (NYBERG et al. 1996). Above 75% of striatal D_2 receptor occupancy patients suffer from parkinsonian-like side-effects.

Table 5 shows prescribed clinical dose ranges of the compounds. Comparing these with the relative D_2 receptor affinity indicates marked discrepancies. For haloperidol, the low dose of 5mg is still relatively high, and for quetiapine the 250mg dose is probably insufficient.

Table 5. Receptor profile and D_2 affinity potency ranking of antipsychotics

	Clinical dose range (mg/day)	Ratio in affinity for D_2 receptors compared to haloperidol	Ratio in affinity for 5-HT$_{2A}$ versus D_2 receptors	Ranking of receptors according to the binding affinity of the drug[a]
Haloperidol	5–20	1	0.006	$D_2 \sim D_3$[b]
Risperidone	4–6	2	10	5HT$_{2A}$>5HT$_7$>α_1~D_2~α_{2C}~α_{2B}
9-OH-Risperidone		2	5.3	5HT$_{2A}$~5HT$_7$>α_1~D_2~D_3~α_{2C}~α_{2B}
Ziprasidone	80–160	3.3	3.2	5HT$_7$~5HT$_{2A}$~5HT$_{2C}$~5HT$_{1D}$~5HT$_{2B}$~5HT$_{1A}$~D_2~5HT$_{1B}$~α_1
Zotepine	150–340	6.2	4.7	H$_1$~5HT$_{2B}$>5HT$_{2C}$~5HT$_{2A}$~α_1~NET~D_3~α_{2B}~5HT$_6$~D_2~5HT$_7$~D_4
Olanzapine	12.5–17.5	30	27	H$_1$~5HT$_{2A}$>5HT$_6$~5HT$_{2B}$>D_4~D_3~5HT$_7$~5HT$_{2C}$~mACh~α_1~D_2~D_1
Clozapine	300–600	87	28	H$_1$>5HT$_{2A}$~α_{2C}~5HT$_{2B}$~5HT$_6$~5HT$_{2C}$~5HT$_7$~α_{2B}~α_1~mACh~α_{2A}~D_4~5HT$_3$~D_2~5ht$_{1F}$~5HT$_{1A}$~D_3~D_1
Pipamperone	80–360	95	30	5HT$_{2A}$~D_4>5HT$_{2B}$~5HT$_{2C}$~α_1~α_{2B}>D_2~D_3
Quetiapine	250–750	355	2.6	H$_1$~α_1~α_{2B}>5HT$_{2B}$~5HT$_{2A}$~α_{2C}~5HT$_{1A}$~5HT$_7$~D_2~D_3~mACh~5HT$_6$~5ht$_{1F}$~5ht$_{1E}$

[a] Potency difference of more or equal to 2.0-fold; ~, potency difference of less than 2.0-fold; receptors are indicated for which the drug has higher or equal affinity than for the D_2 receptors.
[b] Haloperidol has equally high affinity for σ-sites.

Risperidone, for which the early clinical studies indicated 6mg, is now being used at lower dose: 4mg or less is recommended. Surprising is the high dose range of 80–160mg of ziprasidone in view of its high D_2 receptor affinity. A relatively fast metabolism could be a reason for using a higher dose range. However, a dose–D_2 receptor occupancy PET study in volunteers showed that 40mg ziprasidone produced over 75% of D_2 receptor occupancy; this compound should best be used at lower dose (BENCH et al. 1993).

Similar dose–D_2 receptor occupancy studies using PET in humans would be useful for all existing and new antipsychotics to indicate a preferred median dose.

Relatively potent D_1 receptor interaction is only seen with olanzapine and clozapine. Recent studies have revealed a role for D_1 receptors in cognitive functions; extensive blockade of D_1 receptors may cause cognitive impairment (WILLIAMS and GOLDMAN-RAKIC 1995).

In a clinical study with a selective D_1 antagonist, no antipsychotic activity was seen; on the contrary, a worsening of symptoms occurred (KARLSSON et al. 1995).

D. Interaction with 5-HT$_2$ and Other 5-HT Receptors

5-HT$_2$ receptor blockade, in particular 5-HT$_{2A}$ receptors, is believed to add to the treatment of the negative symptoms (DUINKERKE et al. 1993; DAVIS and JANICAK 1996; KING 1998). 5-HT$_{2A}$ receptors are densely present in the frontal cortex and the accumbens and were demonstrated to be localised on cortical GABA interneurons and on apical dendrites of glutamatergic neurons (JAKAB and GOLDMAN-RAKIC 1998). 5-HT$_{2A}$ receptors appear to have a role in the regulation of glutamatergic transmission and 5-HT$_{2A}$ antagonists were shown to antagonise behavioural effects of glutamate NMDA antagonists (SORENSEN et al. 1993; SCHMIDT et al. 1995). Since defective glutamatergic transmission is thought to be involved in schizophrenia, the therapeutic effects of 5-HT$_{2A}$ antagonists may be explained in this way (CARLSSON et al. 1997; LEYSEN 2000). Studies using risperidone have shown that the benefits of 5-HT$_{2A}$ receptor blockade are only apparent when there is no over-blockade of central D_2 receptors. It was suggested that the 5-HT$_{2A}$ and D_2 receptor affinity should differ by at least one order of magnitude (MELTZER and NASH 1991). Table 5 shows that this is achieved for risperidone, olanzapine, clozapine and pipamperone. 9-OH risperidone, ziprasidone, zotepine and quetiapine still bind with higher affinity to 5-HT$_{2A}$ than to D_2 receptors, but with a potency difference of less than 10-fold. Haloperidol has a more than 100-fold lower affinity for 5-HT$_{2A}$ than for D_2 receptors. Except for risperidone, 9-OH risperidone and haloperidol, the antipsychotics have, in addition, a relatively high affinity for 5-HT$_{2C}$ and 5-HT$_{2B}$ receptors. Studies on receptor signalling revealed that all the compounds are full antagonists at 5-HT$_{2A}$, 5-HT$_{2B}$ and 5-HT$_{2C}$ receptors. 5-HT$_{2C}$ antagonism may confer anxiolytic properties but can also contribute to weight gain (TECOTT et al. 1996; BROMIDGE et al. 1997). 5-HT$_{2B}$ receptors are scarcely found in the brain and its central role is still enigmatic.

All new antipsychotics have a relatively high affinity for 5-HT$_7$ receptors, at which they probably act as antagonists. 5-HT$_7$ receptors are excitatory receptors with high concentration in the thalamic nuclei, dentate gyrus, cortex and amygdala. The regional distribution of this receptor in the brain suggests that its blockade may be of importance for the treatment of psychotic or mood disorders, yet its particular function is still to be elucidated.

The tricyclic antipsychotics zotepine, olanzapine, clozapine and quetiapine have a relative high affinity for 5-HT$_6$ receptors. Their effect on 5-HT$_6$ receptor signalling is still to be investigated. Recent studies with selective 5-HT$_6$ antagonists have shown an improvement of cognitive functions (SLEIGHT et al. 1997; SLEIGHT et al. 1998).

Ziprasidone, clozapine and quetiapine interact with 5-HT$_1$ receptors, e.g. with 5-HT$_{1A}$ receptors. In general, it is seen, that 5-HT$_{1A}$ receptors are more readily stimulated than blocked by ligands. An investigation of the effect of the antipsychotics on 5-HT$_{1A}$ receptor signalling is required in order to assess possible functional consequences. Stimulation of 5-HT$_{1A}$ receptors can amplify effects produced by 5-HT$_{2A}$ receptor blockade and 5-HT$_{1A}$ agonism may add anxiolytic activity; as such 5-HT$_{1A}$ agonism is expected to confer a beneficial effect (ASHBY et al. 1994). The central role of 5-HT$_{1D}$, 5-ht$_{1E}$ and 5-ht$_{1F}$ receptor is as yet unknown.

E. Interaction with Various Biogenic Amine Receptors

All the antipsychotics, except haloperidol, are relatively potent α_1 adrenoceptor blockers; clozapine, pipamperone and quetiapine are more potent blockers of α_1-adrenoceptors than of D$_2$ receptors. Although some authors have suggested that α_1-adrenoceptor blockade may confer antipsychotic effects, its major effect will be sedation, orthostatic hypotension and reflex tachycardia. The relative potent α_2 receptor blockade, observed with risperidone, 9-OH risperidone, clozapine and quetiapine, and the blockade of the norepinephrine transporter by zotepine could contribute to certain antidepressant actions such as improved motivation and drive.

The tricyclic compounds zotepine, olanzapine, clozapine and quetiapine all exert their most potent action at H$_1$ receptors. As a consequence these compounds are highly sedative. H$_1$ receptor blockade may also lead to substantial weight gain, which may still be aggravated by the concomitant 5-HT$_{2C}$ receptor blockade.

Olanzapine, quetiapine and in particular clozapine are noted for their muscarinic cholinergic receptor interaction. Blockade of these receptors may mask certain effects of D$_2$ receptor blockade such as EPS. However, muscarinic cholinergic receptor blockade may cause confusion, hallucinations, impairment of cognitive functions and induce various peripheral side-effects, such as dry mouth, constipation and urinary retention. Recent studies on muscarinic cholinergic receptor subtypes have revealed that the relative potencies of the compounds at the different subtypes differ (see Table 4).

Moreover, clozapine appeared to be an agonist at m4 receptors, whereas it was an antagonist at the other subtypes (BOLDEN et al. 1993). The functional consequence of differential interactions with the different muscarinic cholinergic receptor subtypes is not understood (for review see CAULFIELD 1993). Distribution studies showed that exocrine glands exclusively contain the excitatory m1 and m3 subtype, whereas in the heart only the inhibitory m2 subtype is found. Other peripheral tissues contain several subtypes. All subtypes occur in the brain and only some regional differences are noted.

F. Future Antipsychotics

In spite of the large number of antipsychotics on the market and almost half a century of experience in the study and use of the drugs, there still is ample room for improvement in the treatment of psychotic disorders: improved and broader therapeutic efficacy, faster onset of action, treatment of resistant patients, treatment of residual symptoms, fewer or no side-effects.

Several compounds are still in clinical development with as basic activity $5\text{-HT}_{2A}/D_2$ antagonism. These compounds are likely to show antipsychotic action but will probably not bring much improvement in therapeutic efficacy. A number of compounds are under study which are D_2 antagonist/5-HT_{1A} agonists, e.g. S-16924 (recently stopped because of cardiac side-effects) and sarizotan. As discussed above, the 5-HT_{1A} agonistic component may add beneficial effects. Sarizotan appears to have less of the unwanted effects than, for instance, α_1 adrenoceptor, H_1 receptor and cholinergic muscarinic receptor blockade (BARTOSZYK et al. 1997). This is an advantage for improving the side-effect profile such as reducing sedation, risk of weight gain, and avoiding muscarinic antagonist side-effects.

Aripiprazole, is a dopamine autoreceptor partial agonist in phase III clinical study in schizophrenia. Through its autoreceptor agonism the drug reduces dopaminergic neuronal activity.

Compounds which interact with the so-called sigma sites, recently identified as enzymes in the sterol synthesis pathway (sigma 1 sites: sterol 7-reductase; sigma 2 sites: sterol delta 8-7 isomerase) (MOEBIUS et al. 1998) are being proposed as potential antipsychotics. Sigma site interaction is a property of many compounds with widely different structures. Compounds on the market, such as haloperidol, ifenprodil, emopamil, have high affinity for sigma sites in addition to their diverse primary pharmacological effects (LESAGE et al. 1995). For over 20 years sigma ligands have been proposed as potential therapeutic agents, but none have reached the market yet.

MDL 100907, the most selective 5-HT_{2A} antagonist known thus far, went into phase III clinical study as a stand-alone treatment for schizophrenia

(SORENSEN et al. 1993; SCHMIDT et al. 1995). No published reports are available; it is said that certain beneficial effects have been observed, but this may not be sufficient as sole treatment.

Sanofi started in 1999 phase II clinical studies with SR142801, a NK3 antagonist, SR-141716A, a cannabinoid antagonist and SR-142948, a neurotensin antagonist. NK3 antagonism and cannabinoid antagonism are expected to mitigate dopaminergic neurotransmission. Neurotensin is localised with dopamine in the mesolimbic dopamine pathway and seems to have a role in the regulation of mesolimbic dopaminergic transmission. Arguments have been put forward for potential therapeutic actions of both neurotensin agonists and antagonists in schizophrenia.

Today, discovery research is much focussed on the glutamatergic system. Mostly based on the observation of symptoms produced by phencyclidine (a glutamate NMDA antagonist), which mimic both positive and negative symptoms of schizophrenia, it has been hypothesised that signalling at the NMDA receptor is impaired in schizophrenia.

Direct NMDA receptor stimulation may rapidly lead to neurotoxicity; therefore, strategies for correcting NMDA receptor signalling indirectly are being explored. Ways to activate the glycine site at the NMDA receptor are being investigated. This can be achieved directly by agonists for this site, e.g. D-cycloserine, or by inhibitors of the glycine transporter type I. Since D-serine has relatively high affinity for the glycine site, activation of serine racemase has been proposed as a possible approach.

AMPAkines are compounds which prolong glutamate signalling at the AMPA receptor by slowing down the rapid desensitisation of this receptor. AMPAkines recently went into early clinical development for schizophrenia.

Metabotropic glutamate receptors can also regulate glutamatergic transmission. mGluR2 agonists are active in certain animal models of psychosis; proof-of-principle studies started with a prototypic compound LY341495 (SCHOEPP et al. 2000).

For an extensive review on future antipsychotics see STAHL and SHAYEGAN (2000).

G. Conclusions

Haloperidol, the most typical antipsychotic, is a highly selective and highly potent D_2 antagonist. Clozapine, the prototype atypical antipsychotic, is the compound with the broadest receptor interaction, hitting at least 22 monoamine receptor subtypes with relevant potency.

Risperidone and 9-OH risperidone are the relatively most "pure" and potent $5\text{-HT}_{2A}/D_2$ antagonist. These compounds clearly show an atypical profile when used at an appropriate low dose. Compounds like olanzapine and quetiapine were designed to match the profile of clozapine. However, it is not clear which are the most relevant properties of clozapine with relation to its "atypical antipsychotic activity". The broader the profile, the less clear the picture and the more likely that various types of side-effects will occur.

Future antipsychotics in clinical development are either still designed on the 5-HT$_{2A}$/D$_2$ antagonist or on the 5-HT$_{1A}$ agonist/D$_2$ antagonist principle. Other approaches aim at modulating dopaminergic transmission by agonism at the dopamine autoreceptor or by interfering with receptors for neuropeptides that affect dopaminergic transmission, such as neurokinins or neurotensin, or by blocking cannabinoid receptors.

Various ways of interfering with glutamatergic transmission are being explored. The clinical demonstration of antipsychotic activity with these new approaches is awaited.

Abbreviations

5-HT/5-ht	5-hydroxytryptamine or (S) serotonin
ACh	acetylcholine
AMPA	α-amino-3-hydroxy-5-methyl-4-isoxazole-4-propionic acid
cAMP	cyclic adenosine monophosphate
D	dopamine
EPS	extrapyramidal symptoms
GABA	γ-aminobutyric acid
H	histamine
h	human
K$_i$	equilibrium inhibition constant
m	muscarinic
mGluR	metabotropic glutamate receptor
NK	neurokinin
NMDA	N-methyl-D-aspartate
n.	nucleus
PET	positron emission tomography
r	rat

References

Ashby CR Jr, Edwards E, Wang RY (1994) Electrophysiological evidence for a functional interaction between 5-HT$_{1A}$ and 5-HT$_{2A}$ receptors in the rat medial prefrontal cortex: an iontophoretic study. Synapse 17:173–181

Arnt J, Skarsfeldt T (1998) Do novel antipsychotics have similar pharmacological characteristics? A review of the evidence. Neuropsychopharmacology 18:63–101

Barnes NM, Sharp T (1999) A review on central 5-HT receptors and their function. Neuropharmacology 38:1083–1152

Bartoszyk GD, Greiner HE, Seyfried CA (1997) Pharmacological profile of EMD 128130: a putative atypical antipsychotic with dopamine D$_2$ antagonist and serotonin 5-HT$_{1A}$ agonistic properties. Soc Neurosci Abstr 23 (part 1):530

Beijsterveldt van LEC, Geerts RJF, Leysen JE, Megens AAHP, Van den Eynde HMJ, Meuldermans WEG, Heykants JJP (1994) Regional brain distribution of risperidone and its active metabolite 9-hydroxy-risperidone in the rat. Psychopharmacology 114:53–62

Bench CJ, Lammertsma AA, Dolan RJ, Grasby PM, Warrington SJ, Gunn K, Cuddigan M, Turton DJ, Osman S, Frackowiak RSJ (1993) Dose dependent occupancy of central dopamine D$_2$ receptors by the novel neuroleptic CP-88,059-01: a study using positron emission tomography and ^{11}C-raclopride. Psychopharmacology 112: 208–314

Bergson C, Mrzljak L, Smiley JF, Pappy M, Levenson R, Goldman-Rakic PS (1995) Regional, cellular and subcellular variations in the distribution of D_1 and D_5 dopamine receptors in primate brain. J Neurosci 15:7821–7836

Bolden C, Cusack B, Richelson E (1991) Antagonism by antimuscarinic and neuroleptic compounds at the five cloned human muscarinic cholinergic receptors expressed in Chinese hamster ovary cells. J Pharmacol Exp Ther 260:576–580

Bristow LJ, Kramer MS, Kulagowski J, Patel S, Ragan CI, Seabrook GR (1997) Schizophrenia and L-745,870, a novel dopamine D_4 receptor antagonist. Trends Pharmacol Sci 18:186–188

Bromidge SM, Duckworth M, Forbes IT, Ham P, King FD, Thewlis KM, Blaney FE, Naylor CB, Blackburn TP, Kennett GA, Wood MD, Clarke SE (1997) 6-Chloro-5-methyl-1-[[2-[(2-methyl-3-pyridyl)oxy]-5-pyridyl]carbamoyl]-indoline (SB-242084): the first selective and brain penetrant 5-HT$_{2C}$ receptor antagonist. J Med Chem 40:3494–3496

Bymaster FP, Calligaro DO, Falcone JF, Marsh RD, Moore NA, Tye NC, Seeman P, Wong DT (1996) Radioreceptor binding profile of the atypical antipsychotic olanzapine. Neuropsychopharmacology 14:87–96

Caulfield MP (1993) Muscarinic receptors – Characterization, coupling and function. Pharmac Ther 58:319–379

Ciliax BJ, Nash N, Heilman C, Sunahara R, Hartney A, Tiberi M, Rye DB, Caron MG, Niznik HB, Levey AI (2000) Dopamine D_5 receptor immulocalization in rat and monkey brain. Synapse 37:125–145

Creese I, Burt DR, Snyder SH (1976) Dopamine receptor binding predicts clinical and pharmacological potencies of antischizophrenic drugs. Science 192:481–483

Carlsson A, Lindqvist M (1963) Effect of chlorpromazine or haloperidol on formation of 3-methoxytryramine and normetanephrine in mouse brain. Acta Pharmacol Toxicol 20:140–144

Carlsson A, Hansson LO, Waters N, Carlsson ML (1997) Neurotransmitter aberrations in schizophrenia: new perspectives and therapeutic implications. Life Sci 61:75–94

Carpenter WT (1995) Serotonin–dopamine antagonists and treatment of negative symptoms. Clin Psychopharmacol 16 [suppl 1]:30S–35S

Davis JM, Janicak PG (1996) Risperidone: a new, novel (and better?) antipsychotic. Psychiatr Ann 26:78–87

Delay J, Deniker P, Harl JM (1952) Traitement des états d'excitation et d'agitation par une méthode médicamenteuse dérivée de l'hibernothérapie. Ann Méd Psychol 110:267–273

Divry P, Bobon J, Collard J (1958) Le R01625 nouvelle thérapeutique symptomatique de l'agitation psychomotrice. Acta Neurol Psychiat Belg 58:878–888

Docherty JR (1998) Subtypes of functional α_1- and α_2-adrenoceptors. Eur J Pharmacol 361:1–15

Duinkerke SJ, Botter PA, Jansen AAI, Van Dongen PA, Van Haaften AJ, Boom AJ, Van Laarhoven JH, Busard HL (1993) Ritanserin, a selective 5-HT$_{2A/2C}$ antagonist, and negative symptoms in schizophrenia: a placebo-controlled double-blind trial. Br J Psychiatry 163:451–455

Hoyer D, Martin GR (1996) Classification and nomenclature of 5-HT receptors: a comment on current issues. Behav Brain Res 73:263–268

Hoyer D, Martin G (1997) 5-HT receptor classification and nomenclature: towards a harmonisation with the human genome. Neuropharmacology 36:419–428

Jakab RL, Goldman-Rakic P (1998) 5-hydroxyhyptamine 2 A serotonin receptors in the primate cerebral cortex: possible site of action of hallucinogenic and antipsychotic drugs in pyramidal cell apical dentrites. Proc Natl Acad Science USA 95:735–740

Janssen PAJ, Van de Westeringh C, Jageneau AHM, Demoen PJA, Hermans BKP, Vandaele GHP, Schellekens KHL, Van der Eycken CAM, Niemegeers CJE (1959) Chemistry and pharmacology of CNS depressants related to 4-(4-hydroxy-4-phenylpiperidino) butyrophenone. Part I. Synthesis and screening data in mice. J Med Pharmac Chem 1:281–297

Janssen PAJ, Niemegeers CJE, Awouters F, Schellekens KH, Megens AA, Meert TF (1988) Pharmacology of risperidone (R64766), a new antipsychotic with serotonin-S_2 and dopamine-D_2 antagonistic properties. J Pharmacol Exp Ther 244: 685–693
Jones H (1997) Risperidone: A review of its pharmacology and use in the treatment of schizophrenia. J Serotonin Res 4:17–28
Karlsson P, Smith L, Farde L, Harwyd C, Sedvall G, Wiesel FA (1995) Lack of apparent antipsychotic effect of the D_1-dopamine receptor antagonist SCH 39166 in acutely ill schizophrenic patients. Psychopharmacology 121:309–316
King DJ (1998) Drug treatment of the negative symptoms of schizophrenia. Eur Neuropsychopharmacol 8:33–42
Kuenstler U, Juhnhold U, Knapp WH, Gertz H-J (1999) Positive correlation between reduction of handwriting area and D_2 dopamine receptor occupancy during treatment with neuroleptic drugs. Psychiat Res: Neuroimaging Section 90:31–39
Lahti RA, Roberts RC, Cochrane EV, Primus RJ, Gallager DW, Tamminga CA (1998) Dopamine D_4 receptors in human postmortem brain tissue of normal and schizophrenic subjects. An [^{3}]NGD-94-1 study. Schizophr Res 29:93
Lesage A, De Loore KL, Peeters L, Leysen JE (1995) Neuroprotective sigma ligands interfere with the glutamate-activated NOS pathway in hippocampal cell culture. Synapse 20:156–164
Leysen JE (1984) Receptors for neuroleptic drugs. In: Burrows GD, Werry JS (eds) Advances in Human Psychopharmacology, vol 3, JAI Press Inc, pp 315–356
Leysen JE (2000) Receptor profile of antipsychotics. In: Ellenbroek BA, Cools AR (eds) Atypical antipsychotics, Birkhäuser Verlag, Basel Boston Berlin, pp 57–81
Leysen JE, Janssen PMF, Schotte A, Luyten WHML, Megens AAHP (1993) Interaction of antipsychotic drugs with neurotransmitter receptor sites in vitro and in vivo in relation to pharmacological and clinical effects: role of $5HT_2$ receptors. Psychopharmacology 112: S40–S54
Leysen JE, Gommeren W, Eens A, De Chaffoy de Courcelles D, Stoof JC, Janssen PAJ (1988) The biochemical profile of risperidone, a new antipsychotic. J Pharmacol Exp Ther 247:661–670
Leysen JE, Gommeren W, Van Gompel P, Wynants J, Janssen PFM, Laduron PM (1985) Receptor binding properties in vitro and in vivo of ritanserin: A very potent and long acting serotonin-S_2 antagonist. Mol Pharmacol 27:600–611
Leysen JE, Niemegeers (1985) Neuroleptics. In: Lajtha A (ed) Handbook of Neurochemistry, vol 9, Plenum Publishing Corporation, pp 331–361
Leysen JE, Niemegeers CJE, Tollenaere JP, Laduron PM (1978) Serotonergic component of neuroleptic receptors. Nature 272:168–171
Lindstrom LH (1989) A retrospective study on the long-term efficacy of clozapine in 96 schizophrenic and schizoaffective patients during a 13-year period. Psychopharmacology 99 [suppl]:84–86
Meltzer HY, Nash JF (1991) Effects of antipsychotic drugs on serotonin receptors. Pharmacol Rev 43:587–604
Migler BM, Warawa EJ, Malick JB (1993) Seroquel: behavioral effects in conventional and novel tests for atypical antipsychotic drug. Psychopharmacology 112:299–307
Moebius FF, Reiter RJ, Bermoser K, Glossman H, Cho SY, Paik YK (1998) Pharmacological analysis of sterol delta-8-delta-7 isomerase proteins with [^{3}H]ifenprodil. Mol Pharmacol 54:591–598
Moore NA, Tye NC, Axton MS, Risius FC (1992) The behavioral pharmacology of olanzapine, a novel "atypical" antipsychotic agent. J Pharmacol Exp Ther 262:545–551
Needham PL, Atkinson J, Skill MJ, Heal DJ (1996) Zotepine: preclinical tests predict antipsychotic efficacy and an atypical profile. Psychopharmacol Bull 32:123–128
Nyberg S, Nakashima Y, Nordström AL, Halldin C, Farde L (1996) Positron emission tomography of in-vivo binding characteristics of atypical antipsychotic drugs. Review of D_2 and $5-HT_2$ receptor occupancy studies and clinical response. Br J Psychiatry 168 [suppl 29]:40–44

Roth BL, Tandra S, Burgess LH, Sibley DR, Meltzer HY (1995) D-4 Dopamine receptor binding affinity does not distinguish between typical and atypical antipsychotic drugs. Psychopharmacology 120:365–368

Reyntjens A, Gelders YG, Hoppenbrouwers M-LJA, Vanden Bussche G (1986) Thymostenic effects of ritanserin (R55667), a centrally acting serotonin-S_2 receptor blocker. Drug Dev Res 8:205–211

Saller CF, Salama AI (1993) Seroquel: biochemical profile of a potential atypical antipsychotic. Psychopharmacology 112:285–292

Schmidt CJ, Sorensen SM, Kehne JH, Carr AA, Palfreyman MG (1995) The role of 5-HT_{2A} receptors in antipsychotic activity. Life Sci 56:2209–2222

Schoepp DD, Cartmell J, Ornstein P, Monn JA (2000) *In vivo* pharmacology of group II metabotropic glutamate receptor agonists: novel agents for psychiatric disorders. Neuropsychopharmacology 23:S2, S86

Schotte A, Janssen PFM, Gommeren W, Luyten WH, Van Gompel P, Lesage AS, De Loore K, Leysen JE (1996) Risperidone compared with new and reference antipsychotic drugs: in vitro and in vivo receptor binding. Psychopharmacology 124:57–73

Seeger TF, Seymour PA, Schmidt AW, Zorn SH, Schulz DW, Lebel LA, McLean S, Guanowsky V, Howard HR, Lowe JA III, Heym J (1995) Ziprasidone (CP-88,059): a new antipsychotic with combined dopamine and serotonin receptor antagonist activity. J Pharmacol Exp Ther 275:101–113

Seeman P (1995) Dopamine receptors and psychosis. Sci Am 273:28–37

Seeman P, Lee T, Chau-Wong M, Wong K (1976) Antipsychotic drug doses and neuroleptic/dopamine receptors. Nature 261:717–719

Sirviö J, MacDonald E (1999) Central α_1-adrenoceptors: their role in the modulation of attention and memory formation. Pharmacol Ther 83:49–65

Sleight AJ, Boess FG, Bos M, Levettrafit B, Riemer C, Bourson A (1998) Characterization of Ro 04-6790 and Ro 63-0563 – potent and selective antagonists at human and rat 5-HT_6 receptors. Br J Pharmacol 124:556–562

Sleight AJ, Boess FG, Bourson A, Sibley DR, Monsma FJ Jr (1997) 5-HT_6 and 5-HT_7 receptors: molecular biology, functional correlates and possible therapeutic indications. DN & P 10:214–224

Sorensen SM, Kehne JH, Fadayel GM, Humphreys TM, Ketteler HJ, Sullivan CK, Taylor VL, Schmidt CJ (1993) Characterization of the 5-HT_2 receptor antagonist MDL 100907 as a putative atypical antipsychotic: behavioral, electrophysiological and neurochemical studies. J Pharmacol Exp Ther 266:684–691

Stahl SM, Shayegan DK (2000) New discoveries in the development of antipsychotics with novel mechanisms of action: beyond the atypical antipsychotics with serotonin dopamine antagonism. In: Ellenbroek BA, Cools AR (eds) Atypical antipsychotics, Birkhäuser Verlag, Basel Boston Berlin, pp 215–232

Tecott LH, Sun LM, Akana SF, Strack AM, Lowenstein DH, Dallman MF, Julius D (1996) Eating disorder and epilepsy in mice lacking 5-HT_{2C} serotonin receptors. Nature 374:542–546

Vanhauwe JFM, Ercken M, Van de Wiel D, Jurzak M, Leysen JE (2000) Effects of recent and reference antipsychotic agents at human dopamine D_2 and D_3 receptor signaling in Chinese hamster ovary cells. Psychopharmacology 150:383–390

Van Rossum JM (1966) The significance of dopamine receptor blockade for the mechanism of action of neuroleptic drugs. Arch Int Pharmacodyn Ther 160:492–494

Van Wielinck PS, Leysen JE (1983) Choice of neuroleptics based on in vitro pharmacology. J Drug Res 8:1984–1997

Williams GV, Goldman-Rakic PS (1995) Modulation of memory fields by dopamine D_1 receptors in prefrontal cortex. Nature 376:572–575

CHAPTER 27
Sleep and Wake Cycle

J. BIERBRAUER and L. HILWERLING

A. Introduction

At the end of the second decade of the twentieth century, Berger conducted the electroencephalograph (JOUVET 1972). BREMER (1935) carried out the first investigations by conduction transection experiments in cats. A preparation done in the intercollicular level (cerveau isolé) created a persistent slow-wave state in the midbrain, while an incision in the medulla oblongata (encephale isolé) did not alter the sleep–wake cycle. Because of this, he assumed that sleep is a passive state, reversible to deafferentation (BREMER 1938). MORUZZZI and MAGOUN (1949) demonstrated the effect of the ascending activating reticular system by causing arousals in the cerveau isolé preparation. In further transection experiments, NAUTA (1946) observed a long-lasting insomnia when setting a lesion in the rostral part of the hypothalamus. Lesions in the mammillary bodies induced sleep, while combined lesions had no effect.

A landmark in modern sleep diagnosis was the discovery of rapid eye movement (REM) sleep by ASERINSKY and KLEITMANN (1953), introducing the dichotomous classification of REM sleep and slow-wave sleep (SWS).

JOUVET (1962) presented data pointing to the pontomedullary region as the trigger zone for REM sleep. Activation is mediated by the cholinergic system from pedunculopontine tegmental and laterodorsal tegmental nuclei through the monoaminergic system of the raphe dorsalis (RD) and locus coeruleus (LC) via the thalamus to the cerebral cortex (JONES 1990). Injections of acetylcholine (ACh) into the medial forebrain bundle, connecting the preoptic area, the lateral hypothalamus and the limbic system produce SWS (VELLUTI 1963). The descending system and the ascending part arising from the spinal cord meet in the pons cerebri (HERNÁNDEZ-PEÓN 1965). Priming the REM sleep was thought to be a function of the caudal raphe system, which interacts with the LC. In particular the caudal two thirds of the LC are thought to be responsible for executing REM, the cranial third is involved in the control of the muscle tone, the medial third communicates with pontine pacemakers of the ponto-geniculo-occipital (PGO) activity. STERIADE et al. (1990) demonstrated the reduction of this activation during SWS in the thalamus cells. SWS sleep is admitted by the serotonergic (5-HT) activation of the RD.

Inhibition of the tryptophan hydroxylation reduces SWS and REM sleep by producing insomnia (Mouret 1968). Giving dopa to cats that have been treated with reserpine reduces REM sleep latency (Matsumoto 1964), alpha-methyl-dopa dislocates norepinephrine (NE) and suppresses REM (Dusan-Peyrenthon and Jouvet 1968). Disulfiram, a NE synthetase blocker, attenuates the quantity of REM (Dusan-Peyrenthon 1968).

There are controversial theories concerning the monoaminergic control of sleep mechanisms. Hobson and Wyzinski (1975) named the cells of the RD and the LC "REM-off" cells because they show quietness during REM and a maximum discharge while awakening. The neurons of the pontine reticular formation (PRF) discharge during REM (REM-on cells). The cells from the LC and the RD are silent in REM (REM-off cells). It is suggested that the arrest of REM-off-cells disinhibits REM-on cells. While auto-excitatory loops to REM-on cells are described, REM-off cells seem to be modulated by auto-inhibitory loops.

A REM period is terminated when the excitatory influence of the excitatory effect of the PRF reaches the monoaminergic activity. The present model reflects on the existence of cell groups rather than of the dichotomy of REM-off and REM-on cells. On the other hand, there are connections from the suprachiasmatic nucleus (SNC) to the RD and to the LC as well, so there is evidence that the circadian system controls the REM-off cells (McCarley 1986).

Lesions in selective parts of the RD or in the LC cause permanent PGO activity but do not purge SWS and REM sleep (Laguzzi and Pujol 1987; Froment and Bertrand 1974). After eliminating 5-HT, cats had a normal sleep–wake cycle 5 days later (Dement and Henriksen 1972). At the least, findings show a lowered 5-HT level during sleep (Puizillout and Daszuta 1979).

B. Dopaminergic Action in Sleep

Initial data suggested that dopamine is only involved in wakefulness. Today, data support the theory that there is involvement in REM sleep. In the following description of various drug actions, we even use the term REM for paradoxical sleep in non-human species and SWS for slow-wave sleep in animals and humans. Because of the small number of published studies, we have to refer to D_1- and D_2-like agonists and antagonists. Only a summary of more specific studies will be given.

I. D_2 Antagonists

There is rich information available on the action of haloperidol on sleep. The kind of action varies from species to species. When SWS is increased in cats (Monti 1968), the amount of REM sleep declines (Takeuchi 1973). In dogs

SWS and REM both increase (WAUQUIR and NIEMEGEERS 1980), in rats the quantity of SWS was lowered, but the component of REM expanded (MONTI 1979; STILLE 1974; TSUCHIYA and FOKUSHIMA 1979).

Not only the variation of species changed the findings, but also the compound used altered the results. In investigations with flupenthixol in rats, SWS was increased but REM remained unchanged (FORNAL and RADULOVACKI 1982). When loxapine was used, a REM suppression was observed in cats (TSUCHIYA and FOKUSHIMA 1979; SCHMIDEK et al. 1974).

These discrepancies have not been explained. Interesting is the finding that the action is dose-related: lower doses of pimozide decreased REM and SWS, intermediate doses did not alter the quantity of REM and SWS, whereas higher doses lowered the parts of SWS and REM (FORNAL and RADULOVACKI 1982). The sensitivity of the dopaminergic neurons for D_2-like antagonists seems to be highest in phases of lowest dopamine turnover, for example in wakefulness (GESSA et al. 1985; TRAMPUS 1990).

II. D_2 Agonists

Apomorphine, a D_2 agonist with weak action on D_1, is one of the best studied drugs in sleep experiments. In rats, wakening phases were prolonged (KAFI 1976). The application of low doses was followed by a significant increase in sleep time (MEREU et al. 1979). The prolonged time was due to the extended SWS phase. There was no correlation to sleep quality and quantity in prior baseline nights (WAUQUIR 1985).

Therefore, the effect of apomorphine is a biphasic action: low doses reduce motility and are hypnotic, while a higher dose leads to a reduction in sleep. In a study by MONTI et al. (1988), similar results for other D_2-like agonists were observed. Only pergolide attenuated REM sleep. When haloperidol was given additionally, the effects of the low-dose application were reversed.

In another study MONTI and FERNANDEZ (1989) explored the response of rats after applying the selective agonist quinpirole. To differentiate the effect of postsynaptic and presynaptic antagonists, apomorphine was combined with one of the latter substances. While blocking the presynaptic sites, they could reverse the suppression of the hypnotic effect. When the postsynaptic receptor sites were blocked, REM sleep was not enlarged to the former baseline amount. So, there is a better correlation of triggering SWS and REM via the presynaptic sites.

In dogs, a dose-related decrease of REM and SWS was observed. In a series of examinations, WAUQUIR and JANSSEN (1980) viewed the interaction of apomorphine and the peripheral-acting D_2 antagonist domperidone. In addition, the interaction of pimozide and domperidone was tested. Only the emetic effect was prevented; the sleep pattern was unchanged. When pimozide was applied, REM decline mediated through low doses of apomorphine was antagonised; in higher doses of apomorphine there was only partial antagonism.

III. D_1 Antagonists

Best data are available for the SCH23390 compound. Synchronisation of the electroencephalogram (EEG) occurred when SCH 23390 was given to rats (GESSA et al. 1985). In lower doses TRAMPUS and ORGINI found an increased amount of SWS and REM (TRAMPUS 1990).

TAKEUCHI (1973) suggested that REM sleep is unrelated to the appearance of SWS because of the inflated SWS/REM relation after the application of SCH23390 and decreased relation after haloperidol. In further studies, MONTI found no altered latency in REM, but a significant decline of wake state and REM sleep was observed; SWS was significantly augmented (MONTI and JANTOS 1990a).

IV. D_1 Agonists

SKF38393 was given in doses of 0.1–4.0mg/kg i.p. to rats. REM was only decreased at the highest doses (MONTI 1968). This overall effect was only weak because of ineffective penetration into brain. On the other hand SKF38393, in combination with the application of SCH23390, depressed REM sleep. SWS sleep decrease could not be found when SKF38393 and SCH23390 were again applied together.

Instead of a decrease in REM sleep an increase was found (TRAMPUS 1990). This might be due to the doses used. They worked with doses ranging from 0.1 to 2.0mg/kg given i.v., while MONTI and JANTOS (1990a) used doses up to 0.003mg/kg.

V. More Specific Studies

In rats, pramipexole, an agonist with a high affinity for D_3 receptors, had an biphasic action (ABERCROMBIE and DEBOER 1997). At lower doses of 30µg/kg REM and NREM sleep were increased, while wakefulness was reduced. In doses of 500µg/kg wakefulness was increased, REM and NREM sleep were reduced. YM-09151-2, a mixed D_2 and D_3 receptor antagonist, prevented the increase of REM initiated by pramipexole. In higher doses (500–1000µg/kg) the antagonist reversed reduction of SWS induced by the 500µg/kg dose of the pramipexole. WIN35428, a potent antagonist of the dopamine transporter, showed biphasic effects upon REM and locomotor activity with low doses increasing and high doses decreasing REM (DE SAINT et al. 1995).

The effect of the dopamine autoreceptor antagonist (–)DS121 in rat was observed by OLIVE et al. (1998). In rats entrained to a light-dark cycle, (–)DS121 dose-dependently increased wakefulness, locomotor activity and body temperature, and decreased both NREM and REM sleep during the first 4h post-treatment. REM interference lasted up to 3h longer than NREM. Low doses of (–)DS121 (0.5 and 1.0mg/kg) produced little waking that was not followed by significant compensatory sleep responses. In contrast, higher doses (5.0 and 10.0 mg/kg) produced compensatory hypersomnolence.

VI. Catecholaminergic Pathway Modulation

Depleting substances like reserpine led to suppression of cortical activation. Confirmatory studies in rats and rabbits revealed that there is a reduction in REM and SWS (GOTTESMANN 1966; TABUSHI 1969). REITE et al. (1969) evaluated an increase in REM. After injections of reserpine, the enlarged amount of SWS was strongly correlated with the homovanillic acid (HVA) (BUCKINGHAM 1976). PGO waves were induced in cats after reserpine treatment, although they were induced without other signs of REM (DELORME and JOUVET 1965).

Disulfiram raised dopaminergic levels by inhibiting of the dopamine decarboxylase. In cats, an increase in SWS and a reduction in PGO waves and REM appeared (PEYRETHON-DUSAN 1968). There are confusing results from studies with inhibition of the tyrosine hydroxylase. TORDA (1968) published data with a decrease in REM. On the other hand, an increased amount of REM was also observed (TORDA 1968; KAFI and CONSTANTINIDIS 1977).

VII. Temperature Regulation

Many studies have been carried out on changes of metabolism and body temperature control and sleep (BACH et al. 1994; BERGER and PHILLIPS 1988; FRIEDMAN et al. 1994). For example, STOHERS and WARNER (1984) observed that the metabolism rate at neutral temperature revealed higher rates in REM sleep than in NREM. In a cooler situation the differences were potentiated.

The core temperature in humans is strongly correlated to the timing and the duration of sleep. In studies with environments free of time cues, the bedtime is near the nadir of temperature (CZEISLER et al. 1980a). At the minimum of the core temperature, the REM sleep latency is at its shortest and the amount of REM sleep is at its longest (CZEISLER et al. 1980b).

There are two main hypotheses for the regulation of NREM sleep and the way temperature is mediated. While NREM is found highest when the body temperature is downregulated, so NREM might be a primary function of energy conservation (BERGER and PHILLIPS 1990). In another hypothesis, a counter of built heat loads is favoured (McGINTY and SZYMUSIAK 1990). So in conclusion, there is clear evidence that sleep is regulated by temperature and vice versa.

There are data for the D_2 agonist lisuride and pergolide causing temperature changes (SÁNCHEZ and ARNT 1992; ZARRINDAST and TABATABAI 1992). These agonists are consistent with the data that apomorphine induces hypothermia in mice. Partial agonists like 3-(4-(4-phenyl-1,2,3,6-tetrahydropyridil)-(L))-butyl)-indole were inactive by themselves, but apomorphine induced hypothermia (HJORTH et al. 1985). While there is only little support for the argument of mediating temperature through D_2 agonists, the data for D_1 agonists are less clear. Most studies failed to demonstrate a decrease in body temperature after injection of SKF38393 (FAUNT and CROCKER 1987). Indeed, in several publications an increase in rodent body temperature was seen (SÁNCHEZ 1989).

After applying reserpine, hypothermia is observed (HJORTH et al. 1985). D_1 agonists and D_2 agonists produce increases in body temperature in these pretreated rodents. Nevertheless the problem is complex because of the interaction of NE and 5-HT.

C. Pharmacological Interactions

As was already mentioned, the monoaminergic system is closely linked to the regulation of sleep. There are broad known interactions with the dopaminergic system. Interactions with other systems are discussed, but there are little data available.

I. Serotonin

The dopamine and 5-HT interactions are evident at the neuroanatomical level. 5-HT neurons project from the RD and median raphe nucleus and ascend to the ventral tegmental area (VTA) and substantia nigra (SN) and further to the dopaminergic neuron projection fields in the forebrain (HERVÉ et al. 1979).

There are two mechanisms described for release of 5-HT in the brain in the sleep–wake cycle. During waking, the axonal release in the hypothalamus is increased, while a reduction is seen in SWS and in REM (CESPUGLIO and JOUVET 1988). The local release of 5-HT is mediated via dendritic connectivities of the RD. Hereby there is a decrease of 5-HT level examined in a waking state, and an increase while SWS or REM is occurring. It was postulated, therefore, that serotonin is autoinhibitory.

There is an important regulation of the serotonergic system by dopamine and vice versa (BERGER and STRICKER 1985). Serotonin inhibition after stimulation of D_1 receptors has been shown (WHITAKER-AZMITIA and SHEMER 1990).

Input from catecholaminergic neurons to the RD are known (TANAKA et al. 1994; PEYRON et al. 1996). In the SN, the serotonergic fibres are more frequent in the pars reticulata than in the pars compacta (MORI and YAMADA 1985; MORI et al. 1985). Direct synaptic junction between serotonin and dopamine are known (LIPOSITS and PAULL 1987).

A significant proportion of the neurons of dorsal raphe complex contain substance P (SP) (BAKER et al. 1991). KHAN et al. (1998) demonstrated the modulation of dopamine by substance P. Because of the finding that SP depresses REM sleep, studies in patients with a narcolepsy were carried out. The narcolepsy is a REM-associated disorder with imperative sleepiness and cataplexies and expression of a weak muscle tone in emotionally triggered situations. In conclusion, with other groups we found decreased levels of SP in the cerebrospinal fluid (CSF), but HVA was increased (STRITTMATTER et al. 1996). These findings might show supersensitivity of the dopaminergic system in narcolepsy, even receptor density was not changed in positron emission tomography (RINNE et al. 1996; STAEDT et al. 1996). Overall, measured monoaminergic levels in CSF are only poor correlated with the integrity of the sleep–wake cycle (STERN 1973).

II. Adrenergic System

NE is converted from dopamine by the dopamine-β-hydroxylase. The LC is thought to be a trigger for REM sleep (vide supra).

The sleep of two patients with central and peripheral dopamine beta-hydroxylase deficiency was studied (TULEN et al. 1991). Untreated, they had an enlarged amount of REM and a decreased pattern of SWS sleep. After restoring the norepinephrine production with D,L-threo-3,4-dihydroxyphenylserine, REM sleep was facilitated. In cats, prazosin, an α_1 antagonist, expanded significantly REM when given in lower doses (HILAKIVI and PUTKONEN 1980; HILAKIVI 1984). Higher doses cut SWS sleep down. In experiments with cats, when prindamine (an NE uptake blocker) and prazosin were given, REM latency was drawn out and SWS lessened. It was speculated that inhibition of NE uptake leads to reduced REM and α_1 antagonism returns REM augmentation.

There are only few data on α_1 agonism. PICKWORTH and NOZAKI injected dogs (1977). REM was completely eliminated in higher doses for a minimum of 2 h. These effects could be resolved by the administration of phenoxybenzamine, an α_1 antagonist.

Phentolamine, a weak α_2 antagonist, led to an increase in REM in cats (PUTKONEN and 1977a). Antagonistic effects after applying of α-methyldopa were also observed (LEPPÄVOURI 1978).

Overall, there are no data for the selective α_2 agonist effect on sleep. Clonidine was used in several studies; even this drug is now thought to mediate action via the sigma receptor. In rats, SWS and REM was decreased, but wakening was reduced (KLEINLOGEL and SAYERS 1975). The same effect, a lowering of SWS and REM pattern, was noticed in other studies (PUTKONEN and STENBERG 1977b; LEPPÄVOURI 1980).

The induced SWS was revered by α_2 antagonism, while phenoxybenzamine did not have any significant force in preventing this action (FLORIO and LONGO 1975). The action of β antagonist propranolol on sleep appears to be mediated via β_1 receptors. The REM sleep was repressed when this drug was given to rats (MENDELSON et al. 1980). When applied in a normal phase of darkness, REM latency was prolonged and SWS was increased. Isoproterenol in rats reversed REM decrease derived by propranolol. Insomnia induced by propranolol could not be antagonist by clenbuterol and salbutamol, two β_1 agonists.

III. Acetylcholine

Cholinergic brainstem projections to the thalamus and midbrain dopamine neurons affect basic arousal processes (for example sleep–wake cycle) and behavioural activation (EVERITT and ROBBINS 1997; MOORE and 1993). Moreover, cholinergic neurons of the pedunculopontine nucleus (Ch5) and laterodorsal tegmental nucleus (Ch6) activate dopamine neurons of the substantia nigra, zona compacta (A9) and ventral tegmental area (A10) via muscarinic and nicotinic receptors (FUTAMI et al. 1995). These pathways activate the pontine reticular formation and induce REM sleep. In patients with chronic schizophrenia early-onset REM sleep that can result from cholinergic Ch5 and Ch6 activation can be found (YEOMANS 1995). The LC in rats has multiple afferent projections arising from neurons containing ACh and dopamine. Further, there are differences in afferent projections to the noradrenergic and cholinergic regions of the LC (SAKAI 1991). The cholinergic system projects to basal forebrain, hypothalamus, brainstem and spinal cord (MOORE 1993). The caudate nucleus is suggested to participate in regulation of the sleep–wakefulness regulation through modulation of thalamo-cortical and hypothalamopaleocortical integration during SWS in Wistar rats (OGANESIAN et al. 1997). After REM sleep deprivation, rat striatum shows alterations in cholinergic and dopaminergic mechanism. Dopamine levels are increased by to 133% and ACh 28% after 10 days (GHOSH et al. 1976). Although dopaminergic–cholinergic interaction in striatum is well described (CONSOLO et al. 1987, 1996; LOGIN and HARRISON 1996; ABERCROMBIE and DEBOER 1997), further investigations are needed to understand the role of striatum in regulation of sleep–wakefulness.

The model of vigilance-controlling apparatus (VCA) suggests that dopaminergic and cholinergic systems upregulate vigilance through enhancing reactivity in the neuronal networks that subserve the organisation of behavioural components. Additional ACh pathways upregulate the vigilance of higher functions, whereas dopaminergic pathways regulate the reactivity of various motor systems (KOELLA 1984).

The interaction of eserine and reserpine let REM sleep appear in normal quantity, while only the treatment with reserpine led to a lessened REM (KARCZMAR and SCOTTI 1970).

IV. Histamine

Neurophysiological, neurochemical and neuropharmacological evidence indicates that cerebral histamine is an important regulator of wakefulness. Histamine neurons play a role in the regulation of vigilance during waking state (HILAKIVI 1987). During a cat's wakefulness, histaminergic neurons display regular discharge of up to 2.3 spikes per second. When the cat enters SWS, the discharging rate decreases up to 0.43 spikes per second (YOSHIMOTO et al. 1989). During deep SWS and REM sleep, all the neurons become silent, like noradrenergic and serotonergic REM-off cells (SAKAI et al. 1990).

1. H_1 Receptor

The central administration of histamines stimulates mesolimbic H_1 receptors but has no effect upon the activity of nigrostriatal dopaminergic neurons (FLECKENSTEIN et al. 1994a). Histamine increases dopaminergic neuron activity projecting to the suprachiasmatic, caudal periventricular and paraventricular hypothalamic nuclei. These led to a decrease in vigilance (FLECKENSTEIN et al. 1994b). The oral application of H_1 receptor antagonists like, promethazine or diphenhydramine lead to a decrease in vigilance (KUDO and KURIHARA 1990). These substances attenuate the activity of the dopaminergic neurons in the brain. The same effect can be reached with D_2 antagonists. These findings suggests that H_1 receptor downregulates the activity of the H_2 receptors (MONTI et al. 1986). On the other hand, there are no dopaminergic presynaptic receptors modulating the histamine release in rabbits (NOWAK 1985).

2. H_2 Receptor

To date there have been no results or studies suggesting that H_2 receptors are involved in sleep–wake regulation (FLECKENSTEIN et al. 1994b). The H_2 receptor antagonist zolantidine may activate the mesolimbic dopaminergic system, but there only a few cases of sleep–wake changes after administration of zolantidine described (MONTI et al. 1990b). However, the oral application of cimetidine given to healthy volunteers increased SWS and number of movements during sleep (NICHOLSON et al. 1985b; NICHOLSON 1985a).

3. H_3 Receptor

The dopaminergic nerve terminals in the mouse striatum are endowed with presynaptic H_3 receptors. Through these presynaptic H_3 receptors histamine inhibits the dopamine release in mouse striatum. Simultaneous blockage of dopamine autoreceptors increases the extent of H_3 receptor-mediated inhibition of dopamine release (SCHLICKER et al. 1993). Exogenous histamine injected intracerebroventricularly induced a biphasic effect: initial transitory hypoactivity followed by hyperactivity expressed by locomotion frequency.

The hypoactivity response is probably due to activation of H_3 receptors as heteroreceptors reducing the activity of the striatal dopaminergic system. The hyperactivity is induced by H_1 receptor activation. Both effects can overlap (CHIAVEGATTO et al. 1998). This biphasic effect may also occur in sleep–wake cycles.

V. GABAergic System

A major function of γ-aminobutyric acid (GABA) neurons and receptors is the regulation of the nigrostriatal dopamine pathway and the expression of dopamine receptor-mediated events. This modulation occurs via three mechanisms: first, through a tonic inhibition of dopamine neuron activity regulating the dopamine synthesis turnover and release; second, via long-term modulation controlling striatal dopamine receptor numbers; and third by modification of the expression of dopaminergic transmission distal to the dopaminergic synapse (LLOYD et al. 1985; VILA et al. 1996). Moreover, GABA receptor agonists such as progabide decrease dopamine turnover in basal ganglia (BARTHOLINI 1985). On the other hand, activation of the presynaptic D_1 receptors led to a stimulation of GABA release in basal ganglia of rats (ACEVES et al. 1991).

In current working REM-sleep models, the central hypothesis is the hyperpolarisation of cholinergic pedunculopontine (PPN) neurons by serotonin. It is suggested that modulation of REM sleep by PPN involves adjacent glutaminergic neurons and alternates afferent neurotransmitters. Dopamine-sensitive GABAergic pathways appear to be the most promising and most likely to be clinically relevant. These pathways excite the main output nuclei of the basal ganglia and the adjacent forebrain nuclei. The GABAergic pathways are ideally sited to modulate the hallmarks of REM sleep. Each originates from a functionally unique forebrain circuit and terminates in a unique pattern upon brainstem neurons. Sleep disorders with changes in quantity, timing and quality of REM sleep are often associated with changes in responsiveness of the cells in the PPN region controlled by these afferents (RYE 1997).

In rats, the activity of dopamine neurons in the VTA and sleep–wake cycles after injection of 25.0μg GABA were studied. The wakefulness time of the free-moving rats was decreased. Furthermore, the time of wakefulness was enhanced by an injection of dopamine (10.0μg). These data suggest that the GABA injection exerts an inhibitory action on dopamine neurons in VTA, mediating sleep–wakefulness through the mesolimbic system (WANG and LIN 1997).

However, little information is available on interaction between dopaminergic and GABAergic system in the sleep–wake cycle.

D. Summary

The present model of sleep reflects on the existence of cell groups in the brain stem, which trigger REM and SWS sleep. Critical structures are the RD and the LC. A high number of connections arise from the RD to the substantia nigra and the ventral structures of the hypothalamus, the generator of slow spindles. Also the cortex is connected through the medial forebrain bundle to the preoptic area, the lateral hypothalamus and the limbic system.

Applying D_2 dopaminergic substances, it can be concluded that effects are moderated in a dose-depended manner: low doses of D_2 agonists and high doses of antagonists lead to a reduction in wake state and produce an increase in the amount of REM and SWS. On the other hand, the confusion is exceeded by the finding that high doses of D_2 agonists and low doses of antagonists exaggerate the wake state and reduce REM and SWS. The data for compounds acting on D_1 receptors are even less clear than studies with more specific substances (for example D_3 or D_4).

Finally, even though there are only few data available on the regulation of temperature mediated by dopamine modulation, the effects of changes in REM and NREM sleep via temperature regulation changes by applying special agonists and antagonists cannot be ruled out. So, the mechanism of action on sleep might not be triggered on critical structures for sleep but instead maybe on structures responsible for thermoregulation.

Care has to be taken in interpreting data from receptor studies by considering the biological rhythms as well. After all, receptor density is modified by the light and dark cycle and even by age (HALL et al. 1996).

References

Abercrombie ED, DeBoer P (1997) Substantia nigra D1 receptors and stimulation of striatal cholinergic interneurons by dopamine: a proposed circuit mechanism. J Neurosci 17:8498–8505

Aceves J, Floran B, Martinez-Fong D, Sierra A, Hernandez S, Mariscal S (1991) L-dopa stimulates the release of [3H]gamma-aminobutyric acid in the basal ganglia of 6-hydroxydopamine lesioned rats. Neurosci Lett 121:223–226

Aserinsky E (1953) Regular occurring periods of eye motility and concomitant phenomena during sleep. Science 118:273–274

Bach V, Bouferrache B, Kremp O, Maingourd Y, Libert JP (1994) Regulation of sleep and body temperature in response to exposure to cool and warm environments in neonates. Pediatrics 93:789–796

Baker KG, Hornung J-P, Cotton RH (1991) Distribution, Morphology and number of monoamine-synthesizing and substance P-containing neurons in the human dorsal raphe nucleus. Neuroscience 42:757–775

Bartholini G (1985) GABA receptor agonists: pharmacological spectrum and therapeutic actions. Med Res Rev 5:55–75

Berger RJ, Phillips NH (1988) Comparative aspects of energy metabolism, body temperature and sleep. Acta Physiol Scand Suppl 574:21–27

Berger RJ, Phillips NH (1990) Comparative physiology of sleep, thermoregulation and metabolism from the perspective of energy conservation. Prog Clin Biol Res 345:41–42

Berger TW, Stricker EM (1985) Hyperinnervation of the striatum by dorsal raphe afferents after dopamine-depleting brain lesions in neonatal rats. Brain Res 336:454–358

Bremer F (1935) Cerveau "isolé" et physiologie du sommeil. CR soc Biol (Paris) 118:1235–1241

Bremer F (1938) L'activité électrique de l'ecorce cérébrale et le probléme physiologique du sommeil. Boll Soc Ital Bioö Sper 13:271–290

Buckingham RC (1976) The effects of reserpine, L-dopa and 5-hydroxytryptophan on 5-hydroxyindole acetic and homovanillic acids in cerebrospinal fluid, behavior and EEG in cats. Neuropharmacology 15:383–392

Cespuglio R, Jouvet M (1988) Opposite variations of 5-hydroxyindoleacetic acid (5-HIAA) extracellular concentrations, measured with voltammetry either in the axonal nerve endings or in the cell bodies of the nucleus raphe dorsalis, throughout the sleep–waking cycle. CR soc Biol (Paris) 307:817–823

Chiavegatto S, Nasello AG, Bernardi M (1998) Histamine and spontaneous motor activity: biphasic changes, receptors involved and participation of the striatal dopamine system. Life Sci 62:1875–1888

Consolo S, Baronio P, Guidi G, Di Chiara G (1996) Role of the parafascicular thalamic nucleus and N-methyl-D-aspartate transmission in the D1 dependent control of in vivo acetylcholine release in rat striatum. Neuroscience 71:157–165

Consolo S, Wu CF, Fusi R (1987) D-1 receptor-linked mechanism modulates cholinergic neurotransmission in rat striatum. J Pharmacol Exp Ther 242:300–305

Czeisler CA, Weitzman E, Moore-Ede MC, Zimmerman JC, Knauer RS (1980a) Human sleep: its duration and organization depend on its circadian phase. Science 210:1264–1267

Czeisler CA, Zimmerman JC, Ronda JM, Moore-Ede MC, Weitzman ED (1980b) Timing of REM sleep is coupled to the circadian rhythm of body temperature in man. Sleep 2:329–346

de Saint H, Python A, Blanc G, Charnay Y, Gaillard JM (1995) Effects of WIN 35,428 a potent antagonist of dopamine transporter on sleep and locomotor activity in rats. Neuroreport 6:2182–2186

Delorme F, Jouvet M (1965) Effects remarquables de la réserpine sur l'activité EEG phasique ponto-géniculo-occipitale. C.R.soc Biol (Paris) 159:900–903

Dement WC, Henriksen SL (1972) Sleep changes during chronic administration of parachlorophenylalanine. Rev Can Biol 31:239–246

Dusan-Peyrenthon D (1968) Effets du disulfiram sur les états de sommeil chez le chat. C.R.soc Biol (Paris) 162:2144–2145

Dusan-Peyrenthon D, Jouvet M (1968) Supression sélective du sommeil paradoxal chez le chat par alpha méthil-dopa. C.R.soc Biol (Paris) 162:116–118

Everitt BJ, Robbins TW (1997) Central cholinergic systems and cognition. Annu Rev Psychol 48:649–684

Faunt JE, Crocker AD (1987) The effects of selective dopamine receptor agonists and antagonists on body temperature in rats. Eur J Pharmacol 133:243–247

Fleckenstein AE, Lookingland KJ, Moore KE (1994a) Activation of noradrenergic neurons projecting to the diencephalon following central administration of histamine is mediated by H1 receptors. Brain Res 638:243–247

Fleckenstein AE, Lookingland KJ, Moore KE (1994b) Differential effects of histamine on the activity of hypothalamic dopaminergic neurons in the rat. J Pharmacol Exp Ther 268:270–276

Florio V, Longo VG (1975) A study of the central effects of sympathomimetic drugs: EEG and behavioral investigations on clonidine and naphazoline. Neuropharmacology 14:707–714

Fornal C, Radulovacki M (1982) alpha-Flupentixol increases slow-wave sleep in rats: effect of dopamine receptor blockade. Neuropharmacology 21:323–325

Friedman TC, Garcia-Borreguero D, Hardwick D, Akuete CN, Stambuk MK, Dorn LD, Starkman MN, Loh YP, Chrousos GP (1994) Diurnal rhythm of plasma delta-sleep-inducing peptide in humans: evidence for positive correlation with body temperature and negative correlation with rapid eye movement and slow wave sleep. J Clin Endocrinol Metab 78:1085–1089

Froment JL, Bertrand N (1974) Effects del'injection intracérébrale de 5,6-hydroxytryptamine sur la monamines cérebrals et les étatsde sommeils du chat. Brain Res 67:405–409

Futami T, Takakusaki K, Kitai ST (1995) Glutamatergic and cholinergic inputs from the pedunculopontine tegmental nucleus to dopamine neurons in the substantia nigra pars compacta. Neurosci Res 21:331–342

Gessa GL, Collu M, Serra M-O, Baggio G (1985) Sedation and sleep induced by high dodes of apomorphine after blockade of D-1 receptors by SCH 23390. Eur J Pharmakol 109:269–274

Ghosh PK, Hrdina PD, Ling GM (1976) Effects of REMS deprivation on striatal dopamine and acetylcholine in rats. Pharmacol Biochem Behav 4:401–405

Gottesmann C (1966) Réserpine et vigilance chez le rat. C.R.soc Biol (Paris) 160:2056–2061

Hall H, Halldin C, Dijkstra D, Wikström H, Wise LD, Pugsley TA, Sokoloff P, Pauli S, Farde L, Sedvall G (1996) Autoradiographic localisation of D3 dopamine receptors in the human brain using the selective D3-dopamine receptor agonist (+)-[3H]PD 128907. Psychopharmacology (Berl) 128:240–247

Hernández-Peón R (1965) A cholinergic hypnogenic limbic forebrain-hindbrain circuit. In: Jouvet M (ed) Aspects anatomo fonctionells du sommeil. Centre National de la Recherche Scientifique, Paris

Hervé D, Simon H, Blanc G, Lisoprawski A, Le Moal M, Glowinski J, Tassin JP (1979) Increased utilization of dopamine in the nucleus accumbens but not in the cerebral cortex after dorsal raphe lesion in the rat. Neurosci Lett 15:127–133

Hilakivi I (1984) Effects of methoxamine, an alpha-1-adrenoceptor agonist and prazosin, an alpha-1-antagonist, on the stages of the sleep-waking cycle in the cat. Acta Physiol Scand 120:363–372

Hilakivi I, Putkonen PS (1980) Prazosin increases paradoxical sleep. Eur J Pharmacol 65:417–420

Hilakivi I (1987) Biogenic amines in the regulation of wakefulness and sleep. Med Biol 65:97–104

Hjorth S, Carlsson A, Clark D, Svensson K, Sanchez D (1985) Dopamine receptor-mediated hypothermia induced in rats by (+)-, but not by (–)-3-PPP. Eur J Pharmacol 107:299–304

Hobson JA, Freedman R (1974) Time-course of discharge rate changes in rat pontine brainstem neurons during the sleep cycle. J Neurophysiol 37:1297–1309

Hobson JA, Wyzinski PW (1975) Sleep cycle oscillation: reciprocal discharge by two brainstem neuronal groups. Science 189:55–58

Jones BE (1990) Influence of the brain stem reticular formation, including intrinsic monoaminergic and cholinergic neurons, on forebrain mechanisms of sleep and waking. In: Mancia M, Marini G (eds) The Diencephalon and Sleep, Raven, New York

Jouvet M (1972) Neurophysiology and neurochemistry of sleep and wakefulness. Springer, New York

Jouvet M (1962) Recherches sur les structures nerveuses et les méchanismes responsables des différentes phases du sommeil physiologique. Arch Ital Biol 100:125–206

Kafi S, Constantinidis J-G (1977) Paradoxical sleep and brain catecholamines in the rat after single and reported administration of alpha-methyl-paratyrosine. Brain Res 135:123–134

Kafi S (1976) Brain dopamine receptors and sleep in the rat: effects of stimulation and blockade. Eur J Pharmacol 38:357–364

Karczmar AG, Scotti DC (1970) A pharmacological model of paradoxical sleep. The role of cholinergic and monoamine systems. Physiol Behav 5:175–182

Khan S, Sandhu J, Whelpton R, Michael-Titus AT (1998) Substance P fragments and striatal endogenous dopamine outflow: interaction with substance P. Neuropeptides 32:519–526

Kleinlogel H, Sayers AC (1975) Effects of clonidine and BS-100–141 on the EEG sleep patterns in rats. Eur J Pharmacol 33:159–163

Koella WP (1984) Biochemistry and pharmacology of vigilance: role of neurotransmitters within the framework of vigilance control. EEG EMG Z Elektroenzephalogr Verwandte Geb 15:180–189

Kudo Y, Kurihara M (1990) Clinical evaluation of diphenhydramine hydrochloride for the treatment of insomnia in psychiatric patients: a double-blind study. J Clin Pharmacol 30:1041–1048

Laguzzi R, Pujol JF (1987) Effets de l'injection intraventriculaire de 6-hydroxydopamine. II. Sur le cycle veille-sommeil du chat. Brain Res 48:295–310

Leppävouri A (1978) Evidence for central alpha adrenoceptor stimulation as the basis of paradoxical sleep suppression by alpha-methyldopa. Neurosci Lett 9:37–43

Leppävouri A (1980) Alpha-adrenoreceptive influences on the control of the sleep-waking-cycle in the cat. Brain Res 193:95–115

Liposits Z, Paull WK (1987) Synaptic interaction of serotonergic axons and corticotropin releasing factor (CRF) synthesizing neurons in the hypothalamic paraventricular nucleus of the rat: a light and electron microscopic immunocytochemical study. Histochemistry 86:541–549

Lloyd KG, Perrault G, Zivkovic B (1985) Implications of GABAergic synapses in neuropsychiatry. J Pharmacol 16 Suppl 2:5–27

Login IS, Harrison MB (1996) A D1 dopamine agonist stimulates acetylcholine release from dissociated striatal cholinergic neurons. Brain Res 727:162–168

Matsumoto J (1964) Effects de réserpine, DOPA et 5HTP sue les 2 états de sommeil. CR soc Biol (Paris) 158:2137–2140

McCarley RW (1986) A limit cycle mathematical model of the REM sleep oscillator system. Am J Physiol 251:1011–1029

McGinty D, Szymusiak R (1990) Keeping cool: a hypothesis about the mechanisms and functions of slow-wave sleep. Trends Neurosci 13:480–487

Mendelson WB, Dawson SD, Wyatt RJ (1980) Effects of melatonin and propanolol on sleep of the rat. Brain Res 201:240–244

Mereu GP, Paglietti E-P, Chessa P, Gessa GL (1979) Sleep induced by low doses of apomorphine in rats. Electroencephalogr Clin Neurophysiol 46:214–219

Monti JM (1968) The effects of haloperidol on the sleep cycle of the cat. Experientia 24:1143

Monti JM (1979) The effects of neuroleptics with central dopamine and noradrenaline receptor blocking properties in the L-dopa and (+) amphetamine-induced waking EEG in the rat. Br J Pharmacology (Berl) 67:87–91

Monti JM, Jantos H, Fernandez M (1988) Biphasic effects of dopamine D-2 receptor agonists on sleep and wakefulness in the rat. Psychopharmacology (Berl) 95:395–400

Monti JM, Jantos H (1990a) Sleep during acute dopamine D1 agonist SKF 38393 or D1 agonist SCH 23390 administration in rats. neuropsychopharmakology. 3:153–162

Monti JM, Orellana C, Boussard M, Jantos H, Olivera S (1990b) Sleep variables are unaltered by zolantidine in rats: are histamine H2-receptors not involved in sleep regulation? Brain Res Bull 25:229–231

Monti JM, Pellejero T, Jantos H (1986) Effects of H1- and H2-histamine receptor agonists and antagonists on sleep and wakefulness in the rat. J Neural Transm 66:1–11

Monti JM, Fernandez M (1989) Effects of the selective D-2 receptor against quinpirole on sleep and wakefulness in the rat. Eur J Pharmacol 169:61–66

Moore RY (1993) Principles of synaptic transmission. Ann NY Acad Sci 695:1–9

Mori S, Yamada H (1985) Immunohistochemical demonstration of serotonin nerve fibers in the subthalamic nucleus of the rat, cat and monkey. Neurosci Lett 62:305–309

Mori S, Yamada H, Sano Y (1985) Immunohistochemical demonstration of serotonin nerve fibers in the corpus striatum of the rat, cat and monkey. Anat Embryol (Berl.) 173:1–5

Moruzzi G-M (1949) Brain stem reticular formation and activation of the EEG. Electroencephalogr. Clin Neurophysiol 1:455–473

Mouret JR (1968) Insomnia following parachlorophenylalanine in the rat. Eur J Pharmacol 5:17–22

Nauta WH (1946) Hypothalamic regulation of sleep in rats. Experimental study. J Neurophysiol 9:285–316

Nicholson AN (1985a) Central effects of H1 and H2 antihistamines. Aviat Space Environ Med 56:293–298

Nicholson AN, Pascoe PA, Stone BM (1985b) Histaminergic systems and sleep. Studies in man with H1 and H2 antagonists. Neuropharmacology 24:245–250

Nowak JZ (1985) Depolarisation-evoked release of dopamine and histamine from brain tissue and studies on presynaptic dopamine-histamine interaction. Pol J Pharmacol Pharm 37:359–381

Oganesian GA, Vataev SI, Titkov ES (1997) Corticostriatal relations in the waking-sleep cycle of normal rats and in pathology. Ross Fiziol Zh Im I M Sechenova 83:37–46

Olive MF, Seidel WF, Edgar DM (1998) Compensatory sleep responses to wakefulness induced by the dopamine autoreceptor antagonist (–)DS121. J Pharmacol Exp Ther 285:1073–1083

Peyron C, Fort P-R, Jouvet M (1996) Lower brainstem catecholamine afferents to the rat dorsal raphe nucleus. J Comp Neurol 364:402–413

Pickworth WB, Nozaki M (1977) Sleep suppression induced by intravenous and intraventricular infusions of methoxamine in the dog. Exp Neurol 57:999–1011

Puizillout JJ, Daszuta A (1979) Release of endogenous serotonin from "encephale isolé" cats. II. Correlations with raphe neuronal activity and sleep and wakefulness. J Physiol (Paris) 75:531–535

Putkonen PS (1977a) Increase in paradoxical sleep in the cat after phentolamine and alpha-adrenoceptor antagonist. Acta Physiol Scand 100:488–490

Putkonen PS, Stenberg D (1977b) Paradoxical sleep inhibition by central alpha-adrenoceptor stimulant clonidine antagonized by alpha-receptor blocker yohimbine. Life Sci 21:1059–1066

Reite M, Stephens LM, Lewis OL (1969) The effect of reserpine and monoamine oxidase inhibitors on paradoxical sleep in the monkey. Psychopharmacologia. 14:12–17

Rinne JO, Partinen M, Nagren K, Ruotsalainen U (1996) Striatal dopamine D1 receptors in narcolepsy: a PET study with [^{11}C]NNC 756. J Sleep Res 5:262–264

Rye DB (1997) Contributions of the pedunculopontine region to normal and altered REM sleep. Sleep 20:757–788

Sakai K (1991) Physiological properties and afferent connections of the locus coeruleus and adjacent tegmental neurons involved in the generation of paradoxical sleep in the cat. Prog Brain Res 88:31–45

Sakai K, Yoshimoto Y, Luppi PH, Fort P, el Mansari M, Salvert D, Jouvet M (1990) Lower brainstem afferents to the cat posterior hypothalamus: a double-labeling study. Brain Res Bull 24:437–455

Sánchez C (1989) The effects of dopamine D-1 and D-2 receptor agonists on body temperature in male mice. Eur J Pharmacol 171:201–206

Sánchez C, Arnt J (1992) Effects on body temperature in mice differentiate between dopamine D2 receptor agonists with high and low efficacies. Eur J Pharmacol 211:9–14

Schlicker E, Fink K, Detzner M, Göthert M (1993) Histamine inhibits dopamine release in the mouse striatum via presynaptic H3 receptors. J Neural Transm Gen Sect 93:1–10

Schmidek WR, Schmidek M-K, Alves MR (1974) Influence of Loxapine on the sleep–wakefulness cycle of the rat. Pharmacol Biochem Behav 2:747–751

Staedt J-S, Kogler A, Hajak G-R, Mayer G, Munz DL, Ruther E (1996) IBZM SPET analysis of dopamine D2 receptor occupancy in narcoleptic patients in the course of treatment. Biol Psychiatry 39:107–111

Steriade M, Paré D, Curró DR (1990) Neuronal activities in brainstem cholinergic nuclei related to tonic activation processes in thalamo-cortical systems. J Neurosci 10:2541–2559

Stern WC (1973) Effects of reserpine on sleep and brain biogenic amine levels in the cat. Psychopharmacologia. 28:275–286

Stille G-L (1974) Die Wirkung von Neuroleptika im chronischen pharmakologischen Experiment. Arzneimittelforschung. 24:1292–1294

Stothers JK, Warner RM (1984) Thermal balance and sleep state in the newborn. Early Hum Dev 9:313–322

Strittmatter M, Grauer MT, Schimrigk K (1996) CFS substance P somatostatin and monoaminergic transmitter metabolites in patients with narcolepsy. Neurosci Lett 218:99–102

Tabushi K (1969) The acute effects of reserpine on the sleep–wakefulness cycle in rabbits. Psychopharmacologia 16:240–252

Takeuchi O (1973) Influences of psychotropic drugs (chlorpromazine, imipramine and haloperidol) on the sleep wakefulness mechanisms in cats. Psychiatr Neurol Jpn 75:424–459

Tanaka M, Tamada Y, Tanaka Y (1994) Catecholaminergic input to spinally projecting neurons in the rostral ventromedial medulla oblongata of the rat. Brain Res Bull 35:23–30

Torda C (1968) Effects of changes of brain norepinephrine content on sleep cycle in rat. Brain Res 10:200–207

Trampus M (1990) The D1 dopamine receptor antagonist SCH 23390 enhances REM sleep in the rat. Neuropharmakology 29:889–893

Tsuchiya T, Fokushima H (1979) Analysis of the dissociation between the neocortical and hypocortical EEG activity induced by neuroleptics. Psychopharmacology (Berl) 63:179–185

Tulen JH, Dzoljic MR, Moleman P (1991) Sleeping with and without norepinephrine: effects of metoclopramide and D,L-threo-3,4-dihydroxyphenylserine on sleep in dopamine beta-hydroxylase deficiency. Sleep 14:32–38

Velluti R (1963) Atropine blockade within a cholinergic hypnogenic circuit. Exp Neurol 8:20–29

Vila M, Herrero MT, Levy R, Faucheux B, Ruberg M, Guillen J, Luquin MR, Guridi J, Javoy-Agid F, Agid Y, Obeso JA, Hirsch EC (1996) Consequences of nigrostriatal denervation on the gamma-aminobutyric acidic neurons of substantia nigra pars reticulata and superior colliculus in parkinsonian syndromes. Neurology 46:802–809

Wang Z, Lin YL (1997) Regulation of dopamine neurons in ventral tegmental area on sleep–wakefulness. Sheng Li Hsueh Pao 49:135–140

Wauquir A, Niemegeers CE (1980) On the antagonist effects of pimozide and domperidone on apomorphine-disturbed sleep–wakefulness in dogs. In: Koella WP (ed) Sleep Karger., Basel 279–282

Wauquir A (1985) Active and permissive roles of dopamine in sleep–wakefulness regulation. In: Wauquir A, Gaillard JM, Monti JM, Radulovacki M (eds) Sleep: neurotransmitters and neuromodulators. Raven., New York. 107–120

Wauquir A, Janssen PJ (1980) Biphasic effects of pimozide on sleep–wakefulness in dogs. Life Sci 27:1469–1475

Whitaker-Azmitia PM, Shemer AV (1990) Prenatal treatment with SKF 38393, a selective D1-receptor agonist: long-term consequences on H-paroxetine binding and on dopamine and serotonin receptor sensitivity. Dev Brain Res 57:181–187

Yeomans JS (1995) Role of tegmental cholinergic neurons in dopaminergic activation, antimuscarinic psychosis and schizophrenia. Neuropsychopharmacology 12:3–16

Yoshimoto Y, Sakai K, Panula P, Salvert D, Stuart M, Jouvet M (1989) Cells of origin of histaminergic afferents to the cat median eminence. Brain Res 504:149–153

Zarrindast MR, Tabatabai SA (1992) Involvement of dopamine receptor subtypes in mouse thermoregulation. Psychopharmacology (Berl) 107:341–346

Subject Index

acetaldehyde 334, 354
N-acetylaspartate 422
acetylcholine (ACh) 63, 65, 72, 89, 160, 397, 491
acetylcholinesterase 19
N-acetylcysteine 330
achievement 447
acoustic startle response 184
activation 218
active avoidance responses 292
active avoidance test 251
adenosine diphosphate (ADP) 347
adenosine triphosphate (ATP) 135, 331
S-adenosylhomocysteine 135
adenylate cyclase 121, 142
adenylyl cyclase 63
adipsia 228
adrenalectomy 180
adrenoceptor, alpha (α)1 485
adrenoreceptor, alpha (α) 22, 392, 476
adrenoreceptor, beta (β) 22
adrenoreceptor, alpha (α)2 33, 327
agitation 390
akathisia 352
akinesia 154, 397
alcohol see ethanol
alogia 445
alpha (α)-amino-3-hydroxy-5-methyl-4-isoxazolepropionic acid (AMPA) 64
alpha (α)-flupenthixol 183
amineptine 395
2-amino-5-phosphonopentanoic acid (AP-5) 339
gamma (γ)-aminobutyric acid (GABA) 294, 328, 424, 499
4-aminopyridine (4-AP) 8, 66
amisulpride 242, 302
amitryptiline 399
amphetamine 1, 29, 91, 145, 179, 219, 252, 265, 285, 303, 355, 393, 418

amygdala 26, 182, 271, 388, 484
anesthesia 6
anesthetics 35
anhedonia 387, 426, 444
anhedonia hypothesis 273, 289, 305
antidepressants 274, 350, 394
antidepressants, tricyclic 403
antidromic activation 6
anti-hypertensives 350
antipsychotic drugs 2, 33, 161, 173, 195, 251, 274, 418, 421, 451
antipsychotic treatment see antipsychotic drugs
antipsychotics see antipsychotic drugs
antipsychotics, tricyclic 484
anxiety 246
aotoreceptor, terminal 25
apamin 8, 19
apathy 407, 444
aphagia 228
apoferritin 344
apomorphine 1, 22, 35, 183, 193, 229, 242, 265, 357, 392, 493
appetitive phase 278
arachidonic acid 64
aripiprazole 485
arousal 103
articulatory loop 187
L-ascorbic acid 330
aspartate 327
associative thought process 199
astrocytes 68
atropine 74
attention 103, 184
attention deficit hyperactivity disorder (ADHD) 185
attribution process 294
autoreceptor 13
autoreceptor, axon terminal 28
autoreceptor, presynaptic 398
autoreceptor, somatodendritic 28

Subject Index

avolition 445
azaperone 283

baclofen 14, 164
basal ganglia 85, 151, 224, 294, 322, 432
– circuitry 158
bed nucleus (stria terminalis) 26
behavioral control 426
behavioral expression 437
behaviour, appetitive 404
behaviour, goal-directed 173, 192, 284, 417, 428, 446
behaviour, instrumental 273, 300
behaviour, routine 438
behaviour, stereotypic 446
benzodiazepine 155, 303
beta (β)-blockers 350
biasing, cortical 432
bicuculline 10, 31, 73, 163
bipolar disorder 391
blood-brain barrier 342
body temperature 495
bradykinesia 305, 321, 342, 425
bradyphrenia 389
brain microdialysis 87
brain-derived neurotrophic factor (BDNF) 340
bromocriptine 141, 188, 193, 241, 394, 407
bufuralol-hydroxylase 349
kappa (κ)-bungarotoxin 20, 99
alpha-(α)-bungarotoxin 102
bupropion 395
burst firing 1
butyrophenone 473

caffeine 139, 143
calcium 335, 345
 calcium channels, L-type 124
calcium conductance 8
calcium currents, N-type 95
cannabinoids 36
cannabis 42
carbachol 69
(S)-4-carboxy-phenylglycine 15
beta (β)-carotene 330
caspase-3 338
catalepsy 139, 144, 243, 255
cataplexy 497
catecholamine neurons mapping 1
caudate nucleus 64
caudate putamen 151
caudate-putamen nucleus 339
cerebellum 324
cerebral cortex 92

chemical axotomy 337
chloral hydrate 6
chlorpromazine 30, 287, 296, 302, 473
cholecystokinin-8 (CKK-8) 31
cholinergic afferents 14
chronic mild stress (CMS) 403
ciliary neurotrophic factor (CTNF) 340
cimetidine 499
cingulate gyrus, anterior 402
clenbuterol 497
clomipramine 395
clonidine 22, 334, 497
clozapine 33, 195, 230, 284, 419, 474, 481
cocaine 36, 155, 219, 265
competence, interpersonal 443
conditioned taste aversion 297
conditioning, associative 174
conditioning, classical 308
conduction velocity 1
w-conozoxin 29
consummatory phase 278
continuous reinforcement (CRF) 286
Contursi family 361
convulsant agents 342
corpus striatum see striatum 321
cortex 118
cortex, entorhinal 420
cortex, orbitofrontal 427
cortex, prefrontal 286, 388, 420, 427
cortical cup technique 104
craving 266
6-cyano-7-nitroquioxaline-2,3-dione (CNQX) 339
cyclic AMP (cAMP) 121
cyclooxygenase 69
cyclosporine A 343, 347
cyclothiazide 66
CYP system 348
cytochrome C 347
cytochrome C oxidase 338
cytochrome P450 69

D1 (dopamine 1) receptor family 23
D_1 (dopamine$_1$) receptors 437
– antagonist 88 391
D2 (dopamine 2) receptor family 23, 391
D_2 (dopamine$_2$) receptors 437
– antagonists 450
– gene polymorphism 423
D_4 (dopamine$_4$) receptors 481
DA-receptor blockers 272
debrisoquine 330, 355
decision-making 429
delusion 390, 406, 417, 444

Subject Index

dementia 324
dentate gyrus 484
depersonalization 450
depolarization block 34
L-deprenyl 330
depression 305, 324, 426
depression, major 387, 392
depression, old age 394
desipramine 405
desmethylimipramine 405
dextrometorphan 350
diagonal band of Broca 103
diazepam 197
diethyldithiocarbamate (DDC) 333
disulfiram 492
domperidone 493
dopamine (dopaminergic) system, mesocortical 2, 420
dopamine (dopaminergic) system, mesocorticolimbic 229, 388, 408
dopamine (dopaminergic) system, nigrostriatal 2, 229
dopamine (dopaminergic) system, prefrontal 163
dopamine beta (β)-hydroxylase 227
dopamine blockade 435
dopamine decarboxylase 495
dopamine function, phasic 421
dopamine function, tonic 421
dopamine overflow 27
dopamine quinone 357
dopamine receptor antagonists 297
dopamine receptor blockade 272, 281
dopamine transporter (DAT) 323, 337, 357, 225
dopamine transporter (DAT) knockout 186, 213
dopamine-reuptake 102
dopaminergic cell bodies 1
dopaminergic neurons
– patterns of firing 4
dorsal raphé nucleus 20
drive reduction 271
drug abuse 2
drug abuse therapy 240
drug addiction 38, 127, 265
drug reward 218, 265
drug withdrawal 40
D-serine 486
dynorphin 73, 152
dyskinesia 144, 157, 161,
dyskinesia, orofacial 155
dyskinesia, tardive 352
dysphoria 407
dysthymia 305
dystonia 353

ecto-5' nucleotidase 135
electrical stimulation 93
electron transport chain 348
emotions 443
endonuclease 346
enkephalin 87, 122, 152, 424
entopeduncular nucleus 74, 154, 158
epidermal growth factor (EGF) 340
epilepsy 137
erection, penile 241
eserine 498
ethanol 35, 40, 224, 241, 354
euphoria 266, 305, 393
excitement 447
experience 432, 447
extrapyramidal functions 272
extrapyramidal motor function 305
extrapyramidal motor system 218
extrapyramidal side effects 33, 242, 418
eye blinking 242

fatigue 284, 407
fenfluramine 198
ferritin 344
FGF-2 340
firing mode pacemaker, random, bursty 5
firing rate 36
fluoxetine 350, 395
flupenthixol 282, 285, 302, 493
forebrain 218, 323
forskolin 125
Fos 419
free radical scavengers 331
free radicals 329
frontal cortical stimulation 16

gamma (γ)-aminobutyric acid (GABA) 8, 68, 85, 122
– antagonists 153
– receptors, A and B 10, 153
gamma-hydrohybutyric acid (GHBA) 31
ganciclovir 214
gene linkage 216
gene targeting 215
genetic recombination 216
glial cell line-derived neurotrophic factor (GDNF) 340
glial cells 67
globus pallidus 8, 11, 76, 155, 294, 424, 432
glutamate 14, 64, 92, 229, 327, 357, 437, 483
glutamate input, cortical 125
glutamate receptor agonist 340

Subject Index

glutamatergic afferents 14
glutamic acid 328, 340
glutathione 361
glycine 31
goal selection 440
G-protein 23, 32
Gray's Type II symmetrical synapses 2
grooming 220
growth hormone (GH) 392
G$_s$ proteins 121
guanylyl cyclase 67

habit responding 271
hallucinations 104, 417, 450
haloperidol 21, 30, 144, 154, 183, 190, 243, 283, 302, 398, 419, 482, 492
hedonia 266
heparin 32
heteroreceptors 63
hippocampus 181, 324, 388, 420
histamine 499
homovanillic acid (HVA) 389, 418
Huntington's disease 197
hyaline inclusions 321
6-hydroxydopamine (6-OHDA) 154, 179, 279, 296, 323
3-hydroxypiridine 354
2-hydroxysaclofen 12
5-hydroxytryptamine (5-HT) see serotonin
5-hydroxytryptophan 21
hyperdopaminergia 418
hyperprolactinaemia 240
hypersomnolence 495
hypodopaminergia 43
hypofrontality 420, 427
hypoglycemia 137
hypokinesia 425
hypomania 393
hypotension, orthostatic 484
hypothalamus 26, 297, 302, 491
hypothermia 225, 244, 496
hypoxia 137

ifenprodil 337
imipramine 395, 399, 401
in situ 3'-end labelling (ISEL) 346
in vivo dialysis 176
incentive arousal 295
indoleamine 361
inflammation 67
information storage 202
inhibitory postsynaptic potential (IPSP) 10
initial segment-somatodendritic (IS-SD) break 6

insomnia 491
instrumental action 269
instrumental responding 281
insulin-like growth factor (IGF)-1 and -2 340
interleukin (IL)-6 340
intracranial self-stimulation (ICSS) 265
iodosulpride 240
iron 344

c-Jun N-terminal kinase (JNK) 359

kainate 67
kassinin see substance K
ketamine 6, 337
kynurenate 17

lactoferrin 344
lactotroph cells (pituitary gland) 26
latent inhibition (LI) 183
L-dopa 154, 228, 407, 417
L-dopa therapy 344
learning 103, 222, 431, 438
learning, associative 174, 435
learning, incentive 270
learning, instrumental 296, 308
learning, stimulus-reward 174
lesion studies 216
Lewy bodies 321, 361
ligand binding affinity 98
limbic split circuit 439, 451
limbic system 70
lipid hydroperoxide 343
lipid peroxidation 331
lipoxygenase 69
lisuride 495
lithium 297, 303
locomotor activity 162, 220, 243, 246, 292, 435
locus coeruleus (LC) 22, 323, 491
loxapine 493
lymphocytes 136

magnesium block 65, 121
mainserin 399
malondialdehyde 343
mamillary bodies 491
mania 305, 390, 407
marijuana 37
mast cells 136
mazindol 69
MDL 100907 485
mecamylamine 102
medial forebrain bundle 286
mediodorsal nucleus 161
medium-spiny neurons 126

Subject Index

melancholia 389
membrane conductance 92
memory 103, 175
memory, emotional 182
memory, procedural 199
memory, reference 186
memory, semantic 199
memory, spatial 181
mepacrine 69
mesencephalon dopaminergic pathways 2
mesolimbic system 144, 151
metamphetamine 230, 325, 329, 356
metergoline 21
1-methyl-4-phenyl-1,2,3,6-tetrahydropyridine (MPTP) 189, 321
1-methyl-4-phenylpyridinium (MPP+) 326
methyllycaconitine 101
methylphenidate (Ritalin) 185, 195, 393, 397, 417
alpha-(α)methyl-p-tyrosine (αMPT) 336, 404
alpha-(α)methyltyrosine 140
metoclopramide 302
metyrapone 69
microcatalepsy 291
microdialysis 176
microsomal ethanol oxidizing system (MEOS) 351
midbrain 323
minaprine 395
mitogenesis 241
MK-801 334
modafinil 163
molindone 285
monoamine oxydase B (MAO-B) 330
– inhibitor 333
morphine 36, 303, 349,
motivation 265, 287, 433, 444
motivation, incentive 387
motivational arousal function 273
motivational impact 276
motivational process 218
motivational stimuli 267
motor behavior 141, 222
motor behavior, catatonic 449
motor capacity 287
motor deficit 278, 285
motor dysfunction 248
motor impairment 281
motor performance 272, 308
motor program 439, 448
movement disorders 127, 425
movement initiation 298

movement, control 161
mRNA 98
muscimol 6
myoclonus 353
myocytes, cardiac 136

NADH dehydrogenase 338
NADPH cytochrome P450 reductase 348
nafadotride 243, 247
naloxone 38, 73, 303
narcolepsy 496
neostigmine 95
neostriatum 10, 297
N-ethylcarboxamide adenosine (NECA) 139
neural networks 175
neurofibrillary tangles 324
neurokinin NK3 32
neuroleptics see antipsychotic drugs
neuromedin N 32
neuronal plasticity 202
neurotensin 32, 163, 397
neurotoxins 322
neurotrophins (NT)-3 340
neurturin 340
neutrophils 136
nicotinamide adenine nucleotide (NAD) 331
nicotine 19, 36, 65, 334, 341
nicotinic antagonists 102
nitric oxide (NO) 67
nitric oxide synthase (NOS) 345
7-nitroindazole (7-NI) 345
N-methyl-D-aspartate (NMDA) 14, 64, 72, 86, 153, 296, 328
nomifensine 69, 154, 395
non-reinforcement 282
noradrenaline see norepinephrine
norepinephrine 22, 188, 226, 323, 388
norepinephrine transporter 484
nor-fluoxetine 353
nucleus accumbens 25, 38, 99, 126, 145, 161, 285, 388, 399, 402
– core 4
– shell 4, 162, 177, 297
nucleus basalis magnocellularis 103

olanzapine 284, 476, 482
olfactory tubercle 4
opioid peptides 72
opioids 224
oxotremorine 65

pain modulation 138
pallidum 440

para-chlorphenylalanine 21
parkin 322
parkinsonism 321
Parkinson's disease 2, 78, 138, 144, 154, 218, 224, 321, 399
– early onset 360
paroxetin 350
parvalbumin 152
patch-clamp recordings 8
Pavlovian (incentive) learning 280, 291, 295, 300
Pavlovian aversive conditioning 178
Pavlovian principle 268
pedunculopontine stimulation 17
pempidine 74
pentobarbital 90
pergolide 193, 197, 495
peroxynitrite 345
perseveration 449
pertussis toxin 23, 30
phencyclidine 252, 337
phenobarbital 349
phenoxybenzamine 497
phospholipase A2 68
picrotoxin 10, 31, 303
pimozide 179, 267, 274, 283, 296, 302, 493
pipamperone 283, 474, 482
piperazine acceptor 350
pirenzepine 20
piribedil 394
planning 429
plasminogen 340
plasticity, synaptic 434
platelets 135
pontomesencephalic cell groups 98
Porsolt swim test 403
potassium 29
potassium channels 63
potassium conductance 7, 89
pramipexole 395, 494
prazosin 497
preoptic area 491
preparatory phase 278
preprodynorphin 89
preproenkephalin 89
prepulse inhibition (PPI) 254
pre-pulse-inhibition 145
prindamine 497
prodynorphin 121
proenkephalin 122
prolactin 392, 473
propranolol 497
protein kinase A 122
protein kinase C 69, 126
pseudodepression 396

psychomotor stimulants 173
psychosis 418, 426
psychostimulant drugs 38, 219, 231, 274, 295
psychostimulant reward 266
psychostimulants see psychostimulant drugs

quetiapine 474, 482
quinelorane 244
quinine 274
quinone 343
quinpirole 25, 96, 141, 179, 249, 400, 493

raclopride 197, 275, 283, 302, 422
raphe dorsalis (RD) 491
rapid eye movement (REM) 104, 491
reaction, aversive 279
reactive oxygen species (ROS) 338
receptor, glutamate 117
receptor, histamine 476
receptor, ionotropic 92, 118
receptor, kainate 118
receptor, metabotropic 15, 118
receptor, muscarinic 20, 65, 68, 190, 476
receptor, neurotensin 67
receptor, nicotinic 37, 65, 101
receptor, opioid 36, 38
receptors, serotonin, overview 477
reinforcement 274, 282, 305, 431
reinforcement deficit 288
reinforcement learning 292
release, dendritic 445
reserpine 139, 247, 397, 407, 492, 495
respiratory chain inhibition 334
responding, instrumental 306
responding, perseverative 191
retardation, psychomotor 389, 397, 426
retina 2
retrorubral field 4
reward 433
rigidity 342
riluzole 64
risperidone 267, 474, 481
Ritalin see methylphenidate
ritanserin 474
rotorod test 222
roxindole 396

2-OH-saclofen 13
salbutamol 497
SCH 23390 185, 190, 255, 276, 494
SCH 39166 303
schizophrenia 2, 104, 127, 144, 194, 232, 396, 399, 406, 474
– prevalence 417

Subject Index

scopolamine 142, 156, 190
seasonal affective disorder (SAD) 392
sedation 139, 484
selective serotonin reuptake inhibitors (SSRI) 186–350, 360, 393
semiquinone 343
semiquinone radicals 345
sensorimotor gating 162
sensorimotor impairment 279
sensorimotor integration 173
sensory-motor network 71
serotonin 21, 63, 186, 226, 388
sexual activity 302
sexual reward 293
single-pulse stimulus 10
SKF 38393 249, 494
sleep 241
sleep mechanisms 138
smokers 99
smooth muscle cells 136
social relations 443
social skills 443
sodium pump, electrogenic 15
somatostatin 67, 397
spinal cord 324
spinophilin 125
spiroperidol 285, 296
spontaneous discharge 1
SR 141716 A 486
SR 142801 486
SR 142948 486
startle reflex 242
sterol synthesis 485
stimulus, appetitive 174, 433
stimulus, arbitrary 300
stimulus, aversive 174
stimulus, conditioned 177, 436
stimulus, environmental 177
stimulus, gustatory 280
stimulus-secretion coupling 103
stress 164
stria terminalis
– bed nucleus 100
striatal modulatory network 86
striatonigral pathway 10
striatum 8, 25, 67, 85, 197, 218, 293, 321
striatum, associative 433
striatum, dorsal 401
striatum, limbic 433, 439
striosomes 70
strychnine 31
substance K (kassinin) 32
substance P 32, 76, 89, 122, 152, 424
substantia nigra substantia nigra 7, 117, 120, 294, 496
– pars compacta 2, 85, 151, 159, 321,
– pars reticulata 8, 74, 159, 424
subthalamic nucleus
subthalamic nucleus 339, 424
– stimulation 16
suicide 389, 401
sulpiride 25, 30, 66, 74, 181, 190, 275, 288, 302, 395, 403
sultopride 276
superoxide dismutase 353
surprise 447
synaptic plasticity 127
alpha-(α)synuclein 322, 361

tachycardia 484
tachykinin 72, 76
tegmental nucleus 19, 103
tetraethylammonium (TEA) 7
Δ^9-tetrahydrocannabinol (THC) 37
tetrodotoxin 63, 65, 72, 99, 165
TGF-beta(β)1 340
thalamic motor nuclei 158
thalamic nucleus 427
thalamus 11, 85, 92, 118, 424, 432
theophylline 139
thymidine kinase gene 214
tissue dopamine 27
alpha-(α)tocopherole 330
trans-1-aminocycolpentane-1,3-dicarboxylate (1S,3R-ACPD) 15
transcription factors 89
transforming growth factor (TGF)-alpha 340
transgene 216
tremor 321, 353
tryptamine 355, 361
tryptophan hydroxylase 357
tryptophan hydroxylation 492
tuberoinfundibular system 240
tyramine 361
tyrosine 24
tyrosine hydroxylase (TH) 17, 98, 143, 163, 227, 326, 336, 346, 350
tyrosine hydroxylase knockout 231
tyrosine hydroxylation 24

unconditioned response (UR) 269
urethane 6

ventral tegmental area (VTA) 4, 102, 117, 161, 218, 405, 496
veratridine 65
vesicular monoamine transporter (VMAT) 326, 337, 360
vesicular monoamine transporter 2 (VMAT-2) 230
vigilance 196

vigilance-controlling apparatus 498
viral agents 322
visuospatial sketchpad 187
visuospatial working memory 198
voltammetry 176

WIN 35428 494
Wisconsin Card Sorting Test
 (WCST) 196
working memory 164, 424

xanthine derivates 138

yawning 241
yohimbine 327

ziprasidone 476, 482
zotepine 476, 482

Printed in the USA
CPSIA information can be obtained
at www.ICGtesting.com
LVHW010911161023
761203LV00003B/6